ANATOMY, PHYSIOLOGY & DISEASE
FOR THE HEALTH PROFESSIONS

Please Return To:

Kelly Kohl
Education Advocate
Diversity and Inclusion Services

920-735-4825
920-851-8269 Cell
920-735-4874 Fax
kohlk@fvtc.edu
www.fvtc.edu/diversity

Appleton Campus
1825 N. Bluemound Dr. • P.O. Box 2277 • Appleton, WI 54912-2277

ANATOMY, PHYSIOLOGY & DISEASE
FOR THE HEALTH PROFESSIONS
Third Edition

Kathryn A. Booth, RN-BSN, RMA (AMT), CPT, CPhT, MS

Military to Medicine, INOVA Health System
Falls Church, VA

J. Virgil Stoia, DC

College of DuPage
Glen Ellyn, IL

ANATOMY, PHYSIOLOGY, AND DISEASE FOR THE HEALTH PROFESSIONS
Published by McGraw-Hill, a business unit of The McGraw-Hill Companies, Inc., 1221 Avenue of the Americas, New York, NY, 10020.
Copyright © 2013 by The McGraw-Hill Companies, Inc. All rights reserved. Printed in the United States of America. Previous editions © 2008 and 2009. No part of this publication may be reproduced or distributed in any form or by any means, or stored in a database or retrieval system, without the prior written consent of The McGraw-Hill Companies, Inc., including, but not limited to, in any network or other electronic storage or transmission, or broadcast for distance learning.

Some ancillaries, including electronic and print components, may not be available to customers outside the United States.

This book is printed on acid-free paper.

1 2 3 4 5 6 7 8 9 0 QVR/QVR 10 9 8 7 6 5 4 3 2

ISBN 978-0-07-340222-2
MHID 0-07-340222-2

Editorial director: *Michael Ledbetter*
Director, digital products: *Crystal Szewczyk*
Managing development editor: *Christine Scheid*
Digital development editor: *Katherine Ward*
Director, Editing/Design/Production: *Jess Ann Kosic*
Lead project manager: *Susan Trentacosti*
Senior buyer: *Michael R. McCormick*
Senior designer: *Srdjan Savanovic*
Lead photo research coordinator: *Carrie K. Burger*
Photo researcher: *Mary Reeg*

Manager, digital production: *Janean A. Utley*
Media project manager: *Cathy L. Tepper*
Outside development house: *Amber Allen, Integra Software Services, Inc.*
Cover design: *Nathan Kirkman*
Interior design: *Maureen McCutcheon*
Typeface: *11/13 ITC Garamond Std Light*
Compositor: *Laserwords Private Limited*
Printer: *Quad/Graphics*
Cover credit: *Illustration Copyright© 2011 Nucleus Medical Media, All rights reserved.*

Credits: The credits section for this book begins on page 710 and is considered an extension of the copyright page.

Library of Congress Cataloging-in-Publication Data

Booth, Kathryn A., 1957-
    Anatomy, physiology, and disease for the health professions / Kathryn. A. Booth, J. Virgil
Stoia.—3rd ed.
      p. ; cm.
    Includes index.
    Rev. ed. of: Anatomy, physiology, and pathophysiology for allied health / Kathryn A. Booth,
Terri D. Wyman. 2nd ed. c2009.
    ISBN 978-0-07-340222-2 (alk. paper)—ISBN 0-07-340222-2 (alk. paper)
    I. Stoia, J. Virgil. II. Booth, Kathryn A., 1957- Anatomy, physiology, and pathophysiology for
allied health. III. Title.
    [DNLM: 1. Anatomy. 2. Health Personnel. 3. Pathology. 4. Physiology. QS 4]
    612—dc23
                                                                          2011047647

WARNING NOTICE: The clinical procedures, medicines, dosages, and other matters described in this publication are based upon research of current literature and consultation with knowledgeable persons in the field. The procedures and matters described in this text reflect currently accepted clinical practice. However, this information cannot and should not be relied upon as necessarily applicable to a given individual's case. Accordingly, each person must be separately diagnosed to discern the patient's unique circumstances. Likewise, the manufacturer's package insert for current drug product information should be consulted before administering any drug. Publisher disclaims all liability for any inaccuracies, omissions, misuse, or misunderstanding of the information contained in this publication. Publisher cautions that this publication is not intended as a substitute for the professional judgment of trained medical personnel.

The Internet addresses listed in the text were accurate at the time of publication. The inclusion of a Web site does not indicate an endorsement by the authors or McGraw-Hill, and McGraw-Hill does not guarantee the accuracy of the information presented at these sites.

www.mhhe.com

To my students and all students using this program: May you find the human body as fascinating as it is and discover the importance of treating it properly to maintain wellness for yourself and those who you care for.

Kathy Booth

I would like to dedicate *Anatomy, Physiology, and Disease for the Health Professions* to all those students that have chosen health care as their life's vocation.

Virgil Stoia

brief contents

Unit 1: The Human Body and Disease

1. Concepts of the Human Body *2*
2. Concepts of Chemistry *18*
3. Concepts of Cells and Tissues *28*
4. Concepts of Disease *52*
5. Concepts of Microbiology *74*
6. Concepts of Fluid, Electrolyte, and Acid–Base Balance *104*

Unit 2: Concepts of Common Illness by System

7. The Integumentary System *124*
8. The Skeletal System *152*
9. The Muscular System *192*
10. Blood and Circulation *226*
11. The Cardiovascular System *270*
12. The Lymphatic and Immune Systems *300*
13. The Respiratory System *330*
14. The Nervous System *372*
15. The Urinary System *426*
16. The Male Reproductive System *450*
17. The Female Reproductive System *470*
18. Human Development and Genetics *498*
19. The Digestive System *524*
20. Metabolic Function and Nutrition *558*
21. The Endocrine System *586*
22. The Special Senses *614*

Appendixes

I Diseases and Disorders   *656*

II Prefixes, Suffixes, and Word Roots in Commonly Used Medical Terms   *674*

III Abbreviations and Symbols Commonly Used in Medical Notations   *677*

Glossary   *680*
Credits   *710*
Index   *718*

Connect

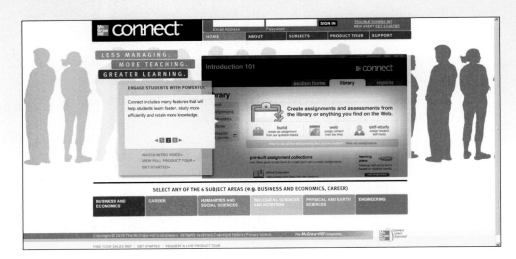

Connect for Anatomy, Physiology, and Disease for the Health Professions

connect.mcgraw-hill.com

McGraw-Hill's new *Connect* and *Connect Plus* are revolutionary online assignment and assessment solutions, providing instructors with tools and students with resources to maximize their success. The information provided below is written to assist the instructor.

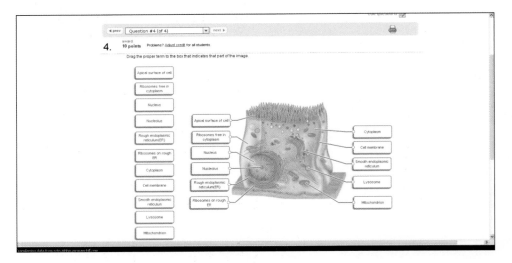

Simple Assignment Management

With *Connect,* creating assignments is easier than ever, so you can spend more time teaching and less time managing. The assignment management function enables you to:

- Create and deliver assignments easily with selectable interactives, vocabulary exercises, end-of-chapter questions, and test bank items.

- Streamline lesson planning, student progress reporting, and assignment grading to make classroom management more efficient than ever.
- Go paperless with *Connect Plus,* which includes an eBook and online submission and grading of student assignments.

Smart Grading

When it comes to studying, time is precious. *Connect* helps students learn more efficiently by providing feedback and practice material when they need it, where they need it. When it comes to teaching, your time also is precious. The grading function enables you to:

- Have assignments scored automatically, giving students immediate feedback on their work and side-by-side comparisons with correct answers.
- Access and review each response; manually change grades or leave comments for students to review.
- Reinforce classroom concepts with practice tests and instant quizzes.

Instructor Library

The *Connect* Instructor Library is your repository for additional resources to improve student engagement in and out of class. You can select and use any asset that enhances your lecture. The *Connect* Instructor Library includes:

- Instructor's Manual
- Test Bank
- PowerPoint presentation
- Body ANIMAT3D
- eBook
  - Fully integrated, allowing for anytime, anywhere access to the textbook.
  - Dynamic links between the exercises or questions you assign to your students and the location in the eBook where that exercise or question is covered.
  - A powerful search function to pinpoint and connect key concepts in a snap.

Student Progress Tracking

Connect keeps instructors informed about how each student, section, and class is performing, allowing for more productive use of lecture and office hours. The progress-tracking function enables you to:

- View scored work immediately and track individual or group performance with assignment and grade reports.
- Access an instant view of student or class performance relative to learning objectives.
- Collect data and generate reports required by many accreditation organizations, such as CAAHEP or ABHES.

Lecture Capture

Increase the attention paid to lecture discussion by decreasing the attention paid to note taking. For an additional charge, Lecture Capture offers new ways for students to focus on the in-class discussion, knowing they can revisit important topics later. Lecture Capture enables you to:

- Record and distribute your lecture with a click of a button.
- Record and index PowerPoint presentations and anything shown on your computer so it is easily searchable, frame by frame.

- Offer access to lectures anytime and anywhere by computer, iPod, or mobile device.
- Increase intent listening and class participation by easing students' concerns about note taking. Lecture Capture will make it more likely you will see students' faces, not the tops of their heads.

Body ANIMAT3D

This 3D animation series brings to life the most complex concepts in anatomy and physiology, offering students an additional resource that visually reinforces selected topics from the text. Throughout the textbook, an animation icon is located next to any content that is further supported by an animation. These also feature pre- and post-animation assessment questions tied to the Revised Bloom's Taxonomy.

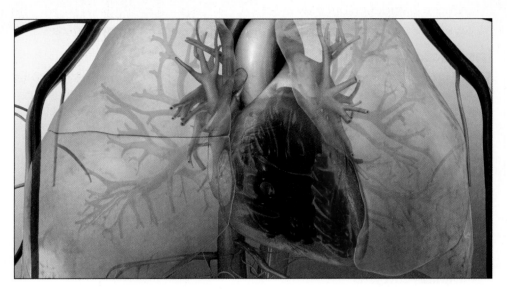

"I am very impressed by the Body ANIMAT3D videos that can be utilized to reinforce our lectures and as additional student resources."

—Angela LeuVoy, *Fortis College–Columbus*

Digital Coloring Book

Coloring Book images are available as digital activities in *Connect*. By filling in different various structures as directed in the activity, students can test their knowledge of human anatomy in different views and learn how each structure relates to neighboring structures.

Digital Atlas of Anatomy

These drag-and-drop activities allow students to identify and label structures on illustrated images for additional exercises to master their knowledge of human anatomy.

Digital Dissection Manual

This drag-and-drop activity allows students to identify and label structures from actual cadaver images, preparing them for working with real bodies.

Coding Corner Example Codes: CPT, ICD-9, and ICD-10

Chapter 1 Concepts of the Human Body
Disease: 50 y/o Male Physical **Body System Affected:** General/GI

Diagnostic Tests	CPT Codes	ICD-9 Diagnosis Descriptions	ICD-9 Codes	ICD-10 Diagnosis Descriptions	ICD-10 Codes
Physical exam; new patient, age 40–64	99386	General health exam	V70.0	Examination, general adult medical exam, without abnormal findings	Z00.00
Screening colonoscopy	45378	Colonoscopy, screening	V76.51	Screening, digestive tract; lower GI	Z13.811

Chapter 2 Concepts of Chemistry
Disease: Phenylketonuria (PKU) **Body System Affected:** Multiple

Diagnostic Tests	CPT Codes	ICD-9 Diagnosis Descriptions	ICD-9 Codes	ICD-10 Diagnosis Descriptions	ICD-10 Codes
Phenylalanine	84030	Phenylketonuria (PKU)	270.1	Phenylketonuria, classical	E70.0
Phenylketones (qualitative)	84035			Phenylketonuria, maternal	E70.1
				Phenylketonuria, unspecified site	C74.10

Coding Corner

Review and exercises as part of the *Connect Plus* resources introduce students to common codes used for the diseases and disorders presented in each chapter. These include diagnosis descriptions for ICD-9 and the NEW ICD-10 codes, common diagnostic tests for each disease and disorder, and the CPT codes used. These exercises serve as an introduction to future billers and coders or any health care professional who is involved with ensuring that insurance claims are complete and accurate, and that patient care is paid for. You can find the code information on the inside back cover of this text, and the exercises online with *Connect Plus*.

about the authors

Kathryn A. Booth (Kathy) is a registered nurse (RN) with a master's degree in education as well as certifications in phlebotomy, pharmacy technician, and medical assisting. She is an author, educator, and consultant for Total Care Programming, Inc., a multimedia software development company. She has over 30 years of teaching, nursing, and health care work experience that spans five states. Her current focus is to develop up-to-date, dynamic health care educational materials to assist herself and other educators and promote health care professions. Kathy volunteers at a free health care clinic in her hometown of Palm Coast, Florida, and is teaching online with INOVA Health System's Military to Medicine program.

Virgil Stoia received his bachelor of science in experimental biology from the University of Michigan. He received his doctor of chiropractic degree from the National College of Chiropractic and has pursued graduate studies in pathology at the University of Illinois at Chicago. He has taught courses in Anatomy and Physiology, Head and Neck Anatomy, Microbiology, Histology, Pathology, and Medical Terminology. He lives with his wife Sue in Chicago, Illinois, and they are parents of a daughter and have one grandchild with another on the way.

preface

Human anatomy and physiology is wonderfully complex and fascinating. The study of the human body is rewarding on both a personal and professional level. Students will learn material that is not only applicable to their chosen health care profession, but that also will allow them to live more rewarding lives. An understanding of the human body and how it functions will shed light on how diseases occur and what can be done to prevent them.

In writing this book, our aim as authors was to make the material interesting, applicable, and manageable for students who may have little or no experience with anatomy, physiology, or pathophysiology (disease). Our intent was to ensure accurate, up-to-date content that stimulates the learner to want to know more.

From inception to publishing, there were multiple levels of reviews and accuracy checks. During this development process we always kept the needs of instructors and students foremost in our minds. That is why we took extra efforts to get feedback from those individuals who will be using this book. The following efforts were made during the process:

- No less than five levels of rigorous market review and accuracy checking, from first-draft manuscript through the final round of page proofs.
- Reading-level assessments completed throughout the development of the textbook to ensure readability.
- The mapping of every piece of content to specific learning outcomes, giving instructors the option to choose which learning outcomes they want to cover and easily assign only the content that addresses those learning outcomes.
- The creation and incorporation of a brand-new art program, with many new figures enlarged to accommodate visual learners.

Reviewer comments include:

It is important that students have the option of following a real case approach and see how it progresses. Students grasp disease and pathophysiologic concepts much better.

—Dr. Scott M. Sofferman, *Nevada State College*

I like the book because it uses such layman's terms yet gives all the information needed.

—Linda Porterfield, *Fortis College–Cincinnati*

I like the amount of practice that is located at the end of each chapter. The questions allow them much needed review.

—Dr. Gerald Heins, *Northeast Wisconsin Technical College*

Walkthrough

Pedagogical Features

As educators, we need tools to help with our instruction. The following elements are built into this text to facilitate student learning.

Specific Learning Outcomes: Specific learning outcomes give the students concrete expectations. Case studies, illustrations, and tables always attempt to keep the focus on the learning outcomes.

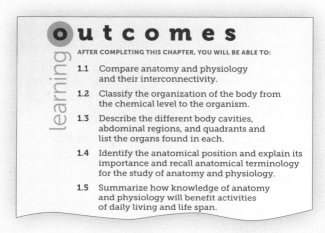

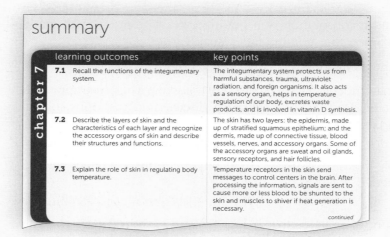

Summary: Chapter summaries that revisit the learning outcomes are in tabular form to make it easier for the students to self-check their understanding. All case studies, questions, and exercises at the end of the chapter further reinforce the learning outcomes.

CASE STUDY 2 *Hypothyroidism*

Tara McDonald, a 28-year-old nurse's assistant, has complained about feeling cold all the time. She also has felt sluggish and is concerned that she has not been able to conceive, although she and her husband have been trying to have a family for some time.

1. Which of the following is *not* typical of hypothyroidism?
   a. Problems with conception
   b. Cold intolerance
   c. Lethargy (sluggishness)
   d. Rapid heart rate
2. Which of the following hormones is *not* secreted by the thyroid gland?
   a. Thymosin
   b. Triiodothyronine

Anatomy, Physiology, and Pathophysiology in Context: The concepts of anatomy, physiology, and pathophysiology are presented in a manner to make the concepts applicable to the real world. Homeostasis is an essential concept that is visited throughout the text. Students learn the connection between health and disease. In addition, through the idea of life span, the effects of aging are looked at in each chapter.

Focus on Clinical Applications: Focus on Clinical Applications makes the connection between anatomy, physiology, and disease concepts and applications in the health professions.

focus on Clinical Applications: Standard Precautions

Health care workers are taught to treat all patients with respect regardless of their status in society, their ability to pay for care, or their diagnosis. At the same time, all health care workers who provide patient contact must do all they can to reduce the spread of potentially contagious diseases and conditions. Standard precautions (formerly referred to as universal precautions) were developed to provide basic protection for both the patient and the caregiver. Standard precautions include the wearing of personal protective equipment (PPE) by caregivers such as gloves, gowns, masks, and goggles when there is the potential for exposure to "moist body substances." These substances include blood, all body fluids and secretions except perspiration, nonintact skin, and mucous membranes. Although breast milk is not included in this list, because it has been shown that HIV can be transmitted from mother to child via breast milk, it is generally treated in the same manner as other body fluids. Standard precautions are an important measure in preventing the transmission of disease in the health care setting and are used in all health care facilities.

focus on Genetics: Recombinant DNA

DNA keeps all of the information needed to recreate an organism. Nearly every cell in the body has some DNA. Most DNA is located in the nucleus of the cell. The infomation in DNA is stored as a code. The code is made up of four different chemical bases: adenine (A), thymine (T), guanine (G), and cytosine (C). The nitrogen bases are found in pairs, with A & T and G & C paired together. The sequence of the bases can be arranged in infinite ways. This sequence and the number of bases is what creates diversity. Each base is also attached to a sugar molecule and a phosphate molecule. Together, a base, sugar, and phosphate are called a nucleotide. Nucleotides are arranged in two long strands that form a spiral called a double helix as seen in Figure 2-4.

   DNA does not actually make the organism, it only makes proteins. The DNA is transcribed into messenger RNA (mRNA). Then mRNA is translated into protein. The protein then forms the organism. By changing the DNA sequence, the way in which the protein is

Focus on Genetics: Focus on Genetics relates genetics to anatomy, physiology, and pathophysiology.

Life Span: The Life Span section concludes each chapter's text discussion and helps students understand the ever-increasing aging patient population.

 learning outcome

Discuss the effects of aging on the muscular system and the steps that can be taken to promote muscle health.

Life Span

As humans age, we lose muscle mass and flexibility. By the time individuals are in their eighties, they have only about half the strength they had in their twenties. Aging causes a decline in the speed and strength of muscle contractions even though the actual endurance of muscle fibers changes very little. Elderly patients often have increasing difficulty with dexterity and gripping ability. Both aerobic exercises and strength training can slow the loss of muscle strength and mobility.

Excretion

Water and salts (electrolytes) are lost through the skin during perspiration. When water loss is excessive, it can be life threatening (see Chapter 6, "Concepts of Fluid, Electrolyte, and Acid–Base Balance"). It also acts to eliminate urea and uric acid as waste products.

U check
The skin is essential to the synthesis of what vitamin?

from the perspective of . . .

A VOCATIONAL OR PRACTICAL NURSE A vocational or practical nurse provides care for patients under the supervision of a physician or registered nurse. How will knowing the functions of the skin help you better treat the patients placed in your care?

From the Perspective of . . . : The From the Perspective of . . . feature applies the content to the skills needed in a related health care profession.

focus on Wellness: Normal Flora

Normal flora is the term used for the microorganisms that inhabit the skin and mucous membranes of healthy individuals. Although some bacteria are considered normal flora, viruses are never considered normal flora. Normal flora function to keep pathogenic organisms from invading and causing disease and may also produce substances needed by our bodies. For example, bacteria present in the gastrointestinal tract are responsible for synthesizing vitamin K. Normal flora may cause disease if they are introduced to "foreign" locations in the body or if they multiply in large numbers. An example is *Neisseria meningitidis* which is normal flora in up to 15 percent of the population. However, under some circumstances the organism can cause meningitis. Normal flora is an example of good bacteria that helps maintain our health and wellness.

Focus on Wellness: Focus on Wellness relates the study of anatomy, physiology, and disease to the learner's personal health.

Pathophysiology: Diseases and disorders common to each system are included in each system chapter to help students understand the relationship of abnormal anatomy and physiology to pathology. Pathophysiology has been expanded in the current edition of the text and makes the anatomy and physiology presented more relevant.

PATHOPHYSIOLOGY

Endocarditis

Endocarditis is an inflammation of the innermost lining of the heart and the heart valves.

Etiology: Bacterial infections are the most common cause of endocarditis. Patients are more susceptible to this condition if they have abnormal heart valves or an underlying heart condition. Intravenous drug abuse and systemic lupus erythematosus (SLE) are two possible causes of endocarditis.

Signs and Symptoms: Common signs and symptoms include weakness, fever, excessive sweating, general body aches, difficulty breathing, and blood in the urine.

Treatment: The treatment addresses the underlying cause. Therefore, if endocarditis is caused by a bacterial infection, the treatment for the condition is intravenous antibiotics followed by oral antibiotics for up to six weeks.

Myocarditis

Myocarditis is an inflammation of the muscular layer of the heart. It is relatively uncommon, but very serious when it does occur because it leads to weakening of the heart wall.

Bone Markings

Every bump, groove, and hole on a bone has a name. These are known as bone markings. There are two main types of bone markings—those that grow out are known as projections, and those that indent the bone are known as depressions. These markings found on bones and the specific terms used to describe skeletal structures are provided in Table 8-1. The table also directs you to the figures throughout the chapter which show these structures.

U check
What is the name for the shaft of long bone?

U check: U check questions are found throughout the text to allow the students to check their learning of the concepts presented.

Study Tips: Study tips are presented at the beginning of each chapter. These tips provide ideas to assist the student while learning the content.

> **study tips**
> 1. Draw a skeleton labeling the axial and appendicular skeletons.
> 2. Make flash cards with the anatomical and common names. For example, write "scapula" on one side of a card and "shoulder blade" on the opposite side.
> 3. Write a sample quiz with answers and page numbers from the text where to find the correct answer.
> 4. You can remember the number of vertebrae in the cervical, thoracic, and lumbar regions by remembering you eat breakfast at 7 (7 cervical vertebrae), lunch at 12 (12 thoracic vertebrae), and dinner at 5 (5 lumbar vertebrae).

> **periosteum** A membrane surrounding the shaft of a bone.

A membrane called the **periosteum** (*peri* = around; *oste* = bone) surrounds the diaphysis. It is made up of dense fibrous connective tissue and has nutrient blood vessels and nerves penetrating it. It is the periosteum that is responsible for "growing pains" during adolescence.

Essential Terms: Essential terms with phonetic pronunciations are presented at the beginning of each chapter. In the chapter, the essential terms are bolded and their definitions are given in the margin. In addition, other key terms are italicized and their definitions and phonetic pronunciations are found in the glossary.

Critical Thinking Questions: Critical thinking questions are included as part of the chapter review questions in every chapter. Students are challenged with critical thinking questions, which avoids mere memorization that often is cumbersome for students.

> **critical thinking questions**
> 1. Andy Finn, a 17-year-old youth, was severely burned while playing around a bonfire with his friends. He was admitted to the emergency department where you are working part time while you get your medical assisting certificate. Andy has second-degree burns on his anterior thorax and third-degree burns on both the front and back of his left leg. Answer the following questions based on the knowledge you have mastered in this chapter.
>   a. Using the rule of nines, estimate the total percentage of Andy's body surface that was affected by this burn.

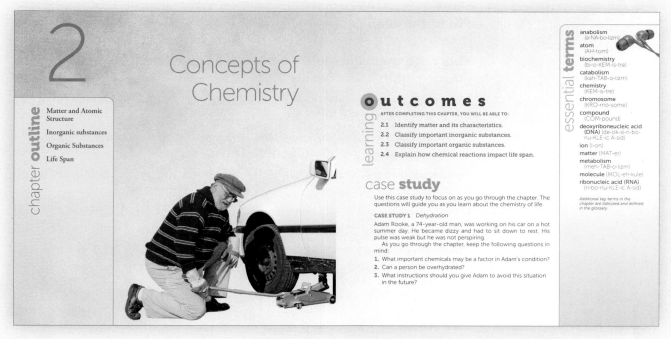

Case Study: Case studies are presented at the beginning of each chapter and are revisited in the text. Appropriate questions about the chapter case study are then included in the review at the end of the chapter.

WALKTHROUGH   *xvii*

Instructor Resources

The Online Learning Center (www.mhhe.com/boothapd3e) that comes included with this text places a host of teaching resources at your disposal:

1. **Instructor's Manual** featuring a lesson plan for each chapter detailing all of the learning features and opportunities for that chapter, lecture outlines, talking points, group activities, discussion topics, figure numbers, and written assignments for every learning outcome. This IM also includes one to three quizzes per chapter, and answer keys for all questions and activities in the text and workbook.
2. **EZTest Online** (www.eztestonline.com) containing nearly 1,500 test questions correlated to chapter learning outcomes, level of difficulty, Revised Bloom's Taxonomy, and relevant CAAHEP/ABHES competencies.
3. **PowerPoint presentations** with images and photos and instructor talking points.
4. **Image Library** containing all art and photos in the text. Every labeled figure is available both with and without labels filled in.

Workbook

A full-color workbook is available for purchase for additional practice. This workbook includes the following features to reinforce the lessons that students learn in each chapter of the text.

Vocabulary Review Matching matches key terms from the text to the correct definitions.

Multiple Choice questions review the content of the chapter by completing the statement or answering the question with the correct choice.

Fill in the Blanks allows students to write the word or phrase that best completes each sentence as another form of content review.

Short Answer questions give students the chance to answer the questions with brief statements.

Labeling provides one or more illustrations on which students can apply the correct labels for further content review.

Critical Thinking questions allow students to apply their knowledge to answer questions with brief statements.

Applications ask students to follow directions and apply their knowledge accordingly.

Case Studies give students time to consider a brief case study and answer questions using the knowledge they've learned from the chapter.

Pathophysiology section provides a variety of disease-related questions for students to answer.

acknowledgments

For this latest edition, vital feedback was received from faculty and students throughout the country, and was relied on to enhance and improve this book and ancillary products. Each person who has offered assistance, comments, and suggestions has our heartfelt thanks. Among these people are the reviewers and consultants who pointed out areas of concern, cited areas of strength, and made recommendations for change. In this regard, the reviewers listed below provided feedback that was enormously helpful in preparing this book. Of course, many others deserve heartfelt thanks as well because once all the information is compiled it takes a lot more people to make it happen. Recognition should go to McGraw-Hill including Ken Kasee, Christine Scheid, and Susan Trentacosti; Amber Allen, Ingrid Benson, Carol Field, and Eric Arima at Integra–Chicago for their development services; and our fellow colleagues who helped us with our ancillary and Connect materials, including Jill Tall of Youngstown State University, Eric Bourassa of Northwest Missouri State University, Jan Sesser of Anthem Education Group, Eldis Rivera of Sanford Brown Institute, and Paula Bostwick. Also, a special thanks to Jody James, Lynn Egler, and Terri D. Wyman, CPC, CMRS who were all of great assistance during the creation and development of this project.

Kathy

In addition to all those at McGraw-Hill that were very helpful and always available to guide the process of making this book a reality, I would like to thank my wife Sue for her help and patience during this exciting adventure and Dr. Bill Militello for his continuous encouragement.

Virgil

Reviewers

Diana Alagna
Stone Academy

Khaliff Ali
Fortis College

Tamra J. Ashley, RN, MSN, CFNP, IBCLC, RMA
Davenport University

Wilfredo Barreto
Florida Technical College

Barry N. Bates
Atlanta Technical College

Stephanie Bernard
Sanford-Brown Institute

Drew Case, RN, MSN, APRN
Southeast Community College–Lincoln

Bonnie J. Crist, MHeD, CMA (AAMA)
Harrison College–Indianapolis

George M. Dalich, PhD
Lake Washington Technical College

Melissa DeMayo
Salter College

Sharon J. Fugate
Madisonville Community College

Souzan Habashy, MD
Keiser University

Monica L. Hall-Woods, PhD
Saint Charles Community College

Jessica Hanzel
Herzing University

Gerald Heins, DC
Northeast Wisconsin Technical College

Kebret T. Kebede, MD
Nevada State College

Anne LaGrange Loving
Passaic County Community College

Terry E. Lancaster
East Tennessee State University

Beth Laurenz, MBA, BS, AAS, CMA (AAMA)
National College

Penny Lee
Medtech College

Angela LeuVoy
Fortis College–Columbus

Melyssa Munch
Star Career Academy

Fred R. Pearson
Brigham Young University–Idaho

Linda Porterfield
Fortis College–Cincinnati

Wanda F. Ragland
Macomb Community College

Jana Rybolt
Harrison College–Fort Wayne

Scott M. Sofferman
Nevada State College

Pennie Tilkens
Northeast Wisconsin Technical College

Rao Veeramachaneni
Colorado State University

Vanja Velickovska
Nevada State College

Terri Wusterbarth
Northeast Wisconsin Technical College

Fadi Zaher
Gateway Technical College

Lisa Zepeda, MS
San Joaquin Valley College–Visalia

Susan Zolvinski
Brown Mackie College–Michigan City

contents

Unit 1: The Human Body and Disease

1 Concepts of the Human Body 2

Introduction 4
The Study of the Human Body 5
Organization of the Human Body 7
Body Cavities, Regions, and Quadrants 9
Anatomical Terminology 12
Life Span 16

2 Concepts of Chemistry 18

Introduction 20
Matter and Atomic Structure 21
Inorganic Substances 22
Organic Substances 24
Life Span 26

3 Concepts of Cells and Tissues 28

Introduction 30
Cell Components 31
Cell Transport 35
The Cell Cycle 37
Stem Cells 39
Cell Death 40
Tissue Types 40
Life Span 49

4 Concepts of Disease 52

Introduction 54
Cell Injury and Death 54
The Body's Response to Injury 57
Immunopathology 58
Neoplasia 59
Developmental and Genetic Diseases 60
Hemodynamic Disorders 61
Environmental and Nutritional Pathology 63
Infectious and Parasitic Diseases 66
Disease Diagnosis and Prognosis 66
Prevention and Treatment of Diseases 68
Life Span 69

5 Concepts of Microbiology 74

Introduction 76
Microscopes 76
Bacterial Cell Structure 78
Bacteria Classification, Growth, and Reproduction 80
Shapes and Arrangements of Bacteria 82
Pathogenicity and Virulence 84
Bacteria and Disease 84
Virology 89
Mycology 90
Parasitology 92
Sexually Transmitted Infections 94
Life Span 99

6 Concepts of Fluid, Electrolyte, and Acid–Base Balance 104

Introduction 106
Fluid Compartments 106
Regulation of Water 107
Electrolytes 110
Acid–Base Balance 116
Acid–Base Imbalances 118
Life Span 118

xxii CONTENTS

Unit 2: Concepts of Common Illness by System

7 The Integumentary System 124

Introduction 126
Functions of the Integumentary System 126
Structures of the Integumentary System 128
Regulation of Body Temperature 132
Skin Color 133
Skin Wounds and Healing 134
Skin Lesions and Disorders 135
Burns 141
Skin Cancer 144
Life Span 146

8 The Skeletal System 152

Introduction 154
Functions of the Skeletal System 154
Bone Structure 155
Bone Growth 157
Bone Classification, Parts, and Markings 158
Axial Skeleton 162
Appendicular Skeleton 173
Joints 178
Bone Fractures 180
Life Span 187

9 The Muscular System 192

Introduction 194
Types and Functions of Muscle 194
Structure and Classification of Muscle 196
Energy Production for Muscle and Muscle Contraction 198
Attachments and Actions of Skeletal Muscles 200
Body Movements 201
Major Skeletal Muscles 207
Strains and Sprains 219
Life Span 222

10 Blood and Circulation 226

Introduction 228
Hematopoiesis 228
Plasma 230
Formed Elements 232
Hemodynamics 241
Hemostasis 244
Blood Typing 247
Blood Vessels 250
Pulmonary and Systemic Circulation 255
Arterial and Venous Systems 256
Life Span 265

11 The Cardiovascular System 270

Introduction 272
Heart Anatomy 272
The Cardiac Cycle 281
Heart Sounds 283
Cardiac Output 285
The Conduction System 286
Electrocardiogram 287
Chest Pain 291
Life Span 296

12 The Lymphatic and Immune Systems 300

Introduction 302
Lymphatic System Components and Functions 302
Disease Defenses 310
Immune System Responses and Acquired (Specific) Immunities 318
Transplantation and Tissue Rejection 319
Major Immune System Disorders 320
Life Span 326

13 The Respiratory System 330

Introduction *332*
Respiratory System Components and Functions *332*
Respiration *358*
Respiratory Volumes and Capacities *361*
Factors That Control Breathing *362*
Gas Exchange *364*
Gas Transport *366*
Life Span *367*

14 The Nervous System 372

Introduction *374*
Functions of the Nervous System *374*
Cells of the Nervous System *375*
The Synapse and Nerve Transmission *381*
Central Nervous System Components and Their Functions *384*
Meninges, Ventricles, and Cerebrospinal Fluid *400*
Peripheral Nervous System Components and Their Functions *404*
Neurological Testing *420*
Life Span *421*

15 The Urinary System 426

Introduction *428*
Kidneys *429*
Urine Formation *439*
Urine Elimination *441*
Life Span *446*

16 The Male Reproductive System 450

Introduction *452*
External Male Reproductive Structures *452*
Internal Male Reproductive Structures *456*
Male Reproductive Hormones *460*
Spermatogenesis *461*
Erection, Orgasm, and Ejaculation *464*
Life Span *465*

17 The Female Reproductive System 470

- Introduction 472
- External Female Reproductive Structures 473
- Internal Female Reproductive Structures 477
- Female Reproductive Hormones 484
- Female Reproductive Cycle 485
- Fertilization 490
- Birth Control and Infertility 490
- Life Span 493

18 Human Development and Genetics 498

- Introduction 500
- Human Development and Inheritance 500
- Hormonal Changes During Pregnancy 512
- The Birth Process 512
- The Postnatal Period 514
- Human Genetics 515
- Life Span 521

19 The Digestive System 524

- Introduction 526
- Overview of Digestive System Functions 526
- Digestive System Organs and Their Functions 528
- Accessory Organs and Their Functions 546
- Phases of Digestion 553
- Life Span 553

20 Metabolic Function and Nutrition 558

- Introduction 560
- Carbohydrate Metabolism 561
- Lipid Metabolism 565
- Protein Metabolism 567
- Homeostasis 570
- Nutrition 570
- Malnutrition 580
- Life Span 582

21 The Endocrine System 586

Introduction 588
Major Endocrine Glands and Hormones 589
Additional Endocrine Glands and Tissues 609
Regulatory Mechanisms 609
Life Span 610

22 The Special Senses 614

Introduction 616
Olfaction 616
Taste 618
Vision 620
Hearing and Equilibrium 641
Life Span 651

Appendixes

I   Diseases and Disorders 656
II Prefixes, Suffixes, and Word Roots in Commonly Used Medical Terms 674
III Abbreviations and Symbols Commonly Used in Medical Notations 677

Glossary 680
Credits 710
Index 718

ANATOMY, PHYSIOLOGY & DISEASE

FOR THE HEALTH PROFESSIONS

UNIT 1 The Human Body and Disease

1 Concepts of the Human Body

chapter outline

The Study of the Human Body

Organization of the Human Body

Body Cavities, Regions, and Quadrants

Anatomical Terminology

Life Span

learning outcomes

AFTER COMPLETING THIS CHAPTER, YOU WILL BE ABLE TO:

1.1 Compare anatomy and physiology and their interconnectivity.

1.2 Classify the organization of the body from the chemical level to the organism.

1.3 Describe the different body cavities, abdominal regions, and quadrants and list the organs found in each.

1.4 Identify the anatomical position and explain its importance and recall anatomical terminology for the study of anatomy and physiology.

1.5 Summarize how knowledge of anatomy and physiology will benefit activities of daily living and life span.

essential terms

anatomical (ana-TOM-ical) **position**
anterior (an-TER-e-or)
caudal (KAW-dal)
cranial (KRAY-nee-al)
deep
distal (DISS-tal)
dorsal (DOOR-sal)
homeostasis (ho-me-o-STA-sis)
lateral (LAT-er-al)
medial (MEE-dee-al)
oblique (o-BLEK)
posterior (pos-TER-e-or)
proximal (PROX-im-al)
sagittal (SAJ-it-al)
superficial (soop-er-FISH-al)
tissue (TISH-oo)
transverse (trans-VERSE)
ventral (VEN-tral)

Additional key terms in the chapter are italicized and defined in the glossary.

case study

Use the case study to focus on as you go through the chapter. The questions will guide you as you learn anatomy and physiology and understand the pathology associated with each body system.

CASE STUDY 1 *Aspiring Medical Assistant*

Ellen Besler is an aspiring medical assistant as well as a 38-year-old wife and mother of two. She has always had an interest in medicine. She would love to become a medical assistant and help others. Ellen has been encouraged by her family to pursue her dream. She really does not know much about anatomy or physiology. She wonders why she has to take the course, and what she can do to increase her chances of success.

As you go through the chapter, keep the following questions in mind:

1. What is anatomy? And what are the different categories of anatomy?
2. What is physiology?
3. How are physiology and anatomy interconnected?
4. What study techniques can you suggest to help Ellen succeed?

study tips

1. Choose a quiet place that has everything you need to study effectively: textbooks, reference books, paper, pencils, note cards, and a computer.

2. Write out a schedule for each week and chapter. Be specific. Most authorities say for every hour of class time you should be spending two to three hours of study time. Do not try to do too much at one time. Divide your time into manageable units and include scheduled breaks.

3. Preview, read, and review the chapter.

4. Make flash cards for the essential terms of the chapter.

5. Look at illustrations and tables and read the captions.

6. Outline the chapter. After each section, ask yourself what you just read.

7. Write down one to three questions to ask your instructor.

8. Meet with a study group and review and quiz each other.

9. Answer the questions at the end of the chapter.

10. What other ideas do you have that will help you be the best anatomist and physiologist you can?

Introduction

The study of anatomy and physiology is one of the most fascinating topics you can undertake. The human body is a complex machine; it is amazing how so many different cells, tissues, and organs, each with a specific purpose, work together to produce a highly efficient organism. Your study will begin at the chemical level and work all the way up to the organism (Figure 1-1). You will also be learning a new language—the language of science and the body. Whether you are learning anatomy and physiology because it is required for the profession you have chosen or strictly for personal reasons, we are confident you will be delighted with the knowledge and understanding you will achieve.

U check
Why have you chosen to study anatomy and physiology?

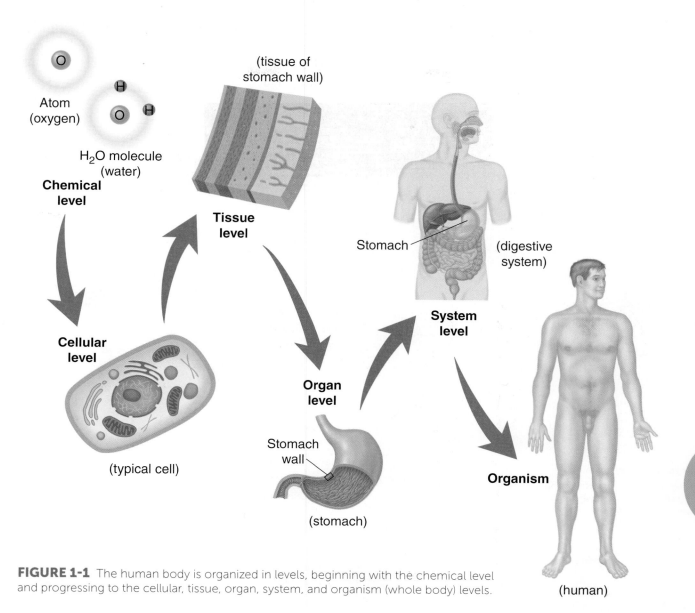

FIGURE 1-1 The human body is organized in levels, beginning with the chemical level and progressing to the cellular, tissue, organ, system, and organism (whole body) levels.

The Study of the Human Body

Anatomy (*ana* = up; *tomy* = to cut) is the science of the study of body structures. There are several branches of anatomy. Gross anatomy is the study of the body at a macroscopic level (unaided by a microscope). Dissection is often used for this study. In this textbook, we will devote our studies to gross anatomy. Physiology is the study of the function of the body's organs. We study anatomy and physiology together because they are interconnected. Essentially the shape or structure of cells, tissues, organs, and the organism (the person) as a whole will determine the function. By having a thorough understanding of normal anatomy and physiology, it is much easier to recognize abnormal situations when we encounter them. This knowledge will help you grasp the meaning of diagnostic and procedural codes if you are going into billing. It can also help you understand the clinical procedures you will perform as a medical assistant, x-ray technician, nurse, or other health care professional. It will be easier to see how and why certain diseases develop.

1.1 learning outcome

Compare anatomy and physiology and their interconnectivity.

homeostasis Relative consistency of the body's internal environment.

Homeostasis is defined as the relative consistency of the body's internal environment. Body conditions that must remain within a stable range include body temperature, blood pressure, and the concentration of various chemicals within the blood. Individual cells must also maintain homeostasis. We will learn how the different organ systems work together to maintain homeostasis and help us remain healthy (Figure 1-2).

> **U check**
> What is the effect on the organism when homeostasis is disrupted?

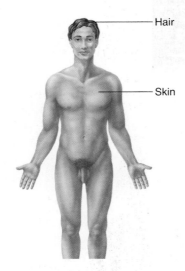

Integumentary System

Serves as a sense organ for the body, provides protection, regulates temperature, prevents water loss, and produces vitamin D precursors. Consists of skin, hair, nails, and sweat glands.

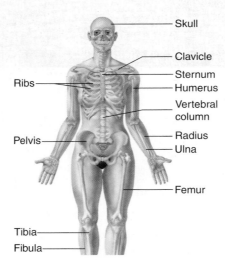

Skeletal System

Provides protection and support, allows body movements, produces blood cells, and stores minerals and fat. Consists of bones, associated cartilages, ligaments, and joints.

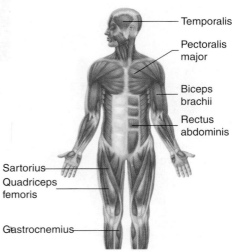

Muscular System

Produces body movements, maintains posture, and produces body heat. Consists of muscles attached to the skeleton by tendons.

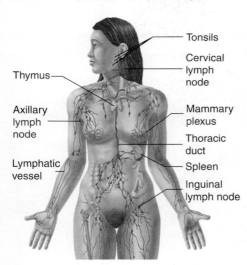

Lymphatic System

Removes foreign substances from the blood and lymph, combats disease, maintains tissue fluid balance, and absorbs fats from the digestive tract. Consists of the lymphatic vessels, lymph nodes, and other lymphatic organs.

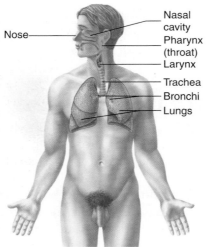

Respiratory System

Exchanges oxygen and carbon dioxide between the blood and air and regulates blood pH. Consists of the lungs and respiratory passages.

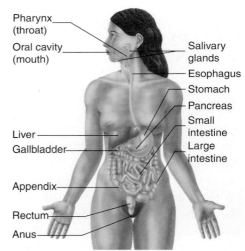

Digestive System

Performs the mechanical and chemical processes of digestion, absorption of nutrients, and elimination of wastes. Consists of the mouth, esophagus, stomach, intestines, and accessory organs.

FIGURE 1-2 Organ systems of the body.

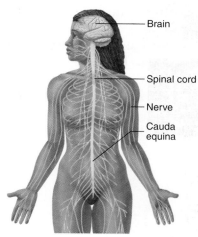

Nervous System
A major regulatory system that detects sensations and controls movements, physiologic processes, and intellectual functions. Consists of the brain, spinal cord, nerves, and sensory receptors.

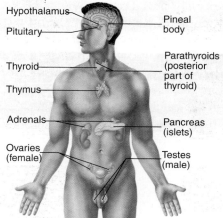

Endocrine System
A major regulatory system that influences metabolism, growth, reproduction, and many other functions. Consists of glands, such as the pituitary, that secrete hormones.

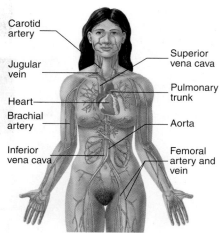

Cardiovascular System
Transports nutrients, waste products, gases, and hormones throughout the body; plays a role in the immune response and the regulation of body temperature. Consists of the heart, blood vessels, and blood.

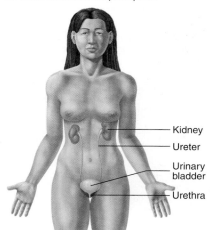

Urinary System
Removes waste products from the blood and regulates blood pH, ion balance, and water balance. Consists of the kidneys, urinary bladder, and ducts that carry urine.

FIGURE 1-2 (concluded)

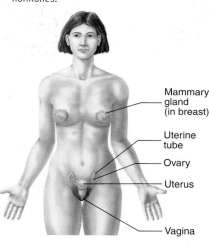

Female Reproductive System
Produces oocytes and is the site of fertilization and fetal development; produces milk for the newborn; produces hormones that influence sexual function and behaviors. Consists of the ovaries, vagina, uterus, mammary glands, and associated structures.

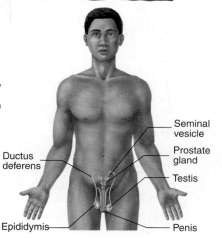

Male Reproductive System
Produces and transfers sperm cells to the female and produces hormones that influence sexual functions and behaviors. Consists of the testes, accessory structures, ducts, and penis.

Organization of the Human Body

The structure of the body can be divided into six different levels of organization with increasing complexity: chemical, cellular, tissue, organ, organ system, and organism (whole body).

The chemical level is the most basic level and is made up of atoms and molecules (two or more atoms joined together by chemical bonds). *Atoms* are the simplest units of all matter. *Matter* is anything that takes up space and has weight. Certain elements or atoms are required for life. These include carbon (C), hydrogen (H), oxygen (O), nitrogen (N), phosphorus (P), calcium (Ca), and sulfur (S). These essential atoms combine to make the essential molecules needed for life including water, glucose, proteins, and

1.2 learning outcome

Classify the organization of the body from the chemical level to the organism.

Basic Chemistry (Organic Molecules)

tissue A group of similar cells that combine to perform a specific function.

deoxyribonucleic acid (DNA). We will look at the chemistry of life more closely in Chapter 2, Concepts of Chemistry.

The next level of organization, the basic structural and functional unit of life, is the cell. Individual cells require a microscope to be looked at closely. When cells act together to perform a specific function, the next level of organization, they are classified as **tissue.** The four basic types of tissues in the body are epithelial, connective, muscle, and nervous tissue (Figure 1-3). Chapter 3, Concepts of Cells and Tissues, will introduce you to the world of organelles, cells, and tissues.

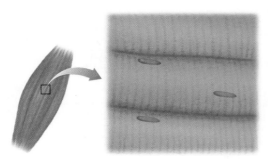

A Skeletal muscle: Elongated cylindrical cells with striations and several nuclei

Locations: Throughout the body where voluntary movement takes place

Functions: Movement of body parts such as the extremities, head and neck, and spine

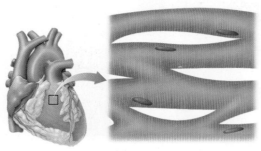

B Cardiac muscle: Short, branching cells with striations and a single nucleus; has intercalated discs between cells for intercellular communication

Location: Heart

Function: Contraction of the heart for blood circulation

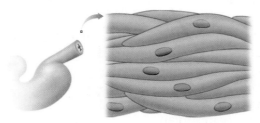

C Smooth muscle: Short tapered cells, not striated with a single nucleus

Locations: Walls of blood vessels and walls of hollow organs such as the stomach and uterus

Functions: Maintains blood vessel diameter; controls movement of food through the digestive tract, as well as urine in the urinary system and the egg and sperm in the reproductive tract

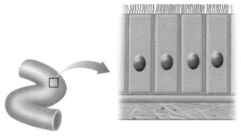

D Epithelium: Cells can be squamous (flat), cuboidal, columnar, or transitional; cells can be arranged in a single layer (simple), stratified, or pseudostratified

Locations: Epithelium lines body surfaces and cavities; for example, it makes up the skin and lining of the digestive tract

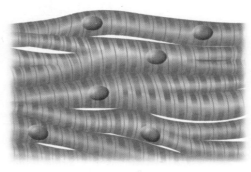

E Connective tissue: The most abundant and varied of the four tissue types; consists of cells and extracellular matrix

Locations: Throughout the body; bone, cartilage, blood, and collagen are examples of connective tissue

Functions: Movement, storage of minerals, transport of oxygen and carbon dioxide, a source of energy, protection, and support

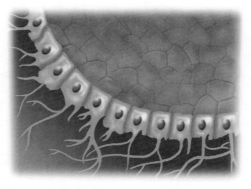

F Nervous tissue: Consists of neurons and neuroglia (supporting cells)

Locations: Brain, spinal cord, and nerves

Functions: Receives, integrates, and responds to various internal and external stimuli

FIGURE 1-3 Four basic tissue types.

> **focus on Wellness:** The Human Body
>
> The human body is an amazing machine. Each organ system works in sync with other organ systems. Each level of the body organization is built on simpler levels of organization. Understanding how the body works will provide you with a better understanding of how to keep it working and maintain the most optimal state of health and wellness.

Two or more tissue types combine to form organs, and organs that perform a common function are called organ systems—the next two organizational levels of the human body. For example, the heart is made of cardiac muscle tissue, connective tissue, and epithelial tissue. The heart and blood vessels unite to form the cardiovascular system. The function of the cardiovascular system is to supply oxygen and nutrients to the cells and tissues of the body and remove carbon dioxide and waste products. The organism consists of all the organ systems working together to function as a unit or living individual.

> **U check**
> What are the four basic tissue types?

Body Cavities, Regions, and Quadrants

Body cavities are spaces that help protect and support organs (Figure 1-4). Two major body cavities are the **dorsal** cavity, located on the posterior aspect of the body, and the **ventral** cavity, located on the front of the body. The dorsal cavity is divided into the **cranial** cavity and the *spinal cavity (vertebral canal)*. The cranial cavity contains the brain and the spinal cavity contains the spinal cord. The organs of the dorsal cavity are well protected because of the skull and the vertebral column. The ventral cavity is divided into the *thoracic cavity* (chest cavity) and the *abdominopelvic cavity*. The *diaphragm* separates the thoracic and abdominopelvic cavities. The thoracic cavity is divided into two pleural cavities, pericardial cavity, and mediastinum. The lungs are found in the *pleural cavities*. The heart is located in the *pericardial cavity* which is found in the mediastinum. The *mediastinum* is a space located between the two lungs laterally and the sternum anterior and the vertebral column posterior. It runs from the first rib superiorly to the diaphragm inferiorly. The abdominopelvic cavity is divided into a superior abdominal cavity and an inferior pelvic cavity. The stomach, small and large intestines, gallbladder, liver, spleen, kidneys, and pancreas are located in the abdominal cavity. The bladder and internal reproductive organs are located in the pelvic cavity. The body cavities with the exception of the abdominopelvic cavity provide protection to the internal organs because of the surrounding skeletal structures. Also, the cavities are lined with protective connective tissue membranes and small amounts of lubricating fluids.

1.3 learning outcome

Describe the different body cavities, abdominal regions, and quadrants and list the organs found in each.

dorsal Toward the back of the body.

ventral Toward the front of the body.

cranial Above or close to the head.

CHAPTER 1 Concepts of the Human Body

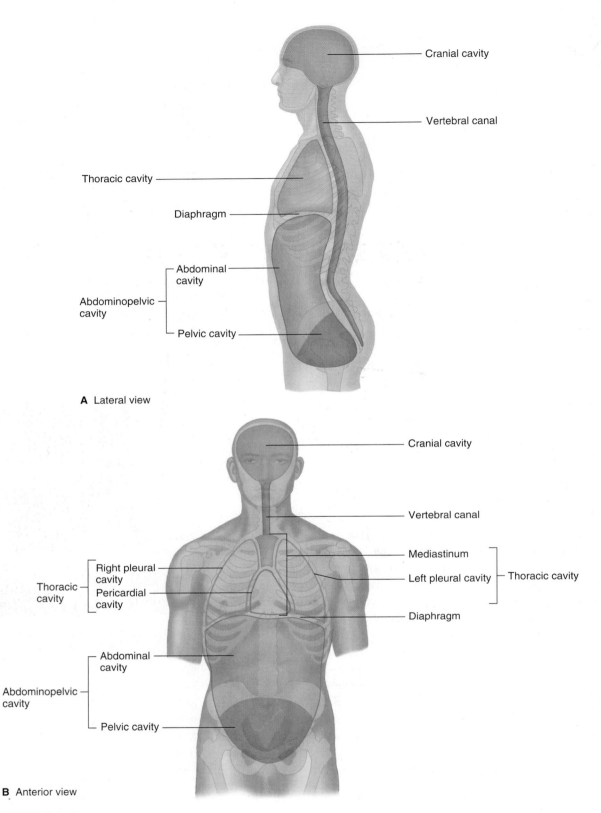

FIGURE 1-4 Major body cavities.

The abdominal area is further divided into nine regions or four quadrants (Figure 1-5). The regions and quadrants help physicians and other medical personnel such as x-ray technicians and medical assistants easily locate organs and vital structures (Figure 1-6). When using the nine-region method,

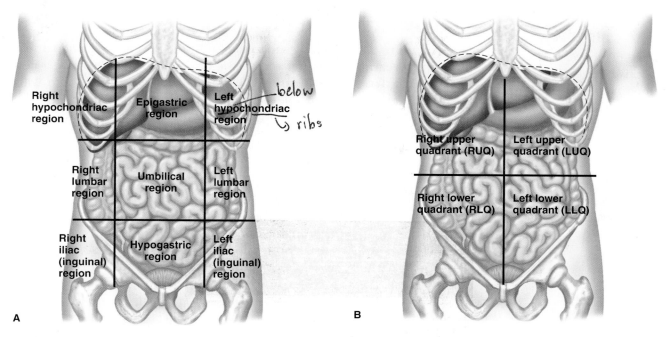

FIGURE 1-5 (A) The abdominal area divided into nine regions and (B) the abdominal area divided into four quadrants.

a tic-tac-toe grid is drawn. Two horizontal lines and two vertical lines are drawn. The top horizontal line is drawn just below the rib cage and the lower horizontal line is drawn just below the tops of the hip bones. The two vertical lines are drawn just medial to the nipples through the middle of the clavicles. The nine regions are named right hypochondriac, epigastric, left hypochondriac, and right lumbar, umbilical, left lumbar, right inguinal, hypogastric, and left inguinal. A simpler method of locating structures is the quadrant method. To form the quadrants, a horizontal line and a vertical line are drawn through the umbilicus. The four quadrants are named the right upper quadrant, left upper quadrant, right lower quadrant, and left lower quadrant.

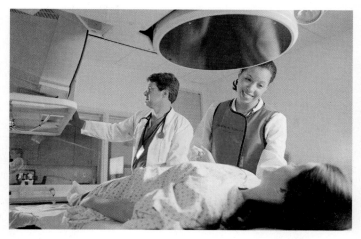

FIGURE 1-6 Knowledge of anatomy is essential to x-ray technicians.

U check
List all the cavities that the heart is found in.

from the perspective of...

AN ADMINISTRATIVE MEDICAL ASSISTANT An administrative medical assistant updates and maintains patients' medical records, fills out insurance forms, and arranges for hospital admissions and laboratory services as well as other duties depending on where he or she works. How will learning the different directional terms help you communicate more effectively with other health care providers?

1.4 learning outcome

Identify the anatomical position and explain its importance and recall anatomical terminology for the study of anatomy and physiology.

anatomical position The body is standing upright, facing forward, with the arms at the sides and the palms of the hands also facing forward.

sagittal A plane that divides the body into left and right portions.

transverse A plane that is also described as horizontal and divides the body into upper and lower portions.

FIGURE 1-7
Anatomical position: The individual is facing forward with the arms at the sides and the palms of the hands facing forward as well.

REMEMBER ELLEN, our aspiring medical assistant? Why do you think it is important for Ellen to have a thorough understanding of the anatomical position?

Anatomical Terminology

Anatomical terms are used to describe the location of body parts and various body regions. Learning anatomy and physiology terms is like learning a new language—one that all health care professionals must speak. To begin with, you must understand the concept of **anatomical position** (Figure 1-7). This is described as a body standing upright and facing forward with the arms at the sides and the palms of the hands facing forward. A person lying on the stomach is said to be in the *prone* position, and someone lying on the back is in the *supine* position.

Planes and Sections

Flat surfaces or planes passing through the body are useful in identifying structures. There are several planes or sections you should become familiar with (Figure 1-8). Medical professionals often use the following terms to describe how the body is divided into sections: sagittal, transverse, and frontal (coronal). A **sagittal** plane divides the body into left and right portions. A *midsagittal* plane runs lengthwise down the midline of the body and divides it into equal left and right halves. A **transverse** (horizontal) plane divides the body into *superior* (upper) and *inferior* (lower)

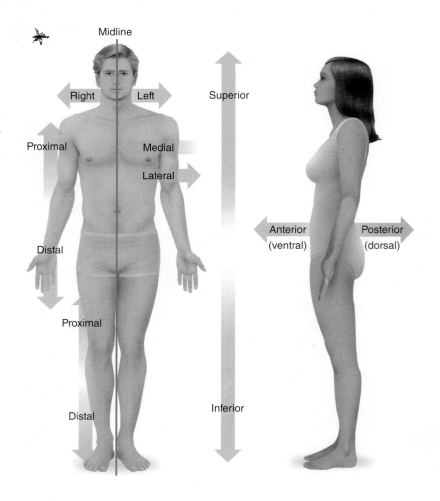

12   UNIT 1   The Human Body and Disease

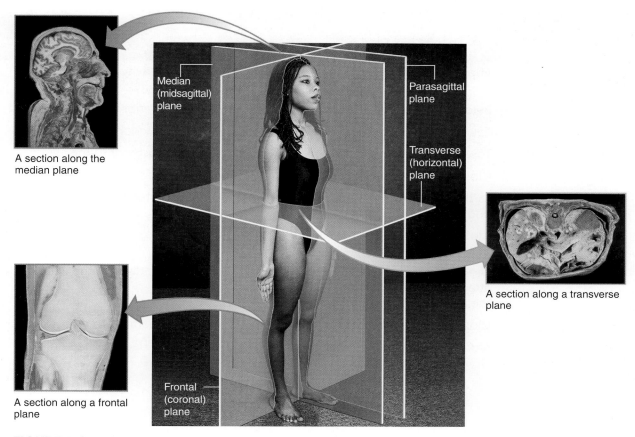

FIGURE 1-8 Sectioning the body along various planes allows observation and identification of internal structures.

portions. A *frontal,* or *coronal,* plane divides the body into **anterior** (frontal) and **posterior** (rear) portions. Although not a main body plane, an **oblique** plane is one that runs at an angle other than perpendicular to a sagittal, horizontal, or coronal plane. It is used in radiology, describing x-ray views.

> **U check**
> List the planes that would run perpendicular to a sagittal plane.

Directional Anatomical Terms

Directional anatomical terms are used to identify the position of body structures relative to other body structures (Figure 1-9 and Table 1-1). The directional anatomical terms are cranial, cephalad, caudal, inferior, ventral, anterior, dorsal, posterior, medial, lateral, proximal, distal, superficial, and deep. A structure that is described as *cranial* would be approaching or close to the head or skull. Cephalad and superior are often used interchangeably with cranial. **Caudal** or inferior refers to away from the head. As stated previously, *ventral* or *anterior* means the front of the body and *dorsal* or *posterior,* the back. **Medial** refers to being comparatively closer to the midline of the body. **Lateral** refers to relatively farther from the midline. For example, the eyes are medial to the ears but lateral to the nose. **Proximal** refers to a structure

anterior Toward the front of the body when in anatomical position or in front of another structure.

posterior Toward the back of the body when in anatomical position or in behind another structure.

oblique At an angle other than perpendicular to a sagittal, horizontal, or coronal plane.

caudal Away from the head.

medial Near the midline of the body.

lateral Away from the midline of the body.

proximal Nearer to the attachment of an extremity to the trunk or nearer to the point of attachment or origin.

TABLE 1-1 Directional Anatomical Terms

Term	Definition	Example
Superior (cranial or cephalad)	Above or close to the head	The thoracic cavity is superior to the abdominal cavity.
Inferior (caudal)	Below or close to the feet	The neck is inferior to the head.
Anterior (ventral)	Toward the front of the body	The nose is anterior to the ears.
Posterior (dorsal)	Toward the back of the body	The brain is posterior to the eyes.
Medial	Close to the midline of the body	The nose is medial to the ear.
Lateral	Farther away from the midline of the body	The ears are lateral to the nose.
Proximal	Close to a point of attachment or to the trunk of the body	The knee is proximal to the ankle.
Distal	Farther away from a point of attachment or from the trunk of the body	The fingers are distal to the wrist.
Superficial	Close to the surface of the body	The skin is superficial to muscle.
Deep	More internal	The bones are deep to the skin.

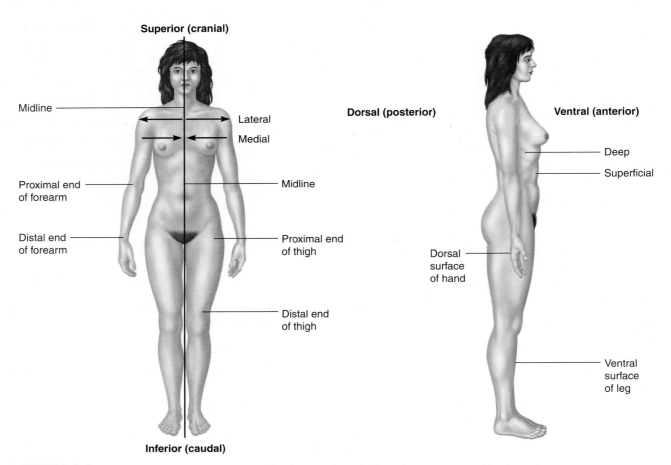

FIGURE 1-9 Directional terms provide mapping instructions for locating organs and body parts.

being closer to the trunk or a specified part. **Distal** is farther from the trunk or specified part. For example, the wrist is more proximal to the elbow than are the fingers. The fingers would be more distal to the elbow than the wrist. **Superficial** is closer to the surface of the body. For example, the skin is more superficial than the heart. **Deep** would be farther from the surface of the body.

distal Farther from the attachment of an extremity to the trunk or farther from the point of attachment or origin.

superficial Located on or near the surface of the body or organ.

deep Away from the surface of the body or organ.

> **U check**
> Is the right shoulder or right ankle more proximal to the right knee?

Anatomical Terms Used to Describe Body Parts

Many other anatomical terms are used to describe different regions or parts of the body (Figure 1-10). For example, the term *brachium* refers to the arm and the term *femoral* refers to the thigh.

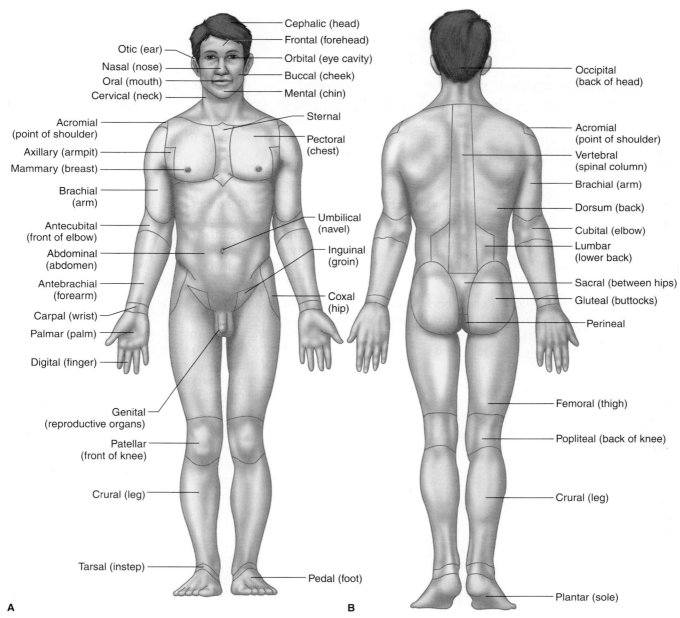

FIGURE 1-10 Numerous anatomical terms are used to describe regions of the body: (A) anterior view and (B) posterior view.

CHAPTER 1 Concepts of the Human Body

1.5 learning outcome

Summarize how knowledge of anatomy and physiology will benefit activities of daily living and life span.

Life Span

From the moment we are conceived we start to age. As early as our twenties and thirties we may start to see subtle changes. We may not have the endurance we did when we were in our teens or early twenties. Skin changes occur with loss of elasticity, loss of subcutaneous fat, and presence of "age spots." Metabolism also starts to slow down, which may affect diet as well as drug metabolism. We also become more susceptible to cancers as our body is not as quick to repair DNA damage. By becoming knowledgeable of anatomy and physiology, you will gain valuable insights into how you can assist your patients in living longer, healthier lives.

Let's take a moment to see how Ellen, our medical assistant student, is doing. Do you think she now sees the importance of learning anatomy and physiology? What are your thoughts?

summary

chapter 1

learning outcomes	key points
1.1 Compare anatomy and physiology and their interconnecti...	Anatomy is the study of structures of the body... the study of the function of... each has a dramatic impact... at is why we often study...
1.2 Classify the organi... the chemical level...	...animals, are classified by... mplex levels of organization. ...hemical level and becomes ...e move through the ...n, organ system, and finally ...e body) level.
1.3 Describe the differ... abdominal regions... the organs found i...	...ivided into a thoracic and ...ity. The dorsal cavity is divided ...nal cavity. You should be able ...ns and structures found in ...s. The abdominal area can be ...ions or four quadrants. This ...cation of structures.
1.4 Identify the anatomical position and explain its importance and recall anatomical terminology for the study of anatomy and physiology.	In the anatomical position, the individual is standing upright, facing forward with the arms at the sides and the palms also facing forward. This is an important concept to prevent misunderstanding when describing body structures. Knowing the correct terminology will allow you to communicate effectively with other health care providers.
1.5 Summarize how knowledge of anatomy and physiology will benefit activities of daily living and life span.	As soon as we are born we start to age. Endurance decreases, skin loses elasticity, and metabolism changes. We become more susceptible to cancers and other illnesses. Understanding normal anatomy and physiology will allow you to understand your own body and its needs as well as quickly recognize and appreciate pathology when it is encountered in your patients.

Handwritten note:

Issues that arise w/ old age:
1) low endurance
2) loss of skin elasticity
3) loss of subcutaneous fat
4) age spots
5) metabolism slows down
6) more susceptable to cancers
7) DNA does not repair as quickly.

chapter 1 review

case study 1 questions
Can you answer the following questions that pertain to Ellen's case study presented earlier in this chapter?
1. How are anatomy and physiology related to each other?
2. Why is it important to understand normal anatomy and physiology before learning what is abnormal?
3. What would you recommend to help make learning effective and fun for Ellen?

review questions
1. The heart is located in which of the following body cavities (there may be more than one correct answer)?
    a. Dorsal cavity   b. Cranial cavity   c. Ventral cavity   d. Thoracic cavity
2. Which level of organization is immediately above the cellular level?
    a. Chemical   b. Tissue   c. Organism (whole body)   d. Organ
3. Which one of the following statements is correct?
    a. The skin is superficial to the wrist.   c. The spine is posterior to the abdomen.
    b. The head is ventral to the knee.   d. The ankle is proximal to the abdomen.
4. Which one of the following is *not* a basic tissue type?
    a. Epithelial   b. Muscle   c. Nervous   d. Connective   e. Bone

critical thinking questions
1. Discuss how a problem at the cellular level would impact homeostasis of the organism.
2. Discuss the organs you would find in the various ventral cavities of the human body.
3. Discuss the organs you would find in the right upper quadrant.

patient education
You are asked to give a short presentation to a class of high school juniors on the benefits of learning human anatomy and physiology. What would be the main points of your talk? Include both personal and professional aspects.

applying what you know
Your best friend has been having some pain on her right side just above her hip. The physician suspects it may be her appendix. Using your knowledge of the quadrants and regions of the abdomen, answer the following questions.
1. Which region of the abdomen would the physician describe the pain as being located?
    a. Right lumbar region   c. Right hypochondriac region
    b. Right iliac (inguinal) region   d. Hypogastric region
2. In what quadrant of the abdomen would the physician describe the pain as being located?
    a. RUQ   b. LUQ   c. LLQ   d. RLQ

CASE STUDY 2 *Football Injury*

Mark Buchholz, a 22-year-old, plays left tackle for his college football team. He was hurt when he made a tackle that prevented the other team from winning the game. The sports physician suspects he may have injured his head and/or neck.
1. What body cavities may be involved?
2. What planes or sections would you want to look at using x-rays?

CHAPTER 1   Concepts of the Human Body

2

Concepts of Chemistry

chapter outline

Matter and Atomic Structure

Inorganic substances

Organic Substances

Life Span

learning outcomes

AFTER COMPLETING THIS CHAPTER, YOU WILL BE ABLE TO:

2.1 Identify matter and its characteristics.
2.2 Classify important inorganic substances.
2.3 Classify important organic substances.
2.4 Explain how chemical reactions impact life span.

case study

Use this case study to focus on as you go through the chapter. The questions will guide you as you learn about the chemistry of life.

CASE STUDY 1 *Dehydration*

Adam Rooke, a 74-year-old man, was working on his car on a hot summer day. He became dizzy and had to sit down to rest. His pulse was weak but he was not perspiring.

As you go through the chapter, keep the following questions in mind:

1. What important chemicals may be a factor in Adam's condition?
2. Can a person be overhydrated?
3. What instructions should you give Adam to avoid this situation in the future?

essential terms

anabolism (a-NA-bo-lizm)
atom (AH-tom)
biochemistry (bi-o-KEM-is-tre)
catabolism (kah-TAB-o-lizm)
chemistry (KEM-is-tre)
chromosome (KRO-mo-some)
compound (COM-pound)
deoxyriboneucleic acid (DNA) (de-ok-si-ri-bo-nu-KLE-ic A-sid)
ion (I-on)
matter (MAT-er)
metabolism (meh-TAB-o-lizm)
molecule (MOL-eh-kule)
ribonucleic acid (RNA) (ri-bo-nu-KLE-ic A-sid)

Additional key terms in the chapter are italicized and defined in the glossary.

study tips

1. Review the essential terms of the chapter.
2. Draw and label a typical atom with its nucleus and orbitals.
3. Make flash cards comparing inorganic and organic compounds.

chemistry The study of matter and how it undergoes change.

biochemistry A branch of chemistry dealing with the chemistry of life.

atom Unit of matter that makes up a chemical element.

molecule The combination of two or more atoms of the same element.

compound The combination of two or more atoms of more than one element.

metabolism The sum of all chemical reactions within an organism.

anabolism Chemical reactions requiring energy where smaller molecules are used to build larger, more complex molecules.

catabolism The chemical breakdown of complex molecules into simpler molecules with the release of energy.

Introduction

Chemistry is the study of matter and how it undergoes change. Regardless of the field you have chosen to go into, an understanding of chemistry can make life more satisfying. Everything we come into contact with is made up of chemicals. Even our bodies are made up of chemicals.

Biochemistry is a branch of chemistry that deals with the chemistry of life or biological chemistry. When two or more **atoms** are chemically combined, a molecule is formed. **Molecules** are the basic units of compounds. A **compound** is formed when two or more atoms of more than one element are combined. An example of a molecule is water, which is composed of two hydrogen atoms and one oxygen atom. Water is also an example of a compound because its molecules are made up of atoms of two different elements—hydrogen and oxygen. Atoms are held together by *chemical bonds*. Energy is stored in chemical bonds. When the bonds are broken, energy is released that can be used by the body. When bonds are formed, energy is required. Two of the more important chemical bonds we see in cells are covalent bonds and hydrogen bonds (Figure 2-1). *Covalent bonds* are formed when two atoms share the same electrons. These are relatively strong bonds. A *hydrogen bond* is a bond that results when a hydrogen atom in a molecule is attracted to an electronegative atom, usually in another molecule. An electronegative atom is one in which the overall charge is negative, such as oxygen, fluorine, or nitrogen. There are other types of bonds, but they do not have much biological relevance and are not discussed here.

Metabolism is the sum of all the chemical reactions that take place in our body. The two processes of metabolism are **anabolism** and **catabolism.** In

focus on Wellness: Chemicals Affect the Body

Everything in our environment is made up of chemicals. Some chemicals are harmful; for example, benzene is a chemical that is used as a solvent and in the synthesis of other substances. It has been proven to be carcinogenic. Of course, other chemicals may be beneficial to our health. Calcium phosphate is used to treat certain bone diseases such as osteoporosis (excessive bone loss). Stay aware of all chemicals you take into your body or are exposed to in order to avoid harm and stay well.

anabolism, small molecules combine to form larger ones (for example, when amino acids combine to form proteins). In catabolism, larger molecules are broken down into smaller ones (for example, when stored glycogen is converted to glucose molecules for energy). The human body is made up of water, proteins, carbohydrates, lipids, and nucleic acids. Therefore, we will explore these substances and their importance to life.

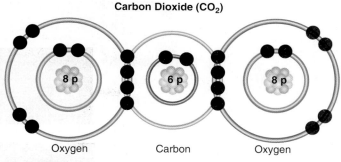

FIGURE 2-1 Two oxygen atoms and one carbon atom sharing electrons to form a covalent bond—carbon dioxide.

U check
What are the two processes that make up metabolism?

Matter and Atomic Structure

Matter is anything that takes up space. It also has weight or the ability to exert a force. Matter can be a gas, liquid, or solid and is made up of substances called elements that consist of the same atom. Table 2-1 lists some of the major elements in the human body. For example, calcium is an element. The terms *atom* and *element* are sometimes used interchangeably. An atom consists of a *nucleus* that is at the center of the atom. The nucleus may contain a number of neutrally charged particles called *neutrons* and positively charged particles called *protons*. A hydrogen atom contains a proton, but does not contain any neutrons. Neutrons and protons have approximately the same mass. Surrounding the nucleus, we find a number of negatively charged particles called *electrons*. Their mass is so small compared to protons and neutrons we say they have no mass (Figure 2-2). Sometimes chemistry is defined as the behavior of electrons because electrons are the most important particles in chemical reactions. Electrons farthest from the nucleus are called valence electrons. Valence electrons in the outermost shells will determine the chemical properties of the atoms by gaining or losing electrons, affecting molecular stability. They may also be shared, as in covalent bonds. The *atomic number* of an element is the number of protons in the element. The *atomic weight* is the sum of the number of protons plus neutrons. *Isotopes* are atoms with the same atomic number, but different atomic weights (see Figure 2-3). This means the number of protons are the same, but the number of neutrons differs. Isotopes have the same chemical behavior, but different physical properties. In some isotopes, the nuclei decay, causing the isotope to emit energy waves or particles. These isotopes are said to be *radioactive*. This is important because some radioisotopes are used to diagnose and treat diseases. Matter can be divided into two large categories—organic and inorganic matter. *Organic* matter contains carbon and hydrogen, and tends to form large molecules. *Inorganic* compounds may contain carbon, but they do not form either hydrogen bonds or covalent bonds. These molecules tend to be smaller.

2.1 learning outcome
Identify matter and its characteristics.

matter Anything that takes up space.

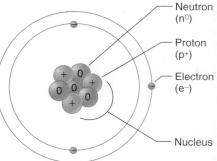

Lithium (Li)

FIGURE 2-2
An atom consists of subatomic particles: neutrons, protons, and electrons. An atom of lithium is shown with its four neutrons, three protons, and three electrons.

U check
What is an isotope?

CHAPTER 2 Concepts of Chemistry

FIGURE 2-3
Isotopes: Atoms with the same atomic number but different atomic weights are isotopes of an element.

TABLE 2-1 Major Elements in the Human Body

Major Elements	Symbol	Approximate Percentage of the Human Body (by Weight)
Oxygen	O	65.0%
Carbon	C	18.5
Hydrogen	H	9.5
Nitrogen	N	3.2
Calcium	Ca	1.5
Phosphorus	P	1.0
Potassium	K	0.4
Sulfur	S	0.3
Chlorine	Cl	0.2
Sodium	Na	0.2
Magnesium	Mg	0.1
		Total 99.9%
Trace Elements		
Chromium	Cr	
Cobalt	Co	
Copper	Cu	
Fluorine	F	Together less than 0.1%
Iodine	I	
Iron	Fe	
Manganese	Mn	
Zinc	Zn	

2.2 learning outcome

Classify important inorganic substances.

LET'S SEE HOW
Adam is doing. What inorganic substances do you think are depleted in him because of the heat?

Inorganic Substances

Examples of inorganic substances are water, oxygen, carbon dioxide, and salts such as sodium chloride (Table 2-2). Water is the most abundant inorganic compound in the body accounting for more than 60% of our body weight. It is an essential component of the cells, the blood, and many other body fluids. It is also called the universal solvent because so many substances dissolve in water. Because it takes a lot of energy to change the temperature of water, it helps regulate body temperature.

 check
What is the most abundant inorganic compound in the human body?

Oxygen is another essential inorganic molecule. We inhale oxygen and then it is transported through our blood attached to hemoglobin within red blood cells (erythrocytes) to all the cells and tissues of our body. It helps

TABLE 2-2 Inorganic Substances Common in Cells

Substance	Symbol or Formula	Functions
I. Inorganic Molecules		
Water	H_2O	Major component of body fluids; medium in which most biochemical reactions occur; transports chemicals; helps regulate body temperature
Oxygen	O_2	Used in energy release from glucose and other organic molecules
Carbon dioxide	CO_2	Waste product that results from metabolism; reacts with water to form carbonic acid
II. Inorganic Ions		
Bicarbonate ions	HCO_3^-	Helps maintain acid-base balance
Calcium ions	Ca^{+2}	Necessary for bone development, muscle contraction, and blood clotting
Carbonate ions	CO_3^{-2}	Component of bone tissue
Chloride ions	Cl^-	Has a role in acid-base balance of the blood, water balance, and the production of hydrogen chloride (HCl) in the stomach
Magnesium ions	Mg^{+2}	Component of bone tissue; required for certain metabolic processes
Phosphate ions	PO_4^{-3}	Required for synthesis of ATP, nucleic acids, and other vital substances; component of bone tissue
Potassium ions	K^+	Required for polarization of cell membranes
Sodium ions	Na^+	Required for polarization of cell membranes; essential for water balance; important in action potential conduction
Sulfate ions	SO_4^{-2}	Helps maintain the health of cartilage

convert glucose to needed energy. Other molecules such as fats and amino acids can also be converted into energy. Some cells such as neurons or nerve cells can go for only a very short period without oxygen before dying. Other cells such as fibroblasts that make connective tissue can last much longer without oxygen before dying. *Carbon dioxide* is exchanged in the cells for oxygen and then is exhaled. Although carbon dioxide is considered a waste product for humans and other animals, it is essential for most plants and is used to make oxygen.

Inorganic salts are the fourth major category of inorganic substances that are important to life. They consist of **ions** (an atom or group of atoms called cations with a positive charge or anions that have a negative charge) such as sodium (Na^+), chloride (Cl^-), phosphate (PO_4^{-3}), potassium (K^+), calcium (Ca^{+2}), magnesium (Mg^{+2}), carbonate (CO_3^{-2}), sulfate (SO_4^{-2}), and bicarbonate (HCO_3^-). They are involved in many important metabolic functions. Some participate in bone growth, blood clotting, and pH regulation to name

Oxygen transport/gas exchange

ion An atom or group of atoms with a net positive or net negative charge.

a few functions. We will look at inorganic salts more closely in Chapter 6, Concepts of Fluid Electrolyte and Acid Base Balance.

> **U check**
> What is the difference between an anion and a cation?

2.3 learning outcome
Classify important organic substances.

Basic chemistry (organic molecules)

Organic Substances

The four major classes of organic matter in the body are carbohydrates, lipids, proteins, and nucleic acids. Each of these organic substances is needed for some aspect of human anatomy and physiology. Balanced diets for healthy living include lipids, carbohydrates, and proteins.

Carbohydrates

Carbohydrates are our body's main source of energy. Sugars are one category of carbohydrate. Sugars can be classified as simple or complex depending on their size. The most common carbohydrate is *glucose* which contains 6 carbon atoms, 12 hydrogen atoms, and 6 oxygen atoms in every molecule. When we have excess glucose, it is stored as glycogen in the liver and skeletal muscles. Starch is one type of carbohydrate found in potatoes, pastas, and breads, which when eaten is broken down into glucose by the body and used as needed. Besides being an energy source, carbohydrates are used to build some structural units. For example, deoxyribose, a type of sugar, is used to make DNA. We will talk more about carbohydrates as well as lipids and proteins in Chapter 20, Metabolic Function and Nutrition.

Lipids

Lipids are fats and are insoluble in water. The three types of lipids found in the body are triglycerides, phospholipids, and steroids. *Triglycerides* are used to store energy for cells. Each gram of fat can provide over twice the energy as a gram of carbohydrate or protein. A major function of *phospholipids* is to make cell membranes. You will learn more about them in Chapter 3, Concepts of Cells and Tissues. Butter and oils are composed of triglycerides, and the body stores these molecules in adipose tissue (fat). *Steroids* are very large lipid molecules used to make cell membranes and some hormones. Cholesterol is an example of an essential steroid for body cells. Cholesterol is used to make estrogen, progesterone, and testosterone to name a few substances.

Protein synthesis

deoxyribonucleic acid (DNA) A nucleic acid consisting of the sugar deoxyribose; the bases adenine, guanine, cytosine, and thymine; and phosphate.

ribonucleic acid (RNA) Is used to synthesize proteins.

Proteins

Proteins have many functions in the body. Many proteins act as structural materials for the building of solid body parts, such as muscle. An example is actin and myosin, contractile proteins that are part of muscle tissue. Other proteins act as hormones, enzymes, receptors, and antibodies.

Nucleic Acids

Deoxyribonucleic acid (DNA) and **ribonucleic acid (RNA)** are two examples of nucleic acids. DNA contains genes—the genetic information of

focus on Genetics: Recombinant DNA

DNA keeps all of the information needed to recreate an organism. Nearly every cell in the body has some DNA. Most DNA is located in the nucleus of the cell. The infomation in DNA is stored as a code. The code is made up of four different chemical bases: adenine (A), thymine (T), guanine (G), and cytosine (C). The nitrogen bases are found in pairs, with A & T and G & C paired together. The sequence of the bases can be arranged in infinite ways. This sequence and the number of bases is what creates diversity. Each base is also attached to a sugar molecule and a phosphate molecule. Together, a base, sugar, and phosphate are called a nucleotide. Nucleotides are arranged in two long strands that form a spiral called a double helix as seen in Figure 2-4.

DNA does not actually make the organism, it only makes proteins. The DNA is transcribed into messenger RNA (mRNA). Then mRNA is translated into protein. The protein then forms the organism. By changing the DNA sequence, the way in which the protein is formed changes.

Recombinant DNA, or rDNA, is taking a piece of one DNA and combining it with another strand of DNA (thus, the name recombinant!). Recombinant DNA is also sometimes referred to as "chimera." By combining two or more different strands of DNA, scientists are able to create a new strand of DNA. This new strand of rDNA has a new code. The most common recombinant process involves combining the DNA of two different organisms. The rDNA works when it is placed in a host and the host cells create proteins using the newly created rDNA information.

This rDNA process has led to the discovery of important drugs. rDNA techniques are being used to create vaccines; prevent and cure sickle cell anemia and cystic fibrosis; and produce human insulin, clotting factors, growth hormone, and other phamaceuticals. These are just the beginning. Many substances are now produced by biotechnological means and may one day be useful in treating other diseases such as Parkinson's or Alzheimer's as well.

cells—and RNA is used to synthesize proteins. DNA may be dispersed in the cytoplasm or condensed to form **chromosomes.** DNA consists of the sugar deoxyribose; the bases adenine, guanine, cytosine, and thymine; and phosphate (see Figure 2-4). Nucleic acids are large, complex molecules made up of carbon, hydrogen, oxygen, nitrogen, and phosphorus. They will be covered in more detail in Chapter 3.

Pharmacology

chromosome A small threadlike structure in the nucleus of the cell made of DNA and containing the genetic information of the cell.

from the perspective of...

A LABORATORY TECHNICIAN A laboratory technician examines and analyzes body fluids and cells. Why is it important to understand the functions of the various types of proteins?

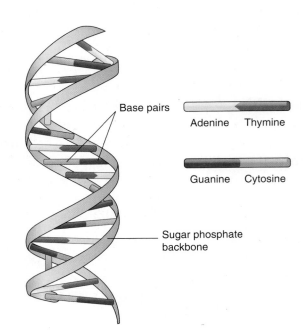

FIGURE 2-4 The nucleic acid DNA stores the information needed to recreate an organism.

2.4 learning outcome

Explain how chemical reactions impact life span.

Life Span

When chemical reactions in the body result in too much or too little of a substance, it can adversely affect life. For example, if a person cannot synthesize the enzyme responsible for converting phenylalanine to tyrosine, excess phenylalanine will accumulate in the body. Although phenylalanine is an essential amino acid, too much can cause harm. Organ damage, mental retardation, and even death are possible. The condition is called *phenylketonuria (PKU)* and most newborns are tested for it as recommended by the U.S. Preventive Services Task Force. Prenatal diagnosis is also available if there is suspicion of it occurring in the infant. When PKU is diagnosed early, special diets have been successful in allowing the patient to have a typical, normal life span.

summary

learning outcomes	key points
2.1 Identify matter and its characteristics.	Matter is anything that takes up space and has mass.
2.2 Classify important inorganic substances.	The more important inorganic substances include water, oxygen, carbon dioxide, and salts such as sodium chloride.
2.3. Classify important organic substances.	Organic compounds include carbohydrates, lipids, proteins, and nucleic acids.
2.4 Explain how chemical reactions impact life span.	Abnormal chemical reactions in the body can affect our health and sometimes even our life span.

case study 1 questions
Can you answer the following questions that pertain to Adam's case study presented earlier in this chapter?
1. What important chemicals may be a factor in Adam's situation?
2. What percentage of the human body is water?
   a. 20 percent
   b. 40 percent
   c. 60 percent
   d. 80 percent
3. What instructions should you give Adam to avoid dehydration and possible heat exhaustion or heat stroke?

review questions
1. What is the most abundant inorganic molecule in the human body?
   a. Glucose
   b. Sodium chloride
   c. Water
   d. Carbon dioxide
2. Genes are made up of
   a. Nucleic acids
   b. Cholesterol
   c. Carbon molecules
   d. Carbohydrates
3. Which of the following are *not* functions of water in the body?
   a. It prevents dehydration.
   b. It can act as a solvent.
   c. It is a source of ATP.
   d. It helps regulate body temperature
4. Cholesterol is necessary for all of the following except
   a. Synthesis of testosterone
   b. Synthesis of estrogen
   c. Synthesis of cell membranes
   d. Synthesis of ATP

critical thinking questions
1. If a person has too little cholesterol in his or her body, how might this affect the person's reproductive capability?
2. Discuss the particles that are found in atoms.
3. Discuss fats, proteins, and carbohydrates, including some of their functions.

patient education
Why do patients need to know about the organic substances related to their diet and health?

applying what you know
1. What inorganic substance accounts for the greatest percentage of mass in the body?
2. What are the three main types of lipids in the body?
3. Which organic substance has the greatest amount of energy per gram of substance?

3 Concepts of Cells and Tissues

chapter outline

Cell Components
Cell Transport
The Cell Cycle
Stem Cells
Cell Death
Tissue Types
Life Span

learning outcomes

AFTER COMPLETING THIS CHAPTER, YOU WILL BE ABLE TO:

3.1 Explain the structure and function of the cell nucleus, the cell membrane, and the cytoplasm and the other cell organelles.

3.2 Compare passive and active transport.

3.3 Discuss the cell cycle including mitosis and meiosis.

3.4 Describe the difference between embryonic and adult stem cells.

3.5 Identify pathological and "normal" cell death and how this relates to the pathophysiology of the organism.

3.6 Contrast the structure and function of the four major tissues.

3.7 Infer how the cell cycle impacts life span and explain how patients can improve their general health by addressing issues at the cellular level.

essential terms

active transport (AK-tiv TRANS-port)
anaphase (AN-ah-faze)
chromosome (KRO-mo-some)
cytokinesis (si-to-kin-E-sis)
cytoplasm (SI-to-plazm)
diffusion (dih-FU-zhun)
filtration (fil-TRA-shun)
interphase (IN-ter-faze)
meiosis (mi-O-sis)
metaphase (MET-ah-faze)
mitosis (mi-TO-sis)
nucleus (NU-kle-us)
organelle (or-gan-EL)
osmosis (os-MO-sis)
prophase (PRO-faze)
telophase (TEL-o-faze)
zygote (ZI-got)

Additional key terms in the chapter are italicized and defined in the glossary.

case study

Use the case study to focus on as you go through the chapter. The questions will guide you as you learn the anatomy, physiology, and pathology of cells and tissues.

CASE STUDY 1 *Fibromyalgia*

Annette Acosta is a 22-year-old dancer who has just been invited to audition for Chicago's Joffrey Ballet. She has been experiencing pain in her right inner thigh area. She has been told that she has a connective tissue disease. She thinks the name of the disease the doctor mentioned was "fibromyalgia." Annette is concerned that this may keep her from dancing.

As you go through the chapter, keep the following questions in mind:

1. What is connective tissue?
2. Are connective tissue diseases more common in either gender?
3. Will Annette be able to pursue her dream of dancing with the Joffrey Ballet?

study tips

1. Draw and label a typical cell. Identify the nucleus, cytoplasm, and cell membrane.

2. Outline the stages of the cell cycle. Use the mnemonic "Play on the **MAT**" for the phases of mitosis (prophase, metaphase, anaphase, and telophase).

3. Make a table comparing the three types of muscle: skeletal, smooth, and cardiac. Include headings for fiber shape, number of nuclei, voluntary or involuntary, and location.

Introduction

The *cell* is the basic unit of life (Figure 3-1). The branch of anatomy that studies cells is *cytology* and the branch dealing with the study of tissues is called *histology*. Some organisms such as bacteria are single-cell organisms. Other organisms, including humans, are multicellular organisms. The human body is composed of trillions of cells of over 250 variations. Each type of a cell has a specific function. Cells with similar functions will form tissues. It is amazing that all the information for these variations is housed in one cell, the fertilized egg or **zygote** which is formed from the joining of a sperm cell and an ovum or egg (Figure 3-2). From this one cell, other cells will be produced that have specific structures and functions. This process of cells becoming specialized is called *differentiation*. It is one of the characteristics that distinguishes mature cells from immature cells. Mature cells are usually more specialized or differentiated than immature cells.

zygote The fertilized egg formed from the union of a sperm and an ovum.

U check
What is the term given to a fertilized ovum?

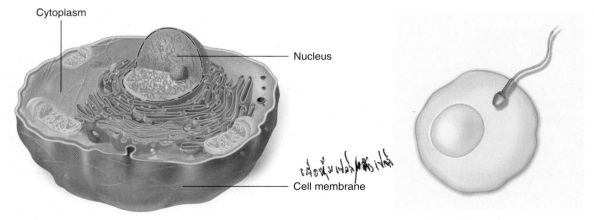

FIGURE 3-1 The cell is the basic unit of life. The study of cells is known as cytology.

FIGURE 3-2 Male sperm and female ovum.

Cell Components

Human cells vary dramatically in size. The largest cells include the fertilized egg (120 to 140 micrometers in diameter) and the anterior horn cell of the spinal cord (approximately 135 micrometers in diameter). Smooth muscle cells can be as long as 500 micrometers in length. Although erythrocytes (red blood cells) are only about 7.5 micrometers in diameter and human sperm is about 25 micrometers in length, granule cells of the cerebellum are as small as 4 micrometers in diameter. The typical mature cell has three components: the nucleus, cytoplasm, and cell membrane (Figure 3-3). A major exception to this is the mature red blood cell or erythrocyte. Immature red blood cells have a nucleus, but it is lost by the time it has matured.

3.1 learning outcome

Explain the structure and function of the cell nucleus, the cell membrane, and the cytoplasm and the other cell organelles.

Cells and tissues

Nucleus

Organisms that have cells with nuclei are called *eukaryotic* cells. The **nucleus** is sometimes called the "brain" of the cell. This is because the nucleus contains deoxyribonucleic acid (DNA) or the genetic material of the cell. DNA is responsible for directing all the activities of the cell. The nucleus is a large spherical organelle that has a double membrane with holes or pores in it that allow the passage of substances into and out of the nucleus. In the nucleus is a region with no surrounding membrane called the *nucleolus*. Some cells may have multiple nucleoli. The purpose of this structure is the production of ribosomes which we will talk about shortly. Also in the nucleus is chromatin which is dispersed or uncoiled DNA and protein called *histones*. When chromatin condenses, **chromosomes** are formed and are held together by the histones. In humans, all cells except the gametes (the egg and sperm cells) have 23 pairs or 46 chromosomes. The fluid environment within the nucleus is the nucleoplasm.

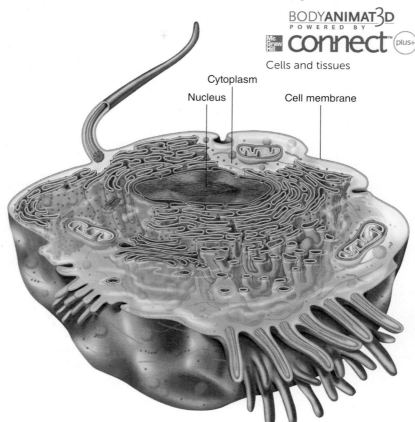

FIGURE 3-3 A typical cell has a nucleus, cytoplasm, and a cell membrane.

nucleus A spherical organelle within the cell that contains genes, the hereditary factors of the cell.

chromosome A small threadlike structure in the nucleus of the cell made of DNA and containing the genetic information of the cell.

cytoplasm Portion of the cell between the cell membrane and the nucleus.

organelle Small, permanent structure within the cytoplasm that serves specific functions.

✓ check
What is the function of the nucleolus?

Cytoplasm and Its Organelles

The **cytoplasm** of a cell is the "inside" of the cell. The cytoplasm is made up of a fluid portion called the *cytosol* and small structures called **organelles**

TABLE 3-1 Organelles and Their Functions

Organelle/Structure	Function
Nucleus	Houses DNA, the hereditary material of the cell
Mitochondria	The "powerhouse" of the cell, synthesis of ATP
Ribosomes	The site of protein synthesis
Endoplasmic reticulum	Intracellular transport, support, storage, synthesis, and packaging of molecules
Golgi apparatus	Processing, sorting, packaging, and delivering proteins and lipids to the cell membrane
Vesicles	A sac that transports substances within and out of the cell
Lysosomes	Contain enzymes for the destruction and degradation of harmful substances
Peroxisomes	Contain enzymes to oxidize organic compounds
Cilia	Move substances along the surface of the cell
Flagella	Allow the sperm cell to move
Centrioles	Involved in pulling the chromosomes apart during mitosis

("small organs"). The cytosol is mostly made up of water, proteins, ions, and nutrients. There are many organelles and other cellular components that perform a multitude of functions for the cell and therefore for the organ system and the organism (Table 3-1). These organelles and structures include the nucleus, mitochondria, ribosomes, endoplasmic reticulum (ER), Golgi apparatus, vesicles, lysosomes, peroxisomes, cilia, flagella, and centrioles.

U check
What are the two components of the cytoplasm?

Mitochondria: We have already talked about the nucleus, so let's direct our attention to the *mitochondrion (*plural *mitochondria)*. This organelle is sometimes called the "powerhouse" of the cell because it is responsible for producing the energy for the cell. It has a double membrane, with the inner membrane thrown into folds called *cristae*. It is on the inner membrane where the energy in the form of molecules called *adenosine triphosphate (ATP)* is produced. ATP is called the "energy currency" of the cell. Cells that require a lot of energy such as cardiac muscle will have several mitochondria, whereas other cells may have few mitochondria. Mature red blood cells have no mitochondria. Mitochondria are very unique organelles because they have some DNA and therefore can direct the synthesis of some proteins.

U check
What is the function of mitochondria? How do they differ from other organelles?

Ribosomes: Ribosomes are responsible for the production of proteins. They are made up of ribosomal ribonucleic acid (RNA), a single-stranded nucleic acid made of ribose (a sugar), nitrogenous bases (adenine, guanine, cytosine, and uracil), and phosphate groups, and composed of two subunits. Ribosomes do not have a membrane and are assembled in the nucleolus. Ribosomes may be found "free" in the cytoplasm or bound to *endoplasmic reticulum*.

Endoplasmic Reticulum: The endoplasmic reticulum (ER) is a network of channels that has a single membrane. The ER is widely distributed in the cytoplasm and is connected to the nuclear envelope, the cell membrane, and various organelles. There are two forms of ER: *rough endoplasmic reticulum (rER)* and *smooth endoplasmic reticulum (sER)*. Rough ER has ribosomes attached to it giving it a rough appearance. The function of rER is to synthesize and process proteins and transport them to the Golgi apparatus for further processing. The processed proteins eventually will leave the cell for use elsewhere. Rough endoplasmic reticulum is abundant in endocrine glands that secrete protein-type hormones. Smooth endoplasmic reticulum lacks *ribosomes* (hence smooth) and is involved in the synthesis of lipids and the detoxification of substances such as drugs. Smooth ER is abundant in the liver.

> **U check**
> How does rough endoplasmic reticulum differ from smooth endoplasmic reticulum?

Golgi Apparatus: The *Golgi apparatus* is a system of six or so stacked membranous sacs called *cisternae* that have an appearance of stacked bowls of decreasing size. In reality, the sacs are connected, but only seem to be disconnected because of the section that was made through the cell for microscopic examination. The function of the Golgi apparatus is to process, package, and transport proteins that were synthesized by the rER. When the processing is done, a part of the Golgi apparatus buds off creating a vesicle containing the protein. The vesicle transports the protein to the cell membrane where it is pushed out of the cell by a process called *exocytosis*.

Vesicles: The *vesicles* ("small sacs") are small membrane-bound sacs that contain and transport various substances in the cell as well as to the cell membrane for export out of the cell.

Lysosomes: *Lysosomes* are relatively small, membrane-bound sacs that contain lytic enzymes capable of destroying and digesting proteins, carbohydrates, nucleic acids, and foreign particles.

Peroxisomes: *Peroxisomes* are similar to lysosomes, but contain peroxidases. These enzymes are involved in breaking down peroxide and other harmful substances to help detoxify the body. They are abundant in the liver.

Cilia: Many cells contain hairlike projections on the outside of the cell membrane called *cilia*. Cilia assist with propelling matter throughout the body tracts, such as within the respiratory system. Cells with cilia are often found in the mucus membranes. Cilia are sometimes damaged in certain individuals such as cigarette smokers. Consequently, particles cannot be moved up the respiratory tract by the cilia and the person develops a cough to assist in this function.

Flagella: *Flagella* are similar to cilia, but are longer. In humans, one is found on each sperm cell (flagellum) and is responsible for its motility. Flagella are also found in many bacteria and parasites, and will be discussed further in Chapter 5, Concepts of Microbiology.

Centrioles: The *centrioles* are two cylindrical organelles near the nucleus. They are essential to cell division because spindles are attached to them and the chromosomes to pull apart the chromosomes during mitosis and meiosis.

> **U check**
> How does the structure and function of cilia differ from that of flagella?

Cell Membrane (Plasma Membrane)

The cell membrane is the boundary between the internal environment of the cell and the external environment (Figure 3-4). If the cell membrane is damaged, the cell may not be able to survive and will die. The cell membrane is made up of two layers of phospholipids, with hydrophilic (attracted to water) "heads" pointing outward from the membrane and hydrophobic (repels water) "tails" pointing inward from the membrane. The plasma membrane is called selectively or semi-permeable, which means that it allows some substances to pass through it while preventing other substances from doing so. The cell membrane consists of proteins, lipids, and some carbohydrates. Proteins in the cell membrane have different functions including acting as channels for certain substances and as receptors for hormones and drugs.

> **U check**
> Why is the cell membrane referred to as "selectively permeable"?

FIGURE 3-4
The cell membrane is a complex semipermeable membrane that is essential to the survival of the cell.

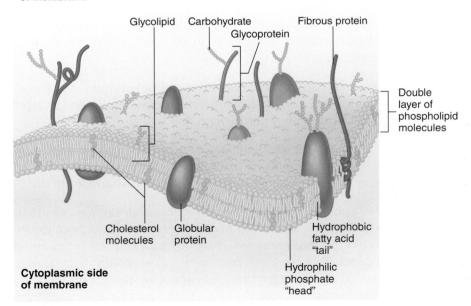

PATHOPHYSIOLOGY

Crenation

When a person drinks salt water, water will leave the cell in an attempt to dilute the hypertonic external environment. If the situation is not corrected, then the cell will shrivel up and die. This process of the cell shriveling up and dying is called crenation.

Cell Transport

3.2 learning outcome

Compare passive and active transport.

Substances are transported within the cell as well as into and out of the cell. Movement of substances in the cytoplasm is achieved through microtubules and concentration gradients. Movement of substances across the semipermeable cell membrane occurs through passive and active transport.

Passive Transport

Passive transport requires no energy and typically molecules move from an area of high concentration to an area of low concentration to achieve equilibrium.

Simple Diffusion: Atoms or molecules of gases or liquids will move from an area of high concentration (more molecules) to an area of low concentration (fewer molecules), referred to as *simple diffusion* (Figure 3-5). The difference between the concentrations of two areas is called a concentration gradient. Molecules tend to move down a concentration gradient. When equilibrium is achieved, it does not mean movement has stopped, but only that the movement in both directions is equal. No energy is required for simple diffusion. Lipid soluble substances such as oxygen, carbon dioxide, and steroids and some drugs cross the cell membrane by simple diffusion.

osmosis The net movement of water molecules through a selectively permeable membrane from an area of high water concentration to an area of low water concentration.

Facilitated Diffusion: Some small molecules that are not lipid soluble also can cross the plasma membrane, but need a little help. Molecules such as glucose, potassium and sodium use a carrier molecule to facilitate ("make easy") movement across the membrane. *Facilitated diffusion* is similar to simple diffusion in that movement is down the concentration gradient and no energy is required.

Osmosis is simple diffusion of water from an area of low solute

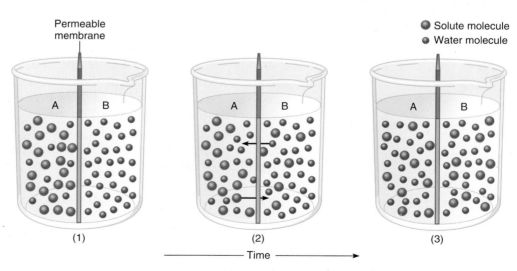

FIGURE 3-5 Simple diffusion is a passive movement of particles from an area of high concentration to an area of low concentration until equilibrium is reached.

CHAPTER 3 Concepts of Cells and Tissues 35

A. Passive transport: No energy is required

B. Active transport: Energy is required

FIGURE 3-6 Passive versus active transport: Passive transport is like rolling a ball down a hill because it requires no energy; active transport is like pushing a ball up a hill because it does require energy.

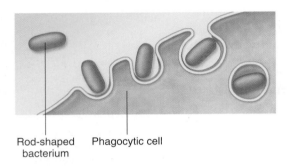

Rod-shaped bacterium | Phagocytic cell

FIGURE 3-7 A cell participating in phagocytosis of bacteria.

filtration The flow of a liquid through a filter due to hydrostatic pressure; no energy is required.

active transport The movement across cell membranes against a concentration gradient and requiring energy (ATP).

concentration to an area of high solute concentration. In this case, the molecules of the solid are too large to pass across the membrane. Instead, water is diffusing from an area of high concentration to an area of low concentration. When there are more particles of a solid, referred to as the solute, inside the cell than outside, the inside is said to be hypertonic to the outside. The outside of the cell is said to be hypotonic to the inside. If the concentrations inside and outside the cell are equal, the external and internal environments are isotonic to each other. Since osmosis is simple diffusion of water, it is a passive process and requires no energy.

Filtration is a passive process, meaning no energy is required. Molecules are moved across a membrane or a filter due to greater pressure on the side where particles are leaving. This concept will be discussed further in Chapter 15, The Urinary System.

Active Transport

Active transport requires energy and typically molecules move from an area of low concentration to an area of high concentration to achieve equilibrium. Figure 3-6 illustrates the difference between passive and active transport.

Endocytosis is a process where energy is required. Part of the cell membrane encloses a substance outside the cell, forming a vesicle, and then invaginates, bringing the substance into the cell interior.

Phagocytosis is a form of endocytosis where larger molecules are engulfed and brought into the cell (Figure 3-7). *Pinocytosis* is similar to phagocytosis, except it is liquid that is being brought into the cell.

Exocytosis is the opposite of endocytosis. Endocrine glands will secrete protein hormones in this manner. Of course, energy is required for this process.

Transcytosis is a combination of endocytosis and exocytosis going on simultaneously.

Sodium-potassium pump pumps sodium ions out of a cell and pumps potassium ions into the cell, creating a high sodium concentration outside the cell and a high potassium concentration within it. This occurs in nerve cells after passage of a nerve impulse.

> **U check**
> What is the difference between passive and active transport?

The Cell Cycle

The *cell cycle* is the series of events that occur in a cell that result in its replication (Figure 3-8). The cell cycle is divided into interphase, mitosis, and cytokinesis. We will also talk about a special type of cell division called meiosis.

Interphase

About 90 percent of a cell's life is spent in **interphase.** This phase was once called the "resting" phase of the cell cycle. This term is no longer used because rather than resting, the cell is very active. During part of interphase, the cell is growing. At other times, it is replicating DNA and getting the cell ready for mitosis.

Mitosis

The remainder (about 10 percent) of the cell cycle is spent in **mitosis** (Figure 3-9). The object or end result of mitosis is the production of two identical "daughter" cells with the same number and kind of chromosomes as the original nucleus. Mitosis is divided into four phases.

1. *Prophase:* In the first phase of mitosis, or **prophase,** several events are taking place. The chromatin is condensing to form chromosomes. Each chromosome consists of two identical sister chromatids joined near the center by a *centromere. Centrioles* (specialized microtubules) move to opposite ends or poles of the cell and spindle fibers form that attach the centrioles to the centromeres of each of the chromosomes.

3.3 learning outcome

Discuss the cell cycle including mitosis and meiosis.

interphase The period between cell divisions when the cell is involved in growth, metabolism, and preparation for cell division.

mitosis The division of the cell nucleus that results in two nuclei with the same number and kind of chromosomes as the original nucleus.

prophase The first phase of mitosis during which the nuclear envelope breaks apart and chromatin condenses to form chromosomes.

Meiosis versus mitosis

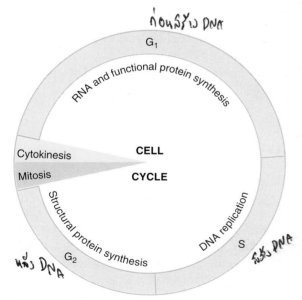

FIGURE 3-8 During the cell cycle, a cell duplicates its contents and divides into two identical daughter cells.

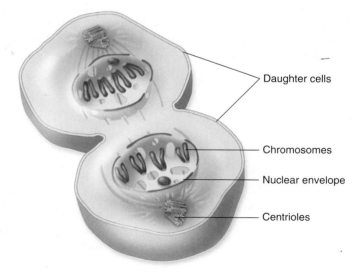

FIGURE 3-9 Cell division includes mitosis or division of the nucleus followed by cytokinesis or division of the cytoplasm.

metaphase The second phase of mitosis in which chromatids line up along the center of the cell.

anaphase The third phase of mitosis where sister chromatids have separated and moved to opposites poles of the cell.

telophase The fourth and final phase of mitosis, where the cell is ready to divide into tow identical daughter cells.

cytokinesis Distribution of cytoplasm into two separate cells during cell division.

2. *Metaphase:* During the **metaphase,** the chromosomes line up at the middle of the cell along an imaginary line called the metaphase plate or equatorial plate.

3. *Anaphase:* In the third phase of mitosis, the **anaphase,** the spindle fibers that attach the centromere to the centrioles shorten, dragging the sister chromatids to opposite ends of the cell.

4. *Telophase:* In the final phase of mitosis, the **telophase,** the nuclear envelope starts to form around the chromosomes that start to unwind to form the thread-like chromatin. The cell is now ready to divide into two identical daughter cells.

Cytokinesis

The actual division of the cytoplasm and the cell is called **cytokinesis.** It is not a part of mitosis, but it begins during late anaphase and continues through telophase. Formation of a cleavage furrow, which is a slight indentation in the cell membrane, initiates cytokinesis. The cleavage furrow appears between the centrosomes and extends around the periphery of the cell. Microfilaments made of actin pull the cell membrane inward until the cell is pinched in two. At this point, a set of chromosomes is in each daughter cell and interphase begins.

> **U check**
> What are the phases of mitosis and what are the major events occurring in each?

Meiosis versus mitosis

meiosis A type of cell division that takes place during gamete production.

Meiosis

A specialized type of cell division that is necessary for sexual reproduction and produces *sperm* and *ova* (eggs) is called **meiosis.** The process actually involves two cell divisions. Rather than resulting in two identical daughter cells, in men it results in four functional sperm cells and in women one functional egg and three nonfunctional *polar bodies* that end up disintegrating. Since the fertilized egg has to supply nutrition for the zygote for a period of time, it gets most of the cytoplasm with very little going to the polar bodies.

> **U check**
> What is the purpose of meiosis?

Regulation

Cells are capable of dividing only a certain number of times before they lose that ability. The number of times a cell can divide depends on the cell type and ranges from roughly 50 to 75 times. Some cells called *labile cells* have the ability to constantly divide (Figure 3-10). These would include skin cells, cells lining the gastrointestinal tract, and blood cells in the bone marrow. *Stable cells* are capable of dividing when necessary. For example, *hepatocytes* (liver cells) have the ability to multiply when the liver is injured. *Permanent cells* do not have the ability to multiply.

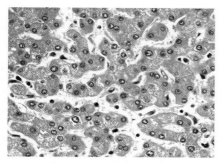
A Labile: Glandular epithelium

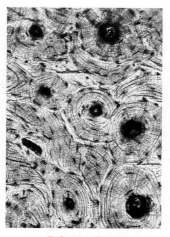
B Stable: Bone

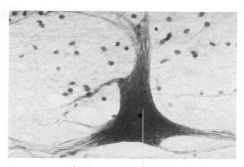
C Pemanent: Neuron

FIGURE 3-10 Cells are classified as (A) labile, (B) stable, or (C) permanent based on their ability to undergo mitosis.

A *neuron* (nerve cell) is an example of this type of cell. Interestingly, research is showing that perhaps even "permanent" cells may have some potential to replace themselves. If this is true, it offers hope to individuals suffering from conditions such as cerebrovascular accident (stroke) or Alzheimer's disease.

Stem Cells

Stem cells are cells that are capable of becoming a variety of different cell types and of self-renewal or production of other stem cells. There are two categories of stem cells—embryonic stem cells and adult stem cells. Embryonic stem cells are derived from embryos that develop from eggs that have been fertilized *in vitro* (in a laboratory) (Figure 3-11). They are not taken from eggs fertilized in a woman's body. Adult stem cells are present in most organs and tissues in our body. A major difference between embryonic stem cells and adult stem cells is that the embryonic stem cell is *totipotent*, or capable of developing into virtually any cell type. Adult stem cells are undifferentiated cells found within an organ or tissue and capable of renewing themselves and developing into some of the specialized cells of that organ or tissue.

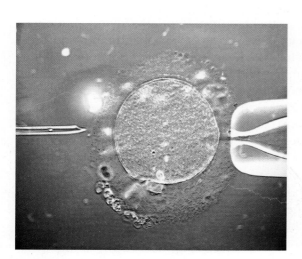
FIGURE 3-11 In vitro fertilization.

3.4 learning outcome

Describe the difference between embryonic and adult stem cells.

> **focus on Genetics:** Stem Cell Research
>
> Stem cell research has been a controversial topic for quite some time and opinions exist on both sides of the issue. It is hoped that agreement can be reached that may open the door to new discoveries that can cure diseases such as Down syndrome, diabetes, and cystic fibrosis.

learning outcome 3.5

Identify pathological and "normal" cell death and how this relates to the pathophysiology of the organism.

Cell Death

Cell death can be a normal occurrence or it may be pathological (abnormal, related to a disease process). *Apoptosis* is referred to as "programmed cell death" and is the reason why most of us do not have webbing between our fingers and toes or a tail (both are seen in the fetus). It is a method of destroying cells that may have mutated and pose the danger of becoming a cancer. Apoptosis also helps delete immune cells that may recognize a person's own body as foreign, causing an autoimmune disease. *Necrosis* is premature cell death that is always pathological. For example, a myocardial infarction (heart attack) results in the death (necrosis) of cardiac muscle cells.

learning outcome 3.6

Contrast the structure and function of the four major tissues.

Tissue Types

Tissues are similar cells grouped together to perform a common function. The four tissue types are epithelial, connective, muscle, and nervous tissue (Figure 3-12). Not only do these tissues differ in their functions, but they also differ in the cell types, cell arrangement, blood supply, connections to other cells, and ability to multiply.

Cells and tissues

Epithelial Tissue

Epithelium or epithelial tissue lines body cavities, organs, and surfaces and makes up glands. For example, epithelium lines the thoracic and abdomino-pelvic cavities. It also lines hollow tubes such as the esophagus and blood vessels as well as hollow organs such as the heart and stomach. Of course, skin is an epithelium that we are all familiar with. Glandular tissue is also epithelial. Glands are classified as exocrine if they secrete their chemical substances into ducts or channels (these are discussed in Chapters 7, The Integumentary System, and 19, The Digestive System). They are referred to as endocrine glands (Chapter 21, The Endocrine System) if they secrete their chemical substances (hormones) directly into the surrounding tissues or blood. Epithelium is avascular and receives its nutrients from blood vessels in the underling connective tissue. It also has a basement membrane which is a thin, sheetlike structure that separates the epithelium from the underlying connective tissue. Epithelium has a nerve supply and is labile (highly mitotic) or capable of dividing constantly. A number of different functions exist that determine the cell shape and arrangement found in the epithelium (Figure 3-13).

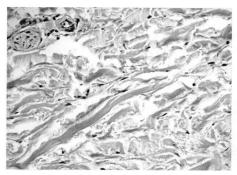
A Dense connective tissue

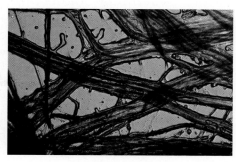
C Nervous tissue

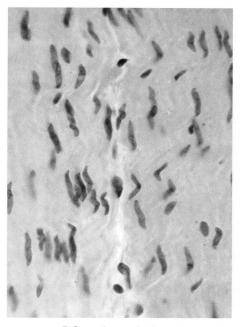
B Smooth muscle tissue

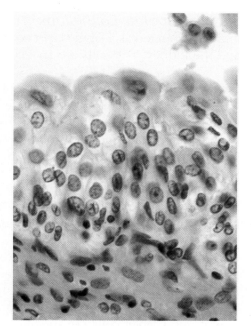
D Transitional epithelium

FIGURE 3-12
The four major tissue types are (A) connective tissue, (B) smooth muscle tissue, (C) nervous tissue, and (D) epithelial tissue.

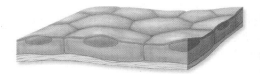
A Simple squamous epithelium

B Simple cuboidal epithelium

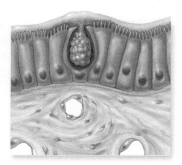
C Columnar epithelium with goblet cell

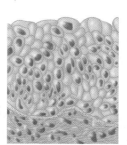
D Transitional epithelium

FIGURE 3-13
Four basic cell shapes are (A) squamous, (B) cuboidal, (C) columnar, and (D) transitional.

CHAPTER 3 Concepts of Cells and Tissues

Simple Squamous Epithelium: The cells in this type of epithelium are flattened (squamous) and only one cell layer in thickness. This will allow easy passage of substances across the membrane. This type of epithelium is found in the lungs for easy exchange of oxygen and carbon dioxide.

Stratified Squamous Epithelium: The skin has a protective function so its cells are flattened or *squamous* in shape and stacked in several layers *(stratified)*. The cells are also tightly packed and held together by specialized cell junctions. Stratified squamous epithelium can also be found the oral cavity, esophagus, vagina, and anus.

Simple Cuboidal Epithelium: These epithelial cells are cuboidal in shape and are arranged as a single layer of cells. They are commonly seen in glands and the kidney. The cells are specialized for secretion and absorption.

Simple Columnar Epithelium: These cells are taller than they are wide. They are arranged as a single layer of cells. The cells are involved in secretion and absorption and are seen in the digestive tract. Some cells have specialized structures such as *cilia* to move substances, *microvilli* to increase surface area for absorption, and *goblet cells* to secrete mucus for protection.

Stratified Columnar Epithelium: These epithelial cells are not common, but can be found in the male reproductive tract and the pharynx.

Pseudostratified Columnar Epithelium: These are columnar cells that give the appearance of being stratified, but are not. This type of epithelium is found in the respiratory tract and has specialized structures such as cilia that can move debris up the respiratory tract that can then be swallowed or spit out. Goblet cells are specific columnar cells that produce the protein mucin, which helps form mucus. Mucus then traps debris, thus protecting the epithelium.

Transitional Epithelium: These epithelial cells are cuboidal with a modification of the *apical* or free surface (the surface facing the lumen). The apical surface is domed or slightly curved to allow stretching or expansion without tearing the epithelium. This epithelium is seen in the urinary bladder where stretching will occur as the bladder fills. It also is seen in the ureters and part of the urethra.

> **U check**
> What type of epithelium is found in the skin?

Connective Tissue

Connective tissue is the most abundant and widely distributed tissue in the body. The definition of connective tissue is tissue made up of living cells and a nonliving matrix. The five major connective tissue types are blood, bone, cartilage, connective tissue proper, and fat. The *matrix* is the material

PATHOPHYSIOLOGY

Thrombocytopenia

When a person has too few platelets, the condition is called *thrombocytopenia* and the person has a risk of increased and dangerous bleeding. Possible causes range from infection to genetics to cancer of the bone marrow. A hematologist is a physician who specializes in blood disorders and will probably be involved with the diagnosis and treatment. Diagnosis begins with a thorough patient history, and then blood tests are run. These include a full blood count, microscopic examination, and bone marrow biopsy. Treatment is dependent on the cause and may include antibiotics, IVIG therapy, plasmapheresis, corticosteroid therapy, and/or platelet replacement.

between the cells and is made up of water, salts, proteins, fibers, and other substances. A variety of cells will be discussed with each connective tissue type. Connective tissue varies in its innervation and the blood supply ranges from none to highly vascular. Connective tissue may be loosely arranged or tightly packed.

Blood: You may not have thought of blood as being a connective tissue, but it is. That is because blood has cells (red and white blood cells) and a nonliving matrix—the *plasma*. In addition to *erythrocytes* (red blood cells) and *leukocytes* (white blood cells), blood also has *thrombocytes (platelets)* that are actually not cells, but are cell fragments (Figure 3-14). Unlike other connective tissue, the matrix (plasma) does not contain fibers. Blood is involved in the transport of gases (oxygen and carbon dioxide) and nutrients and wastes. It also is involved in immunity and blood clotting. Blood and blood cells will be discussed in more detail in Chapter 10, Blood and Circulation, and Chapter 12, The Lymphatic and Immune Systems.

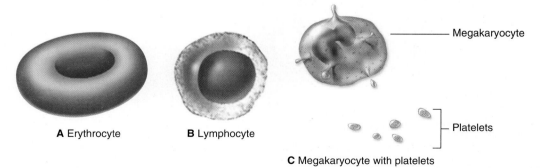

A Erythrocyte **B** Lymphocyte **C** Megakaryocyte with platelets

FIGURE 3-14 Formed blood elements: (A) erythrocyte (red blood cell), (B) lymphocyte (type of white blood cell), and (C) a megakaryocyte producing platelets (thrombocytes).

CHAPTER 3 Concepts of Cells and Tissues

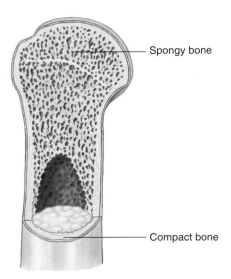

FIGURE 3-15
Compact and spongy bone.

Bone (Osseous) Tissue: Bone tissue is a very hard and rigid connective tissue (Figure 3-15). However, it is not the hardest tissue in the body. (Do you know what the hardest tissue in the body is? The enamel on your teeth!). The extracellular matrix is made up of collagen fibers that give bone its flexibility, and mineral salts such as calcium phosphate that gives bone its hardness. Osteoblasts, osteocytes, and osteoclasts are the bone cells in this tissue. Osteoblasts produce the bone matrix, osteocytes are mature bone cells, and osteoclasts tear down bone as part of the remodeling process. Bone acts as protection, attachment for muscles, hematopoiesis (blood cell production), and storage for certain chemicals such as calcium (over 99 percent of the body's calcium is stored in bone) and phosphorus. For more information on bone, see Chapter 8, The Skeletal System.

Cartilage: Cartilage is another rigid connective tissue (Figure 3-16). It helps support, protect, and act as a framework for bone development. Cartilage is avascular and receives nutrients from nearby blood vessels. Because it has no direct blood supply, healing is slow in cartilage injuries. There are two types of cartilage cells: chondroblasts and chondrocytes. Three types of cartilage are based on the extracellular matrix.

Hyaline Cartilage: Hyaline is the most abundant cartilage in the body. It makes up most of the fetal skeleton which will be replaced by bone, is involved in bone formation in many bones, is the cartilage found in synovial joints, and is the cartilage that attaches the ribs to the sternum (its smooth surface allows for easy gliding of joints). It has collagen fibers in the matrix and has a glassy appearance when examined with a microscope. People sometimes take supplements such as chondroitin sulfate and glucosamine to keep cartilage healthy.

Elastic Cartilage: Elastic cartilage is very flexible because of the high content of elastic fibers. This type of cartilage can be found in the external ear and the epiglottis.

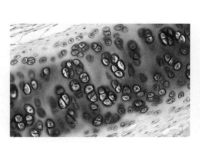

A Hyaline cartilage
(from a synovial joint)

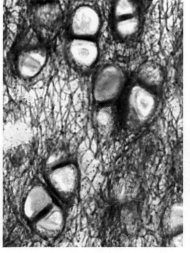

B Elastic cartilage
(from the external ear)

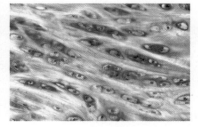

C Fibrocartilage
(from an intervertebral disc)

FIGURE 3-16 Three types of cartilage: (A) hyaline, (B) elastic, and (C) fibrocartilage.

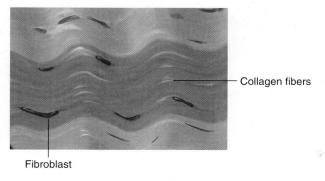

FIGURE 3-17 Fibroblast in connective tissue and collagen fibers.

Fibrocartilage: Fibrocartilage has an abundance of collagen fibers and is a great shock absorber. We see this cartilage in the intervertebral discs between the vertebra and the pubic symphysis. Many diseases affect the cartilage and joints. These will discussed in detail in the respective chapters.

Connective Tissue Proper: Bone, cartilage, and blood are considered specialized connective tissues whereas connective tissue proper is a name given to fibrous connective tissue. This includes loose and dense connective tissue. Connective tissue proper has different types of fibers and *fibroblasts* (immature cells)—cells responsible for making the fibers and matrix (Figure 3-17). *Fibrocytes* (mature cells) are a less active form of cell.

Loose Connective Tissue: Areolar tissue, fat (adipose tissue), and reticular connective tissue are forms of loose connective tissue.

- Areolar tissue forms thin membranes and is distributed throughout the body including in the digestive, respiratory, reproductive, and urinary systems. It surrounds blood vessels, and epithelium rests on areolar tissue. The abundance of loose space in areolar tissue allows leukocytes to easily move, find, and destroy pathogens.
- Fat (adipose tissue) is made of up of *adipocytes* (fat cells) and fat globules. Adipose tissue is found throughout the body; it cushions and protects organs, insulates, and provides a source of energy when other sources such as protein and carbohydrates are not available.
- Reticular connective tissue is made up of collagen fibers and is found as a framework in organs such as the spleen and liver as well as in the thymus and lymph nodes.

focus on Wellness: Fat Cells

Fat cells are a type of connective tissue that act as storage sites for energy in addition to cushioning and protecting organs. They also help insulate the body. An excess of fat cells can result in obesity, which is a body weight exceeding 20 percent of a desirable standard because of too much adipose tissue. Obesity is a risk factor for cardiovascular disease, hypertension, lung disease, cancer, diabetes, and arthritis. A proper diet and a regular exercise program can help control obesity and improve the health of an individual.

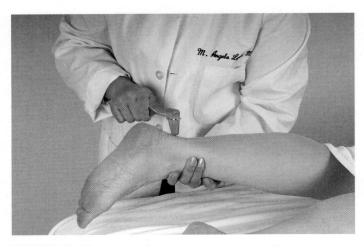

FIGURE 3-18 An examiner performing an Achilles' tendon reflex on a patient.

Dense Connective Tissue: Dense regular connective tissue and dense irregular connective tissue are both forms of dense connective tissue.

- Dense regular connective tissue is made of collagen fibers that are tightly packed and arranged in a parallel pattern. There may be some elastic fibers as well. This type of connective tissue is poorly vascularized which means healing is a slow process. Dense connective tissue is found in *tendons* (connect muscle to bone) and *ligaments* (connect bone to bone); see Figure 3-18.
- Dense irregular connective tissue is more random in arrangement and is able to resist forces that are multidirectional. The dermis has dense irregular connective tissue.

Elastic connective tissue is made up mostly of elastic fibers that are yellow in color. There may be some collagen fibers and fibroblasts. It is found in the ligamentum flava ("yellow ligament") in the vertebral column and the walls of larger arteries, the heart, and the bronchi.

> **REMEMBER ANNETTE,** the ballerina who was diagnosed with fibromyalgia? What type of connective tissue is involved in this disease?

U check
What are the two major features of all connective tissue?

Muscle contraction

Muscle Tissue

Muscles can do only one thing—contract. When they contract, the muscle cells shorten and thicken. Muscle cells are also called muscle fibers or *myocytes.* Contraction and relaxation of muscles will cause movement in the body; examples might be movement of the arms and legs, propulsion of food along the gastrointestinal tract, or pumping of blood through the body. Three types of muscle tissue are skeletal, smooth, and cardiac (Figure 3-19). We will talk briefly about them here and discuss them in more detail in Chapter 9, The Muscular System, and Chapter 11, The Cardiovascular System.

PATHOPHYSIOLOGY

Fibromyalgia

Fibromyalgia is chronic muscle and connective tissue pain. The patient may suffer from symptoms such as sleep disturbance, fatigue, and joint stiffness. The symptoms can vary widely from individual to individual. Females are nine times as likely to experience fibromyalgia as males. Diagnosis of fibromyalgia is controversial, and not all doctors consider it a true medical disease because of the absence of objective diagnostic tests. There may also be a genetic predisposition for the disorder. Various hormonal (dopamine, serotonin, and growth hormone) imbalances may play a role and trauma may be a trigger in some cases. Treatment is mostly symptomatic, including the use of pharmaceuticals that may give some relief. Unfortunately, most patients do not experience long-term relief, nor can they expect a cure from the disorder.

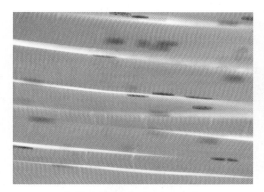
A Skeletal (voluntary) muscle

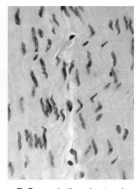
B Smooth (involuntary) muscle

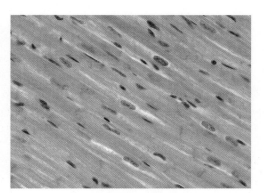
C Cardiac (heart) muscle

FIGURE 3-19 The three muscle types are (A) skeletal, (B) smooth, and (C) cardiac.

Skeletal Muscle: Skeletal muscle is also called *striated muscle* because of alternating light and dark bands in the muscle fiber. The fibers are long and narrow. The cells also have multiple nuclei (multinucleate). Skeletal muscle is also called *voluntary muscle* because it only contracts when innervated and we can consciously control its contraction. Skeletal muscle is attached to bones by tendons and causes the bones to move. An example would be the biceps brachii muscle that can rotate the forearm and flex the elbow.

Smooth Muscle: Smooth muscle is called *visceral* or *involuntary muscle*. It does not have striations or bands, and is associated with the walls of hollow organs such as the gastrointestinal tract and uterus. We also cannot consciously control its contractions because it is under the control of the autonomic nervous system. Its spindle-shaped fibers are shorter than those of skeletal muscle and the fibers have a single, centrally located nucleus.

Cardiac Muscle: Cardiac muscle is found only in the heart. It has striations like skeletal muscle but contracts involuntarily like smooth muscle. Cardiac muscle cells branch, with a single nucleus. Cardiac muscle also has a specialized structure, the *intercalated disc,* that helps the muscle cells contract as a single unit.

> **U check**
> How is cardiac muscle similar to and different from smooth muscle and skeletal muscle?

Nervous Tissue

Nervous tissue is located in the brain, spinal cord, and peripheral nerves of the body. Two categories of cells are found in the nervous system: the neuron and *neuroglial cells* (*neuroglia*). Neurons are the structural and functional unit of the nervous system (Figure 3-20). The human brain contains over 100 billion neurons (the octopus brain contains about 170 million).

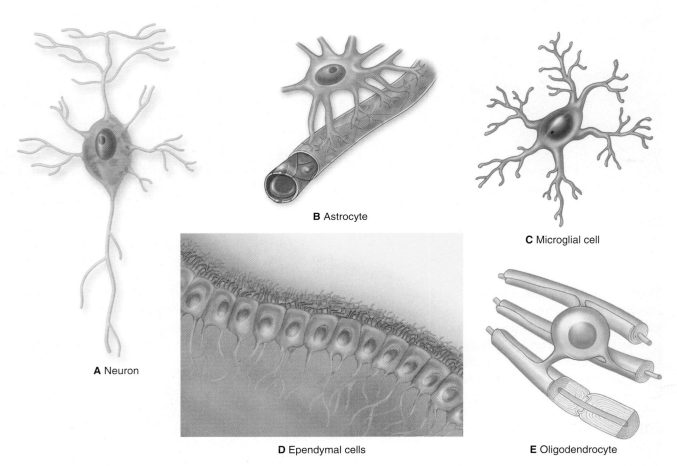

FIGURE 3-20 Two categories of nervous system cells are neurons (A) and neuroglial cells: (B) astrocyte, (C) microglial cell, (D) ependymal cells, and (E) oligodendrocyte.

UNIT 1   The Human Body and Disease

Neurons conduct electrical impulses, or *action potentials*. Neuroglia (*glia* means glue) are supportive cells, and there are four different types in the central nervous system with specific functions. Astrocytes help to maintain the chemical environment around neurons. Microglial cells perform a phagocytic function. Ependymal cells produce cerebrospinal fluid and oligodendrocytes produce the myelin sheath around axons. Neuroglia do not transmit action potentials but can multiply, whereas neurons are not capable of multiplying. We will discuss the nervous system in greater detail in Chapter 14, The Nervous System.

> **U check**
> What is the structural and functional unit of the nervous system?

Life Span

Cells have a limited number of divisions or "multiplications" within them. This means that the organism has a limited life span. Aging then is a normal process. In addition, Rudolf Virchow, known as the "father of pathology," said that all pathology or disease starts at the cellular level. This means anything that affects the cell will have a positive or negative impact on the individual. Glucose also can affect life span by forming cross links with proteins. This can result in the loss of elasticity in tissues. In diabetes, atherosclerosis is accelerated, often leading to early heart disease and even early death of the individual.

3.7 learning outcome

Infer how the cell cycle impacts life span and explain how patients can improve their general health by addressing issues at the cellular level.

summary

	learning outcomes	key points
3.1	Explain the structure and function of the cell nucleus, the cell membrane, and the cytoplasm and the other cell organelles.	Cells are made up of a nucleus, cytoplasm, and a cell membrane with a couple of exceptions. Each has a special role in maintaining the life of the individual. A number of organelles have specific functions. For example, mitochondria are the powerhouses of the cell that produce ATP. Ribosomes are involved in protein synthesis, and lysosomes contain lytic enzymes that can kill foreign organisms.
3.2	Compare passive and active transport.	Passive transport requires no energy and includes simple diffusion and osmosis. Active transport does require energy and includes phagocytosis and pinocytosis.
3.3	Discuss the cell cycle including mitosis and meiosis.	The cell cycle consists of interphase, mitosis which has four phases (prophase, metaphase, anaphase, and telophase), and cytokinesis. Mitosis is the method by which most cells reproduce. Meiosis is a type of cell division involving the gametes or sex cells.

chapter 3

learning outcomes	key points
3.4 Describe the difference between embryonic and adult stem cells.	Stem cells are undifferentiated cells that have the potential to become different types of cells.
3.5 Identify pathological and "normal" cell death and how this relates to the pathophysiology of the organism.	"Normal" cell death includes the natural aging process at the cellular level, whereas pathological cell death is termed necrosis.
3.6 Contrast the structure and function of the four major tissues.	The four basic tissue types are epithelial, connective, muscle, and nervous tissue. Each has many variations and is structured for its unique functions.
3.7 Infer how the cell cycle impacts life span and explain how patients can improve their general health by addressing issues at the cellular level.	Cells are capable of reproducing only a finite number of times. Once that number has been reached the cell and, according to some theories on aging, the individual will die. Rudolf Virchow, the "father of pathology," said all disease begins at the cellular level. Health also begins there.

chapter 3 review

case study 1 **questions**

Can you answer the following questions that pertain to Annette, the ballerina from the first case study presented in this chapter?

1. What is connective tissue?
2. What connective tissue is most likely involved in Annette's case?
3. Does this connective tissue have a good or poor blood supply? How will this affect healing?
4. Are connective tissue diseases more common in either gender?
5. How could this diagnosis affect Annette's dream of dancing with the Joffrey Ballet?

review **questions**

1. Which type of connective tissue is avascular?
   - **a.** Fat
   - **b.** Cartilage
   - **c.** Bone
   - **d.** Blood
2. What are the two components of connective tissue?
   - **a.** A nonliving matrix
   - **b.** Living cells
   - **c.** Actin
   - **d.** Myosin
3. What is the structural and functional unit of the nervous system?
   - **a.** Neuron
   - **b.** Platelet
   - **c.** Thrombocyte
   - **d.** Erythrocyte
4. Hydrophilic heads and hydrophobic tails are part of the
   - **a.** Nucleus
   - **b.** Cell membrane
   - **c.** Ribosome
   - **d.** Golgi apparatus

critical **thinking questions**

1. Mitochondria are sensitive to decreases in oxygen levels to the cell. If a person is hypoxic (inadequate oxygen to the tissues), explain how may this impact cellular function?
2. Relate how the structure of each of the three components of a "typical" cell is uniquely suited to its function.
3. Differentiate between the characteristics of bone and cartilage. Determine which is more important and explain why.

patient **education**

What impact do you think nutrition and healthy habits have on your cells? What are some unhealthy habits that affect your cells?

applying **what you know**

1. What organelle is responsible for protein synthesis?
2. What are gametes?
3. What are the three types of bone cells?
4. Is "visceral" muscle voluntary or involuntary muscle?
5. Where do we find intercalated discs?
6. Why is hyaline cartilage well suited for synovial joints?
7. Is epithelium vascular or avascular?

CASE STUDY 2 *Juvenile Diabetes*

Stacey Kaiser has type 1 (juvenile) diabetes. She has heard that stem cell transplantation may help with her disease, but she does not know much about the topic.

1. Can you explain what stem cells are?
2. Can you differentiate between embryonic and adult stem cells?

4 Concepts of Disease

chapter outline

- Cell Injury and Death
- The Body's Response to Injury
- Immunopathology
- Neoplasia
- Developmental and Genetic Diseases
- Hemodynamic Disorders
- Environmental and Nutritional Pathology
- Infectious and Parasitic Diseases
- Disease Diagnosis and Prognosis
- Prevention and Treatment of Diseases
- Life Span

learning outcomes

AFTER COMPLETING THIS CHAPTER, YOU WILL BE ABLE TO:

4.1 Compare cell injury and cell death.

4.2 Distinguish among inflammation, repair, regeneration, and fibrosis.

4.3 Describe immunopathology.

4.4 Explain neoplasia.

4.5 Identify causes of developmental and genetic diseases.

4.6 Explain hemodynamic diseases.

4.7 Summarize common environmental and nutritional disorders.

4.8 Describe infectious and parasitic diseases.

4.9 Distinguish among signs, symptoms, diagnosis and prognosis, or disease.

4.10 Explain prevention and treatment of diseases.

4.11 Explain the effects of disease on human life span.

case study

Use the case study to focus on as you go through the chapter. The questions will guide you as you learn the concepts of pathology including signs and symptoms, diagnosis, treatment, and prognosis.

CASE STUDY 1 *Lung Disease*

Eugene Kivett is a 47-year-old lab technician. He works in a hospital preparing specimens for examination by the pathologist. He has smoked two packs of cigarettes per day for the last 25 years. He has had a persistent cough for the last three weeks, and he has noticed some blood occasionally when coughing and is also experiencing shortness of breath.

As you go through the chapter, keep the following questions in mind:

1. What signs of disease is Eugene showing?
2. What are the symptoms he is experiencing?
3. What are possible diagnoses that must be considered?
4. What are the treatments for the various diagnoses you have listed?
5. What are the prognoses for the various diagnoses you have listed?

essential terms

apoptosis (AP-op-TOE-sis)

atrophy (AT-ro-fee)

benign (be-NINE)

carcinoma (kar-sih-NO-mah)

differential diagnosis (dif-er-EN-shul die-ag-NO-sis)

embolism (EM-bo-lizm)

hyperchromatic (hi-per-kro-MA-tik)

hyperplasia (hi-per-PLAY-zhee-a)

hypertrophy (hi-PER-tro-fee)

iatrogenic (i-at-ro-JEN-ik)

malignant (mah-LIG-nant)

metaplasia (met-a-PLAY-zhee-a)

metastasis (meh-TAS-tah-sis)

necrosis (neh-KRO-sis)

neoplasm (NE-oh-plazm)

pathogenic (path-oh-JEN-ik)

prognosis (prog-NO-sis)

sarcoma (sar-KO-mah)

teratogens (ter-AT-o-jen)

Additional key terms in the chapter are italicized and defined in the glossary.

study tips

1. Make flash cards with one side labeled "sign or symptoms" and list an example on the reserve side. Example: Write "sign" on the front and write "fever" on the reverse side.
2. Write a quiz that deals with the following concepts: atrophy, hypertrophy, hyperplasia, metaplasia, dysplasia, and anaplasia.
3. Write down one to three questions to ask your instructor.
4. Meet with a study group and review and quiz each other.

Introduction

The term *pathology* comes from "pathos" (suffering) and "ology" (the study of); thus it is defined as the study of suffering. Pathology is the study of disease. Disease is any structural or physiological change that disrupts homeostasis. At one time, people thought disease and illness were caused by supernatural forces. In the 1800s, Rudolf Virchow, who is considered the "father of pathology," stated that all disease originated at the cellular level. Pathology then concerns itself with abnormal structural and physiologicalal changes of organs and organ systems. To appreciate the abnormal, we must have a firm grasp of normal. That is one of the main reasons for studying anatomy and physiology. In this chapter, we will explore some of the concepts underlying this fascinating science. Diseases will be presented in the chapters in which the specific organs or organ systems affected are discussed.

4.1 learning outcome

Compare cell injury and death.

atrophy The reduction in size or wasting of an organ.

Cell Injury and Death

When a cell is injured by toxins, trauma, infection, cancer, or other agents, it may adapt or die. The major ways a cell can adapt are atrophy, hypertrophy, hyperplasia, metaplasia, dysplasia, and intracellular storage, as well as calcification and hyaline changes.

Atrophy is the reduction in the size of an organ (Figure 4-1). Atrophy may be physiological or pathological. For example, as we get older we start to lose body mass. This is a normal occurrence and therefore is a physiological event. Pathological atrophy refers to a change that is caused by a disease or disorder. Because there are

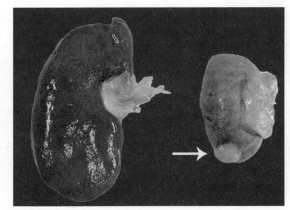

FIGURE 4-1 This kidney (right) has undergone pathological atrophy to become much smaller than normal.

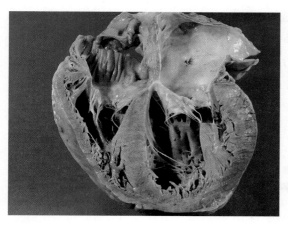

FIGURE 4-2 A heart from a cadaver showing hypertrophy of the heart wall.

several causes of pathological atrophy, it may be reversible in some instances and irreversible in others. For instance, if someone is immobilized for some reason, he or she will see muscles waste or atrophy due to disuse. In most cases, once the individual is up and moving around again, he or she will regain muscle mass and strength. Inadequate oxygenation (*hypoxia*) or inadequate nutrients and paralysis, such as after a stroke, will result in atrophy that may be a permanent condition.

Hypertrophy is the opposite of atrophy. It is defined as an increase in the size of an organ and may be physiological or pathological. For instance, when someone works out with weights, you expect to see an increase in the size of the person's muscles. This results in an increase in muscle function and is generally considered to be a positive outcome. When the heart works harder because of resistance in blood vessels, the heart will get larger due to hypertrophy of the heart muscle (*myocardium*). This is an abnormal condition, known as myocardial hypertrophy, which can put a person at risk for serious illness and eventually death (Figure 4-2).

Hyperplasia is an increase in the number of cells in an organ or tissue. It is not the same as hypertrophy, but if the hyperplasia is significant, hypertrophy will also take place. Hyperplasia may result from physiological or pathological conditions. For example, the endometrium of the uterus responds to estrogen to produce more cells, in preparation for a possible pregnancy. Hyperplasia is also seen in cancers as well as other pathologies and allows for the cancer or other pathology to spread more quickly. Hyperplasia of the uterine cervix may be due to chronic irritation and should be monitored (Figure 4-3). Chronic trauma, such as irritation of the foot from poorly fitting shoes, may cause callous formation, which is another example of hyperplasia.

hypertrophy An increase in the size of an organ with physiological or pathological origin.

hyperplasia An increase in the number of cells in an organ or tissue.

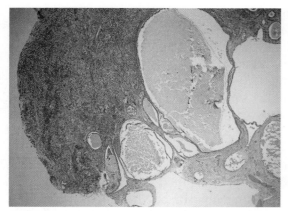

FIGURE 4-3 Photomicrograph of the uterine cervix showing increased cellularity as seen in hyperplasia.

U check
What is the difference between hypertrophy and hyperplasia?

CHAPTER 4 Concepts of Disease

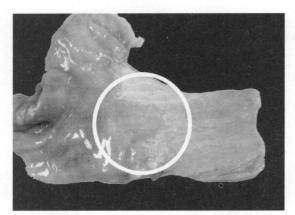

FIGURE 4-4 A specimen taken from the esophagus demonstrating metaplasia (Barrett's esophagus).

metaplasia The change of one adult cell from one type to another.

apoptosis Programmed cell death or cell suicide.

necrosis Pathological cell death.

Metaplasia is the change of one adult cell type to another. An example would be columnar epithelium changing to squamous epithelium, which is the disease called Barrett's esophagus, a condition that can lead to esophageal cancer (Figure 4-4). Metaplasia may also be a response to chronic irritation. For example, with gastro-esophageal reflux disorder (GERD), the chronic irritation will cause the columnar epithelium in the esophagus to change to squamous epithelium.

Dysplasia is an abnormal, disordered growth and maturation of cells. Dysplastic cells do not look like normal cells. They may have a variety of shapes and sizes, nuclei that are darker than normal, and a disorderly arrangement of cells. Although dysplasia may reverse spontaneously, it is considered a precancerous condition and must be watched. In the uterus, dysplasia may be seen on examination of the cervix during a colposcopy or from cell examination during a Pap smear (Figure 4-5). Even though it is considered a precancerous situation, dysplasia does not always progress to cancer.

Intracellular storage of substances may be normal or abnormal and the substances stored may be harmful or harmless. Fat normally accumulates in the body, providing a reserve of energy when needed. In some individuals,

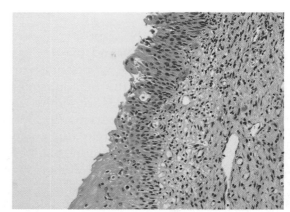

FIGURE 4-5 Photomicrograph of cervical epithelial cells showing precancerous dysplasia.

there can be an abnormal accumulation of fat. This commonly occurs in the liver with diabetes and alcoholism; the medical term for this is fatty liver. The heart, kidney, and skeletal muscle are other organs that may accumulate increased fat. Other substances that can accumulate in cells are glycogen, mucopolysaccharides, cholesterol and abnormal proteins, pigments, iron, and melanin. All of these can interfere with normal functioning of the cell when the amount of the accumulated substance becomes excessive. Lipofuschin (sometimes called liver spots when appearing on the skin) is a stored brownish-purple pigment that is associated with aging. It does not appear to interfere with cell function.

Cell death can be described as apoptosis or necrosis. **Apoptosis** is called programmed cell death or cell suicide. As explained in Chapter 3, apoptosis is the reason we do not have webbing between our fingers and toes and why we do not have a tail. Both of these exist in the embryo or fetus, but the cells are "programmed" to die so that these structures do not persist after birth. This is normal cell death. **Necrosis** is pathological cell death, which occurs when cells suffer injury or insult and are not able to adapt and so die. For example, when the toes of a patient receive little or no oxygen and nutrients due to poor circulation in conditions such as diabetes, necrosis causing gangrene can occur (Figure 4-6).

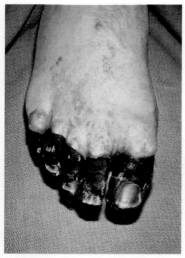

FIGURE 4-6 Necrosis of the foot resulting in gangrene.

> **U check**
> Does dysplasia always lead to cancer?

The Body's Response to Injury

When cell injury occurs, the body has an amazing ability to respond. It responds through the process of inflammation, repair, regeneration, and fibrosis.

4.2 learning outcome
Distinguish among inflammation, repair, regeneration, and fibrosis.

Inflammation

Inflammation is a vascular response of tissue to an injury, infection, allergy, or autoimmune disease. Inflammation is normally a protective response to destroy or limit both the injurious agent and the amount of damaged tissue. At times it can get out of control and cause even more damage—for example, rheumatoid arthritis. Inflammation can be classified as acute or chronic based on the length of the inflammation.

Acute inflammation usually lasts from one to two days. This is the type of inflammation that might occur from a fall. The four cardinal signs of inflammation as described by the ancient Romans are (1) *rubor* (redness), (2) *calor* (heat), (3) *tumor* (swelling), and (4) *dolor* (pain). A fifth sign, (5) loss of function, was added much later but is now considered a cardinal sign as well. Certain white blood cells are associated with acute inflammation. Neutrophils and monocytes that are transformed to macrophages are the most common. The typical outcome of acute inflammation as the result of an injury is repair (healing), regeneration, or *fibrosis* (scar formation).

> **U check**
> What are the cardinal signs of acute inflammation?

Chronic inflammation usually is inflammation that lasts days, weeks, or even months. Chronic osteoarthritis and rheumatoid arthritis are an examples of chronic inflammation (Figure 4-7).

Repair, Regeneration, and Fibrosis

Repair involves the replacement of the damaged tissue with connective tissue cells and fibers and new blood vessels. This combination of connective tissue and blood vessels is called granulation tissue. Eventually, the granulation tissue will be replaced with functional cells. If the injury is large or severe, a scar may form.

FIGURE 4-7
Inflammation in an arthritic hand. A person with arthritis may experience pain, heat, and swelling of the joint as well as redness of the overlying skin.

Regeneration is the renewal of damaged tissue or an appendage. In humans, we do not see regeneration of appendages (arms, legs), but we do see regeneration in the liver, bone, and to some extent skeletal muscle.

Fibrosis is the production of connective tissue at the site of injury. Scarring is a type of fibrosis and can occur if the injury is extensive or if healing is not complete. Excessive scar formation, called keloids (or proud flesh), can occur in anyone but are more common in dark-skinned individuals (Figure 4-8).

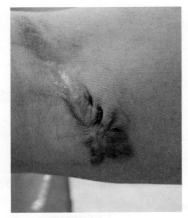

FIGURE 4-8 Keloid (excessive scar tissue formation) on an elbow.

4.3 learning outcome

Describe immunopathology.

Immunopathology

Immunopathology is pathology of the immune system and can involve acquired or congenital diseases. A more detailed discussion of the immune system is presented in Chapter 12. We will discuss a few specific situations here.

Immune reaction to transplanted tissues is commonly known as *transplant rejection*. On the surface of cells in our body are proteins called *major histocompatibility complex* or *human leukocyte antigens* (*HLAs*). HLAs received their name because they were first found on white blood cells (leukocytes). We now know that most cells in our body have these antigens on their cell surface. Mature red blood cells do not have HLAs. If there is not an appropriate match of HLA between the donor and the recipient of an organ transplant, the recipient's immune system will reject the transplant. The most likely donors (in descending order) for a successful transplant are an identical twin, sibling, or parent.

Immunodeficiency diseases can be classified as congenital (present at birth) or acquired (developed after birth). An example of congenital immunodeficiency disease is *primary antibody deficiency.* In this condition, the patient does not develop antibodies to protect from recurrent bacterial or viral infections as well as other problems.

> **U check**
> What are human leukocyte antigens?

HIV/AIDS

Human Immunodeficiency Virus (HIV)/Acquired Immunodeficiency Syndrome (AIDS) is the most well-known acquired immunodeficiency disease. The viruses HIV-1 and to a lesser extent HIV-2 are responsible for this disease. They are transmitted though body fluids such as blood, semen, vaginal fluid, and breast milk. Many cases of HIV infection are spread as a venereal disease (sexually transmitted infection, or STI), both homosexually and heterosexually. The virus is also passed through sharing of intravenous needles and from mother to fetus or mother to infant. There have also been relatively rare cases of transmission from patient to caregiver, mostly through blood exposure. The virus attacks the immune system by targeting T-lymphocytes. When the numbers of T-lymphocytes fall below certain levels, the patient becomes susceptible to a host of opportunistic infections that can make the individual ill and potentially be fatal. One of these diseases, Kaposi's sarcoma, is a cancer of the connective tissue around blood vessels. It appears as multiple lesions, typically on the lower extremities, but can also appear elsewhere (Figure 4-9). The increase of antiviral drugs to treat HIV infection has greatly increased the life span and quality of life of individuals infected with HIV.

Autoimmunity and autoimmune diseases occur when the body's immune system does not recognize itself and attacks its own tissues and organs. Basically, there is an inability of the body to distinguish self from non-self. Autoimmune diseases are more common in women than men and there appears to be a rise in the number of diagnoses. The reason for this is not clear. Autoimmune diseases can be very debilitating and even cause death. Some examples of autoimmune diseases are *systemic lupus erythematosus (SLE), rheumatoid scleroderma,* and *Sjögren syndrome.* SLE is a chronic autoimmune disease that may involve any organ, but it most often involves the kidneys, joints, serous membranes, and skin. *Scleroderma* is a progressive disease affecting blood vessels and excessive deposition of collagen in the skin and organs such as the lungs, gastrointestinal tract, heart, and kidneys. It occurs four times as often in women as men. *Sjögren syndrome* is an autoimmune disease characterized by dry eyes (*keratoconjunctivitis sicca*) and dry mouth (*xerostomia*). It is more common in individuals of Nordic ancestry.

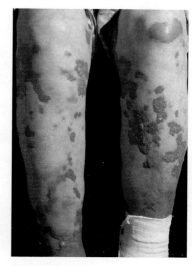

FIGURE 4-9
Typical lesions seen in Kaposi's sarcoma associated with AIDs.

U check
What cells are targeted in HIV/AIDS?

Neoplasia

Neoplasia ("new growth" in Greek) is the new and abnormal development of cells. **Neoplasms** are tissues that grow without the control of the usual restraints on cell proliferation. They are classified as benign or malignant. Because solid neoplasms are space-occupying lesions they are referred to as *tumors.* If the tumor stays localized in the tissue, it is designated as **benign.** If the tumor has the ability to spread to distant sites, it is referred to as **malignant** or *cancer*.

By definition, all cancers are malignant. Benign tumors often have the ending "oma" to their name. For example, a chondroma is a benign tumor of cartilage cells. Benign tumors typically stay localized, grow slowly, and have a connective tissue capsule surrounding them. Malignant tumors often have the ending **"carcinoma"** or **"sarcoma"** depending on whether the cancer is of epithelial or *mesenchymal* (mesodermal tissue that forms connective tissue and blood and smooth muscles) origin, respectively (Figure 4-10). For example, breast cancers are classified as carcinomas because they originate from epithelial cells. A cancer of bone is called an osteosarcoma because bone arises from mesenchyme. Cancers are usually faster growing, and because they do not have a fibrous connective tissue capsule they can invade and metastasize (spread).

U check
What is the difference between a carcinoma and a sarcoma?

neoplasm Tissues that grow without the control of the usual restraints on cell proliferation.

4.4 learning outcome
Explain neoplasia.

benign A tumor that stays localized in the tissue.

malignant Most tumors have the ability to spread to other sites of the body.

carcinoma A malignant tumor of epithelial origin.

sarcoma A malignant tumor of mesenchymal origin.

REMEMBER OUR lab technician, Eugene? Is he at risk for any neoplasms? What is the difference between a benign and malignant tumor?

metastasis The spread of tumor cells.

hyperchromatic Having a nucleus with a color that is darker than normal.

Invasion and **metastasis** involve destruction of localized tissue and the spread to distant sites. The spread (metastasis) to a distant site can be through the lymphatics (lymphatogenous) or through the blood (hematogenous). Carcinomas most often metastasize lymphatogenously and sarcomas hematogenously.

Grading and staging of cancers is how we determine the aggressiveness of a cancer. It is also used to develop a treatment plan and to give a prognosis. *Grading* is where the nucleus of the cancer cell is examined to see how it differs from a normal cell. Several changes distinguish a malignant cell from a normal cell. A malignant cell has a larger nucleus that takes up much more of the interior of a cell than does the nucleus of a healthy cell. The nucleus also is darker in color; the term for this is **hyperchromatic**. It indicates that more mitotic activity is going on in the malignant cell as it replicates itself many times over. The malignant cell nucleus will also have abnormal shapes and mitotic figures in it. Cancers are graded I through IV according to their degree of differentiation. *Staging* is used to determine how far the tumor has spread. It may show no spread, may involve only local lymph nodes, or may reveal spread to distant nodes and organs. Staging often uses a "TNM" classification, where T represents the primary tumor, N represents nodal involvement, and M stands for metastasis. Staging is more important than grading in predicting the outcome of the cancer.

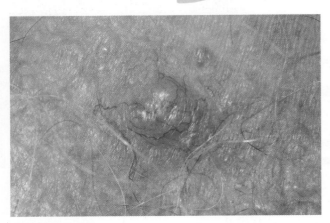

FIGURE 4-10 Lesion of basal cell carcinoma, the most common form of skin cancer.

> **U check**
> What is the difference between grading and staging of a cancer?

Developmental and Genetic Diseases

4.5 learning outcome

Identify causes of developmental and genetic diseases.

Developmental and genetic diseases are those that originate in utero or are passed on from our parents. Over 250,000 documented birth defects occur in newborns in the United States each year. Approximately 20 percent of these are due to heredity. About 6 percent are due to maternal or uterine factors, and over 70 percent are of unknown cause.

Teratology (Greek "teraton," for monster) is the study of developmental anomalies. **Teratogens** are chemical, physical, and biological agents that can cause birth defects. Teratogens affect each individual to a different extent. For example, fetal alcohol syndrome (Figure 4-11) involves a spectrum of physical and mental disabilities and malformations caused by intrauterine exposure to alcohol. Damages related to exposure to teratogens is also dependent on the developmental stage of the fetus. For instance, rubella (German measles) exposure causes problems for the fetus only during the first trimester of development. Teratogens, such as drugs, alcohol, and cigarettes, are also dose dependent. However, it is impossible to predict a safe dose of most substances for pregnant women. Therefore, pregnant women should avoid all teratogens.

teratogens Chemical, physical, and biological agents that can cause birth defects.

Chromosomal abnormalities are studied in *cytogenetics*. The normal number of chromosomes in somatic cells is 46—44 *autosomes* and 2 *sex chromosomes*. Abnormalities may include differences in the chromosome number, chromosomal deletions, and translocations. *Trisomy 21* or *Down syndrome* is an example of a syndrome that results from an abnormal number of chromosomes (Figure 4-12). The individual is born with three #21 chromosomes rather than the normal two. Down syndrome is the single most common cause of mental retardation. The risk of having a Down syndrome child increases significantly when a woman is older than 35.

In *multifactorial inheritance* both genetic and environmental factors play a role in the onset of the disease. Some examples of multifactorial inheritance include cleft palate, heart disease, and Alzheimer's disease.

Sex-linked disorders are those that are passed on through the sex chromosomes. The most common are those that are passed on through the mother's X chromosome. Hemophilia is an example of an X-linked disorder. Because the Y chromosome is very small, there are few genes located on it and there is a less of a chance for Y-linked disorders. An example of a Y-linked disorder is retinitis pigmentosa.

FIGURE 4-11 Fetal alcohol syndrome is a disorder seen in children who have been exposed to alcohol in utero.

Screening of genetic disorders has become more common and may involve screening of potential parents or prenatal screening during pregnancy. Screening is more often done in women over the age of 35, in families with a history of a child with a chromosomal abnormality or with a family history of X-linked disorders, and for known carriers of a genetic disorder.

FIGURE 4-12 Many people with Down syndrome lead productive and fulfilling lives.

U check
What is a teratogen?

Hemodynamic Disorders

4.6 learning outcome

Explain hemodynamic diseases.

Hemodynamics is the study of blood flow through the heart and blood vessels. We will give an overview of some conditions in this chapter, but more detail is given in Chapter 11, The Cardiovascular System.

Thrombosis is the formation of a thrombus (clot), which is an aggregate of coagulated blood containing platelets, fibrin, and other elements within a blood vessel. Our blood vessels develop thrombosis as result of atherosclerosis

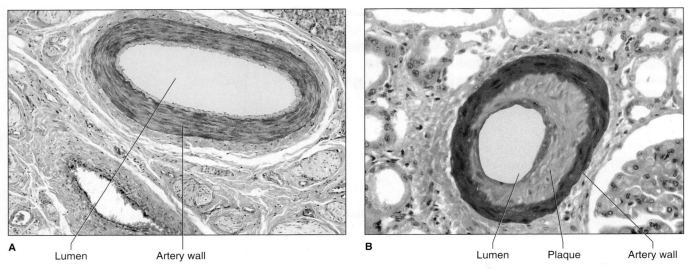

FIGURE 4-13 Atherosclerosis will cause the formation of plaque and thrombi on the inner wall of arteries: (A) light micrograph of a normal artery (50x); (B) the inner wall of an artery narrows as a result of atherosclerosis (100x).

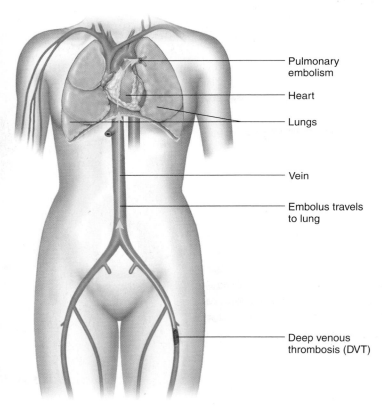

FIGURE 4-14 A pulmonary embolus; the majority of pulmonary emboli orginate from a deep venous thrombosis (DVT) in the lower extremities.

(Figure 4-13). The most common vessels to experience thrombosis are the coronary, cerebral, mesenteric, and renal arteries as well as the deep veins of the lower extremities (deep venous thrombosis, or DVT). The events of thrombosis are endothelial injury, alterations in blood flow, and increased coagulability of the blood. Once a thrombus is formed, several outcomes are possible: The thrombus may *lyse* or break up; it may *propagate* or increase in size; it may organize, which is the invasion of connective tissue into the thrombus; it may *canalize*, which is the formation of a new lumen or channel through the thrombus; and it may break off or undergo *embolization*.

Embolism is the movement of any substance through the arterial or venous circulations causing an obstruction (Figure 4-14). An embolus may be fat, bone, gas, tumors, amniotic fluid, or coagulated blood. The most common cause of an embolus is a thrombus that has broken loose in part or in whole from where it was formed.

Infarction is the death or necrosis of tissue distal to the occlusion of an artery. When you hear the term infarction, myocardial infarction (heart attack) is what probably comes to mind. But infarctions can occur in other organs such as the brain (cerebrovascular accident or stroke), intestines, kidneys, and lungs. Infarctions can cause serious illness, injury, and in some cases, death.

embolism An obstruction caused by the movement of a substance through arterial or venous circulations.

U check

What is the difference between a thrombus and an embolus?

Edema is the accumulation of excess fluid in the *interstitium*, or spaces between cells. Edema may be localized (in one area) or generalized (throughout the body). Examples of causes of localized edema are burns, lymphatic obstruction, immune reactions, and hives (Figure 4-15). Causes of generalized edema include heart failure, kidney failure, cirrhosis or other diseases of the liver, and starvation.

Fluid loss and overload can have serious effects by disturbing fluid compartments, electrolyte balance and hemodynamics. One condition related to fluid loss and fluid overload is *shock*, the inability of the circulatory system to adequately perfuse organs and tissues. Types of shock include:

- *Cardiogenic shock* is caused by pump or heart failure. Any condition such as a myocardial infarction or myocarditis that interferes with the pumping efficiency of the heart can cause cardiogenic shock.
- *Hypovolemic shock* occurs when there is a dramatic decrease in blood volume. Hemorrhage, fluid loss from severe burns, and severe diarrhea are some causes of hypovolemic shock.
- *Septic shock* occurs when there is a dangerously severe drop in blood pressure. It is caused by severe infections, especially those involving Gram-negative organisms.
- *Anaphylactic shock* is an extreme hypersensitivity (allergic) reaction that leads to widespread vasodilation.
- *Neurogenic shock* results from brain or spinal cord injury.

Fluid overload, an increase in volume of fluid within the body, also may result from congestive heart failure.

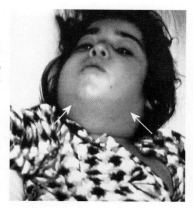

FIGURE 4-15
A young girl with localized edema of the pharyngeal area.

Immune response: Hypersensitivity

U check
What is the definition of *shock*?

Environmental and Nutritional Pathology

Environmental and nutritional pathology involve all aspects of our environment and diet that cause disease. In recent years, this has become a serious issue for health care providers and government leaders.

Smoking is a major health hazard that affects millions of people and has direct health costs that run in the billions of dollars each year. Smoking tobacco is the single largest factor for preventable deaths in America (Figure 4-16). As much as 30 percent of all cancer deaths are attributable to smoking and up to 30 percent of cardiovascular deaths are due to smoking tobacco. In addition, almost a third of lung disease deaths are due to smoking and 20 to 30 percent of low-birth-weight babies are probably due to smoking. The amount and duration of cigarette smoking is given in "*pack years*." A person who smokes 2 packs per day for 10 years would be a 20 pack year smoker ($2 \times 10 = 20$). How many pack years has Eugene, our lab technician, smoked?

Cigarette smoking is an independent risk factor for heart disease. Smoking also works together with hypertension and high serum cholesterol to magnify the negative impact on the heart. Smoking accounts for almost 90 percent

4.7 learning outcome

Summarize common environmental and nutritional disorders.

of lung cancer deaths. Lung cancer is the single most common cause of death due to cancer in both men and women.

Alcoholism is an addiction that has characteristics of dependence and withdrawal symptoms. There are about 10 million alcoholics in the United States (Figure 4-17). Men are more likely to be diagnosed with alcoholism, but the number of women being diagnosed is growing as is the number of young people who are being diagnosed with alcoholism. Certain ethnic groups have higher than average rates of alcoholism (for example, Native Americans) while others have lower rates (for example, Chinese and Jews). Alcoholism affects many organs, including the brain and nervous system. Individuals may experience brain atrophy, cerebellar degeneration, amnesia, and polyneuropathy, to name just a few disturbances. The liver typically will be affected. The person may first develop fatty liver, then alcoholic hepatitis, and finally cirrhosis. Other organ systems affected include the heart, pancreas, endocrine system, gastrointestinal tract, and bone.

FIGURE 4-16 Smoking is the number-one health hazard in the United States, responsible for more deaths and illnesses than any other single factor.

A

B

FIGURE 4-17 (A) Alcohol is the most abused substance in the United States; over 10 million people suffer from alcoholism; (B) One of the possible complications of alcoholism is cirrhosis of the liver.

Drug abuse is the use of any substance that is not according to accepted medical, social, or legal practices. Most abused drugs are taken to alter an individual's mood or perception. Some of the more commonly abused drugs are heroin, alcohol, marijuana, cocaine, barbiturates, psychedelic drugs (such as LSD), and various inhalants such as paint sprays, gasoline, and hair spray.

Iatrogenic injuries are the unintended injuries caused by physicians and other health care workers and by prescribed drugs or therapies. An example may be something as simple as an infection caused by health care professionals not washing their hands when treating patients. Another example is when there is excessive radiation exposure to patients who are not given a lead apron to protect their ovaries or testes when having dental x-rays taken.

Nosocomial injuries are those caused in a hospital setting that were not present in the individual before entering the hospital. For example, a *Candida* infection is a common nosocomial infection caused by the use of catheters. *Physical* injuries include a spectrum of injuries including physical and chemical trauma, electrocution, irradiation, and thermal injury. Physical injuries are often related to the age of a person as well as his or her occupation. For example, young adults have an increase in physical injuries when they begin to drive.

Nutritional pathology includes deficiencies and excesses. Deficiencies in calories and/or proteins may result in starvation. Starvation consists of many organ system changes. There is decreased body mass and diminished skeletal muscle, loss of subcutaneous fat, neurological changes, dermatological effects, and impaired immunity. Two examples of nutritional diseases that also have psychological components are anorexia nervosa (an unhealthy misconception of one's body weight resulting in self-starvation) and bulimia (binge and purge eating patterns). On a milder level, an individual may have vitamin or mineral deficiencies. On the other end of the spectrum is nutritional excess. Although there is hunger and even starvation in the United States, the more widespread problem is dietary excesses and obesity. *Body mass index (BMI)* is calculated as weight in kilograms per height^2. A BMI of 25 to 30 is classified as overweight, 30 to 40 as obese, and over 40 as morbidly obese (Figure 4-18).

Many diseases are attributable at least in part to obesity. These include type 2 diabetes mellitus, heart disease, and joint problems. Obesity is on the rise and is considered an epidemic in the United States.

iatrogenic Unintended injuries caused by physicians and other health care workers and by prescribed drugs or therapies.

FIGURE 4-18 Morbid obesity is of epidemic proportions in the United States. It contributes directly or indirectly to many diseases such as heart disease, cancer, and diabetes.

Obesity

U check
What is iatrogenic injury?

CHAPTER 4 Concepts of Disease

learning outcome 4.8

Describe infectious and parasitic diseases.

pathogenic Having the ability to cause disease.

Infectious and Parasitic Diseases

Worldwide, infections are the most common cause of illness and death. Infectious organisms include bacteria, viruses, fungi, and parasites. The study of these organisms constitutes the discipline of microbiology. We will further explore this discipline in Chapter 5, Concepts of Microbiology.

The terms *pathogenicity* and *virulence* mean the ability to cause disease and the strength of its pathogenicity, respectively. Not all organisms are **pathogenic** or have the ability to cause disease. The majority of bacteria are actually nonpathogenic. If two organisms can cause the same disease and one requires fewer organisms to do so, it is more *virulent* than the other organism. To be infectious, an organism must enter the body, avoid host defenses, grow, and utilize the host's resources.

Host factors have a role in infection. The host has defense mechanisms for protection. These include *physical barriers* such as the skin and mucosa. Physical barriers also include chemicals like lysozyme found in tears. *Inflammation* and our *immune system* also protect us from a daily onslaught of organisms in the environment. Other host factors make individuals more or less susceptible to infection such as heritable differences. Some individuals lack a cell receptor that makes them less vulnerable to some forms of malaria. Age is also an important consideration. At the two extremes of age, the very young and the very old are subject to certain diseases. Viral and bacterial disease may be more harmful in infants than in older children and adults. Pneumonias are more common in elderly people. Certain behaviors can put people at risk for disease. For example, intravenous drug abuse puts the individual at risk for diseases such as hepatitis B and C as well as many STIs. Compromised host defenses whether congenital or acquired also put people at risk for infections.

learning outcome 4.9

Distinguish among signs, symptoms, diagnosis and prognosis, or disease.

Disease Diagnosis and Prognosis

As you can see, diseases and disorders have a multitude of causes. The body responds in a variety of ways. The cause of the disease or disorder and the response of the body is what causes signs and symptoms. The diagnosis or discovery of the disease or disorder is based on these signs and symptoms. The prognosis or outcome of the disease or disorder is based upon many factors.

Signs and Symptoms of Diseases

Signs and symptoms are used by the clinician to make a diagnosis of a disease. *Signs* are objective findings, characteristics that are evident to someone besides the patient, and can be detected through the senses. Therefore, problems such as a heart murmur (hearing), a skin rash (sight), a fever (touch), and halitosis (smell) are all signs. *Symptoms* are subjective findings that for the most part are obvious only to the individual. Pain is an example of a symptom. Putting together the signs and symptoms a patient is experiencing along with the rest of his or her medical history is the beginning of the process of making a diagnosis.

> **U check**
> What is the difference between a sign and a symptom?

Etiology and Diagnoses of Diseases

Etiology is the cause of a disease. For example, it may be a bacterial infection for impetigo, smoking for lung cancer, or physical trauma for an abrasion. A *diagnosis* is the identification of a specific disease or condition. A combination of a medical history, physical examination, and diagnostic tests are used to make a diagnosis.

A patient may come in with a complaint of cough. The physician may think of a common cold as a *tentative diagnosis*—that is, a "temporary" diagnosis until more information is gathered. Once the patient history is taken, including a family medical history and the signs and symptoms of the patient complaint, the physician may revise the diagnosis. The physician will then most likely do a physical examination. If the patient relates that his cough has been ongoing for three weeks and that he has seen blood occasionally in his sputum and that he had no previous cold and no known exposure to anyone with tuberculosis, the physician may come up with a differential diagnosis. A **differential diagnosis** is a list of possible diagnoses ranked based on the current information available. In this case, the physician may come up with lung cancer, pneumonia, and tuberculosis as possibilities. She will then order specific tests to rule in or rule out the different possible diagnoses. In this case, a chest x-ray would be one of the first tests to be done. The physician may also order blood work or other lab tests. If after these tests are completed the physician is confident of her findings, she will make a *definitive diagnosis*—the diagnosis of which she is most certain. If her findings are inconclusive or if the condition does not respond to treatment, then she will order additional tests as she rethinks the diagnosis.

differential diagnosis A ranked list of possible diagnoses based on the information available.

Depending on the disease suspected, the physician may order a number of tests. Effectiveness and cost of the procedure(s) must be considered, however, so the physician needs to be selective in her choice of tests. The best test to diagnose a specific problem is called the *gold standard* for that disease. For example, a biopsy is the gold standard for most cancers. Besides biopsy, diagnostic tests include imaging: x-rays, CT (computerized tomography) scans, magnetic resonance imaging (MRI), angiography, and ultrasound. Blood tests and urinalyses are often done on healthy individuals to determine a baseline for the person. Auscultation with a stethoscope is a common technique to make a diagnosis (Figure 4-19). These are some of the more common tests, but more are being added as new technology becomes available.

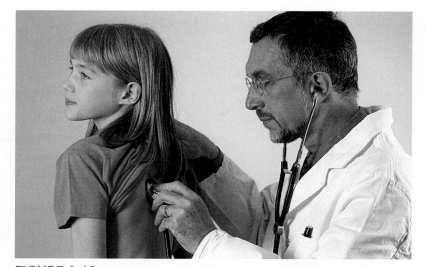

FIGURE 4-19 A health care provider uses as variety of methods such as auscultation to make a diagnosis.

U check

What is meant by a "gold standard" test?

from the perspective of...

A HEALTH INFORMATION TECHNOLOGIST A health information technologist manages the flow of information between health care providers and consumers. Why is it important to keep the patient information confidential?

Prognosis

Prognosis is the likely outcome of a disease. It may range from excellent with no complications up to and including lasting effects or death. The prognosis is based not only on the given disease, but also on several other considerations. The age and general health of the individual, how advanced the disease was at the time of the diagnosis, and the available treatment options all factor into the prognosis.

prognosis The likely outcome of a disease.

4.10 learning outcome

Explain prevention and treatment of diseases.

Prevention and Treatment of Diseases

"An ounce of prevention is worth a pound of cure" is very true for many diseases. A healthy lifestyle can prevent or reduce the progression of many diseases. Individuals should stay active, both physically and mentally; do low-impact exercises several times a week without overdoing it; stay mentally alert by challenging the brain with puzzles and games, and maintaining social contact with friends and people with similar interests; eat a balanced diet with plenty of fresh fruits and vegetables; stay hydrated and take vitamins and supplements if there is a risk of deficiency; get plenty of rest; see a physician on a regular basis for blood pressure checks; and be alert to any signs of cancer or other diseases such as diabetes.

The two main categories of treating diseases are the *allopathic* approach and the *complementary and alternative medicine (CAM)* approach.

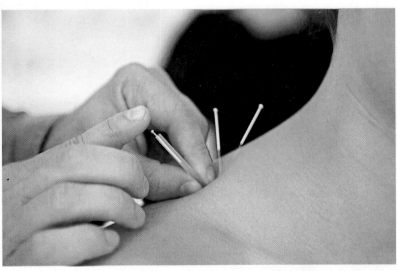

A   B

FIGURE 4-20 Herbal medicine (A) and acupuncture (B) are forms of complementary and alternative medicine.

Allopathic medicine is practiced by medical doctors (MDs) and doctors of osteopathy (DOs). Basically, this kind of medicine treats disease by the use of remedies that produce effects different from those produced by the disease itself. It is also called *conventional medicine*. It uses chemicals (antibiotics and chemotherapy), surgery, and radiation in much of its practice. CAM uses herbal medicines, chiropractic, aromatherapy, and a variety of Eastern medicine practices such as Ayurvedic medicine and traditional Chinese medicine (Figure 4-20). *Integrative medicine* is the approach that attempts to combine aspects of conventional medicine with CAM.

> **U check**
> What are three things you can do to help prevent disease?

Life Span

We are all aware that life expectancy has changed dramatically over the last 100 or so years. In 1900, the average life expectancy for men in the United States was 43.6 years. The 2010 projection has risen to 75.7 years. The average life expectancy for women in 1900 was 48.3 years and is projected to be 80.8 years in 2010. Factors that have impacted these statistics are women surviving childbirth, fewer infectious diseases, better access to health care, safer work environments, and higher awareness of healthy lifestyles. Of course disease still has a powerful effect on life expectancy, but with new research and genetic engineering it will not be surprising to see life expectancy rise further in the future. In the United States, however, concern still exists that our sedentary lifestyle will continue to increase the occurrence of childhood obesity, type 2 diabetes, hypertension, and other diseases.

4.11 learning outcome

Explain the effects of disease on human life span.

summary

	learning outcomes	key points
4.1	Compare cell injury and cell death.	Cell injury occurs when there is some type of insult to the cell. The cell can adapt or it will suffer cell death.
4.2	Distinguish among inflammation, repair, regeneration, and fibrosis.	Inflammation is a protective responsive that may sometimes become uncontrolled. It can be classified as acute (short term) or chronic (long term). Repair occurs when there is cell injury that is not too extensive. Regeneration is limited in humans. Fibrosis or scarring takes place when the injury is more severe or larger.
4.3	Describe immunopathology.	Immunopathology is the study of diseases of the immune system. These can be congenital, acquired, or autoimmune.

continued

learning outcomes	key points
4.4 Explain neoplasia.	*Neoplasia* means new growth and is the new and abnormal development of cells. Neoplasms are classified as benign or malignant (cancer).
4.5 Identify causes of developmental and genetic diseases.	Developmental and genetic diseases can occur because of exposure to teratogenic agents or chromosomal abnormalities and mutations.
4.6 Explain hemodynamic diseases.	Hemodynamic diseases have to do with any interference in blood flow. These include infarctions and emboli among other disorders such as shock.
4.7 Summarize common environmental and nutritional disorders.	These disorders include addictions such as tobacco use and alcoholism, and starvation and obesity.
4.8 Describe infectious and parasitic diseases.	Infectious diseases are some of the most common diseases in the world. They include bacterial, viral, fungal, and parasitic diseases.
4.9 Distinguish among signs, symptoms, diagnosis and prognosis, or diseases.	Signs are objective findings and symptoms are subjective findings. A variety of tools including the patient history, the physical examination, and laboratory tests are used to diagnose disease. Prognosis is the likely outcome of a disease.
4.10 Explain prevention and treatment of diseases.	Prevention entails having a healthy lifestyle with a balanced diet, exercise, and the avoidance of risk factors. Treatment can be by conventional medicine (allopathic approach), complementary and alternative medicine, or a combination of both (integrative medicine).
4.11 Explain the effects of disease on human life span.	Although science and medicine have made great strides in improving health, disease continues to have a dramatic effect on the quality of life and life expectancy.

chapter 4 review

case study 1 questions
Can you answer the following questions that pertain to Eugene, our lab technician?
1. What signs of disease is Eugene showing?
2. What symptoms is he experiencing?
3. What are possible diagnoses that must be considered?
4. What are the treatments for the various diagnoses you have listed?
5. Based on Eugene's signs and symptoms, which of the following may be the most helpful procedure to make a diagnosis?
   a. Urinalysis
   b. CBC
   c. Chest x-ray
   d. Antibody test
6. Based on the information you have, which one of the following is *least* likely to be Eugene's diagnosis?
   a. Pneumonia
   b. Tuberculosis
   c. Lung cancer
   d. Sjögren syndrome

review questions
1. Which of the following means the same as a tissue organ getting smaller?
   a. Hypertrophy
   b. Atrophy
   c. Metaplasia
   d. Dysplasia
2. The disordered growth and maturation of cells considered a precancerous condition is called
   a. Hypertrophy
   b. Atrophy
   c. Metaplasia
   d. Dysplasia
3. Which type of tumor stays localized in the tissue and is considered harmless, not cancerous?
   a. Malignant
   b. Sarcoma
   c. Benign
   d. Carcinoma
4. Harm or illness caused by a medical professional is referred to as
   a. Idiopathic
   b. Metastatic
   c. Iatrogenic
   d. Nosocomial
5. What is the term for pathological cell death?
   a. Apoptosis
   b. Metastasis
   c. Invasion
   d. Necrosis

critical thinking questions
1. A 17-year-old boy is experiencing fatigue and has a lesion on his scrotum. Based on what you have learned in this chapter, what are some appropriate tests that would help make a diagnosis?
2. What is the difference between apoptosis and necrosis?

patient education
As humans age, we are more susceptible to a variety of diseases. However, there are things that we can do to stay healthy—can you name a few?

applying what you know

1. Which of the following is the correct order of events in developing a treatment plan for a patient?
    a. Medical history, differential diagnosis, physical exam, definitive diagnosis, lab tests and diagnostic procedures
    b. Physical exam, definitive diagnosis, lab tests and diagnostic procedures, medical history, differential diagnosis
    c. Medical history, differential diagnosis, definitive diagnosis, lab tests and diagnostic procedures, physical exam
    d. Medical history, differential diagnosis, physical exam, lab tests and diagnostic procedures, definitive diagnosis

CASE STUDY 2   Down Syndrome

A couple in their late 40s are considering starting a family. They have heard about Down syndrome, but really do not understand what it is. They have come to you with some questions.

1. Who is at increased risk of having a child with Down syndrome?
    a. Women younger than 30 have an increased risk of having a child with Down syndrome
    b. Women with multiple pregnancies are at an increased risk of having a child with Down syndrome
    c. Women older than 35 have an increased risk of having a child with Down syndrome
    d. Women who drink alcohol during their pregnancy have an increased risk of having a Down syndrome child
2. Down syndrome is also known as
    a. Trisomy 15   c. Trisomy 21
    b. Trisomy 18   d. Trisomy 24
3. Down syndrome is the most common cause of
    a. Childhood deafness   c. Diabetes mellitus
    b. Mental retardation   d. Deafness

5
Concepts of Microbiology

chapter outline

- Microscopes
- Bacterial Cell Structure
- Bacteria Classification, Growth, and Reproduction
- Shapes and Arrangements of Bacteria
- Pathogenicity and Virulence
- Bacteria and Disease
- Virology
- Mycology
- Parasitology
- Sexually Transmitted Infections
- Life Span

outcomes

AFTER COMPLETING THIS CHAPTER, YOU WILL BE ABLE TO:

5.1 Identify the common types of microscopes used in microbiology and contrast magnification and resolution.

5.2 Describe the general structure of bacteria and explain the purpose of staining bacteria.

5.3 Explain bacteria and how it grows and reproduces.

5.4 Illustrate the shapes and arrangements of bacteria.

5.5 Explain the difference between pathogenicity and virulence.

5.6 Identify types of bacteria and the diseases they cause.

5.7 Infer why viruses are not considered living organisms, and compare hepatitis A and B.

5.8 Identify the different types of mycoses.

5.9 Identify types of parasites and the diseases they can produce.

5.10 Describe the major sexually transmitted infections.

5.11 Model how microbiology impacts life span.

case study

Use the case study to focus on as you go through the chapter. The questions will guide you as you learn the concepts of microbiology.

CASE STUDY 1 *Gonorrhea*

Michelle Lopane is a sexually active 17-year-old. She has heard from a friend that the boy she is seeing may have gonorrhea. Michelle is very concerned and has approached you because she knows you are a medical assistant. She knows that the disease is bad and even heard that she may not be able to have children because she has the disease. She hopes you can answer her questions.

As you go through the chapter, keep the following questions in mind:

1. What type of STI is gonorrhea?
2. What can Michelle do to prevent infection of gonorrhea?
3. What signs and symptoms she should watch for?
4. What are the complications of gonorrhea?

essential terms

aerobic (air-O-bik)
amphitrichous (am-fee-TRIK-us)
anaerobic (an-air-O-bik)
bacillus (buh-SIL-us)
bactericidal (bak-te-re-SI-dal)
bacteriostatic (bak-te-re-o-STA-tik)
coccus (KOK-us)
conjugation (kon-joo-GAY-shun)
decline phase (dee-KLINE faze)
endemic (en-DEM-ik)
eukaryotes (you-KARE-ee-otes)
exponential phase (eks-po-NEN-shul faze)
flagella (flah-JEL-lah)
lag phase (lag faze)
lophotrichous (lo-fo-TRIK-us)
monotrichous (mah-no-TRIK-us)
mycology (my-KOL-o-je)
pathogenicity (path-o-gen-IH-sit-e)
peritrichous (per-ee-TRIK-us)
pilus (PIE-lus)
prokaryotes (pro-KARE-ee-otes)
stationary phase (STAY-shun-air-ee faze)
virulence (VIR-yoo-lens)

Additional key terms in the chapter are italicized and defined in the glossary.

study tips

1. Make flash cards for the organisms discussed in the chapter.
2. Draw and label the bacterial growth curve.
3. Draw and label the different shapes and arrangements of bacteria and flagella.

Introduction

Microbiology is the science that studies organisms that are too small to study without the use of a microscope. Although this is the standard definition of microbiology, certain organisms (some parasites and fungi) can be studied without the use of a microscope. The four groups of organisms we will focus on are bacteria, viruses, fungi, and parasites. Each of these groups has members that have medical significance to humans. There are organisms that cause serious diseases. Tuberculosis, influenza, and anthrax are just three examples of potentially deadly diseases caused by microorganisms. However, others are harmless or even beneficial. For example, some organisms are used to make antibiotics and chemotherapeutic drugs.

The history of microbiology dates back at least to the mid-1300s when there was widespread use of quarantine to control the spread of bubonic plague. In the fifteenth century, Peruvian Indians used the bark of a willow tree (which contains the active ingredient quinine) to treat malaria. In the nineteenth century, Louis Pasteur used his knowledge of yeast to save the French wine industry and he also developed the vaccine for rabies. These are just some of the fascinating stories that surround the study of microbiology.

5.1 learning outcome

Identify the common types of microscopes used in microbiology and contrast magnification and resolution.

Microscopes

Studying the majority of organisms in this field requires the use of microscopes. In 1676, Antonie van Leeuwenhoek, a Dutch linen maker, developed the first simple microscope to observe defects in the cloth he purchased. He then used his microscope to inspect drops of water. He observed "little animals" that we know now were bacteria and protozoa (Figure 5-1).

The *light microscope* is the basic microscope that most students are familiar with (Figure 5-2A). It uses a light source to illuminate a specimen that is prepared and placed on a slide for examination. The two characteristics that we are concerned with for this type of microscope are magnification and resolution. *Magnification* is how greatly we can enlarge an object. The maximum magnifying power of a light microscope is approximately 1,000x (one thousand times the actual size of the object). *Resolution* is the ability to distinguish two objects as being separate and distinct objects. The maximum resolving power of the light microscope is about 200 nanometers (nm); there are 1 billion nanometers in 1 meter. Magnifying power and resolution are inversely related. As magnification increases, resolution decreases and vice versa. You can therefore recognize that at some point we are limited as to how much magnification we can get with a light microscope and still get a clear view of the object.

> **U check**
> What is the difference between magnification and resolution?

The *electron microscope* uses electrons rather than visible light to view objects. Electron microscopes have high resolving power which allows us to have greater magnification (Figure 5-2B). There are two types of electron microscopes: the *transmission electron microscope (TEM)* and the *scanning electron microscope (SEM)*. The TEM is similar in many respects to a light microscope except it uses an electron gun that shoots electrons at a specimen instead of light reflection. Some electrons pass through the specimen and some are diffracted. An image is then made on photographic film. The SEM has a lower resolving power than the TEM, but has the advantage of giving a three-dimensional image of the object.

Darkfield illumination is a type of viewing that does not use light. It is performed on the same type of microscope that bright field (light) microscopy is performed. A special condenser is used to block the direct light rays and light is deflected off a mirror at an oblique angle. Darkfield illumination technique is especially

FIGURE 5-1 Antonie van Leeuwenhoek, the inventor of the light microscope, observing protozoa in 1676.

A

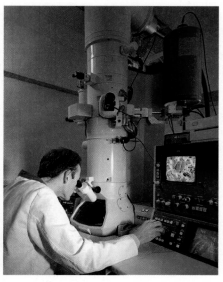

B

FIGURE 5-2 Two of the basic microscopes used in the study of microbiology are the light microscope (A) and the electron microscope (B).

CHAPTER 5 Concepts of Microbiology

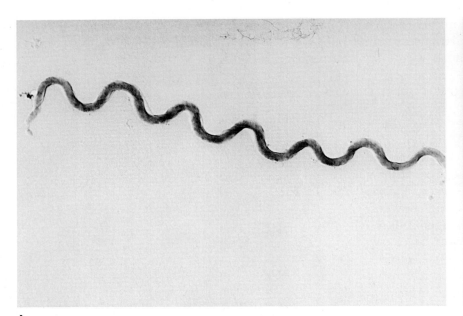
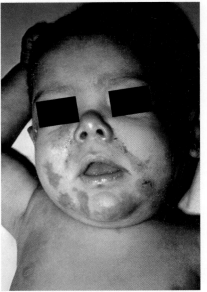

A   B

FIGURE 5-3
(A) *Treponema pallidum*, the organism that causes syphilis, is shaped like a corkscrew (or spirochete). (B) A baby with congenital syphilis.

useful for viewing organisms such as *Treponema pallidum* (the organism that causes syphilis). As seen in Figure 5-3, this organism can cause disease if transferred to a child at birth. We will discuss the organism and disease in more detail later in this chapter.

A third type of microscope, the *phase microscope,* uses light waves passing through transparent objects like cells in different phases based on the makeup of the materials they pass through. This microscope is particularly good at viewing the internal structures of living cells. A fourth type, a *confocal microscope,* uses laser light and computer-assisted image enhancement to give a three-dimensional image.

5.2 learning outcome

Describe the general structure of bacteria and explain the purpose of staining bacteria.

eukaroyote An organism that contains cells that have nuclei.

prokaryotes One-celled organisms that do not have a nucleus.

Bacterial Cell Structure

For a general discussion on cell structure, refer to Chapter 3, Concepts of Cells and Tissues. This section will focus on some of the unique or different features of bacteria.

Humans, animals, and plants contain cells that have nuclei and therefore are referred to as **eukaryotes.** Bacteria do not have a nucleus and are classified as **prokaryotes.** They are unicellular (one-celled) organisms. Bacteria also do not have organelles such as mitochondria or endoplasmic reticulum, but they do have a cell membrane that is similar to that in humans. Bacteria have a protective cell wall as do plants, but humans and other animals do not (Figure 5-4). This wall, which surrounds the entire bacterial cell, responds to certain stains that help in identifying the bacteria.

Two of the more common stains are the *Gram stain* and the acid fast stain. When using a Gram stain for identification, the bacteria is categorized depending on whether or not the cell wall retains the stain. Bacteria are classified as Gram-positive if the stain is retained or Gram-negative if the stain is not retained. Gram-positive bacteria have a cell wall structure that consists of two or three layers. Gram-negative organisms have a very complex cell wall with multiple layers. The staining characteristics of bacteria are important because some antibiotics work better against Gram-positive organisms while other antibiotics are more effective against Gram-negative bacteria.

Another feature of some bacteria is the presence of a *capsule* that surrounds the bacterium. The capsule consists of sugars as well as other components that make the capsule sticky and contribute to the invasiveness of some bacteria. An example is *Streptococcus mutans* that has a sticky capsule that allows it to adhere to tooth enamel and cause caries (tooth decay).

Another feature of some bacteria is flagella (singular, flagellum). **Flagella** are used for locomotion of the bacterium. In the human, the only cell that has a flagellum is the mature sperm cell for motility. In bacteria that have flagella, there are four possible arrangements (Figure 5-5). One arrangement is **monotrichous,** where there is a single flagellum on one end of the organism. A second arrangement is **lophotrichous,** where there are multiple flagella on one end of the cell. A third arrangement is **peritrichous,** where the flagella are distributed over the entire perimeter of the cell. A fourth type of flagellar arrangement is **amphitrichous,** where there is a flagellum on each of the opposite poles of the bacterial cell.

An additional special feature of bacteria is a **pilus** (Latin "hair," plural, pili). This small, hairlike structure is shorter than a flagellum but is made out of protein like a flagellum is (Figure 5-6). Pili have two functions: to attach to other cells, making the bacterium pathogenic; and to transmit

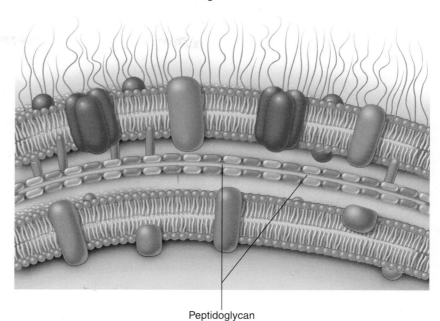

FIGURE 5-4 Bacteria, like plants but unlike humans and other animals, have a cell wall that provides additional protection to the organism.

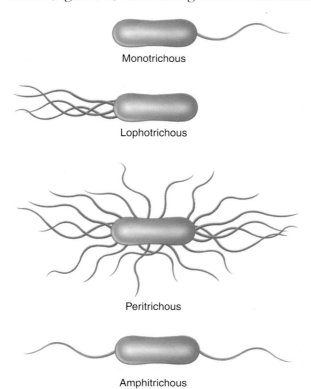

FIGURE 5-5 In bacteria that have flagella, there are four possible arrangements: monotrichous, lophotrichous, peritrichous, and amphitrichous.

flagella Structures on microorganisms used for locomotion.

monotrichous Having a single flagellum on one end of the organism.

lophotrichous Having multiple flagella on one end of the cell.

peritrichous Having flagella distributed over the entire perimeter of the cell.

amphitrichous Having flagella on opposite ends of a bacterial cell.

pilus A small, hairlike structure shorter than a flagellum but made of protein like a flagellum.

CHAPTER 5 Concepts of Microbiology

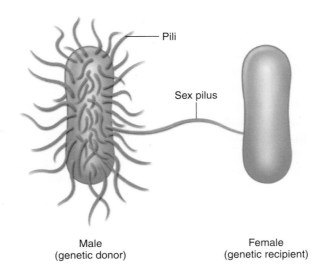

FIGURE 5-6
A pilus is a small hairlike structure that attaches to other cells to transmit disease or to exchange genetic materials with similar cells.

genetic material from one bacterial cell to a second bacterial cell, in which case it is referred to as a sex pilus.

> **U check**
> What is the difference between a eukaryote and a prokaryote?

Bacteria Classification, Growth, and Reproduction

LO 5.3

learning outcome
Explain bacteria and how it grows and reproduces.

Bacteria is organized and classified in order to better understand its positive and negative effects on the body. It grows and reproduces in a very unique fashion.

Bacteria Classification

Classification, nomenclature, and identification are parts of taxonomy. Bacteria are grouped or classified based on similarities between different cells. Bacteria, like all living things, are placed in taxonomic ranks from kingdom (for example, humans belong in the animal kingdom) through division, class, order, family, genus, and species (see Figure 5-7). Each rank is more restrictive than the previous rank.

We will be concerned with *genus* and *species,* referred to as binomial nomenclature ("two names") since that is how most organisms are known. For example, humans are also known by the term *Homo sapiens,* which means "wise man." The genus name is always listed first and is capitalized. The species name comes second and is not capitalized; both names are italicized, except when using them as common names. Therefore, *Homo* is the genus name and *sapiens* the species name for humankind.

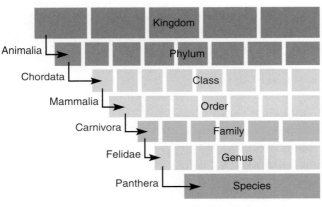

FIGURE 5-7 Taxonomy is a system for classifying and naming organisms.

> **U check**
> What is meant by binomial classification?

Growth and Death of Microorganisms

Growth is the orderly increase in the components of an organism, and so bacterial cell multiplication is a part of growth. Bacteria often are measured in terms of concentration—the number of *viable* (living) cells per unit volume of culture medium (Figure 5-8). Microorganisms can be cultured or grown on *media* that contain the nutrients and other factors necessary for their multiplication. Different types of media are based on the needs of the given bacteria. A common medium is agar, which is a gel derived from red algae. When bacterial cells are given unlimited nutrients and there are no negative factors present, bacteria grow exponentially. This means they will double in number in a given period of time. For example, if the doubling time for a certain bacterium is 20 minutes, its numbers will double every 20 minutes. If you start with 20 bacterial cells, after 60 minutes (that would be 3 doublings; one doubling every 20 minutes) you will have 40 (20 × 2), 80 (40 × 2), and finally 160 (80 × 2) bacterial cells after 1 hour.

The growth curve of bacteria has four phases (see Figure 5-9). The first phase is the **lag phase** where there is no evident growth in numbers. During this time, the bacteria are making enzymes and other products that will promote growth. The second phase is the **exponential phase** where there is very rapid growth. This continues until either nutrients are depleted and/or waste products accumulate in the environment interfering with growth. The third phase is the **stationary phase** where growth and death balance each other and there is neither growth nor decline. The final phase is the **decline phase** (or death phase). In this phase, death of bacteria exceeds the production of new cells.

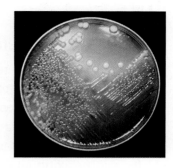

FIGURE 5-8 A medium is a special collection of nutritional factors for culturing or growing bacteria. In this image, spherical bacteria are seen growing on a cultured plate.

lag phase The first phase of the growth curve of bacteria in which there is no evident growth in numbers.

exponential phase The second phase of the growth curve of bacteria in which there is very rapid growth.

stationary phase The third phase of the growth curve of bacteria in which growth and death balance each other and there is neither growth nor decline.

decline phase The last phase of the growth curve of bacteria caused by the death of cells exceeding the production of cells.

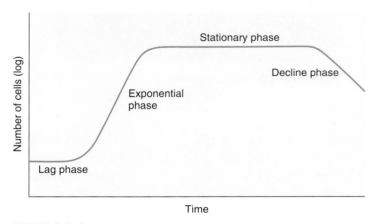

FIGURE 5-9 Bacterial growth curve showing its four phases.

Binary fission is an asexual process where a bacterial cell divides in half to produce two identical daughter cells. Some bacterial cells are capable of **conjugation,** where there is transmission of genetic information from one bacterium to a second. It is considered by some to be a primitive type of sexual reproduction.

Besides the normal factors influencing growth and death of bacteria, antimicrobial agents can influence the cell population. If an agent is

conjugation Transmission of genetic information from one bacterium to a second.

> **focus on** Wellness: Normal Flora
>
> *Normal flora* is the term used for the microorganisms that inhabit the skin and mucous membranes of healthy individuals. Although some bacteria are considered normal flora, viruses are never considered normal flora. Normal flora function to keep pathogenic organisms from invading and causing disease and may also produce substances needed by our bodies. For example, bacteria present in the gastrointestinal tract are responsible for synthesizing vitamin K. Normal flora may cause disease if they are introduced to "foreign" locations in the body or if they multiply in large numbers. An example is *Neisseria meningitidis* which is normal flora in up to 15 percent of the population. However, under some circumstances the organism can cause meningitis. Normal flora is an example of good bacteria that helps maintain our health and wellness.

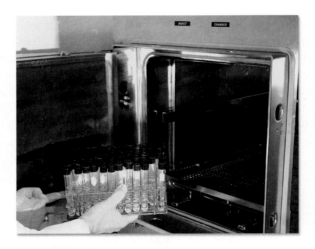

FIGURE 5-10 An autoclave is used to sterilize media and instruments (it is bactericidal).

bacteriostatic Unable to kill an organism but able to inhibit its multiplication.

bactericidal Able to kill bacteria.

coccus A bacterium that is spherical or oval in shape.

bacillus A bacterium that is longer than it is wide.

bacteriostatic, it does not kill the organism, but inhibits its multiplication. Agents that kill bacteria are called **bactericidal.** Examples would be sterilization (instruments in an autoclave, as shown in Figure 5-10), disinfectants (1:10 bleach solution), antiseptics, and some antibiotics.

> **U check**
> What are the four phases of the bacterial growth curve?

5.4 learning outcome

Illustrate the shapes and arrangements of bacteria.

Shapes and Arrangements of Bacteria

Bacteria have four basic shapes. A cell is described as a **coccus** (plural, cocci) if it is spherical or oval in shape. If it is longer than it is wide, it is called a rod or **bacillus** (plural, bacilli). A bacterium that is curved or has a

focus on Clinical Applications: Standard Precautions

Health care workers are taught to treat all patients with respect regardless of their status in society, their ability to pay for care, or their diagnosis. At the same time, all health care workers who provide patient contact must do all they can to reduce the spread of potentially contagious diseases and conditions. Standard precautions (formerly referred to as universal precautions) were developed to provide basic protection for both the patient and the caregiver. Standard precautions include the wearing of personal protective equipment (PPE) by caregivers such as gloves, gowns, masks, and goggles when there is the potential for exposure to "moist body substances." These substances include blood, all body fluids and secretions except perspiration, nonintact skin, and mucous membranes. Although breast milk is not included in this list, because it has been shown that HIV can be transmitted from mother to child via breast milk, it is generally treated in the same manner as other body fluids. Standard precautions are an important measure in preventing the transmission of disease in the health care setting and are used in all health care facilities.

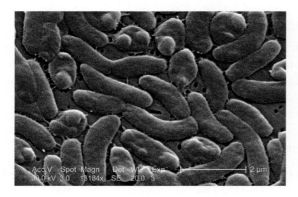

FIGURE 5-11 A scanning electron micrograph of *Vibrio vulnificus* bacteria, a comma-shaped organism similar to *Vibrio cholerae,* an organism responsible for many deaths throughout history.

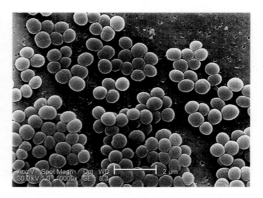

FIGURE 5-12 *Staphylococcus aureus* is a bacterium that is responsible for a variety of diseases and is arranged in grapelike clusters.

corkscrew appearance is a spirillum or *spirochete*. A curved, comma-shaped organism is called a *vibrio*, as is shown in Figure 5-11. A variety of arrangements is most common among the cocci. Cocci that appear in pairs have the prefix *diplo–*. If they appear in groups of four, they are referred to as "tetrads" or have the prefix *tetro–*. A group or "grapelike" cluster of cocci is denoted with the prefix *staphylo–* (see Figure 5-12), and cocci arranged like a chain or belt have the prefix *strepto–*. We will look at examples of each when we discuss specific bacteria later in the chapter.

U check

What does the prefix *staphylo–* mean?

CHAPTER 5 Concepts of Microbiology

5.5 learning outcome

Explain the difference between pathogenicity and virulence.

pathogenicity The ability to cause disease.

virulence The degree of pathogenicity of an organism that has the ability to cause disease.

Pathogenicity and Virulence

Pathogenicity refers to the ability to cause disease. Most bacteria are not pathogenic. Our discussion in this chapter will mostly address pathogenic bacteria, those that can cause disease. **Virulence** is the degree of pathogenicity of an organism that has the ability to cause disease. If you have two different species of bacteria that are capable of causing disease they are both termed pathogenic. If it takes a few hundred of one of the species to cause the disease, but several thousand of the second, then the first organism is described as being more virulent than the second organism.

> **U check**
> What is the difference between pathogenicity and virulence?

5.6 learning outcome

Identify types of bacteria and the diseases they cause.

aerobic Requiring oxygen.

Bacteria and Disease

There are numerous bacteria that cause disease. Understanding these bacteria can provide you a further understanding of the diseases they cause. The following sections discuss bacteria that cause disease.

Gram-Positive *Bacillus* and *Clostridium* Species

The various *bacillus* species includes large, **aerobic** (requiring oxygen) Gram-positive rods.

Bacillus Anthracis: *B. anthracis* causes anthrax. It commonly infects animals, but can infect humans. It enters the body through the skin (cutaneous anthrax), the lungs (inhalation anthrax), and rarely through the gastrointestinal mucosa (gastrointestinal anthrax). Ninety-five percent of cases are cutaneous. It is highly fatal, but can be treated with antibiotics if treatment is started early. This organism has taken on new importance in recent years because of the threat of terrorism.

Corynebacterium Species: *Corynebacterium* species are Gram-positive bacilli. Some members of this species are found as normal flora of the mouth and mucous membranes. *C. diphtheria* is the most important and well-known bacterium in this group. It causes diphtheria (whooping cough), which presents with a sore throat, inflammation in the respiratory tract, and fever and can be fatal. It is treated with antitoxins and antimicrobial drugs, and a vaccine (often in the form of diphtheria, pertussis, and tetanus [DPT]) is also available to prevent infection.

anaerobic Bacteria that do not require oxygen to live.

Clostridium Species: *Clostridium* species are large **anaerobic** (do not require oxygen to live) Gram-positive rods. Rather than cause disease themselves, these organisms produce toxins that actually cause the disease. The three species to be familiar with are *C. botulinum,* which causes botulism, a type of food poisoning associated with (damaged) canned or improperly prepared home-canned foods; *C. tetani,* which causes tetany; and *C. perfringens,* which causes a type of gangrene. Treatment includes antitoxins, antibiotics, wound debridement, and hyperbaric oxygen therapy.

> **U check**
> What is the name and type of organism that causes diphtheria?

PATHOPHYSIOLOGY

Clostridium Tetanus

C. tetani is a rod-shaped, anaerobic bacterium of the genus *Clostridium*. Like other *Clostridium* species, it is Gram-positive. It causes *tetanus,* which is commonly called lockjaw (*trismus*). This disease has a high mortality rate, especially in infants. Immediate treatment is necessary to prevent death or long-lasting effects. This disease is completely preventable through regular vaccinations and booster shots.

Etiology: A neurotoxin produced by the bacterium *C. tetani,* which lives naturally in soil and water, causes tetanus. A saprophyte is a bacterium or fungus that feeds on dead or decaying organic matter and in the case of *C. tetani* lives in soil. People most commonly acquire this bacterium through open wounds caused by objects contaminated with soil.

Signs and Symptoms: Symptoms usually appear between 5 and 10 days after infection. Muscle spasms in the jaw, neck, and facial muscles are usually the first signs. Other signs and symptoms include worsening of the muscle spasms that spread to other body locations and may cause bone fractures; dyspnea (breathing difficulties); irritability; fever; profuse sweating; and drooling. The diagnosis is usually based on the type of wound and the characteristic signs and symptoms of the disease. Tetanus antibody tests can also be used in making a diagnosis, but cultures of the wound site often produce false-negative findings.

Treatment: Administering antitoxin and antibiotics is the specified treatment. Additional treatments include wound cleaning, muscle relaxants, sedation, and bed rest. The insertion of an endotracheal tube and mechanical ventilation may be needed for patients with severe breathing difficulties.

Clostridium Species

Botulism is usually thought of as a disease that affects the gastrointestinal tract, but it can also affect various muscle groups. This disease most commonly affects infants. Although a person can survive this disease, its effects may be long-lasting. Interestingly, Botox, the drug used to erase wrinkles and treat some eye disorders such as strabismus, is derived from the neurotoxin produced by this organism.

Etiology: This disease is a rare but very serious disorder caused by the bacterium *C. botulinum,* which normally lives in soil and water. If this bacterium gets on food, it produces a toxin that can lead to a type of food poisoning. The foods most likely to contain

(continued)

Clostridium Species *(concluded)*

C. botulinum are (damaged or improperly prepared) canned vegetables, cured pork, raw fish, honey, and corn syrup. A person can also acquire this bacterium through open wounds that are not cleaned properly.

Signs and Symptoms: This disease causes many symptoms, including dysphagia (difficulty swallowing), paralysis, muscle weakness, nausea and vomiting, abdominal cramps, double vision, dyspnea (difficulty breathing), poor feeding and suckling in infants, the inability to urinate, the absence of reflexes, and constipation. The signs and symptoms usually appear 8 to 40 hours after the toxin is ingested. The definitive diagnosis is made by either a blood test to identify the toxin or an analysis of the suspected food.

Treatment: Treatment includes emergency hospitalization; intubation to maintain airways, including mechanical ventilation if respiratory muscles are impaired; intravenous fluids or nasogastric (NG) feeding if swallowing is impaired; and the administration of an antitoxin.

Prevention Tips: You can instruct patients to prevent botulism by observing the following guidelines:

- Never give honey or corn syrup to infants.
- Sterilize home-canned foods properly (250°F for 35 minutes).
- Do not use foods from bent or bulging cans (a bulging can is an indication that gases are being produced by the organism).
- Never eat foods that smell as if they may have spoiled.
- Cook and store foods properly.

U check
What is meant by an organism being a saprophyte?

Gram-Positive Cocci

Of the *Staphylococcus* species, *S. aureus* is the best known. It is a Gram-positive spherical organism that is arranged in grapelike clusters, from which it gets its name (staphylo). It also gives off a golden color when cultured, which is what "aureus" refers to (the element gold has the chemical symbol Au for this reason). Infections of the skin and mucous membranes are commonly caused by *S. aureus*. It is found on the skin and in the noses of up to 25 percent of healthy humans, and is also responsible for food poisoning. A common way to get this type of food poisoning is through contact with food prepared by food workers who carry the bacteria or through contaminated milk and cheeses. *Staphylococcus* is salt tolerant and can grow in salty foods like ham. Staphylococcal toxins are resistant to heat and cannot be destroyed by cooking. Foods at highest risk of contamination with *Staphylococcus* include sliced meat, puddings, some pastries, and sandwiches.

from the perspective of...

A SURGICAL TECHNICIAN A surgical technician assists in surgical operations under the supervision of surgeons or registered nurses. It is important to keep in mind the many pathological organisms that may cause illness in a hospital setting.

Streptococcus species are Gram-positive bacteria that form pairs or chains. *S. pyogenes* is responsible for "strep" throat. It also causes impetigo, a skin disease that is relatively common in children. The term *pyogenes* means "pus producing." Streptococcal infections are treated with antibiotics. *S. pneumonia* is another species that is also termed *Diplococcus pneumoniae* because the organism is found in pairs.

> **U check**
> What is the shape and arrangement of *S. aureus*?

Gram-Negative Rods

Enterobacteria are a group of Gram-negative bacteria that are rod shaped and live in the intestinal tract of animals and humans. They are among the most common pathogenic organisms, responsible for causing diseases including pneumonia, meningitis, diarrhea, food poisoning, and urinary tract infections. *Escherichia coli* (*E. coli*) is a part of the normal flora of the GI tract and yet is responsible for approximately 90 percent of first urinary tract infections in women. Infections caused by *E. coli* are usually treated successfully with a variety of antibacterial medications, but the infection can also be deadly.

> **U check**
> What organism is responsible for the majority of first time UTIs in women?

***Vibrios* and *Helicobacter*:** Vibrios are comma-shaped bacteria that are very common in surface waters. *Vibrio cholerae* is associated with contaminated water. It causes cholera, a disease that presents with nausea and a profuse diarrhea that is described as "rice water stools" because of its very watery consistency. Cholera epidemics have been responsible for thousands of deaths and the mortality rate without treatment is between 25 and 50 percent. Treatment is antibiotics and replacement of fluid and electrolytes lost due to the profuse diarrhea.

H. pylori is a spiral-shaped Gram-negative rod associated with the formation of duodenal (intestinal) and gastric (stomach) ulcers and even gastric cancer. It is treated with antibiotics and acid-suppressing agents.

Haemophilus Influenzae: *Haemophilus influenzae* is a rod-shaped Gram-negative bacterium found on the mucous membranes of the upper respiratory tract. It causes respiratory infections in children and adults and is a frequent cause of meningitis in children. Untreated, it can cause death in up to 90 percent of infected individuals. It is treated most frequently with the antibiotic ampicillin.

> **U check**
> What organism has been shown to be the cause of duodenal ulcers?

Mycobacteria

Mycobacteria are rod-shaped aerobic bacteria. They usually are identified through the acid fast stain. *Mycobacterium tuberculosis* is the organism that causes tuberculosis, or TB (Figure 5-13). TB most commonly involves the lungs, but can infect other organs such as the stomach, kidneys, and even the spine (TB of the spine is called Pott's disease). Approximately 90 percent of individuals who are infected with a primary case of TB have no symptoms. Secondary cases often include symptoms of cough with blood-tinged sputum. Reactivation can have serious consequences, including death.

TB can be difficult to treat and involves a combination of drugs. For example, isoniazid and rifampin are two major drugs used to treat the disease. The World Health Organization (WHO) estimates there were 9.4 million cases of TB in the world in 2008. We are also seeing more strains of TB that are resistant to many of the conventional medicines. *M. leprae* is another species of *mycobacterium* and is responsible for causing leprosy.

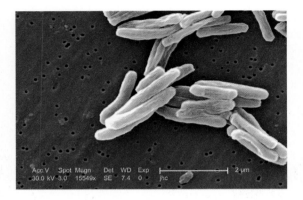

FIGURE 5-13 A scanning electron micrograph of *M. tuberculosis*, the rod-shaped organism responsible for tuberculosis.

> **U check**
> With what symptoms do the majority of individuals with primary tuberculosis present?

Focus on Clinical Applications: Tuberculosis

M. tuberculosis is a Gram-positive aerobic rod that causes tuberculosis (TB). It is a very slow-growing organism and requires a special medium in order to be cultured. It is spread through respiratory droplets and is more common in crowded conditions. Because TB is aerobic, it "prefers" an environment where oxygen concentrations are high. This is why the lungs are the primary organs affected. However, TB can infect any organ including the brain, kidneys, and spine. TB of the spine is called Pott's disease. Over 90 percent of individuals with primary TB are asymptomatic. Secondary TB is usually due to reactivation of a dormant lesion. Signs of secondary TB include a chronic cough with possible blood-tinged sputum, low-grade fever, night sweats, fatigue, anorexia, and weight loss. Treatment of TB includes a lengthy regimen of antibiotics with drugs such as isoniazid, rifampin, and ethambutol. Drug-resistant strains of TB have developed that are very difficult to treat. A vaccine is available for individuals who are at higher risk of contracting the disease.

Spirochetes

Spirochetes are spiral-shaped organisms that are very motile. *Treponema pallidum* is a spirochete that causes syphilis. The first symptom is usually a hard (indurated), painless lesion on the genitalia called a chancre. Syphilis has three stages, with the first two being contagious. It can have serious consequences if it progresses to the third stage, especially when involving the cardiovascular system or central nervous system. It is identified through darkfield microscopy. The treatment for syphilis is penicillin.

> **U check**
> What stages of syphilis are contagious?

Virology

Viruses are not considered living organisms, but are infectious particles. Their genetic material is either DNA or RNA. The infectious part of a virus is called a *virion*. Viruses infect bacteria and algae as well as plants and animals. To replicate, viruses need a host cell. This makes them very difficult to treat, because the host cell will suffer injury or death when the virus is treated. Viruses are spread through a variety of means and cause a wide spectrum of diseases including measles, mumps, polio, rabies, viral hepatitis, herpes (including shingles), and HIV/AIDS. Some viruses even cause cancer. The human papillomavirus (HPV) can cause cervical cancer of the uterus or cancer of the colon. Diagnosis is usually based on serology or blood tests. Treatment includes a variety of antiviral drugs and immunoglobulins. Although for many viral infections there is no cure, preventative vaccines are being developed. For some viral infections, such as chicken pox, the individual may develop immunity after an infection.

> **5.7 learning outcome**
> Infer why viruses are not considered living organisms, and compare hepatitis A and B.

> **U check**
> Why are viruses not considered living organisms?

PATHOPHYSIOLOGY

Viral Hepatitis

Hepatitis A, B, and C are the more common types of viral hepatitis. Hepatitis A and C have RNA for their genetic material, and hepatitis B has DNA for its genetic material.

Etiology: Hepatitis A (HAV) is spread through the fecal-oral route which means drinking water or food is contaminated with the virus. Rarely, it may be spread though sexual practices. Hepatitis B (HBV) is spread through blood and other body fluids, so it can be spread sexually. A percentage of those with acute HBV will go on to develop chronic HBV. Worldwide, it is estimated that approximately one-third of the population—nearly 2 billion people—have been

(continued)

Viral Hepatitis (concluded)

infected with HBV. Over 200 million of those individuals have chronic HBV. Hepatitis C (HCV) is spread through blood products. Some people with HCV have a past history of intravenous drug abuse. HCV can become chronic, and both HBC and HCV can progress to cirrhosis of the liver and, in some cases, liver cancer.

Signs and Symptoms: HAV causes a short, mild, flulike illness with nausea, vomiting, diarrhea, and a loss of appetite accompanied by weight loss. The patient typically shows signs of jaundice (yellow skin and whites of eyes, darker yellow urine, and pale feces), itchy skin (pruritus), and abdominal pain.

The symptoms of HBV usually appear about two to three months after infection and range from mild to severe. Symptoms are abdominal pain, dark urine, arthralgia (joint pain), loss of appetite, and nausea and vomiting.

Often individuals with HCV are asymptomatic. If symptoms are present, they are usually mild, nonspecific, and intermittent. Symptoms may include fatigue, mild right-upper-quadrant abdominal discomfort or tenderness, nausea, poor appetite, or muscle and joint pains.

Treatment: Hepatitis A has no specific treatment. Most people return to normal health within a few months. Patients should avoid alcohol and fatty foods because these can be hard for the liver to process. Infected individuals should get plenty of rest and follow a healthy diet. Hepatitis B is treated with antiviral medications, whereas HCV is often treated with alpha interferon.

U check
Which type of viral hepatitis is spread through the fecal-oral route?

5.8 learning outcome

Identify the different types of mycoses.

mycology The study of fungi.

Mycology

Mycology is the study of fungi (singular, fungus) of which there are more than 50,000 species. Fungi break down and recycle organic matter. Most species are harmless or even beneficial. For example, some fungi are responsible for the production of certain foods and wines. Other fungi are used to make antibiotics and immunosuppressive drugs. Like humans, fungi are eukaryotes—their cells possess a nucleus. Most fungi are obligate or facultative aerobes, which means they either need or can survive in oxygen. Fungal infections are called mycoses and can be difficult to treat. Candidiasis (thrush and vaginal infections) and dermatophytosis (ringworm) are the mycoses with the highest incidence of disease. Mycoses typically are

> **Focus on Wellness:** Preventing Infections
>
> Staying well by preventing infections and other illnesses is a goal of almost everyone. Where to start? The first "habit" is one we were all taught as small children: Wash your hands! Ask any infection control expert and these words will be the first to be uttered. Our hands, which allow us to accomplish so many tasks, also readily spread germs to others. Medical professionals are taught to wash their hands before and after treating or examining any patient, but what about in our everyday lives? Of course, we wash our hands after using the bathroom, but is it being done correctly? Towels, paper towel dispenser handles, faucets, and door handles all are touched hundreds, maybe even thousands, of times a day. When using a paper towel dispenser with a handle, dispense the required length of paper before washing. Turn on the faucet and dispense soap; wash your hands and leave the water running while you dry your hands. Now use the towel to turn off the faucet. You may also want to use the towel to open the bathroom door—there is no guarantee that everyone uses proper technique or even if everyone washed their hands after using the bathroom.
>
> Other infection prevention ideas include the following: Cough or sneeze into the bend of your arm, not your hand; receive a flu shot; keep tetanus and other immunizations up-to-date; get enough sleep each night (seven to eight hours for most people); and maintain a healthy diet with plenty of fruits and vegetables.

classified as superficial, cutaneous, subcutaneous, systemic, and opportunistic. We will look at one or two examples of each.

> **U check**
> What is the term given for a fungal infection?

Superficial Mycoses

Black piedra is an infection of the hair shaft caused by the fungus *Piedraia hortai*. Axillary, pubic, beard, and scalp hair may be affected. It is treated with antifungal medications such as imidazole (many antifungal medications can be recognized by their *–azole* suffix, or ending).

Cutaneous Mycoses

Cutaneous mycoses infect the superficial keratinized tissue such as skin, hair, and nails. Among the most common infections of this type are the tineas. These are better known to many as ringworm because of their appearance of having a red, serpentinelike border and a clear center. They are named after the body part affected—*tinea pedis* (athlete's foot), *tinea cruris* (jock itch), *tinea capitis* (head, see photo in Figure 5-14 for ringworm of the scalp), *tinea unguium* (nails), and *tinea barbae* (beard). Peeling and cracking of the skin with pruritus (itchiness) and pain are common symptoms.

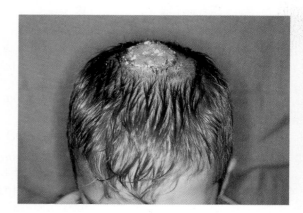

FIGURE 5-14 *Tinea capitis* (ringworm of the scalp) is a contagious fungal infection.

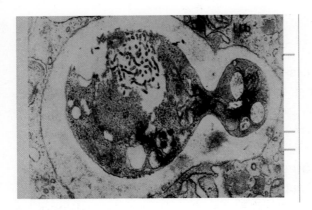

FIGURE 5-15 *Histoplasma capsulatum* is the most common cause of fungal pulmonary infections in humans and animals.

endemic Very common.

Subcutaneous Mycoses

Sporothrix schenckii is a fungus associated with horticultural plants such as roses. People who work in the horticultural industry or who work frequently in gardens are more susceptible to this infection. This organism can cause a chronic infection that may be limited to the subcutaneous tissues or, in rare cases, becomes systemic and life threatening.

Systemic Mycoses

Histoplasma capsulatum is a saprophyte and the most common cause of fungal pulmonary infections in humans and animals. Histoplasmosis is contracted through the inhalation of dried bird droppings, especially pigeons and chickens (Figure 5-15 shows a photo of *Histoplasma capsulatum*). It is **endemic** (very common) in the Ohio and Mississippi river valleys of North America.

Opportunistic Mycoses

Individuals with poorly developed or compromised immune systems are at risk for infections. This means that newborns, diabetics, and people with HIV/AIDS would be more susceptible to these infections. Opportunistic mycoses are fungi that are often ubiquitous (found everywhere), but have little impact on healthy people. Perhaps the best example is *Candida albicans,* which is a type of yeast infection. *Candida* is common in moist parts of the body such as the oral mucosa, vagina, axilla, and groin and where the skin overlaps due to obesity. Unfortunately, *Candida* infections are quite common in hospital settings due to improper hygiene by hospital personnel.

5.9 learning outcome

Identify types of parasites and the diseases they can produce.

Parasitology

A *parasite* is an organism that needs a host to survive. A "successful" parasite is one that does not kill its host. If the host is killed, the parasite must look for a new "home." A parasite grows, feeds, and is sheltered by the host while providing nothing of benefit to the host. The major categories of parasites are protozoa, helminthes ("worms"), and arthropods.

Protozoa and Helminths

Giardia lamblia is a common pathogenic protozoan found in the intestinal tract of humans. It can cause watery, greasy, foul-smelling stools. The individual may also experience malaise, weakness and weight loss, cramping, distention, and flatulence. It is treated with metronidiazole (Flagyl®) which is effective in over 90 percent of infections. Helminths are parasitic worms such as tapeworms, roundworms, and hookworms.

> **U check**
> What is the definition of a parasite?

UNIT 1 The Human Body and Disease

PATHOPHYSIOLOGY

Trichinella Spiralis

Trichinosis is an infection caused by *Trichinella spiralis*.

Etiology: This disease is caused by worms that are usually ingested by eating undercooked meat, especially pork, bear, and other wild game animals (Figure 5-16). Once ingested, the worms can leave the digestive tract and infect skeletal muscles, the heart, lungs, and brain. Proper cooking of all meat will prevent trichinosis.

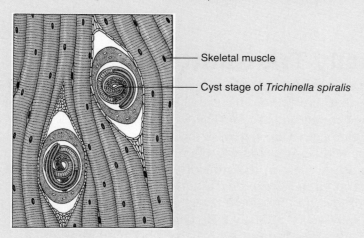

FIGURE 5-16 Diagram of trichinella *spiralis,* the organism that causes trichinosis. The organism has an affinity for skeletal muscle and is seen here in the cyst stage.

Signs and Symptoms: Common symptoms include abdominal pain, diarrhea, muscle pain, fever, and pneumonia. In more serious cases, arrhythmias (irregular heart rhythms), heart failure, and encephalitis (swelling of the brain) can result. The diagnosis is usually based on the patient's symptoms. When people are infected with parasites, there is typically an increase in their eosinophil (a type of white blood cell) count. Therefore, a blood test is done to determine if there is an increase in eosinophils. Sometimes, a muscle biopsy reveals the presence of the worms.

Treatment: Patients with trichinosis are treated with medications such as mebendazole to kill worms in the digestive tract and with anti-inflammatory drugs to reduce muscle pain and swelling. There is no cure for this disease once the worms leave the digestive tract and infect other tissues.

Arthropods

Lice are an example of a parasitic arthropod. Three types of lice have significance to human health: *Pediculus humanus capitis* (head louse), *P. humanus*

FIGURE 5-17 Light micrograph of a female human body louse.

humanus (body louse, as shown in Figure 5-17), and *P. thirus pubis* (crab louse). Lice (singular is louse) feed on the blood of their host. In the developed world, up to 10 percent of children are infested with head lice. Infestations cause itching of the scalp and may also result in fever, aches, and secondary infections. Diagnosis is based on finding live lice or empty egg shells (nits) attached to hair. Shampoos containing 1 percent benzene hexachloride (Kwell©) should be used for most cases of infestation. Sheets, clothing (including hats), and personal care items such as combs and brushes must also be treated to avoid reinfection.

5.10 learning outcome

Compare the major sexually transmitted infections.

Sexually Transmitted Infections

Sexually transmitted infections diseases (STIs) have been around for centuries. Note that they were once called sexually transmitted diseases. Herpes was around in ancient Greece. The term *herpes* in Greek means "to creep." This is a reference to the spread of the lesions on the skin. Syphilis was common in the Middle Ages and was often treated with mercury, which gave rise to the phrase "a night in the arms of Venus leads to a lifetime on Mercury." Another term for STI is *venereal disease; venereal* is derived from Venus, the goddess of love. In the United States, prior to 1960, syphilis and gonorrhea were the only major STIs. In 1976, chlamydia was first recognized; in 1981, AIDS was identified; and in 1982, herpes became prevalent. In 1992, pelvic inflammatory disease (PID) was recognized as being related to infections with certain STIs, and in 1996 human papillomavirus (HPV) was recognized as the cause of up to 90 percent of cervical cancers. National reporting of gonorrhea and syphilis began in 1941 and national reporting for chlamydia began in 1984. STIs have had, and continue to have, a dramatic impact on the health and lives of millions of individuals. Women continue to bear the brunt of most of the long-term consequences of many of the STIs.

PATHOPHYSIOLOGY

Chlamydia Trachomatis

Chlamydia is the most commonly nationally reported STI in the United States. Chlamydia may be grossly underreported because, for women, there are often no symptoms until the disease has spread. Its incidence, like gonorrhea and syphilis, is on the rise in this country.

Etiology: Chlamydia is caused by the bacterium *C. trachomatis*.

Signs and Symptoms: For females, there may be no symptoms. If they do occur, it will be two to three weeks after exposure. Symptoms may include abnormal vaginal discharge and burning on urination. As the disease progresses to other reproductive organs (pelvic inflammatory disease, or PID), there may be abdominal pain, pain during intercourse, and intermenstrual bleeding. Men may have a penile discharge and pain on urination.

In women, PID is a common cause of infertility; men seldom have serious complications related to chlamydia.

Treatment: It is important for both partners to be treated to avoid re-infection. Effective antibiotics against chlamydia include azithromycin and doxycycline. Patients must complete the medication cycle and avoid sexual contact until both partners are cured. Because of the high rate of reinfection, women should be retested three to four months after treatment.

Neisseria Gonorrhoeae

N. gonorrhoeae is a Gram-negative coccus that is a common sexually transmitted infection. Gonorrhea is the second most nationally reported STI in the United States. Over 1 million cases were reported each year from 1976 to 1980. Gonorrhea cases then declined steadily, but have started to rise again in the past two years. It is a highly infectious organism with a 20 to 30 percent chance of infection in men from a single exposure. In women the risk is even greater. The infected female is often asymptomatic, but gonorrhea can cause infertility by damaging the fallopian tubes. In the past, gonorrhea has been treated successfully with penicillin, but resistance has developed. Therefore, other antibiotics may have to be used for treatment.

Etiology: Gonorrhea is caused by *N. gonorrhoeae*, a bacterium that thrives in the warm, moist areas of the reproductive tract, urethral tract, mouth, throat, eyes, and anus.

Signs and Symptoms: Men often have no symptoms. If they do appear, it will be two to three days after exposure, when the patient will experience burning on urination and/or white, yellow, or greenish penile discharge. Women may also be asymptomatic, but common symptoms include dysuria, increased vaginal discharge, and intermenstrual bleeding. Gonorrheal infections, like chlamydia, may lead to PID in women. In men, they may lead to epididymitis.

Treatment: Antibiotics are effective against gonorrhea and both partners must be treated. However, drug-resistant strains of gonorrhea have developed, especially since the Vietnam War, making successful treatment more difficult. It is not unusual for a patient with gonorrhea to also be diagnosed with chlamydia or other STIs, which are usually tested for when testing for gonorrhea is taking place. In addition, gonorrhea lives well and actively in the throat,

(continued)

> **YOU MAY REMEMBER** that Michelle, in the case study, was concerned about her fertility.

Neisseria Gonorrhoeae (concluded)

even if there is not an active genitourinary infection, which can lead to additional coinfection.

Treponema Pallidum

Syphilis rates, once on the verge of elimination, rose for the seventh consecutive year in 2009 according to the Centers for Disease Control and Prevention (CDC). It is increasing most rapidly in males, especially in those who have sex with other males. Experts say because some of these men also have sex with women, it is inevitable that we will start to see an increase in syphilis rates in women as well.

Etiology: Syphilis is caused by the bacterium *T. pallidum,* a spirochete.

Signs and Symptoms: Primary-stage syphilis usually involves the appearance of a hard, painless sore known as a *chancre,* which appears 10 to 90 days after exposure. It will remain for three to six weeks, disappearing even without treatment. If untreated, the disease lies dormant, progressing to the second stage, which is characterized by a nonpruritic rash and lesions in the mucus membranes. The rash may be associated with flulike symptoms. Again, all symptoms disappear without treatment. After a long latent period without symptoms, often years later, the third stage becomes apparent with damage to the brain, eyes, heart, blood vessels, liver, and bones. This damage may lead to muscular incoordination, paralysis, numbness, blindness, dementia, and, finally death. The first two stages are contagious; the third is not.

Treatment: The cure for syphilis in its early stage is a single dose of intramuscular (IM) penicillin. Additional doses are needed for disease that has been present longer than a year. Other antibiotics are also effective for patients allergic to penicillin. Sexual contact must be avoided until treatment is completed to avoid further spread of the disease.

> **U check**
> Which STIs are required to be reported nationally?

Herpes Simplex Virus

Herpes simplex virus infections include those caused by both herpes simplex 1 (HSV-1) and herpes simplex 2 (HSV-2).

Etiology: Herpes viruses cause both infections. In most cases, HSV-1 causes oral blisters or vesicles known as cold sores (see

Figure 5-18). HSV-2 causes what is commonly known as genital herpes—although HSV-1 has also been known to cause the genital type of infection. Herpes infection may also be passed from an infected mother to her child during pregnancy and birth with potentially fatal outcomes.

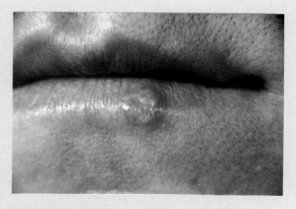

FIGURE 5-18 A cold sore caused by herpes simplex virus 1.

Signs and Symptoms: Many infected persons have minimal or no symptoms. Typical symptoms of genital herpes include vesicles (blisters) on or around the genitals or rectum. These blisters break, leaving tender ulcers for two to four weeks. The number and severity of outbreaks tend to decrease over a period of years.

Treatment: There is no treatment to rid the patient of the herpes virus; however, antiviral medications such as Acyclovir® may shorten outbreaks when they occur. Daily suppressive therapies with medications like Valtrex® may reduce the risk of transmission. Pregnant women with active outbreaks should deliver the child via cesarean (C) section to avoid exposing the child to an infected birth canal.

Trichomoniasis

Trichomoniasis (also known as "trich") is a common, curable STI.

Etiology: *Trichomonas* infection is caused by the protozoan parasite *T. vaginalis*. It has a short, undulating flagellum on one end of its body and four flagella on its anterior aspect. It moves by a rotating motion.

Signs and Symptoms: The vagina is the most common site of infection for females; the urethra for males. Male patients may have penile irritation, dysuria, or a mild penile discharge, but more often than not, men have no symptoms. Females often have a frothy yellow-green vaginal discharge with a strong "fishy" odor to it. Itching and irritation of the vulva is also common.

Treatment: Oral metronidiazole (Flagyl) is the treatment of choice. Both partners, even those who are asymptomatic, must be treated to avoid reinfection. Sexual contact should be avoided until after treatment is completed.

(continued)

Condyloma Acuminatum

Condyloma acuminatum, also known as genital warts, is one of the most common STIs in the world. It is important to note that not everyone infected with human papillomavirus (HPV) has symptoms. HPV has been implicated in an increased risk of cervical cancer in women and colon cancer in homosexual men.

Etiology: Human papillomavirus, or HPV.

Signs and Symptoms: Genital warts appear weeks or months after infection in the vulva, vagina, and cervix in women. In men, warts appear on the scrotum and penis. They have also been known to appear around the anus and on the thighs and groin area. It is important to note that as a viral infection, HPV can be spread even if the patient has no outward signs of the disease.

Treatment: Imiquimod cream, 20 percent podophyllin antimiotic solutions, and TCA (trichloroacetic acid) may be used to remove warts. Cryosurgery, cautery, and laser surgery may also be used to remove the warts; however, the virus remains in the patient's body. Currently, there is no treatment to rid the body of the virus. There are vaccines to prevent infection with HPV. Gardasil® is an HPV vaccine that helps protect against four types of HPV. In females ages 9 to 26 the vaccine helps protect against two types of HPV that are responsible for up to 75 percent of cervical cancers. It is also effective against two types of HPV that cause 90 percent of genital warts. In males ages 9 to 26 the vaccine gives protection in 90 percent of genital warts cases.

> **U check**
> What is the cause of the majority of cervical cancer cases?

HIV/AIDS

AIDS (acquired immunodeficiency syndrome) is caused by HIV (human immunodeficiency virus). It may take 10 or more years for a person to progress from asymptomatic HIV infection to AIDS, and this time is lengthening because of improved testing methods and improved antiviral medications. During the time the virus is destroying T-lymphocytes, an infected person may be spreading the infection before a diagnosis is made. The majority of adults with the disease are between the ages of 25 and 45.

The greatest risk factor for contracting the virus is unprotected sexual activity.

Homosexual men constitute a large percentage of individuals living with AIDS in the United States, but virtually no one is immune. African American men and women as well as Hispanics/Latinos make up a disproportionate percentage of people infected with HIV when compared to their numbers in the population as a whole. According to the CDC, in 2006, AIDS was the third leading cause of death for black males and black females aged 35 to 44, and the fourth leading cause of death for Hispanic/Latino males and females in the same age range.

Etiology: HIV is contracted through infection by the human immunodeficiency virus. It is spread through sexual contact, intravenous drug abuse, and body fluids.

Signs and Symptoms: Numerous signs and symptoms include decreased T-cell count, flulike symptoms, and a host of opportunistic infections such as *Pneumocystis carinii* pneumonia (PCP), Kaposi's sarcoma (KS), and cytomegalovirus (CMV) infection.

Treatment: Antiviral medications have been successful in decreasing the viral load and maintaining T-cell counts in some patients, but there are many side effects to these drugs and a strict medicine regime. Other medications include drugs to support the immune system and to treat opportunistic infections as they arise.

U check
What is the difference between HIV and AIDS?

Life Span

Infection in the Unites States is not the problem it was a century ago. However, with the emergence of drug-resistant strains of bacteria such as methicillin-resistant *S. aureus* (MRSA) and *M. tuberculosis*, infection once again is taking on increased significance. Infection also remains a major cause of morbidity and mortality in other parts of the world. Diseases such as malaria and polio still have an impact on third-world countries. This is why it is necessary to have an understanding of microbiology and the steps that can be taken to reduce the transmission and spread of infection.

5.11 learning outcome

Model how microbiology impacts life span.

CHAPTER 5 Concepts of Microbiology

summary

chapter 5

learning outcomes	key points
5.1 Identify the common types of microscopes used in microbiology and contrast magnification and resolution.	The most commonly used microscopes are the light microscope, electron microscopes, phase microscope. Magnification is how large an object can be made with a microscope. Resolution is the ability to distinguish two objects as being separate and distinct objects.
5.2 Describe the general structure of bacteria and explain the purpose of staining bacteria.	Flagella are hairlike structures that are used for locomotion. A monotrichous arrangement is where there is a single hair on one end of the bacterium; lophotrichous is a tuft of flagella at one end, peritrichous is where flagella are distributed over the entire perimeter of the bacterium, and amphitricous includes flagella on each end. Stains are used in the identification of bacteria. Two of the more commonly used stains are the Gram stain and the acid fast stain.
5.3 Explain bacteria and how it grows and reproduces.	There are four phases to the growth of bacteria: the lag phase is flat, showing no growth but where nutrients are being produced and the cell is getting ready for rapid growth. The exponential phase follows and is a period of very rapid growth. The stationary phase is flat, indicating a "balance" between the production and death of cells. The last phase is the decline phase and is caused by the death of cells exceeding the production of cells. Bacteria reproduce asexually by binary fission and a type of reproduction known as conjugation.
5.4 Illustrate the shapes and arrangements of bacteria.	The major shapes of bacteria are spheres (cocci), rods (bacilli), spirochetes, and vibrios. They can be arranged individually or in pairs, groups of four, clusters, and chains.
5.5 Explain the difference between pathogenicity and virulence.	Pathogenicity is the ability of an organism to cause disease. Virulence is how strongly pathogenic an organism is.
5.6 Identify types of bacteria and the diseases they cause.	*Clostridium C. tetani* is a Gram-positive anaerobic bacillus. It is the organism that causes tetanus. *Staphylococcus S. aureus* is a Gram-positive spherical organism that is arranged in "grape-like" clusters. It is part of the normal flora for many people and also can cause many skin infections, food poisoning, and more serious systemic infections. *Mycobacterium M. tuberculosis* is the aerobic, Gram-positive bacillus that is responsible for causing tuberculosis.

chapter 5

learning outcomes	key points
5.7 Infer why viruses are not considered living organisms, and compare hepatitis A and hepatitis B.	Viruses are considered "particles" and not living organisms because they must rely on a host cell for their reproduction. Hepatitis A is an RNA virus that is spread through the fecal oral route and the disease it produces is typically self-limiting. Hepatitis B is a DNA virus and is spread through body fluids. It has both an acute and chronic phase and can lead to cirrhosis and even liver cancer.
5.8 Identify the different types of mycoses.	Tinea is ringworm, a fungal infection. It is named after the body part infected. For example, *T. capitis* is ringworm of the head, and *T. pedis* is ringworm of the feet (athlete's foot).
5.9 Identify types of parasites and the diseases they can produce.	Parasites are organisms that require a host to survive. There are three major categories of parasite; protozoa that are common pathogens, helminthes (worms), and arthropods that include organisms such as lice.
5.10 Compare the major sexually transmitted infections.	Syphilis is caused by *T. pallidum*. A hard, painless lesion is seen in the first stage. The second stage may see flulike symptoms and a rash, or there may be no symptoms. The third stage may see serious damage to organ systems, including the nervous system and the cardiovascular system. The first two stages are contagious; the third stage is not.
5.11 Model how microbiology impacts life span.	The organisms that are studied in microbiology have both improved and endangered our lives. They are responsible for foods such cheeses, breads, and wines. They are used in the manufacture of many of our antibiotics and they are also the reason for millions of illnesses and deaths over the centuries.

chapter 5 review

case study 1 questions
Can you answer the following questions that pertain to Michelle's case study at the beginning of this chapter?

1. What type of STI is gonorrhea?
   - a. Bacterial infection
   - b. Viral infection
   - c. Parasitic infection
   - d. Fungal infection
2. Which of the following is the surest way of preventing a gonorrhea infection?
   - a. Shower after intercourse
   - b. Use a condom
   - c. Use spermicide
   - d. Practice abstinence
3. All of the following are possible symptoms of a gonorrheal infection *except*
   - a. Pain on intercourse
   - b. Yellow or bloody vaginal discharge
   - c. Otitis media
   - d. Pain or a burning sensation when urinating
4. All of the following are possible complications of a gonorrheal infection *except*
   - a. Epididymitis
   - b. PID
   - c. Miscarriage
   - d. Endometriosis

review questions

1. The ability to distinguish two objects as being separate and distinct is called
   - a. Resolution
   - b. Pathogenicity
   - c. Magnification
   - d. Virulence
2. Which of the following is used to diagnose syphilis?
   - a. Brightfield illumination
   - b. Darkfield illumination
   - c. Scanning electron microscopy
   - d. Transmission electron microscopy
3. A bacterium that has flagella arranged as a tuft at one end of the bacterium has what type of flagella arrangement?
   - a. Lophotrichous
   - b. Monotrichous
   - c. Polytrichous
   - d. Peritrichous
4. Which phase of the bacterial growth curve has the most rapid bacterial growth?
   - a. Lag phase
   - b. Exponential phase
   - c. Stationary phase
   - d. Decline phase
5. Which of the following organisms is involved in causing tooth decay?
   - a. *S. mutans*
   - b. *S. aureus*
   - c. *C. botulinum*
   - d. *C. perfringens*
6. Which of the following causes a profuse diarrhea described as "rice water" stools?
   - a. *V. cholera*
   - b. *M. tuberculosis*
   - c. *C. botulinum*
   - d. *C. perfringens*
7. Which of the following has been implicated in the cause of cervical cancer?
   - a. HIV
   - b. HPV
   - c. HVA
   - d. HBV
8. Fungal infections are considered
   - a. Opportunistic infections
   - b. Marginal infections
   - c. Gram-negative infections
   - d. Gram-positive infections

critical **thinking questions**

1. If a person has gonorrhea, the individual
   a. Will have immunity to other STIs
   b. Is treated with antiviral medications
   c. Is diagnosed with a bacterial disease
   d. Will have a hard, painless lesion called a chancre
2. How may a case of pharyngitis evolve into bacterial meningitis?

patient **education**

Review the cause, symptoms, treatment, and prevention of tuberculosis or another disease discussed in this chapter. Practice educating a patient about the disease you select. Role-play with a friend or classmate.

applying **what you know**

A 24-year-old female hiker stepped on a rusty nail, receiving a puncture wound on the bottom of her foot.

1. Which of the following is the most likely infection she may get from this injury?
   a. *S. aureus*
   b. *C. tetani*
   c. *S. schenckii*
   d. *S. pyogenes*
2. Which of the following is a likely complication if the wound is not treated properly?
   a. Trismus
   b. Chancre
   c. Vesicles
   d. Warts
3. Which of the following is *not* part of the treatment for this wound?
   a. Antitoxin
   b. Antibiotics
   c. Wound cleansing
   d. Alpha interferon

CASE STUDY 2   *Hepatitis*

A 52-year-old man has recently returned from a vacation in Mexico. He has a fever and complains of being nauseous and having pain in the right upper quadrant of his abdomen. He is monogamous and does not use illicit drugs.

1. Of the following, which one is he most likely to have?
   a. Hepatitis A
   b. Hepatitis B
   c. Hepatitis C
   d. Hepatitis D
2. Based on his probable diagnosis, which one of the following is true?
   a. He is at increased risk for liver cancer
   b. He is at increased risk for liver failure
   c. The disease is typically self-limiting
   d. A liver transplant will be necessary
3. He most likely contracted his illness
   a. Through contaminated food or water
   b. Through respiratory droplets
   c. By touching someone with the disease
   d. By a parasite infection

6
Concepts of Fluid, Electrolyte, and Acid–Base Balance

chapter outline

Fluid Compartments

Regulation of Water

Electrolytes

Acid–Base Balance

Acid–Base Imbalances

Life Span

learning outcomes

AFTER COMPLETING THIS CHAPTER, YOU WILL BE ABLE TO:

6.1 List the major fluid body compartments.

6.2 Summarize the different mechanisms by which the body gains or loses water and explain how water moves between the various body compartments.

6.3 Describe the electrolyte composition and the function of the major electrolytes of the body compartments.

6.4 Describe a buffer system and list examples.

6.5 Relate how acid–base imbalances affect homeostasis.

6.6 Infer how fluids, electrolytes, and acid–base balance impact life span.

essential terms

acidosis
(as-ih-DO-sis)

aldosterone
(al-DOS-ter-one)

alkalosis
(al-kah-LO-sis)

angiotensin
(an-je-oh-TEN-sin)

atrial natriuretic peptide (ANP)
(AY-tree-uhl na-tre-u-RET-ik PEP-tide)

hyperkalemia
(HI-per-ka-LEE-me-uh)

hyperphosphatemia
(HI-per-fos-fay-TEE-me-uh)

hypokalemia
(HI-po-ka-LEE-me-uh)

vasopressin
(VAS-oh-press-in)

Additional key terms in the chapter are italicized and defined in the glossary.

case study

Use the case study to focus on as you go through the chapter. The questions will guide you as you learn the physiology and pathology of the fluid compartments and acid–base balance.

CASE STUDY 1 *Electrolyte Imbalance*

Mike Ambrosino, a 72-year-old man, was diagnosed with type 2 diabetes 11 years ago. He has developed cataracts and is now experiencing kidney failure.

As you go through the chapter, keep the following questions in mind:

1. What role do the kidneys play in water regulation?
2. How is sodium regulated by the kidneys?
3. Would you expect Mike to be experiencing hypernatremia or hyponatremia?
4. What are possible treatments for Mike?

> **study tips**
> 1. Draw a picture of the various body fluid compartments and list the percentages contained by each.
> 2. Draw and label a flow diagram of the renin-angiotensin system and its effects on the body.
> 3. Make a two-column table and label the heads *acidosis* and *alkalosis*. Under each heading, identify factors that contribute to each condition.

Introduction

Body fluids, which are mostly water and the solids found within them, play an essential role in maintaining homeostasis in the body. The body fluids are found in cells, blood vessels, and in the intercellular spaces (spaces between cells). In addition, many of the solids found in the body fluids are capable of conducting an electrical impulse and therefore are called electrolytes. Understanding the regulation of fluid loss and gain as well as fluid movement is necessary to appreciating homeostasis. The connection between electrolyte regulation and health and disease will become evident as you move through the chapter.

6.1 learning outcome

List the major fluid body compartments.

Fluid Compartments

We are all aware that a large percentage of our body is made up of water. In women, the amount of water is about 55 percent of the total body mass, and in men it is about 60 percent. However, did you know that water is found in different compartments and that movement between compartments plays an important role in homeostasis? The two major compartments are intracellular fluid and extracellular fluid (Figure 6-1).

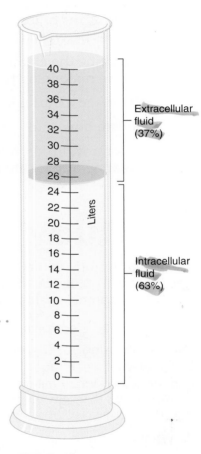

FIGURE 6-1 Of the 40 liters of water in the body of an average adult male, about two-thirds is intracellular and one-third is extracellular.

✓ check
What are the two major fluid compartments?

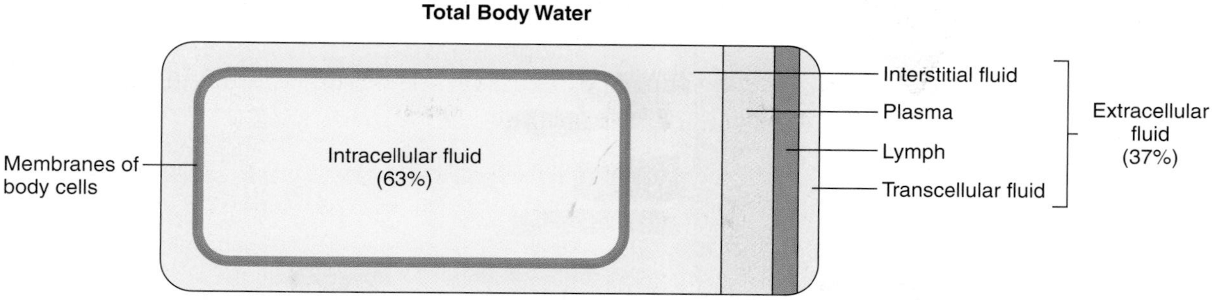

FIGURE 6-2 Cell membranes separate fluid in the intracellular compartment from fluid in the extracellular compartment. Approximately two-thirds of the water in the body is inside cells.

Intracellular Fluid

The fluid that is found in cells is called intracellular (*intra* = within) fluid. Since our body is composed of trillions of cells, it is not surprising that the majority of water is found in them. About two-thirds of the body fluid is found in cells as the cytosol. If you remember from Chapter 3, Concepts of Cells and Tissues, the cytosol is the liquid component of the cytoplasm. The remaining one-third of the body fluid is found outside the cell in the extracellular (*extra* = outside) fluid. The cell membrane or plasma membrane is the semipermeable barrier that separates the intracellular fluid from the interstitial fluid, discussed in the following section.

Extracellular Fluid

The extracellular fluid is further broken down into two components: interstitial fluid and plasma. Interstitial fluid is found between cells and makes up 80 percent of the extracellular fluid, or about 24 to 25 percent of the total body fluid. Plasma makes up 20 percent of the extracellular fluid, or about 6 to 8 percent of total body water. The endothelium of blood vessels separates plasma from the interstitial fluid in the surrounding environment (Figure 6-2).

Regulation of Water

To be in balance, we must remember an obvious observation: Body water gain must equal body water loss (Figure 6-3). Each day, we gain approximately 2½ liters (L) of water, and each day we lose about 2½ L.

6.2 learning outcome
Summarize the different mechanisms by which the body gains or loses water and explain how water moves between the various body compartments.

Water Gain

The two mechanisms by which we gain water are ingestion and metabolism. Ingestion includes everything we drink and eat. About 1,600 milliliters (mL) of our gain is from the water and other liquids we drink. The food we eat supplies almost 600 mL of our daily water intake. The remaining gain is mainly from metabolic reactions, especially cellular respiration where water is a byproduct as the cells produce energy in the form of adenosine triphosphate (ATP). Metabolism accounts for around 200 mL of our daily water intake. Together, ingestion and metabolism are responsible for 2,500 mL (2.5 L) of water gain every day.

Water Loss

The human body loses water totaling about 2,500 mL daily through four mechanisms. The first way we lose water is through the kidneys. The kidneys

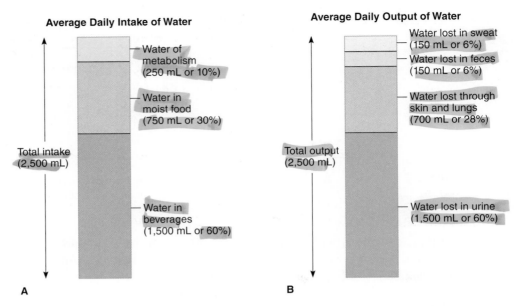

FIGURE 6-3 Water balance: (A) Major sources of body water and (B) routes by which the body loses water. Urine production is most important in the regulation of water balance.

(see Chapter 15, The Urinary System) excrete 1,500 mL of urine each day in a process called micturition (urination). Another 600 mL is lost through the skin. About 200 mL of fluid loss through our skin is perceived as perspiration (sweat). Four hundred milliliters of the loss is not perceived, and so it is referred to as *insensible perspiration*. When we breathe, we exhale 300 mL each day and we defecate another 100 mL. A woman's menses or period is also a way of losing fluid.

> **U check**
> How much water is gained and lost each day?

Water Regulation and Movement

We have looked at how water intake must equal water loss to maintain balance in the body. There is an area in the hypothalamus (see Chapter 14, The Nervous System) that is the thirst center for the body. This thirst center affects water balance by either increasing or decreasing our thirst. The hypothalamus is stimulated in several ways to increase the desire to drink. If we lose too much water we experience dehydration which can be life threatening, especially in children and elderly people who are more susceptible to dehydration and its effects. With dehydration, blood volume drops and hormones such as renin and angiotensin II are released that will cause several things to happen, including stimulation of the hypothalamus. With a decrease in body water, nerves in the mouth will stimulate thirst as will baroreceptors in the heart and blood vessels.

Another very important way water is regulated is by actions of the kidneys. Sodium as part of sodium chloride, or salt (NaCl), is the major determinant of this regulatory mechanism. It is said that where sodium goes, water follows. This means that when we excrete more sodium, we

angiotensin A hormone involved in the regulation of water and sodium.

also excrete more urine (water). The opposite is also true. If we ingest a large amount of sodium, we develop increased thirst, drink more fluid, and often retain more of that fluid causing edema (swelling) in the tissues. Several hormones, such as **angiotensin,** influence the reabsorption of sodium which will cause more water to be retained by the body. Angiotensin II is a hormone that is converted from angiotensin I by renin, a hormone secreted by the kidneys. Angiotensin II causes vasoconstriction, but it also targets the adrenal cortex to secrete **aldosterone.** Aldosterone will impact renal tubules to reabsorb more sodium and consequently more water is also reabsorbed. **Atrial natriuretic peptide (ANP)** is a hormone secreted by the right atrium. ANP is secreted when blood volume (body fluid) increases. ANP will cause increased excretion of sodium and increased water loss. Antidiuretic hormone (also known as **vasopressin**) is produced by the hypothalamus and stored in the posterior pituitary. It is released when there is a decrease in fluid volume. It will cause the kidneys to reabsorb more water, thereby increasing fluid volume.

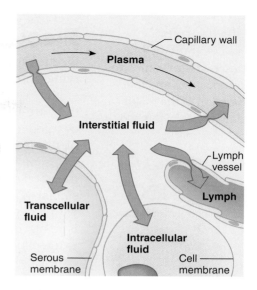

FIGURE 6-4 Movement of fluid between compartments if affected by osmolarity.

Movement of water between compartments is affected by osmolarity (see Chapter 2, Concepts of Chemistry) of the various compartments. For example, if you drink salt water, the osmolarity of the interstitial fluid will increase. This will move water from the cells into the interstitial fluid, causing cells to shrink. On the other hand, if you drink too much water and your kidneys are not able to excrete it fast enough, water will move into your cells and the cells will swell and possibly rupture (Figure 6-4).

aldosterone A hormone secreted by the adrenal cortex for the absorption of sodium by the kidneys.

atrial natriuretic peptide (ANP) A hormone secreted by the right cardiac atrium to increase sodium and water excretion when blood volume increases.

vasopressin Antidiuretic hormone.

> ## Ʊ check
> What effect does increased secretion of atrial natriuretic peptide have on water balance in the body?

---

focus on Wellness: Hydration in Elderly People

Older adults must be made aware that they may have lost some of the compensatory mechanisms for maintaining the right amount of water in their bodies. While working outside on hot, humid days, older adults should dress in loose clothing and keep their head shielded with a hat. They should take frequent breaks even if not feeling tired, and drink small amounts of cool water on a regular basis to avoid becoming dehydrated, even if they are not feeling thirsty. This is because as we age we do not have the same thirst drive that a younger person has. Older adults may not perspire as much as a younger person, which allows heat to remain and build up in the body. To combat this, they may want to use a mist spray bottle to cool themselves from time to time. If they (or others) notice any changes such as confusion, tremors, or weakness, they should take a break immediately, stay in the shade, and drink a cool (nonalcoholic or decaffeinated) beverage. If an older adult is not feeling better after a short rest, medical attention should be sought.

6.3 learning outcome

Describe the electrolyte composition and the function of the major electrolytes of the body compartments.

Fluid/electrolyte imbalance

Electrolytes

Electrolytes are substances that dissolve in solution and are capable of carrying an electrical current (Figure 6-5 and Table 6-1). This allows the generation of action potentials (see Chapter 14, The Nervous System). When they are in solution, electrolytes form ions that have a charge. A positively charged ion is called a cation and a negatively charged ion is an anion. Some of the ions create a concentration gradient due to their osmolarity, which helps move water between fluid compartments. Other ions participate in metabolic reactions. The characteristics of plasma and interstitial fluid are similar. One major difference, however, is that plasma has many proteins with a negative charge (anions) and interstitial fluid has relatively few. There is a big difference in the concentration of ions between intracellular and extracellular fluid. Intracellular fluid has potassium (K^+) as the most abundant cation, and proteins and phosphates (HPO_4^{2-}) are the most abundant anions. In the extracellular fluid, sodium (Na^+) is the most abundant cation and chloride (Cl^-) the most abundant anion.

Sodium

Sodium accounts for 90 percent of the extracellular cations. The normal plasma concentration of sodium is approximately 140 milliequivalent per liter (mEq/L). The normal plasma concentration of extracellular fluid is 300 milliosmole per liter (mOsm/L) and sodium is responsible for about half of it. Many of the foods we eat are high in sodium content. Because sodium causes the body to retain fluid, this extra fluid can contribute to or worsen a person's hypertension. As mentioned earlier, sodium levels are regulated by aldosterone, ADH (antidiuretic hormone), and ANP (atrial natriuretic peptide). A sodium concentration below normal is called hyponatremia (*hypo* = below) and a level above normal is called hypernatremia (*hyper* = excessive). By the way, the chemical name for sodium is natrium because it is so abundantly found in nature.

Hyponatremia: Hyponatremia (*natri* = sodium) is a sodium plasma concentration less than 135 mEq/L. The causes of hyponatremia include vomiting, diarrhea, diuretics, excessive water ingestion, and a deficiency of aldosterone. Blood chemistry tests are used to diagnose hyponatremnia.

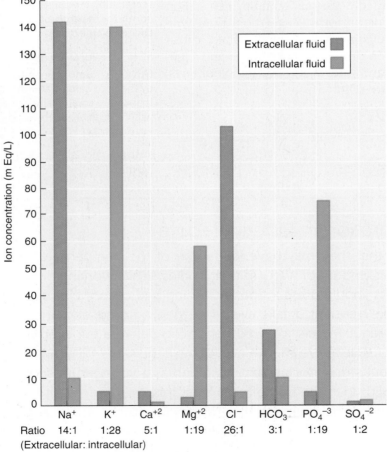

FIGURE 6-5 Extracellular fluids have relatively high concentrations of sodium (Na^+), calcium (Ca^{2+}), chloride (Cl^-), and bicarbonate (HCO_3^-) ions. Intracellular fluid has relatively high concentrations of potassium (K^+), magnesium (Mg^{2+}), phosphate (PO_4^{3-}) and sulfate (SO_4^{2-}) ions.

TABLE 6-1 Understanding Electrolytes

Electrolyte	Function		
Bicarbonate ion	Maintains blood pH.		
Condition	**Causes**	**Signs and Symptoms**	**Treatment**
High or low bicarbonate levels	Conditions that affect respiratory function, kidney disease and metabolic disorders	There are no specific signs, but if bicarbonate levels are too low, acidosis may occur and hyperventilation observed; when bicarbonate levels are too high, alkalosis may occur and the patient may have apathy, confusion, and stupor	Treatment for low bicarbonate levels is to administer oral alkali and when levels are too high, saline solution is given to enhance renal excretion of bicarbonate ions
Electrolyte	**Function**		
Chloride	Easily crosses the plasma membrane and helps balance the anions in the fluid compartments.		
Condition	**Causes**	**Signs and Symptoms**	**Treatment**
Low; hypochloremia	Overhydration, vomiting, congestive heart failure, aldosterone deficiency, and diuretics such as furosemide	Shallow breathing, hypotension, and tetany (muscle spasms)	Correct cause of imbalance and provide IV therapy
High; hyperchloremia	Dehydration, kidney failure, high chloride intake, some drugs, and hyperaldosteronism	Lethargy, weakness, metabolic acidosis, and rapid deep breathing	Correct cause of imbalance and provide IV therapy
Electrolyte	**Function**		
Sodium	Accounts for 90 percent of the extracellular cations. Helps regulation of water balance and the generation of electrical impulses.		
Condition	**Causes**	**Signs and Symptoms**	**Treatment**
Low; hyponatremia	Decreased sodium intake, overhydration, vomiting, diarrhea, low aldosterone levels, and diuretics	Muscle weakness, confusion, headaches, shallow breathing, hypotension, mental confusion, and coma	Correct cause of imbalance and provide IV therapy (saline solution)
High; hypernatremia	Dehydration, excessive sodium in the diet, and certain drugs	Increased thirst, edema, hypertension, and convulsions	Correct cause of imbalance and provide IV therapy to dilute the sodium concentration and excrete the excess
Electrolyte	**Function**		
Potassium	The most abundant cation in the intracellular fluid. It is essential in generating an action potential and regulating pH in body fluids.		
Condition	**Causes**	**Signs and Symptoms**	**Treatment**
Low; hypokalemia	Vomiting, diarrhea, kidney disease, decreased potassium intake, and some diuretics	Muscle fatigue, shallow breathing, confusion, abnormal ECG changes, and increased urine output	Correct cause of imbalance and provide IV therapy

continued

TABLE 6-1 (concluded)

Condition	Causes	Signs and Symptoms	Treatment
High; hyperkalemia	Kidney failure, high potassium intake, kidney failure, crushing injuries, and aldosterone deficiency	Nausea, vomiting, diarrhea, muscle weakness, ventricular fibrillation, and even death	Correct cause of imbalance and provide IV therapy

Electrolyte	Function
Calcium	The most abundant mineral in the body with over 98 percent stored in bone and teeth. Involved in blood clotting, transmission of nerve impulses, and excitability of muscle.

Condition	Causes	Signs and Symptoms	Treatment
Low; hypocalcemia	Increased calcium loss, decreased calcium intake, increased phosphate levels, and hypoparathyroidism	Numbness and tingling in digits, hyperactive reflexes, muscle cramps, and bone fractures	Correct cause of imbalance and provide IV therapy
High; hypercalcemia	Hyperparathyroidism, excessive intake of calcium or vitamin D, and some cancers	Lethargy, weakness, nausea, vomiting, increased urination, bone pain, mental changes, and coma	Correct cause of imbalance and provide IV therapy

Electrolyte	Function
Phosphate	The majority (85 percent) is found in bone and teeth as calcium phosphate salts. Helps maintain acid–base balance.

Condition	Causes	Signs and Symptoms	Treatment
Low; hypophosphatemia	Loss through increased urination or decreased intestinal absorption	Mental changes, myalgia, tingling of the digits, and decreased coordination	Correct cause of imbalance and provide IV therapy
High; hyperphosphatemia	Dehydration, kidney failure, cellular destruction, and increased intake of phosphates	Nausea, vomiting, muscle weakness, hyperactive reflexes, tetany, and tachycardia	Correct cause of imbalance and provide IV therapy

Electrolyte	Function
Magnesium	Over 50 percent is found in bone; most of the remainder is located in intracellular fluid. Acts as a cofactor in many metabolic reactions.

Condition	Causes	Signs and Symptoms	Treatment
Low; hypomagnesemia	Decreased intake or increased loss through the feces or urine; seen in diabetes, alcoholism, and starvation	Irritability, tetany, mental changes, anorexia, nausea, vomiting, and heart abnormalities	Correct cause of imbalance and provide IV therapy
High; hypermagnesemia	Kidney failure, increased intake, and aldosterone deficiency	Nausea, vomiting, muscle weakness, hypotension, and altered mental status	Correct cause of imbalance and provide IV therapy

The signs and symptoms include muscle weakness, confusion, hypotension, headache, and coma. To treat hyponatremia, correct the cause of the sodium loss or deficiency and give saline solution to replace the sodium lost.

Hypernatremia is a plasma sodium concentration greater than normal. Dehydration or excessive sodium in the diet as well as some drugs can result in hypernatremia. The patient may exhibit increased thirst, edema, or convulsions. As with hyponatremia, blood chemistry tests are used to diagnose hypernatremia. To correct the hypernatremia, treat the underlying cause of the excessive sodium and administer intravenous (IV) fluids to dilute the sodium concentration and cause the excess to be excreted.

> **REMEMBER MIKE,** our type 2 diabetic who is experiencing kidney failure? Would you expect him to be experiencing hypernatremia or hyponatremia?

> **U check**
> What is the most abundant cation in the extracellular fluid?

Chloride

The most abundant anion in the extracellular fluid is chloride. The normal plasma concentration is about 100 mEq/L. Because chloride can easily cross the plasma membrane, it helps balance the anions in the fluid compartments. ADH helps regulate chloride concentrations by regulating water balance.

Hypochloremia: Hypochloremia (*chlor* = chloride) is a plasma chloride concentration less than normal. Overhydration, vomiting, congestive heart failure, and diuretics can cause hypochloremia. Shallow breathing, hypotension, and tetany (muscle spasms) are some of the signs and symptoms of this condition. Blood chemistry tests are used to diagnose hypochloremia. Treatment includes correcting the underlying cause of the imbalance and giving IV fluids including chloride.

Hyperchloremia: Hyperchloremia is a plasma chloride concentration greater than normal. Hyperchloremia can be caused by dehydration, kidney failure, high chloride intake, drugs, and increased aldosterone secretion. Lethargy, weakness, and rapid deep breathing are some of the signs and symptoms of hyperchloremia. Blood chemistry tests are used to diagnose the disorder. Treatment includes treating the underlying cause of the excess chloride and administration of IV fluids.

Potassium

The normal concentration of potassium ions in the intracellular fluid is about 140 mEq/L and it is the most abundant cation within cells. Potassium is important in maintaining the volume of cells, essential in generating an action potential, and is involved in regulating the pH (the hydrogen ion concentration in solution) of the body fluids. The concentration of potassium in plasma is much less—4.0 to 5.0 mEq/L because it is under the control of aldosterone. A higher than normal plasma concentration of potassium is called hyperkalemia (*kali* = potassium).

hypokalemia A plasma potassium concentration less than normal.

Hypokalemia: **Hypokalemia** is a plasma potassium concentration less than normal. Vomiting, diarrhea, kidney disease, and some diuretics can cause hypokalemia. Muscle fatigue, confusion, shallow breathing, abnormal electrocardiogram (ECG) changes, and increased urine output are signs and symptoms of hypokalemia. Blood chemistry tests are used to diagnose hypokalemia, and it is treated by correcting the underlying cause of the potassium loss and administering IV fluids.

hyperkalemia A plasma potassium concentration that is greater than normal.

Hyperkalemia: **Hyperkalemia** is a plasma potassium concentration greater than normal. Hyperkalemia can be caused by kidney failure, excessive potassium intake, crushing injuries, and aldosterone deficiency. Nausea, vomiting, diarrhea, muscle weakness, ventricular fibrillation, and even death can be seen with this condition. Blood chemistry tests are used to diagnose the condition. The underlying cause should be addressed and the patient given IV fluids.

Calcium

The majority (almost 99 percent) of the body's calcium is stored in bones and teeth. The concentration of calcium in the plasma is about 10 mEq/L with half attached to plasma proteins and the other half circulating unattached in the blood. Calcium accounts for the hardness of bone and tooth enamel. It is also a factor in blood clotting, release of neurotransmitters, and the contraction of muscle.

Two hormones regulate plasma concentrations of calcium. Calcitonin (see Chapter 21, The Endocrine System) is secreted by the thyroid gland and moves calcium into bone from the plasma. Parathyroid hormone (PTH) is secreted by the parathyroid gland and is the antagonist to calcitonin. It targets the osteoclast to resorb bone, releasing calcium and causing it to move into the plasma. PTH also causes the kidneys to reabsorb calcium and stimulates calcitriol (vitamin D) to absorb calcium from food in the digestive tract. So calcitonin increases the amount of calcium in the bone and PTH increases the amount of calcium in the plasma.

Food absorption

> **U check**
> What two hormones regulate calcium balance?

Hypocalcemia: *Hypocalcemia* (*calci* = calcium) is a plasma calcium potassium concentration less than normal. Decreased calcium intake, increased calcium loss, and hypoparathyroidism can cause hypocalcemia. Numbness and tingling of the fingers, hyperreflexia, fractures, muscle cramps, tetany, and even death may result from hypocalcemia. Blood chemistry tests are used to diagnose hypocalcemia. The underlying cause of the low calcium level should be treated and the individual given IV fluids.

Hypercalcemia: *Hypercalcemia* is a plasma calcium concentration greater than normal. Causes include hyperparathyroidism, some cancers, bone disease, and excessive vitamin D levels. Lethargy, anorexia, weakness, bone pain, mental status changes, and coma are some of the signs and symptoms of hypercalcemia. Blood chemistry tests are used to diagnose hypocalcemia. Treatment includes addressing the underlying cause of the hypercalcemia and administration of IV fluids.

Phosphate

The majority (approximately 85 percent) of phosphate is found in bones and teeth as part of calcium phosphate salts. The remaining 15 percent is found in ionized form in the body. Approximately 100 mEq/L is found in the intracellular fluid and about 2.0 mEq/L is found in the plasma. The levels of phosphate ions in the plasma are regulated by calcitriol and PTH. Calcitriol increases absorption of phosphates from the gastrointestinal tract. As stated earlier, PTH stimulates osteoclast activity and the breakdown of bone. This raises plasma levels of phosphates. PTH also inhibits the reabsorption of phosphate from the tubules.

Hypophosphatemia: *Hypophosphatemia* is a plasma phosphate concentration less than normal. Increased phosphate loss and decreased phosphate absorption can cause hypophosphatemia. Confusion, decreased coordination, lethargy, seizures, and coma are seen in patients with hypophosphatemia. Blood chemistry tests are used to diagnose hypophosphatemia. Treatment includes correcting the underlying pathology and giving IV fluids.

Hyperphosphatemia: **Hyperphosphatemia** is a plasma phosphate concentration greater than normal. Kidney failure and destruction of cells can lead to hyperphosphatemia. Muscle weakness, increased reflexes, anorexia, vomiting, tetany, and tachycardia are some of the signs that are seen in hyperphosphatemia. Blood chemistry tests help diagnose the disorder, and the underlying cause must be treated and then IV fluids given.

hyperphosphatemia
A plasma phosphate concentration that is greater than normal.

Magnesium

Almost 55 percent of the body's magnesium is found in bone as magnesium salts. The remaining 45 percent is found in the ionized form with almost all of it located within cells (intracellular fluid). Only a very small amount is found in the extracellular fluid. Intracellular concentration of magnesium is 35 mEq/L. In the plasma, the concentration is only about 1.5 to 2.0 mEq/L. Magnesium is important in nerve transmission, muscular activity including that of the heart, secretion of PTH, and many metabolic reactions.

Hypomagnesemia: *Hypomagnesemia* is a plasma magnesium concentration less than normal. Alcoholism, malnutrition, diabetes mellitus, and diuretics cause hypomagnesemia. Weakness, tetany, convulsions, anorexia, nausea, and heart abnormalities are signs and symptoms seen with hypomagnesemia. Blood chemistry tests make the diagnosis of hypomagnesemia. The underlying cause should be treated and IV fluids given.

Hypermagnesemia: *Hypermagnesemia* is a plasma magnesium concentration greater than normal. Kidney failure, excessive antacid use, hypothyroidism, and hypoaldosteronism cause hypermagnesemia. Hypotension, muscle weakness, changed mental status, and vomiting are seen in hypermagnesemia. To make the diagnosis, blood chemistry tests are given. To correct hypermagnesemia, treat the underlying cause of the condition and give IV fluids.

> **U check**
> Where is the majority of ionized magnesium found?

Bicarbonate

Bicarbonate ions (HCO_3^-) are found in high concentrations (24 mEq/L) in the plasma. The concentration in the veins is slightly higher than that in the arteries. The reason is that the carbon dioxide concentration is greater in the veins and lowers the pH of the plasma. Bicarbonate ions act as a buffer to raise the plasma pH. The kidneys regulate the amount of bicarbonate by synthesizing or excreting the ions as the body's need changes.

> **U check**
> Why is the bicarbonate concentration greater in veins than in arteries?

6.4 learning outcome
Describe a buffer system and list examples.

Acid–Base Balance

The normal pH of arterial blood is 7.35 to 7.45. This is equivalent to 40 mEq/L of hydrogen ion (H^+) concentration in the plasma. Survival is dependent on keeping the pH of our body within this very narrow range. Three mechanisms handle situations where there is excess acidity (excess H^+ ions): buffer systems, carbon dioxide exhalation, and kidney excretion.

Buffers

Buffers convert strong acids and bases to weak acids and bases (Figure 6-6), which keeps the pH within the desired range. Three major buffer systems include the protein buffer system, the carbonic acid–bicarbonate buffer system, and the phosphate buffer system. Proteins can act as both acid and base buffers as needed. In red blood cells, hemoglobin is the major protein buffer and albumin is the major protein buffer in the plasma.

Carbonic acid–bicarbonate can buffer acids or bases because the bicarbonate ion can act as either a weak base or a weak acid. However, it is not capable of buffering respiratory pH due to changes in the levels of carbon dioxide. If the pH of arterial blood drops to 6.8 or rises to 8.0 for more than a few hours, a person usually cannot survive (Figure 6-6).

The phosphate buffer system is most active in the cytoplasm of cells (intracellular fluid) because of the high concentrations of phosphate ions found there. To a lesser extent, it also buffers the pH of urine (Table 6-2).

Carbon Dioxide

Carbon dioxide can make the plasma more acidic. When we exhale and eliminate carbon dioxide, the pH of plasma rises. The rate and depth of breathing therefore have an important effect on the plasma pH. When we breathe more rapidly and deeply, more CO_2 is exhaled and the pH increases. When breathing is slower and more shallow, the pH decreases and the plasma becomes more acidic. Changes in plasma pH are detected by chemoreceptors in the medulla oblongata of the brain stem, the aorta, and the carotid arteries. If the plasma is too acidic, the chemoreceptors stimulate breathing and the blowing off of carbon dioxide. When the plasma is more acidic, the chemoreceptors inhibit breathing.

BODYANIMAT3D
POWERED BY
connect plus+
Acid-base balance: Acidosis
Acid-base balance: Alkalosis

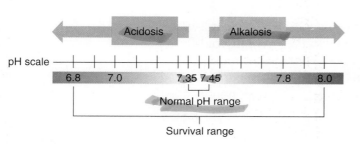

FIGURE 6-6 If the pH of arterial blood drops to 6.8 or rises to 8.0 for more than a few hours, a person usually cannot survive.

> ### focus on Clinical Applications: IV Therapy
>
> Anyone who has ever been in a hospital or emergency department has seen patients receiving intravenous (IV) therapy. IV fluids are solutions, including medications, that are delivered directly into the bloodstream through a vein. The most common reasons for IV therapy include:
>
> - To replace and maintain fluid and electrolyte balance.
> - To administer medications, including chemotherapeutic agents, intravenous anesthetics, and diagnostic reagents.
> - To transfuse blood and blood products.
> - To deliver nutrients and nutritional supplements.
>
> IV fluids are given to patients for fluids lost or depleted due to hemorrhage, vomiting, or diarrhea. IV fluids such as normal saline are given during and after surgery to maintain fluid balance. Some IVs provide access to the vascular system for emergency situations. For patients who are dehydrated or at risk for dehydration, malnutrition, and/or electrolyte imbalance, IV therapy is the fastest route to replace or build up the missing elements.

Hypercapnia: *Hypercapnia* is an arterial blood carbon dioxide concentration greater than normal. Seizures and chronic obstructive pulmonary diseases (COPDs) such as emphysema, asthma, sleep apnea, and hypoventilation can lead to high carbon dioxide levels in the blood. Headache, drowsiness, inability to think clearly, rapid breathing, tachycardia, and increased blood pressure are signs and symptoms of hypercapnia. Arterial blood gases are drawn and analyzed to diagnose hypercapnia. Treatment is correcting the underlying cause of the excessive carbon dioxide and administering ventilation.

TABLE 6-2 Chemical Acid–Base Buffer Systems

Buffer System	Constituents	Actions
Protein system (and amino acids)	Bicarbonate ion (HCO_3^-)	Converts a strong acid to a weak acid
	Carbonic acid (H_2CO_3)	Converts a strong base to a weak base
Carbonic acid–bicarbonate system	Monohydrogen phosphate ion (HPO_4^{-2})	Converts a strong acid to a weak acid
	Dihydrogen phosphate ($H_2PO_4^-$)	Converts a strong base to a weak base
Phosphate system	$-NH_3^+$ group of an amino acid or protein	Releases a hydrogen ion in the presence of excess base
	$-COO^-$ group of an amino acid or protein	Accepts a hydrogen ion in the presence of excess acid

Kidneys

Metabolic reactions produce large amounts of metabolic acid byproducts in too large a quantity to be handled by any of the previously discussed mechanisms. The kidneys, therefore, are responsible for excreting a great amount of the excess hydrogen ions from these reactions. Although the lungs participate in acid–base balance, the greater amount of work is done by the kidneys. That is why when we see renal failure; there is often an accompanying imbalance in pH.

> **U check**
> Why are proteins important in acid–base balance?

Relate how acid–base imbalances affect homeostasis.

acidosis A condition in which the pH of the blood plasma is less than 7.35.

alkalosis A condition in which the pH of the blood plasma is greater than 7.45.

Homeostasis

Acid–Base Imbalances

When acid–base balance is disrupted, either **acidosis** (plasma pH less than 7.35) or **alkalosis** (plasma pH greater than 7.45) is the result (Figures 6-7). Both can cause serious consequences and even death, but by different means. Acidosis depresses the central nervous system and the person may enter a coma and die. In alkalosis, there is increased excitability of the central nervous system and peripheral nerves. The person may experience muscle spasms and convulsions, coma, and eventually death.

Respiratory acidosis occurs when carbon dioxide is not removed from plasma by respiration (breathing); see Figure 6-8. A value above 45 millimeters of mercury (mm Hg) in the arterial blood is considered respiratory acidosis. This is common in chronic obstructive lung diseases such as emphysema as well as after trauma to the medulla oblongata.

Respiratory alkalosis happens when the carbon dioxide level in arterial blood is less than 35 mm Hg (Figure 6-9). This is often the result of hyperventilation (e.g., from an anxiety attack) or as the result of a cerebrovascular accident (CVA, or stroke). Metabolic acidosis is when the arterial bicarbonate concentration is below 22 mEq/L (Figure 6-10). This may occur with diarrhea, ketoacidosis, and inability of the kidneys to excrete the excess hydrogen ions from protein metabolism.

Metabolic alkalosis is a bicarbonate concentration in the arterial blood greater than 26 mEq/L. This can be caused by vomiting, excessive use of antacids, and dehydration.

> **U check**
> What is the difference between respiratory and metabolic acidosis?

Infer how fluids, electrolytes, and acid–base balance impact life span.

Life Span

Problems with fluid balance and acid–base homeostasis are seen at the two extremes of life. The water content in infants is about 75 percent of their body mass and is even higher in premature infants. In addition, the amount of water in the extracellular fluid is much greater than in the intracellular fluid. Therefore, infants are very sensitive to changes in fluid intake and elimination. Infants have a very high rate of metabolism and yet their kidneys are not fully developed. This means that any excess hydrogen ions from protein metabolism may not be efficiently excreted, resulting in acidosis.

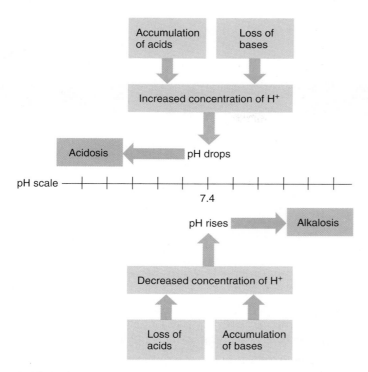

FIGURE 6-7 Acidosis results from accumulation of acids or loss of bases. Alkalosis results from loss of acids or accumulation of bases.

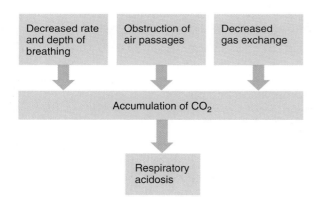

FIGURE 6-8 Some of the factors that lead to respiratory acidosis.

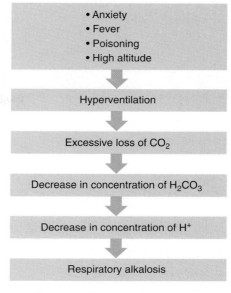

FIGURE 6-9 Some of the factors that lead to respiratory alkalosis.

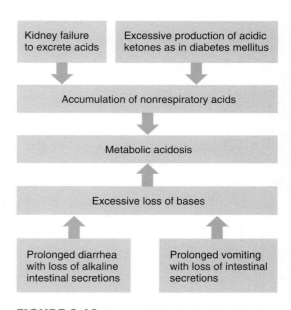

FIGURE 6-10 Some of the factors that lead to metabolic acidosis.

A different set of physiological changes in elderly people present numerous potentially dangerous situations. With aging, there is a decrease in intracellular fluids, decrease in muscle mass (resulting in total body potassium decrease), respiratory and renal function decline, and inefficiency in body temperature regulation. Older individuals also commonly lose the drive for thirst that is more readily apparent in younger people, so it is very common to see dehydration in elderly individuals. They may experience hypernatremia or even hyponatremia, as well as acidosis.

summary

chapter 6

learning outcomes	key points
6.1 List the major fluid body compartments.	The major body fluid compartments are the extracellular and intracellular fluid compartments. The extracellular compartment is composed of interstitial fluid and plasma.
6.2 Describe the different mechanisms by which the body gains or loses water and explain how water moves between the various body compartments.	The body gains water by ingestion of fluids as well as foods and metabolic reactions in the body. The body loses fluids through micturition (urination), breathing, sweating, and defecation.
	Water moves between the various body compartments through the action of hormones and electrolytes.
6.3 Describe the electrolyte composition and the function of the major electrolytes of the body compartments.	The electrolyte composition of the body compartments consists of cations such as sodium and anions such as chloride.
	Electrolytes allow the generation of action potentials. Some ions create a concentration gradient due to their osmolarity. This helps move water between fluid compartments. Other ions participate in metabolic reactions.
6.4 Describe a buffer system and list examples.	Buffer systems maintain the pH of the body around a very narrow range to protect homeostasis and life. The protein buffer, carbonic acid–bicarbonate buffer, and the phosphate buffer are the major buffer systems.
6.5 Relate how acid–base imbalances affect homeostasis.	Acid–base imbalances include respiratory acidosis and alkalosis as well as metabolic acidosis and alkalosis. Any of these will disrupt homeostasis and put the person at risk of serious illness and death.
6.6 Infer how fluids, electrolytes, and acid–base balance impact life span.	Very young and very old individuals are most affected by variations in fluid and electrolyte imbalances. These can often have serious implications on their health.

case study 1 questions

Remember Mike, our 72-year-old man who is experiencing kidney failure because of his diabetes? Can you answer these questions regarding the case study?

1. What role do the kidneys play in water regulation?
2. How is sodium regulated by the kidneys?
3. Would you expect Mike to be experiencing hypernatremia or hyponatremia?
4. What are possible treatments for Mike?

review questions

1. The right atrium of the heart secretes which of the following?
   a. Angiotensin II
   b. Angiotensin I
   c. Aldosterone
   d. Atrial natriuretic peptide
2. The amount of water a typical adult normally takes in through the fluids he drinks each day is closest to
   a. 200 mL
   b. 600 mL
   c. 1,600 mL
   d. 4,000 mL
3. What is the normal body fluid percentage in an adult female?
   a. 25 percent
   b. 35 percent
   c. 45 percent
   d. 55 percent
4. Plasma is
   a. Interstitial fluid
   b. Extracellular fluid
   c. Intracellular fluid
   d. Cytosol
5. Secretion of aldosterone by the adrenal cortex will
   a. Increase absorption of calcium by the kidneys
   b. Increase absorption of sodium by the kidneys
   c. Increase vasoconstriction of the arterioles
   d. Increase excretion of sodium by the kidneys
6. What is the most abundant cation in extracellular fluid?
   a. Potassium
   b. Chloride
   c. Calcium
   d. Sodium
7. Where is the thirst control center located?
   a. Cerebellum
   b. Thymus
   c. Hypothalamus
   d. Pituitary gland
8. The majority of the body's calcium is found in
   a. Plasma
   b. Red blood cells
   c. Interstitial fluid
   d. Bone
9. What is hypokalemia?
   a. An increase of sodium in the blood
   b. A decrease of sodium in the blood
   c. An increase of potassium in the blood
   d. A decrease of potassium in the blood

critical **thinking questions**

1. A patient has been a cigarette smoker for 35 years. He has just been diagnosed with emphysema. How will this COPD (chronic obstructive pulmonary disease) affect his acid–base balance?
2. Illustrate how the kidneys are involved in the regulation of pH in the body.
3. Compare the difference between respiratory and metabolic acidosis.

patient **education**

When individuals are ill with nausea, vomiting, and/or diarrhea, this can have an effect on their fluid balance. Using what you know about the body's water balance, what should you teach patients about their intake and output during illness?

applying **what you know**

A four-year-old boy has been vomiting most of the night. He has a slight fever and feels achy. His mother thinks he has the flu.

1. In what fluid compartment is the majority of water found in young children?
2. Why do children have a high risk of electrolyte imbalance?
3. What are the signs and symptoms the mother should watch for?

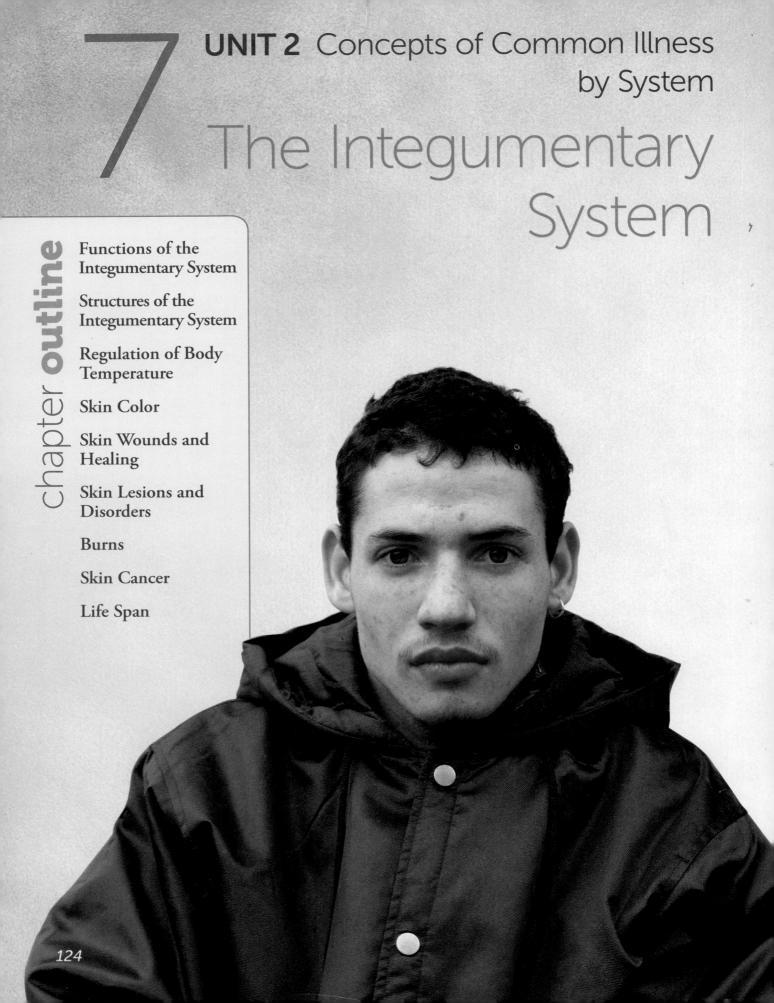

7 The Integumentary System

UNIT 2 Concepts of Common Illness by System

chapter outline

- Functions of the Integumentary System
- Structures of the Integumentary System
- Regulation of Body Temperature
- Skin Color
- Skin Wounds and Healing
- Skin Lesions and Disorders
- Burns
- Skin Cancer
- Life Span

learning outcomes

AFTER COMPLETING THIS CHAPTER, YOU WILL BE ABLE TO:

7.1 Recall the functions of the integumentary system.

7.2 Describe the layers of skin and the characteristics of each layer and recognize the accessory organs of skin and describe their structures and functions.

7.3 Explain the role of skin in regulating body temperature.

7.4 Explain the factors that affect skin color.

7.5 Explain the process of skin healing, including scar production.

7.6 Recognize and describe common skin lesions and summarize the etiology, signs and symptoms, and treatments of common skin disorders.

7.7 Relate the degrees of burns to their appearance and treatment.

7.8 Summarize the etiology, signs and symptoms, and treatments of the various types of skin cancer.

7.9 Describe the effects of aging on skin.

essential terms

apocrine (APP-o-crihn)
arrector pili (ah-REK-tor PI-li)
carotene (KARE-o-teen)
cyanosis (si-ah-NO-sis)
dermis (DER-mis)
eccrine (EK-krin)
epidermis (ep-ih-DER-mis)
hypodermis (hi-po-DERM-is)
lunula (LU-nu-la)
melanin (MEL-ah-nin)
sebaceous glands (se-BAY-shus GLANDS)
stratum basale (STRA-tum bas-A-le)
stratum corneum (STRA-tum KOR-ne-um)
stratum germinativum (STRA-tum jer-mi-ni-TIE-vum)
stratum granulosum (STRA-tum gran-you-LO-sum)
stratum lucidum (STRA-tum LU-si-dum)
stratum spinosum (STRA-tum spy-NO-sum)
subcutaneous (sub-ku-TAY-nee-us)
sudoriferous (soo-dor-RIF-or-us)

Additional key terms in the chapter are italicized and defined in the glossary.

case study

Use the case study to focus on as you go through the chapter. The questions will guide you as you learn the anatomy, physiology, and pathology of the integumentary system.

CASE STUDY 1 *Acne Vulgaris*

Adam Boyle is a 17-year-old high school senior. He is concerned because he has a severe case of acne that is causing some scarring. He has had the problem for the last two years. Adam remembers that his older brother had the same problem.

As you go through the chapter, keep the following questions in mind:

1. What is the cause of acne vulgaris?
2. How is the diagnosis made? Is it a clinical diagnosis or are specific tests used?
3. What are some treatments for people with this condition?
4. How is this similar to, but different from, rosacea?
5. What will be your patient teaching regarding this condition?
6. What is the prognosis for people who have acne?

study tips

Here are some tips that can help you learn about the integumentary system.

1. Draw the dermis and epidermis; label the structures you would find in each.
2. Draw a body and indicate the parts that make up the "rule of nines" for burns.
3. Make flash cards for the major types of skin cancers; include the ABCs of melanoma.

Introduction

When people think of the integumentary system, they almost always immediately think of the skin. Although this is correct, it is not "the whole story." The integumentary system also consists of accessory organs such as sebaceous (oil) glands and temperature receptors, to mention just two of many structures found in this amazing system.

The skin is considered a dry epithelial membrane that covers and protects the organs of the body. It is also the largest organ in the body by weight, and its unique structure gives it both strength and flexibility. The skin is a highly regenerative organ; it replaces itself approximately every 30 days! More than likely you have had a minor cut or other skin injury sometime in the past. It is likely that it healed without scarring and with very little intervention from you. Of course, with larger wounds at least some scarring may have occurred, even with medical intervention. In this chapter we will also look at the factors that affect skin healing.

U check
Why is the integumentary system much more than just skin?

7.1 learning outcome

Recall the functions of the integumentary system.

Functions of the Integumentary System

Our skin is not just a covering. The skin and accessory organs of the integumentary system have several functions. These include protection, body temperature regulation, vitamin D production, sensory perception, and excretion.

Protection

The skin acts a physical barrier. If the skin is not broken and there is no inflammation, it serves as our first line of defense against foreign organisms such as bacteria (Figure 7-1). In fact, we normally have some bacteria on the surface of our skin, but as long as the skin is not "compromised" (broken), these bacteria are relatively harmless. The skin is considered a nonspecific defense mechanism because it protects without discrimination (we will discuss nonspecific and specific defense mechanisms in Chapter 12,

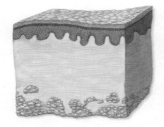

FIGURE 7-1
Intact skin acts as a physical barrier to protect the body.

The Lymphatic and Immune Systems). The skin protects against both chemical (substances) and mechanical (trauma) damage. The skin also protects underlying structures (our organs and vessels) from ultraviolet (UV) radiation and dehydration.

Body Temperature Regulation

Skin plays a key role in body temperature regulation. This temperature regulation is part of homeostasis (Figure 7-2). One way this is done is by having more or less blood shunted to the surface of the skin. By shunting blood to the skin, the body can cool down and when blood is shunted away from the skin, blood is directed to the core of the body for warming.

So if a person is too hot and needs to lower his or her body temperature, several things happen. Blood flow is increased to the skin, and blood vessels located in the dermis dilate. This allows heat to dissipate (be lost to the environment). The individual's skin will take on a pinkish hue and be warmer to the touch because of the increased blood flow. Sweat glands are activated to increase perspiration, further cooling the body. On the other hand, when a person becomes cold, the dermal blood vessels constrict and more blood is shunted to the core of the body which is warmer than the skin. This prevents heat loss at the skin surface and produces warming of the blood as it circulates through the core of the body.

Vitamin D Synthesis

Exposure of skin to ultraviolet rays from the sun initiates the series of reactions that leads to the production of active vitamin D. The absorption of calcium is dependent on the body having adequate stores of vitamin D. Calcium helps with bone health. People in more northern climates often require vitamin D supplements in the winter months due to lack of sun exposure.

Sensory Perception

The dermal layer of the skin has a variety of receptors for light touch, pain, pressure, and temperature. Each receptor responds to a certain stimuli but not others. Receptors also vary in terms of abundance relative to each other. For example, there are far more pain receptors than cold receptors in the body. In addition, receptors vary in terms of the concentration of their distribution over the surface of the body, such as the fingertips having far more touch receptors than the skin of the back. Receptors are specialized nerve cells or fibers.

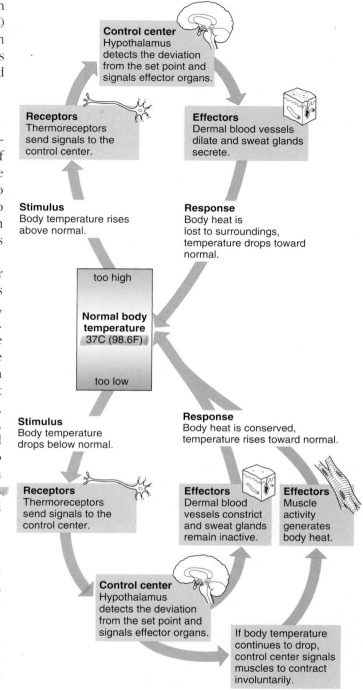

FIGURE 7-2 Body temperature regulation is an example of homeostasis.

CHAPTER 7 The Integumentary System

Excretion

Water and salts (electrolytes) are lost through the skin during perspiration. When water loss is excessive, it can be life threatening (see Chapter 6, "Concepts of Fluid, Electrolyte, and Acid–Base Balance"). It also acts to eliminate urea and uric acid as waste products.

> **U check**
> The skin is essential to the synthesis of what vitamin?

from the perspective of . . .

A VOCATIONAL OR PRACTICAL NURSE A vocational or practical nurse provides care for patients under the supervision of a physician or registered nurse. How will knowing the functions of the skin help you better treat the patients placed in your care?

7.2 learning outcome

Describe the layers of skin and the characteristics of each layer and recognize the accessory organs of skin and describe their structures and functions.

epidermis The top layer of the skin.

stratum basale (stratum germinativum) the deepest layer of the epidermis, comprised of a single row of cells that are producing millions of new skin cells daily.

stratum germinativum Also called the stratum basale; the deepest layer of the epidermis, comprised of a single row of cells that are producing millions of new skin cells daily.

stratum spinosum The epidermal layer above the stratum basale containing several rows of slowly dividing cells and pre-keratin protein.

Structures of the Integumentary System

The skin is a relatively thin but complex organ. It is composed of a top layer called the epidermis (*epi* = above; *derm* = skin) and a bottom layer known as the dermis (Figure 7-3). Beneath the dermis is the subcutaneous (*sub* = below; *cutane* = skin) or hypodermis (*hypo* = below or beneath) layer, which is a layer of support tissue.

Epidermis

The **epidermis** is the most superficial layer of the skin. It is composed of four to five distinct layers depending on whether we are talking about thin or thick skin, respectively. It is *avascular* (without a blood supply) and is made up of at least four different cell types. The deepest layer is called the **stratum basale** (*basale* = bottom) or the **stratum germinativum** (*germinate* means "cause to grow"). This layer is made of a single row of cells that are actively dividing (mitosis) to literally produce millions of new skin cells daily. Melanocytes are found here and they contain special organelles called melanosomes that produce melanin—a pigment responsible in part for skin color; it also protects us from dangerous UV rays of the sun (Figure 7-4). Up to 25 percent of the cells in this layer are melanocytes and an accumulation of this pigment in one spot is a freckle. Just above the stratum basale is the **stratum spinosum** (spiny layer) which is so named because the cells have extensions or spikes. In this layer, which is several cells thick, pre-keratin protein is produced by keratinocytes, the most numerous cell type in the epidermis. The cells

128 UNIT 2 Concepts of Common Illness by System

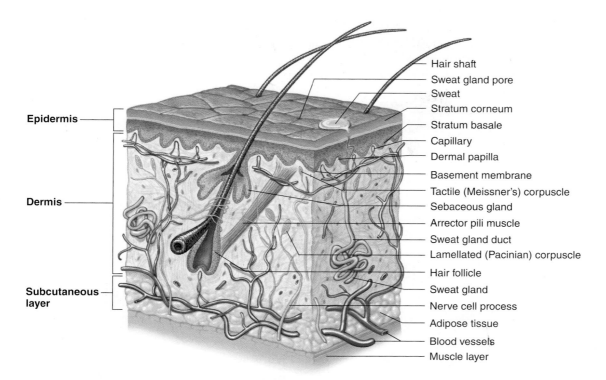

FIGURE 7-3 Section of skin and the underlying subcutaneous layer.

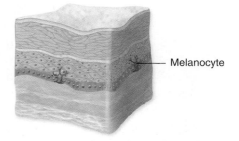

FIGURE 7-4 Melanocytes, cells in the epidermis responsible for the production of melanin.

here do divide, but not as actively as do the cells in the stratum basale because these layers do not get adequate nutrition from the underlying dermis and the cells die as they move to the surface of the skin.

The next layer of the epidermis is the **stratum granulosum** (granular layer) which is composed of a thin layer of cells producing two types of granules. The first type is a glycolipid that has a waterproofing quality. The other granule type helps make keratin fibers. The next layer, the **stratum lucidum** (clear layer), is very thin and translucent and consists of dead keratinocytes. This layer is found only in thick skin such as the soles of feet. The most superficial (top) layer of the epidermis is the **stratum corneum** (horny layer) and accounts for the thickness of the epidermis since it is composed of 20 to 30 cell layers. The cells in this layer are dead, flat (squamous), and fully keratinized. These cells are constantly being shed and replaced by cells from the layers below. The keratin (from the keratinocytes) in this layer helps the skin protect underlying organs and structures from bacteria and viruses and helps prevent dehydration from fluid loss.

stratum granulosum The middle layer of the epidermis, comprised of a thin layer of cells that produce granules to provide waterproofing and make keratin fibers.

stratum lucidum The epidermal layer just below the stratum corneum consisting of thin, translucent dead keratinocytes found only in thick skin.

stratum corneum The most superficial layer of the epidermis, made of 20 to 30 layers of dead, squamous, fully keratinized cells.

dermis The bottom layer of the skin.

Dermis

The **dermis** lies immediately below the epidermis. It is much thicker and more complex than the epidermis and contains all four of the major types of tissues: epithelial, connective, muscle, and nervous tissue. Two main regions of the dermis have an abundance of connective tissue—the papillary and reticular layers. The *papillary layer* is the more superficial of the two, composed of areolar or loose connective tissue with upward projections—called *dermal* *papillae*—responsible for fingerprints. Capillaries are numerous in this layer, providing nutrients to the avascular epidermis and permitting heat to dissipate from the skin surface. Pain and touch receptors can also be found here. The *reticular layer* is the thickest part of the skin and is made up of dense, irregular connective tissue. This layer has many arteries and veins as well as sudoriferous (sweat) and sebaceous (oil) glands, pressure receptors, and nerves. Hair follicles, arrector pili muscles, lymphatic vessels, collagen and elastic fibers, and adipose cells are also found here. The dermis attaches the epidermis to the underlying subcutaneous layer.

> **U check**
> What are the five layers of the epidermis in thick skin?

PATHOPHYSIOLOGY
Decubitus Ulcers

When there is a restriction of the blood supply to the skin, cell death and even ulcers can occur. *Decubitus ulcers,* sometimes referred to as bedsores or pressure sores, are seen in patients who are bedridden or wheelchair bound and are not turned or repositioned regularly (Figure 7-5). The ulcers typically are seen over bony prominences (hips, heels, elbows) where the body's weight puts increased pressure on the skin. Patients who lie in bed for long periods should be turned and moved frequently to prevent bedsores.

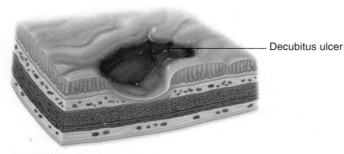

FIGURE 7-5 A decubitus ulcer (bedsore).

subcutaneous (hypodermis) A layer of support tissue below the dermis.

Subcutaneous Layer

The **subcutaneous** layer, also known as the **hypodermis,** is technically not a layer of the skin; it is a layer of support tissue. It lies immediately below the dermis and consists mostly of adipose (fatty) tissue and loose connective tissue. The fat in this layer acts as a reservoir for energy as well as an insulator and cushion against injury. The amount of fat varies from one body region to another and from person to person. The subcutaneous (hypodermis) layer, like the dermis, also contains blood vessels and nerves.

Accessory Organs

The accessory organs of the skin include hair, hair follicles, arrector pili muscles, sebaceous (oil) glands, nails, and sudoriferous (sweat) glands.

Hair is found over the entire body except for on thick skin (palms of hands and soles of feet), areas of the external genitalia, the nipples, and the lips. Hairs are enclosed in *follicles* that are formed from both the epidermis and dermis. Most of the follicle is made up of keratinocytes. As new keratinocytes are produced, old ones are pushed toward the surface of the skin. The old keratinocytes stick together to produce a hair. The part of the hair below the skin surface is the root, and the part extending above the surface is the shaft. Melanocytes located in the hair follicles produce the pigment that gives hair its color. As we age, melanocytes produce less pigment and the hair turns gray or white. When a hair follicle goes into a resting cycle, the hair falls out. Typically, the hair follicle will enter a new growing cycle and produce a new hair. Sometimes hair follicles completely die, and *alopecia* (baldness) develops.

Arrector pili muscles are made of smooth muscle that connects hair follicles to the papillary layer of the dermis. When the muscles contract, they cause a dimpling of the skin known as goosebumps. This happens when a person is cold or nervous. Because of their close proximity and relative position to **sebaceous** *(oil)* **glands,** when arrector pili muscles contract, *sebum* (oil) is released onto the hairs to keep them soft, moist, and pliable. Sebum is deposited onto skin to inhibit the growth of bacteria and keep it soft as well. Sebaceous glands are found on most areas of the skin except for the palms of the hands and the soles of the feet. They become more active for both sexes at puberty, when there is an increase of hormones called androgens.

> **U check**
> What is the function of the arrector pili muscles?

Nails also are derived from the epidermis and are composed of three main parts (Figure 7-6): the nail body, which is the majority of the nail; the free edge, or part that grows out from the body; and the cuticle, which is attached to the skin. The **lunula** is the crescent-shaped lighter colored area of the nail that is seen just above the cuticle.

The part of the nail embedded in the skin is called the nail root. The nail root contains active keratinocytes that produce nail growth. Most of the body appears pink because of the rich blood supply to the underlying dermis. Beneath each nail is the nail bed, which holds the nail down to underlying skin and provides nutrients to the nail. The nail plate acts as a protective shield for the delicate tissues of the underlying nail bed (Figure 7-7). Nails function to protect the ends of the fingers and toes. They are formed by epithelial cells with hard keratin, which is more permanent than the softer keratin found in skin. When a person is not getting and distributing enough oxygen to the peripheral area of the body, the nails may develop **cyanosis,** or become bluish in color.

> **U check**
> From what layer of the skin are nails derived?

Most **sudoriferous** *(sweat)* **glands** are located in the dermis with their ducts opening onto the epidermis. Two types of sweat glands include **eccrine** *(merocrine)* and **apocrine** (Figure 7-8). The most numerous of the sudoriferous glands are the eccrine glands. They are widely distributed over the body, with their greatest concentration on the forehead, neck, and back. They secrete perspiration made up of water, sodium chloride, and urea. Eccrine glands respond to high temperatures and act to cool the body when necessary. Apocrine glands produce a thicker type of sweat than eccrine glands. In addition to water, salt, and urea, their secretion contains a milky white protein and fat. Apocrine glands are mostly found in the axilla (armpit) and

arrector pili Smooth muscles that connect hair follicles to the papillary layer of the dermis.

sebaceous gland An exocrine gland in the dermis or subcutaneous layer that produces perspiration.

lunula The crescent-shaped area of the nail just above the cuticle.

cyanosis A bluish discoloration.

sudoriferous glands Sweat glands.

eccrine A type of sweat gland that is widely distributed in most areas of the body.

apocrine A type of sweat gland found in the axilla, groin, areola of the breasts, and bearded region of the face of males.

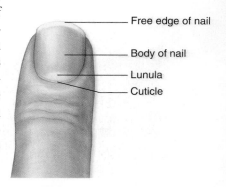

FIGURE 7-6 Superior view of a nail.

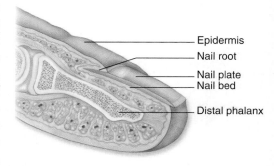

FIGURE 7-7 Cross-section of a nail.

CHAPTER 7   The Integumentary System

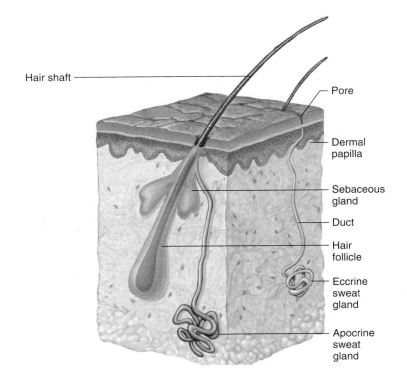

FIGURE 7-8
Location of eccrine and apocrine sweat glands and their ducts.

inguinal (groin) regions. In women, these glands will also enlarge at specific times of the menstrual cycle. Apocrine glands are primarily activated by nervousness or stress, producing what is known as a "cold sweat," but can also be activated by heat. Bacteria often break down and digest the proteins in the sweat produced by apocrine glands, resulting in body odor.

> **U check**
> What is the difference between an eccrine gland and an apocrine gland?

7.3 learning outcome

Explain the role of skin in regulating body temperature.

Regulation of Body Temperature

The skin has warm and cold *thermoreceptors* that are made up of free nerve endings. The warm receptors function optimally at temperatures of 25°C (77°F) to 45°C (113°F). Cold receptors work best between 10°C (50°F) and 20°C (68°F). Pain is felt at temperatures below 10°C (50°F) and above 45°C (113°F). If a person feels too warm or cold, thermoreceptors will send a message to the hypothalamus in the brain. Among its functions, the hypothalamus detects any temperature deviation from the "normal." If the body temperature is too high, blood vessels in the skin will dilate and sweat glands will secrete a watery perspiration, which causes heat to dissipate and the body temperature to fall back toward normal. If the body temperature is too low, blood vessels in the skin will constrict and blood will be shunted to the core of the body where temperatures are higher and the blood will be warmed. In addition, shivering may occur which will assist in warming the body.

> **U check**
> What body mechanisms cause us to cool down and to heat up?

Skin Color

7.4 learning outcome

Explain the factors that affect skin color.

Melanin, carotene, and hemoglobin are the three factors that determine skin color (Figure 7-9). **Melanin** (pigment) ranges in color from yellow to brownish black. The more melanin individuals have in their skin, the darker their skin color. Regardless of skin color, all people have about the same number of melanocytes. What does vary is the number of melanosomes, the organelles that actually produce the melanin. Darker-skinned individuals have more melanosomes than do lighter-skinned persons.

Carotene is a yellow pigment found mainly in the stratum corneum and the fat cells of the hypodermis. It is derived from foods such as carrots. The third contributor to skin color is *hemoglobin*. This pigment is responsible for carrying oxygen in the blood. Well-oxygenated blood will give the skin a "ruddy" or pinkish complexion.

FIGURE 7-9 Variation in skin coloration.

U check
What three factors are responsible for skin color?

melanin Yellow to brownish black pigment found in some parts of the body such as the skin, retina, and hair.

carotene A yellow pigment found mainly in the stratum corneum and the fat cells of the hypodermis.

PATHOPHYSIOLOGY

Jaundice

With some conditions the skin as well as the whites of the eyes may take on a yellow cast that is referred to as *jaundice* (Figure 7-10). Jaundice occurs when there is too much bilirubin—a waste product—in the blood. This excessive bilirubin can be caused by too much bilirubin being produced from the breakdown of red blood cells, a liver disorder that prevents the liver from removing the extra bilirubin, or a blockage of the ducts that secrete bilirubin into the intestines. For adults, jaundice is typically due to liver disease but sometimes caused by gallbladder or pancreas disorders. Newborns sometimes get jaundice due to the breakdown of red blood cells after birth.

Sometimes the skin may take on other unusual colors. For example, if the blood is poorly oxygenated or there is insufficient circulation, the individual's skin may take on a bluish or cyanotic appearance. In still another example, a bronzing of the skin may be an indication of a hypoactive adrenal gland (Addison's disease).

CHAPTER 7 The Integumentary System

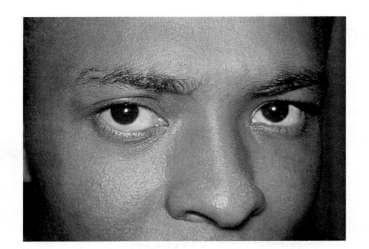

FIGURE 7-10 Yellowing of the sclera (whites of the eye) in an individual with jaundice.

learning outcome

Explain the process of skin healing, including scar production.

Skin Wounds and Healing

When the skin is injured (Figure 7-11) it often becomes inflamed, which makes the skin look red. This is due to dilation of blood vessels in the area which allows more blood at the site of injury. The dilated blood vessels become more permeable ("leaky"), permitting fluid to leave the blood vessels and enter the spaces between cells. This causes swelling in the area, and there may also be pain because of the activation of pain receptors. The overall process in most cases will promote healing as the increased blood flow will bring more oxygen and nutrients to the injured site.

In addition to the fluid that leaves the circulatory system to assist in healing, there are various white blood cells that help fight infections. (These will be discussed in Chapter 12, The Lymphatic and Immune Systems.) When there is a minor skin wound with sharp, well-defined margins, the healing is referred to as healing by primary intention. If the wound is large with irregular margins, it is referred to as healing by secondary intention. If the wound becomes infected, it is referred to as healing by tertiary intention. When the dermis is injured, a blood clot will form and eventually a scab will replace the clot. In most cases, if the initial wound was not large or severe, the skin will heal without permanent scarring. However, if the injury is more extensive, a scar may form. Collagen fibers are white in color and make up the greatest component of the scar. Although a scar may appear tough, it

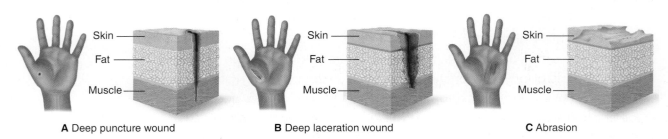

FIGURE 7-11 (A) Deep puncture wound, (B) deep laceration wound, and (C) abrasion.

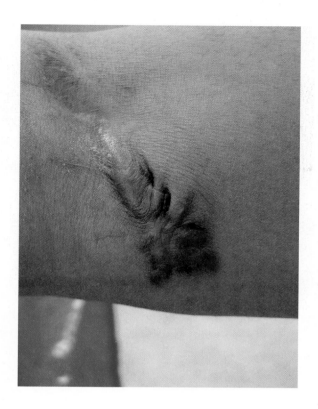

FIGURE 7-12
Keloid formation due to excessive scar formation.

actually is not as strong as the original skin that it has replaced. Sometimes scarring can be excessive. These are called *keloids* or "proud flesh" because of their appearance (Figure 7-12). Keloids can occur in anyone, but are more common in darker-skinned individuals.

Skin Lesions and Disorders

Skin lesions and skin disorders are two types of changes to the skin. Lesions are any variation in the skin. Skin disorders are abnormalities of the skin. The medical specialty that deals with the numerous skin lesions and disorders that exist is dermatology. The medical professional who specializes in these disorders is a dermatologist.

Skin Lesions

Lesions may be as normal as freckles or moles, which are increased concentrations of melanin in a specific area, or they may be as serious as skin ulcers or tumors. These variations in the skin or lesions can help the dermatologist or other physician determine the underlying cause or problem. Numerous types of lesions occur and can be classified as primary, secondary, or vascular (Figure 7-13). We will discuss a few here. Table 7-1 lists additional types of lesions.

A *macule* is a flat lesion that is not raised above the surface of the skin. It may be the same or a different color than the adjacent skin. A freckle is an example of a macule. A *vesicle* is a small blister; a cold sore is an example. A *bulla* is a larger blister or a collection of small vesicles, and a *pustule* is a pus-filled lesion such as a pimple. An ulcer is an example of *erosion*, where there is an excavation of the skin. An *excoriation*, an

7.6 learning outcome

Recognize and describe common skin lesions and summarize the causes, signs and symptoms, and treatments of common skin disorders.

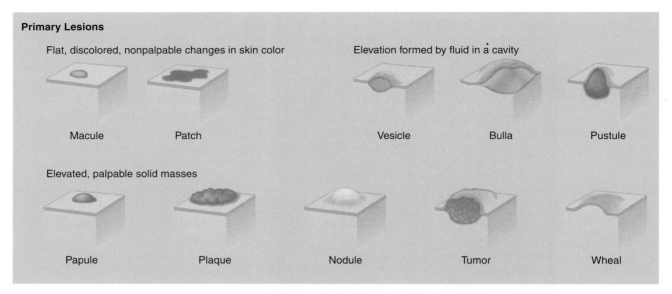
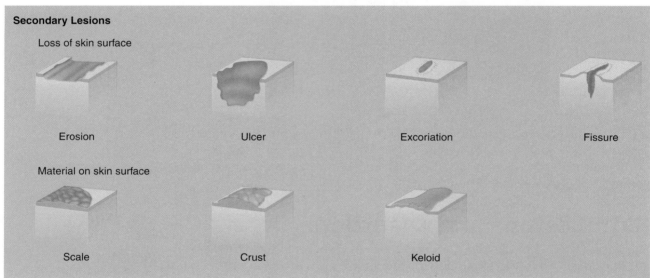
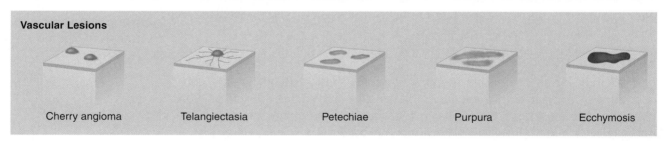

FIGURE 7-13 Types of skin lesions.

abrasion of the skin, may result when an individual constantly scratches himself or herself vigorously. *Petechia* are pinpoint hemorrhages that show up as small red dots on the skin. They can result from conditions such as platelet deficiencies.

TABLE 7-1 Common Skin Lesions and Descriptions

Lesion Name	Description
Bulla	A large blister or cluster of blisters
Cicatrix	A scar, usually inside a wound or tissue
Crust	Dried blood or pus on the skin
Ecchymosis	A black and blue mark or bruise
Erosion	A shallow area of skin worn away by friction or pressure
Excoriation	A scratch; may be covered with dried blood
Fissure	A crack in the skin's surface
Keloid	An overgrowth of scar tissue
Macule	A flat skin discoloration, such as a freckle or a flat mole
Nodule	A large pimple or small node (larger than 6 cm)
Papule	An elevated mass similar to but smaller than a nodule
Petechiae	Pinpoint skin hemorrhages that result from bleeding disorders
Plaque	A small, flat scaly area of skin
Purpura	Purple-red bruises usually the result of clotting abnormalities
Pustule	An elevated (infected) lesion containing pus
Scale	Thin plaques of epithelial tissue on skin's surface
Tumor	A swelling of abnormal tissue growth
Ulcer	A wound that results from tissue loss
Vesicle	A blister
Wheal	Another term for *hive*

Skin Disorders

Skin disorders may range from a serious condition to just a change in appearance. As mentioned earlier, *alopecia* is the absence or loss of hair, especially on the head (baldness). It is can be caused by a variety of factors including genetics, aging, chemotherapy, and infection. Fungal infection is probably the most common cause, but alopecia can also be caused by bacterial infections. The treatment for alopecia depends on the cause. Hair transplants are becoming quite common and are often successful, and drugs such as Rogaine may slow down hair loss. Some hair loss is temporary, such as that caused by chemotherapy. Note that although the hair may grow back when chemotherapy is completed, it is not unusual for the hair to grow back a different color or texture.

Cellulitis is an inflammation of the connective tissue in skin and is most often seen on the face and legs. A bacterial infection by *Staphylococcus aureus* or *Streptococcus* is the most common cause. The skin may be

Inflammation

> ### focus on Wellness: Preventing Acne
>
> Acne vulgaris, commonly known as acne, is an inflammatory condition of the skin follicles and sebaceous glands. It results in comedoes (blackheads and whiteheads) as well as papules and pustules. They occur mainly on the face, but they can also occur on the neck, back, and chest. Acne often appears in adolescence relating to the increase in sex hormones, especially testosterone which will stimulate increased production of sebum by the sebaceous glands. Pores become blocked by excess sebum and dead epidermal cells and then are infected by certain bacteria (propionibacteria). The result is a pimple. Treatments include washing the face regularly with noncomedogenic (does not contribute to comedo formation) soaps, keeping the hands away from the face, and washing the hair frequently to keep oil from the scalp away from the face. Makeup should be removed daily and makeup or lotion with sunscreen should be used. In severe cases, a dermatologist may recommend oral antibiotics such as tetracycline or other medications such as retinol, which is rich in vitamin A and helps moisturize and detoxify the skin.

erythematous (red), feel tight and hot, and possibly be painful. The individual may also experience a fever due to the infection. Because it is caused by bacteria, cellulitis is treated with antibiotics. In more serious cases, the antibiotics may have to be administered intravenously.

Dermatitis is a general term defining any inflammation of the skin that can be caused by a wide range of disorders. *Comedos* are collections of bacteria, dead epithelial cells, and dried sebum. They are also known as blackheads or whiteheads. When there is an infection of the sebaceous gland(s) the result is *acne vulgaris,* which causes pustules, papules, and scarring of the affected skin. Testosterone may contribute to the formation of acne. The treatment consists of a combination of reducing oil production, prescribing antibiotics or anti-inflammatory medications, and increasing the rate of cell turnover.

Eczema is a very common chronic dermatitis that often has acute phases or flare-ups followed by periods of remission. Eczema often first appears in childhood, but is seen with equal and increasing frequency in adults. The most common cause of eczema is an allergic hypersensitivity (often to an unknown allergen) and it is often associated with other allergies such as asthma and hay fever. Environmental irritants, stress, diet, and medications can bring about exacerbations (outbreaks) of this disease. The rash of eczema can appear anywhere on the body and appears as a red, scaly pruritic (itchy) rash that may even be painful. Anti-inflammatory drugs such as steroids and nonsteroidal anti-inflammatory drugs (NSAIDS) are used to treat many skin disorders including eczema. If there is a secondary bacterial infection, antibiotics will be prescribed. Avoiding known triggers may also help reduce the frequency of outbreaks.

Folliculitis is inflammation of hair follicles. When it involves a single hair follicle the condition is called a *furuncle*. When more than one hair follicle is involved it is a *carbuncle* (Figure 7-14). A common cause is skin

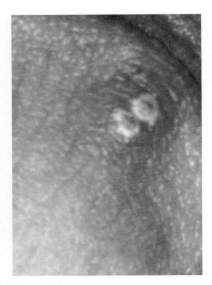

FIGURE 7-14 A carbuncle (an infection of a hair follicle).

RECALL ADAM BOYLE. What do you think is most likely the cause of his acne?

irritation from shaving or excess rubbing of skin areas which then are infected by *S. aureus* in most cases. The hair follicles become red, itchy, or painful and filled with pus. Treatment includes prescribing antibiotics to treat the infection and avoiding irritation of the affected skin.

Herpes simplex virus type 1 and 2 (HSV-1, HSV-2) are two of the more common herpes viruses. Typically HSV-1 causes cold sores, is very contagious, and is spread through contact with infected saliva. HSV-2, known as genital herpes, is sexually transmitted. However, note that there can be crossing over, with HSV-1 infecting the genitals and HSV-2 infecting the mouth. It is estimated that more than 40 million Americans are infected with genital herpes. Signs and symptoms include painful or itchy vesicles or blisters in the genital area. Although there is no cure for the infection, many medical and alternative treatments are available. Antiviral drugs such as acyclovir may decrease the frequency of outbreaks. Also, lysine (an amino acid) may lessen the severity and duration of an outbreak. Getting sufficient rest and adequate nutrition and controlling stress can help reduce and shorten the severity of outbreaks.

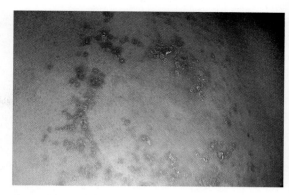

FIGURE 7-15 The varicella-zoster virus (shingles).

Herpes zoster, more commonly known as *shingles* (Figure 7-15), is an inflammatory condition of the skin caused by the varicella-zoster virus which is the same herpes virus responsible for causing chicken pox. After a person has chicken pox, the virus becomes dormant in the dorsal root ganglion located just off the spinal cord. Stress or illness can reactivate the virus and cause shingles, sometimes decades after the individual had chicken pox. The patient may experience a painful burning sensation prior to the appearance of the herpes simplex lesions. Typically the lesions appear as painful vesicles on the skin distributed along the path of a single spinal nerve. The skin that is innervated by that nerve and where the vesicles are seen is called a dermatome. Treatment may include antiviral medications to shorten the duration of the disease and pain medications to control the pain. It usually takes several weeks for the lesions to heal and recurrences are possible. Sometimes the patient may continue to have pain long after the lesions have disappeared, which is known as postherpetic neuralgia (nerve pain). A vaccine is now available to individuals over 60 years of age that can prevent herpes zoster.

Impetigo is a condition most often seen in children and is caused by a *Staphylococcus* or *Streptococcus* infection. The lesions are sometimes described as "honey crusted" because their appearance is similar to that of dried honey (Figure 7-16). Impetigo most commonly is seen on the face or lower leg and is highly contagious when contact is made with the lesions. This condition is treated with antibiotics. Instructing the patient to wash the lesions two or three times a day with soap and water will help remove the *exudate* (fluid from the lesions) and decrease the spread to other skin areas.

Pediculosis, more commonly known as lice, comes in three forms: head lice (pediculosis capitis), body lice (pediculosis corporis), and pubic lice (pediculosis pubis). All forms are caused by parasitic lice and are associated with overcrowded conditions and often with poor hygiene. Pubic lice are also spread by sexual contact. The skin is pruritic (itchy) and excoriations may occur

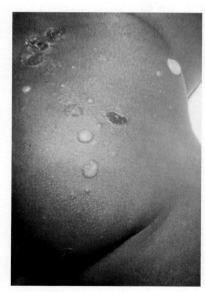

FIGURE 7-16
Impetigo, an infection caused by *S. aureus* or *S. pyogenes.*

CHAPTER 7 The Integumentary System 139

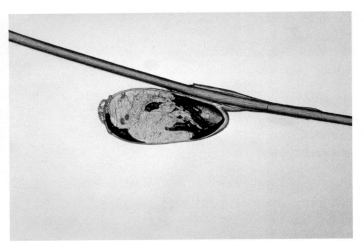

FIGURE 7-17 A nit (egg) of *Pediculosis humanus,* the cause of lice.

because of aggressive scratching. Lice are often identified by the eggs, or *nits,* that can be seen in the hair of the affected body area. *Pediculosis humanus* is the nit that causes lice in humans (Figure 7-17). Prescription medications and shampoos are often necessary, but over-the-counter (OTC) treatments such as Nix are often effective in ridding the hair of lice. Individuals should avoid sharing combs, hats, and towels with anyone diagnosed with lice. All bedding, hats, and clothing exposed to the parasite will also need to be treated thoroughly to avoid reinfection.

Psoriasis is a common chronic inflammatory skin condition that has an autoimmune basis. Patients are diagnosed by the appearance of characteristic silvery, scaly lesions anywhere on the body, but commonly on extensor surfaces such as the elbows and knees. Psoriasis is usually a lifelong condition with periods of remission and exacerbation. Sometimes joint pain known as psoriatic arthritis may also occur with this disorder. Many medications are used to treat psoriasis, including anti-inflammatory drugs and therapeutic ointments such as creams with vitamins A and D, hydrocortisone creams, and retinoids. Some patients also get relief with controlled UV ray (PUVA) treatments; severe cases may require hospitalization to get exacerbations under control.

Rosacea is a skin disorder that commonly appears as facial redness, predominantly over the cheeks and nose. It is often considered an adult form of acne. Rosacea results from dilation of small facial blood vessels and probably has an autoimmune connection. It is seen most often in fair-skinned individuals and is not curable, but is managed fairly well with topical cortisone creams. In severe cases, electrolysis may be useful in destroying large or dilated blood vessels.

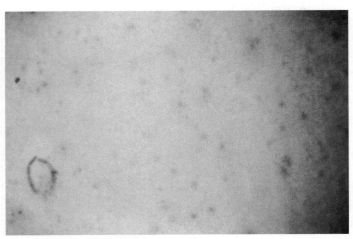

FIGURE 7-18 Scabies infection.

Scabies is a highly contagious skin condition caused by the mite, *Sarcoptes scabei.* As the mite burrows beneath the skin it defecates, causing an inflammatory reaction seen as red lines on the surface of the skin (Figure 7-18). The skin often is extremely itchy, especially at night. Most cases are easily treated with prescription medications. Because scabies is contagious, it is wise to treat an entire family if one member is infected.

Tinea (ringworm) is a fungal skin infection that gets its name because it appears serpentine like a worm (Figure 7-19). Names representing the body location affected are given by adding to –*tinea* the part of the body affected. For example, tinea corporis means the location is on the trunk or body, tinea capitis affects the scalp, and tinea pedis refers to the feet and is commonly known

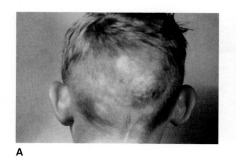 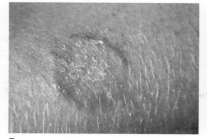 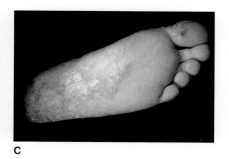

FIGURE 7-19 Ringworm: (A) tinea capitis, (B) tinea corporis, and (C) tinea pedis.

as athlete's foot. Ringworm is caused by fungi called dermatophytes. Topical and oral antifungal agents are used to treat tinea. It is highly contagious so there should be no sharing of sheets, towels, and other personal care items. Wrestlers and other athletes often get tinea because of their contact with others.

Verrucae (warts) are harmless skin growths (Figure 7-20) caused by the HPV (human papillomavirus). They appear most often on the hands, feet, and face. Warts can be smooth, flat, rough, raised, dark, small, or large. Warts are often removed with OTC medications, but can also be treated through surgery, laser surgery, cryosurgery (freezing), or burning.

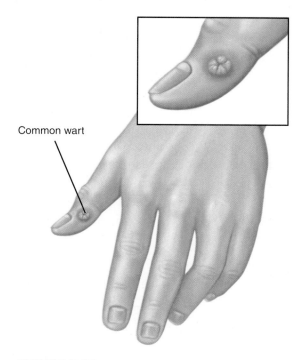

FIGURE 7-20 Wart on a hand.

Burns

After motor vehicle accidents, burn injuries are the most common cause of accidental death in the United States. Although there are different causes of burns, such as chemical, boiling water, and steam, the most common cause is fire. In 2007, over one-half million burn and fire victims required medical care. In the same time period more than 4,000 deaths resulted from burns and fire (it is not always easy to tell if the victim died from burns related to the fire or smoke inhalation). The number of burn units has dropped from 132 in 2004 to 127 in 2010. Some experts feel this is putting our nation at risk in a post–September 11, 2011, world.

On a more positive note, research is being conducted to come up with new and better artificial skin for burn victims. Burns are often classified based on the amount of body surface or *total body surface area (TBSA)* involved and the severity of the burn. These are the two greatest

7.7 learning outcome

Relate the degrees of burns to their appearance and treatment.

Burns

> **focus on** Clinical Applications: The Rule of Nines
>
> A quick way to estimate the extent of body surface area affected by burns is the rule of nines (Figure 7-21). This method divides the body into 11 areas, each accounting for 9 percent of the total body surface. The 11 body areas of the rule of nines with their percentages are as follows:
>
> - Head, 9 percent (front and back, 4.5 percent each)
> - Right arm, 9 percent (front and back, 4.5 percent each); left arm, 9 percent (front and back, 4.5 percent each)
> - Right leg (front), 9 percent; left leg (front), 9 percent
> - Right leg (back), 9 percent; left leg (back), 9 percent
> - Trunk (front), 18 percent ; trunk (back), 18 percent
> - Genital area, 1 percent
>
>   Total body area = 100 percent.
>
>   The rule of nines applies only to adults. For infants a modified rule of nines is used and for children the Lund and Browder chart is used.

predictors for the risk of death due to burns because of fluid loss and infection. The Focus on Clinical Applications feature discusses the rule of nines to estimate the amount of body surface area affected by a burn or burns.

Burn severity, or the severity of burns, indicates the depth of the injury. Burns can be classified as first degree, second degree, or third degree. A first-degree burn is considered superficial and involves only the epidermis. It is characterized by pain, redness, and swelling. Unless they are extensive, first-degree burns do not require medical attention and usually heal well. A mild sunburn would be an example of a first-degree burn. A second-degree is called a partial thickness burn and both the epidermis and dermis are affected. In addition to pain, redness, and swelling, there will also be blistering. If a partial-thickness burn affects 1 percent or more of the body surface, medical attention should be sought. A person's hand covers about 1 percent of total body surface. If a partial-thickness burn affects 9 percent or more of the body surface, shock may develop which can be life threatening. Medical attention should always be sought for a person who has experienced shock. A third-degree or full-thickness burn involves the epidermis, dermis, hypodermis, and often underlying structures such as muscle and bone. The skin often looks black or charred, a condition known as an eschar. Regardless of the amount of surface area affected, full-thickness burns always require immediate medical attention. See Figure 7-22 on page 144 for examples of the degrees of burn severity.

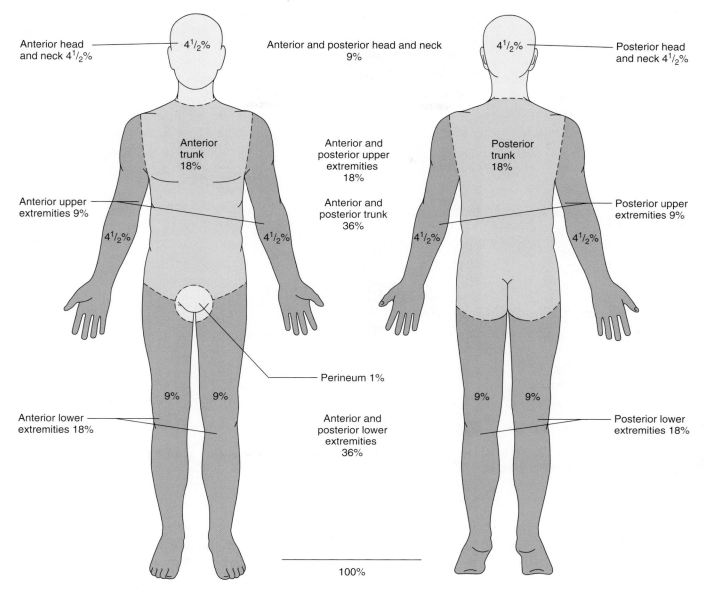

FIGURE 7-21 The rule of nines is used to estimate the extent of damage caused by burns. The body is divided into regions, each representing 9 percent (or some multiple of 9 percent) of the total skin surface area.

Infection and loss of fluids and electrolytes are major concerns with severe burns. Some general guidelines for treating burns are as follows:

- Anything sticking to the burn should be left in place.
- Butter, lotions, or ointments should not be applied to a burn because they may cause an infection to develop. Only ointments prescribed by a doctor or recommended by a pharmacist should be used.
- The burn should be cooled with large amounts of cool water. Ice or extremely cold water should be avoided.
- The burn should be covered with a sterile sheet or plastic bag. Burns to the face, however, should not be covered.

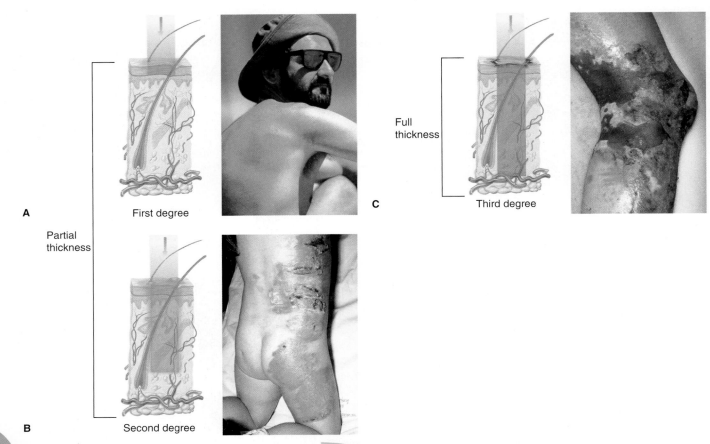

FIGURE 7-22 The degrees of burn severity include (A) superficial (first-degree) burns, (B) partial-thickness (second-degree) burns, and (C) full-thickness (third-degree) burns.

- Emergency medical personnel should be contacted for serious burns.
- In burns affecting the mouth and throat, the airways should be checked to see if there is any swelling that could interfere with breathing.
- Burns to the head are always more serious than burns to other body parts and almost always require emergency medical treatment.

U check
What are the two major ways to classify burns?

7.8 learning outcome

Summarize the causes, signs and symptoms, and treatments of the various types of skin cancer.

Skin Cancer

The three types of skin cancer are *basal cell carcinoma, squamous cell carcinoma,* and *malignant melanoma* (Figure 7-23). Excessive exposure to sunlight is the greatest single risk factor for skin cancer and tanning beds are implicated as well. Basal cell carcinoma originates from the basal layer of the epidermis, whereas squamous cell carcinoma arises from the upper cells of the epidermis. Melanomas arise from melanocytes. Basal cell carcinoma rarely *metastasizes* (spreads) whereas squamous cell carcinoma and melanoma often do. Basal cell carcinoma accounts for over 90 percent of all skin

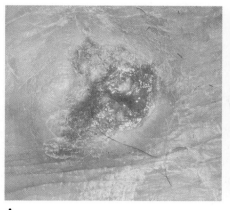  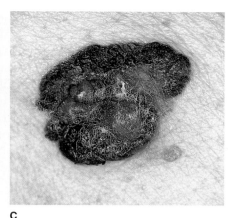

A　　　　　　　　　　　　B　　　　　　　　　　　　C

FIGURE 7-23 Skin cancer types: (A) basal cell carcinoma, (B) squamous cell carcinoma, and (C) malignant melanoma.

cancers, squamous cell carcinoma accounts for approximately 2,500 deaths each year. Although malignant melanoma accounts for less than five percent of skin cancers, it is responsible for over 75 percent of deaths due to skin cancer.

Signs and symptoms of basal cell and squamous cell carcinoma include changes to the affected skin area, including a new growth or sore on the skin that does not heal. The appearance may be waxy, smooth, red, pale, flat, or lumpy, and it may or may not bleed. Several forms of treatment are available for these two cancers: In *curettage,* a sharp instrument is used to scoop out the cancerous spot. *Electrodessication* uses electrical currents to minimize bleeding as well as to kill any remaining cancer cells. In Mohs surgery, the cancerous lesion is shaved off one layer at a time. *Cryosurgery* uses freezing to kill cancer cells, whereas laser therapy uses a laser beam of light to destroy the cancer cells.

Malignant melanoma is the most aggressive of the three skin cancers. People with fair skin, blonde or red hair, and blue or green eyes are at increased

---

focus on Wellness: The ABCDEs of Malignant Melanoma

Knowing the "ABCDE" rule will help you recognize a possible malignant melanoma.

A: **Asymmetry.** The lesion is not symmetrical in shape.

B: **Border.** The border is irregular and not well demarcated, with the edges blending into the surrounding skin.

C: **Color.** It is not uniform in color (brown or black), but contains a mixture of colors including red and blue.

D: **Diameter.** The lesion is larger than 6 mm or approximately the diameter of a pencil eraser.

E: **Evolution.** If an existing mole or lesion has any changes in size, shape, or color or begins to itch or bleed, it should be checked out. This is the newest addition to the ABCD acronym.

risk of contracting malignant melanoma. Melanoma is more common in females, but appears to have a worse prognosis when seen in males. It is also one of the few cancers for which we are seeing an increase in incidence. Melanoma can occur anywhere on the body, but most often appears on the trunk, head, and neck in men and on the arms and legs in women. The Focus on Wellness feature discusses the ABCDEs of malignant melanoma.

The staging (how far it has spread) of the cancer will determine the treatment. Available treatments include surgery to remove the melanoma and lymph node biopsy to determine if the cancer has spread, with removal of cancerous lymph nodes. For advanced cancers, chemotherapy and radiation therapy may be used. Immunotherapy also may be used to boost the immune system. Malignant melanoma has five different stages, which are described from the least to the most serious:

Stage 0. Malignancy is found only in the epidermis.
Stage I. Malignancy has spread to the epidermis and dermis, and has a thickness of 1 to 2 mm.
Stage II. Malignancy has a thickness of 2 to 4 mm and may have ulceration.
Stage III. Malignancy has spread to one or more nearby lymph nodes.
Stage IV. Malignancy has spread to other body organs or other lymph nodes far away from the original melanoma.

> **U check**
> Which type of skin cancer is responsible for the most deaths?

7.9 learning outcome

Describe the effects of aging on skin

Life Span

As we age, our skin marks the passage of time (Figure 7-24). A loss of elastic and collagen fibers occurs, the amount of underlying fat decreases, and the dermis becomes thinner. The circulation to the skin also decreases. All of these changes result in loss of firmness and wrinkling of the skin. The skin in elderly individuals also may become more transparent and pale. It is common to see *lentigos*—more commonly called "liver spots" or age spots—that are caused not by the liver but are the result of excessive melanin production because of overexposure to sunlight (UV rays). Melanin decreases in the hair follicles causing the hair to turn gray or white. The hair of the scalp

---

focus on Wellness: Skin Cancer Warning Signs

Although the thought of getting any cancer is frightening, early detection can make a significant difference in the prognosis. The most common warning sign of skin cancer is a change in the appearance of the skin. Therefore, if you have any suspicious change to your skin such as a lesion that does not seem to heal, or that changes in any way, you should seek the professional care of a physician.

as well as that on the rest of the body becomes thinner. Older people also have a reduced ability to tolerate temperature changes. This is due in part to a decrease in the number of sudoriferous glands, and with less perspiration it is more difficult for the body to adjust to high temperatures. At the same time, the loss of adipose tissue and decreased circulation results in a lessened ability to retain heat, leading to increased sensitivity to cold.

FIGURE 7-24
Skin changes seen in aging.

summary

chapter 7

learning outcomes	key points
7.1 Recall the functions of the integumentary system.	The integumentary system protects us from harmful substances, trauma, ultraviolet radiation, and foreign organisms. It also acts as a sensory organ, helps in temperature regulation of our body, excretes waste products, and is involved in vitamin D synthesis.
7.2 Describe the layers of skin and the characteristics of each layer and recognize the accessory organs of skin and describe their structures and functions.	The skin has two layers: the epidermis, made up of stratified squamous epithelium; and the dermis, made up of connective tissue, blood vessels, nerves, and accessory organs. Some of the accessory organs are sweat and oil glands, sensory receptors, and hair follicles.
7.3 Explain the role of skin in regulating body temperature.	Temperature receptors in the skin send messages to control centers in the brain. After processing the information, signals are sent to cause more or less blood to be shunted to the skin and muscles to shiver if heat generation is necessary.

continued

chapter 7

learning outcomes	key points
7.4 Explain the factors that affect skin color.	Melanin, carotene, and hemoglobin are the three pigments responsible for giving the skin its color.
7.5 Explain the process of skin healing, including scar production.	If a wound is minor, healing may occur without scarring. In more extensive injuries, permanent scarring may result.
7.6 Recognize and describe common skin lesions and summarize the causes, signs and symptoms, and treatments of common skin disorders.	Numerous types of skin lesions exist, from the common acne vulgaris to the deadly skin cancers. The skin has a great capacity for healing. Dermatological disorders may be caused by infection, autoimmune diseases, exposure to sunlight or chemicals, and trauma. The signs and symptoms usually include some type of rash and possibly itching, burning, or pain.
7.7 Relate the degrees of burns to their appearance and treatment.	A first-degree burn involves only the epidermis and has some reddening and discomfort. A second-degree burn involves the epidermis and dermis. Reddening, blisters, and pain may occur. Third-degree burns involve the epidermis, dermis, and subcutaneous layers and may even destroy bone and muscle. The skin is charred in appearance.
7.8 Summarize the causes, signs and symptoms, and treatments of the various types of skin cancer.	Basal cell carcinoma typically has an ulcerated appearance. Squamous cell carcinoma may display as a new growth or sore on the skin that does not heal. The appearance may be waxy, smooth, red, pale, flat, or lumpy. Malignant melanoma is the most deadly form of skin cancer. A melanoma has a unique appearance with variation of colors often part of the lesion. Most skin cancers are due to exposure to UV radiation, but chemicals and other factors may play a role. Most skin cancers are treated with surgery or chemotherapy.
7.9 Describe the effects of aging on skin.	As we age, our skin loses collagen and elastic fibers. The underlying subcutaneous layer loses fat. Circulation to the skin may lessen and the skin becomes wrinkled and subject to a greater number of skin conditions.

case study 1 questions

Can you answer the following questions that pertain to Adam's case presented earlier in this chapter?

1. What is a cause of acne vulgaris?
   a. Bacterial infections of sebaceous glands
   b. Autoimmune disease
   c. High levels of estrogen
   d. Too much protein in the diet
2. How is the diagnosis made?
   a. Clinically
   b. Blood tests
   c. Urinalysis
   d. Radiographically
3. What are some treatments for people with this condition? (There may be more than one correct answer.)
   a. Antibiotics
   b. Anti-inflammatory medications
   c. Special cleansers
   d. Antiviral medications
4. Which of the following has an earlier age of onset?
   a. Rosacea
   b. Acne vulgaris

review questions

1. Which one of the following is the most superficial layer of the epidermis?
   a. Stratum corneum
   b. Stratum basale
   c. Stratum granulosum
   d. Stratum spinosum
2. Which one of the following is found only in thick skin?
   a. Stratum corneum
   b. Stratum lucidum
   c. Stratum granulosum
   d. Stratum spinosum
3. Which one of the following is not a determinant of skin color?
   a. Carotene
   b. Melanin
   c. Hemoglobin
   d. Keratin
4. A flat lesion that is not raised above the surface of the skin is a(n)
   a. Macule
   b. Papule
   c. Pustule
   d. Excoriation

critical thinking questions

1. Andy Finn, a 17-year-old youth, was severely burned while playing around a bonfire with his friends. He was admitted to the emergency department where you are working part time while you get your medical assisting certificate. Andy has second-degree burns on his anterior thorax and third-degree burns on both the front and back of his left leg. Answer the following questions based on the knowledge you have mastered in this chapter.
   a. Using the rule of nines, estimate the total percentage of Andy's body surface that was affected by this burn.

b. What percentage of his body was affected by partial-thickness (second-degree) burns?
c. What percentage was affected by full-thickness (third-degree) burns?
d. What layers of skin have been affected by each of the burn types?
e. What functions of the skin are lost by each of these injuries?
f. What types of treatments does each burn require?

patient **education**

Create a patient educational brochure that teaches a patient how to prevent skin cancer as well as helps the patient learn the signs of skin cancer.

applying **what you know**

1. Why do people with darker skin tones have a lower incidence of skin cancer than those who have lighter complexions?
2. What are the accessory organs of the skin? Describe their structures and functions.
3. What are the two types of sweat glands? Where is each located, and how do their secretions differ?

CASE STUDY 2 *Impetigo*

Guy Ashley is a nine-year-old who is always coming down with colds and runny noses. His mother says that the "crust" around his nose is different this time and that he also has a red area on his left cheek. She is concerned that this is more than a cold.

1. What is the cause of Guy's condition?
   a. *S. aureus*
   b. Varicella-zoster virus
   c. Mumps
   d. Diabetes mellitus
2. Who is most at risk for this condition?
   a. Older males
   b. Older individuals of either gender
   c. Children
   d. Women of reproductive age
3. Which of the following would you expect the doctor to prescribe?
   a. An antibiotic
   b. An antiviral medication
   c. Corticosteroids

CASE STUDY 3 *Basal Cell Carcinoma*

Beverly Byrd is a 34-year-old hair stylist who likes to spend her free time participating in outdoor activities. In the winter she likes to visit the tanning salon to keep that great tan she got in the summer. Recently, Beverly noticed an "ulcer" on her left cheek. She has been hearing quite a bit on the radio about spending too much time in the sun or tanning salons and is afraid she may have skin cancer.

1. Which one type of skin cancer is the most likely to metastasize?
    a. Basal cell carcinoma
    b. Squamous cell carcinoma
    c. Malignant melanoma
2. How will the diagnosis be made?
    a. Biopsy
    b. Blood tests
    c. Urinalysis
    d. X-rays

8 The Skeletal System

chapter outline

- Functions of the Skeletal System
- Bone Structure
- Bone Growth
- Bone Classification, Parts, and Markings
- Axial Skeleton
- Appendicular Skeleton
- Joints
- Bone Fractures
- Life Span

outcomes

AFTER COMPLETING THIS CHAPTER, YOU WILL BE ABLE TO:

8.1 Recall the general structure and functions of the skeletal system.

8.2 Describe the general structure of bone.

8.3 Compare the two major types of bone growth.

8.4 Summarize how bones are classified, their parts, and markings.

8.5 Recall the bones of the axial skeleton.

8.6 Recall the bones of the appendicular skeleton.

8.7 Relate the major types of joints to their location and function.

8.8 Describe the major types of bone fractures and diseases of the skeletal system.

8.9 Explain the effects of aging on the skeletal system.

case study

Use the case study to focus on as you go through the chapter. The questions will guide you as you learn the anatomy, physiology, and pathology of the skeletal system.

CASE STUDY 1 *Fracture*

Twelve-year-old Tony Hodge is a running back for his middle school football team. On the last play of the last game of the season he scored the winning touchdown for his school. As he crossed the goal line he fell on his outstretched arm. Tony heard a snap and thinks he broke his "collarbone."

As you go through the chapter, keep the following questions in mind:

1. What is the "collarbone"?
2. What is the difference between a simple (closed) and compound (open) fracture?
3. How is a fracture diagnosed?
4. What is the healing process when a bone is fractured?
5. What advice would you give to Tony?

essential terms

appendicular skeleton (ap-en-DIK-you-lar SKEL-uh-ton)

axial skeleton (AK-cee-al SKEL-uh-ton)

canaliculi (kan-ah-LIK-yoo-lie)

cancellous (KAN-cell-us)

diaphysis (die-AH-fih-sis)

endochondral (en-doe-KON-dral)

endosteum (en-DOS-tee-um)

epiphysis (eh-PIH-fih-sis)

hematopoiesis (hee-ma-toh-poh-EE-sis)

lacunae (lah-KU-ni)

lamella (la-MELL-ah)

osteoblast (OS-tee-oh-blast)

osteoclast (OS-tee-oh-clast)

osteocyte (OS-tee-oh-site)

osteon (OS-tee-ohn)

osteoporosis (os-tee-oh-po-RO-sis)

periosteum (per-ee-OS-tee-um)

scoliosis (sko-lee-OH-sis)

suture (SU-chure)

synovial (sin-OH-vee-al)

Additional key terms in the chapter are italicized and defined in the glossary.

study tips

1. Draw a skeleton labeling the axial and appendicular skeletons.

2. Make flash cards with the anatomical and common names. For example, write "scapula" on one side of a card and "shoulder blade" on the opposite side.

3. Write a sample quiz with answers and page numbers from the text where to find the correct answer.

4. You can remember the number of vertebrae in the cervical, thoracic, and lumbar regions by remembering you eat breakfast at 7 (7 cervical vertebrae), lunch at 12 (12 thoracic vertebrae), and dinner at 5 (5 lumbar vertebrae).

Introduction

The skeletal system is made up of bones, joints, and related connective tissues. There are normally 206 bones in the adult skeleton. The skeleton has two major divisions: the axial skeleton and the appendicular skeleton (Figure 8-1). The **axial skeleton** has 80 bones that include the bones of the skull, hyoid bone, auditory ossicles, vertebral column, sternum, and ribs. The **appendicular skeleton** has 126 bones and includes the pectoral and pelvic girdles and the upper and lower extremities. The upper extremities (the arms) are attached to the axial skeleton by the pectoral girdle while the lower extremities (the legs) are attached to the axial skeleton by the pelvic girdle.

axial skeleton Division of the adult skeleton made up of the skull, hyoid bone, auditory ossicles, vertebral column, sternum, and ribs.

appendicular skeleton Division of the adult skeleton including the pectoral and pelvic girdles and the upper and lower extremities.

LO 8.1

U check
What are the two divisions of the skeletal system?

8.1 learning outcome

Recall the general structure and functions of the skeletal system.

Functions of the Skeletal System

Perhaps the first thing you think of when it comes to the skeletal system is protection and support. These are two very important functions. The skull protects the brain from injury and the ribs and vertebra protect viscera, such as the heart and lungs. The skeletal system also allows us to sit, stand, and move by the attachment of muscles to bones. However, the skeletal system has several other functions essential for life. Calcium is an essential mineral for our bodies and bone acts as a reservoir for over 99 percent of the calcium in our body. Both white and red blood cells are produced by red bone marrow contained within bone.

U check
The skeletal system acts as a reservoir for what mineral?

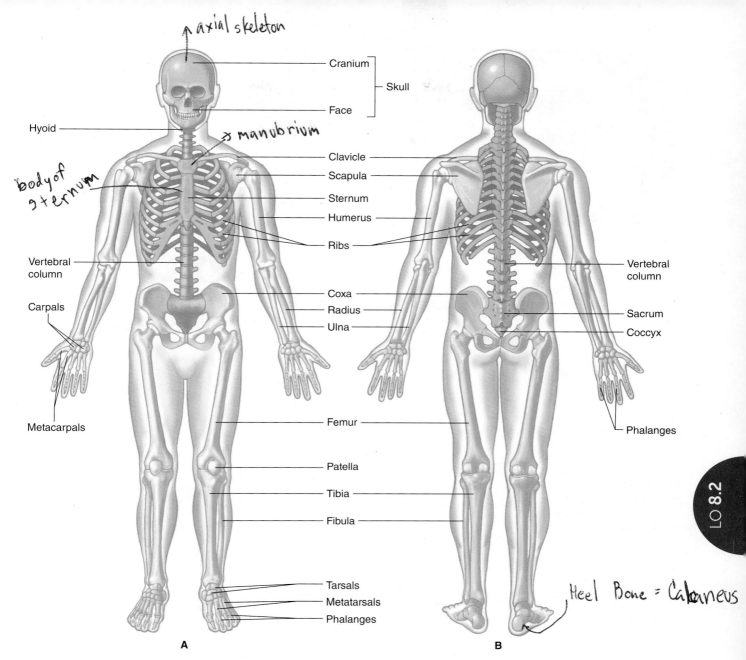

FIGURE 8-1 Major bones of the skeleton: (A) anterior view and (B) posterior view. The axial skeleton is shown in orange and the appendicular skeleton is shown in yellow.

Bone Structure

Bone is a living connective tissue that contains various types of cells, blood vessels, and nerves. The two types of bone tissue include **cancellous** (spongy) bone which has spaces filled with red bone marrow and *compact* bone which is denser, as shown in Figure 8-2. Compact bone has a unique arrangement of cells and canals. **Osteons** (Haversian systems) are the fundamental unit of compact bone (Figure 8-3). Each osteon is a long,

8.2 learning outcome

Describe the general structure of bone.

cancellous Spongy bone tissue that contains red bone marrow.

CHAPTER 8 The Skeletal System 155

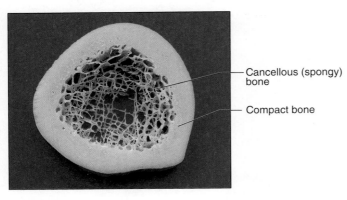

FIGURE 8-2 Cross section of bone showing compact and cancellous (spongy) bone.

osteon Also called a Haversian system; the basic unit of structure in adult compact bone consisting of a central canal and concentrically arranged lamellae, lacunae, osteocytes, and canaliculi.

lamellae Layers of bone that surround the central canal.

lacunae Spaces within the lamellae containing osteocytes.

endosteum A membrane lining the central canal of a bone and the holes of cancellous bone containing the bone-forming osteoblasts.

canaliculi Small canals in the bone that are perpendicular to the central canal and carry blood vessels and nerves to the lamellae and lacunae.

osteoblast An immature bone cell that causes mineralization of the bone matrix.

osteocyte A mature bone cell.

osteoclast A cell that breaks down bone for remodeling.

cylindrical tube consisting of concentric layers of bone called **lamellae** that surround a central canal. The lamellae have an appearance similar to the growth rings of a tree. Within the lamellae are **lacunae** (lakes) where osteocytes can be found. Blood vessels and nerves are contained within the central canal which is lined with a membrane called the **endosteum** (*endo* = within; *oste* = bone). The endosteum is also found within the holes of cancellous bone and contains the bone-forming osteoblasts. Smaller canals, called **canaliculi,** run perpendicular to the central canal and carry blood vessels and nerves to the lamellae and lacunae. The three types of bone cells are osteoblasts, osteocytes, and osteoclasts. Immature bone cells are called **osteoblasts** and are responsible for mineralizing the bone matrix (Figure 8-4). **Osteocytes** are mature bone cells, and **osteoclasts** are cells that tear down bone for remodeling and cause the release of calcium. There has to be a balance

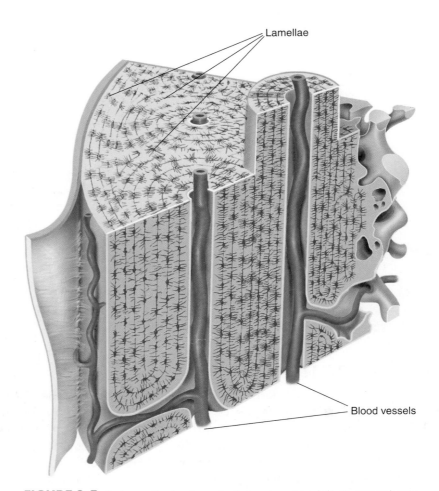

FIGURE 8-3 Osteon: the basic unit of structure in adult compact bone.

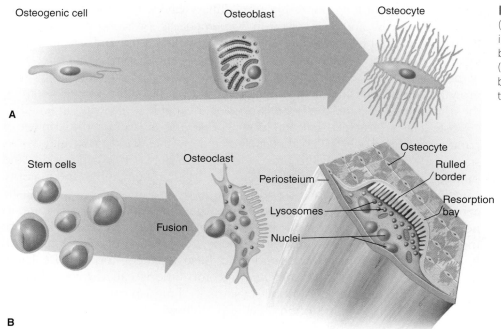

FIGURE 8-4
(A) Osteoblasts mature into osteocytes, the bone-forming cells; (B) osteoclasts are the bone-resorbing cells of the skeleton.

(homeostasis) between the building up and the tearing down of bone. This process is called *remodeling* and it is a lifelong process. Too much of either building up or tearing down can result in poor health.

> **U check**
> What is a Haversian system?

Bone Growth

Bones grow through a process called *ossification* or *osteogenesis*. There are two types of ossification: intramembranous and endochondral. In intramembranous ossification, bones begin as tough, fibrous membranes that are replaced by a bone matrix that is produced by osteoblasts. Calcium and other minerals are deposited within the matrix and then it calcifies. Bone is second only to enamel as the hardest tissue in our body. Bone is a connective tissue which means that it has a nonliving matrix and living cells. The matrix is made up of calcium, phosphate, collagen fibers, and water. The collagen fibers give bone its flexibility and the minerals give bone its strength. Other factors essential to bone growth include vitamin A for stimulation of osteoblasts, vitamin C for collagen synthesis, as well as growth hormone, insulin, and thyroid hormone. Except for the mandible (lower jaw), the bones of the skull are formed by intramembranous ossification.

In **endochondral** ossification, bones start out as cartilage models made of hyaline cartilage. This is common in long bones such as the femur. Eventually, the osteoblasts form a bone "collar" around the **diaphysis** (shaft) of the cartilage model and eventually will replace the cartilage model. The area where ossification first begins is called the primary ossification center. Each end or **epiphysis** of the long bone will begin to ossify later in the process and so are called secondary ossification centers. Some bones, such as the vertebrae, have more than two ossification centers.

learning outcome 8.3
Compare the two major types of bone growth.

endochondral A type of ossification in which bones start out as cartilage models.

diaphysis Shaft of a long bone.

epiphysis The end of a long bone.

In the center of long bones is the medullary cavity. It is initially filled with red bone marrow and is involved in *hematopoiesis,* or blood cell formation. Red marrow will be replaced by yellow marrow (fat tissue) and lose its capacity to produce blood cells. Between the epiphysis and diaphysis of a long bone is the epiphyseal plate or growth plate. Between 18 and 25 years of age, the growth plate which is made of hyaline cartilage is replaced by bone, becoming the epiphyseal line, and the bone is no longer capable of growing in length.

Osteoclasts are cells that tear down bone and are under the control of parathyroid hormone (PTH) that is secreted by the parathyroid glands. As stated earlier, throughout life there is a balance between tearing down and building up bone—a process known as remodeling. Remember, the majority of calcium is stored in bone so when the body requires more, osteoclastic activity increases to release calcium into the bloodstream. When there is not as great a demand for calcium elsewhere in the body, osteoblasts are more active, keeping calcium within bone. As we discussed in Chapter 6 (Concepts of Fluid, Electrolyte, and Acid–Base Balance), too much calcium in the blood is called *hypercalcemia* and too little is called *hypocalcemia*. As we age, the number of osteoblasts starts to diminish causing less calcium to be found in bones and often leaving people who are elderly, particularly women, at increased risk for fractures.

> **U check**
> What are the two types of ossification?

Bone Classification, Parts, and Markings

8.4 learning outcome
Summarize how bones are classified, their parts, and markings.

Bone is classified based on its histology (makeup) and individual bones are classified based on their shape. When classifying bones by their shape, the following terms are used:

- Flat bones are thin and consist of a layer of spongy bone between two layers of compact bone. Flat bones include the sternum, ribs, scapulae, and cranial bones.
- Long bones are greater in length than width. Long bones are mostly made up of compact bone, but their epiphyses or ends are mostly spongy bone. Long bones include the femur, tibia, fibula, humerus, radius, ulna, and phalanges (digits).
- Short bones are cuboidal in shape. They consist mostly of spongy bone with just a thin outer layer of compact bone. Short bones include the carpals (wrist bones) except for the pisiform, and tarsal (ankle) bones except for the calcaneus (heel bone).
- Irregular bones have a variety of shapes and include the vertebrae, coxal (hip bones), some facial bones, and the calcaneus.
- Sesamoid bones have their name because of their resemblance to sesame seeds. The patella and pisiform are examples of sesamoid bones.

Figure 8-5 shows the five shape classifications: flat bone, long bone, short bone, irregular bone, and sesamoid bone.

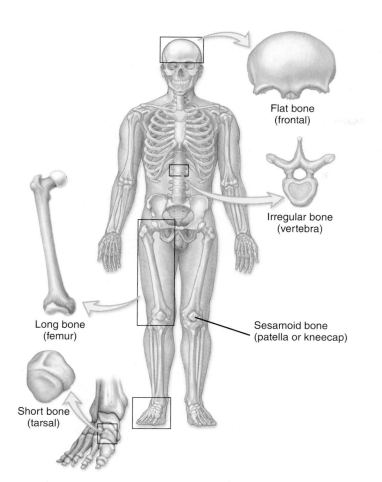

FIGURE 8-5
Classification of bone by shape; five different shapes are recognized—flat, long, short, irregular and sesamoid.

Long Bones

The five basic bone shapes are long, short, flat, irregular, and sesamoid. Long bones are classified as such because they are longer than they are wide. The *femur* (the longest and heaviest bone in the body) and the *humerus* are two bones that quickly come to mind. Additional examples of long bones are the tibia and fibula (Figure 8-6). The *phalanges* (singular, *phalanx*), the digits of the hands and feet, are also long bones because they are longer than they are wide.

Long bones have several regions, as displayed in Figure 8-7. The diaphysis (shaft) makes up the majority of the length of the long bone. It is a tubular structure with a thick collar of compact bone surrounding a space called the medullary cavity. In children, this is the location of red bone marrow and is responsible for **hematopoiesis.** Hematopoiesis is the production of blood cells, including 10 to 15 million erythrocytes (red blood cells) per second. In adults, red bone marrow is replaced by yellow bone marrow except in a few locations such as the skull, ribs, vertebra, and pelvis. At each end of a long bone there is an epiphysis (plural, epiphyses) which is an expanded area consisting of a thin layer of compact bone surrounding more abundant cancellous bone. Covering each epiphysis is articular cartilage of the hyaline type. It is smooth and aids in the almost frictionless movement of joints.

hematopoiesis The production of blood cells.

CHAPTER 8 The Skeletal System

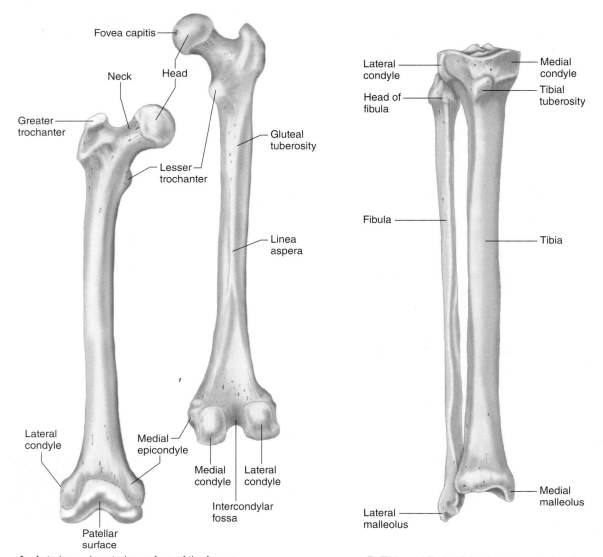

A Anterior and posterior surface of the femur.

B Tibia and fibula of the right leg, anterior view.

FIGURE 8-6 (A) Bone structure of the femur and (B) bone structures of the tibia and fibula.

periosteum A membrane surrounding the shaft of a bone.

A membrane called the **periosteum** (*peri* = around; *oste* = bone) surrounds the diaphysis. It is made up of dense fibrous connective tissue and has nutrient blood vessels and nerves penetrating it. It is the periosteum that is responsible for "growing pains" during adolescence.

Bone Markings

Every bump, groove, and hole on a bone has a name. These are known as bone markings. There are two main types of bone markings—those that grow out are known as projections, and those that indent the bone are known as depressions. These markings found on bones and the specific terms used to describe skeletal structures are provided in Table 8-1. The table also directs you to the figures throughout the chapter which show these structures.

> **U check**
> What is the name for the shaft of long bone?

UNIT 2   Concepts of Common Illness by System

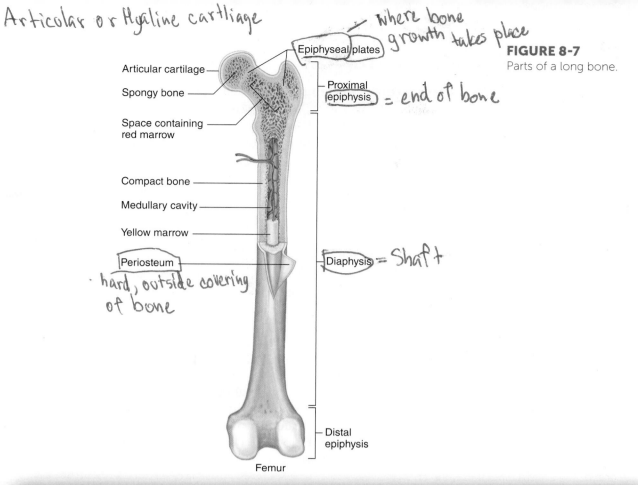

FIGURE 8-7
Parts of a long bone.

TABLE 8-1	Terms Used to Describe Skeletal Structures	
Term	**Definition**	**Examples**
Condyle	A rounded process that usually articulates with another bone	Medial and lateral condyles of the femur (refer to Figure 8-6A)
Crest	A narrow, ridge-like projection	Iliac crest of the ilium (refer to Figure 8-25)
Epicondyle	A projection situated above a condyle	Medial epicondyle of the femur (refer to Figure 8-6A)
Foramen	An opening through a bone that is usually a passageway for blood vessels, nerves, or ligaments	Mental foramen of the mandible (refer to Figure 8-8)
Fossa	A relatively deep pit or depression	Olecrannon fossa of the humerus (refer to Figure 8-24C)
Head	An enlargement on the end of a bone	Head of the femur (refer to Figure 8-6A)
Process	A prominent projection on a bone	Mastoid process of the temporal bone (refer to Figure 8-9)
Suture	An interlocking line of union between bones	Lambdoidal suture between the occipital and parietal bones (refer to Figure 8-9)
Trochanter	A relatively large process	Greater trochanter of the femur (refer to Figure 8-6A)
Tubercle	A small, knoblike process	Greater tubercle of the humerus (refer to Figure 8-24A)
Tuberosity	A knoblike process usually larger than a tubercle	Tibial tuberosity of the tibia (refer to Figure 8-6B)

CHAPTER 8 The Skeletal System

learning outcome 8.5

Recall the bones of the axial skeleton.

Axial Skeleton

The axial skeleton is made up of the skull, ribs, sternum, and vertebra. The axial skeleton consists of 80 bones that lie along the longitudinal axis of the body. The bones of the axial skeleton are important because of their role in protecting internal organs and attachment of many muscles.

The Skull

Skull bones are divided into two types: cranial bones and facial bones. Figures 8-8 through 8-12 show various views of the skull. Eight cranial bones form the top, sides, and back of the skull and 14 facial bones form the face. In addition, 3 ossicles (the smallest bones in the body) are found in each middle ear. The skull bones of an infant are not completely fused at the time of birth (Figure 8-13). Not all the sutures have been formed, creating "soft spots" which are tough membranes that have not yet ossified felt on an infant's skull. These are the *fontanelles* (little fountains), named because they appear to pulsate like a fountain. The fontanelles allow the bones of the infant's skull to be "molded" as the infant moves through the birth canal.

There are 6 fontanelles in the infant skull: one anterior, one posterior, two anterolateral, and two posterolateral in location. The anterior fontanelle is located midline between the two parietal bones and the frontal bone. It is the largest of the fontanelles and ossifies 18 to 24 months after birth. The posterior fontanelle is also located in the midline and is seen between the two parietal bones and the occipital bone. The posterior fontanelle is the first to close or ossify and this about 2 months after birth. By about 2 years of age,

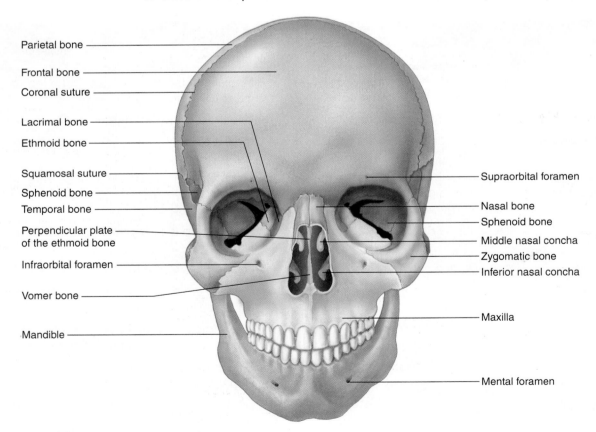

FIGURE 8-8 Anterior view of the skull.

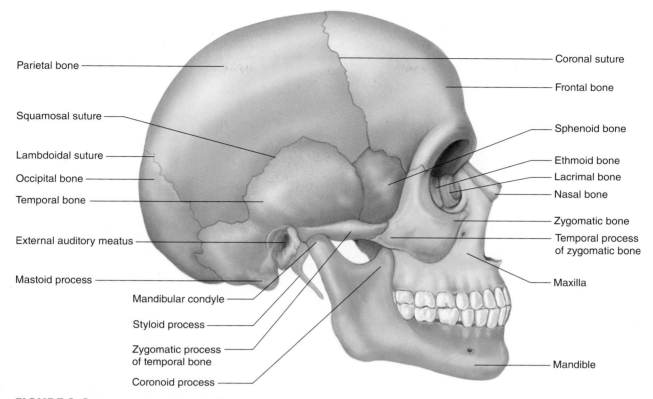

FIGURE 8-9 Lateral view of the skull.

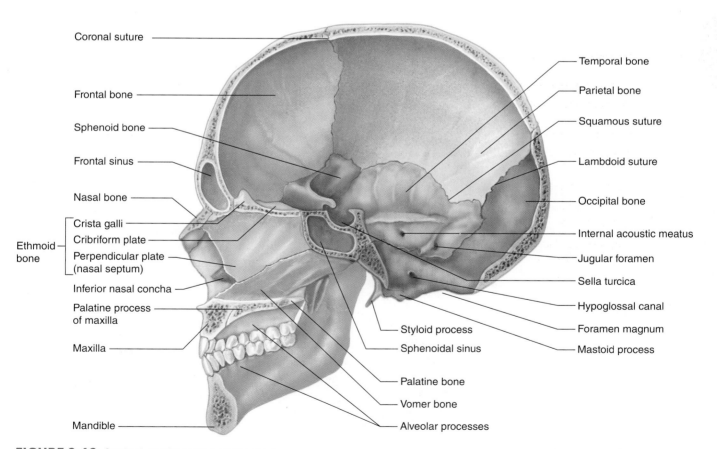

FIGURE 8-10 Sagittal section view of the skull.

CHAPTER 8 The Skeletal System 163

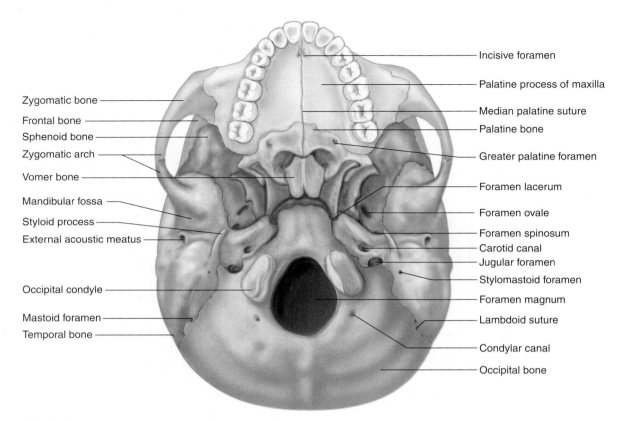

FIGURE 8-11 Inferior view of the skull.

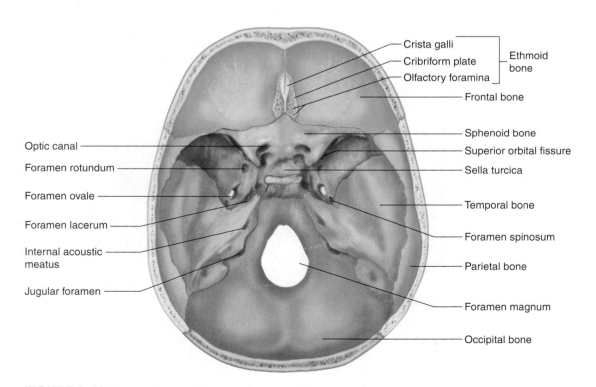

FIGURE 8-12 Floor of the cranial cavity, viewed from above.

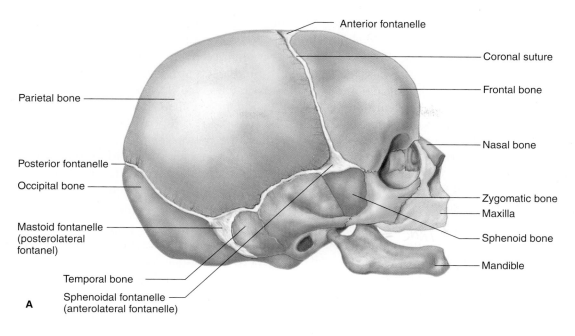

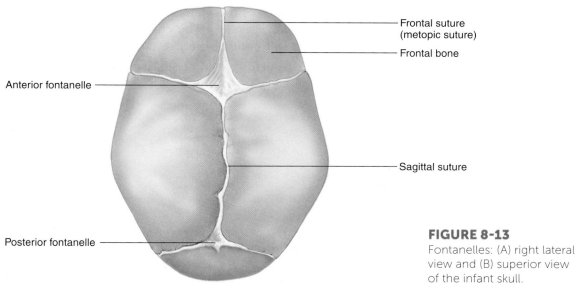

FIGURE 8-13
Fontanelles: (A) right lateral view and (B) superior view of the infant skull.

all the fontanelles have closed to form **sutures** that are immovable joints between bones of the skull.

The major cranial bones include the following:

- *Frontal* bone is a single bone that forms the anterior portion of the cranium. It is the bone in the forehead.
- *Parietal* bones are paired bones that form most of the top and sides of the skull. They come together at the sagittal suture.
- *Occipital* bone is a single bone that forms the back of the skull. The large opening at the base of the occipital bone is called the *foramen magnum,* and the spinal cord runs through it to allow for attachment to the brain stem. Two rounded elevations on the base of the occipital bone are called the occipital condyles on either side of the foramen magnum. They articulate with the first vertebra. This joint between the skull and the first vertebra (*atlas*) allows a person to nod the head yes.

suture An immovable joint between the bones of the skull.

- *Temporal* bones are paired bones on each side of the head that form the lower sides of the skull. The word *temporal* refers to "time." It is called this because this is where a person's hair often first starts to turn gray indicating the passage of time. Through each temporal bone runs a canal called the external auditory meatus or ear canal. Just behind each ear on the temporal bone is a bony prominence called the mastoid process. Because of its dense structure, the part of the temporal bone where the mastoid process is found is called the petrous portion. This mastoid process is the site of attachment of the sternocleidomastoid muscle which is a very important muscle involved in the movement of the head. You will learn about muscles in Chapter 9, The Muscular System.
- *Sphenoid* bone is an unpaired bone that is the most complex bone of the skull. It is shaped somewhat like a butterfly and forms part of the floor of the cranium (skull). In the center of this bone is a deep depression called the hypophyseal fossa, which is where the pituitary gland is located. The sphenoid bone can be seen from either the left or right side of the skull and it also helps form the two orbits in which the eyes are located.
- *Ethmoid* bone is an unpaired bone located between the sphenoid bone and the nasal bones. Together, the ethmoid bone and sphenoid bone form part of the floor of the cranium. With the *vomer* bone, the ethmoid bone forms the *nasal septum*. The ethmoid bone is superior to the vomer. The ethmoid bone also forms part of the boney orbit.
- *Auditory ossicles* (small bones) are the smallest bones of the body and not usually considered part of the skull: the *malleus* (hammer), *incus* (anvil), and *stapes* (stirrup). They are found in the middle ear, are involved in hearing, and are connected to each other by synovial joints (Figure 8-14). We will discuss these important bones in more detail in Chapter 22, The Special Senses.

U check
The mastoid process is part of what bone?

FIGURE 8-14
Auditory ossicles of the middle ear.

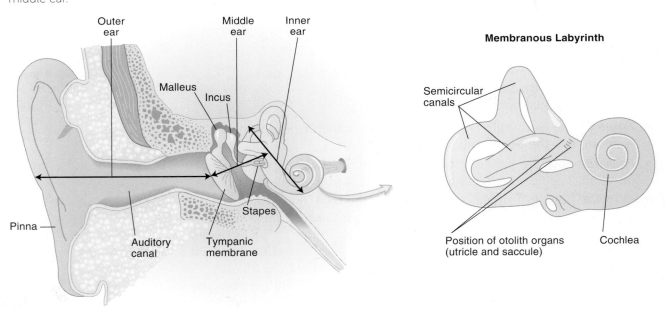

Facial Bones

The 14 facial bones include the following:

- The *mandible* is a single bone; it is the lower jawbone and the largest and strongest facial bone. The mandible is the only movable bone in the skull other than the ossicles (Figure 8-15). It articulates with the temporal bone in front of the external auditory meatus. With the temporal bone, it forms the *temporomandibular joint (TMJ)* which is a synovial joint. The mandible forms the chin, and the lower teeth are embedded in it.
- *Maxillae* are paired bones that have fused to form the upper jawbone of the facial skeleton. The upper teeth are embedded in the maxillae. In addition, the maxillae help form the bony orbit of the eye and the anterior aspect of the hard palate or roof of the mouth. The maxillae articulate with every bone of the face except the mandible.
- *Zygomatic* bones are paired bones that are more commonly called the cheekbones. They also help form the bony orbit, or the eye socket. Along with the zygomatic process of the temporal bone, the zygomatic bone forms the zygomatic arch.
- *Nasal* bones are paired small bones that form the bridge of the nose. The rest of the nose is cartilage.
- *Palatine* bones are paired bones that form the posterior aspect of the hard palate as well as the lateral wall of the nasal cavity and a small part of the floor of the orbit.
- The *vomer* ("plow" because of its shape) is a single bone that forms the inferior part of the nasal septum.
- *Lacrimal* (*lacri* = tear/cry) bones are paired bones that help form the bony orbit or the eye socket. They are the smallest bones of the face; they have a groove on their surface called the lacrimal fossa that collects tears and passes them into the nasal cavity.

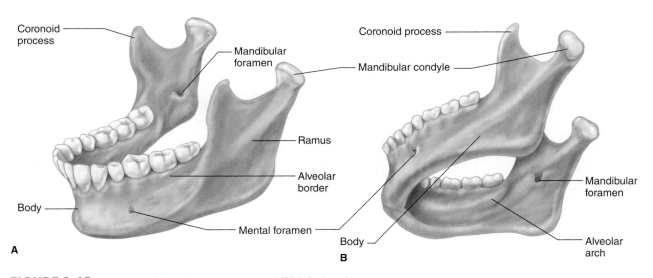

FIGURE 8-15 Mandible: (A) left lateral view and (B) inferior view.

CHAPTER 8 The Skeletal System

- *Inferior nasal conchae* are paired bones that are separate from the ethmoid bone and its *middle nasal conchae*. The nasal conchae increase the surface area in the nose to filter and warm the air that is inhaled.

> **U check**
> What is the only freely movable joint in the skull?

The Rib Cage

The rib cage is formed by the sternum and 12 pairs of ribs. The sternum is also called the breastbone and is actually made up of three bones. From most superior to inferior they are the *manubrium*, the *body*, and the *xiphoid process*. Anteriorly, the clavicles (collar bones) and most of the ribs attach to the sternum. Posteriorly, the ribs articulate with the thoracic vertebrae. The first 7 pairs of ribs are called *true ribs* because they attach directly to the sternum through pieces of cartilage called *costal* (*costo* = rib) cartilage (Figure 8-16). The next 5 pairs of ribs are called *false ribs* because their costal cartilages attach indirectly to the sternum or do not attach at all. Rib pairs 11 and 12 are called *floating ribs* because their costal cartilages do not attach to the sternum anteriorly or to any other structure.

> **U check**
> What are the three parts of the sternum?

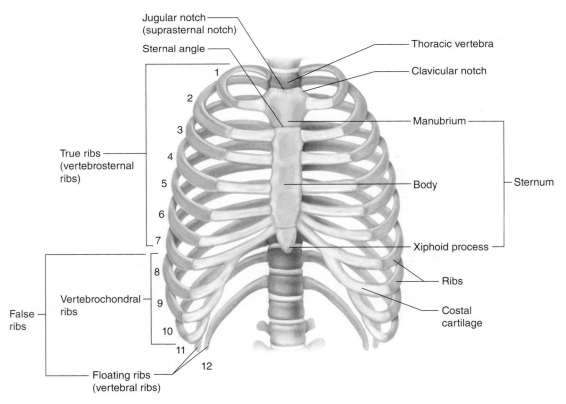

FIGURE 8-16 The thoracic cage includes the thoracic vertebrae, the ribs, the sternum, and costal cartilages.

The Spinal Column

The vertebral or spinal column consists of 7 cervical (*cervix* = neck) vertebrae, 12 thoracic (*thorax* = chest; they are across from it) vertebrae, 5 lumbar (*lumbo* = low back) vertebrae, a sacrum, and a coccyx. (*Hint:* To remember the number of vertebrae think breakfast at 7, lunch at 12, and dinner at 5.) There are four major curvatures to the spine (Figure 8-17). Relative to the anterior surface of the body, the thoracic and sacral curves are concave and are called primary curvatures because we are born with them. The cervical

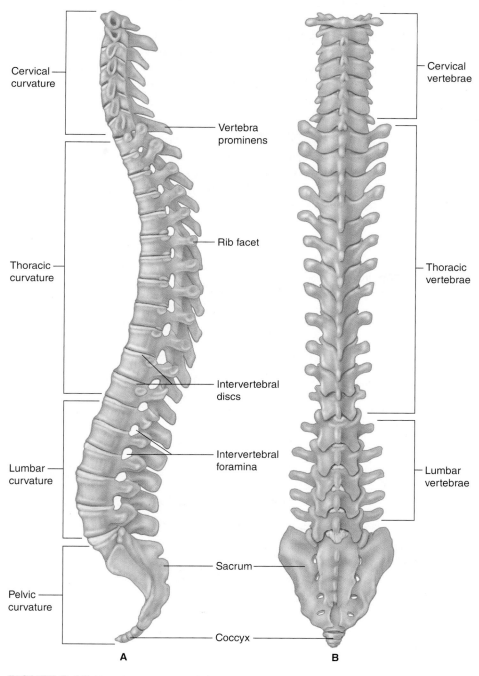

FIGURE 8-17 Vertebral column: (A) lateral view and (B) posterior view.

CHAPTER 8 The Skeletal System

and lumbar curves are convex relative to the anterior surface of the body. They are called secondary curvatures because we are not born with them, but instead they are developmental. The cervical curve begins to develop as the infant begins to lift and hold its head up. The lumbar curve starts to develop when the infant begins to stand upright.

- *Cervical vertebrae,* located in the neck, are the smallest and lightest of all the vertebrae. The first cervical vertebra is called the *atlas* or *C1* and the second is called the *axis* or *C2* (Figure 8-18). When you turn your head from side to side (saying "no"), your atlas is pivoting around your axis. The other 5 cervical vertebrae are noted as C3 to C7. All cervical vertebrae have three foramina—two transverse foramina and a single vertebral foramen (Figure 8-19). The vertebral artery and the corresponding vein and nerve fibers travel through the transverse foramina. The spinal cord passes through the vertebral foramen, which is quite large in this region. The seventh cervical vertebra has a very large spinous process that is a projection from the posterior aspect of the vertebra, and so is called *vertebra prominens*.
- *Thoracic vertebrae* are the posterior attachments for the 12 pairs of ribs. The thoracic vertebrae have long, sharp, spinous processes that you can feel when you run your finger down someone's spine. Many muscles of the back are attached to them. The thoracic vertebrae are denoted as T1 to T12. See Figures 8-19 and 8-20.

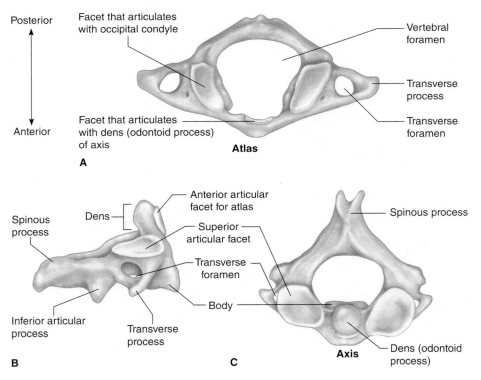

FIGURE 8-18 Atlas and axis: (A) superior view of the atlas, (B) right lateral view of the axis, and (C) superior view of the axis.

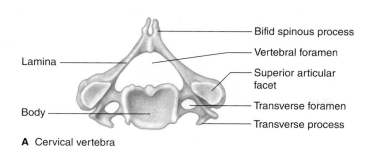
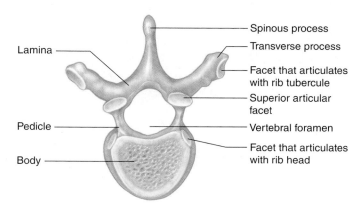
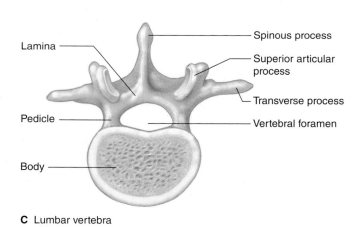

FIGURE 8-19
Superior view of (A) a cervical vertebra, (B) a thoracic vertebra, and (C) a lumbar vertebra.

- *Lumbar vertebrae* are the most massive and sturdy of the vertebrae because they support much of the weight of the body (Figure 8-19). They are charted using the abbreviations L1 to L5.
- *Sacrum* is a triangular-shaped bone that actually consists of five fused vertebrae (Figure 8-21). These vertebrae are noted as S1 through S5.
- *Coccyx* (the tailbone) is a small, triangular-shaped bone made up of three to five fused vertebrae (Figure 8-21).

U check

How many thoracic vertebrae do humans have?

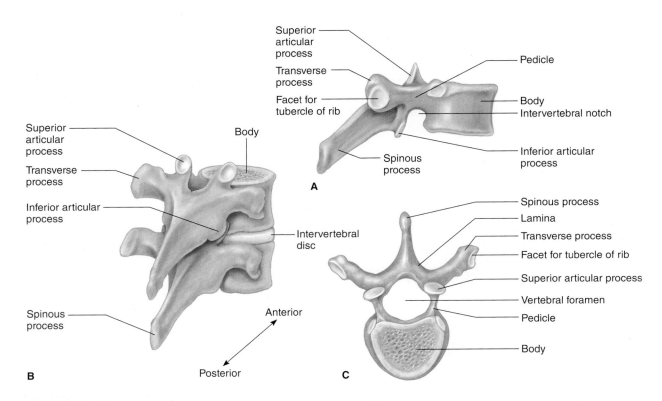

FIGURE 8-20 Typical thoracic vertebra: (A) right lateral view, (B) adjacent vertebrae joined at their articular processes, and (C) superior view.

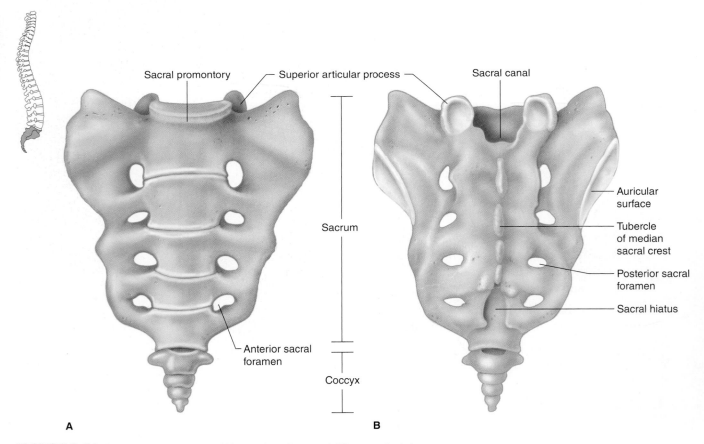

FIGURE 8-21 Sacrum and coccyx: (A) anterior view and (B) posterior view.

Appendicular Skeleton

The appendicular skeleton consists of 126 bones and is made up of the pectoral (*pect* = chest) and pelvic girdles as well as the upper and lower extremities.

Pectoral Girdle

The bones of the shoulders are called the pectoral girdle and include the clavicles and the scapulae (Figures 8-22 and 8-23). They attach the upper extremities, or arms, to the axial skeleton. The clavicles are commonly known as the collarbones. They are slender in shape and each articulates with the sternum and scapula. The scapulae, or shoulder blades, are thin, triangular-shaped flat bones located on the dorsal surface of the rib cage. Each scapula joins with the head of a humerus and a clavicle. The upper limb, or arm bones, includes the humerus, radius, and ulna. The humerus is located in the upper part of the arm. Proximally it articulates with the scapula and distally with the radius and the ulna. The head of the humerus articulates with the glenoid fossa of the scapula. Because this fossa is rather shallow, the humerus can become easily dislocated.

8.6 learning outcome

Recall the bones of the appendicular skeleton.

RECALL TONY HODGE—he thinks he broke his collarbone. Now that you know where the collarbone is, how do you think he did it?

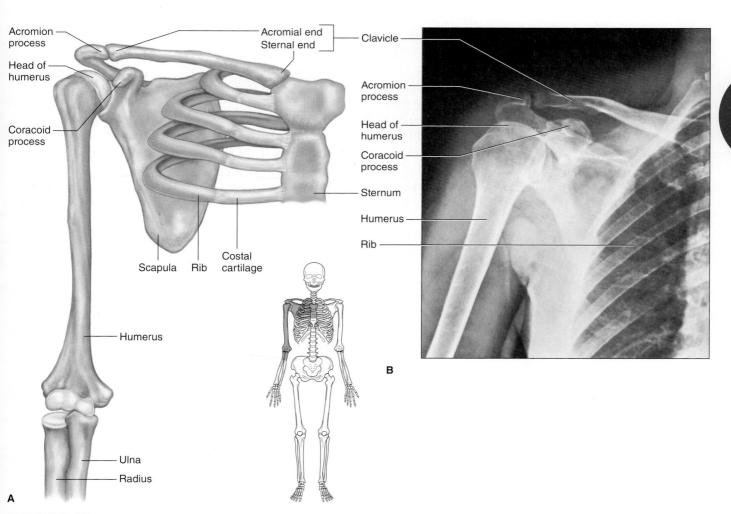

FIGURE 8-22 Pectoral girdle: each of the two pectoral girdles consists of a scapula and clavicle.

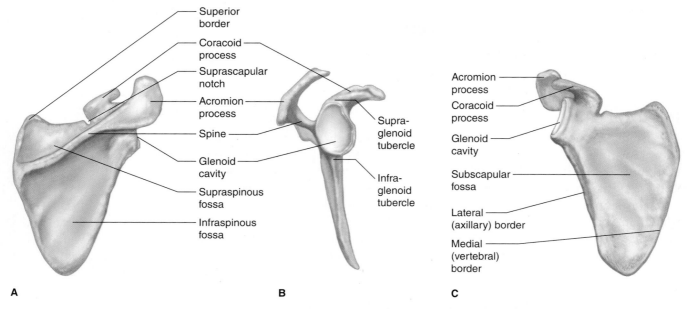

FIGURE 8-23 Right scapula: (A) posterior surface; (B) lateral view showing the glenoid cavity that articulates with the head of the humerus.

The radius is the lateral bone of the forearm and when the body is in the anatomical position, it is on the same side of the arm as the thumb. Proximally it joins with the humerus and the ulna and distally with the carpal (wrist) bones. The ulna is the medial bone of the lower arm. The proximal end of the ulna joins with the humerus to form the elbow joint. Distally, it also joins with the radius and some of the carpal bones of the wrist.

The bones of the hand include carpals (wrist), metacarpals (*meta* = beyond; the hand is beyond the wrist), and phalanges (fingers); see Figure 8-24. The 8 carpal bones make up the wrist. Each hand has 5 metacarpal bones that make up the palm of the hand. There are 14 phalanges, the bones of the fingers, in each hand—3 for each finger and 2 per thumb. The joints between the phalangeal bones are the proximal and distal interphalangeal (PIP and DIP) joints. The joints that join the phalanges to the metacarpals are called the metacarpophalangeal (MCP) joints, or knuckles.

> **U check**
> What are the bones of the pectoral girdle?

Pelvic Girdle

The pelvic girdle consists of the coxal (hip) bones and the lower extremities (Figures 8-25 and 8-26). The coxal bones attach the legs to the axial skeleton and help protect the pelvic organs. Each coxal bone has three parts: the ilium, the ischium, and the pubis. The ilium is the most superior part of a coxal bone and when you put your hands on your hips, you are touching the ilium. The hip projection or ridge is the iliac crest. The ischium forms the lower part of a coxal bone and when an individual is sitting, he or she is sitting on the ischial tuberosities. The two pubic bones form the front of the coxal bone and come together at the pubic symphysis, which is a fibrocartilaginous joint. This joint becomes more lax in response to hormones when a woman who is pregnant approaches the time to give birth, allowing easier passage of the fetus through the birth canal.

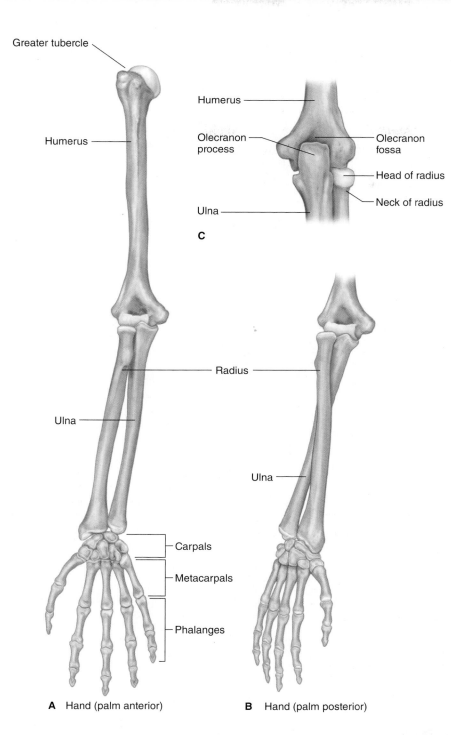

FIGURE 8-24
Right upper limb: (A) anterior view with the hand and palm facing anterior and (B) with the hand and palm facing posterior; and (C) posterior view of the right elbow.

Lower Extremities

The bones of the lower limb, or leg, include the femur, patella, tibia, fibula, tarsals, metatarsals, and phalanges (see Figure 8-27). The femur is the thigh bone and the longest and heaviest bone in the body. Its proximal end joins with the hipbone at the hip socket or *acetabulum* (acetabulum means "wine or vinegar cup" because of its resemblance to the cup used by Romans to hold wine). The head of the femur is held in place by ligaments and muscles. The distal end of the femur attaches to the tibia and the patella (little dish) or knee-cap. The tibia is the medial bone of the lower leg and is commonly called the shinbone. Its proximal end joins with the femur and fibula and its distal end

FIGURE 8-25 Pelvic girdle: (A) anterior view and (B) posterior view.

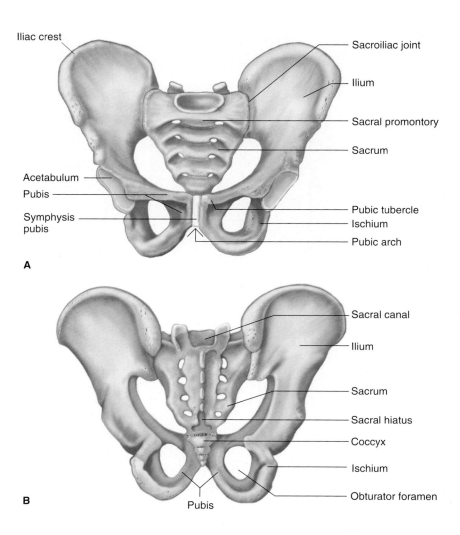

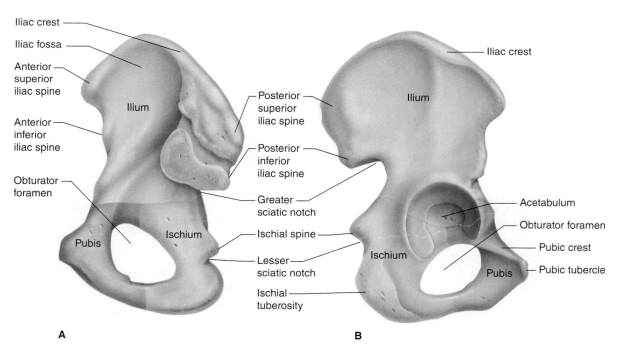

FIGURE 8-26 Right coxal (hip) bone: (A) medial view and (B) lateral view.

176 UNIT 2 Concepts of Common Illness by System

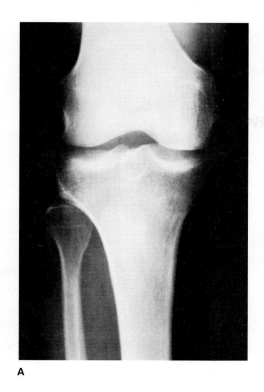
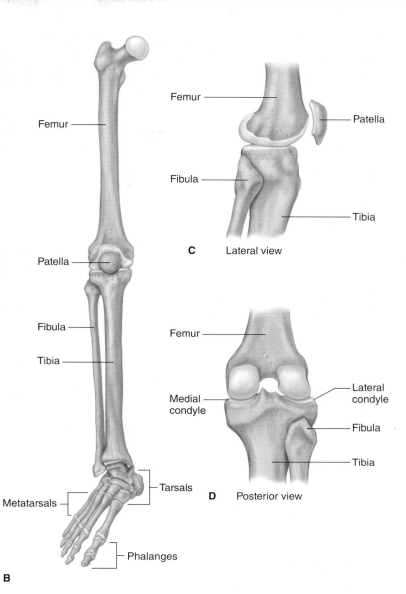

FIGURE 8-27
Lower limb: (A) radiograph of anterior view of the knee showing the femur, tibia, and fibula; (B) anterior view of the lower limb; (C) lateral view of the knee; (D) posterior view of the knee.

attaches to the ankle bones. The fibula is the thinner, lateral bone of the lower leg that joins with the ankle bones at its distal end. The patella, or kneecap, is considered a sesamoid bone because of its resemblance to a sesame seed.

The bones of the foot include the tarsals (ankle), the metatarsals (foot; "after" the ankle), and the phalanges (toes). The calcaneus, or heel bone, is the largest tarsal bone. The talus bone sits on top of calcaneus and articulates with the tibia and fibula. Five additional tarsal bones help make up the posterior part of the foot. Five metatarsal bones form the front of the foot. The bones of the toes are called phalanges. Each foot contains 14 phalanges—2 in each big toe and 3 in each of the other toes (just like in the hand and thumb). The joints between the phalanges of the foot are also called interphalangeal joints like those of the fingers. The joints that join the toes to the foot are called metatarsophalangeal (MTP) joints.

The male and female skeletons vary a bit. Table 8-2 illustrates some of the differences between the male and female skeletons.

U check

The ilium, ischium, and pubis make up what bone?

TABLE 8-2 Differences between the Male and Female Skeletons

Part	Differences
Skull	Male skull is larger and heavier, with more conspicuous muscular attachments. Male forehead is shorter, facial area is less round, jaw is larger, and mastoid processes are more prominent than those of a female.
Pelvis	Male pelvic bones are heavier and thicker, and have more obvious muscular attachments. The obturator foramina and the acetabula are larger and closer together than those of a female.
Pelvic cavity	Male pelvic cavity is narrower in all diameters, and is longer, less roomy, and more funnel-shaped. The distances between the ischial spines and between the ischial tuberosities are less than in a female.
Sacrum	Male sacrum is narrower, sacral promontory projects forward to a greater degree, and sacral curvature is bent less sharply posteriorly than in a female.
Coccyx	Male coccyx is less movable than that of a female.

8.7 learning outcome

Relate the major types of joints to their location and function.

synovial The type of joint that is the most movable.

Joints

Joints are the junctions, or articulations, between bones. Based on their structure, joints can be classified as fibrous, cartilaginous, or synovial. The bones of *fibrous joints* are connected by short connective tissue fibers abundant with collagen that do not allow much, if any, movement. Examples of fibrous joints are found between cranial bones and facial bones. As mentioned earlier, fibrous joints in the skull are called sutures. The bones connected by cartilaginous joints are connected by a disc of cartilage and are slightly movable. The joints between vertebrae are cartilaginous joints. The most numerous type of joint in the body, and the most movable one, is the **synovial** joint. A synovial joint is covered with hyaline cartilage and is surrounded by a fibrous joint capsule lined with a synovial membrane (Figure 8-28). The synovial membrane secretes a slippery fluid called

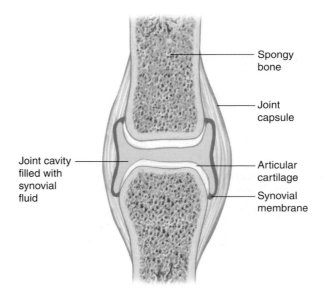

FIGURE 8-28
Structure of a synovial joint.

> **focus on Wellness:** Maintaining Bone Health Throughout Life
>
> Many factors influence bone health, including diet, exercise, and a person's overall lifestyle. You can help patients improve or maintain their bone health by teaching them about behaviors that will support bone health.
>
> - Bone-Healthy Diet: Good nutrition is essential for proper bone growth during childhood and the teen years. It is equally important in adulthood to maintain healthy bones. Bone-building nutrients are found in dairy products, broccoli, kale, spinach, salmon, sardines, egg yolks, whole grains, and fruits, especially bananas and oranges. Calcium and vitamin D are particularly important for healthy bones. Without vitamin D, the body cannot absorb calcium from the digestive tract. Without calcium, bone tissue will slowly wear away. Supplements should be taken if a person's diet does not include adequate amounts of calcium and vitamin D.
>
> - Bone-Healthy Exercises: Weight-bearing and strength-training exercises are best for bone health. When your muscles contract, they pull on your bones. This tension stimulates bones to thicken and strengthen. Lifting weights is an effective way to increase the tension on bones. Other activities such as jogging, walking briskly, or playing a sport regularly will also stimulate bones to increase in density.
>
> - Bone-Healthy Lifestyle: A person with a bone-healthy lifestyle avoids smoking and alcohol. Smoking rids the body of calcium, which is necessary for bone growth. Alcohol prevents calcium absorption in the digestive tract. People who smoke are almost twice as likely to develop osteoporosis as nonsmokers.

synovial fluid, which acts as a lubricant to allow the bones to move easily across each other. Bones are also attached to each other by dense connective tissue fibers called *ligamets*. Examples of synovial joints are those of the elbows, knees, shoulders, hips, and knuckles.

> **U check**
> What is the most numerous type of joint in the body?

from the perspective of . . .

A RADIOLOGY TECHNICIAN A radiology technician operates x-ray and other diagnostic imaging equipment as directed by a physician to help diagnose various disorders. How will knowing the anatomy of the human skeleton help you perform this job effectively and efficiently?

8.8 learning outcome

Describe the major types of bone fractures and diseases of the skeletal system.

Bone Fractures

Although several different types of bone fractures occur, we will mention only a few of the more common ones. A *compound* or *open* fracture is one where the skin is broken (Figure 8-29). A *simple* or *closed* fracture is one where the skin remains intact. *Greenstick* fractures, commonly seen in children, occur in bones that are not completely ossified so there is a bending rather than a complete breaking of bone. *Impacted* fractures occur when one end of the fractured bone is driven into the interior of the other. This is seen in falls from great heights. Fractures are commonly diagnosed based on the clinical presentation and confirmed by x-rays, as shown in Figure 8-30. Computerized tomography (CT) scans and magnetic resonance imagining (MRI) are special tests that are used to more clearly assess the damaged area. Fractured bones heal through a series of steps.

1. A hematoma (blood clot) forms at the site of the fracture.
2. A fibrocartilaginous callus forms.
3. A bony callus forms.
4. Remodeling occurs when dead cells are removed and new bone cells are laid down.

REMEMBER TONY, our high school football player? What sorts of tests were done to confirm the fracture?

Complete healing of fractures, particularly those involving complicated breaks and surgical intervention, may take several months.

U check
What is the difference between a compound fracture and a simple fracture?

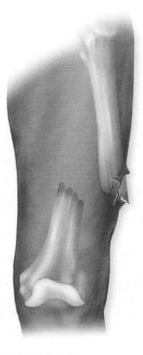

FIGURE 8-29 Open fracture showing penetration of the skin.

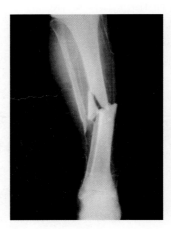

FIGURE 8-30 X-ray of a fractured femur.

focus on Clinical Applications: Evaluating Bones

Several diagnostic procedures are available to evaluate bones. Health care professionals may need to determine if a bone is broken. This is usually done with an x-ray. Bone scans help the health care professional diagnose the causes of bone pain, arthritis, bone infections, and bone cancers. During a bone scan, radioactive dyes are injected into the patient and concentrate in bone tissue to allow abnormalities to be seen. Bone-density tests are painless procedures used to determine the density of a person's bones. Because osteoporosis shows no symptoms in early stages, these tests are necessary to determine if a patient is developing osteopenia (less than normal bone mineral density) or osteoporosis (bone thinning and low bone mineral density).

PATHOPHYSIOLOGY

Arthritis and Osteoarthritis

Osteoporosis

Osteoarthritis versus rheumatoid arthritis

Arthritis is a general term meaning joint inflammation. More than 100 types of arthritis exist; we will discuss the most common types. *Osteoarthritis (OA)* is also known as degenerative joint disease (DJD), or "wear and tear" arthritis. It primarily affects the weight-bearing joints of the hips and knees. The cartilage between the bones and the bones themselves begin to break down.

Etiology: It may be due to trauma, but more frequently OA occurs in elderly individuals due to the aging process.

Signs and Symptoms: Joint stiffness, aching, and pain, especially with weather changes due to pressure changes affecting the joints. Often fluid occurs around the joint, and joint movement creates grating noises that indicate joint destruction.

Treatment: Anti-inflammatory drugs, including aspirin and NSAIDS (nonsteroidal anti-inflammatory drugs) such as naproxen, may be used. Intra-articular steroid injections may be tried for severe cases. In some cases, a series of injections of medications containing hyaluronic acid are used when other treatments do not work. These injections serve as joint fluid replacement. Some success has been found with transplanting harvested cartilage cells from the patient's healthy knee cartilage, which are then grown in the lab and reinjected into the patient's affected joint. Surgical scraping of the joint may also be done to remove deteriorated bone fragments. As a last resort, joint replacement may be recommended.

(continued)

Osteoarthritis versus rheumatoid arthritis

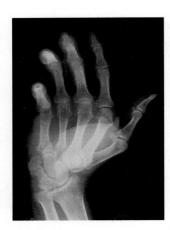

FIGURE 8-31
A hand with rheumatoid arthritis showing typical deformity.

Rheumatoid Arthritis

Rheumatoid arthritis (RA) is a chronic systemic inflammatory disease that attacks the smaller joints such as those in the hands and feet (Figure 8-31). The surrounding soft tissue is also often involved. Periods of exacerbation (increased pain) are often followed by periods of remission (decreased pain). RA is more common in women than in men.

Etiology: RA is believed to have an autoimmune component, although infection may play a role.

Signs and Symptoms: In this disease, the synovial membrane is destroyed, causing edema and congestion. Tissue becomes granular and thick, eventually destroying the joint capsule and bone. Scar tissue forms, bones atrophy, and visible deformities become apparent due to the bone misalignment and immobility. Patients have moderate to severe pain in the affected joints.

Treatment: Treatment includes anti-inflammatory drugs, heat or cold treatments, and cortisone injections. Immunosuppressive drugs may help. Low-impact aerobic exercise may be helpful and some patients find warm water exercises beneficial.

U check
What is the cause of rheumatoid arthritis?

Gout

Gout (gouty arthritis) is a type of arthritis associated with high uric acid levels in the blood with crystalline deposits in the joints, kidneys, and various soft tissues.

Etiology: Gout is caused by deposits of uric acid crystals in the joints. People with gout do not break down uric acid properly and cannot remove it from their bloodstream. It appears to have a genetic component.

Signs and Symptoms: Symptoms include acute or chronic joint pain, commonly in the great toe because the crystals come out of solution at lower temperatures such as those seen in the distal joints.

Treatment: Treatments include pain medications, anti-inflammatory drugs, and dietary modifications. Patients should eliminate from their diet certain foods such as red meat, fish, beer, cheese, and red wine which contribute to the formation of uric acid.

Bursitis

Bursitis is inflammation of a bursa, which is the fluid-filled sac that cushions tendons (tendons attach muscles to bone). Bursitis occurs most commonly in the elbow, knee, shoulder, and hip.

Bursitis (concluded)

Etiology: Overuse and trauma to joints are the most common causes of bursitis. Bacterial infections can also cause it.

Signs and Symptoms: Signs and symptoms include joint pain, swelling, and tenderness in the structures surrounding the joint.

Treatment: Common treatments are bed rest, pain medications, steroid injections (to control the inflammation), aspiration of excess fluid from the bursa, and antibiotics if caused by infection.

Carpal Tunnel Syndrome

Carpal tunnel syndrome (CTS) occurs when the median nerve in the wrist is excessively compressed by an inflamed tendon (the flexor retinaculum).

Etiology: Overuse of the wrist is a common cause of this syndrome. Repetitive activities such as keyboarding, assembly-line work, painting, and sports such as racquetball put the participants at risk for developing CTS.

Signs and Symptoms: Weakness, numbness, pain, or a tingling sensation in the hand, wrist, or elbow are common complaints.

Treatment: This condition can be treated with wrist splints, pain medications, steroid injections, and alteration in the work (or sport) habits to better position and support the wrists. If these treatments do not improve the patient's condition, surgery to reduce the pressure on the affected nerves may be needed.

Kyphosis

Kyphosis is an abnormally exaggerated curvature of the spine, most often at the thoracic level and often referred to as "humpback" (Figure 8-32). The hunchback of Notre Dame probably suffered from this along with several other skeletal abnormalities.

Etiology: Adolescent kyphosis may result from growth retardation or improper development of the epiphyses as a result of rapid growth. Poor posture may exacerbate the condition. The adult form of kyphosis is most often the result of aging and degenerative disc disease of the intervertebral discs and/or actual pathological (caused by disease) vertebral fracture from underlying osteoporosis.

Signs and Symptoms: In adolescent kyphosis, no symptoms may occur other than the exaggerated back curvature. There may be mild pain, tiredness, or tenderness or stiffness of the thoracic spine. In adult kyphosis, the upper back is rounded and there may be pain, back weakness, and fatigue.

(continued)

Kyphosis (concluded)

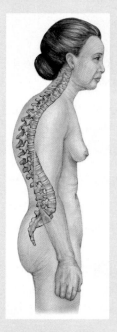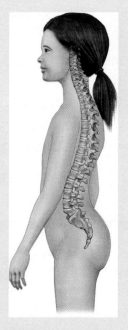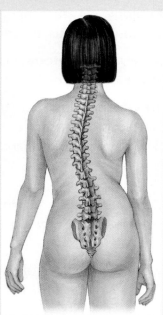

FIGURE 8-32 Abnormal curvatures of the vertebral column: (A) kyphosis, (B) lordosis, and (C) scoliosis.

Treatment: Childhood kyphosis may be treated with exercise, a firm mattress, and a back brace if needed until growth is completed to keep the spine in alignment. Spinal fusion or grafting may be needed in rare cases of neurological damage or disabling pain. Harrington rods may also be surgically inserted to keep the vertebrae aligned.

Lordosis

Lordosis is an exaggerated inward (convex) curvature of the lumbar spine. The condition is sometimes called swayback (Figure 8-32).

Etiology: Obesity, tuberculosis of the spine, poor posture, pregnancy, osteoporosis, and continual wearing of high heels can all cause this condition.

Signs and Symptoms: The main sign of lordosis is visual inward curvature of the lower back accompanied by mild to moderate pain.

Treatment: Prevention is best achieved by preventing or treating the underlying cause.

Scoliosis

Scoliosis is an abnormal, S-shaped lateral curvature of the thoracic or lumbar spine (Figure 8-32).

Etiology: This disorder can develop prenatally when vertebrae do not fuse together. It can also result from diseases that cause

---

scoliosis An abnormal, S-shaped lateral curvature of the thoracic or lumbar spine.

Scoliosis (concluded)

weakness of the muscles that hold vertebrae together. In some cases, there may be a genetic basis.

Signs and Symptoms: A patient with scoliosis usually has a spine that looks bent to one side, with one shoulder or hip appearing to be higher than the other. Patients often experience back pain.

Treatment: Treatment includes different types of back braces, surgery to correct spinal curves, and physical therapy to strengthen the spine.

U check
Which condition has an abnormal lateral curvature?

Osteogenesis Imperfecta

People with *osteogenesis imperfecta* (brittle-bone disease) have decreased amounts of collagen in their bones, which leads to very fragile bones. Four types of this disease occur. Type 1 is the most common, and type 2 is the most severe.

Etiology: The disorder is hereditary and very often runs in families.

Signs and Symptoms:

- Type 1: Fractures, blue sclera, dental problems, hearing loss, triangular face, abnormal spinal curves
- Type 2: Very small stature, small chest
- Type 3: Very small stature, barrel-shaped chest, fractures at birth, loose joints, small muscles
- Type 4: Dental problems, abnormal spinal curves, loose joints

Treatment: Because there are many symptoms of this disease, the list of treatments is extensive and includes fracture repair, surgery to strengthen bones by inserting metal rods into them, dental procedures, physical therapy, braces to prevent bone deformities, wheelchairs and other supportive aids, medications, and counseling. Surgeries may be required to treat lung and heart problems that sometimes occur with this disease.

Osteoporosis

Osteoporosis is a condition in which bones thin (become more porous) over time. It is a very common disorder in the United States and affects women more than men, with Caucasians affected more than other races. The condition occurs with hypocalcemia when bone is broken down to release calcium into the blood and is not replaced in sufficient amounts. A decrease in bone density is the result.

(continued)

osteoporosis A condition in which bones become thin and more porous.

Osteoporosis

Osteoporosis (concluded)

Etiology: The causes of osteoporosis include hormone deficiencies (estrogen in women and testosterone in men), a sedentary lifestyle, a lack of calcium and vitamin D in the diet, bone cancers, corticosteroid excess (usually as a result of endocrine diseases), smoking, excessive alcohol consumption, and the use of steroids.

Signs and Symptoms: There are usually no symptoms in the early stages of this disease. Patients at high risk, including those with a family history of osteoporosis, should undergo bone densitometry studies to identify the condition in the early stages so treatment can be started. Patients in later stages of the disease may experience pathological fractures (usually of the spine, wrists, or hips), back and neck pain, a loss of height over time, and kyphosis.

Treatment: Common treatments include medications to prevent bone loss and relieve bone pain, hormone replacement therapy, lifestyle changes to prevent bone loss (including regular weight-bearing exercise and diets or supplements that include calcium, phosphorous, and vitamin D), moderation or elimination of the use of alcohol, and cessation of smoking.

Osteosarcoma

Osteosarcoma is a type of bone cancer that originates from osteoblasts, the cells that make bone tissue. It is most often seen in children, teens, and young adults and more often in males than females. Usually this type of cancer affects bones of the legs and especially the metaphysis of the long bone.

Etiology: The cause of osteosarcoma is not known.

Signs and Symptoms: Primary symptoms include pain in affected bones (usually the legs), swelling of the soft tissue around affected bones, and an increase in pain with movement of the affected bones.

Treatment: Treatments include surgery, chemotherapy, and radiation therapy. Amputation of the affected limb and fitting of a prosthesis may be necessary to prevent metastasis.

Paget's Disease

Paget's disease causes bones to enlarge and become deformed and weak. It usually affects people older than 40 years of age.

Etiology: Although the etiology is not well understood, Paget's disease may be caused by a childhood viral infection and may have some genetic basis.

Paget's Disease *(concluded)*

Signs and Symptoms: Bone pain, deformed bones, and fractures are common symptoms. Patients may experience headaches and hearing loss if the disease affects skull bones.

Treatment: Treatments include surgery to remodel bones, hip replacements, medications to prevent bone weakening, and physical therapy.

> **U check**
> What type of exercise is beneficial for slowing down the progression of osteoporosis?

Life Span

8.9 learning outcome
Explain the effects of aging on the skeletal system.

Bone mass decreases with age. By 30 years of age, this process has already begun in some people. Bone production declines while bone resorption continues at a normal level. Although men lose bone density at the same rate as women after 65 years of age, men typically start with greater bone mass (see Table 8-2), so in most cases the effects of osteoporosis are not as great in men as in women. Women may lose 8 to 10 percent of their skeletal mass. Certain bones decline at different rates. The epiphyses of long bones, vertebrae, and the jaws lose mass at a greater rate than other bones. Bones become fragile, with spontaneous fractures not uncommon. With vertebral fractures, spinal nerves may be affected causing severe pain.

summary

learning outcomes	key points
8.1 Recall the general structure and functions of the skeletal system.	The functions of the skeletal system include protection, support, movement, and storage of minerals.
8.2 Describe the general structure of bone.	Two types of bone tissue are cancellous (spongy) bone and compact bone. Osteons (Haversian systems) are the fundamental unit of compact bone. Immature bone cells are called osteoblasts. Osteocytes are mature bone cells, and osteoclasts are cells that tear down bone.
8.3 Compare the two major types of bone growth.	Two types of ossification are intramembranous and endochondral. In intramembranous ossification, bones begin as tough, fibrous membranes that are replaced by bone produced by osteoblasts. In endochondral ossification, bones start out as cartilage models.

continued

chapter 8

learning outcomes	key points
8.4 Summarize how bones are classified, their parts, and markings.	Bones are classified by their density; spongy or cancellous bone has spaces filled with red bone marrow, whereas compact bone is much denser. Bones are also classified according to on their shape. There are flat bones, long bones, short bones, irregular bones, and sesamoid bones. The diaphysis (shaft) makes up the majority of the length of the bone. It surrounds a space called the medullary cavity. At each end of a long bone there is an epiphysis.
8.5 Recall the bones of the axial skeleton.	The axial skeleton is made up of the skull, hyoid bone, auditory ossicles, ribs, sternum, and vetebrae.
8.6 Recall the bones of the appendicular skeleton.	The appendicular skeleton consists of 126 bones and is made up of the pectoral and pelvic girdles as well as the upper and lower extremities.
8.7 Relate the major types of joints to their location and function.	Joints are the junctions or articulations between bones. Based on their structure, joints can be classified as fibrous, cartilaginous, or synovial. The most numerous and most movable joint is the synovial joint. The intervertebral disc is an example of a fibrous joint.
8.8 Describe the major types of bone fractures and diseases of the skeletal system.	Several different types of bone fractures occur. These include compound (open), simple (closed), greenstick, and impacted fractures.
8.9 Explain the effects of aging on the skeletal system.	Bone mass decreases with age. This puts people at risk for fractures and other injuries and diseases.

chapter 8 review

case study 1 questions

Can you answer the following questions that pertain to Tony Hodge, our young football player, whose case was presented in the beginning of this chapter?

1. What is the "collarbone"?
   a. The scapula
   b. The humerus
   c. The clavicle
   d. The sternum
2. What is the difference between a simple (closed) and compound (open) fracture?
   a. An open fracture does not break the skin, but a closed fracture does.
   b. Both an open and a closed fracture break the skin.
   c. An open fracture breaks the skin, but a closed fracture does not.
   d. Neither an open nor a closed fracture breaks the skin.
3. How is a fracture diagnosed?
   a. CT scan
   b. MRI
   c. X-ray
   d. Blood tests
4. What is the healing process when a bone is fractured?
   a. Hematoma formation occurs first
   b. Fibrocartilaginous callus formation occurs first
   c. Formation of a bony callus occurs first
   d. Remodeling occurs first

review questions

1. Which one of the following is *not* part of calcium regulation?
   a. Osteoclasts
   b. Calcitonin
   c. Parathyroid hormone
   d. Vitamin A
2. Greenstick fractures are most often seen in
   a. White men
   b. Black men
   c. Elderly people
   d. Young children
3. What vitamin is essential for the absorption of calcium?
   a. Vitamin D
   b. Vitamin E
   c. Vitamin C
   d. Vitamin B
4. Which of the following is *not* part of the orbit?
   a. Maxilla
   b. Frontal
   c. Sphenoid
   d. Mandible
5. Which of the following is *not* part of the pelvic girdle?
   a. Ilium
   b. Tibia
   c. Ischium
   d. Pubic bone
6. What hormone gives some protection against osteoporosis?
   a. Antidiuretic hormone
   b. Estrogen
   c. Follicle-stimulating hormone
   d. Prolactin
7. What is the name for the shaft of a long bone?
   a. Epiphysis
   b. Diaphysis
   c. Metaphysis
   d. Fontanelle
8. What cells secrete the components for bone formation?
   a. Erythrocytes
   b. Osteoblasts
   c. Osteoclasts
   d. Mast cells

critical thinking questions

1. Why are weight-bearing exercises beneficial for bone health?
2. What are the effects of aging on the skeletal system?

patient education

Women, as they age, are at a high risk for osteoporosis (porous bones). Before osteoporosis occurs women get a condition known as osteopenia. Osteopenia is low bone mineral density. This condition can be slowed down or stopped by good bone health practices. Create a teaching plan for a 53-year-old woman who has just been diagnosed with osteopenia.

applying what you know

A 32-year-old man fractured his right ankle playing soccer when he was 17 years old. Within the last 8 to 10 months he has been experiencing pain in the ankle. He has no other joint problems. His physician says he probably has arthritis. Applying what you have learned in this chapter, answer the following questions:

1. What type of arthritis is the man most likely to have and what are some predisposing factors for this disease?

CASE STUDY 2 *Osteoporosis*

Nancy Marr is a 74-year-old retired culinary professional. She has been experiencing some back pain the last several months and is beginning to "hunch over." When she visited her physician, Nancy was told that she probably has osteoporosis and her physician wants to run some tests.

1. What is osteoporosis?
2. What are the risk factors for osteoporosis?
3. What tests will the physician will most likely order to confirm the diagnosis?

CASE STUDY 3 *Scoliosis*

Eleven-year-old Mary Park has been experiencing back pain for the last couple of months. At first her mother thought it may be related to Mary's menstrual periods, which she started this year. However, the pain seems to be constant and not related to her menstrual cycle. Her shoulders also do not seem to be level.

1. What is scoliosis?
2. What are the causes of scoliosis?

9

The Muscular System

chapter outline

- Types and Functions of Muscle
- Structure and Classification of Muscle
- Energy Production for Muscle and Muscle Contraction
- Attachments and Actions of Skeletal Muscles
- Body Movements
- Major Skeletal Muscles
- Strains and Sprains
- Life Span

outcomes

AFTER COMPLETING THIS CHAPTER, YOU WILL BE ABLE TO:

9.1 Relate the types of muscle tissue to their locations and characteristics, and describe how visceral (smooth) muscle produces peristalsis.

9.2 Describe the structure of skeletal muscle.

9.3 Explain how muscle tissue generates energy.

9.4 Define the terms *origin* and *insertion* and their relationship to the skeletal muscles.

9.5 Identify the types of body movements produced by and disorders that affect skeletal muscles.

9.6 Relate the major skeletal muscles of the body to their actions.

9.7 Compare strain and sprain and other muscular injuries.

9.8 Discuss the effects of aging on the muscular system and the steps that can be taken to promote muscle health.

essential terms

agonist (AG-uh-nist)
antagonist (an-TAG-uh-nist)
aponeurosis (ap-o-noo-RO-sis)
endomysium (en-do-MIZ-ee-um)
epimysium (ep-ih-MIZ-ee-um)
fascicle (FAS-ih-kul)
insertion (in-SER-shun)
intercalated disc (in-TER-cah-la-ted disk)
myocytes (MY-uh-sites)
origin (OR-ih-jin)
perimysium (per-ih-MIZ-ee-um)
prime mover (prime MOVE-er)
sarcolemma (sar-koh-LEM-uh)
sarcoplasmic reticulum (sar-koh-PLAZ-mik re-TIK-you-lum)
synergist (SIN-er-jist)
tendon (TEN-dun)

Additional key terms in the chapter are italicized and defined in the glossary.

case study

Use the case study to focus on as you go through the chapter. The questions will guide you as you learn the anatomy, physiology, and pathology of the muscular system.

CASE STUDY 1   *Muscular Dystrophy*

Danny Kolker is a very bright 11-year-old who loves sports. Up until about six months ago he had no problem participating in soccer, baseball, and basketball. In the past few months, Danny has started to become "clumsy" and uncoordinated. He also complains about feeling tired frequently. His calf muscles seem to be somewhat enlarged, but his thighs seem to have lost some muscle mass.

As you go through the chapter, keep the following questions in mind:

1. What is the cause of muscular dystrophy (MD)?
2. Which type of MD does Danny seem to have? Why?
3. Why is it more common in boys? Is there a genetic basis for the disease?
4. What is the treatment for MD?

study tips

1. Write a study schedule for the chapter. Be realistic and reward yourself with a treat when you accomplish what you have set out to do.

2. Flash cards are a great way to learn the muscles. On one side of the card, write the name of the muscle. On the reverse side, write the origin, insertion, and action. Go through your flash cards at least three times each day. This will take you less than 5 minutes each time you review the muscles, but you will be surprised by how much you will remember by the end of this chapter. (Review the flash cards at least once a week for the remainder of the course and you will put them into your permanent memory.)

Introduction

You know that hitting a baseball or taking a walk on a beautiful day involves muscles, but did you know that the movement of food through the gastrointestinal tract involves muscles or that the beating of the heart requires continuous muscular activity? In addition to these activities, muscles participate in heat production and are essential for maintaining posture and body tone. You rarely think about it, but muscles hold the bones tightly together so that the joints remain stable. There are also very small muscles holding the vertebrae together to make the spinal column stable. Muscles can only do one thing: contract. When muscles contract, they move bones, circulate blood, or move food through the digestive system. The human body has more than 600 individual skeletal muscles and whether you are an Olympic weight lifter or a couch potato, the number of muscles you have is the same. Although each muscle is a distinct structure, muscles act in groups to perform particular movements.

> **U check**
> What are three actions that muscles produce?

9.1 learning outcome

Relate the types of muscle tissue to their locations and characteristics, and describe how visceral (smooth) muscle produces peristalsis.

Types and Functions of Muscle

The three types of muscles are smooth, skeletal, and cardiac. Although there are some similarities among the three, differences account for their unique functions. Table 9-1 gives the muscle types, locations, and other features.

Smooth Muscle

There are two types of smooth muscle—multiunit and visceral. *Multiunit smooth muscle* receives its name because when one fiber is stimulated, many adjacent fibers are also stimulated. This type of muscle is responsible for changing pupil diameter to allow more or less light into the eye and focusing the eye by changing lens thickness. It is also found in walls of some blood

TABLE 9-1 Types of Muscle Tissue

Muscle Group	Major Location	Major Function	Striated (Yes/No)	Mode of Control	Rate of Contraction	Intercalated Discs
Skeletal muscle	Attached to bones and skin of the face	Produces body movements and facial expressions	Yes	Voluntary	Fast to contract and relax	No
Smooth muscle	Walls of hollow organs, blood vessels, and iris	Moves contents through organs; vasoconstriction	No	Involuntary	Slow to contract and relax	No
Cardiac muscle	Wall of the heart	Pumps blood through the heart	Yes	Involuntary	Groups of muscle fibers contract as a unit	Yes

vessels to allow expansion and contraction, as well as in the air passages in the lungs. Multiunit smooth muscle contracts in response to neurotransmitters and hormones. *Visceral* (also known as single unit) *smooth muscle* contains sheets of muscle cells that are in close contact with each other. It is found in the walls of hollow organs such as the stomach, intestines, bladder, and uterus as well as the walls of small arteries and veins.

Muscle fibers in visceral smooth muscle respond to neurotransmitters, but they also stimulate each other to contract; therefore, the muscle fibers tend to contract and relax together. Like cardiac muscle, visceral muscle has *autorhythmicity*. The cells are tapered (fusiform shape) on both ends like some cigars, and have a single nucleus.

Visceral muscle fibers have no visible striations, even though they have actin and myosin filaments, because the filaments are more disorderly in their arrangement than in striated muscle. Smooth muscle produces an action called peristalsis. *Peristalsis* is a rhythmic contraction that pushes substances through tubes of the body. For example, peristalsis in the lower two-thirds of the esophagus moves the *bolus* of food through the stomach; peristaltic muscle movements in the Fallopian tubes propel the ovum (egg) through the tubes toward the uterus. Two *neurotransmitters* are involved in smooth muscle contraction—*acetylcholine (ACh)* and *norepinephrine*. Depending on the smooth muscle type, these neurotransmitters cause or inhibit contractions.

Skeletal Muscle

Skeletal muscle is also called striated or voluntary muscle. It has actin and myosin filaments responsible for the striations that are visible under a microscope. Because skeletal muscles are under some conscious control, they are called voluntary. These muscles are responsible for body movement, maintaining posture, and heat generation through shivering. Skeletal muscle fibers respond to the neurotransmitter acetylcholine (ACh) by contracting (Figure 9-1). Once contraction has occurred, skeletal muscles release an enzyme called *acetylcholinesterase*, which breaks down acetylcholine and causes muscle relaxation.

> **U check**
> Which of the muscle types contain actin and myosin filaments?

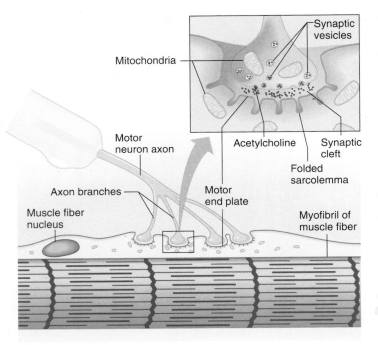

FIGURE 9-1 Neuromuscular junction includes the end of a motor neuron and the motor end plate of a muscle fiber.

intercalated disc Cross bands of cardiac muscle cells that permit the cells to work as a single unit.

Cardiac Muscle

It is probably obvious that cardiac muscle is found in the heart (*cardi* = heart). The heart is a pump that moves blood through the heart and out to the systemic circulation. The heart is discussed in more detail in Chapter 11, The Cardiovascular System.

Cardiac muscle has similarities to both smooth muscle and skeletal muscle. Like smooth muscle, cardiac muscle is an *involuntary* muscle because it contracts "involuntarily"—that is, without any conscious thought. Like skeletal muscle, cardiac muscle has *striations* that appear as "stripes" when examined under a microscope. The striations are created by the overlapping of thick and thin contractile proteins known as filaments. The thin filaments are *actin* and the thick filaments are *myosin*. We find these same proteins in skeletal muscle.

Cardiac muscle cells are sometimes called cardiomyocytes (*cardi* = heart; *myo* = muscle; *cyte* = cell) and typically have one nucleus. The cells also have numerous large mitochondria because of the high energy demand of the heart. Cardiac muscle cells have cross bands called **intercalated discs** (*intercal* = to insert between) that permit the cells to work as a single unit by allowing communication between adjacent cells. (Refer to Figure 3-19 in Chapter 3 to view the intercalated discs.) In addition to being innervated by nerves, the heart is also capable of "self-excitation" or autorhythmicity which means the cells can contract without external innervation. The cardiac muscle cells can also contract for a much longer period of time than can skeletal muscle. Nerves speed up or slow down the contraction rate of the heart. Like smooth muscle, cardiac muscle responds to two neurotransmitters—acetylcholine and norepinephrine. Acetylcholine slows the heart rate, and norepinephrine speeds it up.

> **U check**
> What is the function of intercalated discs?

learning outcome

Describe the structure of skeletal muscle.

myocytes Muscle cells.

sarcolemma The cell membrane of a muscle fiber.

sarcoplasmic reticulum The endoplasmic reticulum of a muscle cell.

Structure and Classification of Muscle

Muscle cells are called **myocytes** or *muscle fibers* because of their long lengths (Figure 9-2). The cell membrane of a muscle fiber is called the **sarcolemma.** The cytoplasm of this cell type is called *sarcoplasm*, and its endoplasmic reticulum is called **sarcoplasmic reticulum.** Most of the sarcoplasm is filled with long structures called *myofibrils*. It is the arrangement of the actin and myosin filaments in myofibrils that produces the striations observed in skeletal and cardiac muscle cells. Muscle fibers are controlled by motor neurons that release chemical substances called neurotransmitters such as acetylcholine, dopamine, and epinephrine onto the fibers.

UNIT 2   Concepts of Common Illness by System

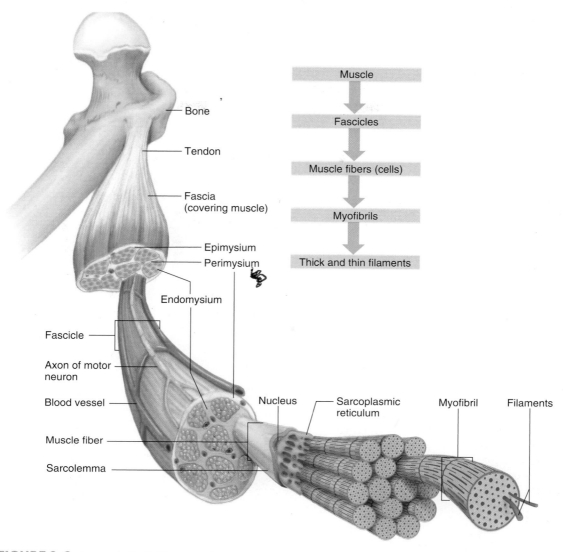

FIGURE 9-2 A muscle is divided into fascicles, fibers, myofibrils, and thick and thin filaments.

Skeletal muscle is wrapped in connective tissue coverings. **Endomysium** is the connective tissue surrounding an individual muscle cell. **Perimysium** is a sheath of connective tissue surrounding a group of 10 to 100 muscle fibers. This grouping of muscle fibers is called a **fascicle**. The **epimysium** is dense, irregular connective tissue that surrounds an entire muscle. *Fascia* is connective tissue located just below the skin that helps support and hold together muscles, bones, nerves, and blood vessels. A **tendon** is a tough, cord-like structure made of fibrous connective tissue that connects a muscle to a bone. A ligament is a tough fibrous connective tissue that binds bone to bone. Chapter 8, The Skeletal System, discusses ligaments in more detail. An **aponeurosis** is a broad, sheet-like structure made of fibrous connective tissue (Figure 9-3) that typically attaches muscles to other muscles. For example, the occipitalis muscle located at the back of the head is connected to the frontalis muscle located at the front of the head by the galea aponeurotica.

endomysium The connective tissue surrounding an individual muscle cell.

perimysium A sheath of connective tissue surrounding a fascicle.

fascicle A group of 10 to 100 muscle fibers.

U check
What is the difference between a ligament and a tendon?

REMEMBER DANNY from the case study? Which types of muscle do you think are causing his problems?

epimysium Dense, irregular connective tissue that surrounds an entire muscle.

tendon A tough, cord-like structure made of fibrous connective tissue that connects a muscle to a bone.

aponeurosis A broad, sheet-like structure made of fibrous connective tissue that typically attaches muscles to other muscles.

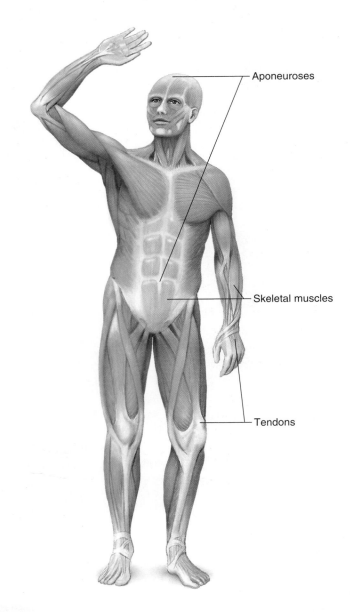

FIGURE 9-3
An aponeurosis consists of fibrous connective tissue and usually connects muscles to other muscles.

> **9.3 learning outcome**
> Explain how muscle tissue generates energy.

Energy Production for Muscle and Muscle Contraction

Adenosine triphosphate (ATP), which is a type of chemical energy, is needed for sustained or repeated muscle contractions. Therefore, muscles must have a ready and plentiful supply of ATP. Three sources for ATP include the following:

1. *Creatine phosphate:* When there is an excess of ATP, creatine phosphate is synthesized. This is a high-energy molecule that is stored in the muscle cell and can provide enough energy for about 15 seconds of muscle activity. Creatine phosphate transfers its high-energy phosphate group to ADP (adenosine diphosphate, a molecule similar to ATP, but having only two high-energy bonds), regenerating new ATP.

2. *Anaerobic respiration:* Glucose is used to generate ATP when the supply of creatine phosphate is depleted. Glucose is derived from the blood and from glycogen stored in muscle fibers and the liver. Glycolysis

breaks down glucose into molecules of pyruvic acid and produces two molecules of ATP. If oxygen levels are low, pyruvic acid is converted to lactic acid which is carried away by the blood. Anaerobic respiration provides enough energy for about 30 to 40 seconds of muscle contraction.

3. *Aerobic respiration*: Aerobic respiration uses the body's store of glucose and oxygen (aerobic) to make ATP. Glucose is broken down into pyruvic acid using oxygen, which is further converted into acetyl coenzyme A, beginning a series of reactions known as the *Krebs cycle* or the citric acid cycle. The oxygen needed for aerobic respiration is derived from a muscle pigment called *myoglobin*, which also gives muscle its pinkish color, or from hemoglobin in the blood. More than 90 percent of the ATP needed for activities lasting more than 10 minutes is provided by aerobic respiration. Pyruvic acid is oxidized in the mitochondria of muscle cells. The products of this oxidation or "burning" is ATP (36 molecules for every molecule of glucose), carbon dioxide, water, and heat.

Oxygen debt occurs when skeletal muscle is used strenuously for several minutes. After strenuous exercise, labored breathing continues and oxygen demand remains high. When pyruvic acid is converted to lactic acid for energy production, the lactic acid builds up in the tissues and causes a burning sensation. The additional oxygen is used to return the muscle fibers to their resting state. This happens in three ways: (1) lactic acid is converted to glycogen, (2) creatine phosphate and ATP are synthesized, and (3) oxygen is replaced in the myoglobin.

Muscle fatigue occurs when a muscle has temporarily lost its ability to contract. Muscle fatigue usually develops during oxygen debt because of the accumulation of lactic acid in the muscle. It can also occur if the blood supply to a muscle is interrupted or if a motor neuron loses its ability to release acetylcholine onto muscle fibers. *Cramps*—which are painful, involuntary contractions of muscles—often accompany muscle fatigue (Figure 9-4).

> **U check**
> What is the process that produces the greatest amount of ATP for muscle activity?

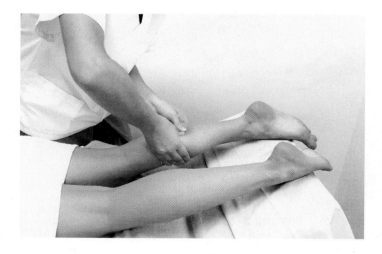

FIGURE 9-4
Cramps are painful, involuntary contractions of muscles that are sometimes treated with massage.

9.4 learning outcome

Define the terms *origin* and *insertion* and their relationship to the skeletal muscles.

Muscle contraction

origin The site of attachment of a muscle to the less movable bone during a muscle contraction.

insertion The site of attachment of a muscle to the more movable bone during muscle contraction.

prime mover The muscle responsible for most of the movement of the involved joint.

agonist A muscle that assists the prime mover in the muscle movement.

Attachments and Actions of Skeletal Muscles

As stated earlier, skeletal muscles produce movement. In order to do this, a muscle must cross a joint and have at least two attachments to bone—one to the bone proximal to the joint and the second distal to the joint. Typically, one of the bones is more moveable than the other when the muscle contracts. The **origin** is the attachment site of a muscle to the less moveable bone during muscle contraction. The **insertion** is the attachment site of a muscle to the more movable bone during muscle contraction (Figure 9-5). For example, the biceps brachii (the muscle on the anterior humerus) attaches to two places on the scapula and to one site on the radius. When the biceps brachii contracts, the radius moves and the arm bends at the elbow. Therefore, the origin of the biceps brachii is where it attaches to the scapula which remains essentially stationary. The insertion site of the biceps brachii is its attachment site on the radius which moves toward the scapula.

Most of the time, body movement is not produced by a single muscle but by a group of muscles. The muscle responsible for most of the movement is called the **agonist,** or **prime mover.** Other muscles called **synergists** help the prime mover by stabilizing joints. The **antagonist** is the muscle that produces a movement opposite to that of the prime mover. When the prime mover contracts, the antagonist must relax in order to produce a smooth body movement. With other muscle movements the roles may switch, with the antagonist becoming the agonist and the agonist becoming the antagonist (Figure 9-6). For example, when you bend your arm at the elbow, the prime mover is the biceps brachii. The synergist muscles are the brachialis and brachioradialis.

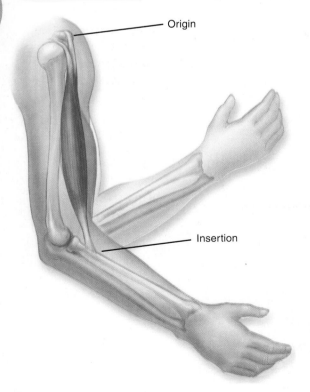

FIGURE 9-5 The biceps brachii muscle has two heads that originate on the scapula. A tendon inserts this muscle on the radius.

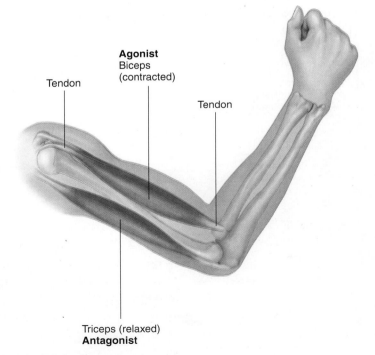

FIGURE 9-6 An agonist is the prime mover for a given action. The antagonist is the muscle that resists the prime mover's action and moves in the opposite direction.

The antagonist is the triceps brachii because its action is to extend the arm at the elbow which is opposite of flexion. When the prime mover and synergists (biceps brachii, brachialis, and brachioradialis) all contract, the antagonist (triceps brachii) relaxes; when the triceps brachii contracts, the biceps brachii, brachialis, and brachioradialis relax. Keep in mind that a muscle in one type of movement may be the prime mover, but in another situation becomes the antagonist. In flexion of the elbow, the biceps brachii is a prime mover, but in extension of the elbow, it becomes an antagonist to the triceps brachii which is now the prime mover when the elbow is being extended.

synergist A muscle that helps the prime mover by stabilizing joints.

antagonist A muscle that produces a movement opposite to that of the prime mover.

> **U check**
> What is the difference between an agonist and an antagonist?

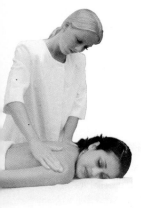

from the perspective of...

A MASSAGE THERAPIST A massage therapist uses massage techniques to help relieve muscle pain due to injury or stress, and in some cases to relieve headache, whiplash, and even some respiratory and cardiovascular conditions. Why is it important to know the origins and insertions of muscles?

Body Movements

9.5 learning outcome

Identify the types of body movements produced by and disorders that affect skeletal muscles.

LO 9.5

Health care professionals need to understand body movements to assist with judging and measuring patient ability to perform range of motion (ROM) exercises when assessing injuries and illnesses (Figures 9-7 through 9-9). Body movements produced by skeletal muscles include the following:

Flexion: bending a body part or decreasing the angle of a joint. For example, when you bend your elbow, you are flexing your arm.

Extension: straightening a body part or increasing the angle of a joint. For example, when you straighten your elbow, you are extending your arm.

Hyperextension: extending a body part past the normal anatomical position; for example, when you arch your body backward.

Dorsiflexion: pointing the toes up.

Plantar flexion: pointing the toes down. (Hint: Think "planting" your toes to help keep the definitions of these straight.)

Abduction: moving a body part away from the midline of the body; for example, moving your arm to the side away from your body. (Hint: Think "abduct," which means to take away.)

Adduction: moving a body part toward the midline of the body; for example, bringing your arm down to rest at the side of your body. (Hint: Think "adding the extremity" back to your body.)

Rotation: twisting or rotating a body part; for example, turning your head from side to side.

CHAPTER 9 The Muscular System 201

Circumduction: moving a body part in a circle; for example, moving your arm in a circular motion.

Pronation: turning the palm of the hand down or lying face down on the abdomen.

Supination: turning the palm of the hand up or lying face up on the back. (Hint: The word *up* is in *supination*.)

Inversion: turning the sole of the foot medially (inward) so that the soles of the feet can touch each other.

Eversion: turning the sole of the foot laterally (outward) so that the soles of the feet are pointing away from each other.

Retraction: moving a body part posteriorly; for example, bringing your lower jaw or mandible backward. (Hint: "Retro" means *going back*.)

Protraction: moving a body part anteriorly; for example, jutting or pushing your mandible forward.

Elevation: lifting a body part; for example, raising your shoulders as in a shrugging expression.

Depression: lowering a body part; for example, lowering your shoulders in depression.

Opposition: bringing together each of the fingers and the thumb. (Hint: The fingers are "opposite" the thumb.)

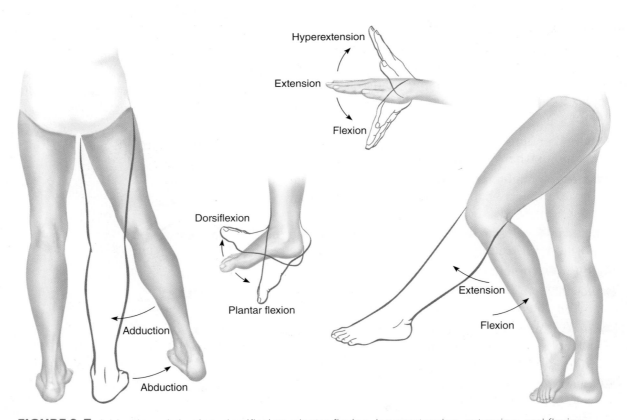

FIGURE 9-7 Adduction, abduction, dorsiflexion, plantar flexion, hyperextension, extension, and flexion.

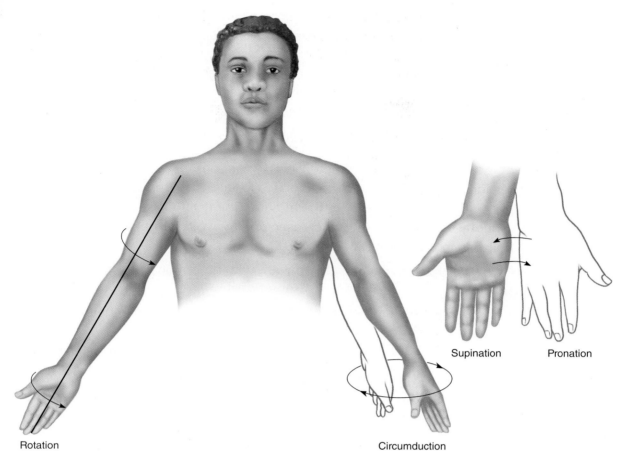

FIGURE 9-8 Rotation, circumduction, supination, and pronation.

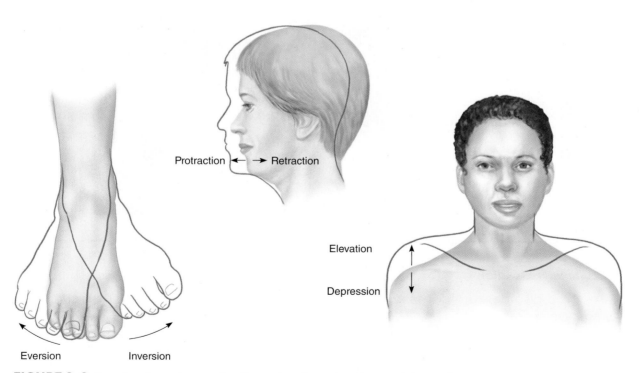

FIGURE 9-9 Eversion, inversion, protraction, retraction, elevation, and depression.

CHAPTER 9 The Muscular System

focus on Clinical Applications: Range of Motion

Assessing joint mobility is accomplished by testing range of motion (ROM), the degree in which a joint is able to move. (*FROM* is the abbreviation commonly used for "full range of motion.") A goniometer is a protractor-like device that is used to measure the amount of joint mobility (Figure 9-10). A health care professional will use the goniometer to measure the amount of movement that can be accomplished by a patient. Measurements are taken when the patient is asked to move the affected joints in various ways.

Once ROM testing is completed, the patient is usually instructed on ROM exercises designed to increase his or her mobility of the affected joint(s). ROM exercises should be done slowly and gently so that pain is not increased and to avoid further injury. ROM exercises are accomplished in three ways:

- Active ROM exercises are performed by the patient with no physical assistance after initial instruction.
- Active assisted ROM exercises are performed by the patient but with the help of another person or with machine assistance.
- Passive ROM exercises are performed on a patient by the health care professional, without the patient's help.

ROM exercises should be performed on patients who are confined to bed to maintain joint function and increase circulation. All the joints including the shoulder, hip, back, wrist, ankle, toes, and fingers are exercised.

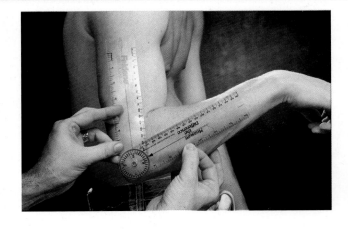

FIGURE 9-10
A goniometer is a protractor-like device that is used to measure the range of motion of a joint.

focus on Genetics: Sex-Linked Disorders

Sex-linked disorders are diseases that are passed along by genes on the X or Y chromosome. Y-linked disorders are extremely rare. X-linked disorders are more common and are passed from a mother who is a carrier for a condition to her son. X-linked disorders typically are recessive. Since males have one X chromosome and one Y chromosome, they need only one of the recessive alleles to have the disease. Hemophilia and muscular dystrophy are examples of X-linked diseases.

PATHOPHYSIOLOGY

Fibromyalgia

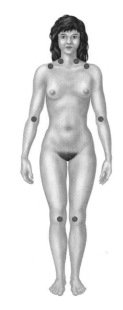

FIGURE 9-11
Tender points seen in fibromyalgia.

Fibromyalgia is a fairly common condition that results in chronic pain primarily in joints, muscles, and tendons (Figure 9-11). It most commonly affects women between the ages of 20 and 50.

Etiology: The causes of this disorder are poorly understood. It may have an autoimmune component and the condition can be aggravated by sleep disturbance, emotional distress, decreased blood flow to muscles, a virus, or any combination of these factors.

Signs and Symptoms: Symptoms of fibromyalgia include fatigue, tenderness in different areas of the body, sleep disturbances, and chronic pain. The diagnosis is usually made by ruling out other possible diseases. It is not normally diagnosed unless a person has muscle and joint pain for at least three months in designated body areas.

Treatment: Treatment is varied and includes antidepressants, anti-inflammatory medications, physical therapy, lifestyle changes to reduce stress, counseling to improve coping skills, reduction or elimination of caffeine to improve sleeping, and diet supplements to improve nutrition.

Muscular Dystrophy

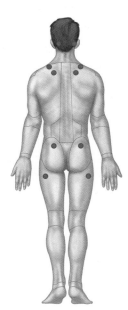

Muscular dystrophy (MD) is a group of inherited disorders characterized by muscle weakness and a loss of muscle tissue. There are at least seven types of muscular dystrophy, and they are distinguished from each other by types of symptoms, the age of symptom onset, and the underlying cause.

Etiology: The cause of this disorder is primarily hereditary, being classified as X-linked disorders (passed on from mother to son). Genetic fetal testing is available for those who request it.

Signs and Symptoms: The signs and symptoms vary widely and depend on the type of MD. Most progress steadily and are eventually fatal. Other types cause mild symptoms, and patients usually have normal life expectancies. Specific signs and symptoms include muscle weakness in various muscle groups, depending on the type of dystrophy; difficulty walking; drooling; a delayed development of motor skills; frequent falls; mental retardation in some types; a curved spine; the formation of a claw hand or clubfoot; a loss of muscle mass; the accumulation of fat or fibrous connective tissue in muscles; and arrhythmias in some types. The progression of the muscular weakness may also include eventual paralysis of the affected muscle groups. The two most common forms of MD are

(continued)

Muscular Dystrophy (concluded)

Duchenne's and Becker's MD. Duchenne's muscular dystrophy is the most common type and progresses more rapidly than Becker's MD.

The diagnosis is primarily made through a muscle biopsy. Other tests include DNA testing; an electromyography (EMG) test, which tests muscle weakness; or an ECG (electrocardiogram), which tests cardiac function.

Treatment: Treatment includes physical therapy to maintain muscle function, the use of braces and wheelchairs, various medications based on the type of MD, and spinal and other surgery on tendons and ligaments to avoid or treat contracture.

Myasthenia Gravis

Myasthenia gravis is a condition in which affected persons experience muscle weakness. In this autoimmune condition, a person produces antibodies that prevent muscles from receiving neurotransmitters from neurons. It most commonly affects young women and older men, especially if they have other underlying autoimmune disorders.

Etiology: Myasthenia gravis is usually considered an autoimmune disorder.

Signs and Symptoms: Symptoms of myasthenia gravis include *diplopia* (double vision); muscle weakness; *ptosis* (drooping of the eyelid); *dysphagia* (difficulty swallowing) and difficulty talking, chewing, lifting, walking or breathing; fatigue; and drooling. The signs and symptoms usually worsen with activity and as the day progresses, but get better with rest. Muscle atrophy is not a common finding in the disease.

Making the diagnosis of myasthenia gravis may be difficult, but a single fiber EMG test is often useful. This test measures the response of a muscle fiber to nervous stimulation. Other diagnostic tests include acetylcholine receptor antibody tests and the Tensilon test. In a positive Tensilon test, muscle activity increases after medication is given that blocks the breakdown of acetylcholine.

Treatment: Treatment includes making lifestyle changes to avoid excessive stress, getting adequate rest, using heat on muscles, using an eye patch to treat double vision, taking medications to improve communication between nerves and muscles, and using medications to suppress the immune system. *Plasmapheresis* may be used to remove harmful antibodies from blood, and in some patients the removal of the thymus may be recommended.

Major Skeletal Muscles

learning outcome

Relate the major skeletal muscles of the body to their actions.

The name of a skeletal muscle often describes it in some way. Its name may indicate the location, size, action, shape, or number of attachments of the muscle. For example, the pectoralis major is named for its large size (major) and its location (pectoral, or chest, region). The sternocleiodmastoid is named for its attachment sites—sterno (sternum), cleido (clavicle), and mastoid (the mastoid process of the temporal bone, located behind the ear). As you study muscles, you will find it easier to remember them if you think about what the name describes. Figure 9-12 shows an anterior view of the superficial skeletal muscles. These include many muscles such as the pectoralis major and minor muscles, the biceps brachii, and the vastus medialis. Figure 9-13 illustrates a posterior view of the superficial skeletal muscles. The following sections illustrate various other muscles, their actions, origins, and insertions.

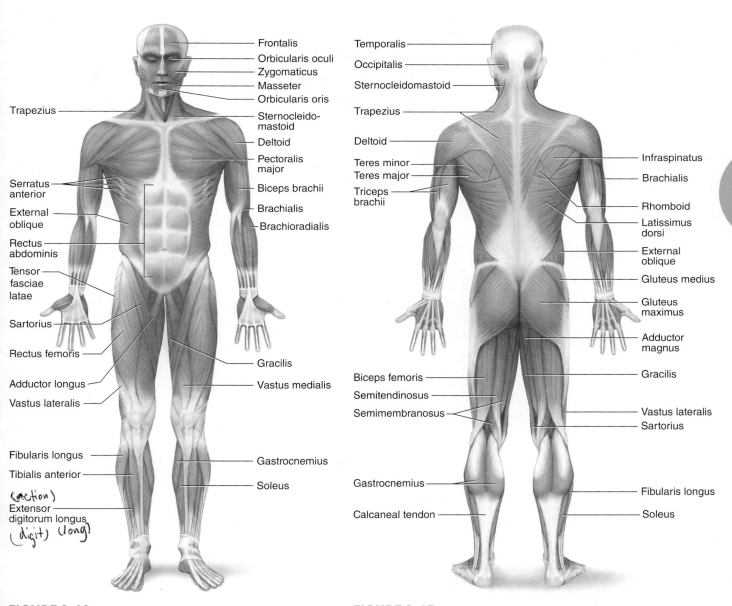

FIGURE 9-12 Anterior view of superficial skeletal muscles.

FIGURE 9-13 Posterior view of superficial skeletal muscles.

CHAPTER 9 The Muscular System

Muscles of the Face and Head

A number of muscles beneath the skin and scalp allow us to make facial expressions and chew our food. These muscles, their actions, origin, and insertions are seen in Figure 9-14 and Table 9-2.

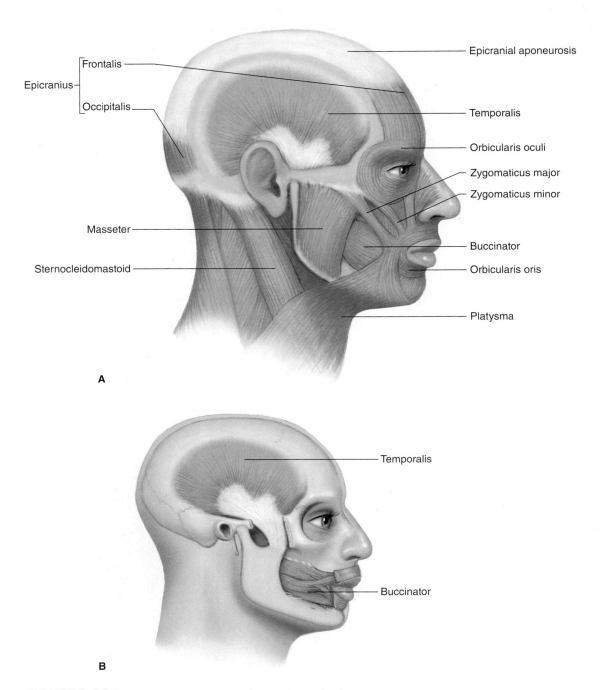

FIGURE 9-14 Muscles of facial expression and mastication.

TABLE 9-2 Muscles of the Face and Head

Muscle	Origin	Insertion	Action
Muscles of Facial Expression			
Frontalis	Frontal bone	Skin of eyebrows	Raises the eyebrows, wrinkles skin of forehead
Orbicularis oris	Muscle fibers surrounding opening of the mouth	Skin around corner of mouth	Closes and protrudes lips as in kissing
Zygomaticus major	Zygomatic bone	Skin at angle of mouth and orbicularis oris	Draws angle of mouth superiorly and laterally as in smiling
Orbicularis oculi	Frontal and maxillary bones	Skin around the eye and eyelid	Closes the eye
Platysma	Fascia of chest and deltoid muscle	Mandible, skin, and muscle at corner of mouth	Tenses skin of neck, pulls down corners of mouth
Muscles of Mastication (Chewing)			
Masseter	Zygomatic arch and maxilla	Angle and ramus of mandible	Closes the jaw
Temporalis	Temporal bone	Coronoid process and ramus of mandible	Elevates and retracts mandible

Muscles of the Upper Body

Without our arms we would be very limited in the types of movements we could do. Although it is possible to adapt to living with no arms, it is obviously not ideal. There are numerous muscles in our upper body. They help us perform a large number of tasks. Figures 9-15 to 9-18 and Table 9-3 illustrate these muscles and the movements they produce, as well as their origin and insertions.

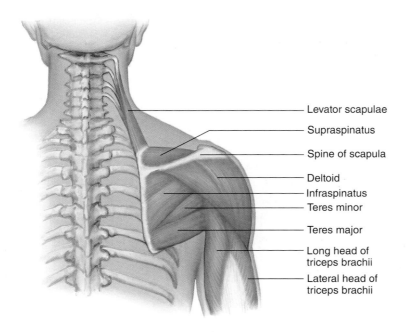

FIGURE 9-15
Muscles of the posterior surface of the scapula and arm.

CHAPTER 9 The Muscular System

FIGURE 9-16 Muscles of the posterior shoulder.

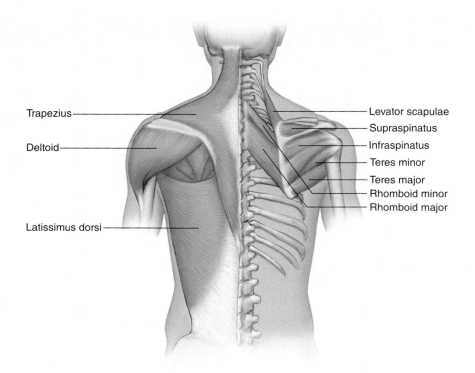

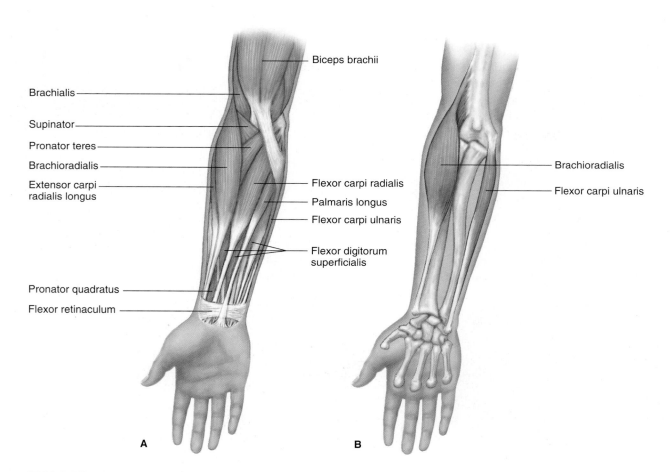

FIGURE 9-17 Muscles of the arm and forearm.

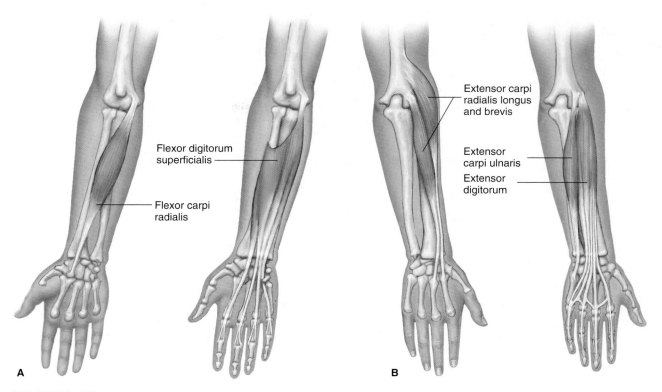

FIGURE 9-18 Muscles of the wrist, hands, and fingers.

TABLE 9-3 Muscles of the Upper Body and Extremities

Muscle	Origin	Insertion	Action
Upper Extremity Muscles			
Pectoralis major	Clavicle, sternum, and costal cartilages of second to sixth ribs	Humerus	Flexes, adducts, and medially rotates arm
Latissimus dorsi	Vertebrae T1-L5, inferior four ribs, sacrum, and iliac crest	Deltoid tuberosity of humerus	Extends, adducts, and medially rotates arm (important muscle in doing the butterfly stroke in swimming)
Deltoid	Clavicle and scapula	Deltoid tuberosity of humerus	Flexes, extends, and rotates humerus
Subscapularis	Scapula	Lesser tubercle of humerus	Medially rotates arm
Infraspinatus	Scapula	Greater tubercle of humerus	Extends arm and rotates it laterally

continued

TABLE 9-3 (concluded)

Muscle	Origin	Insertion	Action
Muscles That Move the Forearm			
Biceps brachii	Scapula	Radius	Flexes and supinates forearm, flexes arm at shoulder joint
Brachialis	Anterior surface of humerus	Ulna	Flexes forearm at elbow
Brachioradialis	Lateral border of humerus	Radius	Flexes forearm at elbow, supinates and pronates forearm
Triceps brachii	Scapula and humerus	Ulna	Extends forearm at elbow, extends arm at shoulder
Supinator	Humerus and ulna	Radius	Supinates forearm
Pronator teres	Medial epicondyle of humerus and ulna	Midlateral surface of radius	Supinates forearm
Muscles of the Wrist, Hands, and Fingers			
Flexor carpi radialis	Humerus	Metacarpals	Flexes and abducts hand (radial deviation)
Flexor carpi ulnaris	Humerus and ulna	Metacarpals	Flexes and adducts hand (ulnar deviation)
Palmaris longus	Humerus	Flexor retinaculum	Weakly flexes hand
Flexor digitorum profundus	Ulna	Phalanges	Flexes fingers (not thumb)
Extensor carpi radialis longus	Humerus	Second metacarpal	Extends and abducts hand at wrist
Extensor carpi radialis brevis	Humerus	Third metacarpal	Extends and abducts hand at wrist
Extensor carpi ulnaris	Humerus and ulna	Fifth metacarpal	Extends and adducts hand at wrist
Extensor digitorum	Humerus	Phalanges	Extends phalanges (not thumb)
Muscles of the Pectoral Girdle			
Trapezius	Occipital bone and vertebrae	Clavicle and scapula	Adduct, depress, and rotate scapula
Pectoralis minor	Ribs	Scapula	Abduct and rotate scapula downward, elevate ribs

Muscles of the Trunk

The abdomen is not protected by bony structures. Instead the walls of the abdomen consist of layers of muscles known as the core. Muscles that help us breathe and the muscles of the abdomen are all located on the torso, or trunk of the body. The most important muscle of respiration is the diaphragm. The respiration process is discussed in more detail in Chapter 13, The Respiratory System. Table 9-4 and Figure 9-19 illustrate the muscles of the torso, their movements, and their origin and insertions.

TABLE 9-4 Respiratory Muscles

Muscle	Origin	Insertion	Action
Respiratory Muscles			
Diaphragm	Sternum, ribs, and vertebrae	Central tendon	Inhales
External intercostals	Ribs	Ribs	Inhale
Internal intercostals	Ribs	Ribs	Exhale
Abdominal Muscles			
External obliques	Ribs	Iliac crest and linea alba	Compress abdomen, flex and rotate vertebral column
Internal obliques	Iliac crest	Ribs and linea alba	Compress abdomen, flex and rotate vertebral column
Transverse abdominis	Iliac crest, ribs	Sternum, linea alba, and pubis	Compresses abdomen
Rectus abdominis	Pubis	Ribs and sternum	Compresses abdomen and flexes vertebral column

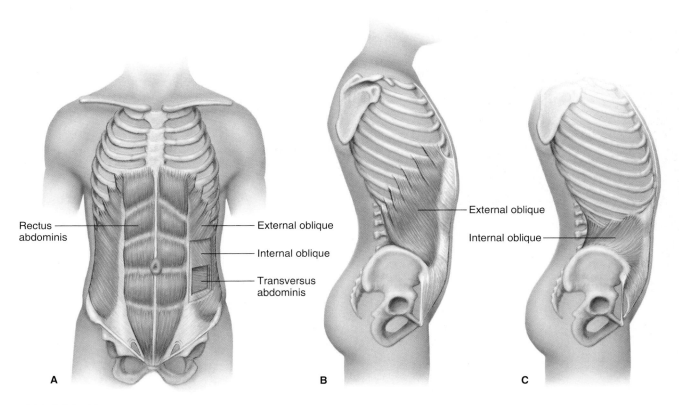

FIGURE 9-19 Muscles of the abdominal wall.

CHAPTER 9 The Muscular System

Muscles of the Lower Extremities

The muscles of the lower extremities help us to walk, run, and jump. They provide us mobility, or the ability to move while we are upright. The lower extremities have some of the strongest muscles in the body. One group of muscles are known as the *quadriceps group.* This muscle group includes the rectus femoris, vastus lateralis, vastus intermedius, and vastus medialis. Figures 9-20 to Figure 9-23 and Table 9-5 illustrate these muscles of the lower extremities, their actions, origins, and insertions.

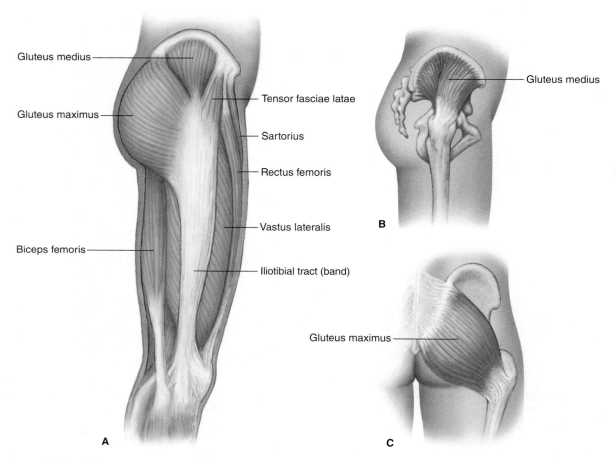

FIGURE 9-20 Muscles of the lateral thigh and leg.

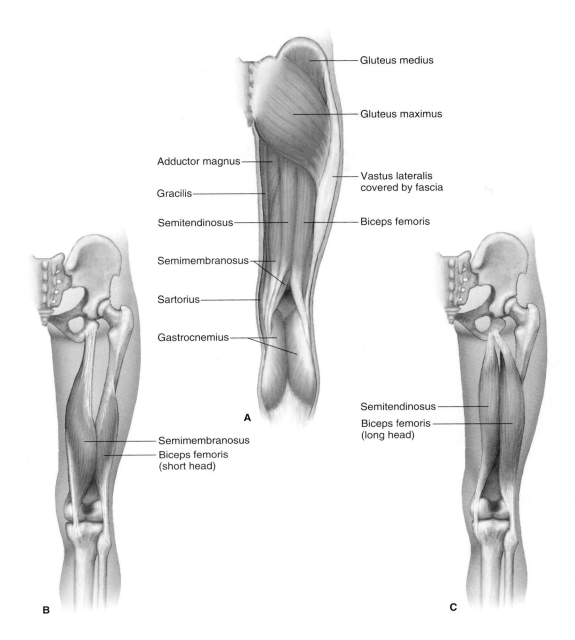

FIGURE 9-21 Muscles of the posterior thigh.

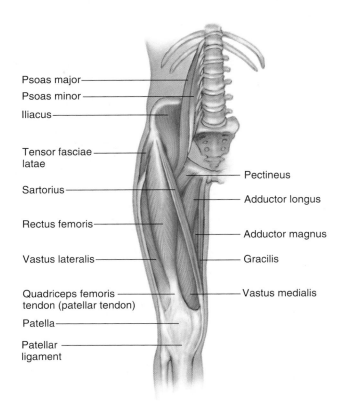

FIGURE 9-22
Muscles of the anterior thigh and leg.

> ### focus on Wellness: Exercise for Muscle and Overall Health
>
> From the time we are quite young we are told of the importance of exercise. It is good for decreasing cholesterol levels and blood pressure, and it helps us keep off or lose unwanted weight. But there is more to the story.
>
> The addition of "strength training" to any exercise program assists in slowing the muscle loss that comes with age. It builds not only muscle strength but also the strength of connective tissues, decreasing the chance of injury such as sprains and strains. Strength or resistance exercise also increases bone density, further decreasing injuries such as fractures that often occur related to osteoporosis. Exercise can also relieve arthritis pain.
>
> Exercise also has been shown to have real benefits for people suffering from depression. Exercise won't replace medication, but it can help. In addition to the above-mentioned benefits, increased energy and improved physical appearance occur. So you can see, the positive side effects of exercise are numerous.

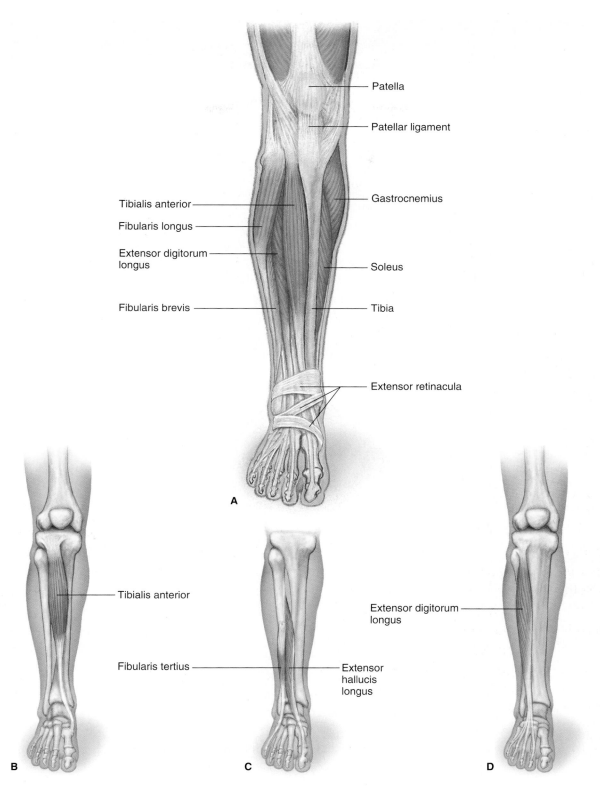

FIGURE 9-23 Muscles of the leg: (A) muscles of the anterior right leg and (B to D) isolated views of muscles associated with the anterior right leg.

CHAPTER 9 The Muscular System

TABLE 9-5 Lower Extremity Muscles

Muscle	Origin	Insertion	Action
Lower Extremity Muscles			
Iliopsoas major	Lumbar vertebrae	Femur	Extends and laterally rotates thigh; flexes trunk
Gluteus maximus	Iliac crest, sacrum, and coccyx	Femur	Extends thigh at hip and laterally rotates thigh
Gluteus medius	Ileum	Femur	Abducts and medially rotates thigh
Gluteus minimus	Ileum	Femur	Abducts and medially rotates thigh
Adductor longus	Pubis	Femur	Adducts, flexes, and rotates thigh
Adductor magnus	Pubis and ischium	Femur	Adducts and rotates thigh
Biceps femoris (Together, the biceps femoris, semitendinosus, and semimembranosus are known as the hamstrings because their tendons are long and are stringlike in the popliteal area behind the knee.)	Ischium and femur	Fibula and tibia	Flexes leg at knee and extends thigh at hip
Semitendinosus	Ischium	Tibia	Flexes leg at knee and extends thigh at hip
Semimembranosus	Ischium	Tibia	Flexes leg at knee and extends thigh at hip
Sartorius (Because the sartorius muscle is involved in sitting cross-legged, it is sometimes called the "tailor's muscle" since that is how tailors used to sit while sewing.)	Anterior superior iliac spine	Medial surface of the body of the tibia	Weakly flexes leg and thigh, abducts and laterally rotates thigh
Rectus femoris	Anterior inferior iliac spine	Patella and tibial tuberosity	Extends leg at knee, flexes thigh at hip joint
Vastus lateralis	Greater trochanter and linea aspera of femur	Patella and tibial tuberosity	Extends leg at knee
Vastus intermedius	Anterior and lateral body of femur	Patella and tibial tuberosity	Extends leg at knee
Vastus medialis	Linea aspera of femur	Patella and tibial tuberosity	Extends leg at knee

TABLE 9-5 (Concluded)

Muscle	Origin	Insertion	Action
Muscles of the Ankle, Foot, and Toes			
Tibialis anterior	Tibia	Tarsal and metatarsals	Dorsiflexes and inverts foot
Extensor digitorum longus	Tibia and fibula	Phalanges of foot	Dorsiflexes foot and extends toes
Gastrocnemius (The gastrocnemius is commonly referred to as the calf muscle; it has a large "belly," hence the name.)	Femur and capsule of knee	Calcaneous	Plantar flexes the foot and flexes leg at knee
Soleus	Fibula and tibia	Calcaneous	Plantar flexes the foot
Flexor digitorum longus	Tibia of middle	Phalanges of foot	Plantar flexes the foot and flexes the middle phalanges of the foot

Strains and Sprains

Strains are caused by stretching or tearing of muscles or tendons. Sprains involve injuries that excessively stretch or tear ligaments at a joint (Figure 9-24). Remember that tendons connect muscle to bone and ligaments connect bone to bone. Ankle and wrist sprains are fairly common, with pain and swelling of the joint the most common symptoms. These types of injuries can be often be prevented. If the patient does not feel that his or her symptoms have improved in several days to a week's time, the patient should contact his or her physician to rule out a more serious injury, such as a torn ligament or muscle, or even a bone fracture.

9.7 learning outcome

Compare strain and sprain and other types of muscular injuries.

FIGURE 9-24
Sprains and strains are usually treated with the RICE protocol (*R* = rest; *I* = ice; *C* = compression; *E* = exercise).

focus on Wellness: Preventing and Treating Strains and Sprains

To help prevent strains and sprains, follow these steps:

- Warmup: Warming up muscles for just a few minutes before intense activity raises muscle temperature, which makes muscles more pliable and less likely to be injured.
- Stretch: Stretching improves muscle performance and should always be done after the warmup and after exercising. A person should never stretch further than he or she can hold for 10 seconds.
- Cooldown: Slowing down the exercise before completely stopping it prevents dizziness and fainting. When a person suddenly stops exercising, blood can pool in the legs which prevents blood from reaching the brain, in turn causing dizziness (vertigo). Cooling down also allows time to remove lactic acid from muscles.

If sprains or strains do occur, immediate treatment using *RICE* protocol is recommended:

R is for rest. Resting minimizes bleeding, further injury, and swelling.

I is for ice. Ice minimizes swelling and pain. A bag that is filled with crushed ice conforms better to a body part than one filled with ice cubes. A bag full of frozen peas or other small vegetables may also be used. The ice should be applied for 10 minutes and then removed for 10 minutes. This should be kept up for about an hour and repeated several times during a 24-hour period. Ice can be applied for a shorter period of time if blood vessels dilate during its application. (Although heat may temporarily make the injury feel better, during the acute phase of the injury, heat should be avoided.)

C is for compression, which minimizes swelling. A bandage should be wrapped loosely around the injured area and the bag of ice. Compression should be applied and removed along with the ice.

E is for elevation. The injured muscle should be elevated, which minimizes swelling. Elevation should be continued as long as swelling is present.

PATHOPHYSIOLOGY

Tendonitis *← information*

Tendonitis is described as the painful inflammation of a tendon as well as of the tendon–muscle attachment to a bone. The most common locations for tendonitis are the elbow, shoulder, hip, heel, and hamstring. Tendonitis may also be associated with bursitis, the inflammation of the bursa located in synovial joints such as the shoulder, elbow, and knee.

Etiology: Tendonitis usually occurs after a sports-related or other repetitive activity that results in injury to the muscle–tendon or tendon-to-bone attachment. For example, tennis and golf can cause tennis elbow and golfer's elbow, respectively. Other musculoskeletal disorders may also cause or exacerbate this condition.

Signs and Symptoms: Pain at the joint or muscle attachment that results in limited range of motion (ROM) of the affected area.

Treatment: During the acute phase of the injury, ice should be used to minimize inflammation. After this initial time period, heat application will assist with relieving joint and muscle pain. Resting the affected area and oral analgesics such as aspirin or acetaminophen may help control the pain.

Tetanus

Tetanus is commonly known as lockjaw.

Etiology: Tetanus is caused by the toxin produced by the bacterium *Clostridium tetani* which is found in soil and water. The usual form of entry is through open wounds.

Signs and Symptoms: Muscle spasms of face, neck, and jaw usually occur 5 to 10 days after infection. Worsening spasms may cause bone fractures, dyspnea, irritability, fever, diaphoresis, and drooling. Diagnosis is usually based on wound type and presenting symptoms. Tetanus antibody tests may also be run.

Treatment: Immediate treatment with antitoxins and antibiotics is needed to prevent death or long-lasting effects. Wound cleaning, muscle relaxants, sedation, endotracheal tubes, and ventilation may be required. Regular vaccinations prevent this disease.

Tetany

Tetany is a condition of (often severe) muscle spasms, cramps, uncontrolled twitching, and sometimes seizures.

Etiology: The cause is low blood calcium.

Signs and Symptoms: Muscle cramps and spasms, uncontrolled twitching, and in severe cases, seizures will occur.

Treatment: Calcium (oral and/or IV) and vitamin D supplements are administered. If the parathyroid is involved, parathyroid hormone (PTH treatments) may also be used.

Torticollis

Torticollis (*tortus* = twisted, *collum* = neck) is also known as wry neck. This condition is due to abnormally contracted neck muscles. The head typically bends toward the side of the contracted muscle while the chin rotates to the opposite side.

(continued)

Torticollis (concluded)

Etiology: Torticollis may be congenital or acquired. Abnormal positioning of the baby's head in utero, birth injury during delivery, and genetics are some congenital factors. Acquired factors include neck injury, infections, poor posture, and head trauma.

Signs and Symptoms: Obvious tilting of the head, headaches, and inflammation and pain of neck muscles associated with restricted range of motion of the neck and head.

Treatment: For the congenital form, passive exercises to stretch the muscles as well as corrected head positioning during sleep (to maintain the straightening accomplished through the exercises) may be helpful. The treatment for acquired torticollis should consist of treating the underlying condition if possible. Heat, cervical traction, neck brace, exercise, massage, and psychotherapy to assist the patient in dealing with the psychological and emotional effects related to the deformity are all treatment options.

9.8 learning outcome

Discuss the effects of aging on the muscular system and the steps that can be taken to promote muscle health.

Life Span

As humans age, we lose muscle mass and flexibility. By the time individuals are in their eighties, they have only about half the strength they had in their twenties. Aging causes a decline in the speed and strength of muscle contractions even though the actual endurance of muscle fibers changes very little. Elderly patients often have increasing difficulty with dexterity and gripping ability. Both aerobic exercises and strength training can slow the loss of muscle strength and mobility.

Summary

	learning outcomes	key points
9.1	Relate the types of muscle tissue to their locations and characteristics, and describe how visceral (smooth) muscle produces peristalsis.	Muscle acts to move things. Examples are skeletal muscle moving joints, cardiac muscle pumping blood, and smooth muscle moving substances through the digestive tract. Skeletal muscle typically is attached to bones and is under voluntary control. Cardiac muscle is found in the heart and is involuntary. Smooth muscle is also involuntary and is found in the walls of hollow organs and the walls of blood vessels.
9.2	Describe the structure of skeletal muscle.	Skeletal muscle is made up of thick (myosin) and thin (actin) filaments that are arranged to produce striations. The fibers (cells) are multinucleated and are cylindrical in shape. They have numerous mitochondria.
9.3	Explain how muscle tissue generates energy.	Energy in the form of ATP (adenosine triphosphate) is needed by muscles for contraction. This energy is generated by creatine phosphate, anaerobic respiration, and aerobic respiration. The greatest amount of energy is derived from aerobic respiration.
9.4	Define the terms *origin* and *insertion* and their relationship to skeletal muscles.	The origin of a muscle is the site of attachment of the skeletal muscle to the less movable bone. The insertion of a skeletal muscle is the site of attachment of the muscle to the more movable bone.
9.5	Identify the types of body movements produced by and disorders that affect skeletal muscles.	Body movements that are produced by skeletal muscles include flexion and extension, abduction and adduction, circumduction, opposition, rotation, inversion and eversion, retraction and protraction, pronation and supination, and depression and elevation.
9.6	Relate the major skeletal muscles of the body to their actions.	There are approximately 600 muscles in the body. In learning them, you need to identify their origin and insertion as well as be able to explain their actions.
9.7	Compare sprain and strain and other muscular injuries.	A sprain is an injury to a ligament. A strain is an injury to a muscle or tendon.
9.8	Discuss the effects of aging on the muscular system and the steps that can be taken to promote muscle health.	Humans lose muscle mass, strength, and flexibility as we age. We can slow the process of aging on muscle and promote general muscle health by eating properly and getting adequate and appropriate age-related activity.

chapter 9 review

case study 1 questions
Can you answer the questions that pertain to Case Study 1 presented in this chapter?
1. Why is muscular dystrophy (MD) more common in males than females?
   a. Males have more muscle mass on average than females, so they are more prone to muscular disease.
   b. Males are more often involved in contact sports.
   c. MD is a sex-linked condition that is passed on from mother to son.
   d. Estrogen has a protective effect against muscular diseases.
2. Which of the following is *not* used to diagnose MD?
   a. Blood enzyme tests
   b. Electromyogram
   c. Muscle biopsy
   d. Stress tests
3. The most severe form of MD is
   a. Duchenne's
   b. Becker's
   c. Smith's
   d. Wilson's

review questions
1. Define the following terms related to muscle structure:
   a. Fascia
   b. Epimysium
   c. Fascicle
   d. Tendon
   e. Aponeurosis
2. Describe the following actions:
   a. Flexion
   b. Extension
   c. Pronation
   d. Supination
   e. Abduction
3. Briefly describe the signs and symptoms of the following muscular diseases:
   a. Fibromyalgia
   b. Muscular dystrophy
   c. Myasthenia gravis
4. Give the origin, insertion, and action of the following muscles:
   a. Sternocleidomastoid
   b. Orbicularis oris
   c. Pectoralis major
   d. Latissimus dorsi
   e. Triceps brachii

critical thinking questions
1. Why is it important to do warmup activities before participating in sporting events?
2. Explain why physical therapy and massage are prescribed for patients in wheelchairs, for those who are bed bound, or even for those in a coma.
3. Explain why cardiac muscle has intercalated discs.

patient education
1. Using the acronym *RICE,* explain how a patient should care for a sprained ankle.
2. Explain to a group of elderly residents at an assisted living facility the types of exercises that will benefit their quality of life. These may include pool exercises and moderate weight training. As a side note, you may want to mention assistive devices such as railings, tub and shower seats, and gripping devices.

applying what you know

1. Relate muscle function to homeostasis.
2. Describe the structure and function of the three muscle types.
3. What are the three processes by which muscles get the energy required for contraction?
4. What is oxygen debt?
5. What type of muscle is found in the walls of hollow organs such as the uterus?
6. Which type of muscle has intercalated discs?
7. Which type of muscle does not show autorhythmicity?

CASE STUDY 2 Poisoning and Muscular Disorders

Five days ago, 40-year-old Mary Carroll came to the doctor's office where you work as a medical assistant. She complained about experiencing generalized muscle weakness as well as "seeing double" and drooping of her right eyelid. She said these symptoms came on suddenly within the last couple of days. The doctor asked Mary to elaborate on her activities the day before she fell ill. She told him that she had sprayed her furniture and carpets with an insecticide to get rid of fleas in her house. She had also dipped her cats and dogs with the same insecticide. The doctor was suspicious of insecticide poisoning so he transferred her to the hospital for treatment. He was concerned that the insecticide blocked acetylcholinesterase in her body.

1. What is the function of acetylcholinesterase?
2. Why is it important for patients to give their doctor a complete account of their activities prior to an illness?

CASE STUDY 3 Muscle Strain

Yesterday, 62-year-old Laura Engler was working in her yard. After 5 hours of raking and planting, she could barely get up and walk.

1. What is the difference between a strain and a sprain?
2. What is the treatment of sprains and strains?

10

Blood and Circulation

chapter outline

- Hematopoiesis
- Plasma
- Formed Elements
- Hemodynamics
- Hemostasis
- Blood Typing
- Blood Vessels
- Pulmonary and Systemic Circulation
- Arterial and Venous Systems
- Life Span

outcomes

AFTER COMPLETING THIS CHAPTER, YOU WILL BE ABLE TO:

10.1 Explain why blood is considered a connective tissue and describe hematopoiesis.

10.2 Describe the components of plasma and their functions.

10.3 Relate the structure and function of erythrocytes, leukocytes, and platelets to anemia, leukemia, and thrombocytopenia, respectively.

10.4 Describe blood pressure and how it is regulated, including a description of essential and secondary hypertension.

10.5 Explain the steps of hemostasis.

10.6 Explain the purpose of blood typing, and describe the ABO blood typing system and the Rh blood typing system.

10.7 Describe types of blood vessels, their functions, and disorders.

10.8 Compare the pulmonary and systemic circuits.

10.9 Identify the major arteries and veins of the body.

10.10 Describe how the circulatory system is affected by age.

essential terms

albumin (al-BYOO-men)
atherosclerosis (ath-er-o-skle-RO-sis)
basophil (BA-so-fil)
embolus (EM-bo-lus)
eosinophil (e-o-SIN-o-fil)
erythrocyte (e-RITH-ro-site)
erythropoiesis (e-rith-ro-po-EE-sis)
hematocrit (he-MAT-o-crit)
hemoglobin (HE-mo-glo-bin)
hemostasis (he-mo-STA-sis)
hypertension (hi-per-TEN-shun)
leukocyte (LOO-ko-site)
lymphocyte (LIM-fo-site)
monocyte (MON-o-site)
neutrophil (NOO-tro-fil)
serum (SEER-um)
thrombocytes (THROM-bo-sites)
thrombus (THROM-bus)

Additional key terms in the chapter are italicized and defined in the glossary.

case study

Use the case study to focus on as you go through the chapter. The questions will guide you as you learn the anatomy, physiology, and pathology of the blood circulatory system.

CASE STUDY 1 *Pernicious Anemia*

Jim Lewis, a 58-year-old stockbroker, has been very fatigued for several months. He also has noticed a tingling sensation in his right hand. Jim says his memory does not seem like it once was and he feels nauseous at times. He used a friend's blood glucose meter to check his blood glucose levels and they were normal. He admits to being an alcoholic and has not been doing well with his addiction for the past year.

As you go through the chapter, keep the following questions in mind:

1. What is the cause of pernicious anemia?
2. Why is it sometimes called megaloblastic anemia?
3. What are the signs and symptoms of pernicious anemia?
4. What is the treatment?
5. What advice would you give to Jim?

study tips

1. Draw the arterial system in parts. Do the head and neck, then the upper extremities, then the chest and abdominal region, then the lower extremities. Doing it in parts will make memorization and learning more manageable. Do the same for the venous system.

2. On flash cards write the name of an artery or vein and on the reverse side write the name of the organ that the blood vessel either supplies (artery) or drains (vein).

3. Write your own case study using the pathology that you learned in this chapter.

Introduction

An average-sized adult body contains approximately 4 to 6 liters of blood, or approximately 8 percent of the total body weight. Blood volume varies from person to person depending on the person's size, the amount of adipose tissue, and the concentrations of certain ions in the blood. Females generally have less blood volume than males. Blood represents life because of the many essential functions it performs. It carries oxygen and nutrients to tissues and carries carbon dioxide and wastes away from tissues. It also acts as the transport mechanism for hormones and helps with heat regulation of the body. The process of transporting blood to and from the body tissues is known as circulation. The heart's powerful muscular pump performs the task of circulation. Circulation of the blood depends on the heart and its ability to contract or beat.

Hematopoiesis

10.1 learning outcome

Explain why blood is considered a connective tissue and describe hematopoiesis.

Homeostasis

erythrocyte A mature red blood cell.

leukocyte A white blood cell.

thrombocyte A cell fragment that is involved in hemostasis (blood clotting).

erythropoiesis The formation of red blood cells.

Remember from Chapter 3, Concepts of Cells and Tissues, that tissues are classified as connective tissue if they have living cells and a nonliving matrix. Blood is classified as a connective tissue because it has a nonliving matrix and living cells. The plasma is the nonliving fluid component of blood. Although it is "nonliving," it has several components that serve essential functions for the body and maintain homeostasis. Formed elements are the cells and cell fragments and include **erythrocytes** (red blood cells, RBCs), **leukocytes** (white blood cells, WBCs), and **thrombocytes** (platelets).

The formation of blood cells from stem cells, both red and white, is called *hematopoiesis*. See Figure 10-1. Recall from Chapter 8, The Skeletal System, that we talked about blood cell formation. During fetal development, erythrocytes are made in the yolk sac, the liver, and the spleen. However, after birth most blood cells are produced in red bone marrow by cells called stem cells or *hemocytoblasts.* The average life span of an RBC is about 120 days, so red bone marrow is constantly making new cells. The hormone *erythropoietin* is responsible for regulating **erythropoiesis,** or the production of RBCs. Erythropoietin is produced by the kidneys and stimulates the red bone marrow to produce new RBCs. The kidneys release this hormone when oxygen

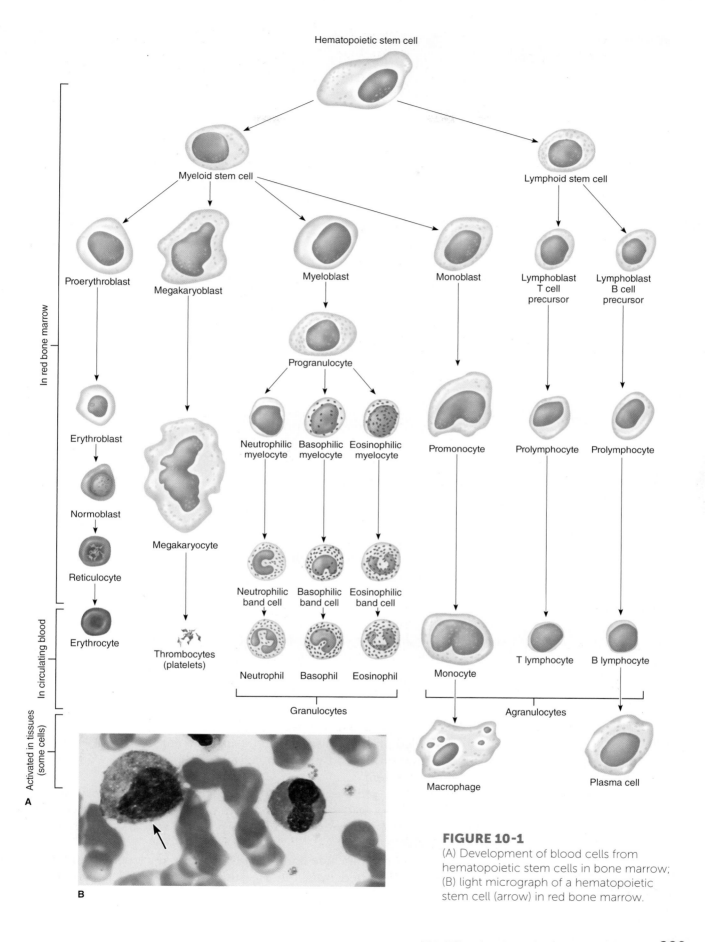

FIGURE 10-1
(A) Development of blood cells from hematopoietic stem cells in bone marrow; (B) light micrograph of a hematopoietic stem cell (arrow) in red bone marrow.

CHAPTER 10 Blood and Circulation

TABLE 10-1 Dietary Factors Affecting Red Blood Cell Production

Substance	Source	Function
Vitamin B_{12} (requires intrinsic factor for absorption via small intestine)	Absorbed from small intestine	DNA synthesis
Iron	Absorbed from small intestine; conserved during red blood cell destruction and made available for reuse	Hemoglobin synthesis
Folic acid	Absorbed from small intestine	DNA synthesis

concentrations in the blood get low. Vitamin B_{12} and folic acid are two dietary factors that affect RBC production. These vitamins are necessary for mitosis, so any actively dividing cell needs them. Iron is also necessary to make hemoglobin. Too few RBCs, too little iron, too little hemoglobin, or insufficient vitamin B_{12} are some causes of anemia. Table 10-1 illustrates some specific dietary factors that affect the production of red blood cells. We will discuss the different types of anemia later in the chapter.

> **U check**
> Why is blood considered a connective tissue?

Plasma

LO 10.2

learning outcome

Describe the components of plasma and their functions.

albumin Smallest but most abundant of the plasma proteins.

serum Blood plasma minus its clotting proteins.

Plasma is the liquid portion of blood. It is made up of approximately 90 percent water and contains various proteins, nutrients, gases, electrolytes, and waste products. The three major types of proteins found in plasma are albumins, globulins, and fibrinogen. Each protein has a unique function.

Albumins are the smallest of the plasma proteins, but are the most abundant, accounting for 60 percent of plasma proteins by weight. Like most proteins in the body, albumins are produced by the liver. They are responsible for the osmotic pressure in blood vessels and, along with hydrostatic pressure (to be discussed later in the chapter), control movement of water between blood vessels and the tissues. Through this mechanism, albumins participate in blood pressure regulation. Albumins also transport hormones, some drugs, free fatty acids, and bilirubin.

Globulins are proteins produced by the liver as well as by specialized blood cells called plasma cells. Some types of globulins are involved in lipid transport. Those produced by plasma cells are antibodies and work as part of the immune system.

Fibrinogen is a protein important in blood clotting. The term **serum** is used for the fluid that is left when all clotting factors are removed from plasma. Table 10-2 illustrates the various proteins and their percentages found in plasma.

Nutrients in plasma include amino acids, glucose, nucleotides, and lipids that have been absorbed from the digestive tract. Because lipids are not water soluble and because plasma is mostly water, lipids must combine with

TABLE 10-2 Plasma Proteins

Protein	Percentage of Total	Origin	Function
Albumin	60%	Liver	Helps maintain colloid osmotic pressure
Globulin	36		
Alpha globulins		Liver	Transport lipids and fat-soluble vitamins
Beta globulins		Liver	Transport lipids and fat-soluble vitamins
Gamma globulins		Lymphatic tissues	Antibodies (immunoglobulins)
Fibrinogen	4	Liver	Plays a key role in blood coagulation

molecules called *lipoproteins* to be transported by plasma. The different types of lipoproteins are *chylomicrons, very low-density lipoproteins* (VLDL), *low-density lipoproteins* (LDL), and *high-density lipoproteins* (HDL). Low-density lipoproteins are sometimes called bad cholesterol and high-density lipoproteins are called good cholesterol. We will look at these more closely later in the chapter when we talk about arteriosclerosis.

focus on Clinical Applications: Understanding Cholesterol Numbers

It is recommended that cholesterol be measured every five years in men from age 20 to 35 and women from age 20 to 45. For males over 35 or females over 45 these numbers should be checked more frequently. The blood tests to measure cholesterol are known as a lipoprotein profile. These tests measure LDL, HDL, and triglycerides. Each of these provides vital information about the heart and blood vessel health. For less risk of heart disease, LDL (low-density) should be kept low and HDL (high-density) should be kept high.

Lipoprotein	Function	Normal Value
LDL (low-density lipoprotein) "bad" cholesterol	Builds up in the arteries and increases the chances of getting heart disease	Less than 70 is ideal. Less than 100 is optimal.
HDL (high-density lipoprotein) "good" cholesterol	Protects against heart disease by taking LDL out of the body and keeping it from building up	Above 60 is optimal. Less than 40 in men or 50 in women increases heart disease risk.
Triglycerides	The chemical form of fat in the body—linked to heart disease	Less than 150 is normal.
Total cholesterol	General indicator of risk for heart disease	Less than 200 is desirable.

Gases dissolved in plasma include oxygen, carbon dioxide, and nitrogen. Electrolytes dissolved in the plasma include sodium, potassium, calcium, magnesium, chloride, bicarbonate, phosphate, and sulfate. Refer to Chapter 6, Concepts of Fluid, Electrolyte, and Acid–Base Balance, for a more thorough discussion on electrolytes. Nonprotein nitrogenous substances include amino acids, urea, and uric acid. Urea and uric acid are waste products produced by cells.

> **U check**
> What is the most abundant protein in the plasma?

10.3 learning outcome

Relate the structure and function of erythrocytes, leukocytes, and platelets to anemia, leukemia, and thrombocytopenia, respectively.

hemoglobin An iron-containing pigment in red blood cells responsible for transporting most of the oxygen and some of the carbon dioxide in blood.

Formed Elements

The formed elements in the blood consist of red and white blood cells and platelets.

Erythrocytes

Erythrocytes, or red blood cells (RBCs), are small cells (about 7.5 micrometers in diameter) that have a biconcave shape (similar to a Life Savers® candy with a depression instead of a hole; see Figure 10-2). Because of their small size, they can easily travel through blood capillaries, the smallest of blood vessels. Mature RBCs do not have a nucleus and consequently cannot reproduce themselves. They contain a pigment called **hemoglobin** whose function is to bind to oxygen and transport it to the tissues where it is needed. Most of the oxygen in the blood is transported attached to hemoglobin. Carbon dioxide also binds to hemoglobin for transport out of the tissues to the lungs for exhalation. Much of the carbon dioxide is transported unattached to hemoglobin. Hemoglobin that carries oxygen is called *oxyhemoglobin* and is bright red in color; hemoglobin that is not carrying oxygen is called *deoxyhemoglobin* and has a darker red color. *Carboxyhemoglobin* refers to hemoglobin that is carrying carbon dioxide. Both oxygen and carbon dioxide levels in the blood can be measured.

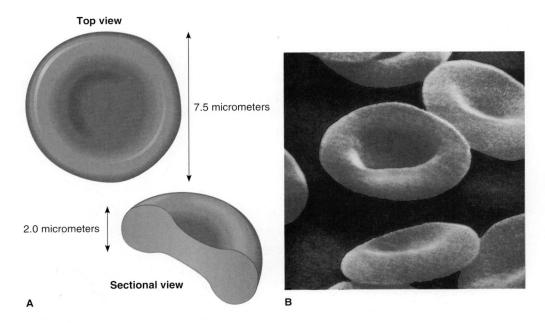

FIGURE 10-2
(A) Red blood cells with characteristic biconcavity; (B) micrograph of RBCs.

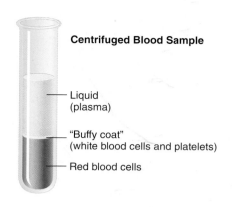
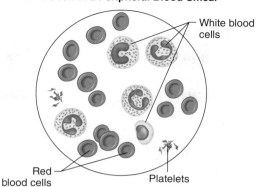

FIGURE 10-3
Centrifuged blood sample and peripheral blood smear showing blood components.

An RBC count is the number of erythrocytes in 1 cubic millimeter (approximately 20 drops of blood) of blood. A normal RBC count in an adult male is about 5.5 million erythrocytes per cubic millimeter of blood and in an adult female about 4.5 million erythrocytes per cubic millimeter of blood. When a sample of blood is centrifuged (spun) in a test tube, the heavier red blood cells settle to the bottom and can thus be measured. The percentage of red blood cells in relationship to the total blood volume is called the **hematocrit,** or packed cell volume (Figure 10-3). A normal hematocrit is about 37 to 43 percent in females and 43 to 49 percent in males. On top of the RBCs is the plasma, the lighter component, which composes approximately 55 percent of the total blood volume. Between the RBCs and the plasma is a thin layer that is whitish in color and is referred to as the "*buffy coat.*" The buffy coat is made up of white blood cells and platelets that make up roughly 1 percent of the total blood volume.

As mentioned previously, the average life span of an erythrocyte is approximately 120 days and old (senescent) RBCs are filtered out by macrophages in the liver and spleen. One of the substances released from destroyed RBCs is a pigment called *biliverdin* that is converted to bilirubin in the liver. Bilirubin is used to make *bile,* a substance that emulsifies or breaks apart fat molecules. Occasionally, something may cause an interruption or stoppage of this process and bilirubin accumulates in the blood, causing a condition known as *jaundice.* Jaundice causes a person's skin and the sclera (white) of their eyes to turn yellow.

hematocrit The percentage of blood made up of red blood cells.

U check
What makes up the buffy coat in a centrifuged blood sample?

PATHOPHYSIOLOGY

Anemias

Anemia is a symptom of an underlying disease process. Anemia occurs when the blood has less than its normal oxygen-carrying capacity. This can occur when the red blood cell count, or hemoglobin concentration, is less than normal. A normal hemoglobin level in men is 13.8 to 17.2 grams per 100 milliliters and in women is 12.1 to 15.1 grams per 100 milliliters.

(continued)

Anemias (concluded)

Anemia is the most common blood disorder in the United States and more women than men are affected by it. Signs and symptoms of all forms of anemia are similar and include tiredness, weakness, pallor (paleness), tachycardia, numbness or coldness of the hands and feet, dizziness, headache, and jaundice.

Etiology: There are many causes of anemia. Some anemias and their individual causes are outlined here.

Iron deficiency anemia: Iron deficiency is the most common cause of anemia. Iron is needed to make hemoglobin, so people with low iron levels often have correspondingly low hemoglobin levels. Pregnant women and women with heavy menstrual cycles are most susceptible to this type of anemia. Others with chronic bleeding conditions such as ulcers and colon cancer are also at risk of developing this type of anemia.

Aplastic anemia: When the bone marrow is destroyed, the result is *aplastic* anemia. Chemotherapy and radiation therapy, as well as some cancers and toxins, can all destroy bone marrow. Because the bone marrow is destroyed, the number of white blood cells produced by the body is also affected.

Hemolytic anemia: In hemolytic anemia red blood cells are destroyed faster than they can be produced. *Sickle cell disease* is an example of a hemolytic anemia. Sickle cell, as it is commonly known, is found most often in African Americans (1 in 500) and in Hispanics (1 in 1,400). Figure 10-4 shows sickling of a red blood cell. Hemolytic anemia is a genetic disorder that is inherited as an autosomal recessive condition. (See Chapter 18 for more information on genetics.) This means that a defective allele (a gene variation) must be inherited from each parent for an individual to have the disease. If only one defective allele is inherited, the individual has *sickle cell trait* and not sickle cell disease. Symptoms are milder with the trait than the disease. Under certain conditions, RBCs that are normal in shape will sickle (become crescent shaped) and this will cause them to get stuck in capillaries. The individual with sickle cell disease may have episodes known as crises that include joint and chest pain, numbness in the hands and feet, jaundice, frequent infections, sores on the skin, delayed growth, stroke, seizures, and breathing difficulties. Retina damage, which causes vision problems, and spleen, liver, or kidney and lung damage may also be seen.

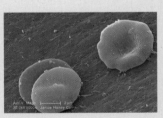

FIGURE 10-4 A sickled red blood cell in sickle cell anemia, a common genetic disorder in African Americans.

Although there is no cure for sickle cell disease, it can be treated or managed with antibiotics, blood transfusions, pain medications, bone marrow transplants, supplemental oxygen, and medications to promote the development of normal hemoglobin.

Pernicious (megaloblastic) anemia: The word *pernicious* means "deadly." Since this anemia is easily treated, today it is less often referred to as pernicious and is more often called megaloblastic anemia. *Megaloblastic* refers to the large red blood cells (*megalo* = large; *blast* = immature cell) that are seen in this condition. Vitamin B_{12} and folic acid are required for the synthesis of erythrocytes, and a deficiency of either causes megaloblastic anemia. *Intrinsic factor (IF)* is required for absorption of vitamin B_{12} by the body. A deficiency of intrinsic factor, a chemical secreted by the parietal cells of the stomach, is responsible for the lack of absorption of vitamin B_{12} by the gastrointestinal (GI) tract.

Chronic illness: Chronic diseases such as AIDS, cancer, rheumatoid arthritis, leukemia, and kidney failure can also be underlying causes of anemia. Table 10-3 lists some types of anemia, their cause, and the defect that is seen.

Treatment: Treatment of the various types of anemia depends on the cause. Treatment of megaloblastic anemia is intramuscular injections of vitamin B_{12}. Treatment must also address factors such as gastrointestinal disease or alcoholism if they are the underlying cause of the anemia.

> **REMEMBER JIM,** our stockbroker friend? Why is he suffering from pernicious (megaloblastic) anemia?

Thalassemia (Inherited Anemia)

Thalassemia is an inherited form of anemia with a defective hemoglobin molecule causing microcytic (small), hypochromic (pale), short-lived RBCs. It is most common among people of Mediterranean descent.

Etiology: Thalassemia is hereditary. Patients of Mediterranean descent are most likely to carry the defective autosomal recessive gene that causes this disease. Like sickle cell, thalassemia comes in two forms: *thalassemia major,* which is the severe form of the disease, and *thalassemia minor,* which causes only minimal symptoms, if any—the latter form is considered to be the "carrier" form of the disease. Also like sickle cell, both parents must contribute the defective allele for the child to be diagnosed with thalassemia major. If only one defective allele is inherited, the result is carrier status (thalassemia minor).

(continued)

Thalassemia (Inherited Anemia) (concluded)

Signs and Symptoms: Thalassemia major is evident in infancy with anemia, fever, failure to thrive, and splenomegaly (enlarged spleen). It is confirmed by characteristic changes in RBCs noted on microscopic examination. As the child matures, splenomegaly may interfere with breathing. In addition, skin becomes freckled or bronzed from the iron deposits created by the rapidly destroyed RBCs. Headache, nausea, and anorexia (loss of appetite) are also common signs and symptoms of this disease.

Treatment: There is no cure for thalassemia. Frequent transfusions are necessary to treat the anemia caused by the destruction of the defective RBCs. A splenectomy (spleen removal) may be recommended to slow the destruction of erythrocytes. Symptoms are otherwise treated as needed, including pain medication for the acute episodes or crises, which are similar to those of sickle cell patients. Patients and their families may require counseling to assist with dealing with the day-to-day difficulties of this disease.

> **U check**
> What is the etiology of aplastic anemia?

Leukocytes

Leukocytes (white blood cells, or WBCs) are categorized into two groups based on whether or not they have granules in their cytoplasm. WBCs known as *granulocytes* have granules in their cytoplasm and include neutrophils, eosinophils, and basophils. WBCs classified as *agranulocytes* do not have granules in their cytoplasm and include monocytes and lymphocytes. Each of the WBC types has a unique appearance and varies in size as well. Unlike RBCs, all white blood cells contain a nucleus.

TABLE 10-3 Types of Anemia

Type	Cause	Defect
Aplastic anemia	Toxic chemicals, radiation	Damaged bone marrow
Hemolytic anemia	Toxic chemicals	Red blood cells destroyed
Iron deficiency anemia	Dietary lack of iron	Hemoglobin deficient
Pernicious anemia	Inability to absorb vitamin B_{12}	Excess of immature cells
Thalassemia	Defective gene	Hemoglobin deficient; red blood cells short-lived

Neutrophils (Figure 10-5) account for about 60 percent of the WBC count; they have a nucleus with three to five lobes and have phagocytic (cell-eating or destructive) qualities. They are important in the destruction of bacteria and their numbers increase in the early stage of acute inflammation. **Eosinophils** (Figure 10-6) account for about 3 percent of all WBCs. They are involved in fighting parasitic infections and are also increased in number in people with acute or active allergies. **Basophils** (Figure 10-7) account for less than 1 percent of all WBCs. They release substances such as histamine, which promotes inflammation, and heparin, which is an anticoagulant. Of the two types of agranulocytes, **monocytes** (Figure 10-8) account for about 8 percent of all WBCs. They become active and quite large (macrophages) when they leave the blood vessel and enter tissues. **Lymphocytes** (Figure 10-9) account for about 30 percent of all WBCs and participate in immunity. Lymphocytes are white blood cells that play an important role in the lymphatic and immune system. They will be discussed in further detail in Chapter 12, The Lymphatic and Immune Systems. Table 10-4 describes the cellular components of blood with their description, numbers, and functions. They will be discussed in further detail in Chapter 12, The Lymphatic and Immune Systems.

neutrophil A type of white blood cell with phagocytic capabilities; characterized by a granular cytoplasm.

eosinophil A type of white blood cell that has a granular cytoplasm and is involved in allergic reactions and parasitic infections.

basophil A type of white blood cell with a granular cytoplasm that is involved in hypersensitivity reactions.

monocyte The largest type of white blood cell with agranular cytoplasm.

lymphocyte A type of white blood cell involved in cell-mediated and antibody-mediated immune responses.

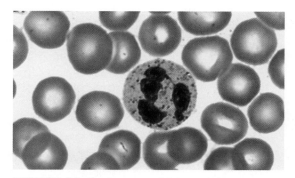

FIGURE 10-5 A neutrophil has a lobed nucleus with three to five components. Neutrophils have lytic enzymes that give them phagocytic capabilities. They are involved in fighting bacterial infections.

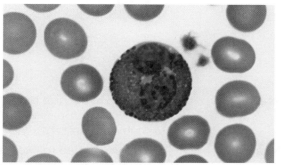

FIGURE 10-6 An eosinophil has granules in the cytoplasm that stain red with acidic dyes. This type of white blood cell can kill some parasites and helps control inflammation in allergic reactions.

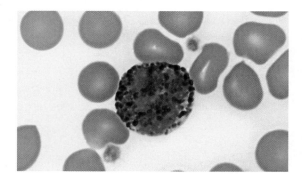

FIGURE 10-7 A basophil has granules that stain deep blue with basic dyes. This type of white blood cell produces heparin to prevent blood clotting and histamine, which increases circulation to injured tissues. Basophils also participate in certain allergic reactions.

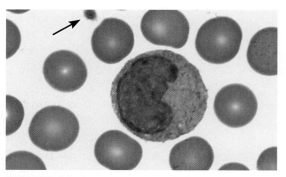

FIGURE 10-8 A monocyte is the largest of blood cells. It leaves the blood vessel to transform into a macrophage, a type of phagocyte.

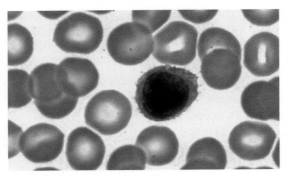

FIGURE 10-9 The lymphocyte is the smallest of the white blood cells and has a large round nucleus. Lymphocytes are active participants in the immune response and are involved in fighting viral infections.

A WBC count is the number of WBCs found in 1 mm^3 of blood. This count is normally between 5,000 and 10,000 cells. A WBC count that is significantly elevated is termed *leukocytosis*. Leukocytosis can be caused by infection, by cancer, and by some drugs. It can also occur after eating a large meal or experiencing stress. A WBC count below normal (less than 3,500 WBCs) is called *leukopenia*. Leukopenia can be caused by viral infections or congenital bone marrow disorders, cancer, immune disorders, and some drugs.

A differential WBC count lists the percentages of the different types of leukocytes in a sample of blood. This is a useful test because the numbers of different WBCs change in certain diseases. For example, the number

TABLE 10-4 Cellular Components of Blood

Component	Description	Number Present	Function
Red blood cell (erythrocyte)	Biconcave disc without a nucleus, about one-third hemoglobin.	4,200,000 to 6,200,000 per microliter	Transports oxygen and carbon dioxide
White blood cell (leukocyte)		4,500 to 10,000 per microliter	Destroys pathogenic microorganisms and parasites and removes worn cells
Granulocytes	About twice the size of red blood cells; cytoplasmic granules are present.		
Neutrophil	Nucleus with two to five lobes; cytoplasmic granules stain light purple in combined acid and base stains.	54–62% of white blood cells	Phagocytizes small particles
Eosinophil	Nucleus is bilobed; cytoplasmic granules stain red in acid stain.	1–3% of white blood cells	Kills parasites and moderates allergic reactions
Basophil	Nucleus is lobed; cytoplasmic granules stain blue in basic stain.	Less than 1% of white blood cells	Releases heparin and histamine
Agranulocytes	Cytoplasmic granules are absent.		
Monocyte	Two to three times larger than a red blood cell; nuclear shape varies from spherical to lobed.	3–9% of white blood cells	Phagocytizes large particles
Lymphocyte	Only slightly larger than a red blood cell; its nucleus nearly fills the cell.	25–33% of white blood cells	Provides immunity
Platelet (thrombocyte)	Cytoplasmic fragment.	130,000 to 360,000 per microliter	Helps control blood loss from broken vessels

of neutrophils increases at the beginning of a bacterial infection or in acute inflammation. Monocytes increase in number about two weeks after a bacterial infection. Eosinophil numbers increase during parasitic infections as well as with allergic reactions. Lymphocytes increase in number with viral infections and chronic illness as well as in leukemia. In AIDS, lymphocyte numbers (particularly T lymphocytes) eventually fall, leaving the patient in an immunocompromised state. Basophils, like eosinophils, increase during allergic reactions. To perform their function, WBCs must typically leave the blood vessels and enter the interstitium (the spaces between the cells). The process of WBCs crossing the blood vessel wall into the interstitium is called *diapedesis* which means "walking across." Figure 10-10 shows the process of diapedesis.

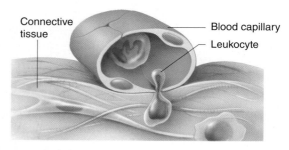

FIGURE 10-10 In diapedesis, leukocytes move across the capillary wall and enter the tissue space outside the blood vessel.

> **U check**
> Which leukocytes are classified as agranulocytes?

PATHOPHYSIOLOGY

Leukemia

Leukemia is a neoplastic condition in which the bone marrow produces a large number of WBCs that are not normal. Figure 10-11 shows a typical increase in the number of leukocytes in a patient who has leukemia. These abnormal cells prevent normal WBCs from carrying out their defensive functions. This disorder is sometimes referred to as cancer of the WBCs. Several different kinds of leukemia exist: acute lymphocytic leukemia, acute myelogenous leukemia, chronic lymphocytic leukemia, and chronic myelogenous leukemia. Lymphoblastic leukemias are caused by a proliferation of lymphocytes, and myelogneous leukemias are caused by a proliferation of cells that give rise to granulocytes.

Etiology: Causes of leukemia include mutations (changes) in WBCs, chemotherapy for the treatment of other cancers, and genetic factors (for example, the inheritance of abnormal genes). Exposure to environmental and chemical agents are sometimes the cause of the mutations in the WBCs.

Signs and Symptoms: The signs and symptoms include fatigue, dyspnea on exertion (DOE), hepatomegaly (an enlarged liver) or splenomegaly (enlarged spleen) or both (hepatosplenomegaly), swollen (nontender) lymph nodes, abnormal bruising, wounds that heal slowly, frequent infections, nosebleeds, bleeding gums, chronic fever, unexplained weight loss, and excessive sweating.

(continued)

Leukemia (concluded)

A patient history, physical examination, loss, and excessive sweating. A patient history, physical examination, CBC, blood smear examination, and imaging are some tools used to diagnose leukemia.

Treatment: Options include chemotherapy, radiation therapy, and medications to strengthen the immune system; antibodies to destroy mutated WBCs; bone marrow transplant; and stem cell transplant.

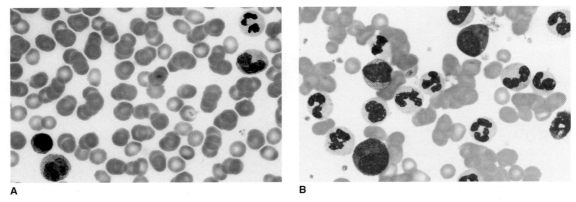

FIGURE 10-11 (A) Normal blood cells. (B) Blood cells from an individual with granulocytic leukemia—note the increased number of leukocytes.

Thrombocytes

Thrombocytes, also known as platelets, are cell fragments that are essential to blood clotting. Platelets are derived from very large cells located in the bone marrow called *megakaryocytes*. These cells fragment (break apart) producing hundreds or even thousands of platelets. Platelets do not have a nucleus. In normal, healthy individuals there are between 130,000 and 360,000 platelets per cubic millimeter of blood.

Thrombocytopenia is a condition in which there are too few platelets, causing abnormal bleeding. This can be caused by leukemia, medications, or even idiopathic (unknown) reasons. The bleeding may be mild or life threatening. The condition opposite of thrombocytopenia is *thrombocythemia*, in which there is an increase in the platelet count. The cause may be idiopathic or secondary to cancer, hemorrhage, and infection, to name a few. Although in most cases the situation is benign, thrombocythemia can cause serious thrombus (clot) or embolus formation, both of which can be limb or life threatening. In those cases, treatment may include platelet lowering medications as well as surgery

U check
What is the difference between thrombocytopenia and thrombocythemia?

focus on Genetics: Hemophilia

Hemophilia is the name given to a group of inherited X-linked blood disorders with symptoms ranging from severe to mild. X-linked disorders are those that are attached to the X chromosome received from the mother by the child. Because X-linked diseases are "recessive" in nature, boys are primarily affected by this disorder since they have only one X chromosome. Women who are carriers of the gene for hemophilia may be identified with a blood test and prenatal testing can be done to diagnose the fetus *in utero*. Hemophilia is caused by the lack of an intrinsic clotting factor; commonly the missing factor is Factor VIII. There is no cure for hemophilia. Treatment revolves around injections/infusions of the missing factor and symptomatic treatment for the accompanying joint aches and pains from repeated bleeds related to the inability of the patient's blood to clot.

Hemodynamics

Hemodynamics (blood dynamics) is the study of blood flow. Although we briefly discuss cardiac output and blood pressure here, they will be covered in more detail in Chapter 11, The Cardiovascular System. This chapter focuses on hypertension, hemostasis, and shock.

Cardiac Output and Blood Pressure

Blood pressure is defined as the force that blood exerts on the inner walls of blood vessels. Blood pressure is highest in arteries and lowest in veins. In the clinical setting, blood pressure refers to the pressure within arteries.

Arterial blood pressure varies with the contraction and relaxation of the ventricles of the heart. When the ventricles contract, blood pressure is greatest in the arteries. This pressure is called the *systolic pressure,* or systole. When the ventricles relax, blood pressure in arteries is at its lowest point. This pressure is called the *diastolic pressure,* or diastole. Blood pressure is usually reported as the systolic number over the diastolic number in the same format as a fraction. For example, in the blood pressure reading 120/80, 120 denotes the systolic pressure and 80 refers to the diastolic pressure.

Many factors affect blood pressure, the most common being cardiac output, blood volume, vasoconstriction, and blood viscosity (thickness). Cardiac output is the total amount of blood pumped out of the heart in one minute. As cardiac output increases and decreases, blood pressure increases and decreases respectively. When a person loses a large amount of blood, his or her blood pressure significantly decreases. If blood pressure falls too low, vasoconstriction, which is the constriction of blood vessel walls, helps raise blood pressure. In contrast, if blood pressure is too high, vasodilation, which is the widening of blood vessels, decreases the blood pressure. Dehydration can also decrease blood pressure. The body monitors and regulates blood pressure in several ways to maintain homeostasis.

10.4 learning outcome

Describe blood pressure and how it is regulated, including a description of essential and secondary hypertension.

LO 10.4

Hypertension

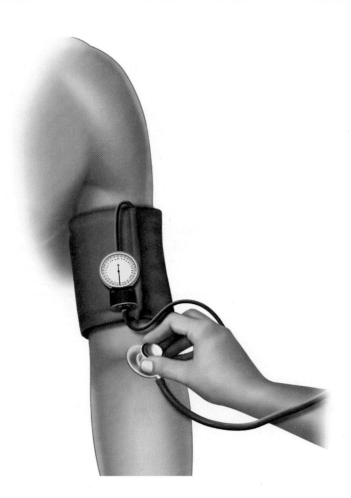

FIGURE 10-12 A sphygmomanometer is used to measure arterial blood pressure.

Baroreceptors are one way blood pressure is regulated. Baroreceptors located in the aorta and carotid arteries detect blood pressure changes and relay the information to the cardiac center in the medulla oblongata. The cardiac center then either increases or decreases the heart rate as needed. Increasing heart rate increases blood pressure and decreasing heart rate lowers blood pressure.

Blood pressure is measured using a sphygmomanometer (blood pressure cuff) and stethoscope. Most health care professionals need to learn how to take a blood pressure reading. Figure 10-12 shows the typical equipment used to measure the arterial blood pressure.

Some people have an artificially elevated blood pressure reading because of anxiety related to visiting the physician. This is referred to as "white coat syndrome." Putting the patient at ease may be effective in lowering this artificially elevated reading.

The pulsation of blood going through the arteries at certain locations can be felt. Figure 10-13 shows the various sites where a pulse is most easily detected. Two common places to feel a pulse

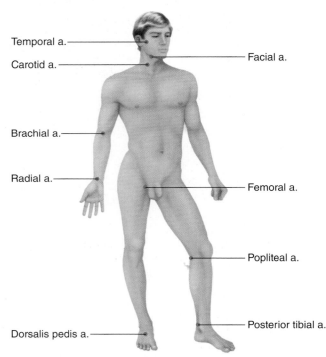

FIGURE 10-13 Sites where an arterial pulse is most easily detected (*a.* stands for artery).

242 UNIT 2 Concepts of Common Illness by System

> **focus on Clinical Applications:** Taking Blood Pressure
>
> One of the most important responsibilities of many health care professionals is that of taking the patient vital signs: temperature, pulse, respiration, and blood pressure. Since all of these are taken so commonly on patients, most people do not realize that it takes time and practice to become proficient in these skills.
>
> To take a blood pressure (BP) two instruments are needed: a sphygmomanometer (BP cuff) and a stethoscope. The BP cuff comes in a range of sizes for children through large adult. Select the appropriate size for the patient whose BP you are taking. It should be about two-thirds the width of the upper arm. Wrap the cuff around the upper arm, just above the antecubital space (bend of the elbow), making it snug enough that it remains in place by itself. The lower edge of the cuff should be about 1 inch above the antecubital fossa, where the stethoscope will be placed so that you can hear the blood flow within the brachial artery.
>
> If the BP cuff includes an aneroid dial, be sure that its placement on the cuff is such that you can read it easily. If a column on the wall is used, be sure it is at eye level.
>
> Close the valve of the pressure bulb of the cuff until it is "finger-tight." Palpate the brachial artery to find its location and place the stethoscope over it, using your fingers, but not your thumb which contains its own pulse. Quickly inflate the cuff to approximately 200 mg Hg. Slowly open the valve of the pressure bulb to deflate the BP cuff, listening carefully for the first repetitive sounds. Note the number of the manometer (dial) when you first hear the sounds—this is the systolic reading. Keep listening until you can no longer hear the sounds and note this number—this is the diastolic reading.
>
> Remove the cuff and replace it. Then recheck the BP a second time to verify the reading. If you have difficulty, ask a coworker to verify your results. Record the BP in the patient's medical record.

are the carotid artery in the neck and radial artery in the wrist. The pulse rate is counted in beats per minute. The pulse rate of a healthy individual is typically between 60 and 90 beats per minute.

> **U check**
> What is the difference between systolic and diastolic pressure?

Hypertension

Commonly known as high blood pressure, **hypertension** is defined as a consistent resting blood pressure measured at 140/90 mm Hg (millimeters of mercury) or higher. Prehypertension is a blood pressure reading between 120 and 139 systolic and 80 and 89 diastolic. Table 10-5 shows blood pressure classifications and their values.

Hypertension is commonly known as the "silent killer" because it increases a person's risk of heart attack, stroke, heart failure, and kidney failure while often presenting no symptoms in the patient. African Americans are twice as likely to have high blood pressure as Caucasians are.

hypertension High blood pressure.

TABLE 10-5 Hypertension

Blood Pressure Classification	Systolic Blood Pressure (mm Hg)	Diastolic Blood Pressure (mm Hg)
Normal	<120	and <80
Prehypertension	120–139	or 80–89
Stage 1 Hypertension	140–159	or 90–99
Stage 2 Hypertension	≥160	or ≥100
Hypertensive crisis	>180	or >110

Over 90 percent of hypertension has no identifiable cause and is referred to as *essential, idiopathic, or primary hypertension.* When a cause has been identified (less than 10 percent of hypertension diagnoses) it is referred to as *secondary hypertension.* Known causes and risk factors for hypertension include narrowing of the arteries, various medications such as oral contraceptives and cold medicines, kidney disease, endocrine disorders, pregnancy, drug use (especially cocaine and amphetamines), sleep apnea, obesity, smoking, a high-sodium diet, excessive alcohol consumption, stress, and diabetes mellitus.

There are usually no symptoms with hypertension. When symptoms do present, they include excessive sweating (diaphoresis), muscle cramps, fatigue, frequent urination, headaches, dizziness, and an irregular heart rate.

The first management tool of hypertension control should be to treat the underlying cause, if known. Other common treatments include a diet low in sodium and fat, weight loss, regular exercise, various medications to slow the heart rate and dilate blood vessels, diuretics to reduce blood volume, and lifestyle changes such as managing stress and smoking cessation. Patient compliance is the key to the successful management of hypertension, which usually cannot be cured, only controlled. Because hypertension itself often has no symptoms and medications have side effects, patients sometimes stop their antihypertensive regimes because "I felt better off the medication." Be sure patients understand that medication compliance is crucial to their treatment and long-term health. They should tell the physician when a medication is not working for them for whatever reason, because there are often many options available for treatment.

> **U check**
> What is the difference between essential and secondary hypertension?

Hemostasis

10.5 learning outcome
Explain the steps of hemostasis.

hemostasis Stoppage of blood flow by natural or artificial means.

Hemostasis refers to the control of bleeding (*hemo* = blood; *stasis* = control). This is important when blood vessels are damaged and bleeding begins. Three processes occur in hemostasis: (1) blood vessel spasm, (2) platelet plug formation, and (3) blood coagulation. If a blood vessel ruptures, the smooth muscle in the damaged vessel contracts and the blood vessel spasms. This spasm reduces the amount of blood lost through the vessel. Platelets begin to stick to the broken area and to each other to form a platelet plug. The platelet plug slows or controls the bleeding temporarily. A blood clot eventually replaces the platelet plug. The formation of a blood clot is called blood *coagulation*. Figure 10-14 shows a scanning electron microscope image of a blood clot showing the very small structures.

Wound healing

From the perspective of . . .

A PHLEBOTOMIST The phlebotomist typically collects blood through a venipuncture (puncture into a vein) or skin puncture for testing. How will knowing the principles of hemostasis help you more confidently perform your duties?

In the coagulation process, the plasma protein fibrinogen is converted to fibrin. Once fibrin forms, it sticks to the damaged area of the blood vessel, creating a mesh web that entraps blood cells and platelets. The resulting mass, the blood clot, stops bleeding until the vessel has repaired itself. Figure 10-15 demonstrates the events involved in hemostasis. The final step in the coagulation process is the dissolution of the clot by fibroblasts.

When a blood vessel is injured, it is normal for a blood clot to form. However, sometimes blood clots form with no known injury; this abnormal blood clot is called

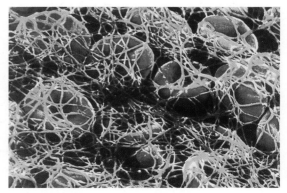

FIGURE 10-14 Scanning electron micrograph of a blood clot. Yellow fibrin threads are covering red blood cells.

FIGURE 10-15 Events of hematopoiesis.

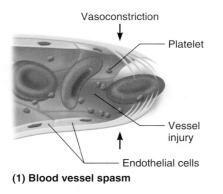

(1) Blood vessel spasm

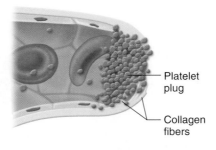

(2) Platelet plug formation

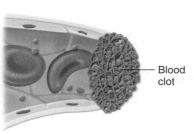

(3) Blood clotting

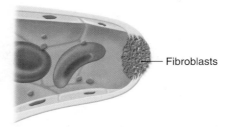

Fibroblasts attracted to the clot, dissolve clot

(4) Fibrinolysis

CHAPTER 10 Blood and Circulation

FIGURE 10-16 Thrombus formation within an artery.

thrombus A stationary clot formed in an unbroken blood vessel.

embolus A blood clot, air bubble, fat, or other foreign material transported by the blood.

a **thrombus** (Figure 10-16). The danger of a thrombus is that if it occludes an artery in the heart or brain it can cause myocardial infarction (MI, or heart attack) or cerebral vascular accident (stroke), respectively. Sometimes a portion of the thrombus can break off and travel through the bloodstream. The traveling portion of the thrombus is called an **embolus.** An embolus is dangerous because it may eventually block a small artery. An embolus that originates in the vein of a leg (known as deep venous thrombosis, or DVT) may travel to the right atrium of the heart (Figure 10-17).

The embolus may then travel through the right ventricle to the lungs. Here the embolus can get stuck in a small artery and cause a pulmonary embolism. If the embolus lodges in a coronary artery, an MI may develop. If the embolus blocks a cerebral artery, the result may be a cerebrovascular accident (CVA). All of these are serious and possibly fatal conditions if not treated or if treatment is not initiated quickly enough.

Shock

Hypoperfusion (too little blood flow) to tissues resulting in decreased oxygen and nutrient delivery to them can result in shock. There are four types of shock: (1) hypovolemic (low volume), (2) cardiogenic (caused by the heart), (3) vascular (pertaining to vessels), and (4) obstructive. *Hypovolemic shock* typically results from trauma that causes severe bleeding and loss of blood volume. *Cardiogenic shock* occurs when the heart is failing and is not pumping an adequate amount of blood to the body. *Vascular shock* is due to vasodilation, or dilation of blood vessels. *Septic shock*, which is caused

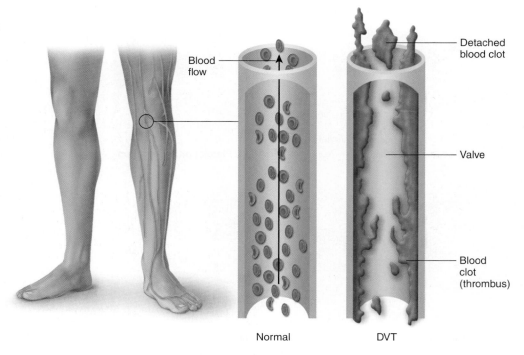

FIGURE 10-17 Deep venous thrombosis (DVT); if a thrombus breaks loose, it becomes an embolus.

by certain bacterial infections, and *neurogenic shock,* which is caused by events such as head trauma, are examples of vascular shock. Septic shock is responsible for tens of thousands of deaths each year and is a major concern in severely burned patients. *Obstructive shock* occurs when there is blockage of an artery causing a decrease in blood flow or perfusion of tissues.

> **U check**
> What are the different types of shock?

Blood Typing

learning outcome

Explain the purpose of blood typing, and describe the ABO blood typing system and the Rh blood typing system.

Blood typing is a way of categorizing blood based on the presence or absence of *antigens* (agglutinogens) on the surface of red blood cells. Blood typing is done to make certain that a person who needs a transfusion will receive blood that is compatible with his or her own. This is to prevent serious and potentially fatal events from happening. Over two dozen different blood grouping systems exist. We will discuss two of them—the ABO and Rh blood groups (Figure 10-18).

ABO Blood Group

The ABO blood group consists of four different blood types: A, B, AB, and O. The blood groups are distinguished from each other in part by their antigens and antibodies. *Agglutination* is the clumping of RBCs that can occur if a patient is given the wrong blood type during a blood transfusion. Clumping can lead to severe anemia and in some cases even death. Agglutination occurs because proteins called *antigens* on the surface of RBCs bind to *antibodies* in plasma. Antibodies in the ABO system are preformed, so

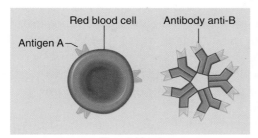
Type A blood

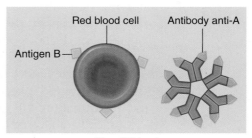
Type B blood

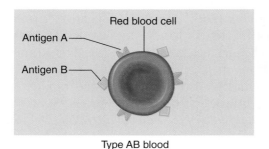
Type AB blood

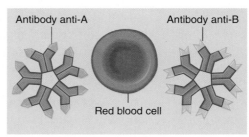
Type O blood

FIGURE 10-18 A, B, AB, and O blood types.

to prevent agglutination, antigens should not be mixed with antibodies that will bind to them. Fortunately, most antibodies do not bind to antigens on blood cells; only very specific ones bind to them.

Let's look at each blood type and its antigen and antibody to better understand this process.

Blood type A: People with type A blood have antigen A on the surface of their RBCs. They have anti-B antibody in their plasma. A person with type A blood should not be given type B blood (and vice versa).

Blood type B: People with type B blood have antigen B on the surface of their RBCs. They have anti-A antibody in their plasma.

Blood type AB: People with type AB blood have both antigen A and antigen B on the surface of their RBCs. They have neither anti-A nor anti-B antibodies in their plasma. People with type AB blood are called *universal recipients* because most of them can receive all ABO blood types. They can receive these blood types because they lack anti-A antibody and anti-B antibody in their plasma, so there is no reaction with antigens A and B of the donor blood.

Blood type O: People with type O blood have neither antigen A nor antigen B on the surface of their RBCs. However, they do have *both* anti-A antibody and anti-B antibody in their plasma. People with type O blood are called *universal donors* because their blood can be given to most people regardless of the recipient's blood type. Type O blood will not agglutinate because it does not have the antigens to bind to antibody A or antibody B.

Table 10-6 shows the different blood types, their antigen and antibody presentation, and the blood types they can receive. Table 10-7 shows the ABO blood type frequencies in the United States.

Rh Blood Group

The *Rh antigen* is a protein first discovered on erythrocytes of the Rhesus monkey, hence the abbreviation Rh. People who are Rh-positive have erythrocytes that contain the Rh antigen. People who are Rh-negative have erythrocytes that do not contain the Rh antigen. If a person who is Rh-negative is given Rh-positive blood, then the Rh-negative person's blood will make antibodies that bind to the Rh antigens. If the Rh-negative person is given Rh-positive blood a second time, the antibodies will bind to the donor cells and agglutination will occur.

Clinically, it is very important for a female to know her Rh type. If an Rh-negative female has a child with an Rh-positive male, there is a 50–50 chance that her fetus will be Rh-positive (Figure 10-19). If the blood of an Rh-positive fetus mixes with the mother's blood that is Rh-negative, the mother will develop antibodies against the fetus's RBCs. Because Rh antibodies are not preformed like the antibodies in the ABO system, the first Rh-positive fetus usually does not suffer from this incompatibility. However, if the mother conceives again and this child is also Rh-positive, the mother's antibodies may cross the placenta and attack and destroy the second fetus's RBCs. This condition is called *hemolytic disease of the newborn (HDN)* or *erythroblastosis fetalis* (Figure 10-20). To prevent this condition, the Rh-negative mother should be given an injection of anti-Rh antibodies (RhoGAM). An injection of RhoGam should be given after each birth, miscarriage, or abortion to prevent the production of these antibodies.

TABLE 10-6 ABO Blood Groups

Blood Type	Antigen Present	Antibody Present	Blood That Can Be Received
A	A	B	A and O
B	B	A	B and O
AB	A and B	None	A, B, AB, and O
O	None	A and B	O

TABLE 10-7 ABO Blood Type Frequencies (%) in the United States

Population	Type O	Type A	Type B	Type AB
Caucasian	45	40	11	4
African American	49	27	20	4
Native American	79	16	4	1
Hispanic	63	14	20	3
Chinese American	42	27	25	6
Japanese American	31	38	21	0
Korean American	32	28	30	10

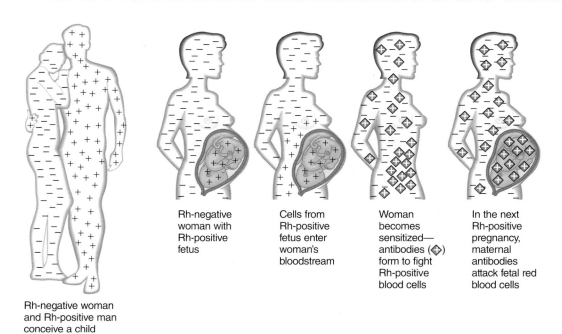

Rh-negative woman and Rh-positive man conceive a child

Rh-negative woman with Rh-positive fetus

Cells from Rh-positive fetus enter woman's bloodstream

Woman becomes sensitized—antibodies (◇) form to fight Rh-positive blood cells

In the next Rh-positive pregnancy, maternal antibodies attack fetal red blood cells

FIGURE 10-19 Development of antibodies in an Rh-negative woman in relation to her Rh-positive fetus.

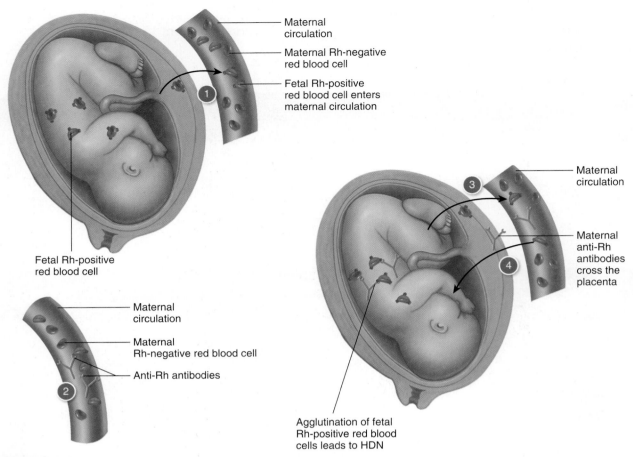

FIGURE 10-20 Infant with erythroblastosis fetalis.

> **U check**
> What is your blood type?

learning outcome

Describe types of blood vessels, their functions, and disorders.

Blood Vessels

Blood vessels are the transport channels through which the blood travels. These vessels include arteries, arterioles, veins, venules, and capillaries. Arteries carry blood away from the heart and the blood in them is oxygenated (an exception is the pulmonary arteries that carry deoxygenated blood to the lungs for oxygenation). Veins carry blood toward the heart and carry deoxygenated blood (the exception is the pulmonary veins that carry oxygenated blood from the lungs back to the heart). Capillaries are the smallest of blood vessels and they connect arterioles to venules.

Arteries and Arterioles

Arteries and arterioles deliver oxygen and nutrients to the tissues. They vary in size from the *aorta* (the largest artery) to the very small *arterioles*. Arteries are high-pressure vessels.

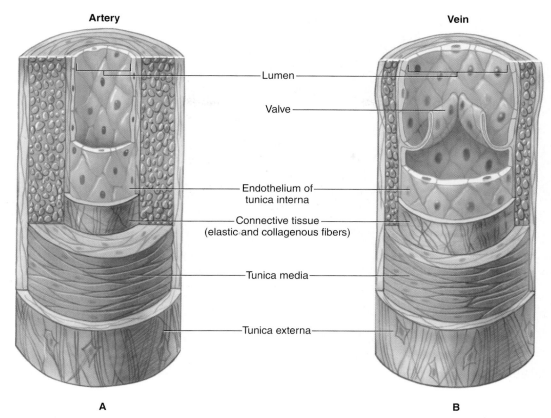

FIGURE 10-21 (A) The wall of an artery; (B) the wall of a vein. The tunics or layers of the arteries and veins differ in thickness.

Arteries have three layers, or tunics ("coats"). Listed in order from the most interior layer adjacent to the lumen to the outermost layer, they are the *tunica intima* (think intimate—within), *tunica media* (media—middle), and *tunica externa* (externa—external). The tunica intima is made up of a single layer of squamous epithelium called endothelium. The tunica media is made up of smooth muscle and is thicker in some arteries than others. The smooth muscle can cause *vasoconstriction* (causing an increase in blood pressure) or *vasodilation* (resulting in a decrease in blood pressure). The tunica externa is made up of connective tissue. Figure 10-21 shows a cross section of an artery and vein and the relative thickness of each of the layers.

Veins and Venules

Blood is under much lower pressure in the veins than in the arteries. Vessels in this low-pressure system range in size from the large superior and inferior vena cavae that return blood to the right atrium of the heart to the very small venules. Like arteries, veins also have a tunica intima, tunica media, and tunica externa. However, the tunica media of veins is thinner, with less smooth muscle to move the blood along its way. Veins depend on two mechanisms to return blood to the heart. One is the pumping action of skeletal muscle such as in the legs. The other is pressure changes in the thoracic cavity that "draw" blood toward the heart. Also, veins have valves that work to prevent blood from "falling down" due to gravity as it moves toward the heart (Figure 10-22). The lumen of veins is also larger than that of the corresponding arteries

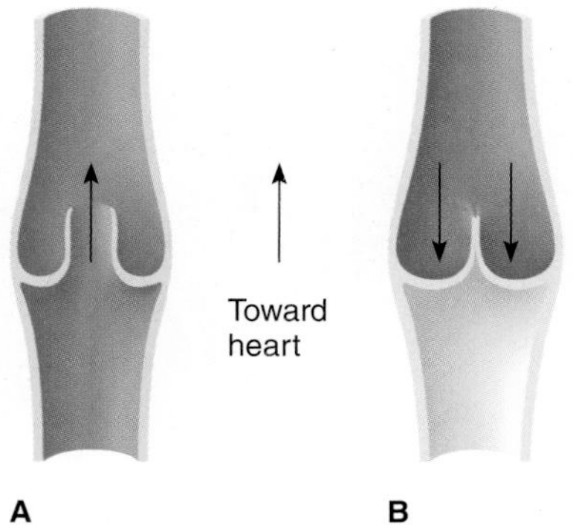

FIGURE 10-22 (A) Valve opens when blood is flowing toward the heart; (B) valve closes to prevent backflow of blood due to gravity.

CHAPTER 10 Blood and Circulation

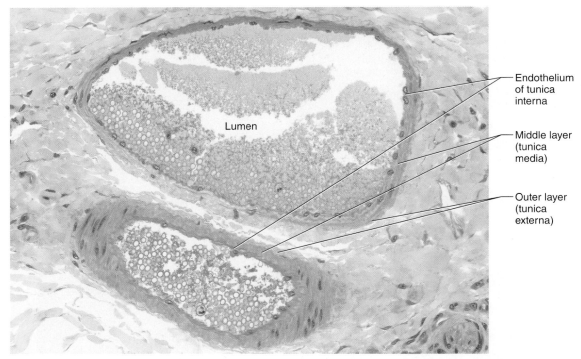

FIGURE 10-23 Note the structural differences in these cross sections of an arteriole (bottom) and venule (top).

(Figure 10-23). The sympathetic nervous system also influences the flow of blood through veins. The sympathetic nervous system causes vein walls to constrict, which forces blood through the veins, but this happens only if blood pressure becomes abnormally low in arteries.

> **U check**
> What are the three layers of an artery or vein?

Capillaries

Capillaries are the smallest of the blood vessels (Figure 10-24). They connect arterioles to venules and have very thin walls—only about one cell layer thick. The thin walls allow substances to easily pass into and out of capillaries. For this reason, they are referred to as *exchange vessels*. Oxygen and nutrients can pass out of a capillary into the tissue fluid of the body, and carbon dioxide and other waste products can pass out of the tissue fluid of the body into a capillary. Tissues that require a lot of oxygen, such as muscle and nervous tissue, have many capillaries. Capillaries have precapillary sphincters that control the amount of blood that flows into them. When the sphincter relaxes, more blood flows into the capillary. The substances that move through the capillary walls (oxygen, carbon dioxide, nutrients, water, and metabolic wastes) do so through diffusion, filtration, and osmosis.

When blood first enters a capillary, it has a high concentration of oxygen and nutrients. The tissue fluid of

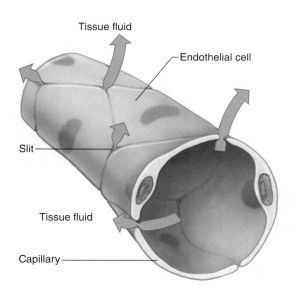

FIGURE 10-24 Substances are exchanged between blood and tissue fluid through openings separating endothelial cells.

UNIT 2 Concepts of Common Illness by System

the body surrounding the capillary usually have low concentrations of oxygen and nutrients but high concentrations of carbon dioxide and other waste products. Substances naturally diffuse from an area of high concentration to an area of low concentration. Therefore, oxygen and nutrients diffuse out of the capillary and into tissue fluid of the body. At the same time, carbon dioxide and waste products diffuse out of the tissue fluid of the body and into the capillary. Because blood is under pressure as it enters the capillary, water is forced through the capillary wall via filtration. This allows water to enter a tissue fluid of the body. By the time blood leaves a capillary, it has a high solid concentration and a low water concentration; water therefore moves back into the capillary through osmosis. Water always moves toward the greater concentration of solids, if possible.

PATHOPHYSIOLOGY

Aneurysm

An *aneurysm* results from a ballooned, weakened arterial wall. The most common locations of aneurysms are the aorta and the arteries in the brain, legs, intestines, and spleen. An aortic aneurysm is a bulge in the wall of the aorta. Most aortic aneurysms occur in the abdominal aorta (abdominal aortic aneurysm, also known as AAA) but some occur in the thoracic aorta. Figure 10-25 shows an abdominal aortic aneurysm. Most aortic aneurysms do not rupture; however, when they do, the resulting hemorrhage is a serious life-threatening emergency.

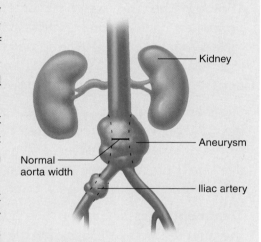

FIGURE 10-25 Abdominal aortic aneurysm (AAA).

Etiology: Most causes of aneurysms are unknown. One identified risk to developing an aneurysm is **atherosclerosis**, which is a hardening of the fatty plaque deposits within the arteries, sometimes associated with a diet high in fat and cholesterol. Smoking and obesity also increase the risk of atherosclerosis. Congenital conditions may cause an aneurysm—some individuals are born with weak aortic walls. A traumatic injury to the chest may also be a risk factor. The risk of developing an aneurysm can be reduced by not smoking, by losing excess weight, and by having a diet low in fats and cholesterol. Periodic screening is an option for patients with a family history of aortic aneurysms.

Signs and Symptoms: There are often no signs and symptoms of an aneurysm. When symptoms do exist, they depend on the blood vessel

(continued)

atherosclerosis An accumulation of cholesterol in the smooth muscle fibers of the tunica media of an artery.

Aneurysm (concluded)

where the aneurysm is located. For example, for abdominal aneurysms, symptoms may include a throbbing pulsation in the abdomen or pain in the back or the side of the abdomen. Severe headaches, dizziness, or loss of consciousness may be associated with a brain aneurysm.

Treatment: The primary treatment is surgery to repair the aneurysm. "Watchful waiting" may also be an option until the aneurysm reaches a size that makes it more dangerous because of increased risk of rupture. At that time, surgery would be undertaken to repair the aneurysm.

Thrombophlebitis

Thrombophlebitis is a condition in which a thrombus and inflammation develop in a vein. It most commonly occurs in the deeper veins of the legs. If a piece of the thrombus breaks loose, as stated earlier, it is called an embolus. The embolus might lodge and obstruct a blood vessel as it moves through the circulation.

Etiology: The causes and risk factors for thrombophlebitis include prolonged inactivity, oral contraceptives, postmenopausal hormone replacement therapy (HRT), some cancers, paralysis in the arms or legs, the presence of a venous catheter, family history of this condition, varicose veins, and trauma to veins.

Signs and Symptoms: The most common symptoms are tenderness and pain in the affected area; redness, swelling, and tenseness of the affected areas; fever; and a positive *Homan's sign*. A positive sign is present when there is pain in the calf or popliteal region when the foot is dorsiflexed with the knee flexed to 90 degrees.

Treatment: This disorder is treated by the application of heat to the affected area, elevation of the legs, anti-inflammatory drugs, anticoagulant medications, the wearing of support stockings, and the removal of varicose veins. Surgery to remove the clot may be needed in some cases.

> **U check**
> What condition can Homan's sign help diagnose?

Varicose Veins

Varicose veins (varices or varicosities) are tortuous or twisted, dilated veins that are usually seen in the legs. They affect women more often than men. When varicose veins occur in the rectum, they are called hemorrhoids. Varicose veins may also occur in the esophagus or the veins surrounding the umbilicus.

Etiology: Varicose veins may be caused by prolonged sitting or standing, damage to valves in the veins, a loss of elasticity in the

Varicose Veins (concluded)

veins, obesity, pregnancy, use of oral contraceptives, or hormone replacement therapy. Family history also seems to play a part in the development of varicose veins. In some cases, varicose veins may be prevented or at least minimized through exercise and elevation of the legs. Esophageal and umbilical varices are often associated with alcoholism.

Signs and Symptoms: Signs and symptoms include discomfort in the legs, discoloration around the ankles, clusters of veins, and enlarged, dark veins that are seen through the skin.

Treatment: The treatment of varicose veins includes the following:

- Sclerotherapy, which is a procedure that prevents blood from flowing through varicose veins.
- Laser surgery to prevent blood from flowing through affected veins.
- Vein stripping, which involves removing the affected veins.
- Insertion of a catheter in the affected veins to destroy them.
- Endoscopic vein surgery to close off affected veins.

focus on Clinical Applications: Phlebotomy

One of the most common procedures performed by healthcare professionals is phlebotomy, or venipuncture—the drawing of blood from a vein to analyze it. Most commonly, phlebotomy is performed using the veins found in the antecubital fossa (the "bend" at the elbow) using one of three veins: the cephalic vein, basilic vein, or medial cubital vein. If the patient has very difficult or heavily scarred veins, veins in the wrist and on the top of the hand may also be used. If only a small amount of blood is required, such as when testing a hematocrit, often a "finger stick" can be performed using capillary blood instead of venous blood. For infants, a heel stick is performed for this purpose. Because there is risk of exposure of body fluids, the phlebotomist or medical assistant protects himself or herself and the patient by using appropriate personal protective equipment (PPE) when drawing blood: gloves, lab coat, and often eye goggles.

Pulmonary and Systemic Circulation

There are basically two circuits in the circulation of blood: the pulmonary circuit and the systemic circuit.

learning outcome 10.8
Compare the pulmonary and systemic circuits.

Pulmonary Circuit

The *pulmonary circuit,* or pulmonary circulation, is referred to as the low-pressure circuit because the blood pressure here is normally only one-eighth

> ### focus on Wellness: Improving Circulation
>
> Like many functions and skills, circulation thrives on a "use it or lose it" concept. To keep the circulatory system functioning well, you have to work it. Exercise is the key. Walking daily, for example, not only gives the heart a workout but also gets the blood pumping and strengthens the circulatory system. Exercise has the added benefit of weight loss. A heavy patient's heart and circulatory systems have to work that much harder to get the blood out to the limbs and then return it to the heart. Lose some weight and the journey becomes that much easier for the cardiovascular and circulatory systems.
>
> Want a reward for your workout? Try a massage. In addition to bringing welcome relief to sore or strained muscles, massage also improves circulation by moving the blood supply throughout tissues.

of that in the systemic circuit. It is called the pulmonary circuit because of the route that blood takes from the heart to the lungs and back to the heart again. The purpose of this circuit is to take deoxygenated blood to the lungs and exchange carbon dioxide for oxygen. The blood returning to the heart is oxygenated blood. The route of the pulmonary circuit can be summarized as follows: right atrium through the tricuspid valve into the right ventricle through the pulmonary semilunar valve into the pulmonary trunk into the pulmonary arteries into the lungs into pulmonary veins and finally into the left atrium.

Systemic Circuit

The *systemic circuit,* or systemic circulation, is referred to as the high-pressure circuit because the pressure is much greater than that seen in the lungs. This is the route blood takes from the heart through the body and back to the heart. The function of this circuit is to deliver oxygen and nutrients to body cells. It also picks up carbon dioxide and waste products from body cells. The systemic circuit can be summarized as follows: left atrium through the bicuspid valve into the left ventricle through the aortic semilunar valve into the aorta to the systemic arteries to arterioles to capillaries to venules to veins to the superior and inferior venae cavae into the right atrium of the heart.

> **U check**
> What is the path a red blood cell will take through the heart?

Identify the major arteries and veins of the body.

Arterial and Venous Systems

Arteries carry blood away from the heart and, with the exception of the pulmonary arteries, have oxygenated blood. Veins carry blood toward the heart and, with the exception of the pulmonary veins, carry deoxygenated blood.

Arterial System

Most of the arteries in the body are paired, meaning that there is a left and a right artery of the same name. There are some major exceptions to this rule.

The aorta comes directly off the left ventricle and is the largest artery in the body. It is unpaired and is arched toward the left. The *brachiocephalic trunk* is a large artery that corresponds to the aorta, but actually comes off it and arches to the right side of the body. Many vessels that supply blood to various parts of the body come off these two arteries. Arteries supply blood to a region, tissue, or structure. When a larger artery divides into smaller arteries, the smaller arteries are called branches.

Parts of the Aorta: The aorta has three parts: the ascending aorta that extends from the left ventricle, the aortic arch, and the descending aorta (made up of the thoracic aorta and abdominal aorta). These are named based on their location or direction the vessel is taking. Table 10-8 identifies the portions of the aorta, their branches, and the regions or organs supplied by the branches. The right and left coronary arteries come off the ascending aorta to supply the heart. The brachiocephalic trunk, the left common carotid artery, and the left subclavian artery come off the aortic arch. The left common carotid artery supplies blood to the head, and neck while the subclavian supplies blood to shoulder, head, and neck. The right common carotid and right subclavian arteries come off the brachiocephalic trunk.

TABLE 10-8 The Aorta and Its Major Branches

Portion of Aorta	Branch	General Regions or Organs Supplied	Portion of Aorta	Branch	General Regions or Organs Supplied
Ascending aorta	Right and left coronary arteries	Heart	Abdominal aorta	Celiac artery	Organs of upper digestive tract
Arch of aorta	Brachiocephalic artery	Right upper limb, right side of head		Phrenic artery	Diaphragm
	Left common carotid artery	Left side of head		Superior mesenteric artery	Portions of small intestines
	Left subclavian artery	Left upper limb		Suprarenal artery	Adrenal gland
Descending aorta				Renal artery	Kidney
				Gonadal artery	Ovary or testis
Thoracic aorta	Bronchial artery	Bronchi		Inferior mesenteric artery	Lower portions of large intestine
	Pericardial artery	Pericardium			
	Esophageal artery	Esophagus		Lumbar artery	Posterior abdominal wall
	Mediastinal artery	Mediastinum		Middle sacral artery	Sacrum and coccyx
	Posterior intercostal artery	Thoracic wall		Common iliac artery	Lower abdominal wall, pelvic organs, and lower limb

Many small branches come off the thoracic aorta to supply organs and structures in the thoracic region. Below the diaphragm, the thoracic aorta becomes the abdominal aorta. It gives off several important vessels. The first of these is the celiac artery, which is one of three unpaired arteries coming off the aorta. It has branches that supply the digestive tract, liver, and spleen. Next are the paired (left and right) phrenic arteries that supply the diaphragm. The superior mesenteric artery is the second unpaired artery coming off the abdominal aorta. It has branches that supply much of the small and large intestines. The third unpaired artery coming off the abdominal aorta is the inferior mesenteric artery that supplies blood to the large intestine and rectum. Other arteries coming off the abdominal aorta are the suprarenal, renal, and gonadal arteries. The abdominal aorta terminates in the pelvic region where it divides into left and right common iliac arteries.

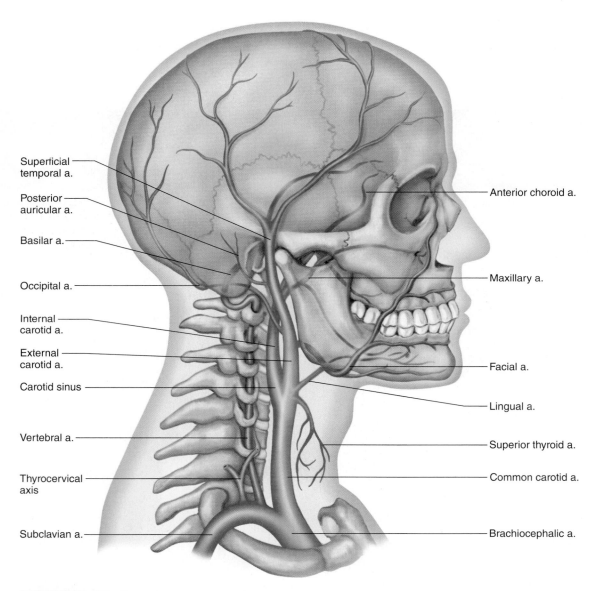

FIGURE 10-26 The major arteries of the head and neck with the clavicle removed (*a.* stands for artery).

Arteries to the Brain, Head, and Neck: Coming off the common carotid arteries are the internal and external carotid arteries. The internal carotid artery is the major blood supply to the brain and has several branches. The external carotid artery has several branches that supply the head, face, and neck region. The left and right vertebral arteries come off the subclavian arteries and supply blood to the vertebra as well as to the brain by combining to form the unpaired basilar artery that enters the skull (Figure 10-26).

Arteries to the Shoulder and Upper Extremity: In the axilla (armpit), the subclavian artery becomes the axillary artery. This artery provides blood for the humerus, shoulder, and chest structures. After leaving the axilla, it becomes the brachial artery supplying structures of the arm. It then divides into a radial artery that is located on the thumb side of the forearm with the body in anatomical position and the ulnar artery on the more medial aspect of the forearm. Figure 10-27 shows the major arteries to the shoulder and upper limb.

Arteries to the Pelvis and Lower Extremity: The left and right common iliac arteries begin where the abdominal aorta ends and then divide into left and right internal and external iliac arteries. The internal iliac arteries supply many organs in the pelvis. The external iliac arteries become the left and right femoral arteries that supply the thigh (Figure 10-28).

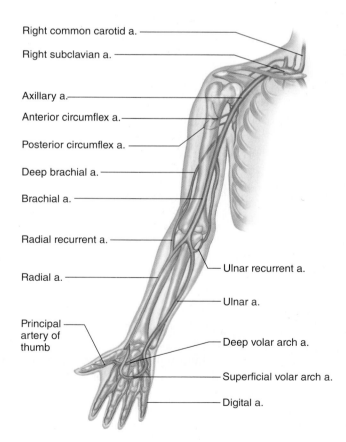

FIGURE 10-27

The major arteries to the shoulder and upper limb (a. stands for artery).

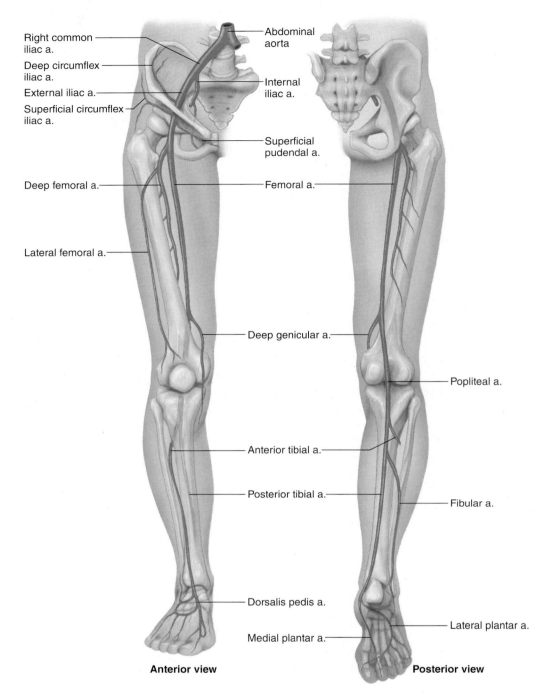

FIGURE 10-28 Major branches of the external iliac artery (a. stands for artery).

The femoral artery becomes the popliteal artery supplying the region behind the knee. It divides into anterior and posterior tibial arteries that supply the leg.

Keep in mind that there are additional named arteries that have not been discussed here and that many unnamed smaller branches come off at various points along the arterial system. Figure 10-29 shows the major vessels of the arterial system. Some major arteries of the body are summarized in Table 10-9.

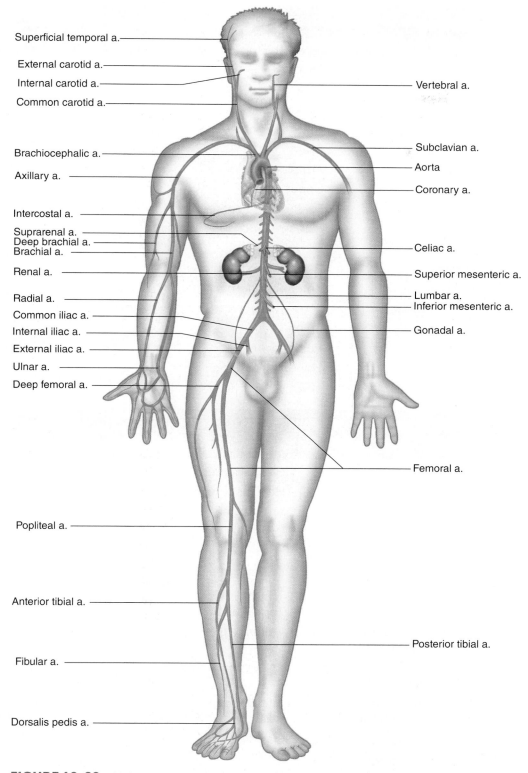

FIGURE 10-29 Major vessels of the arterial system (*a.* stands for artery).

U check
What is the large blood vessel coming off the aortic arch that goes to the right side of the body?

TABLE 10-9 Major Arteries of the Body

Artery	Anatomic Location or Organ Supplied
Lingual	Tongue
Facial	Face
Occipital	Back of scalp and neck
Maxillary	Teeth, jaw, and eyelids
Ophthalmic	Eye
Axillary	Armpit area
Brachial	Upper arm
Ulnar	Forearm and hand
Radial	Forearm and hand
Intercostals	Rib area
Lumbar	Posterior abdominal wall
External iliac	Anterior abdominal wall
Common iliac	Legs, gluteal area, and pelvic organs
Femoral	Thigh
Popliteal	Posterior knee
Tibial	Lower leg and foot

TABLE 10-10 Major Veins of the Body

Vein	Anatomic Location or Organ Drained
Jugular	Head and neck
Brachiocephalic	Head and neck
Axillary	Armpit area
Brachial	Upper Arm
Ulnar	Lower arm and hand
Radial	Lower arm and hand
Intercostal	Thorax
Azygos	Thorax and abdomen
Gastric—part of the hepatic portal system	Stomach to the liver
Splenic—part of the hepatic portal system	Spleen, pancreas, and stomach to the liver
Mesenteric—part of the hepatic portal system	Intestines to the liver
Hepatic portal—part of the hepatic portal system	Gastric, splenic, and mesenteric veins to the liver
Hepatic	Liver to the inferior vena cava
Iliac	Pelvic organs, legs, and gluteal areas
Femoral	Thighs
Popliteal	Knees
Saphenous	Legs

Venous System

As you may remember, veins (with the exception of the pulmonary veins that carry oxygenated blood from the lung to the heart) are blood vessels that carry oxygen-poor or deoxygenated blood toward the heart. Typically, arteries that supply blood to an area have corresponding veins associated with them. Large veins often have the same names as the arteries they run next to. However, there are exceptions to this rule as well. For example, the veins next to carotid arteries are called jugular veins. Smaller veins join to form larger veins. The smaller veins are called "tributaries" to the larger veins. Veins do not "supply" blood, but instead "drain" blood from an area. The largest veins of the body are the superior and inferior vena cavae which empty into the right atrium. The superior vena cava generally collects blood from veins above the heart and the inferior vena cava collects blood from veins below the heart. The major veins of the body are summarized in Table 10-10 and illustrated in Figures 10-30 through 10-34.

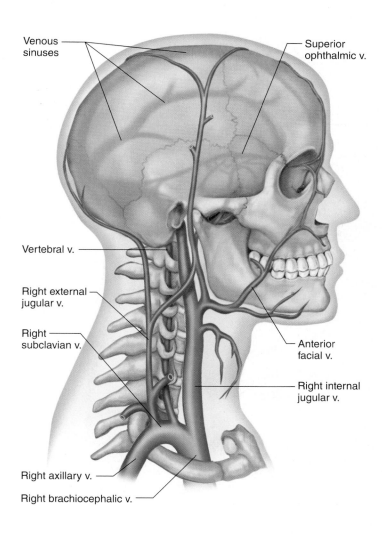

FIGURE 10-30
Major veins of the head, neck, and brain. The clavicle has been removed (v. stands for vein).

CHAPTER 10 Blood and Circulation

FIGURE 10-31 The major veins to the shoulder and upper limb (*v.* stands for vein; *vv.* stands for veins).

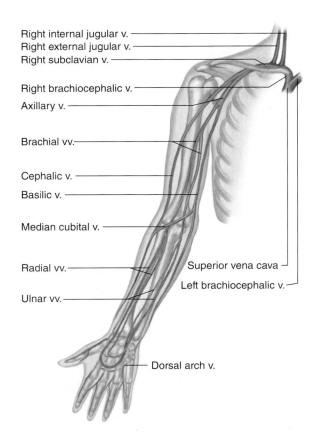

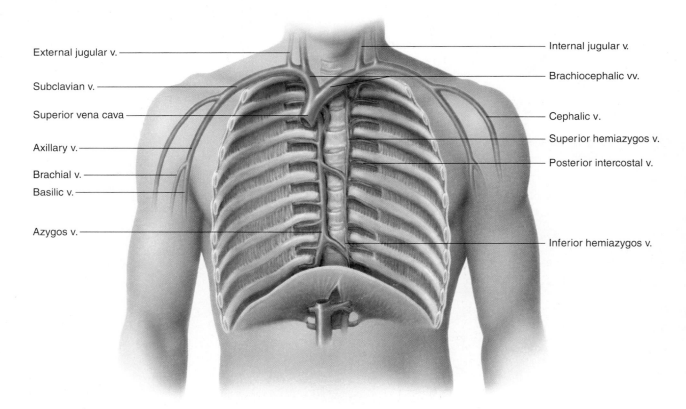

FIGURE 10-32 Veins that drain the thoracic wall (*v.* stands for vein; *vv.* stands for veins).

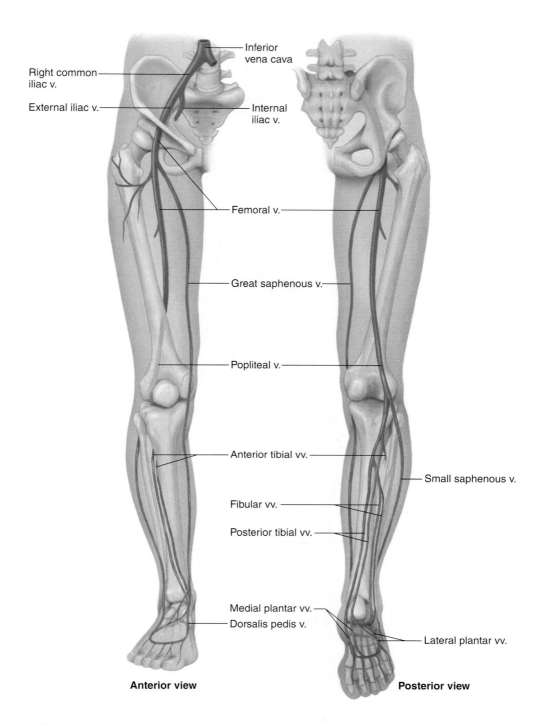

FIGURE 10-33 The major veins of the lower limb and pelvis (v. stands for vein; vv. stands for veins).

Life Span

As we get older, changes occur in our blood vessels. Due to diet, lifestyle, and/or genetics, there may be a buildup of cholesterol and plaque in our arteries that leads to arteriosclerosis, which was discussed earlier in the

learning outcome

Describe how the circulatory system is affected by age.

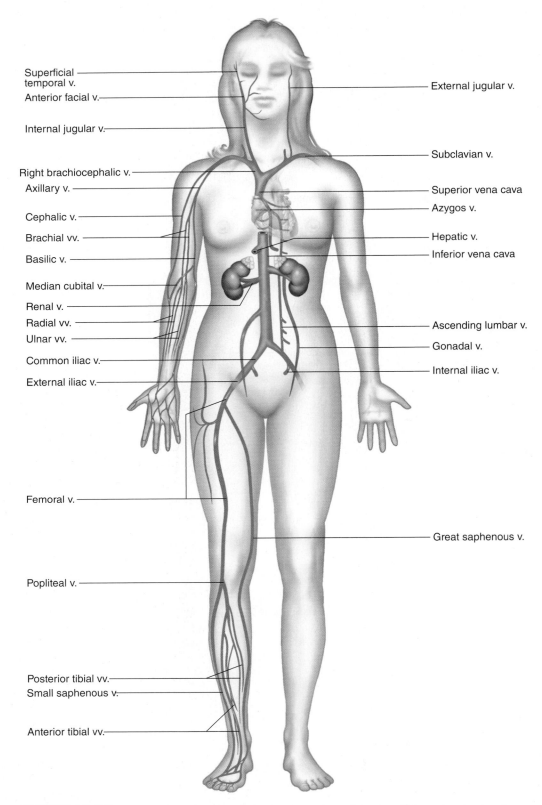

FIGURE 10-34 The major veins that drain the thoracic wall (*v.* stands for vein; *vv.* stands for veins).

chapter. This can lead to hypertension, strokes, and heart attacks. Other changes related to arteriosclerosis are weakening of arterial walls resulting in aneurysms or weakening of venous valves causing varicose veins.

summary

chapter 10

learning outcomes	key points
10.1 Explain why blood is considered a connective tissue and describe hematopoiesis.	Blood is considered a connective tissue because it is composed of living cells (RBCs and WBCs) and a nonliving matrix (plasma).
10.2 Describe the components of plasma and their functions.	Plasma is the liquid portion of blood. It is made up of approximately 90 percent water and contains various proteins, nutrients, gases, electrolytes, and waste products.
10.3 Relate the structure and function of erythrocytes, leukocytes, and platelets to anemia, leukemia, and thrombocytopenia, respectively.	Red blood cells (RBCs, erythrocytes) are small cells that have a biconcave shape. Mature RBCs do not have a nucleus and consequently cannot reproduce themselves. They contain a pigment called hemoglobin whose function is to bind to oxygen. Anemia can result when there is a problem with the oxygen-carrying capacity of red blood cells. Excessive production of lymphocytes can give rise to lymphocytic anemia and excessive production of myelocytes can give rise to myelogenous leukemia. Platelets are cell fragments derived from a megakaryocyte. They are involved in blood clotting and when there are too many or too few platelets, there is disruption of homeostasis.
10.4 Describe blood pressure and how it is regulated, including a description of essential and secondary hypertension.	Blood pressure is the force of the blood on arterial walls as the blood circulates through the body. Contraction of the ventricles causes systolic pressure, and relaxation of the ventricles creates the diastolic pressure. Essential hypertension is high blood pressure of an unidentified or idiopathic cause, and secondary hypertension is high blood pressure with a known cause.
10.5 Explain the steps of hemostasis.	The four steps of hemostasis are (1) blood vessel spasm, (2) platelet plug formation, (3) blood clotting, and (4) fibrinolysis.
10.6 Explain the purpose of blood typing, and describe the ABO blood typing system and the Rh blood typing system.	Blood typing is a way of categorizing blood based on the presence or absence of antigens on the surface of red blood cells. The two most common blood typing systems are the ABO and Rh blood typing systems.
10.7 Describe types of blood vessels, their functions, and disorders.	The major types of blood vessels are arteries, veins, and capillaries. Other than the pulmonary vessels, arteries carry oxygenated blood away from the heart and veins carry deoxygenated blood toward the heart. Capillaries are very small vessels that participate in exchange of oxygen and carbon dioxide as well as

continued

chapter 10

learning outcomes	key points
	nutrients and waste products. Common disorders of blood vessels include aneurysms, thrombophlebitis, and varicose veins.
10.8 Compare the pulmonary and systemic circuits.	The right side of the heart pumps the blood for the pulmonary circuit. Blood is pumped into the pulmonary trunk and pulmonary arteries and then into the lungs. The left side of the heart is the pump for the systemic circuit. The left ventricle pumps blood through the aorta for distribution to the body.
10.9 Identify the major arteries and veins of the body.	The largest artery in the body is the aorta, which travels through the thorax and abdomen. It gives off many arteries and terminates as the right and left common iliac arteries that branch into other arteries. The superior and inferior vena cava are the largest veins in the body and they return blood to the right atrium. Many veins drain blood from the tissues of the body to eventually empty into these two veins.
10.10 Describe how the circulatory system is affected by age.	As we get older, there are changes in our blood vessels. There may be a buildup of cholesterol and plaque leading to arteriosclerosis. This can lead to hypertension, strokes, and heart attacks.

chapter 10 review

case study 1 questions

Can you answer the following questions that pertain to the case study presented earlier in this chapter?

1. Why is pernicious anemia sometimes called megaloblastic anemia?
   a. Because the red blood cells are much larger than normal.
   b. Because the cells easily lyse or rupture.
   c. Because the white blood cells are larger than normal.
   d. Because there is a large increase in the number of red blood cells.
2. What is the cause of pernicious anemia?
   a. Inadequate absorption of vitamin B_{12} or folic acid
   b. Too little iron
   c. A gene mutation
   d. Too little vitamin C
3. What is the treatment for pernicious anemia?
   a. Intramuscular vitamin B_{12} injections
   b. Intramuscular vitamin C injections
   c. Plasmapheresis
   d. Iron pills

review questions

1. What is the largest artery in the body?
   a. Aorta   b. Carotid artery   c. Renal artery   d. Celiac trunk
2. What is the major artery of the thigh?
   a. Femoral artery   b. Celiac trunk   c. Renal artery   d. Popliteal artery

critical thinking questions

1. What danger is there to the circulatory system in a 14-hour plane flight where the passenger remains seated the entire flight?

2. Discuss the difference between the pulmonary and systemic circuits.
3. Discuss the structural and functional differences between arteries and veins.

patient **education**

Your family has a history of heart disease. Develop a plan for teaching your family members how to reduce their risk of heart and vascular disease through changes in lifestyle.

applying **what you know**

1. Michael Huckaby, a 22-year-old man, broke his humerus during a bicycle marathon. What blood vessel is at risk of being damaged with this type of fracture and what is the danger for Michael?

CASE STUDY 2 *Leukemia*

Randall Swift, a four-year-old boy, was diagnosed with acute lymphoblastic leukemia. He has experienced some weight loss and tiredness. Randall also has several swollen lymph nodes in his neck and has had several instances of unexplained fever.

1. What is leukemia?
   a. A cancer of melanocytes
   b. A cancer of white blood cells
   c. A cancer of thrombocytes
   d. A cancer of erythrocytes
2. All of the following are used to diagnose leukemia *except*
   a. CBC (complete blood count)
   b. Blood smear and microscopic examination
   c. Physical examination
   d. Hematocrit
3. All of the following are treatments for leukemia *except*
   a. Chemotherapy
   b. Radiation therapy
   c. Immunotherapy
   d. Thymectomy (removal of the thymus)

CASE STUDY 3 *Rh Incompatibility*

Tessa White, a 28-year-old woman, wants to have a baby. She is concerned because she knows her husband's blood type is Rh-positive and she is Rh-negative. Tessa does not know what this means, but has heard it could be dangerous.

1. What is blood typing?
   a. A procedure to match blood of a potential donor and potential recipient
   b. A procedure to determine paternity
   c. A procedure used to treat sickle cell disease
   d. A procedure to treat pernicious anemia
2. Which of the following is considered the universal donor?
   a. Blood type A
   b. Blood type B
   c. Blood type AB
   d. Blood type O
3. Which of the following is considered the universal recipient?
   a. Blood type A
   b. Blood type B
   c. Blood type AB
   d. Blood type O
4. If the fetus is Rh-positive, which of the following may affect the newborn because of the Rh incompatibility with Tessa?
   a. Hemolytic disease of the newborn
   b. Sickle cell disease
   c. Pernicious anemia
   d. Hepatitis

11 The Cardiovascular System

chapter outline

- Heart Anatomy
- The Cardiac Cycle
- Heart Sounds
- Cardiac Output
- The Conduction System
- Electrocardiogram
- Chest Pain
- Life Span

outcomes

AFTER COMPLETING THIS CHAPTER, YOU WILL BE ABLE TO:

11.1 Identify the location, chambers, valves, and layers of the heart and some disorders that affect them.

11.2 Interpret the cardiac cycle.

11.3 Relate the major heart sounds to their significance.

11.4 Discuss how cardiac output is determined.

11.5 Compare the components of the conduction system of the heart.

11.6 Explain the electrical events in the heart and abnormalities.

11.7 Explain the cardiac and noncardiac causes of chest pain.

11.8 Relate the effects of aging of the heart to life span.

essential terms

aortic semilunar valve (a-OR-tik sem-e-LOO-nar valve)

arrhythmia (a-RITH-me-a)

atrium (plural, atria) (A-tre-um)

atrioventricular (AV) node (a-tre-o-ven-TRIK-u-lar node)

bicuspid valve (bi-KUS-pid valve)

bundle of His (bundle of HISS)

depolarization (de-po-lar-i-ZA-shun)

endocardium (en-do-KAR-de-um)

epicardium (ep-i-KAR-de-um)

mitral valve (MI-tral valve)

myocardial infarction (MI) (mi-o-KAR-de-al in-FARK-shun)

myocardium (mi-o-KAR-de-um)

pericardium (per-i-KAR-de-um)

pulmonary semilunar valve (PUL-mo-nar-ee sem-e-LOO-nar valve)

Purkinje fibers (pur-KIN-je fibers)

repolarization (re-po-lar-i-ZA-shun)

sinoatrial (SA) node (si-no-A-tre-al node)

tricuspid valve (tri-KUS-pid valve)

ventricle (VEN-tri-kul)

Additional key terms in the chapter are italicized and defined in the glossary.

case study

Use the case study to focus on as you go through the chapter. The questions will guide you as you learn the anatomy, physiology, and pathology of the heart.

CASE STUDY 1 *Myocardial Infarction (MI)*

A 55-year-old female investment banker, Krista Judge, was meeting with a major client. During the meeting, the conversation became quite excited and Krista became anxious. She started to sweat profusely and felt a crushing pain in her chest just behind the "breastbone." She also felt a pain going up to her left jaw. Others at the meeting called for an ambulance. The emergency medical personnel administered oxygen and aspirin and rushed Krista to the hospital. Her blood pressure was very low. She said the pain was the worst she had ever experienced. "It's like an elephant standing on my chest," Krista told the physicians. The physicians started a drip of lidocaine for the pain. They also drew blood to check for cardiac enzymes and ran an electrocardiogram. Based on Krista's presentation and the results of the blood tests and ECG, a diagnosis of myocardial infarction (heart attack) was made.

As you go through the chapter, keep the following questions in mind:

1. Is Krista typical of someone you would expect to be at risk for an MI? Why or why not?
2. What were the signs that helped make the diagnosis? Are they different for men?
3. What were the tests that helped make the diagnosis?
4. Could this have been something other than an MI?

study tips

Here are some suggestions that may help you maximize your learning. These are recommendations that others have utilized to help them succeed. Use those that you feel will be beneficial to you.

1. Learn the essential terms for the chapter. Make flash cards and spend 5 to 10 minutes every day going over the terms.

2. Look at illustrations and tables, reading the captions.

3. Draw the heart, labeling the chambers and valves. Use arrows to show the flow of blood through the heart.

4. Make cards to represent all the parts of the cardiovascular system. Shuffle the deck and then sort the cards in the order that blood travels through the cardiovascular system.

Introduction

LO 11.1

The cardiovascular system is fascinating. It is composed of the heart (the vital "pump"), blood, and blood vessels (Figure 11-1). In Chapter 10, Blood and Circulation, we covered the composition of blood and the function of blood vessels. We will focus on the heart in this chapter. Although all organ systems are important, because heart disease is the number one cause of death in the United States the cardiovascular system warrants close attention. Because of the profession you have chosen, you will be working with people who have heart disease. You may also have friends and family members who suffer from various conditions relating to the heart and the circulatory system. Therefore, an understanding of the cardiovascular system is very important.

As we discovered in Chapter 10, Blood and Circulation, the main function of the cardiovascular system is to supply oxygen and nutrients to the tissues of the body and to remove carbon dioxide and waste products (Figure 11-2). Understanding the components and the function of the heart and its associated "pipes" vascular system, or blood vessels, when they are working correctly is key to understanding what can happen when they are not working properly. The heart pumps several thousand gallons of blood in a day and several millions of gallons each year.

learning outcome

Identify the location, chambers, valves, and layers of the heart and some disorders that affect them.

Heart Anatomy

As you study the anatomy of the heart, it is important to keep in mind that it functions as a pump. It is made up of four chambers that are separated by valves that control the flow of blood. It also has three layers, including a thick muscular layer, that allows blood to be pumped into the lungs as well as throughout the body.

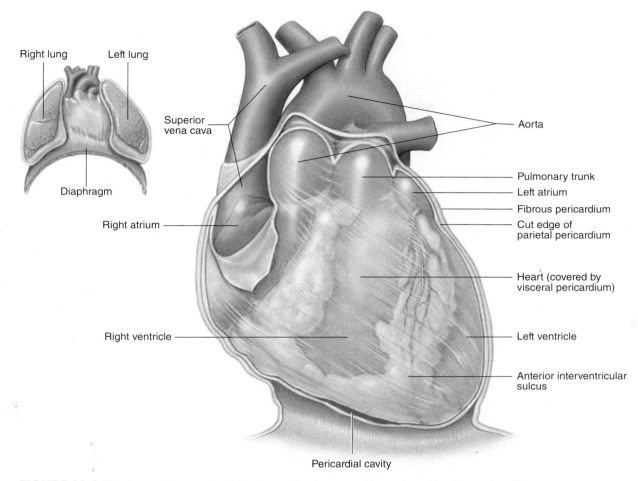

FIGURE 11-1 The heart is located within the mediastinum and is enclosed by the pericardium.

Location of the Heart

Let's first talk about the size, location, and position of the heart. It is about the size of a clenched fist. It weighs 280 to 340 grams (approximately 10 ounces) in men, and slightly less in women (230 to 280 grams, or approximately 8 ounces). It lies posterior to (behind) the sternum just slightly to the left of the midline of the body (Figure 11-3). It occupies a space called the thoracic mediastinum (middle septum) that is located between the right and left lungs. Along with the heart and great vessels, other structures that can be found in the mediastinum include the esophagus, trachea, and thymus. The heart extends from the level of the second rib to about the level of the sixth rib. Inferiorly, the heart rests on the diaphragm (the major muscle of respiration). The heart has four borders (superior, inferior, medial, and lateral) and three surfaces (sternocostal, diaphragmatic, and pulmonary). The superior border is called the *base* and is wider than the inferior surface, which is called the *apex*. The base is formed mainly by the left atrium. The apex is formed by the left ventricle which points inferolaterally.

U check
What is the mediastinum?

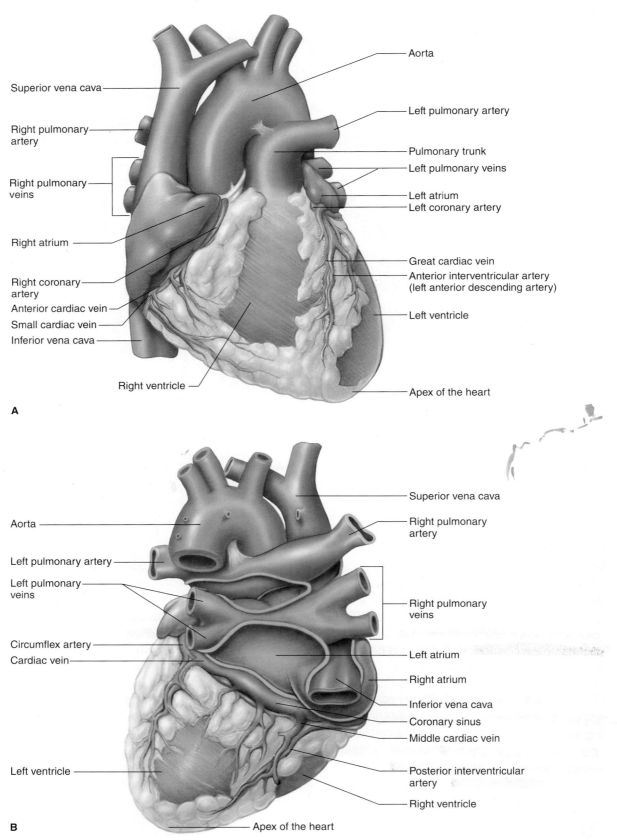

FIGURE 11-2 The blood vessels associated with the surface of the heart: (A) anterior view and (B) posterior view.

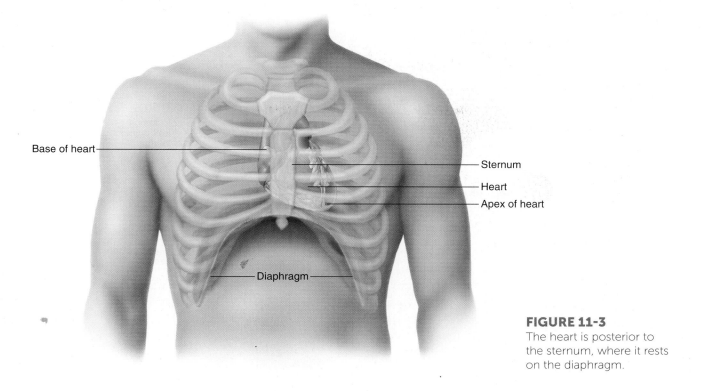

FIGURE 11-3
The heart is posterior to the sternum, where it rests on the diaphragm.

Chambers of the Heart

The inside of the heart consists of four chambers (Figure 11-4). There are two atria (singular, atrium, meaning "antechamber" or "waiting room") that sit on two ventricles (pumping chambers). Singularly, the chambers are known as the right and left **atrium** and right and left **ventricle.** The atria are separated by a septum or wall called the *interatrial septum* and the ventricles are separated by a wall called the *interventricular septum*. There is also a septum that separates the atria from the ventricles. If you said this is called the *atrioventricular septum*, you would be correct. The walls of the ventricles (especially the left ventricle) are thicker than the walls of the atria. Can you guess why?

The right atrium receives blood that is returning from the body to the heart. It forms the right border of the heart. It also receives blood from the *coronary sinus*. Remember that the blood being returned to the heart is deoxygenated or low in oxygen. There is also a depression in the interatrial septum called the *fossa ovalis*. In the fetus, this depression was an actual opening between the two atria called the *foramen ovale* which allows blood to bypass the fetal lungs, as oxygen is being provided by the mother.

The right ventricle receives blood from the right atrium. This chamber forms the anterior surface of the heart, most of the inferior border of the heart, and a part of the diaphragmatic surface. There are three cone-shaped muscles in the ventricle that are attached to the three cusps of the tricuspid valve. Their function is to prevent the valve from "falling" into the right atrium when the pressure builds up in the right ventricle. The right ventricle pumps the deoxygenated blood from the right atrium to the lungs to be oxygenated.

The left atrium forms most of the base of the heart. Four pulmonary veins carrying oxygenated blood from the lungs enter the left atrium. They have no valves so as not to hinder the entry of blood into the chamber. The left ventricle receives blood from the left atrium. The left ventricle works harder

atrium A superior chamber of the heart.

ventricle An inferior chamber of the heart.

CHAPTER 11   The Cardiovascular System

FIGURE 11-4
(A) Frontal section of the heart: connection between the right ventricle and the pulmonary trunk.
(B) Pathway of blood through the heart and lungs and on to other body parts.

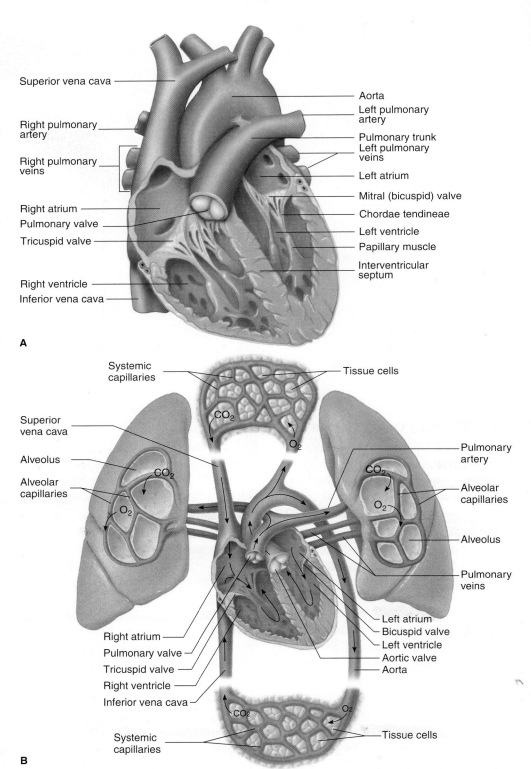

than any other chamber because it pumps oxygenated blood out of the heart via the aorta into the systemic circulation. Its myocardium is twice as thick as that of the right ventricle.

U check

What are the four chambers of the heart?

Valves of the Heart

There are four valves in the heart. These valves act as "gatekeepers" as blood passes from one chamber to the next and as blood leaves the heart to go to the lungs or the systemic circulation of the body. Figure 11-5 is a superior view of the valves of the heart. Figure 11-6 shows the heart valves and the fibrous skeleton that supports them. Interestingly, there are no valves that restrict the entry of blood into the heart.

The first of these heart valves is the **tricuspid valve** (also called the right AV, or atrioventricular valve), which is located between the right atrium and right ventricle. Figure 11-7 shows the tricuspid valve, which is found between the left atrium

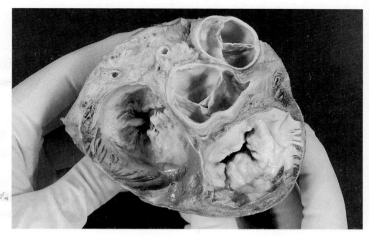

FIGURE 11-5 Photograph of the heart valves (superior view).

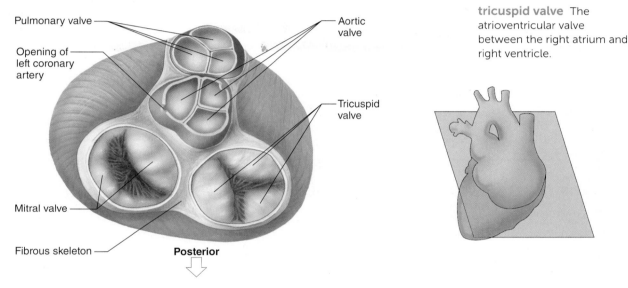

tricuspid valve The atrioventricular valve between the right atrium and right ventricle.

FIGURE 11-6 The valves of the heart and the fibrous skeleton that support them.

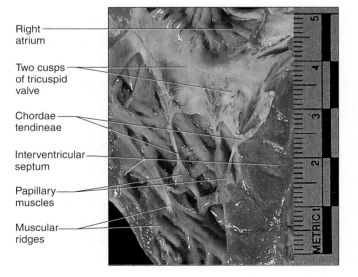

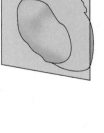

FIGURE 11-7 Photograph of a human tricuspid valve.

CHAPTER 11 The Cardiovascular System

and left ventricle. It is called tricuspid because the valve consists of three cusps or leaflets. These three cusps prevent blood from flowing back into the right atrium when the right ventricle contracts. The cusps of this valve are anchored by cordlike structures called *chordae tendineae* to specialized cardiac muscle called *papillary muscles*. These muscles contract when the ventricles contract. This prevents the valve from "falling" into the right atrium, and keeps blood from regurgitating (flowing back) into the upper chamber.

The next valve is the **pulmonary semilunar valve.** It is situated between the right ventricle and the trunk of the pulmonary arteries, and prevents blood from flowing back into the right ventricle. Because its cusps are shaped like a half moon, this valve is called a semilunar valve.

The **bicuspid valve** has two cusps and is located between the left atrium and the left ventricle. It prevents blood from flowing back into the left atrium when the left ventricle contracts. This valve is also known as the **mitral valve,** or the left AV (atrioventricular) valve. *Mitral* comes from the word *mitre* because of the similarity in shape to the mitres (hats) worn by Catholic bishops. Like the tricuspid valve, the bicuspid valve has chordae tendineae attached to papillary muscles.

The **aortic semilunar valve** is positioned between the left ventricle and the aorta. It prevents blood from flowing back into the left ventricle and is also known as a semilunar valve because of the moonlike shape of its cusps.

pulmonary semilunar valve The valve between the right ventricle and the pulmonary trunk.

bicuspid valve The atrioventricular valve on the left side of the heart.

mitral valve The atrioventricular valve on the left side of the heart, also known as the bicuspid valve.

aortic semilunar valve The valve located between the left ventricle and the aorta.

> **U check**
> What are the names of the four heart valves and where are they located?

focus on Genetics: Congenital Heart Disease

Congenital heart disease refers to a problem with the heart's structure and function due to abnormal heart development before birth. *Congenital* means present at birth. Congenital heart disease is responsible for more deaths in the first year of life than any other birth defect. Two types of congenital heart disease are cyanotic and noncyanotic. Cyanotic heart diseases cause blue discoloration of the skin due to *hypoxia* (lack of oxygen). Some examples of congenital heart disorders are listed below.

Cyanotic Disorders

- Tetralogy of Fallot (TOF)
- Transposition of the great vessels
- Tricuspid atresia
- Truncus arteriosus
- Pulmonary atresia
- Ebstein's anomaly

Noncyanotic Disorders

- Ventricular septal defect (VSD)
- Atrial septal defect (ASD)
- Patent ductus arteriosus (PDA)
- Aortic stenosis
- Pulmonic stenosis
- Coarctation of the aorta

Depending on the defect, the outcome can vary. For example, VSD may never cause problems and coarctation of the aorta may not cause problems for years. Certain defects require careful patient monitoring, whereas others heal over time and still others require specific treatment including possible surgical intervention. TOF is one congenital defect that in most cases requires surgery in infancy.

Layers of the Heart

The heart consists of three layers (Figure 11-8). Table 11-1 lists the layers of the heart, their composition, and their functions. The innermost layer is called the **endocardium**. It is very thin and lined with endothelium. In addition to lining the chambers of the heart, it also lines the heart valves and is continuous with the lining of the large blood vessels that are attached to the heart. The middle layer is the **myocardium,** which is the cardiac muscle. It makes up over 90 percent of the heart. Remember from Chapter 9, The Muscular System, that cardiac muscle is unique to the heart because it has striations like skeletal muscle, but, like smooth muscle, it is under involuntary control. It is the myocardium that is responsible for the pumping action of the heart and it is thickest in the left ventricle. The outermost layer of the heart is the **epicardium**, also known as the visceral pericardium, and it contains fat, which helps cushion the heart if the chest experiences blunt trauma.

endocardium Endothelium and smooth muscle that lines the inner layer of the heart, the heart valves, and the tendons that hold the valves open.

myocardium The middle layer of the heart wall made up of cardiac muscle.

epicardium The thin outer layer of the heart wall also known as the visceral pericardium.

U check
Which layer of the heart is made up of cardiac muscle?

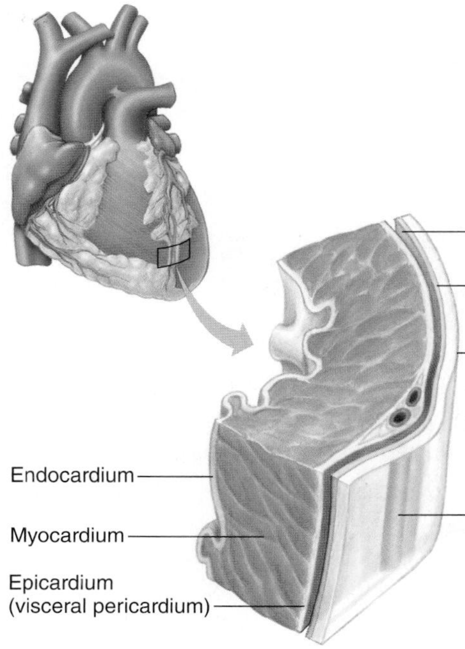

FIGURE 11-8
The heart wall has three layers: an endocardium, a myocardium, and an epicardium.

TABLE 11-1 Layers of the Heart		
Layer	**Composition**	**Function**
Epicardium (visceral pericardium)	Serous membrane of connective tissue covered with epithelium and including blood capillaries, lymph capillaries, and nerve fibers	Forms a protective outer covering; secretes serous fluid
Myocardium	Cardiac muscle tissue separated by connective tissues and including blood capillaries, lymph capillaries, and nerve fibers	Contracts to pump blood from the heart chambers
Endocardium	Membrane of epithelium and underlying connective tissue, including blood vessels and specialized muscle fibers	Forms a protective inner lining of the chambers and valves

Membranes of the Heart

Two membranes surround the heart (a double membrane like this is called a *serosa*). The entire membrane around the heart is called the **pericardium** and it covers the heart and the large blood vessels attached to it. The pericardium consists of an outer fibrous layer called the parietal pericardium and an innermost layer called the visceral pericardium, which lies directly on top of the heart. Together, the visceral and parietal pericardia (plural) form the pericardial sac. A very small amount of *serous* (watery) fluid is found within the sac and acts to reduce friction between the two membranes.

pericardium The membrane that surrounds the heart, made up of a superficial fibrous layer and a deep serous layer.

> **U check**
> What is the difference between the visceral and parietal pericardia?

PATHOPHYSIOLOGY

Endocarditis

Endocarditis is an inflammation of the innermost lining of the heart and the heart valves.

Etiology: Bacterial infections are the most common cause of endocarditis. Patients are more susceptible to this condition if they have abnormal heart valves or an underlying heart condition. Intravenous drug abuse and systemic lupus erythematosus (SLE) are two possible causes of endocarditis.

Signs and Symptoms: Common signs and symptoms include weakness, fever, excessive sweating, general body aches, difficulty breathing, and blood in the urine.

Treatment: The treatment addresses the underlying cause. Therefore, if endocarditis is caused by a bacterial infection, the treatment for the condition is intravenous antibiotics followed by oral antibiotics for up to six weeks.

Myocarditis

Myocarditis is an inflammation of the muscular layer of the heart. It is relatively uncommon, but very serious when it does occur because it leads to weakening of the heart wall.

Etiology: The most common cause of myocarditis in the United States is a viral infection, especially with Coxsackie B viruses. Myocarditis may also be caused by exposure to certain chemicals, allergens, and bacteria.

Signs and Symptoms: Signs and symptoms include unexplained fever as well as chest pain that is described as similar to a heart attack. Dyspnea, decreased urine output, fatigue, and fainting may also accompany myocarditis. Most often, the diagnosis is based on clinical findings.

Treatment: Treatment normally includes steroids to reduce inflammation, bed rest, and a low-sodium diet.

Pericarditis

Pericarditis is inflammation of the pericardium.

Etiology: This condition is most commonly caused by complications of viral or bacterial infections. However, myocardial infarctions and chest injuries can also lead to pericarditis.

Signs and Symptoms: Symptoms include sharp, stabbing chest pains, especially during deep inspirations. Fever, fatigue, and orthopnea (difficulty breathing while lying down) are also common symptoms. On *auscultation* (listening) of the heart, a "friction rub" may be heard which is diagnostic of the condition. A friction rub sounds similar to rubbing the palms of your two hands together.

Treatment: The treatment usually includes analgesics for pain and diuretics to remove excess fluids around the heart. If pericarditis is caused by bacteria, antibiotics are also used. In chronic cases, surgery may be required to remove part of the membranes surrounding the heart.

U check
What are the etiologies of endocarditis?

The Cardiac Cycle

The cardiac cycle is made up of the events from one heart beat to the next heart beat (Figure 11-9). The two atria of the heart contract and then relax almost simultaneously. The two ventricles also contract and relax almost simultaneously. Contraction of atria is called atrial *systole* and of the ventricles is called ventricular systole. Relaxation of the atria is atrial *diastole* and relaxation of the ventricles is ventricular diastole. When the atria are contracting, the ventricles are relaxing and when the atria are relaxing, the ventricles are contracting. The pressure on the left side of the heart is considerably greater

Interpret the cardiac cycle.

Cardiac cycle

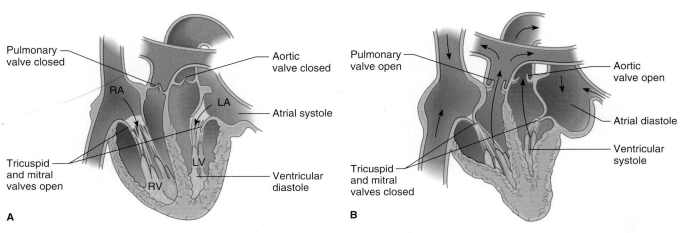

FIGURE 11-9 The atria (A) empty of blood during atrial systole and (B) filled with blood during atrial diastole.

than that on the right side of the heart. When the atria contract, the tricuspid and bicuspid (mitral) valves open, and blood enters the ventricles. When the right ventricle contracts, the tricuspid valve closes. At approximately the same time, the pulmonary semilunar valve opens and blood enters the pulmonary trunk. When the left ventricle contracts, the bicuspid valve closes, the aortic semilunar valve opens, and blood enters the aorta. The number of cardiac cycles per minute is the same as the heart rate (pulse rate).

Several factors influence heart rate:

Exercise. Strenuous exercise increases the heart rate because there is a greater demand for oxygen.

Parasympathetic innervation. The parasympathetic system slows things down. The vagus nerve (Cranial Nerve X) is the parasympathetic innervation to the heart. (The cranial nerves and the parasympathetic system are discussed in more detail in Chapter 14, The Nervous System.) Therefore, when the vagus nerve is stimulated, it slows the heart rate (Figure 11-10).

Sympathetic innervation. Sympathetic innervation increases heart rate. This is part of the "flight or fight" response that prepares the body for increased activity. (This response will be discussed in more detail in Chapter 14.)

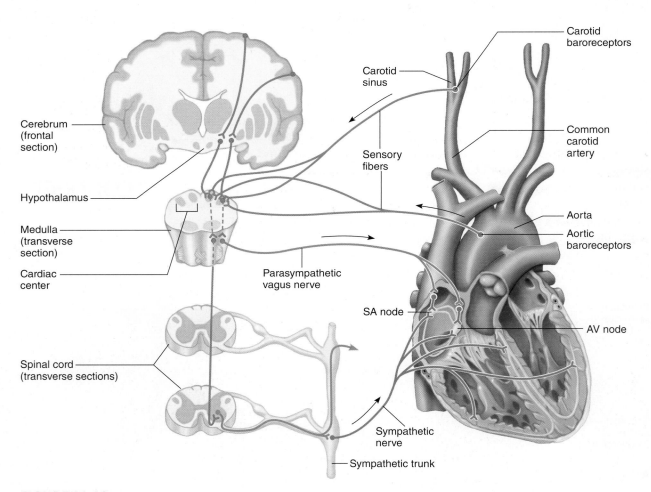

FIGURE 11-10 Autonomic innervation of the heart including the SA and AV nodes.

Cardiac control center: This center is located in the medulla oblongata of the brain. When blood pressure rises, this control center sends impulses to decrease the heart rate. When blood pressure falls, it sends impulses to increase the heart rate. (The functions of the medulla oblongata will also be discussed further in Chapter 14.)

Body temperature. An increase in body temperature usually increases the heart rate. In fact, for every 1 degree Celsius rise in body temperature, the heart rate increases approximately 10 beats per minute. This explains the higher heart rate when a person runs a fever.

Potassium ions. A low concentration of potassium ions in the blood decreases the heart rate, but a high concentration causes an **arrhythmia** (abnormal heart rhythm).

arrhythmia An irregular heart rhythm.

> **U check**
> What are three factors that can increase heart rate?

from the perspective of...

AN ECG TECHNICIAN An ECG technician uses an electrocardiogram to record a tracing of the electrical activity of the heart. This tracing is known as an ECG or EKG. Why is an understanding of the cardiac cycle important to know?

Heart Sounds

Heart sounds are heard through *auscultation* of (listening to) the heart with a stethoscope. The heart sounds are caused by turbulence that is created by the closing of the valves. Heart sounds are closely linked to the events in the cardiac cycle. Although there are four heart sounds associated with a normal heart, there are two heart sounds that you should be familiar with.

The first heart sound (S1) is described as *lubb*. It is the sound made during the closing of the tricuspid and bicuspid valves when the ventricles contract. It is louder and longer in duration than the second heart sound (S2). The second heart sound is described as *dubb* and takes place when the pulmonary and aortic semilunar valves close as the atria contract. Figure 11-11 shows the locations on the chest where heart sounds are best heard using a stethoscope. Sometimes abnormal sounds called murmurs may be heard. This may be an indication of valvular disease and may indicate possible serious heart problems, but often murmurs are harmless.

11.3 learning outcome
Relate the major heart sounds to their significance.

PATHOPHYSIOLOGY

Murmurs

Murmurs are simply defined as abnormal heart sounds. Normally, heart sounds are clear, strong, and smooth as valves close completely and blood flows over the lining of the heart with no resistance. Not all murmurs indicate a heart disorder; some are quite

(continued)

Murmurs (concluded)

benign. Murmurs are graded from 1 to 6, with 1 being barely audible (and the least serious) and 6 being quite loud (and the most serious).

Etiology: The cause of some heart murmurs is not known. In children, the failure of the foramen ovale or ductus arteriosus to close completely after birth can cause murmurs. Other causes of heart murmurs include stress, previous viral or other illnesses affecting the heart, and defective heart valves that do not close completely.

Signs and Symptoms: The signs and symptoms vary considerably depending on the cause and severity of the heart murmur. Severe symptoms include weakness, pallor, edema (fluid retention), and other common signs associated with heart failure. For many people with a mild heart murmur, there may be no symptoms at all.

Treatment: Many times, no treatment is required. Surgery to correct valvular defects or other heart defects may be needed in more serious cases.

> **U check**
> What is the cause of the first heart sound?

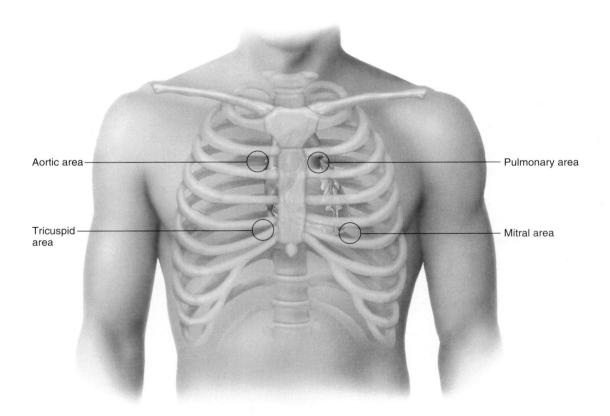

FIGURE 11-11 Thoracic regions where heart sounds of the individual valves are best heard.

Cardiac Output

Cardiac output is the volume of blood that is pumped out of the left or right ventricle each minute. It is determined by the volume of blood ejected with each stroke times the heart rate, or number of beats per minute.

$$CO = SV \times HR \text{ (Cardiac Output = Stroke Volume} \times \text{Heart Rate)}$$

If someone has a resting stroke volume of 74 milliliters and a resting heart rate of 72 beats per minute, then his or her cardiac output will be:

$$74 \text{ mL} \times 72 \text{ bpm} = 5{,}328 \text{ mL/min}$$

Since the typical volume of blood in an adult male is approximately 5,000 mL (5 L), this means that an individual's total blood volume circulates through the body every minute. Anything that affects stroke volume or heart rate will affect cardiac output. Three factors that influence stroke volume are the amount of blood returning to the heart, which is called preload; the contractility or strength of contraction of the heart; and the force that resists ejection of blood from the ventricles, which is called afterload. Heart rate can be influenced by a variety of factors including nervous system regulation, hormones such as epinephrine, age, gender, body temperature, and exercise.

> **11.4 learning outcome**
> Discuss how cardiac output is determined.

U check

What initial effect will a drug that increases heart rate have on cardiac output?

---

focus on Wellness: Heart Disease Prevention and Treatment

The best way to help prevent heart disease is to eat healthy and participate in a regular exercise program. Eating healthy includes eating plenty of fruits and vegetables and limiting fat intake. You may want to talk with your primary care physician or registered nutritionist to develop a plan specifically for you. Aerobic exercise is excellent for maintaining a healthy heart. You want to do exercises that will increase cardiac output and metabolic rate. Exercising 20 minutes at a time three to five times a week will make a difference in the health of your heart. With regular exercise, pumping efficiency improves, hemoglobin concentration increases, and oxygen delivery to tissues increases. Aerobic exercises include activities such as swimming, running, brisk walking, and bicycling. After several weeks of a regular exercise regimen, you should start to see some of the benefits. Check with your physician before beginning an exercise program.

Treatment for heart disease depends on the actual disease and can range from changing your lifestyle to heart transplants. Heart transplants are fairly common today and need surpasses donor availability.

11.5 learning outcome

Compare the components of the conduction system of the heart.

sinoatrial (SA) node
A collection of muscle fiber cells in the posterior wall of the right atrium that acts as the pacemaker of the heart.

atrioventricular (AV) node
A collection of cells located in the interatrial septum that is part of the conduction system of the heart, also known as the secondary pacemaker.

The Conduction System

The conduction system of the heart refers to the electrical activity of the heart that is responsible for the contraction of the atria and ventricles. The heart has its own inherent rhythmic activity that allows the heart not only to beat for a lifetime but also to continue beating for a short time when all nerves to the heart are cut. This "self-excitability" of the heart is called autorhythmicity. Several structures are important in maintaining proper electrical activity and functioning of the heart. These structures are shown in Figures 11-12 and 11-13 and discussed in the following sections.

Sinoatrial Node

A *node* is a collection of specialized cardiac muscle fibers that have a certain rate of activity. The **sinoatrial (SA) node** is located in the posterior wall of the right atrium and generates an impulse that flows to the atrioventricular node. The SA node is known as the primary or "natural pacemaker" because it is responsible for setting the heart rate. A typical heart rate is approximately 70 beats per minute for most adults. The SA node is controlled by the autonomic nervous system (ANS) and is innervated by both the sympathetic division (speeds up heart rate) and parasympathetic division (slows down heart rate). The autonomic nervous system will be explored in more detail in Chapter 14, The Nervous System. A rate greater than 100 beats per minute is called *tachycardia* and a rate less than 60 beats per minute is known as *bradycardia*. The electrical impulse or action potential travels from the SA node to the muscle fibers of both the right and left atria causing them to contract. The action potential then reaches the atrioventricular node.

Atrioventricular Node

The **atrioventricular (AV) node** is located in the interatrial septum, just above the ventricles. The electrical impulses that leave the SA node travel to

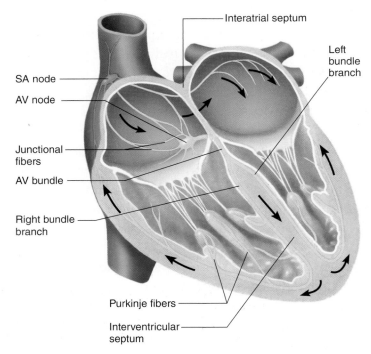

FIGURE 11-12 The cardiac conduction system coordinates the cardiac cycle.

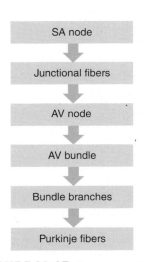

FIGURE 11-13 Components of the cardiac conduction system.

TABLE 11-2 ECG Tracing Components and Pattern

Component	Typical Appearance	Heart Activity
P wave	Upward small curve	Atrial depolarization with resulting atrial contraction
QRS complex	Q, R, and S waves	Ventricular depolarization and resulting ventricular contraction (larger than the P wave); atrial repolarization occurs (not seen)
T wave	Small upward sloping curve	Ventricular repolarization
U wave	Small upward curve	Repolarization of the bundle of His and Purkinje fibers (not always seen)
PR interval	P wave and baseline prior to QRS complex	Beginning of atrial depolarization to the beginning of ventricular depolarization
QT interval	QRS complex, ST segment, and T wave	Period of time from the start of ventricular depolarization to the end of ventricular repolarization
ST segment	End of QRS complex to the beginning of T wave	Time between ventricular depolarization and the beginning of ventricular repolarization

The S-T segment is elevated above the baseline in acute myocardial infarction and depressed when the heart is experiencing hypoxia. The Q-T interval usually extends from the beginning of the QRS to the end of the T wave. This interval may be lengthened by damage to the heart muscle, electrical abnormalities, or myocardial ischemia. There may be irregularities to the conduction system of the heart. For example, ventricular tachycardia is a rapid heart rate involving the ventricles that can progress to ventricular fibrillation, which can be life threatening. Figure 11-15 illustrates ventricular fibrillation. Tachycardia is a heart rate above 100 beats per minute as shown in Figure 11-16. Figure 11-17 shows bradycardia, which is a heart rate below 60 beats per minute. Atrial flutter is an abnormally rapid rate of atrial depolarization as seen in Figure 11-18.

U check
Which wave represents atrial contraction?

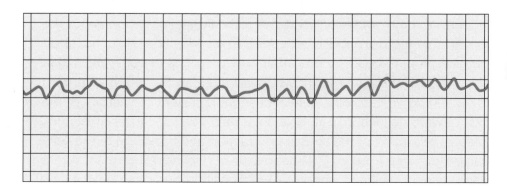

FIGURE 11-15
Ventricular fibrillation is rapid, uncoordinated depolarization of the ventricles.

FIGURE 11-16
Tachycardia is a rapid heart rate.

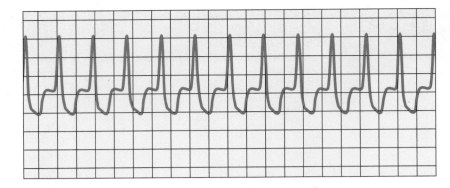

FIGURE 11-17
Bradycardia is a slow heart rate.

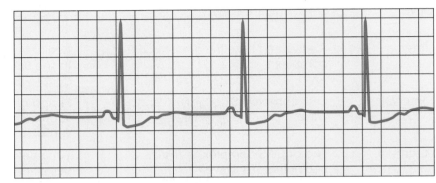

FIGURE 11-18
Atrial flutter is an abnormally rapid rate of atrial depolarization without a 1:1 depolarization of the ventricles.

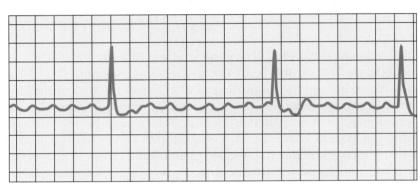

focus on Clinical Applications: Recording an ECG

An ECG is an electrical recording of the heart. A physician may notice some abnormality of the heart while performing a physical examination and recommend that a patient have an ECG, or it may be that the physician wants to get a "baseline" for future reference. Also, when purchasing more life insurance, an insurance company may require that a client have an ECG. The ECG measures the electrical activity of the myocardium and indirectly gives information about the mechanical activity of the heart. To record an ECG, electrodes are placed on the skin of the chest and each of the four extremities. A pen records the heart activity on a strip of paper moving at a known rate. Because the speed of the paper is known, heart rate can be determined from the ECG. The pen moves up and down recording deflections that represent the activity of the heart. For example, the first upward deflection is atrial depolarization which causes atrial contraction. Other waves represent relaxation of the atria or contraction and relaxation of the ventricles. Information that can be gathered from an ECG besides heart rate include past or current myocardial infarction, enlargement of the heart, abnormal heart position, and other heart injury.

Chest Pain

Many people go to the emergency room every year because of chest pain. Chest pain must be taken seriously because it may be a symptom of a life-threatening situation. Typically the causes of chest pain can be cardiac or noncardiac in origin. Following are a few causes of each.

Cardiac Causes

Cardiac causes of chest pain include myocardial infarction (heart attack), angina pectoris, and pericarditis, to name just a few. Some of these may present with merely pain or they may be accompanied by other symptoms. Other symptoms may range from minor limitations to activities of daily living all the way to severe limitations, and in some cases death may result.

- *Myocardial infarction (MI):* Commonly called a heart attack, an MI is caused by blockage of coronary arteries and often causes death of the individual. The pain associated with a heart attack is often described as pressure or fullness in the chest. Sometimes pain also occurs in the back, neck, face (especially the temporomandibular joint), shoulder, or arms (the left arm more than the right). Other signs and symptoms may include shortness of breath (SOB), profuse sweating (diaphoresis), nausea, and dizziness. Remember, someone experiencing an MI may not experience all of the symptoms. Women are noted to have fewer of the "classic" signs of MI and need to be educated regarding the differences between men and women. See Table 11-3 for some of the differences in MI signs between men and women. Diabetics do not always have the severe pain that nondiabetics have.

- *Angina pectoris:* Angina is caused by a narrowing or spasm of coronary arteries and is not immediately life threatening. It may be an indication that a person is at increased risk for an MI. The pain of angina is usually described as a tight feeling in the chest and is often brought about by stress or physical activity, called *stable angina*. This type of chest pain usually goes away after the stress or physical activity stops. Sublingual nitroglycerin acts as a vasodilator to alleviate the chest pain.

11.7 learning outcome

Explain the cardiac and noncardiac causes of chest pain.

TABLE 11-3 Differences in MI Signs and Symptoms between Men and Women

Men	Women
Chest pain or discomfort in the center of the chest; described as a heaviness, tightness, pressure, aching, burning, numbness, fullness or squeezing feeling. It is sometimes thought to be indigestion or heartburn.	Pressure or pain in the center of the chest that may radiate to the jaw pain or localize in the jaw
Areas of pain may include other areas of the upper body such as the arms, left shoulder, back, neck, jaw, or stomach	Upper back or shoulder pain; may radiate down the right or left arm
Lightheadedness, dizziness, weakness, or anxiety	Lightheadedness or dizziness
Diaphoresis (cold sweat)	Unusual fatigue that may last for several days
Indigestion; may include vomiting	
Shortness of breath	
Irregular heartbeat; usually rapid	

- *Pericarditis:* This condition is characterized by inflammation of the sac surrounding the heart. It usually produces a sharp and localized pain in the chest. It is not immediately life threatening, but should be treated. This condition often produces a fever.

Noncardiac Causes

Not all chest pain is cardiac in origin. Patients may experience chest pain due to inflammation of the lungs or costal cartilage (costochondritis), gastric or esophageal irritation (heartburn), gallbladder, dental pain, or even emotional stress. Because there is often uncertainty as to the origin of chest pain, it should always be investigated to rule out the more serious cardiac disorders.

- *Heartburn:* When acid from the stomach enters the esophagus, the individual may experience a burning sensation in the chest. This is known as gastroesophageal reflux disease, or GERD. This pain usually follows a meal and gets worse if a patient bends forward or tries to lie on the back.
- *Panic attacks:* During times of intense stress or fear, chest pains can occur and are often accompanied by increased heart and breathing rates. There may also be excessive sweating.
- *Pleurisy:* This condition occurs when the membranes surrounding the lungs become inflamed. Pleurisy produces a sharp chest pain that usually feels worse when a patient coughs or inhales.
- *Costochondritis:* This condition is also called *Tietze's syndrome* and occurs when there is inflammation at the junction of the rib bone and breastbone (sternum). The pain can mimic that of a heart attack.
- *Pulmonary embolism:* This is caused by a blood clot in an artery in the lungs. The pain is severe, sharp, and increases with deep breaths or coughing. There may be shortness of breath, an increased heart rate, and dizziness, and it can cause sudden death.
- *Myalgia (muscle pain):* Myalgia is caused by strained or injured muscles. It is made worse with movement (especially of the arms) and made better with relaxation.
- *Broken ribs:* Fractured ribs tend to produce sharp and localized chest pain. They are more painful with movement.
- *Inflammation of the gallbladder (cholecystitis) or pancreas (pancreatitis):* Pain associated with these conditions usually begins in the abdomen and spreads to the chest.

Diagnostic Tests

Many diagnostic tests, such as the electrocardiogram, have been around for years. Others, such as nuclear scans, are of more recent origin. Also, some are more invasive (involve entry into the body) than others. Physicians now have an arsenal of tests to help them make a quick and definitive diagnosis of most chest pain.

- *Electrocardiogram (ECG/EKG):* This test can help determine if someone is experiencing or has experienced a heart attack in the past.
- *Stress tests:* Stress tests are ECGs performed while a patient is exercising or has been given drugs to increase the heart rate. Stress tests are useful for determining the health of the coronary arteries.
- *Blood tests:* When heart tissue is damaged such as from an MI, the cardiac muscle releases certain enzymes into the blood. These include creatine phosphokinase (CPK) and lactate dehydrogenase (LDH).

- *Chest x-ray:* X-rays show the size and shape of the lungs and heart and can therefore indicate conditions such as congestive heart failure in which the heart may be enlarged.
- *Nuclear scan:* These scans follow radioactive substances through the blood vessels of the heart and lungs and can reveal narrow or obstructed arteries.
- *Electron beam computerized tomography (EBCT):* This procedure is much like a CT scan of the arteries and is useful for finding narrowed arteries.
- *Coronary catheterization:* This procedure uses a contrast medium that is followed through coronary arteries. It can also show narrowing of the arteries.
- *Echocardiogram:* This procedure uses sound waves (ultrasound) to visualize the shape or defects of the heart.
- *Endoscopy:* This procedure involves inserting a tube with a tiny camera down the throat and into the stomach. It helps diagnose disorders of the stomach or esophagus that might produce chest pain.

PATHOPHYSIOLOGY

Arrhythmias

Arrhythmias, also known as *dysrhythmias,* are abnormal heart rhythms or rates. The most common type of heart arrhythmia is *atrial fibrillation,* which is a rapid but sporadic beating of the atria. The most serious type of heart arrhythmia is *ventricular fibrillation,* which results when the ventricles receive rapid, uncoordinated electrical impulses that cause them to quiver instead of producing a coordinated contraction. Most sudden cardiac deaths are caused by ventricular fibrillation.

Etiology: These abnormal rhythms usually result when the electrical impulses of the cardiac conduction system do not travel correctly through the heart. The list of risk factors and causes is long. Some examples include:

- Certain drugs (such as amphetamines, cocaine, and ephedrine)
- Hypertension
- Previous heart attack
- Coronary artery disease
- Valvular heart disorders
- Thyroid conditions
- Diabetes mellitus
- Sleep apnea
- Smoking
- Excess alcohol and/or caffeine consumption

Signs and Symptoms: Symptoms of arrhythmia include dyspnea (shortness of breath), dizziness or fainting, an uncharacteristically rapid or slow heart rate, a fluttering feeling in the chest, and chest pain.

(continued)

Coronary artery disease

Arrhythmias (concluded)

Treatment: The first management of arrhythmias should be to treat the underlying cause. Other treatment options include the following:

- Pacemakers
- Various medications such as beta blockers and anti-arrhythmics
- Cardiopulmonary resuscitation if there is no evidence of blood flow
- Vagal maneuvers to slow the heart rate; these include holding the breath, straining (having the patient bear down as if he or she is having a bowel movement), or putting the face in cool water
- Electrical shock (defibrillation) to reset heart rhythms
- Radiofrequency catheter ablation, a procedure that destroys a small amount of heart tissue to change the flow of the electrical current through the heart
- Use of an implantable cardioverter defibrillator (ICD), a device that regulates heart rhythms
- Maze procedure, a surgical procedure that forms scars in the atria. This procedure is used to treat atrial fibrillation. The scar tissue will not conduct an electrical impulse and therefore will block the arrhythmia. The scar tissue directs the impulses through a controlled path or "maze" to the ventricles.
- Surgery to correct heart defects such as narrow coronary arteries

Congestive Heart Failure

Heart failure overview

Congestive heart failure (CHF) is failure of the heart to pump effectively. It usually develops over a long period of time. The heart weakens and eventually loses its ability to supply blood to the body. Heart failure is sometimes classified as left-sided or right-sided failure depending on the side of the heart affected.

Etiology: Many risk factors exist for this condition, including smoking, being overweight, a diet high in fats and cholesterol, a lack of exercise, atherosclerosis, a history of MI, hypertension, a damaged heart valve, excessive alcohol consumption, and diabetes mellitus. Congenital heart defects (those present at birth) and drugs that weaken the heart (especially cocaine, heroin, and some antineoplastic drugs for cancer) may also contribute to the development of this disorder.

Left-sided heart failure

Right-sided heart failure

Signs and Symptoms: Signs and symptoms depend on what side of the heart is failing. With left-sided heart failure, there is not enough blood pumping out of the heart to the body. Blood starts to back up into the chest and lungs. Some of the many signs and symptoms of left-sided heart failure include lightheadedness, kidney failure, shortness of breath, pulmonary hypertension, and cough. The signs and symptoms of right-sided heart failure include fluid accumulation and edema (swelling) of the feet, ankles, liver, and abdomen. Additionally, the patient may experience nausea and loss of appetite which can lead to *cachexia* (severe wasting of the body).

Congestive Heart Failure *(concluded)*

Treatment: Common treatment options include medications to slow a rapid heart beat, diuretics to decrease edema and fluid accumulation in the lungs, and medications to reduce blood pressure. If the heart failure has progressed significantly, procedures such as surgery to repair defective heart valves or other heart defects, implantation of a cardiac pacemaker, or a heart transplant may be warranted.

Myocardial Infarction

An "infarction" is death of tissue due to deprivation of oxygen and it can occur in various organs. In the brain, it is called a cerebrovascular accident (CVA, or stroke). Infarction may also occur in joints, the spleen, and the lungs. In the heart it is called **myocardial infarction (MI),** or more commonly, a heart attack. When an MI occurs, the cardiac muscle sustains damage because of the resulting ischemia (*isch* = blockage; *emia* = blood). In the past, most MIs were fatal; however, today more people are surviving heart attacks. One of the reasons for this increased survival is that people are more aware of the signs and symptoms of heart attacks and therefore are seeking earlier treatment. Additionally, more people are trained in cardiopulmonary resuscitation (CPR) and in using automated external defibrillators (AEDs) that are being placed in many public places. Remember, heart tissue that dies in an MI does not regenerate. It is a permanent situation that can affect the efficiency of the heart for the rest of the patient's life.

Etiology: Myocardial infarctions can be caused by obstruction of the coronary arteries as a result of atherosclerosis, a thrombus, or an embolus. Drugs such as cocaine can also cause coronary arteries to spasm. Preventing an MI includes treating or reducing the risk of atherosclerosis, including the use of lipid-lowering agents to lower cholesterol. Risk of MI may be further lowered by controlling blood pressure, not smoking, limiting alcohol intake, having a healthy diet, engaging in regular physical exercise, and avoiding the use of drugs such as cocaine.

Signs and Symptoms: Common symptoms of MI include recurring, squeezing chest pain; pain in the shoulder, arm, back, teeth, or jaw; chronic pain in the upper abdomen; shortness of breath (SOB), especially on exertion (known as dyspnea on exertion, or DOE); diaphoresis (excessive sweating); dizziness or fainting; and nausea or vomiting. It should be noted that MI symptoms in women often do not include typical symptoms such as chest pain, instead often presenting more as indigestion, tiredness, arm pain, and SOB.

Treatment: The first treatment, if possible, is chewing (not swallowing) an aspirin at the onset of symptoms. This allows the aspirin to be absorbed more quickly into the circulation. Aspirin can help

(continued)

---

myocardial infarction (MI) A heart attack.

Myocardial Infarction (concluded)

REMEMBER KRISTA, our investment banker? Which of the symptoms she is having are consistent with a myocardial infarction? Is there anything she could do before emergency personnel arrive?

begin to break up blood clots which will allow essential oxygen to get to the myocardium. In an unconscious patient without a pulse or respiration, CPR should be administered. Other treatment options include the use of an AED (if available) and thrombolytic drugs to destroy the blood clots that block a coronary artery. These drugs are effective only if begun within three hours of the first symptoms, so time is crucial. Anticoagulant medications, such as heparin and warfarin (Coumadin), should be administered to thin the blood, as well as medications that slow the heart rate, such as atenolol. Angioplasty or coronary artery bypass graft (CABG) to open up the coronary arteries are also treatment options.

11.8 learning outcome

Relate the effects of aging of the heart to life span.

Life Span

As we age, changes occur in the vascular system that impact the heart. Because arterial elasticity is much less at age 70 than it is at 20, hypertension is common and the heart has to work harder. Heart rate decreases slightly with age and the cardiac cycle lengthens. The heart muscle may hypertrophy and the heart valves thicken and may calcify. However, with appropriate exercise, the health of the cardiovascular system can be maintained for most individuals.

summary

learning outcomes	key points
11.1 Identify the location, chambers, valves, and layers of the heart and some disorders that affect them.	The heart lies behind the sternum just to the left of midline of the body. It occupies the thoracic mediastinum. The heart contains two atria and two ventricles. The ventricles are more muscular than the atria, and the left ventricle is more muscular than the right ventricle. There are four valves in the heart. The tricuspid valve is located between the right atrium and right ventricle. The pulmonary semilunar valve is located at the junction of the right ventricle and pulmonary trunk. The bicuspid valve is between the left atrium and left ventricle. The aortic semilunar valve is located at the exit of the left ventricle into the aorta. There are three layers to the heart: the endocardium (innermost), myocardium (middle muscular layer), and epicardium (outermost layer).
11.2 Interpret the cardiac cycle.	The cardiac cycle is made up of the events from the beginning of one heartbeat to the beginning of the next heartbeat. The two atria of the heart contract and then relax simultaneously. The two ventricles also contract and relax simultaneously.

chapter 11

learning outcomes	key points
11.3 Relate the major heart sounds to their significance.	The first heart sound is lubb, the closing of the tricuspid and bicuspid valves during ventricular contraction. The second heart sound is dubb and takes place when the pulmonary and aortic semilunar valves close and the atria contract.
11.4 Discuss how cardiac output is determined.	Cardiac output is the volume of blood pumped by the heart each minute. Cardiac output equals stroke volume times heart rate.
11.5 Compare the components of the conduction system of the heart.	The components of the conduction system of the heart are the sinoatrial and atrioventricular nodes, the bundle of His, left and right bundle branches, and Purkinje fibers.
11.6 Explain the electrical events in the heart and abnormalities.	The sinoatrial node is the primary pacemaker. This begins the electrical activity. From here, the electrical impulse goes to the atrioventricular node and then to the bundle of His, the bundle branches, and then to the Purkinje fibers that then excite the heart cells to cause contraction of the heart.
11.7 Explain the cardiac and noncardiac causes of chest pain.	Chest pain can be cardiac in origin (myocardial infarctions, angina pectoris, and pericarditis) or noncardiac in origin (heartburn, costochondritis, and panic attacks).
11.8 Relate the effects of aging of the heart to life span.	As individuals age, structural and physiological changes occur that impact heart health. Cardiovascular disease is the most common cause of death in the United States.

case study 1 **questions**

Can you answer the following questions that pertain to the case presented earlier in this chapter?

1. Which of the following is *not* likely to be a sign of a myocardial infarction?
   a. Chest pain
   b. Diaphoresis
   c. Increased appetite
   d. Referred pain down the left arm

2. How does the pain experienced in angina pectoris compare to the pain of a myocardial infarction?
   a. The pain typically lasts longer with angina pectoris.
   b. The pain typically lasts longer with a myocardial infarction.
   c. The pain in angina pectoris often radiates down the right arm.
   d. Nitroglycerin can relieve the pain of a myocardial infarction in most cases.

3. Which of the following may be a sign of congestive heart failure?
   a. Swelling of the lower extremities
   b. Diaphoresis
   c. Low back pain
   d. Wheezing when breathing

4. Which of the following procedures can help diagnose a myocardial infarction?
   a. Urinalysis
   b. Spirometry
   c. ECG
   d. EEG

review questions

1. What effects will an embolus that has broken loose from a deep vein thrombosis (DVT) cause compared to an embolus that has detached from a thrombus in the wall (mural thrombus) of the left atrium?
   a. An embolus from a DVT is more likely to cause a pulmonary infarction.
   b. An embolus from a DVT is harmless.
   c. An embolus from a DVT is more likely to cause a stroke.
   d. An embolus from a DVT is no different than an embolus from a mural thrombus.

2. A thrombus or blood clot in the anterior interventricular artery (the major artery supplying the left ventricle) is most likely to affect which of the following?
   a. Left ventricle
   b. Brain
   c. Lungs
   d. Right atrium

3. All of the following are tests that may be ordered to make a diagnosis of myocardial infarction *except*
   a. EEG
   b. Cardiac enzymes
   c. ECG
   d. Doppler

critical thinking questions

1. Trace the flow of blood through the heart, mentioning relevant chambers, valves, and blood vessels.
2. Choose a disease mentioned in the chapter that you found particularly interesting. What are some of the signs and symptoms? How is the condition diagnosed and treated? What is the prognosis and what is your patient teaching?
3. You friend's mother is going to have an ECG. Explain to your friend how the ECG is used to diagnose heart problems.
4. What are at least three things that would have a positive effect in reducing the risk for heart disease?

patient education

You want to be able to understand how to live heart-healthy so you can teach the patient how to live heart-healthy. Create a teaching plan that includes the important aspects of patient education for heart-healthy living such as prevention, detection, and treatment of risk factors. Role-play a patient education session with a partner using your teaching plan.

applying what you know

1. With your understanding of the factors that influence cardiac output, which of the following is typically true for well-trained athletes?
   a. Stroke volume is decreased.
   b. Stroke volume is increased.
   c. Resting heart rate is increased.
   d. Resting respiratory rate is increased.

2. Imagine that you are a drop of blood traveling through the heart. For each of the following statements, identify the vessel or structure you are in.
   a. Returning from the brain, you are about ready to enter the heart. What vessel are you in?
   b. After you enter the right atrium, you have to go through a door in order to enter the right ventricle. What is the name of this door?
   c. You made it through the lungs successfully and are now traveling back to the heart. What vessels are you in?

CASE STUDY 2 *Sudden Chest Pain*

Steven Phillips, a 45-year-old man working in car sales, was playing a pickup game of basketball with his son. Although overweight and a heavy smoker, Steven was going to show his son the "old man" still had it. During the game he started to sweat and feel pain in his chest. He immediately stopped and sat down to rest. Within 10 minutes, the pain went away.

1. Severe sweating is known as
   a. Electrophoresis
   b. Diuresis
   c. Meiosis
   d. Diaphoresis
2. What do you suspect has happened to Steven?
   a. He broke a rib.
   b. He was having a heart attack.
   c. He was experiencing angina pectoris.
   d. He was having an asthmatic attack.
3. What tests would the physician run to confirm your suspicions?
   a. EEG
   b. U/A
   c. ECG
   d. Nerve conduction test
4. What treatment would you expect the doctor to prescribe?
   a. Nitroglycerin
   b. Calcium channel blockers
   c. ACE inhibitors
   d. Prozac
5. Which of the following is *not* a risk factor for heart disease?
   a. Smoking
   b. Low-cholesterol diet
   c. A sedentary lifestyle
   d. Family history of heart disease

CASE STUDY 3 *Shortness of Breath*

An 87-year-old retired army gunnery sergeant, Jack Lamborn, has been experiencing shortness of breath and swelling of the ankles. This has been going on for several years, but has gotten much worse in the last eight months. He has no diagnosed history of heart disease. However, Jack does report being a heavy smoker since joining the army at 17 years of age. He also admits to drinking "more than I should."

1. What are Jack's symptoms indicative of?
   a. Congestive heart failure
   b. Stroke
   c. Epilepsy
   d. Angina pectoris
2. What is dyspnea?
   a. Profuse sweating
   b. Ringing of the ears
   c. Difficulty breathing
   d. Chest pain
3. What factor may have contributed to Jack's condition?
   a. Heavy smoking
   b. Drinking too much
   c. Kidney failure
   d. Heart failure

12 The Lymphatic and Immune Systems

chapter outline

Lymphatic System Components and Functions

Disease Defenses

Immune System Responses and Acquired (Specific) Immunities

Transplantation and Tissue Rejection

Major Immune System Disorders

Life Span

outcomes

learning

AFTER COMPLETING THIS CHAPTER, YOU WILL BE ABLE TO:

12.1 Explain the components of the lymphatic system and their functions.

12.2 Compare innate (nonspecific) and acquired (specific) immunity.

12.3 Distinguish between the four types of acquired (specific) immunity.

12.4 Explain the different types of grafts or transplants.

12.5 Describe the major immune system disorders.

12.6 Describe changes to the immune system with aging.

case study

Use the case study to focus on as you go through the chapter. The questions will guide you as you learn the anatomy, physiology, and pathology of the lymphatic and immune systems.

CASE STUDY 1 *Lymphadenopathy*

A few days ago 17-year-old Lisa Haverland came to the doctor's office very concerned that she had a "serious, possibly fatal, illness." She has been running a slight fever for the past week, has been very tired, has tender lymph nodes in her neck, and has been losing weight without dieting. Her chart indicates that she has never been sexually active and has never used intravenous drugs.

As you go through the chapter, keep the following questions in mind:

1. Is it likely that Lisa has a life-threatening illness?
2. What tests should be done to diagnose Lisa's condition?
3. Based on her symptoms, what disease or disorder is Lisa likely to have?
4. Is Lisa's condition contagious? What precautions should she take to avoid spreading her illness?

essential terms

acquired (specific) immunity (a-KWIRD i-MU-ni-te)
allergen (AL-er-jen)
antigen (AN-ti-jen)
autoimmune disease (aw-toe-ih-MUN di-ZEZ)
cytokines (SI-toe-kines)
histamine (HISS-ta-men)
human leukocyte antigens (HLAs) (human LOO-ko-site AN-ti-jens)
humoral immunity (HU-mor-al i-MU-ni-te)
immunoglobulins (Igs) (i-mu-no-GLOB-u-lins)
immunity (i- MU-ni-te)
innate (nonspecific) immunity (in-NATE i- MU-ni-te)
interstitial (tissue) fluid (in-ter-STISH-al FLOO-id)
lymph (LIMF)
lymph node (LIMF node)
plasma cells (PLAZ-ma cells)
spleen (SPLEN)
thymus (THI-mus)
tonsils (TON-sils)

Additional key terms in the chapter are italicized and defined in the glossary.

301

study tips

1. Write a short essay two to four paragraphs in length comparing innate (nonspecific) and acquired (specific) immunity.
2. Draw the human body and locate the major lymph vessels and the major groups of lymph nodes.
3. Make a table comparing T lymphocytes and B lymphocytes. In your table, compare site of origin, site of differentiation, primary locations, and functions.
4. Meet with a study group and review and quiz each other.

Introduction

The *immune system* is responsible for protecting humans against bacteria, viruses, fungi, toxins, parasites, and cancer. It works with the organs of the *lymphatic system*—the thymus, spleen, lymphatic vessels, lymph nodes, and lymphoid tissue—to clear the body of these disease-causing agents (Figure 12-1).

LO 12.1

learning **outcome**

Explain the components of the lymphatic system and their functions.

lymph Fluid collected from the spaces between cells into the lymphatic vessels.

Lymphatic System Components and Functions

The lymphatic system is a network of connecting vessels that collects the interstitial (or tissue) fluid found between cells. These lymphatic vessels then return this fluid, now called **lymph,** to the bloodstream. The lymphatic system also picks up lipids and fat-soluble vitamins (A, D, E, and K) from the digestive tract and transports them to the bloodstream. The third function of the lymphatic system is to protect the body against disease-causing agents.

Interstitial (Tissue) Fluid and Lymph Fluid

Fluid constantly leaks out of blood capillaries into the spaces between cells. This fluid is high in nutrients, oxygen, and small proteins. Most of this fluid is picked up by body cells. However, some of the fluid persists between cells. Increased tissue hydrostatic pressure moves **interstitial (tissue) fluid** into the lymphatic vessels. This fluid is destined to become lymph.

interstitial (tissue) fluid The fluid found between the cells.

> **U check**
> How is lymphatic fluid moved into the lymphatic vessels?

Lymph and lymph node circulation

Lymphatic Vessels and Lymph Circulation

The lymphatic vessels transport lymph and excess fluid away from the interstitial spaces toward the heart. Smaller lymphatic capillaries join to form larger lymphatic vessels.

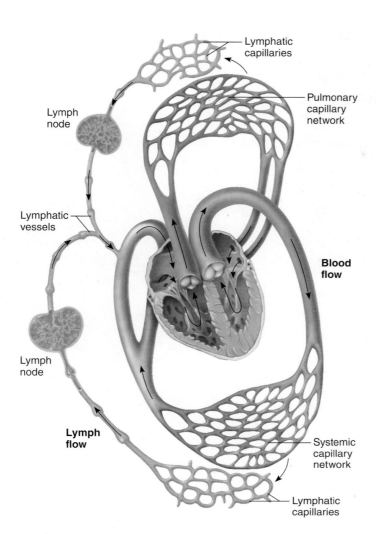

FIGURE 12-1
Schematic representation of lymphatic vessels.

Lymphatic Capillaries: The lymphatic capillaries are similar in structure to blood capillaries except that lymph capillaries are larger in diameter and normally allow interstitial fluid to flow into the lymph capillaries but not out. This is facilitated by the thin, very permeable lymph capillary walls that are lined by a single layer of squamous epithelial cells called endothelium. The epithelial cells of the lymphatic capillary wall overlap, creating flap-like valves that allow fluid to enter the capillary, but do not allow fluid to exit under normal conditions. However, under some conditions, lymph can leak out of the vessels, causing *edema* or fluid buildup in the interstitial spaces (Figures 12-2 and 12-3). Lymph capillaries converge to form lymphatic trunks, which in turn merge to form lymphatic ducts. This is similar to the structure of the venous system, but lymphatic vessels have thinner walls and more valves than do veins.

> **Ü check**
> What is edema?

Lymphatic Trunks and Ducts: Lymphatic trunks are typically named after the region in which they are found. For example, the *jugular trunks* (left and right) drain the head and neck region. The *lumbar trunk* drains

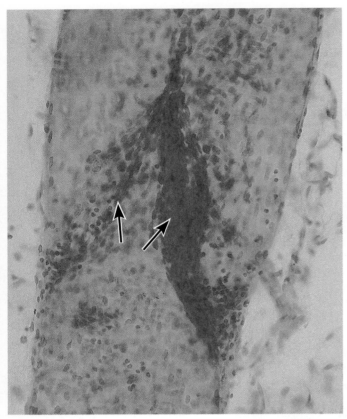

FIGURE 12-2 Light micrograph of the flaplike valve (arrows) within a lymphatic vessel (60x).

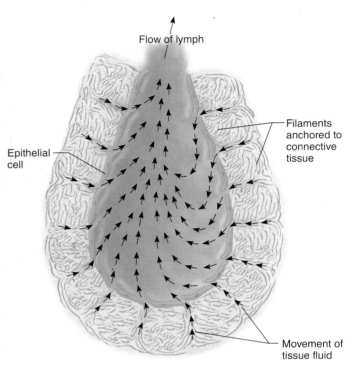

FIGURE 12-3 Interstitial (tissue) fluid enters lymphatic capillaries through flaplike valves between epithelial cells.

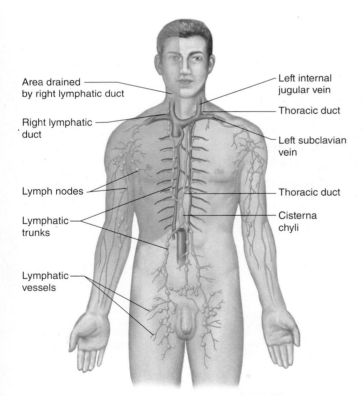

FIGURE 12-4 Areas drained by the right lymphatic duct (shaded) and the thoracic duct (not shaded).

the lymph from the lower extremities, the *subclavian trunk* drains the upper limbs, and the *bronchomediastinal trunk* drains lymph from the thorax. The trunks eventually empty into the two lymphatic ducts. The *right lymphatic duct* receives lymph from the upper-right side of the body. It empties its contents into the right internal jugular and right subclavian veins. These veins return the lymph to the right atrium by way of the superior vena cava.

The other duct is the *thoracic duct* which is the largest lymphatic vessel in the body. It drains lymph from all parts of the body that are not drained by the right lymphatic duct. The thoracic duct begins at the level of the second lumbar vertebra and has a dilated sac or channel called the *cisterna chyli* (Figure 12-4). (Cisterna comes from the word *cistern,* meaning "reservoir," and *chyle* is the milky white fluid that is obtained from dietary lipids.) The thoracic duct passes through the diaphragm beside the aorta. The thoracic duct empties into the junction of the left internal jugular and left subclavian veins.

Lymph is moved along by two pumps and flows in only one direction—toward the heart. The *skeletal muscle pump* utilizes skeletal muscle contractions to move the lymph toward the heart. The other pump is the *respiratory pump* that utilizes pressure changes

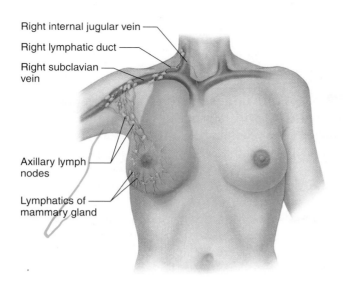

FIGURE 12-5 Lymph drainage of the right breast illustrates a localized function of the lymphatic system. A mastectomy can disrupt this drainage, causing painful edema.

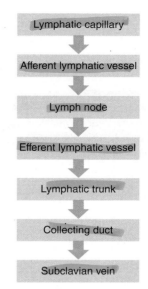

FIGURE 12-6 The lymphatic pathway begins with lymphatic capillaries; lymph moves through nodes and additional vessels, and then drains into the subclavian vein.

in the thorax to assist circulation. Along its way to the venous blood, lymph will pass through more than 600 *lymph nodes* that are found throughout the body (Figures 12-5 and 12-6).

U check
What is interstitial (tissue) fluid?

PATHOPHYSIOLOGY

Lymphedema

Lymphedema is the blockage of lymphatic vessels.

Etiology: This condition may be caused by genetics, parasitic infections, trauma to the vessels, tumors, radiation therapy, cellulitis (a skin infection), and surgeries such as mastectomies and biopsies in which lymph tissues have been removed.

Signs and Symptoms: The common symptom consists of tissue swelling that lasts longer than a few days or increases over time.

Treatment: Treatment options include compression stockings for swelling in the legs or arms, elevation of the affected limb, or surgery to remove abnormal lymphatic tissue. Physical and massage therapy have also been found helpful in the early stages of lymphedema to spread the fluid into surrounding tissues for reabsorption. Patients with lymphedema need to maintain good nutrition, reduce foods high in salt and fat, exercise regularly, avoid infections, and keep the skin clean.

Lymphoid Organs and Tissues

Lymphoid organs and tissues are widespread throughout the body and consist of lymph nodes, thymus, and spleen.

lymph node An oval- or bean-shaped structure that is part of the immune system and is located along the paths of lymphatic vessels.

Lymph Nodes: **Lymph nodes** are very small, glandular structures that usually cannot be felt or palpated very easily (Figure 12-7). They are located along the paths of larger lymphatic vessels and are spread throughout the body. A major exception is that no lymph nodes are found in the nervous system. There is a greater concentration of nodes in certain places in the body such as the cervical (neck), axillary (armpit), inguinal (groin), supratrochlear (medial side of the elbow), pelvic, thoracic, and aortic (thorax) nodes. The indented side of a lymph node is called the hilum. Nerves and blood vessels enter the node through the hilum. The lymphatic vessels that carry lymph to the node and are located opposite the hilum are called afferent lymphatic vessels (meaning "toward"). About four or five afferent vessels are associated with each node. The lymphatic vessels that carry lymph out of a node are called efferent vessels (meaning "away from").

A lymph node usually has only one or two efferent vessels. Because more lymph enters the node than can exit at one time, lymph tends to concentrate in the node and pressure builds up that assists in the filtration process. Lymph nodes are also surrounded by a fibrous capsule of connective tissue. The lymph node is divided into an inner portion called the medulla and an outer portion called the cortex.

Two important cell types are found inside the nodes—macrophages and lymphocytes. Together these form the *lymph nodules,* the functional units of the lymph node. Lymph nodules are found in the cortex of the lymph node. Macrophages digest unwanted pathogens in the lymph and the lymphocytes are part of the immune response against the pathogen. Lymph nodes are also responsible for the generation of some lymphocytes.

When someone has a viral or bacterial infection, he or she may have *lymphadenitis,* an inflammation of the lymph nodes. Any disease of the

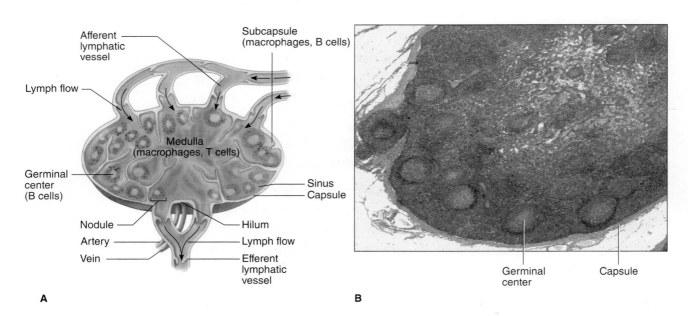

FIGURE 12-7 (A) Section of a lymph node. (B) Light micrograph of a lymph node (20x).

lymph nodes is called *lymphadenopathy*. The terms *lymphadenitis* and *lymphadenopathy* are often used interchangeably by health care professionals, but it is important to remember that the suffix *–itis* refers to inflammation and the suffix *–pathy* refers to disease. Other causes of lymphadenopathy are autoimmune disease and malignancy.

> **U check**
> What is the structure of a lymph node?

REMEMBER LISA, the young woman in the case study experiencing tender lymph nodes? What are some possible reasons for her symptoms?

Thymus The **thymus** is a soft, bilobed organ located behind the sternum, just below the thyroid gland and above the heart in the mediastinum (a space located between the right and left lungs). The thymus is large in the infant and reaches its maximum size of 1 to 2 ounces when the child is about two years of age. After adolescence the thymus starts to atrophy (waste away) or *involute* (turn in on itself). In older adults, the thymus is tiny, almost nonexistent (Figure 12-8). Starvation or acute disease can sometimes accelerate this process. The outer portion of the thymus is called the cortex. This is where T lymphocytes (T cells) that have been produced in the bone marrow proliferate. They then move to a more central portion of the gland called the medulla, where they mature. The thymus also produces the hormone *thymosin* which stimulates the production of mature lymphocytes.

thymus A bilobed organ of the immune system located in the mediastinum, which is responsible for the maturation of T lymphocytes (T cells).

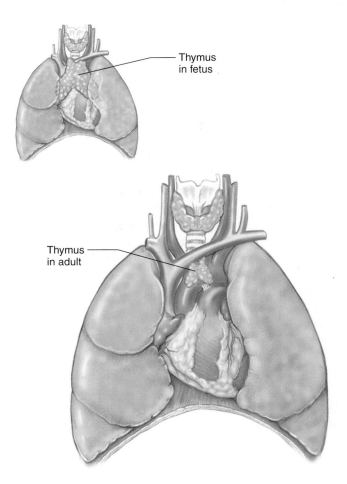

FIGURE 12-8
Size of the thymus relative to other thoracic organs in the fetus and the adult (figure is not drawn to scale).

CHAPTER 12   The Lymphatic and Immune Systems

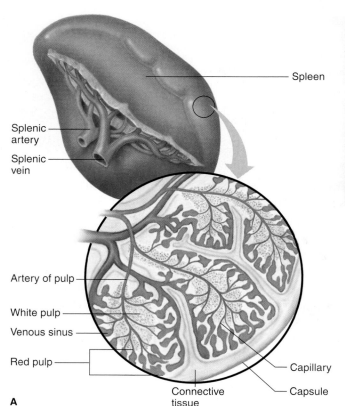
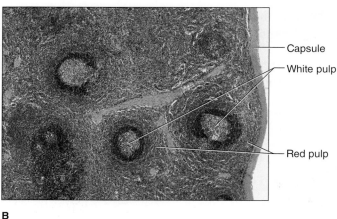

FIGURE 12-9
(A) The spleen resembles a large lymph node.
(B) Light micrograph of the spleen (40x).

spleen Large lymphoid organ located on the left side of the upper abdomen that is involved in blood cell formation in the fetus and filtering of the blood after birth.

Spleen: The **spleen** is the largest lymphatic organ (Figures 12-9 and 12-10). It is located in the upper-left quadrant of the abdominal cavity, just below the diaphragm and behind the stomach. It is protected by the rib cage and normally is not palpable. The spleen is divided into lobules with two types of tissues: white pulp and red pulp. The white pulp is concentrated with lymphocytes similar to those seen in lymph nodes. Red pulp has an abundance of red blood cells, lymphocytes, and macrophages. The spleen filters blood in much the same way that lymph nodes filter lymph. The spleen also removes senescent (old) worn-out red blood cells from the bloodstream. If the spleen is injured or becomes enlarged due to disease, a condition known as *splenomegaly*, the spleen is often removed (a *splenectomy*) to prevent rupture of the spleen if trauma were to occur. When a splenectomy is performed, the patient's liver takes over most of its functions.

Table 12-1 summarizes the characteristics of the major organs of the lymphatic system.

U check
What is the function of the spleen?

tonsils An aggregation of lymphatic nodules located in the throat.

Lymphatic Nodules: Lymphatic nodules are masses of lymphatic tissue not surrounded by a capsule. These are often referred to as mucosa-associated lymphoid tissue (MALT) because they are distributed in the connective tissue of mucosa. The **tonsils** are three sets of lymphoid tissue that form a ring known

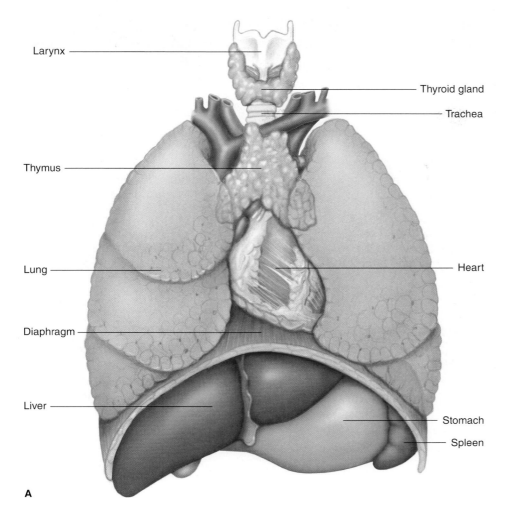
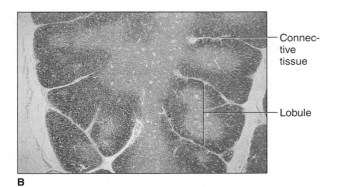

FIGURE 12-10
(A) The thymus is bilobed and is located between the lungs and superior to the heart; the spleen is inferior to the diaphragm and posterior and lateral to the stomach. (B) A cross section of the thymus (15x); note how the thymus is divided into lobules.

as the *Ring of Waldeyer* at the junction of the oral cavity and oropharynx and the junction of the nasal cavity and nasopharynx. These three sets include

- Pharyngeal tonsils or adenoids, located at the junction of the mouth and oropharynx
- Palatine tonsils, located at the junction of the nasal cavity and nasopharynx
- Lingual tonsils, located at the base of the tongue

TABLE 12-1 Major Organs of the Lymphatic System

Organ	Location	Functions
Lymph nodes	In groups or chains along the paths of the larger lymphatic vessels	Filter foreign particles and debris from lymph; produce and house lymphocytes that destroy foreign particles in lymph; house macrophages that engulf and destroy foreign particles and cellular debris carried in lymph
Thymus	In the mediastinum posterior to the upper portion of the body of the sternum	Houses lymphocytes; differentiates thymocytes into T lymphocytes
Spleen	In the upper-left portion of the abdominal cavity inferior to the diaphragm, posterior and lateral to the stomach	Blood reservoir; houses macrophages that remove foreign particles, damaged red blood cells, and cellular debris from the blood; contains lymphocytes

The appendix, or *vermiform appendix* (*appendix* = appendage; *vermiform* = wormlike), is usually located in the lower-right quadrant at the junction of the large and small intestines. It was once thought that the appendix had no function. We now know it is part of the immune system. There are also lymph nodules known as Peyer's patches located in the small intestine.

> **U check**
> What is the Ring of Waldeyer?

Disease Defenses

12.2 learning outcome
Compare innate (nonspecific) and acquired (specific) immunity.

immunity Being resistant to injury and infection by pathogens.

innate (nonspecific) immunity Immunity a person is born with.

acquired (specific) immunity Formation of antibodies and lymphocytes after exposure to an antigen.

Immunity, also known as *unsusceptibility,* can be divided into **innate (nonspecific) immunity** and **acquired (specific) immunity.** *Susceptibility* is the lack of resistance or vulnerability to disease. We are born with innate or nonspecific immunity. Acquired or specific immunity builds up over a period of time after exposure to an antigen; we are not born with it. It involves a specific response to a certain antigen or microbe. Acquired immunity is typically slower to respond than innate immunity, but unlike innate immunity, which does not have memory, acquired immunity does remember dangerous microbes.

Disease may result from injury, infection, or malignancy, to name only three potential causes. An *infection* is the presence of a pathogen in or on the body. A pathogen is a disease-causing agent such as a bacterium, virus, toxin, fungus, or protozoan. Certain species are more susceptible to some infections than others. For example, humans do not get plant diseases or many diseases that affect animals. A disease that is transmissible from an animal to humans is called a *zoonotic disease.*

Innate (Nonspecific) Defenses

Let's discuss the body's defenses in more detail. Innate (nonspecific) defense mechanisms include physical barriers, chemicals, phagocytosis, inflammation, and fever. Table 12-2 outlines the various innate defenses.

TABLE 12-2 Types of Innate (Nonspecific) Defenses

Type	Description
Species resistance	A species is resistant to certain diseases to which other species are susceptible
Mechanical barriers	Unbroken skin and mucous membranes prevent the entrance of some infectious agents; fluids wash away microorganisms before they can firmly attach to tissues
Chemical barriers	Enzymes in various body fluids kill pathogens; pH extremes and high salt concentration also harm pathogens; interferons induce production of other proteins that block reproduction of viruses, stimulate phagocytosis, and enhance the activity of cells to resist infection and the growth of tumors; defensins damage bacterial cell walls and membranes; collectins attach to microbes; complement stimulates inflammation, attracts phagocytes, and enhances phagocytosis
Natural killer cells	Distinct type of lymphocyte that secretes perforins that lyse virus-infected cells and cancer cells
Inflammation	A tissue response to injury that helps prevent the spread of infectious agents into nearby tissues
Phagocytosis	Neutrophils, monocytes, and macrophages engulf and destroy foreign particles and cells
Fever	Elevated body temperature inhibits microbial growth and increases phagocytic activity

Physical Barriers: Skin and the mucosa that line the digestive tract, oral cavity, and other areas act as physical or mechanical barriers that prevent infection from various microorganisms. They act as a *first line of defense* against pathogens. Intact or unbroken skin prevents the entry of most organisms into the body, while mucous membranes (mucosa) permit the entry of a few organisms. Because the epidermis of the skin sloughs off on a regular basis, it also removes superficial bacteria on the body's surface. In addition, the hairs in the nostrils and cilia in the respiratory system help move debris out of the body.

Chemical Barriers: Chemicals and enzymes in body fluids provide barriers that destroy pathogens. For example, acids in the stomach destroy pathogens that are swallowed. Lysozymes in tears destroy pathogens that come in contact with the eye surface. Salt in sweat also kills bacteria, and *interferon* in blood blocks viruses from infecting cells. Chemicals, along with the remaining nonspecific defense mechanisms, form the *second line of defense.*

Phagocytosis: Neutrophils and monocytes are white blood cells that have phagocytic characteristics when they encounter pathogens that may have invaded body tissues. Monocytes, after leaving the circulation, are transformed into macrophages, which are very large and active phagocytes capable of ingesting much larger particles than neutrophils can ingest. Together, neutrophils, monocytes, and macrophages make up what is called the *mononuclear phagocytic system* or the *reticuloendothelial system.*

TABLE 12-3 Major Actions of an Inflammatory Response

Action	Result
Blood vessels dilate; capillary permeability increases and fluid leaks into tissue spaces	Tissues become red, swollen, warm, and painful
White blood cells invade the region	Pus may form as white blood cells, bacterial cells, and cellular debris accumulate
Interstitial (tissue) fluids containing clotting factors seep into the area	A clot containing threads of fibrin may form
Fibroblasts arrive	A connective tissue sac may form around the injured tissues
Phagocytes are active	Bacteria, dead cells, and other debris are removed
Cells divide	Newly formed cells replace injured ones

Inflammation

Natural Killer Cells: *Natural killer (NK) cells* are another type of lymphocyte. They primarily target cancer cells but also protect the body against many types of pathogens. Like cytotoxic T cells, NK cells kill harmful cells on contact. They secrete chemicals called *perforins* that punch holes in the membranes of harmful cells, which cause the cells to lyse or rupture. Unlike B and T cells, NK cells do not have to rely on memory to recognize a specific antigen to start destroying pathogens.

Inflammation: When an area of the body becomes injured or infected with a pathogen, inflammation, which is a vascular response, can result. The five cardinal signs of acute inflammation are *heat, swelling, erythema (redness), pain,* and *loss of function.* Blood vessels in the injured area first constrict, then dilate to allow more blood flow to the site. The increased blood flow is responsible for the heat and redness. As the blood vessels dilate, they also become more permeable or "leaky." As fluid leaves the blood vessels, it accumulates in the interstitial spaces, causing edema.

The edema stimulates pain receptors and also causes the release of chemicals that produce pain. The pain and damaged tissue are responsible for loss of function. Certain white blood cells, such as neutrophils and monocytes, leave the blood vessels to fight the infection and clean up the area of dead bacteria and cellular debris. Although inflammation is a protective mechanism, it can sometimes go out of control and cause harm. For example, arthritis is a chronic disease that has flareups (exacerbations) of inflammation that can be destructive to the affected joints. Table 12-3 summarizes the process of inflammation.

Fever: An elevated body temperature is a fever. Patients experiencing a fever are said to be *febrile* (those without fever are termed *afebrile*). Fever causes the liver and spleen to remove iron from the bloodstream. Because many pathogens need iron to reproduce, this inhibits their growth. Fever also activates phagocytic cells in the body to attack pathogens. Like inflammation, fever is a protective mechanism that can go out of control and cause serious harm and even death.

> **U check**
> What are the cardinal signs of acute inflammation?

Acquired (Specific) Defenses

Specific defense is another name for *immunity*. Immunity is the *third line of defense* and is classified as cellular immunity, or humoral immunity. As we mentioned earlier, a major difference between nonspecific defenses and specific defenses is that nonspecific defenses do not have memory and specific defenses do. Specific immunity protects the body against very specific pathogens. For example, a person who has had chickenpox develops a specific defense that prevents that person from getting chickenpox again. However, this specific defense does not protect the person from other diseases.

Antigens are simply defined as foreign substances in the body. Pathogens have many antigens on their surfaces. The immune system is programmed to recognize antigens in the body. Foreign substances called *haptens* are too small to elicit an immune response by themselves, but haptens are able to combine with larger proteins in the blood. They are then capable of eliciting an immune response. Penicillin is an example of a hapten.

Antibodies and complements are the major proteins involved in specific defenses. Lymphocytes and macrophages are the major WBCs (white blood cells) involved in specific defenses. Lymphocytes and macrophages produce **cytokines,** which have a wide array of functions including regulation of the inflammatory response.

B Cells: There are two major categories of lymphocytes: B cells and T cells (Figures 12-11 and 12-12). Both of these cell types are produced in red bone marrow, but they have different functions. B and T cells are found in lymph nodes, the spleen, the thymus, the lining of digestive organs, and bone marrow. Both B and T cells recognize antigens in the body; however, they respond to antigens in different ways. B cells do not attack antigens directly. Some B cells respond to antigens by becoming **plasma cells,** while others become memory B cells. After exposure to antigens, the plasma cells make antibodies that are directed against a specific antigen. This type of antibody production can take days after the first exposure. The antibodies attach to the antigens in the *humors* (fluids) of the body, which is why this type of immunity is known as **humoral immunity,** or *antibody-mediated response*. Groups of similar B cells recognize only one specific type of antigen. They become activated when a specific antigen binds to receptors on the surface of the B cell. The job of *memory B cells* is to recognize the antigen the next time exposure to the same antigen occurs. The memory B cells launch a stronger and more immediate response to the antigen, usually within hours of exposure. This is much faster than the initial response, which may take several days after the first exposure. B-cell activation usually requires assistance of T cells.

T Cells: As mentioned earlier, T lymphocytes are given their name because they go to the thymus to mature or be processed. They account for 60 to 80 percent of circulating lymphocytes. T cells bind to antigens on cells and attack them directly. This type of response is called a *cell-mediated response*, or *cellular immunity*. T cells also respond to antigens by secreting cytokines called *lymphokines*, which increase T-cell production and directly kill cells that have antigenic potential.

Before a T cell can respond to an antigen, it must be activated. T-cell activation begins when a macrophage ingests and digests a pathogen that has antigens on it. The macrophage then takes some of the antigens from the pathogen and puts them on its cell membrane next to a large protein complex called a *major histocompatibility complex (MHC)*. Every human except

antigen A substance that has the ability to elicit an immune response.

cytokines Chemicals that are made by lymphocytes that aid in the destruction of antigens.

plasma cell A cell derived from B lymphocytes and that produce antibodies.

humoral immunity Immune response in which plasma cells are derived from B lymphocytes.

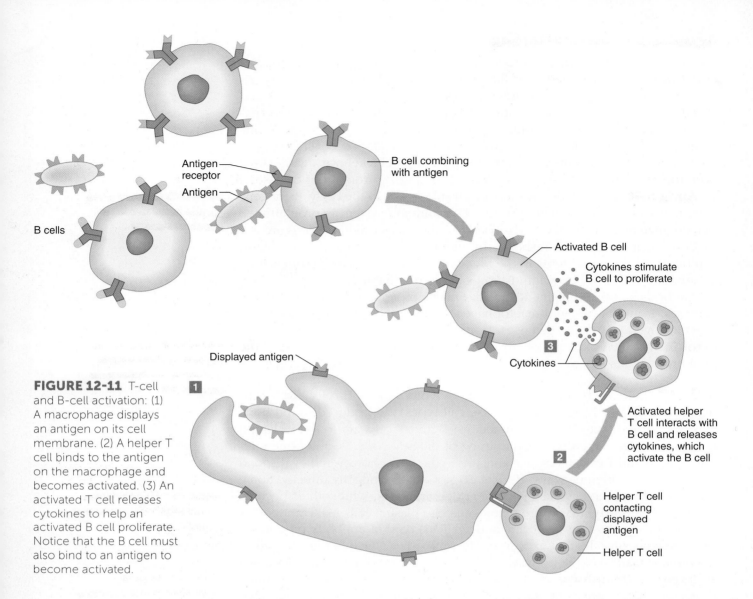

FIGURE 12-11 T-cell and B-cell activation: (1) A macrophage displays an antigen on its cell membrane. (2) A helper T cell binds to the antigen on the macrophage and becomes activated. (3) An activated T cell releases cytokines to help an activated B cell proliferate. Notice that the B cell must also bind to an antigen to become activated.

human leukocyte antigens (HLAs) Surface proteins on white blood cells and other nucleated cells used to type tissues for transplantation.

identical twins has a unique MHC that acts as a "cell identity marker." You can think of it as "internal fingerprints." White blood cells and all other nucleated cells have MHC proteins. Red blood cells, however, do not have a nucleus and therefore do not have MHC proteins. Another name for MHC is **human leukocyte antigens (HLAs)** because they were first identified on human white blood cells. HLAs are actually found on all cells in the body that possess a nucleus. A T cell that has a receptor for the antigen recognizes and binds to the antigen and the MHC on the surface of the macrophage. The T cell is now activated and begins to divide to form other types of T cells. It is important to note that T cells cannot be activated without macrophages and MHC proteins. Some activated T cells become *cytotoxic* T cells. This type of T cell is important in protecting the body against viruses and cancer cells. Other activated T cells become *helper T cells,* which carry out many important roles in immunity. Helper T cells increase antibody formation, memory cell formation, B-cell formation, and phagocytosis. Some activated T cells become memory T cells. Like memory B cells, memory T cells "remember" the pathogen that activated the original T cell. When a person is later exposed to the same pathogen, memory cells trigger an immune response that is more effective than the first

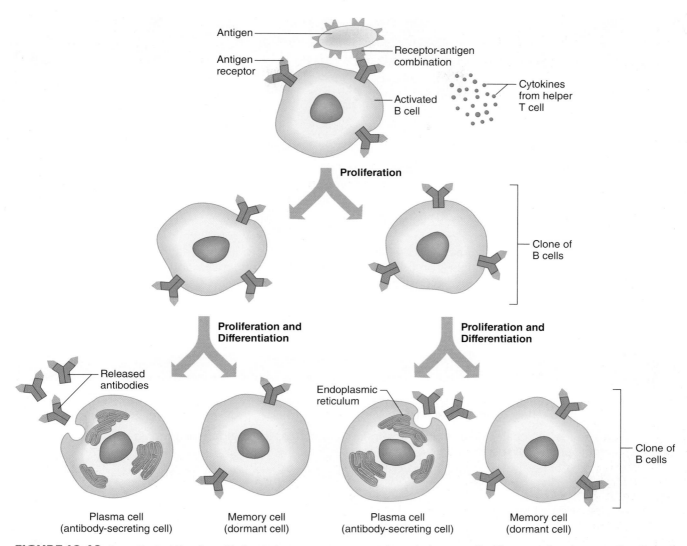

FIGURE 12-12 An activated B cell multiplies to become memory cells and plasma cells. Plasma cells secrete antibodies.

immune response. The production of memory cells prevents a person from experiencing the same disease twice. Some T cells are targeted by certain diseases. For example, a special type of helper cell called a CD4 cell is the target of HIV, the AIDS-causing virus. (CD4 stands for the "cluster of differentiation" antigen it possesses that allows it to recognize a macrophage bearing a foreign antigen.) Table 12-4 compares the characteristics of T and B cells, and Table 12-5 describes the steps in antibody production by T cells and B cells.

Antibodies: Antibodies, also called **immunoglobulins (Ig),** are proteins produced by plasma cells. There are five major classes of antibodies. Their structures are similar in some respects, but differences make them unique in terms of their function. When antibodies bind to antigens, they take one of the following three actions:

immunoglobulin (Ig) An antibody produced by plasma cells in response to exposure to an antigen.

1. They allow phagocytes to recognize and destroy antigens.
2. They make antigens clump together, causing them to be destroyed by macrophages. This is how blood cells are destroyed when an incompatible blood transfusion is made.
3. They cover the toxic portions of antigens to make them harmless.

TABLE 12-4 A Comparison of T Cells and B Cells

Characteristic	T Cells	B Cells
Origin of undifferentiated cell	Red bone marrow	Red bone marrow
Site of differentiation	Thymus	Red bone marrow
Primary locations	Lymphatic tissues, 70 to 80% of the circulating lymphocytes in blood	Lymphatic tissues, 20 to 30% of the circulating lymphocytes in blood
Primary functions	Provide cellular immune response in which T cells interact directly with the antigens or antigen-bearing agents to destroy them	Provide humoral immune response in which B cells interact indirectly, producing antibodies that destroy the antigens or antigen-bearing agents

TABLE 12-5 Steps in Antibody Production

T-Cell Activities

1. Antigen-bearing agents enter tissues.
2. An accessory cell, such as a macrophage, phagocytizes the antigen-bearing agent, and the macrophage's lysosomes digest the agent.
3. Antigens from the digested antigen-bearing agents are displayed on the membrane of the accessory cell.
4. Helper T cell becomes activated when it encounters a displayed antigen that fits its antigen receptors.
5. Activated helper T cell releases cytokines when it encounters a B cell previously combined with an identical antigen-bearing agent.
6. Cytokines stimulate the B cell to proliferate.
7. Some of the newly formed B cells give rise to cells that differentiate into antibody-secreting plasma cells.
8. Antibodies combine with antigen-bearing agents, helping destroy them.

B-Cell Activities

1. Antigen-bearing agents enter tissues.
2. B cell encounters an antigen that fits its antigen receptors.
3. The B cell is activated, either alone or in conjunction with helper T cells. The B cell proliferates, enlarging its clone.
4. Some of the newly formed B cells differentiate further to become plasma cells.
5. Plasma cells synthesize and secrete antibodies whose molecular structure is similar to the activated B cell's antigen receptors.
6. Antibodies combine with antigen-bearing agents, helping destroy them.

TABLE 12-6 Characteristics of Major Immunoglobulins

Type	Occurrence	Major Functions
IgA	Exocrine gland secretions	Defends against bacteria and viruses
IgD	Surface of most B lymphocytes	Unknown; may activate B cells
IgE	Exocrine gland secretions	Promotes inflammation and allergic reactions
IgG	Plasma and tissue fluid	Defends against bacteria, viruses, and toxins; activates complement
IgM	Plasma	Reacts with antigens on some red blood cell membranes following mismatched blood transfusions; activates complement

The following is a list of the different types of immunoglobulins:

- **IgA:** This antibody is found in breast milk, sweat, tears, saliva, and mucus. IgA prevents entry of pathogens into the body. It accounts for 10 to 15 percent of all circulating antibodies and protects the mucosa against bacteria and viruses.
- **IgD:** This antibody was discovered in the 1960s and its role is not completely understood. IgD is found on the cell membranes of B cells and is thought to control the activity of the B cells. It accounts for less than 1 percent of circulating antibodies.
- **IgE:** This antibody is found wherever IgA is located and accounts for less than 1 percent of the antibodies in blood. IgE protects against parasite infections. It is also associated with allergies and hypersensitivity reactions.
- **IgG:** This is the most prevalent antibody in the blood—approximately 80 percent of the total number of antibodies. IgG primarily recognizes bacteria, viruses, and toxins. It can also activate the complement system. This system is made up of over two dozen proteins produced by the liver, which act as a "bridge" between innate and specific immunity. They destroy microorganisms by phagocytosis and the inflammatory response. IgG is the only antibody capable of crossing the placenta; consequently, it protects the fetus.
- **IgM:** This antibody is very large and primarily binds to antigens on food, bacteria, or incompatible blood cells. It also activates the complement system. IgM antibodies make up 5 to 10 percent of the circulating antibodies and are also found in lymph. IgM is the first antibody to be secreted by plasma cells after exposure to an antigen.

Table 12-6 describes the characteristics of the major immunoglobulins.

> **U check**
> What is the difference between an antibody and an antigen?

> **focus on Wellness: Assisting Your Immune System**
>
> There are specific things people can do to boost their immune systems. Getting regular exercise is very important. In addition, getting the proper nutrition is essential. Some vitamins act as antioxidants and have been shown to protect against cancer. These include vitamins A, C, and E. The immune system also needs minerals like zinc (Zn) and selenium (Se). Dark green leafy vegetables and orange vegetables like carrots are high in antioxidant vitamins. These nutrients, especially vitamin C, also are valuable in fighting colds, influenza, and other viral diseases.

learning outcome

Distinguish between the four types of acquired (specific) immunity.

Immune System Responses and Acquired (Specific) Immunities

A primary immune system response occurs the first time a person is exposed to an antigen. This response is slow and takes several weeks to occur. In this response, memory cells are produced. A secondary immune system response occurs the next time a person is exposed to the same antigen. This response is very quick and usually prevents a person from developing a disease from the antigen. Memory cells carry out the secondary immune response.

Humans are born with very little immunity. Immunity continues to develop as long as the immune system is healthy and functioning. There are four different types of acquired (specific) immunity (Table 12-7).

Naturally Acquired Active Immunity

A person develops naturally acquired active immunity by being naturally exposed to an antigen and subsequently making antibodies and memory cells against the antigen. Having an infectious disease caused by pathogens leads to the development of this type of immunity. Naturally acquired active immunity is usually long lasting. For example, when a person has had the

TABLE 12-7 Practical Classification of Acquired (Specific) Immunity

Type	Mechanism	Result
Naturally acquired active immunity	Exposure to live pathogens	Stimulation of an immune response with symptoms of a disease
Artificially acquired active immunity	Exposure to a vaccine containing weakened or dead pathogens or their components	Stimulation of an immune response without the symptoms of a disease
Naturally acquired passive immunity	Antibodies passed to fetus from pregnant woman with active immunity or to newborn through breast milk from a woman with active immunity	Short-term immunity for newborn without stimulating an immune response
Artificially acquired passive immunity	Injection of gamma globulin containing antibodies or antitoxin	Short-term immunity without stimulating an immune response

mumps, he or she will develop resistance to any subsequent infections by the virus that causes the disease.

Artificially Acquired Active Immunity

A person develops artificially acquired active immunity by being injected with a pathogen (inactivated or weakened), which causes the body to make antibodies and memory cells against the pathogen. Immunizations or vaccines cause this type of immunity. The first reported vaccination took place when Edward Jenner, an English physician living in the 18th century, purposely gave a young boy serum taken from a cowpox lesion to give him protection against the more serious smallpox. This type of immunity is usually long lasting. Vaccines are typically given for protection against viral infections. Some examples are the poliomyelitis, measles, tetanus, diphtheria, and hepatitis B vaccines.

Naturally Acquired Passive Immunity

A person gets naturally acquired passive immunity from his or her mother. When a mother breast-feeds, she passes antibodies to her baby through the breast milk. Shortly after birth (and even during pregnancy), *colostrum* is produced which contains important antibodies and nutrients for the infant. A mother also passes antibodies to her baby across the placenta. This type of immunity is short-lived.

Artificially Acquired Passive Immunity

A person is given artifically acquired passive immunity when injected with antibodies. For example, if a snake bites a person, a physician injects the patient with antibodies (antivenom) to neutralize the venom. This type of immunity is short-lived.

> **U check**
> What are the differences between the four types of acquired (specific) immunity?

Transplantation and Tissue Rejection

12.4 learning outcome

Explain the different types of grafts or transplants.

Transplantation is almost commonplace today. Tissues and organs that are transplanted include the cornea, kidney, lung, pancreas, bone marrow, skin, liver, and heart. Transplants are much more successful today but there are still health risks, including risk of rejection. The transplanted tissue may actually cause harm to the host's tissues, which is called graft-versus-host disease (GVHD). The major histocompatibility complex of the host or recipient and the possible donor are examined for a match. This means finding a donor and recipient whose tissues are antigenically similar.

The four major types of transplants include:

1. *Autograft:* tissue taken elsewhere from the person's own body.
2. *Isograft:* tissue or organ taken from an identical twin (the best possible match from someone other than the host himself or herself).
3. *Allograft:* tissue from someone other than the individual or an identical twin.
4. *Xenograft:* tissue taken from another species, such as a pig or other primate.

focus on Wellness: Why Immunize?

When you talk to friends, relatives, and classmates, it is not unusual to hear stories of grandmothers and great-grandmothers who had 5, 6, or even 10 or 12 children, yet often only 2 or 3 survived childhood. Common childhood illnesses that we seldom think about today, such as polio, smallpox, measles, rubella, and whooping cough, were the culprits. These diseases went through families and even entire towns, often killing many children in their path. Today, thanks to physicians and scientists such as Edward Jenner and Drs. Sabin and Salk, these diseases have all but been eradicated in the United States as a result of vaccinations. To prevent resurgence of some of these childhood diseases in schools and colleges, many schools require proof of immunization prior to admittance.

Immunizations against diseases that are not common in this country, but are still prevalent in other countries, are recommended for U.S. citizens traveling to foreign lands. Depending on where you are traveling, you may need immunizations for hepatitis, malaria, cholera, and other diseases. These immunizations are not only for your safety, but also for the safety of the people you could expose on your return to the United States. In general, the side effects of immunizations, as long as you are healthy when receiving the vaccine, are mild and the benefits may include saving your life.

Immunosuppressive drugs are given to help reduce the risk of rejection, but can have serious side effects such as infection, kidney damage, and even cancer. The demand for transplants continues to exceed the supply.

> **Ʉ check**
> Which type of graft is most likely to be rejected by the host?

Major Immune System Disorders

Allergies

12.5 learning outcome
Describe the major immune system disorders.

Immune response—Hypersensitivity

allergen An antigen that elicits a hypersensitivity reaction.

histamine A chemical found in mast cells, basophils, and plasma cells; when released, causes vasodilation, increased vascular permeability, and constriction of bronchioles.

An allergic reaction is an excessive immune response to a (usually harmless) stimulus (antigen). Examples include allergies to pollen or nuts. Most people do not react to these substances, but they can cause severe responses in others.

Substances that trigger allergic responses are called **allergens**. Allergic reactions involve IgE antibodies and mast cells. When IgE antibodies bind to allergens, they cause mast cells to release histamine and heparin. **Histamine** contributes to the inflammatory response and causes constriction of smooth muscle. It is involved in producing many of the symptoms seen in asthma. A patient receiving allergy therapy often is injected with tiny amounts of the known allergen. This causes the body to produce IgG antibodies that prevent IgE antibodies from binding to the allergen. IgG antibodies do not trigger immune responses because they do not activate mast cells.

Most allergies do not cause life-threatening conditions, but some can. One life-threatening condition that can result is *anaphylaxis*. In this condition, blood vessels dilate so quickly that blood pressure drops very suddenly and the body cannot compensate. Without treatment, patients may go into *anaphylactic shock* and die.

Allergy symptoms vary depending on what part of the body is involved or exposed to allergens. Allergens that are inhaled often cause a runny nose, sneezing, coughing, or wheezing. An allergen that is ingested causes nausea, diarrhea, or vomiting. Skin allergens cause rashes, whereas systemic allergens—those that get in the blood, such as penicillin for people who are allergic to it—are often the most life-threatening because they can affect many organ systems.

Avoidance (prevention) of triggers is probably the best treatment for allergies. In addition, many allergies are effectively treated with over-the-counter (OTC) medications called *antihistamines*. Stronger medications are available by prescription. Nasal sprays and decongestants can reduce the symptoms of allergies. When a person experiences anaphylaxis, an injection of *epinephrine* is the treatment of choice. Epinephrine causes vasoconstriction, which increases blood pressure and dilates the bronchi, allowing more air to enter the lungs.

> **U check**
> What is the function of histamine?

Autoimmune Disease

Autoimmune disease is a failure of self-recognition. This means the body does not recognize a part of itself and so considers it "foreign" and begins "fighting it" by attacking its own antigens. Any tissue in the body is capable of being affected by autoimmune disease, but connective tissue is especially affected.

autoimmune disease A disease in which the immune system targets the individual's own tissues.

Autoimmune diseases are much more common in women (up to 75 percent of autoimmune diagnoses). Additionally, in women under 65 years of age, autoimmune disease is the fourth leading cause of disability and it is one of the top 10 causes of death in women of all ages. Unfortunately, individuals who have an autoimmune disease are often called complainers or whiners until a definitive diagnosis is made.

One of the first lines of treatment for autoimmune disease is corticosteroids to control the inflammatory response that is usually a part of these diseases. Autoimmune diseases are on the rise and many diseases are now being found to have an autoimmune component. The National Institutes of Health (NIH) has now identified over 80 autoimmune diseases, and more than 25 million people in the United States are affected by at least one of them. Some examples of autoimmune diseases include scleroderma, rheumatoid arthritis, multiple sclerosis, myasthenia gravis, systemic lupus erythematosus, glomerulonephritis, Crohn's disease, and insulin-dependent (type 1) diabetes mellitus. Many of these diseases will be discussed in further detail in other chapters. For more information on diseases with possible or probable autoimmune components, refer to the appropriate body system pathophysiology section. Table 12-8 lists several autoimmune disorders and two are discussed in more detail in the following text.

Systemic Lupus Erythematosus:
Systemic lupus erythematosus (SLE) is commonly referred to as *lupus* (Latin for "wolf") because the facial rash that often appears creates a wolf-like appearance As with many autoimmune disorders, lupus affects women much more often than men. Lupus may affect only a few organs, but more often it affects many organ systems of the body. The prognosis for SLE has improved dramatically over the last 50 years. The most common causes of death from lupus are renal failure and/or infections.

Less than 5 percent of the time SLE is caused by a reaction to a prescribed drug, but the vast majority of cases are due to autoimmunity. Antinuclear antibodies attack the cells' DNA.

TABLE 12-8 Autoimmune Disorders

Disorder	Symptoms	Antibodies Against
Glomerulonephritis	Lower back pain	Kidney cell antigens that resemble streptococcal bacteria antigens
Graves disease	Restlessness, weight loss, irritability, increased heart rate and blood pressure	Thyroid gland antigens near thyroid-stimulating hormone receptors, causing overactivity
Type I diabetes mellitus	Thirst, hunger, weakness, emaciation	Pancreatic beta cells
Hemolytic anemia	Fatigue and weakness	Red blood cells
Multiple sclerosis	Weakness, incoordination, speech disturbances, visual complaints	Myelin in the white matter of the central nervous system
Myasthenia gravis	Muscle weakness	Receptors for neurotransmitters on skeletal muscle
Pernicious anemia	Fatigue and weakness	Binding site for vitamin B on cells lining the stomach
Rheumatic fever	Weakness, shortness of breath	Heart valve cell antigens that resemble streptococcal bacteria antigens
Rheumatoid arthritis	Joint pain and deformity	Cells lining joints
Systemic lupus erythematosus	Red rash on face, prolonged fever, weakness, kidney damage, joint pain	Connective tissue
Ulcerative colitis	Lower abdominal pain	Colon cells

The signs and symptoms of lupus are extensive and may include any or all of the following: fatigue, general body aches, fever, weight loss, alopecia (hair loss), arthritis (most common finding with over 80 percent of patients suffering from arthritis), numbness of the fingers and toes, "butterfly" rash on the face, sensitivity to sunlight (photophobia), vision problems, nausea, nosebleeds (epistaxis), headaches, mental disorders, seizures, abnormal blood clots, chest pains, inflammation of the heart (carditis), anemia, shortness of breath, fluid accumulation around the lungs, renal failure, and blood in the urine (hematuria).

Treatment options include anti-inflammatory medications, including steroids, as well as protective clothing and creams to prevent damage from sunlight. Dialysis, immunosuppressive medications, and kidney transplants may be necessary for more serious lupus cases.

> **U check**
> What is the most common finding or sign in SLE?

Chronic Fatigue Syndrome: Chronic fatigue syndrome (CFS) is a condition in which a person feels severe tiredness that cannot be relieved by rest and is not attributable to other illness. The causes are primarily unknown, although the Epstein-Barr virus (EBV) is suspected as a possible cause. This condition may also be caused by an autoimmune response against the

focus on Genetics: Autoimmune Diseases

How many of your parents or grandparents have arthritis or type 2 diabetes mellitus? Or do you have a cousin or aunt with Crohn's disease? You may have one of these conditions yourself or you may be concerned you may develop one. Is there a genetic connection for autoimmune diseases? They do seem to run in families. Is our stressful, chemical-laden environment to blame? No one seems to know for sure, but down the road there may be genetic testing for autoimmune diseases just as there is for many other inherited conditions.

nervous system. There is no hard evidence pointing to a single causative factor and it is more likely that a combination of events is responsible.

The most common symptom of CFS is severe fatigue. Other signs and symptoms include mild fever, sore throat, tender lymph nodes in the neck or axilla, general body aches, joint pain, sleep disturbances, and depression.

HIV/AIDS

focus on Clinical Applications: Caring for Patients with HIV/AIDS

HIV infection is a contagious disease transmitted through body fluids. Laws and protocols regarding the diagnosis, management, and control of infectious diseases have been established to help prevent the transmission of infectious diseases. For example, there is the bloodborne pathogen standard that requires health care practitioners to take certain precautions, such as hand washing and wearing gloves, to prevent the spread of disease. Although these vary by state, here are some of the common laws that you should know as a health care professional caring for a patient with HIV/AIDS.

- HIV counseling and testing are typically controlled by the Department of Health (DOH). The DOH must establish protocols for reporting positive antibody test results and reporting these as required to the CDC.
- The DOH should have programs to ensure that individuals who test positive for the HIV antibody are counseled regarding their results and are referred to a health care provider who is clinically competent to treat HIV-infected patients.
- When a positive-HIV antibody test is noted, the patient's sexual and/or needle-sharing partners need to be informed so they can be tested. Partner notification services are established to help patients locate and notify these contacts.
- Health care professionals who treat pregnant women offer HIV testing at certain established points during pregnancy to help prevent the transmission from the mother to child (MTC). Pregnant women have the right to refuse testing.
- Most states allow adolescents to seek care, contraception methods, and treatment for STIs. Depending on the state, when adolescents request HIV testing the parent may or may not need to be notified.
- Health care professionals have the responsibility of maintaining the confidentiality of the patient. This includes keeping confidential any medical record that contains an HIV test result. Health care providers cannot disclose the client's HIV status to anyone without the client's expressed consent.

Since the cause is unclear, treatment includes a variety of therapies. Antiviral drugs, medications to treat the depression associated with this condition, and pain medications are often used.

Lymphedema

Lymphedema is blockage of the lymphatic vessels that drain excess fluid from different areas of the body. This condition may be caused by parasitic infections, trauma to the vessels, tumors, radiation therapy, cellulitis, and surgeries such as mastectomies and biopsies in which lymph tissues have been removed. The common symptom is tissue swelling that lasts longer than a few days or increases over time.

Treatment options for lymphedema include compression stockings for swelling in the legs or arms, elevation of the affected limb, or surgery to remove abnormal lymphatic tissue. Physical and massage therapy have also been found helpful in the early stages of lymphedema to disperse the excess fluid into surrounding tissues for reabsorption.

Mononucleosis

Mononucleosis is also known as *mono*. Because it frequently affects teenagers and is a highly contagious viral infection spread through the saliva of the infected person, it has earned the nickname "the kissing disease." Mono is also spread through coughing and sneezing.

Mononucleosis is caused by either the Epstein-Barr virus (EBV) or the cytomegalovirus (CMV). Unexplained fever, extreme fatigue, and sore throat are common. Other symptoms include weakness, headache, and swollen, tender lymph nodes. Rest, proper nutrition, and antibiotics to prevent secondary infections usually result in recovery from acute symptoms in a week or two, although complete recovery may take a month or longer.

> **U check**
> What is the cause of mononucleosis?

PATHOPHYSIOLOGY

HIV/AIDS

Human immunodeficiency virus/acquired immunodeficiency syndrome (HIV/AIDS) is a viral disease that is spread through body fluids such as semen, blood, or vaginal secretions. About 50 percent of people with HIV infection will progress to AIDS within 10 years.

HIV was first recognized in the San Francisco area in 1981 when it was observed that a number of homosexual men were coming down with a rare form of cancer called *Kaposi's sarcoma* (Figure 4-9, page 59). The Centers for Disease Control and Prevention (CDC) estimates that more than 1 million people are living with HIV in the United States and over 20 percent of those people are not aware they are HIV-positive. The number of new HIV infections in the United States has leveled off to about 55,000 per year, with almost 20,000 individuals dying of AIDS each year.

Etiology: AIDS is caused by the human immunodeficiency virus (HIV), and is spread in three ways: (1) through sexual contact with someone who is HIV-positive, (2) by sharing needles with someone who is HIV-positive, and (3) from mother to fetus or infant through the placenta during the birth process or through breast milk if the mother is HIV-positive. HIV attacks specific lymphocytes called T lymphocytes. This places the individual at risk of contracting many other potentially fatal diseases.

Signs and Symptoms: Initially, the HIV-infected person may present with no symptoms or possibly flulike symptoms. Symptoms of AIDS include T-cell counts below 200 (normal is more than 400); fever, *diaphoresis* (profuse sweating), and weakness; weight loss; frequent infections, including herpetic ulcers of the mouth, skin, and genitals; tuberculosis; yeast infections of the mouth, esophagus, and vagina; meningitis; and encephalitis. Cytomegalovirus (CMV), a specific type of herpetic virus, may also affect the eyes and other internal organs. Kaposi's sarcoma (KS) is a connective tissue cancer commonly seen in AIDS patients; Chapter 4, Concepts of Disease, discusses Kaposi's sarcoma in more detail. Table 12-9 explains how HIV is transmitted.

Treatment: There is no cure for AIDS, but in the United States, treatments are available that significantly delay the progression of the disease for many patients. Treatments include the use of various antiviral drugs, but many of these drugs have serious side effects. Antibiotics are also used to treat infections.

U check
What is Kaposi's sarcoma?

TABLE 12-9 HIV Transmission

How HIV Is Transmitted
Sexual contact, particularly anal intercourse, but also vaginal intercourse and oral sex
Contaminated needles (intravenous drug use, injection of anabolic steroids, accidental needle stick in medical setting)
During birth from infected mother
In breast milk from an infected mother
In infected blood transfusion or other tissue (precautions usually prevent this)
How HIV Is *Not* Transmitted
Casual contact (social kissing, hugging, handshakes)
Objects (toilet seats, deodorant sticks, doorknobs)
Mosquitoes
Sneezing and coughing
Sharing food
Swimming in the same water
Donating blood

12.6 learning outcome
Describe changes to the immune system with aging.

Life Span

As individuals get older, the body starts to "slow down." This is also true of the immune system. The thymus atrophies, causing fewer T lymphocytes to mature. In addition, fewer B lymphocytes are produced. Older people are more susceptible to infections and cancers. To protect elderly individuals, vaccines against common diseases like the flu, pneumonia, and shingles (varicella) are recommended.

summary

learning outcomes	key points
12.1 Explain the components of the lymphatic system and their functions.	The major components of the lymphatic system are lymph nodes and vessels, the thymus, and the spleen. The function of these structures is to provide protection to the body against foreign organisms and other agents that are potentially harmful to the body.
12.2 Compare innate (nonspecific) and acquired (specific) immunity.	Innate (nonspecific) immunity consists of mechanisms such as physical barriers and inflammation. Acquired (specific) immunity consists of T cells and B cells.
12.3 Distinguish between the four types of acquired (specific) immunity.	The four types of acquired (specific) immunity are (1) naturally acquired active immunity, (2) artificially acquired active immunity, (3) naturally acquired passive immunity, and (4) artificially acquired passive immunity.
12.4 Explain the different types of grafts or transplants.	There are four major types of grafts: autograft, isograft, allograft, and xenograft.
12.5 Describe the major immune system disorders.	Autoimmune disorders occur when the body fails to recognize itself and the immune system starts to attack and destroy the tissues. Some autoimmune diseases include rheumatoid arthritis, lupus (SLE), and scleroderma. HIV/AIDS is a viral disease that is transmitted through body fluids. It is treated with a spectrum of drugs, nutrition, and counseling.
12.6 Describe changes to the immune system with aging.	As we age our body slows down and we become more susceptible to immune disorders.

case study 1 questions

Can you answer the following questions that pertain to the case study presented in this chapter?

1. The most likely diagnosis is:
   a. Mononucleosis
   b. Lung cancer
   c. Strep throat
   d. An STI
2. The most likely etiologic agent is:
   a. A virus
   b. A bacterium
   c. The spirochete *Treponema pallidum*
   d. HPV

review questions

1. Proteins found in the blood that are involved in immunity include
   a. Antibodies
   b. Antigens
   c. Haptens
   d. Hemoglobin
2. Antibody-mediated immunity is also called
   a. Humoral immunity
   b. Cellular immunity
   c. Lysogenic immunity
   d. Granulomatous immunity
3. Which one of the following is the most numerous under normal conditions?
   a. Neutrophils
   b. Lymphocytes
   c. Monocytes
   d. Eosinophils
4. Which cells are elevated in parasitic infections?
   a. Eosinophils
   b. Lymphocytes
   c. Neutrophils
   d. Platelets
5. Which one of the following transforms into a macrophage upon leaving the circulation?
   a. Monocytes
   b. Neutrophils
   c. Eosinophils
   d. Platelets

critical thinking questions

1. What can we do while we are young to help maintain a healthy, younger-looking skin as we age?
2. What is the difference between Innate (specific) and acquired (nonspecific) immunity?

patient education

Someone you know has just been diagnosed with HIV. Several of your friends come to you concerned that they may "catch" HIV from this person. Take some time to develop a list of ways your friends can and cannot catch HIV and then role-play, discussing this list with your friends.

applying what you know

1. Why does your skin become red, swollen, and painful when a splinter enters your finger?
2. How do T cells and B cells protect against infection?

3. Why are MHC proteins important considerations in organ transplants?
4. Discuss autoimmune disease and give two examples of autoimmune diseases.
5. What are the different types of antibodies and in what situations does each type dominate?

CASE STUDY 2 *HIV/AIDS*

Dennis Spisak, a 24-year-old man, is frightened because during a night of drinking and partying he had sex with another man. He is concerned that he may have contracted HIV.

1. How is HIV transmitted?
    a. Through touching
    b. Through kissing
    c. Through exchange of body fluids
    d. By sharing combs, hats, or other clothing
2. What are the signs and symptoms of HIV?
    a. Flulike symptoms early in the infection
    b. Paralysis on one side of the body
    c. Tachycardia
    d. Bradycardia

CASE STUDY 3 *Immunity*

Sandy Wolbers, a 26-year-old woman, has just become a first-time mother. She wants to be the best mother she can and wants to protect her child from illness. She wants to know if breast-feeding is really beneficial and also wonders if vaccines are a good thing for her baby as he gets older.

1. Which of the following is secreted by the breast before true milk is produced?
    a. Colostrum
    b. Vasopressin
    c. Follicle stimulating hormone
    d. Oxytocin
2. What type of immunity is being conferred on Sandy's baby by breast-feeding?
    a. Artificially acquired active immunity
    b. Naturally acquired active immunity
    c. Naturally acquired passive immunity
    d. Artificially acquired passive immunity
3. What are immunoglobulins?
    a. Antibodies
    b. Antigens
    c. Interferons
    d. Lipids
4. Which one of the following immunoglobulins can cross the placenta and give the fetus protection?
    a. IgA
    b. IgG
    c. IgM
    d. IgE
    e. IgD

13
The Respiratory System

chapter outline

Respiratory System Components and Functions

Respiration

Respiratory Volumes and Capacities

Factors That Control Breathing

Gas Exchange

Gas Transport

Life Span

outcomes

AFTER COMPLETING THIS CHAPTER, YOU WILL BE ABLE TO:

13.1 Describe the components of the respiratory system and their functions.

13.2 Compare internal and external respiration.

13.3 Describe the respiratory air volumes and capacities.

13.4 Describe the factors that influence breathing.

13.5 Explain how gases are exchanged in the lungs and tissues.

13.6 Describe how oxygen is transported from the lungs to body cells, and how carbon dioxide is transported from body cells to the lungs.

13.7 Explain how aging of the respiratory system affects life span.

essential terms

alveolus (al-VE-o-lus)
asthma (AZ-ma)
bronchi (BRON-ki)
diaphragm (DI-a-fram)
emphysema (em-fi-SE-ma)
exhalation (eks-ha-LA-shun)
glottis (GLOT-is)
inhalation (in-ha-LA-shun)
larynx (LAR-inks)
nasal conchae (NA-zal KONG-kee)
paranasal sinus (par-a-NA-zal SI-nus)
pharynx (FAR-inks)
pleura (PLOO-ra)
sinusitis (si-nyoo-SI-tis)
surfactant (sur-FAK-tant)
thyroid cartilage (THI-royd KAR-tih-lij)
trachea (TRA-ke-a)

Additional key terms in the chapter are italicized and defined in the glossary.

case study

Use the case study to focus on as you go through the chapter. The questions will guide you as you learn the anatomy, physiology, and pathology of the respiratory system.

CASE STUDY 1 *Emphysema*

Emil Mowry is a 62-year-old rancher. He remembers rolling his first cigarette when he was 11 years old. He doesn't remember ever wanting to quit smoking. For the last 5 years he has had increasing difficulty breathing. Emil was diagnosed by Dr. Adams as having emphysema. He was told that he has "COPD." He didn't ask the doctor what this means, but he plans to look it up on the Internet.

As you go through the chapter, keep the following questions in mind:

1. What is emphysema?
2. What is COPD?
3. What is the etiology and treatment of emphysema?
4. What advice would you give to Emil?

study tips

1. Make flash cards for the essential terms of the chapter.
2. Outline the chapter. After each section, ask yourself what you just read.
3. Draw and label the main structures of the respiratory system.
4. Write a sentence or two describing the function of each of the labeled structures in study tip 3.

Gas exchange

Introduction

The respiratory system (Figure 13-1) functions to move gases into and out of the body. These gases are oxygen and carbon dioxide. What may seem like a simple process actually involves several steps. The movement of air into and out of the lungs is called breathing, or ventilation. The exchange of oxygen and carbon dioxide between the alveoli and the capillaries is external respiration. Oxygen and carbon dioxide must be transported between the lungs and virtually every tissue in the body. The exchange of gases at the cellular level is internal respiration, or cellular respiration; it uses oxygen to produce ATP (adenosine triphosphate).

learning outcome 13.1
Describe the components of the respiratory system and their functions.

Respiratory System Components and Functions

The respiratory system is divided into two regions: the upper respiratory tract and the lower respiratory tract (Table 13-1). Each has both physiologic and pathologic significance. The upper respiratory tract consists of the nose, nasal cavity, paranasal sinuses, and pharynx (Figure 13-2). The lower respiratory tract consists of the larynx, trachea, bronchi, and the lungs (Figure 13-3). We will discuss the upper respiratory tract first.

The Upper Respiratory Tract

Nose: The nose is the passageway for the entrance and exit of air. The nose is made up of bone, hyaline cartilage, and adipose tissue covered with skin. The nose serves an important physiologic function and has also been the subject of many jokes and great profiles. Several hundred thousand cosmetic (and medically necessary) surgeries known as *rhinoplasty* are done on noses each year. The openings of the nose are the nostrils, or *external nares*. The hairs in the nose assist in preventing debris and unwanted particles from entering the upper respiratory tract via the nares.

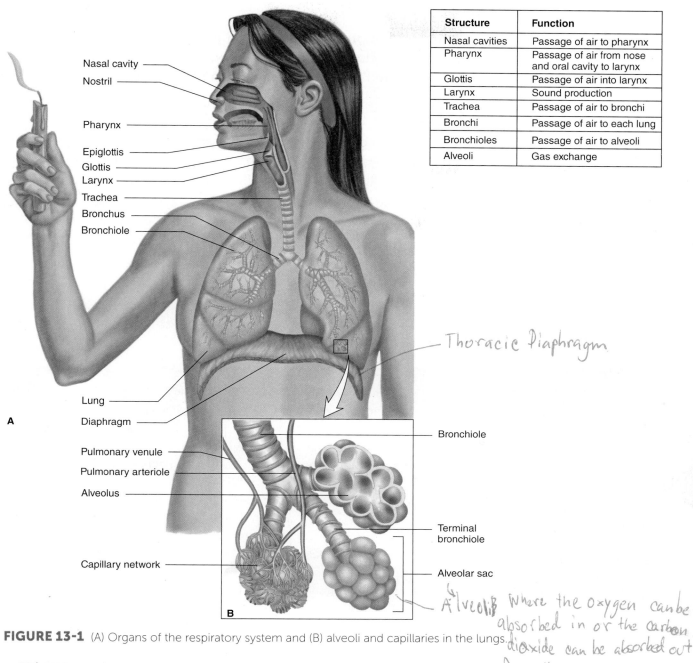

FIGURE 13-1 (A) Organs of the respiratory system and (B) alveoli and capillaries in the lungs.

Alveolis where the oxygen can be absorbed in or the carbon dioxide can be absorbed out from the blood

PATHOPHYSIOLOGY

Allergic Rhinitis

Allergic rhinitis is a hypersensitivity reaction to various airborne allergens.

Etiology: Many causes of allergic rhinitis may be seasonal (such as hay fever) or continual (such as those caused by dust, molds, colognes, cigarette smoke, animal dander, and mites).

(continued)

Immune response—
hypersensitivity

Allergic Rhinitis (concluded)

Signs and Symptoms: There are numerous signs and symptoms, which may include sneezing; itchy, watery eyes; red, swollen eyelids; congested nasal mucous membranes; and nasal discharge.

Treatment: Treatment commonly includes the use of *over-the-counter* (OTC) antihistamines and decongestants. Severe cases may be treated with prescription medications such as Allegra® and Zyrtec®. Patients should also avoid known allergens whenever possible. Air filters and air conditioners will assist in keeping down allergen counts. An allergist may be able to help by giving desensitization injections for long-term management.

> **U check**
> What is rhinoplasty?

TABLE 13-1 Parts of the Respiratory System

Part	Description	Function
Upper Respiratory Tract		
Nose	Part of face centered above the mouth and inferior to the space between the eyes	Nostrils provide entrance to nasal cavity; internal hairs begin to filter incoming air
Nasal cavity	Hollow space behind nose	Conducts air to pharynx; mucous lining filters, warms, and moistens incoming air
Paranasal sinuses	Hollow spaces in various bones of the skull	Reduce weight of the skull; serve as resonant chambers
Pharynx	Chamber posterior to the nasal cavity, oral cavity, and larynx	Passageway for air moving from nasal cavity to larynx and for food moving from oral cavity to esophagus
Lower Respiratory Tract		
Larynx	Enlargement at the top of the trachea	Passageway for air; prevents foreign objects from entering trachea; houses vocal cords
Trachea	Flexible tube that connects larynx with bronchial tree	Passageway for air; mucous lining continues to filter air
Bronchial tree	Branched tubes that lead from the trachea to the alveoli	Conducts air to the alveoli; mucous lining continues to filter incoming air
Lungs	Soft, cone-shaped organs that occupy a large portion of the thoracic cavity	Contain the air passages, alveoli, blood vessels, connective tissues, lymphatic vessels, and nerves of the lower respiratory tract

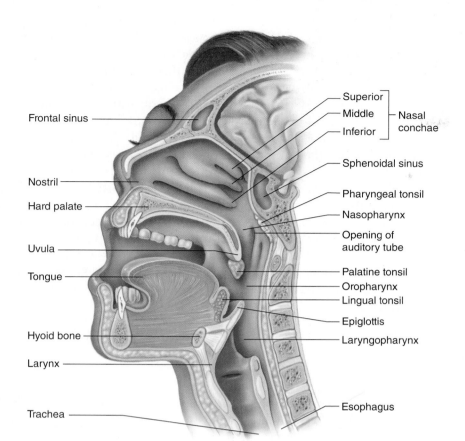

FIGURE 13-2 Major structures associated with the upper respiratory tract in the head and neck.

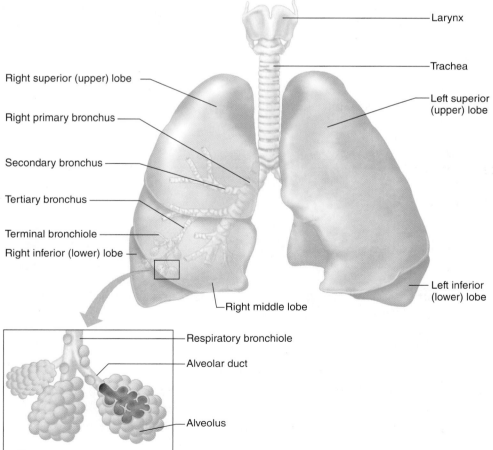

FIGURE 13-3 The bronchial tree consists of the airways that connect the trachea and alveoli. The alveolar ducts and alveoli are enlarged in the inset.

CHAPTER 13 The Respiratory System

nasal conchae Lobed structures on the lateral walls of the nasal cavity that function to increase air turbulence.

Nasal Cavity: The nasal cavity is the space posterior to the nose and is divided into left and right compartments by the *nasal septum*. You may recall from Chapter 8, The Skeletal System, that the ethmoid and vomer bones make up the nasal septum. Three lobelike structures called **nasal conchae** (turbinates) extend from each of the lateral walls of the nasal cavity. The conchae, like most of the nasal cavity, are lined with a mucous membrane that warms and moistens air as it passes through the nasal cavity. The nasal conchae increase the surface area of the nasal cavity. The nasal cavity is also lined with ciliated cells that form a pseudostratified membrane that has mucus-secreting *goblet cells*. The mucus helps trap particles and the cilia help move the particles toward the pharynx to be swallowed, so that they do not get further into the respiratory system. Pathogens are then destroyed by the acidic pH in the stomach. The roof of the nasal cavity is called the *cribriform plate* through which the olfactory nerves pass on their way to the brain. You will learn more about the olfactory nerves in Chapter 14, The Nervous System. The floor of the nasal cavity is formed by the hard and soft palates, which also form the roof of the mouth.

> **U check**
> What bones make up the nasal septum?

paranasal sinuses Mucus-lined air cavities in the skull.

Paranasal Sinuses: The **paranasal sinuses** are made up of the *maxillary, frontal, sphenoid,* and *ethmoid* sinuses. As you may recall from Chapter 9, The Muscular System, they are named after the bones in which they are found. The largest of the paranasal sinuses are the maxillary sinuses. Both the maxillary and frontal sinuses are paired structures (right and left). The sphenoid sinus is unpaired and the ethmoid sinus (sometimes called ethmoid air cells) consists of four to six small cavities. As with the nasal cavity, the sinuses are lined by mucosa, which continues the process of warming and moistening the air we breathe. The sinuses are not present at birth, but develop in the teenage years. They act as resonant chambers that affect the timber or quality of our voices. In addition, they lighten the weight of the skull.

> **U check**
> What structures make up the paranasal sinuses?

sinusitis Inflammation of the paranasal sinuses.

Inflammation

PATHOPHYSIOLOGY

Sinusitis

Sinusitis is an inflammation of the membranes lining the sinuses of the skull and it can be acute or chronic in nature.

Etiology: Bacteria, excess mucus production in the sinuses, blockage of sinus openings, and the destruction of cilia that move mucus out of the sinuses can cause this disorder.

Signs and Symptoms: Fever, cough, headache, pharyngitis, facial pain, nasal congestion, and tooth pain, as well as a cough (often

> **Sinusitis** *(concluded)*
>
> from the draining mucus) are the common signs and symptoms of sinusitis.
>
> **Treatment:** Treatment options include the use of nasal decongestants, nasal steroid sprays, and a humidifier; applications of heat to the face; analgesics for pain; and antibiotics. Sinus *lavage* may be used to clear the sinuses, or surgery may be required to unblock sinus openings.

The Pharynx: The **pharynx** is a "dual" organ of both the respiratory system and the digestive system. The pharynx consists of three regions: the *nasopharynx* (at the junction of the nasal cavity and pharynx), *oropharynx* (the junction of the oral cavity and pharynx), and *laryngopharynx* (the area including the larynx, or voice box) from most superior to most inferior. During inspiration, air flows from the nasal or oral cavity into the pharynx. From the pharynx, air flows into the larynx.

pharynx The throat.

U check
What are the three regions of the pharynx?

---

focus on Wellness: Seasonal Allergies

Does the changing of the season bring you sneezing, wheezing, runny nose, and itchy, watery red eyes? You might be suffering from seasonal allergies. These allergies might also affect you in other ways. Recent studies suggest that having poor control of allergy symptoms can be associated with a negative effect on your activities of daily living as well cause significant fatigue and mood changes. Controlling your allergies to the best of your ability can make you feel better in more ways than one. Here are a few ideas.

- Try exercising to release endorphins, which will improve your feeling of well-being.
- Plan ahead, and follow the weather and pollen forecasts when you expect to spend time outdoors, particularly on sunny, warm, and windy days, when pollen levels are highest.
- Consider using gentle eyelid irrigation and nasal saline sprays to wash and/or dilute pollens that have accumulated.
- If you are outside during a high-pollen day, consider shampooing your hair nightly and change your clothing before entering the bedroom to prevent pollens from being deposited onto your bedding. Hair gel and products can act like a pollen magnet.
- Take OTC or prescription medications as directed.

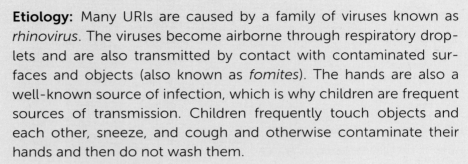

Respiratory tract infection

> ## PATHOPHYSIOLOGY
> ### Upper Respiratory Infection
>
> *Upper respiratory (tract) infection (URI)* is the term often used for the common cold, including pharyngitis (sore throat).
>
> **Etiology:** Many URIs are caused by a family of viruses known as *rhinovirus*. The viruses become airborne through respiratory droplets and are also transmitted by contact with contaminated surfaces and objects (also known as *fomites*). The hands are also a well-known source of infection, which is why children are frequent sources of transmission. Children frequently touch objects and each other, sneeze, and cough and otherwise contaminate their hands and then do not wash them.
>
> **Signs and Symptoms:** A URI is a generally self-limiting condition of approximately one week's duration, following an initial incubation period of two to five days. Symptoms include pharyngitis, nasal congestion, rhinitis (nasal inflammation), *rhinorrhea* (runny nose), headache, fever, and general malaise. There may also be a nonproductive cough, especially at night.
>
> **Treatment:** Because colds are usually self-limiting, care is typically symptomatic and includes *antipyretics* (fever reducer), analgesics (pain reliever), decongestants, and antitussives (cough suppressants). Adequate rest and plenty of fluids to flush the system and stay hydrated are also helpful. Antibiotics are ordinarily prescribed only for patients with an underlying illness or complication such as superimposed bacterial infection.

larynx The voice box.

Larynx and Vocal Cords: The **larynx** is commonly called the voice box because it houses the vocal cords. It sits superior to, and is continuous with, the trachea, or windpipe. Because of its location, the larynx serves the following three important functions:

1. Controls the flow of air during breathing
2. Protects the airway
3. Produces the sounds necessary for speech

thyroid cartilage The largest single cartilage of the larynx.

The larynx (Figures 13-4 and 13-5) is where the common pathway of the pharynx divides to allow air to enter the trachea and food to enter the esophagus. There are three cartilages in the larynx. The largest is the **thyroid cartilage,** and it forms the anterior wall of the larynx. During adolescence, the male starts to produce more testosterone which causes the thyroid cartilage to enlarge producing the "Adam's apple." The smaller elastic *epiglottic cartilage* forms the framework of the epiglottis. The *epiglottis* is the flap-like structure that closes off the larynx during swallowing so that food and liquids do not enter the respiratory system. If someone swallows (aspirates) food or liquid into the lungs, he or she may get "aspiration pneumonia."

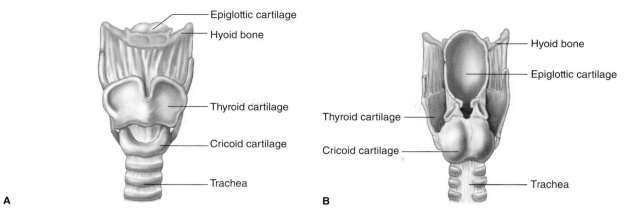

FIGURE 13-4 (A) Anterior view and (B) posterior view of the larynx.

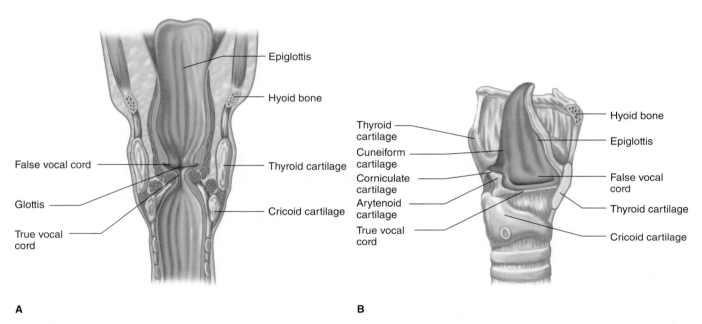

FIGURE 13-5 (A) Frontal section and (B) sagittal section of the larynx.

The third cartilage of the larynx is the *cricoid cartilage*. It forms most of the posterior wall of the larynx and a small part of the anterior wall. The mucous membrane within the larynx forms two pairs of folds called the vestibular folds. The upper vestibular folds are known as the *false vocal cords* because they do not produce sound. The lower vestibular folds form the *true vocal cords,* which do produce sound. The vocal cords stretch between the thyroid cartilage and the cricoid cartilage, and the opening between them is called the **glottis** (Figure 13-6). When the true vocal cords are stretched, the voice becomes higher in pitch. When they are relaxed, the voice becomes lower in pitch. Men have thicker vocal cords, which is why their voices are generally deeper than those of females.

glottis The true vocal cords and the space between them.

U check
What is the function of the epiglottis?

CHAPTER 13 The Respiratory System

FIGURE 13-6
The vocal folds as viewed from above with the glottis (A) closed and (B) open. (C) Photograph of the glottis and vocal folds.

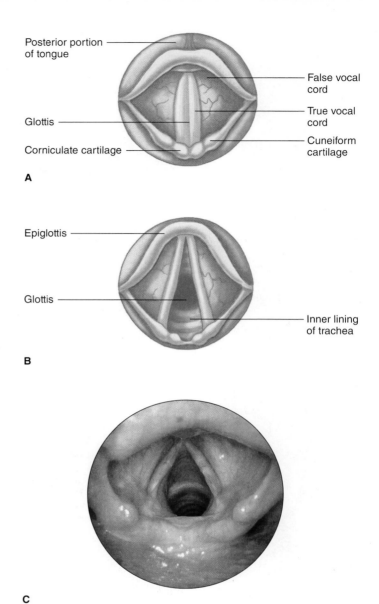

The Lower Respiratory Tract

Trachea: The **trachea** (windpipe) is a tubular organ (Figure 13-7) approximately 5 inches in length, made of about 20 C-shaped rings that posteriorly are open or incomplete. The rings are structured this way to allow a mass or *bolus* of food to pass down the esophagus without being pushed up against the cartilage rings. Smooth muscle and connective tissue fill in the openings of the cartilage rings. The trachea extends from the larynx to the **bronchi** and, like the upper respiratory tract, is lined with mucosa containing goblet cells and ciliated cells that constantly move mucus up to the pharynx where it (and any bacteria, viruses, or other harmful substance trapped in the mucus) is swallowed and destroyed or excreted by the digestive system. The chronic cough seen in many cigarette smokers is due to the destruction or paralysis of the respiratory tract cilia. The only way a smoker can get mucus

trachea The airway extending from the larynx to the main stem bronchus, also known as the windpipe.

bronchi Branches of the respiratory passageway.

> **focus on Clinical Applications:** Caring for Patients Who Snore
>
> When the muscles of the palate, tongue, and throat relax and airflow causes these soft tissues to vibrate, snoring occurs. These vibrating tissues produce the harsh sounds characteristic of snoring. Snoring causes daytime sleepiness and is sometimes associated with a condition known as *obstructive sleep apnea (OSA)*. In OSA, the relaxed throat tissues cause airways to collapse, which prevents a person from breathing. Snoring affects approximately 50 percent of men and 25 percent of women older than the age of 40 years. The common causes of snoring include:
>
> - Enlargement of the tonsils or adenoids.
> - Being overweight.
> - Alcohol consumption.
> - Nasal congestion.
> - A deviated nasal septum.
>
> The severity of snoring varies among people. The Mayo Clinic's Sleep Disorders Center uses the following scale to determine the severity of snoring:
>
> **Grade 1:** Snoring heard from close proximity to the face of the snoring person.
> **Grade 2:** Snoring heard from anywhere in the bedroom.
> **Grade 3:** Snoring heard just outside the bedroom with the door open.
> **Grade 4:** Snoring heard outside the bedroom with the door closed.
>
> As a health care professional you can educate patients about making lifestyle modifications and using aids to help reduce their snoring:
>
> - Lose weight.
> - Change the sleeping position from the back to the side.
> - Avoid the use of alcohol and medications that cause sleepiness.
> - Use nasal strips to widen the nasal passageways.
> - Use dental devices to keep airways open.
>
> In addition, patients may benefit from a *CPAP (continuous positive airway pressure)* machine, which uses a mask attached to a pump that forces air into their passageways while they sleep. If these therapies are not effective, patients may need surgery to trim excess tissues in the throat, such as an *uvulotomy*, or laser surgery to remove a portion of the soft palate.

out of the trachea is to cough because the cilia are not present to assist with the process.

A *tracheostomy* is an emergency procedure performed when the airway is blocked for any reason. An opening is made in the trachea from outside the throat to allow air to enter the lungs, bypassing the blockage. This procedure is usually temporary, but in the case of some pharyngeal or laryngeal cancers, it may be done electively as a permanent procedure.

FIGURE 13-7
The trachea transports air between the larynx and the bronchi.

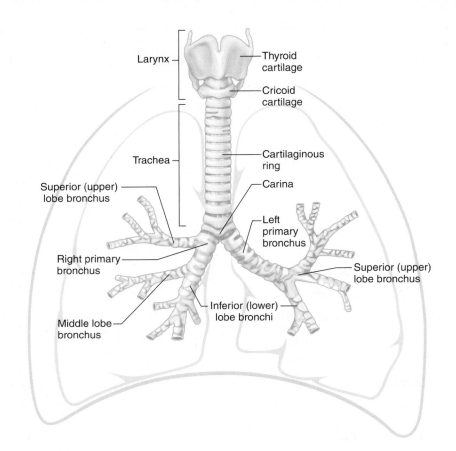

Respiratory tract infection

PATHOPHYSIOLOGY

Influenza

Influenza, commonly called "the flu," is an infection of both the upper and lower respiratory tracts. Infants, elderly individuals, and people with compromised or deficient immune systems, as well as those with chronic respiratory illnesses such as chronic obstructive pulmonary disease (COPD), are at the highest risk of developing influenza. Although the flu, which normally lasts between 5 and 10 days, is self-limiting in most people, it is responsible for thousands of deaths each year. In 1918, a pandemic (worldwide *epidemic*) of the Spanish flu broke out that claimed tens of millions of lives worldwide.

Etiology: Influenza is caused by a number of different viruses that attack the respiratory system. A yearly flu vaccination can help prevent or minimize the severity of the flu. Yearly immunizations are necessary because there are multiple strains of influenza. Therefore, every year the vaccine is created for the known strains that are common that year. This should be explained to patients so they understand that a flu shot is needed each year for that year's specific strains of the virus.

Signs and Symptoms: Common symptoms include rhinorrhea (runny nose), *pharyngitis* (sore throat), sneezing, fever or chills, a dry cough, *myalgias* (muscle pain), fatigue, *anorexia* (loss of appetite), and diarrhea.

Influenza *(concluded)*

Treatment: OTC analgesics and antipyretics can alleviate the aches and pains as well as the fever associated with the flu. Other treatment recommendations include bed rest, fluids, and antiviral medications.

focus on Wellness: Respiratory Hygiene/Cough Etiquette

Because of the potential for an outbreak of respiratory illness, especially certain types of flu, the Centers for Disease Control and Prevention (CDC), an agency of the federal government, has created a new standard. The respiratory hygiene/cough etiquette standard applies to everyone everywhere, in or outside a health care facility. Individuals should cover their cough or sneeze, use the *flu salute,* and clean their hands frequently. See Figure 13-8 for more details and follow these practices at all times.

FIGURE 13-8
The correct method of covering your cough.

PATHOPHYSIOLOGY
Laryngitis

Laryngitis is an acute inflammation of the larynx.

Etiology: The causes of this condition are varied and include viruses; bacteria; polyp formation in the larynx; excessive talking, shouting, or singing; allergies; smoking; frequent heartburn; frequent use of alcohol; damage to nerves that supply the larynx; and stroke (cerebrovascular accident, or CVA) that paralyzes vocal cord muscles.

Signs and Symptoms: Signs and symptoms include *dysphonia* (a hoarse voice), pharyngitis, a dry cough and dry throat, and tickling sensations in the throat.

Treatment: The most common treatment options are antibiotics, management of heartburn, and avoidance of cigarettes and alcohol, as well as voice rest. The treatment of more serious causes includes surgical removal of laryngeal polyps and surgery to tighten the vocal cords.

Bronchial Tree: The distal end of the trachea branches to the left and right. From this point, branching results in the *bronchial tree*. The first branches off the trachea are called *primary* or *main stem bronchi*. The location of this bifurcation is known as the *carina*. The main stem bronchi are lined with ciliated pseudostratified columnar epithelium and goblet cells. The right main stem bronchus is shorter, wider, and more vertically oriented than the left main stem bronchus. It is for this reason that a swallowed foreign object is more likely to go down the right main stem bronchus than the left. The branches of the primary bronchi are called *secondary bronchi* and the secondary bronchi branch into *tertiary bronchi* which then branch into very small *bronchioles*. At the ends of the bronchioles are air sacs called alveoli.

The walls of the airways contain cartilage plates rather than the rings found in the trachea. The amount of cartilage lessens as the airways become smaller until it disappears completely in the smallest passageways. Conversely, smooth muscle becomes more abundant as the airways become smaller to allow for the expansion required during **inhalation** (inspiration) and **exhalation** (expiration).

inhalation Breathing in.

exhalation Breathing out.

> **U check**
> What are the main anatomical differences between the right and left main stem bronchi?

asthma Reaction characterized by smooth muscle spasms in the bronchi and excessive production of mucus.

PATHOPHYSIOLOGY
Asthma

Asthma is a condition in which the tubes of the bronchial tree become obstructed as a result of inflammation and hyperactivity of the smooth muscle in the airways.

Asthma (concluded)

Etiology: The causes can include allergens (pollen, pets, dust mites, etc.), cigarette smoke, pollutants, perfumes, cleaning agents, cold temperatures, and exercise (in susceptible individuals). Often, a specific cause cannot be identified.

Signs and Symptoms: Symptoms include difficulty breathing, tightness in the chest, wheezing, and coughing, all of which can cause a feeling of suffocation and increased anxiety, which in turn worsens the symptoms.

Treatment: Treatment includes avoiding allergens, the use of steroidal and nonsteroidal inhalers such as Advair® and Flovent®, as well as oral medications such as Singulair® and other *bronchodilators* to reduce inflammation. Patients should avoid known allergens and smoky environments, and those who smoke should stop smoking. Strongly scented items such as perfumes, hair products, and cleaning agents should also be avoided.

Asthma

Bronchitis

Bronchitis is inflammation of the bronchi and may be classified as acute or chronic. Acute bronchitis often follows a cold, coming on during the URI itself. Chronic bronchitis is a condition that typically takes decades to develop and is often due to cigarette smoking.

Etiology: Most cases of acute bronchitis are caused by the same viruses that cause the common cold. One noninfectious cause of acute bronchitis is *gastroesophageal reflux disease* (GERD), a condition in which acid from the stomach refluxes into the esophagus and some of the acid gets into the airways. Other noninfectious causes include exposure to cigarette smoke, pollutants, and the fumes of household cleaners.

Signs and Symptoms: The signs and symptoms of acute bronchitis include chills, fever, coughing up yellow-gray or green mucus, tightness in the chest, wheezing, and dyspnea.

Treatment: Acute bronchitis is treated with rest, fluids, nonprescription and prescription cough medicines, and the use of a humidifier.

The Lungs: The lungs are cone-shaped organs that contain the grapelike clusters of alveoli, connective tissue, the bronchial tree, nerves, lymphatic vessels, and many blood vessels. The right lung is larger than the left and is divided into three lobes, known as the right upper, middle, and lower lobes. The left lung is smaller because the heart is also located on the left side of the thoracic cage, taking up some of the space (Figure 13-9). The left lung has only two lobes—the upper and lower lobes. A projection of the upper lobe of the left lung, called the *lingula* (Latin for "tongue"), is comparable to the middle lobe of the right lung and often is the site of a lung biopsy.

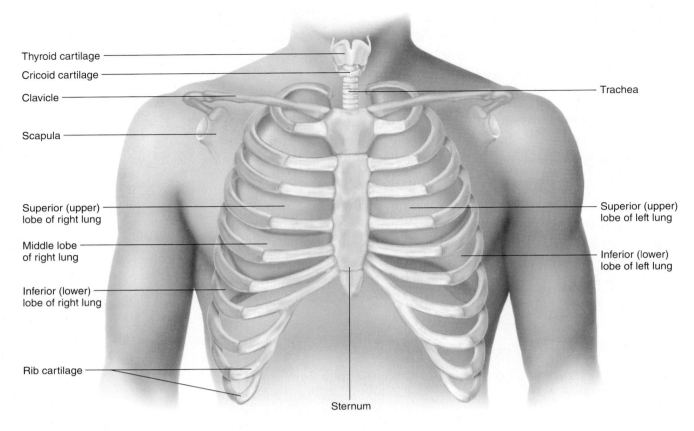

FIGURE 13-9 Location of the lungs in the thoracic cavity.

pleura The serous membrane that covers the lungs and lines the walls of the chest and diaphragm.

A serosa or double-walled membrane called the **pleura** surrounds the lungs (Figure 13-10). The outer membrane is known as the *parietal pleura* and the innermost membrane is the *visceral pleura*. The pleura produces a slippery, serous (watery) fluid called pleural fluid that helps decrease friction as the membranes move against each other during breathing.

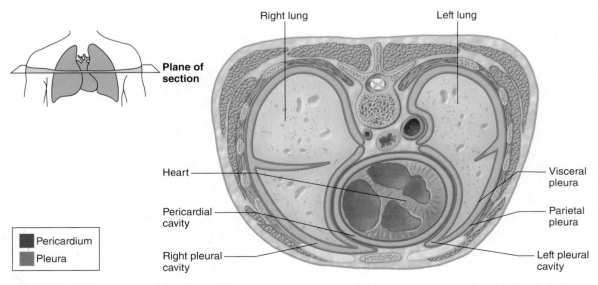

FIGURE 13-10 Transverse section through the heart and lungs illustrating the visceral and parietal pleura.

346   UNIT 2   Concepts of Common Illness by System

PATHOPHYSIOLOGY

Mesothelioma

Mesothelioma is a type of cancer that affects the pleura and is a result of asbestos exposure.

Etiology: Exposure to asbestos fibers in the workplace or home is the cause of mesothelioma.

Signs and Symptoms: Diagnosis of mesothelioma may be missed because the signs and symptoms are similar to those of many other diseases. Almost two thirds of individuals may have low back pain. Shortness of breath, persistent cough, difficulty swallowing, and weight loss are additional symptoms. Some individuals may have blood in their sputum, muscle weakness, and hoarseness.

Treatment: Treatment often consists of a combination of surgery, chemotherapy, and radiation therapy.

Pleurisy

Pleuritis, or *pleurisy,* is a condition in which the pleura become inflamed. This often causes the membranes to stick together or can cause an excess amount of fluid to form between the membranes.

Etiology: Causes include viruses, pneumonia, autoimmune diseases such as lupus or rheumatoid arthritis, tuberculosis, a pulmonary embolism, and trauma to the chest.

Signs and Symptoms: Symptoms include fever or chills; dry cough; shortness of breath; and a sharp, stabbing type of chest pain during respiration. The buildup of fluid between the visceral and parietal pleura can often be seen on a chest x-ray.

Treatment: Analgesics may be prescribed to relieve chest pain. Anti-inflammatory drugs, antibiotics, and the removal of fluid around the lungs by *thoracocentesis* (needle puncture of the thorax to withdraw fluid) are the primary treatment options.

Pleural Effusion

Pleural effusion is a buildup of fluid in the pleural cavity.

Etiology: Effusions are caused by either an overproduction of pleural fluid or an inadequate absorption of the fluid. These often result from an underlying disease, such as congestive heart failure (CHF), cirrhosis, tuberculosis, cancer, lupus, or rheumatoid arthritis.

Signs and Symptoms: As the fluid builds in the pleural space, the lungs begin to compress, reducing the gas exchange of oxygen and carbon dioxide. Infective processes may result in a pus buildup, which is known as *empyema*.

(continued)

Pleural Effusion (concluded)

Treatment: Thoracocentesis is done to remove the fluid and/or pus. *Thoracostomy,* which requires insertion of a tube into the thorax to continually drain the fluid, may be required to maintain drainage of the acute phase of the illness. Oxygen may be administered to increase oxygen concentration in the lungs and antibiotics may be given for any existing bacterial infection.

Tuberculosis

Tuberculosis (TB) is a disease that kills more than 2 million people worldwide each year (Figure 13-11). Although it primarily affects the lungs, it can spread to other parts of the body, such as the kidneys or vertebrae.

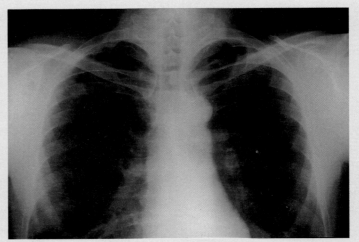
Healthy lungs

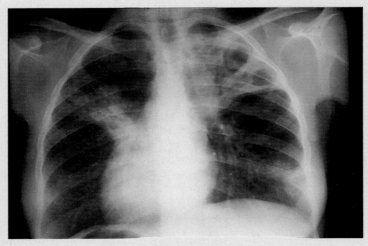
Tuberculosis

FIGURE 13-11 Healthy lungs have a dark and clear appearance on a radiograph. Tubercular lungs have a cloudy appearance because of fibrosis that forms to wall off the infection.

Tuberculosis (concluded)

Etiology: This disease is caused by various strains of the bacterium *Mycobacterium tuberculosis*. The following factors may be contributing to widespread tuberculosis:

- *HIV infection.* HIV infection makes a person more vulnerable to TB.
- *Crowded living conditions.* This factor allows TB to spread easily. Tuberculosis may be found in some prisons and homeless shelters.
- *Poverty.* Poverty prevents some patients with TB from seeking or completing therapy.
- *Drug-resistant bacterium.* Drug-resistant strains of the bacterium that causes TB have increased, which makes treating the disease more difficult. Some countries such as Mexico and India have reported more than the typical number of cases of drug-resistant TB.
- *Long-term therapy.* Current treatments require taking antibiotics for many months. Some patients with TB do not complete the treatment and therefore are noncompliant. This is one of the contributing factors in the drug-resistant strains of TB that are emerging.

Signs and Symptoms: Almost 90 percent of primary cases of TB are asymptomatic. When symptoms do occur, they include a persistent cough that may be blood tinged, unintended weight loss, fever or chills, fatigue, night sweats, pain when breathing or difficulty breathing, and pain in other affected areas.

Treatment: The first step should be TB testing to detect carriers of this disease, who should then be treated. Therapy normally lasts six months to a year. Drug-resistant cases of TB may require years of drug therapy to treat. Isolating the patient during the contagious phase of the disease (usually two to four weeks after treatment begins) is required. Also, during the initial stages of treatment, the patient should be encouraged to receive adequate bed rest and maintain an adequate nutritious diet. Without treatment, 50 percent of patients eventually die.

> **U check**
> What is the etiology of tuberculosis?

Atelectasis

Atelectasis is commonly called collapsed lung. It may occur after abdominal or thoracic surgery or because of pleural effusion, which may consist of blood (*hemothorax*), fluid

(continued)

Atelectasis (concluded)

(*hydrothorax*), air (*pneumothorax*), or pus (*pyothorax*) in the pleural cavity.

Etiology: Acute atelectasis may occur after any injury to the ribs or trauma to the thorax. Postsurgical patients are also susceptible to atelectasis. Athletes sometimes experience spontaneous pneumothorax that is often self-limiting. Cancer patients and those with inflammatory conditions such as pleurisy (pleuritis) may be subject to chronic atelectasis, as may patients with underlying COPD or cystic fibrosis.

Signs and Symptoms: These include *dyspnea* (difficulty breathing), *cyanosis* (a bluish coloration of skin and mucous membranes due to poor oxygenation of tissues), and *diaphoresis* (excessive perspiration due to illness). The individual may also experience anxiety, tachycardia, and intercostal muscle retraction. Depending on the cause of the atelectasis, there may also be chest pain.

Treatment: In acute cases, *thoracocentesis* may be needed to drain the pleural cavity. For chronic atelectasis, treatment may include chest percussion, postural drainage, coughing, deep breathing exercises, and *intermittent positive-pressure breathing (IPPB)*.

check
What is pyothorax?

Chronic Obstructive Pulmonary Disease

Chronic obstructive pulmonary disease (COPD) is a group of lung disorders in which obstruction limits airflow to the lungs. Emphysema and chronic bronchitis are the most common types of COPD. **Emphysema** is a chronic condition that destroys the alveoli of the lungs. That is why it is sometimes called "vanishing lung disease."

Etiology: The most common cause of emphysema is long-term smoking, but it can have a genetic basis that has no connection to smoking. Sometimes exposure to cigarette smoke, pollutants, and the dust from grains, cotton, wood, or coal may play a role.

Signs and Symptoms: Symptoms include shortness of breath that progresses over time, chronic cough, unintended weight loss, and fatigue. Pulmonary function tests and arterial blood gases become progressively more abnormal as the disease progresses. In advanced cases, patients develop characteristic barrel chest that describes the round barrel shape the chest develops over time. These individuals are also called "pink puffers" because their face often has reddish hue and they seem to "puff" when they breathe. A chest x-ray will show hyperinflated lungs.

COPD

emphysema A lung disease in which the alveolar walls are destroyed; often caused by cigarette smoking.

Chronic Obstructive Pulmonary Disease *(concluded)*

Treatment: Smoking cessation and avoidance of triggers such as exposure to cold environments and pollutants should be the first treatment measures. Vaccinations to prevent the flu and pneumonia as well as antibiotics to control the respiratory infections associated with emphysema may also be administered. In addition, patients may be treated with bronchodilators, supplemental oxygen, inhaled steroids, and respiratory therapy. The most serious cases may require either surgery to remove damaged lung tissue or a lung transplant, without which patients will develop respiratory and/or heart failure, resulting in death.

> **U check**
> What is the most common cause of COPD?

> **REMEMBER EMIL,** our rancher with emphysema? What do you think his chest x-ray will show?

Legionnaire's Disease

Legionnaire's disease is an acute type of bacterial pneumonia. It gets its name because it was first identified in 1976 at an American Legion convention in Philadelphia, Pennsylvania.

Etiology: This disease is caused by *Legionnaire bacillus,* a gram-negative organism. It grows in the standing water such as that found in commercial air conditioning, humidifiers, water heaters, and evaporative condensers. It survives chlorination and is not spread by person-to-person contact.

Signs and Symptoms: The symptoms include fever, which may spike as high as 105°F, fatigue, anorexia, dyspnea, frequent coughing, chest pain, muscle aches, and headache. Complications may include hypotension, heart arrhythmias, respiratory and renal failure, as well as shock, which is often fatal.

Treatment: Erythromycin is the antibiotic of choice. In addition, antipyretics and respiratory therapy, including oxygen and ventilator support, may be prescribed. Supportive therapy, such as IV fluids, is also used.

Lung Cancer

Cancer is the number two cause of death in the United States behind heart disease. Lung cancer kills more men and women than any other cancer (Figure 13-12).

Etiology: Lung cancer has been shown unequivocally to be most often caused by cigarette smoking. Smoking accounts for approximately 85 percent of all lung cancer cases. Also implicated in recent years is exposure to secondhand smoke. Additional causes are exposure to radon, asbestos, and industrial carcinogens.

(continued)

Lung Cancer (continued)

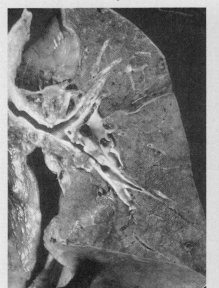

FIGURE 13-12 The lung on the left is healthy; the lung on the right has a cancerous tumor.

Signs and Symptoms: The respiratory symptoms include a cough that worsens over time, *hemoptysis* (coughing up blood), dyspnea, wheezing, shortness of breath, and recurrent bronchitis. Other symptoms are chest pain, dysphonia, unexplained weight loss, and bone pain if the cancer has spread.

Classification: Lung cancer is classified as follows:

- *Small cell lung cancer.* This type occurs almost exclusively in smokers. It is the most aggressive type and spreads readily to other organs. Small cell lung cancer that spreads to other organs is termed "extensive."
- *Squamous cell lung cancer.* This type of lung cancer arises from the epithelial cells that line the tubes of the lungs. It occurs most commonly in men.
- *Adenocarcinoma.* This type arises from the mucous-producing cells of the lungs. It develops most commonly in women and is the most common type of lung cancer in nonsmokers.
- *Large cell carcinoma.* This type of lung cancer arises from the peripheral areas of the lungs.

Cancer Stages: Refer to Chapter 4, Concepts of Disease, for an explanation of staging. Squamous cell lung cancer, adenocarcinoma, and large cell carcinoma are staged as follows:

- **Stage 0:** Cancer is found only in the lining of the tubes of the lungs.
- **Stage 1:** Cancer has spread from the lining of the tubes to lung tissues.

Lung Cancer (concluded)

- **Stage 2:** Cancer has spread to the lymph nodes or the chest wall.
- **Stage 3:** Cancer has spread to the lymph nodes and to other organs within the chest.
- **Stage 4:** Cancer has *metastasized* (spread) to organs outside the chest.

Treatment: Treatment varies depending on the type of cancer and the stage. Common treatment options include chemotherapy and radiation therapy. More serious cases may require the surgical removal of tumors (if they are confined), a *lobectomy* (the removal of a lung lobe or lobes), or a *pneumonectomy* (the removal of an entire lung).

> **U check**
> What is hemoptysis?

Pneumonoconiosis

Pneumonoconiosis is the name given to lung diseases that result from years of exposure to different environmental or occupational types of dust. The specific pneumonoconiosis is named after the dust that is inhaled. All of these disorders have *fibrosis* of the lungs in common. Three of the more prevalent types of pneumonoconiosis are *anthracosis, asbestosis,* and *silicosis.*

Etiology: Anthracosis (black lung disease) results from exposure to coal dusts. Asbestosis results from lung exposure to asbestos, whereas silicosis arises from exposure to silica sand from sand blasting and ceramic manufacture.

Signs and Symptoms: *Tachypnea* (rapid breathing), nonproductive cough, progressive dyspnea on exertion (DOE), pulmonary hypertension, recurrent respiratory infections, and eventual right ventricular hypertrophy are seen. In all cases, fibrous tissue takes over healthy lung tissue, which destroys the alveoli and eventually the bronchioles.

Treatment: Treatment for all types includes avoiding the causative dust, preventing respiratory infections, using bronchodilators, and using supplemental oxygen as needed. Respiratory therapy can also be useful in assisting patients to rid themselves of respiratory secretions. Unfortunately, lung cancer is also commonly seen with many of the pneumoconioses.

Pneumonia

Pneumonia (pneumonitis) is characterized by an inflammation of the lungs. There are over 50 different causes of pneumonia, and

(continued)

Pneumonia (concluded)

they range from mild illnesses to those that are fatal. Double pneumonia refers to inflammation of both lungs.

Etiology: Pneumonia can be caused by bacteria, viruses, fungi, and parasites. It can also be caused by foreign matter that enters the lungs (for example, stomach contents that enter the lungs after vomiting), known as *aspiration pneumonia*.

Signs and Symptoms: Common signs and symptoms include fever or chills, headache, chest or muscle pain, fatigue, dyspnea, and sputum consisting of rust-colored, green, or yellowish mucus.

Treatment: Rest, fluids, OTC pain medications, and antibiotics are the most common treatments. In severe cases, oxygen and ventilator support may be required.

Pulmonary Edema

Pulmonary edema is a condition in which fluids fill the alveoli of the lungs, making gas exchange difficult or even impossible.

Etiology: Pulmonary edema commonly occurs concurrently with left heart failure. The left side of the heart does not pump blood efficiently and blood starts to back up into the lungs, resulting in edema in the alveoli. Other causes for pulmonary edema include myocardial infarction (MI or heart attack), *cardiomyopathy* (heart muscle disease), heart valve disorders, lung infections, allergic reactions, smoke inhalation, drowning, various drugs such as narcotics and heroin, chest injuries, and high altitudes.

Signs and Symptoms: The symptoms of pulmonary edema are shortness of breath; *orthopnea* (difficulty breathing, especially when lying down); a feeling of suffocating; wheezing; a productive cough that produces pink mucus; rapid weight gain; pallor; and diaphoresis (profuse sweating).

Treatment: Treatment includes oxygen therapy, diuretics to eliminate excess fluids, and morphine to reduce anxiety and shortness of breath.

Pulmonary Embolism

Pulmonary embolism is a blocked artery in the lungs.

Etiology: Usually the artery is blocked by an embolus, typically from a blood clot that has broken loose from a vein in the legs (deep vein thrombosis, or DVT). If an artery in the lungs is completely blocked, death can occur suddenly from respiratory failure.

Respiratory failure

Pulmonary Embolism (concluded)

Those at highest risk of developing this condition are persons who have had previous heart attacks, cancer, a fractured hip, or chronic lung diseases. Women who use birth control pills and individuals who have a pacemaker may also be at risk for developing a pulmonary embolism. In addition, long periods of inactivity (such as during long airline flights), increased levels of clotting factors in the blood (usually caused by certain cancers), injury to veins, and a previous CVA with residual paralysis of the arms or legs may also cause this condition. A sedentary lifestyle as well as auto or airplane travel—or any activity that requires prolonged sitting or standing—are also major risk factors for developing a pulmonary embolism. A low-dose aspirin taken daily, as well as plenty of fluids and frequent movement of the arms and legs, may help prevent the development of a pulmonary embolism.

Signs and Symptoms: Symptoms include fainting, a sudden shortness of breath, hemoptysis, wheezing, tachycardia (rapid heartbeat), diaphoresis, and chest pain that may spread to the shoulder, arm, or the face.

Treatment: Support stockings can be used to promote circulation. The patient should rest until the blood clot has dissolved and may be prescribed thrombolytic (clot-dissolving) medications, such as tissue plasminogen activator (TPA). Anticoagulants, typically warfarin (Coumadin), may be used to help prevent formation of new blood clots in the deep veins of the body. Finally, surgery may be used to place a filter in the vena cava to prevent blood clots from reaching the lungs.

U check
What is the most common cause of a pulmonary embolus?

Alveoli: The **alveolus** (Latin, for "cavity"; plural, *alveoli*) is the functional unit of the lungs (Figures 13-13 and 13-14). The alveoli sometimes are referred to as pulmonary *parenchyma*. (Parenchyma means "working tissue" of any organ or organ system.) Through the process of diffusion, red blood cells in the capillaries release carbon dioxide into the alveoli. Conversely, the alveoli release oxygen into the blood through the thin capillary walls. This process of gas exchange in the alveoli is known as *external respiration*.

There are approximately 300 million alveoli in the lungs, consisting of two types. The *Type I alveolus* is more numerous and is composed of a squamous or flat cell. It is the actual site of oxygen and carbon dioxide exchange. The *Type II alveolus* is more cuboidal in shape, and produces a chemical called **surfactant** that is made up of lipoproteins and phospholipids. Surfactant breaks the surface tension of the lungs, allowing them to expand. Surfactant is not produced by the fetus until late in a pregnancy.

alveolus Air sac in the lungs; the functional unit of the lungs.

surfactant A complex molecule of lipoproteins and phospholipids secreted by Type II alveolar cells to reduce the surface tension of the lungs.

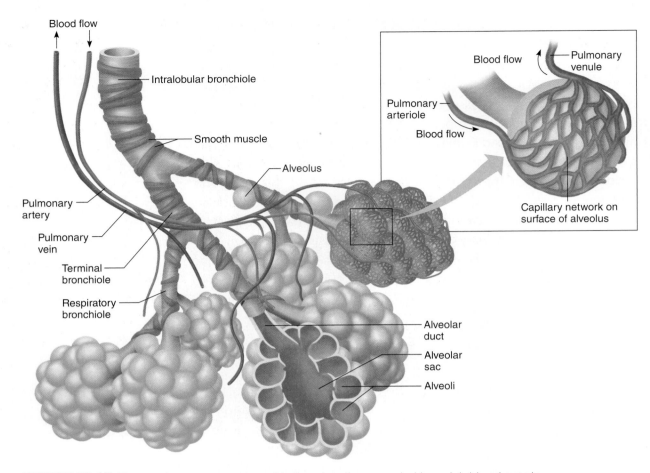

FIGURE 13-13 The respiratory passages end in tiny alveoli surrounded by a rich blood supply.

FIGURE 13-14
Light micrograph of alveoli (250x).

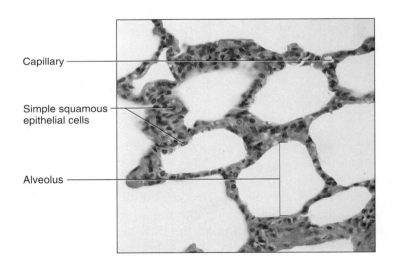

PATHOPHYSIOLOGY
Respiratory Distress Syndrome

Respiratory distress syndrome (RDS), or *hyaline membrane disease,* kills apparently healthy infants. At highest risk are newborns, especially "preemies," to infants eight months of age.

Respiratory Distress Syndrome *(concluded)*

Etiology: The etiology is unknown. The underlying problem is known to be a lack of surfactant in the lungs, which helps prevent the alveoli from totally collapsing on expiration. Without surfactant, the alveoli collapse, resulting in poor oxygenation due to difficulty with reinflation of alveoli.

Signs and Symptoms: RDS is usually diagnosed directly after birth, when the infant's breathing becomes rapid and shallow. The infant's nares flare and the accessory muscles are used to aid in respiration. The infant may also exhibits "grunting" noises in an attempt to breathe.

Treatment: Treatment must be immediate, and preferably in a neonatal intensive care unit (NICU). Oxygen therapy, the insertion of an *endotracheal tube,* ventilator support, and the use of artificial surfactant are all used in an attempt to keep the alveoli inflated. Infants who survive RDS may be at higher risk for respiratory infections later in life, but as their lungs continue to mature, this threat lessens.

Sudden Infant Death Syndrome

Sudden infant death syndrome (SIDS, or crib death) is the sudden and unexpected death of an infant under one year of age, whose body on autopsy shows no explainable cause of death. Usually a baby with this disorder simply goes to sleep and never wakes up. There were 4,895 reported SIDS deaths in 1992 and 2,247 in 2007. This dramatic decline had to do with parent education.

Etiology: The causes of SIDS are unknown, but certain risk factors have been identified:

- Male babies are more likely to die of SIDS.
- Babies between two weeks and six months of age are at greatest risk.
- Premature or low-birth-weight babies are at higher risk.
- A baby with a sibling who died of SIDS is at increased risk.
- Babies who are African American or Native American are at higher risk.
- Babies who were prenatally exposed to alcohol, cocaine, or heroin are at greater risk.
- Babies who sleep on their stomachs are at increased risk.

Treatment: There is no treatment, but certain preventive measures can be taken. Positioning the infant on the back to sleep (put the baby "back to sleep") helps reduce risk. This is especially important for those infants known to be at risk, those with previous *apneic* (lack of breathing) episodes, and those who have lost a sibling to SIDS.

(continued)

> **Sudden Infant Death Syndrome** *(concluded)*
>
> At-risk infants may also be sent home with an apnea monitor that will sound an alarm if breathing ceases. Research into this disease is ongoing. Support groups are available and are suggested for families who have experienced the tragedy of losing a child to SIDS.
>
> **U check**
> What are the risk factors for SIDS?

13.2 learning outcome

Compare internal and external respiration.

Respiration

Breathing, or pulmonary ventilation, consists of two events—inspiration (inhalation) and expiration (exhalation). During inspiration, air, which is 21 percent oxygen, flows from the nasal or oropharynx, through the sinuses into the larynx, trachea, and bronchial tree. It eventually reaches the alveoli of the lungs.

Air flows into the airways during inspiration because the thoracic cavity enlarges, creating a negative pressure in the lungs. The *intercostal muscles* located between the ribs raise the ribs, further enlarging the thoracic cavity. The atmospheric pressure outside the body is then greater than the pressure inside the lungs, and air flows from an area of high pressure to an area of low pressure. The events that enlarge the thoracic cavity and therefore lead to inspiration involve several muscles.

The **diaphragm** is the major muscle of respiration (Figure 13-15). When it contracts, it flattens, which increases the amount of space in the thoracic cavity and creates the negative pressure necessary to draw air into the lungs.

diaphragm The dome-shaped muscle between the thoracic and abdominal cavities; a major muscle of respiration.

U check
What is the major muscle of respiration?

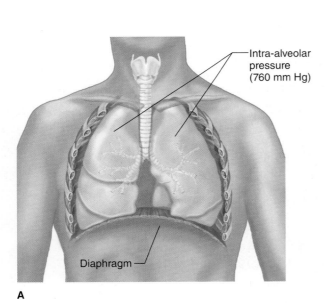
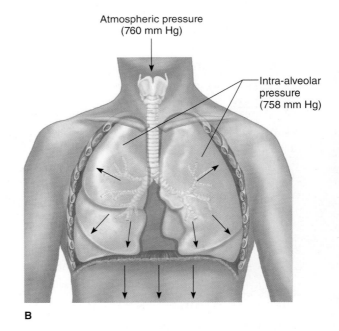

FIGURE 13-15 Normal inspiration: (A) Prior to inspiration, the intra-alveolar pressure is 760 mm Hg. (B) The intra-alveolar pressure decreases to about 758 mm Hg as the thoracic cavity expands and the atmospheric pressure pushes air into the airways.

> **focus on** Wellness: Why and When to Wear a Mask?
>
> Many diseases are spread by airborne droplets through sneezing, coughing, speaking, and even laughing. Wearing a mask when working closely with patients protects them from any respiratory ailment the health care worker may have been exposed to and may protect the health care professional from exposure to any respiratory ailment that the patient may be carrying. Surgeons and their assistants wear masks during surgery. Dental professionals wear masks when working on a patient's mouth. All health care professionals should wear a mask to prevent transmission if there is any chance a patient has a respiratory disease or simply "just in case." During outbreaks of the flu, individuals who have the flu should wear a mask to prevent the spread within their home. Although you may see healthy individuals wearing masks in crowded areas to prevent themselves from getting the flu, research does not support this practice. Most flu appears to spread by large droplets or through hand-to-face contact. Wearing a mask is a good practice, but performing hand hygiene is a better way to prevent the spread of pathogens.

During expiration, or exhalation, air rich with carbon dioxide flows out of the airways. Air flows out because the thoracic cavity becomes smaller, which increases the pressure inside the cavity. When the pressure inside the cavity becomes greater than the atmospheric pressure, air flows out. During expiration, the diaphragm relaxes. When the diaphragm relaxes, it domes up into the thoracic cavity, which decreases the space in the cavity. The intercostal muscles lower the ribs, which further decreases the size of the thoracic cavity. The pressure inside the thoracic cavity is now greater than atmospheric pressure, so air is exhaled.

Normal inspiration is considered an active process and normal expiration is considered a passive process (Figures 13-16 and 13-17). The major events during inspiration include:

1. Nerve impulses travel on phrenic nerves to muscle fibers in the diaphragm, contracting them.
2. As the dome-shaped diaphragm moves downward, the thoracic cavity expands.
3. At the same time, the external intercostal muscles contract, raising the ribs and expanding the thoracic cavity further.
4. The intra-alveolar pressure decreases.
5. Atmospheric pressure, greater on the outside, forces air into the respiratory tract through the air passages.
6. The lungs fill with air.

The major events during expiration include:

1. The diaphragm and external respiratory muscles relax.
2. Elastic tissues of the lungs and thoracic cage, stretched during inspiration, suddenly recoil, and surface tension collapses alveolar walls.
3. Tissues recoiling around the lungs increase the intra-alveolar pressure.
4. Air is squeezed out of the lungs.

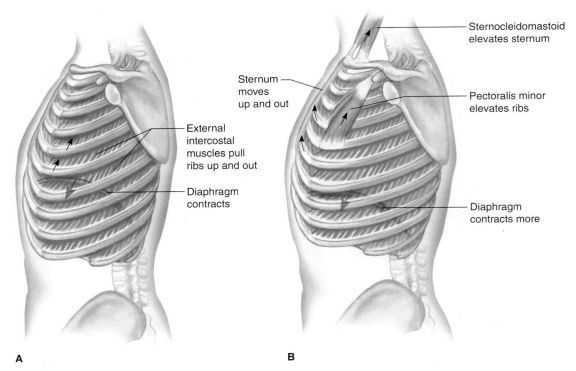

FIGURE 13-16 Maximal inspiration. (A) Shape of the thorax at the end of a normal inspiration. (B) Shape of the thorax at the end of maximal inspiration, aided by contraction of the sternocleidomastoid and pectoralis minor muscles.

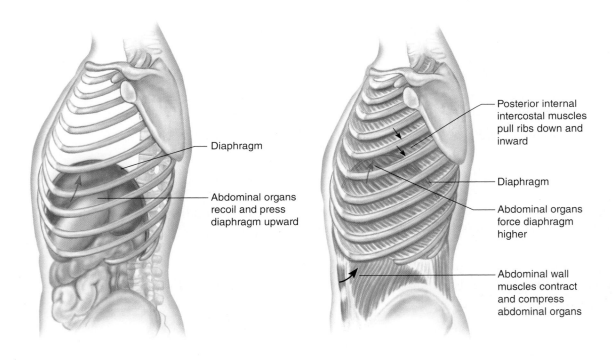

FIGURE 13-17 Expiration. (A) Normal resting expiration is due to elastic recoil of the lung tissues and the abdominal organs. (B) Contraction of the abdominal wall muscles and posterior internal intercostal muscles aids in maximal expiration.

Respiratory Volumes and Capacities

The volume of air entering and leaving the lungs depends on the effort that is used when breathing. These different volumes are called *respiratory volumes* and can be measured to assess the health of the respiratory system. *Respiratory capacities* can be calculated by adding certain respiratory volumes together. The different types of volumes and capacities are described in Table 13-2.

13.3 learning outcome

Describe the respiratory air volumes and capacities.

TABLE 13-2 Respiratory Air Volumes and Capacities

Name	Volume*	Description
Tidal volume (TV)	500 mL	Volume moved in or out of the lungs during a normal respiratory cycle
Inspiratory reserve volume (IRV)	3,000 mL	Volume that can be inhaled during forced breathing in addition to resting tidal volume
Expiratory reserve volume (ERV)	1,100 mL	Volume that can be exhaled during forced breathing in addition to resting tidal volume
Residual volume (RV)	1,200 mL	Volume that remains in the lungs at all times
Inspiratory capacity (IC)	3,500 mL	Maximum volume of air that can be inhaled following exhalation of resting tidal volume: IC = TV + IRV
Functional residual capacity (FRC)	2,300 mL	Volume of air that remains in the lungs following exhalation of resting tidal volume: FRC = ERV + RV
Vital capacity (VC)	4,600 mL	Maximum volume of air that can be exhaled after taking the deepest breath possible: VE = RV + IRV + ERV
Total lung capacity (TLC)	5,800 mL	Total volume of air that the lungs can hold: TLS = VC + RV

*Values are typical for a tall, young adult.

U check

What is tidal volume?

from the perspective of...

A RESPIRATORY THERAPIST A respiratory therapist evaluates, treats, and cares for patients with breathing or other cardiopulmonary disorders. Why is it necessary to understand respiratory volumes and capacities?

Describe the factors that influence breathing.

Factors That Control Breathing

Breathing is controlled by the respiratory center of the brain, which is located in the pons and medulla oblongata (Figure 13-18). The medulla oblongata controls the rhythm and the depth of breathing. The pons controls the rate of breathing. You will learn more about the medulla oblongata and the pons in Chapter 14, The Nervous System.

Other factors that affect breathing are the carbon dioxide levels in the blood and the pH of the blood. The rate and depth of breathing increase when carbon dioxide levels rise in the blood. The rate and depth of breathing also increase when the blood pH drops. Fear and pain also increase the breathing rate. Hyperventilation occurs when someone breathes rapidly and deeply. Hyperventilation decreases the amount of carbon dioxide in the blood. However, it should be noted that the respiratory rates in patients with chronic obstructive pulmonary disease (COPD) are stimulated by decreased oxygen levels. Therefore, giving a patient with COPD a high level of oxygen may actually decrease the "breathing reflex." The "inflation reflex" also helps regulate the depth of breathing. Stretch receptors in pleural membranes are activated when the lungs are stretched beyond a certain point. This triggers the depth of breathing to decrease to prevent overinflation of the lungs.

Normal, everyday situations also alter breathing patterns. For example:

- *Coughing:* A deep inspiration occurs and the glottis is closed. As the air forces the glottis open, a rush of air is forced upward to clear the lower respiratory passages.
- *Sneezing:* The same process occurs as in coughing except that air is moved to the nasal passages by lowering the uvula. This causes a clearing of the upper respiratory passages.
- *Laughing:* A deep breath is expelled in short bursts, expressing happiness or nervousness.
- *Crying:* The same respiratory process occurs as in laughing, but the expression is one of sadness.

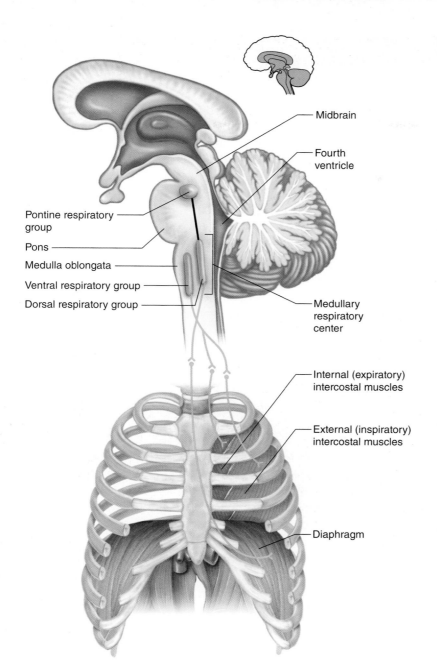

FIGURE 13-18
The respiratory areas are located in the pons and the medulla oblongata.

- *Hiccuping (hiccoughing):* These are spasmodic contractions of the diaphragm against a closed glottis. Interestingly, the purpose for this is not known.
- *Yawning:* This is a deep inspiration that is hypothesized to increase the amount of air brought to the alveoli.
- *Speaking:* Air is forced through the larynx, vibrating the vocal cords. Words are formed by the tongue, lips, and teeth, allowing for verbal communication.

Table 13-3 lists activities that affect breathing.

TABLE 13-3 Nonrespiratory Air Movements

Air Movement	Mechanism	Function
Coughing	Deep breath is taken, glottis is closed, and air is forced against the closure; suddenly the glottis is opened, and a blast of air passes upward	Clears lower respiratory passages
Sneezing	Same as coughing, except air moving upward is directed into the nasal cavity by depressing the uvula	Clears upper respiratory passages
Laughing	Deep breath is released in a series of short expirations	Expresses happiness, amusement, or nervousness
Crying	Same as laughing	Expresses sadness or pain
Hiccuping	Diaphragm contracts spasmodically while glottis is closed	No useful function known
Yawning	Deep breath is taken	Some hypotheses, but no established function
Speaking	Air is forced through the larynx, causing vocal cords to vibrate; actions of lips, tongue, and soft palate form words	Vocal communication

learning outcome

Explain how gases are exchanged in the lungs and tissues.

Oxygen transport and gas exchange

Gas Exchange

Gas exchange occurs in the alveoli of the lungs and at the capillary tissue interface in the body. The exchange in the lungs is called external respiration. Each of the approximately 300 million alveoli in the lungs is a small air sac at the end of respiratory bronchioles. As stated earlier, the alveolus is made up of simple squamous epithelium. The lungs also contain an abundance of capillaries that are also one cell layer in thickness. Gases have to diffuse across only two cells to move back and forth between the capillaries and the alveoli. Because the concentration of oxygen is greater in the lungs than in the blood, it moves down its concentration gradient to enter the blood. Conversely, the concentration of carbon dioxide is greater

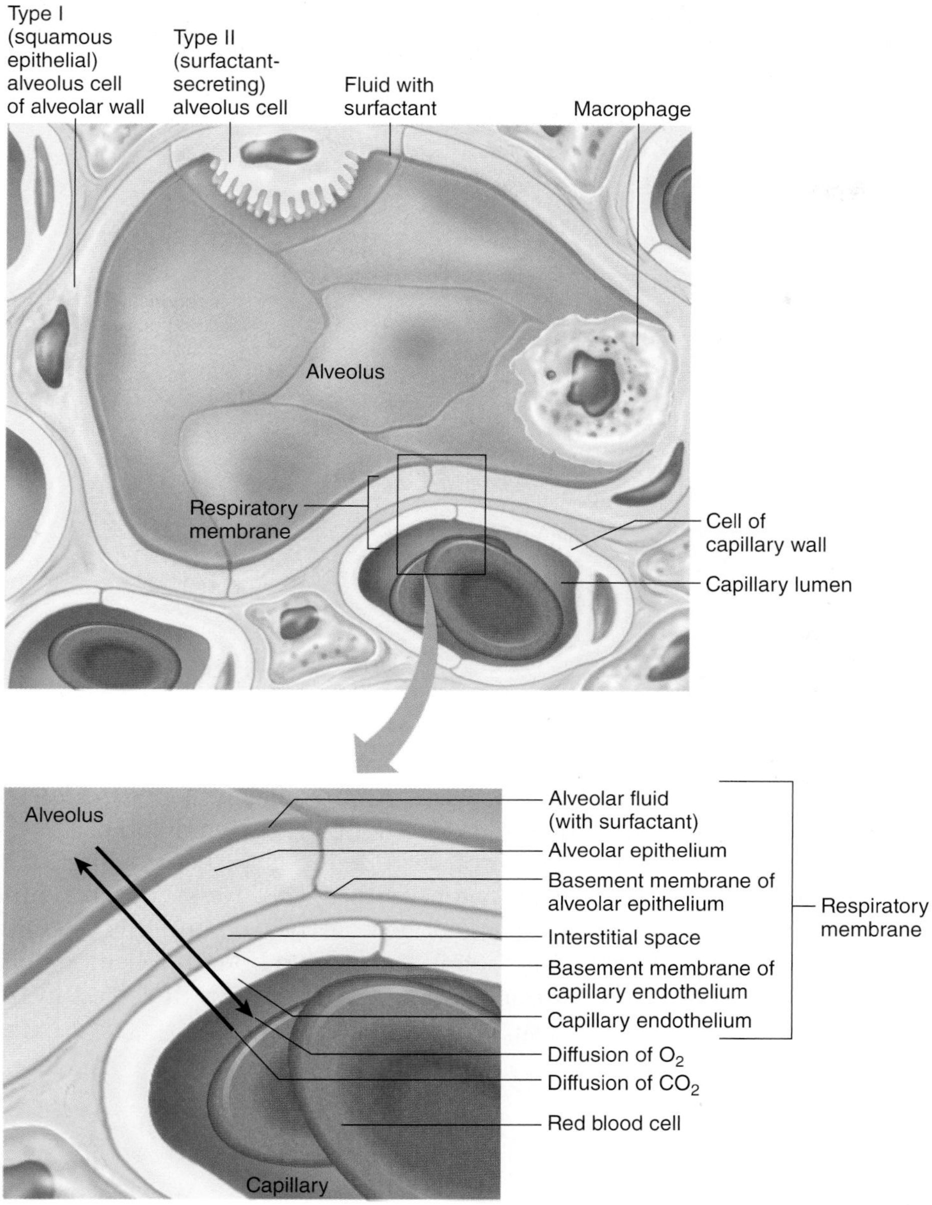

FIGURE 13-19
The respiratory membrane consists of the walls of the alveolus and the capillary.

in the blood than in the lungs. Therefore, carbon dioxide moves out of the capillaries to enter the alveoli and then to be exhaled out of the body (Figure 13-19). In total, there are almost 800 square feet of alveolus surface and approximately 1 liter of blood available in the lungs for gas exchange to occur.

The situation is reversed at the cellular level in the body. The concentration of oxygen is greater in the blood vessels (capillaries) than in the tissues. Consequently, oxygen moves out of the blood to enter the tissues (Figure 13-20). Carbon dioxide concentration is greater in the tissues than the blood, so carbon dioxide moves from the tissues to the blood. This is called internal respiration.

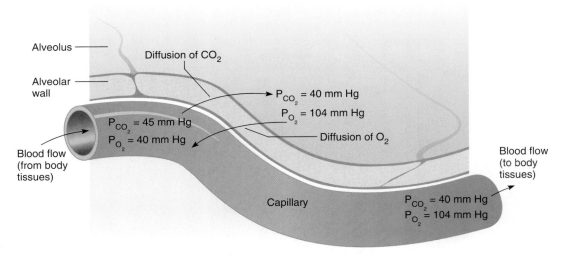

FIGURE 13-20 Gases are exchanged between alveolar air and capillary blood because of differences in partial pressures.

Describe how oxygen is transported from the lungs to body cells, and how carbon dioxide is transported from body cells to the lungs.

Gas transport

Gas Transport

Almost 99 percent of the oxygen entering the blood binds to the hemoglobin in red blood cells (Figure 13-21). Hemoglobin bound to oxygen is called *oxyhemoglobin* and is bright red in color. Approximately 1 percent of the oxygen stays dissolved in plasma and does not bind to hemoglobin.

Under normal circumstances carbon dioxide binds to the protein part of hemoglobin, different from the oxygen binding site. Hemoglobin bound to carbon dioxide is called carbaminohemoglobin (Figure 13-22). Only about 23 percent of the carbon dioxide is transported bound to hemoglobin. Less than 10 percent of the carbon dioxide travels free or dissolved in the plasma, and 70 percent travels as the bicarbonate ion, which acts as a buffer

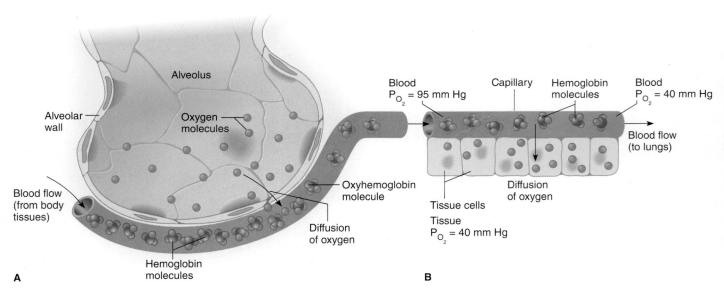

FIGURE 13-21 Blood transports oxygen: (A) Oxygen molecules, entering the blood from the alveolus, bond to hemoglobin. (B) In the tissues, hemoglobin releases the much of the oxygen.

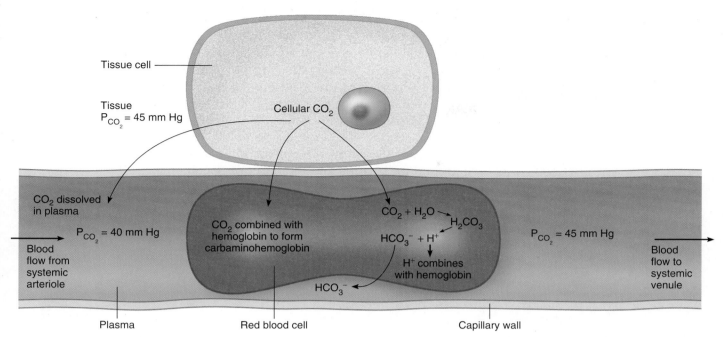

FIGURE 13-22 Carbon dioxide is transported in the blood mostly in the form of bicarbonate ions, but also dissolved in the plasma.

TABLE 13-4	Gases Transported in Blood	
Gas	**Reaction**	**Substance Transported**
Oxygen	Combines with iron atoms of hemoglobin molecules	Oxyhemoglobin
Carbon dioxide	About 7% dissolves in plasma About 23% combines with the amino groups of hemoglobin molecules About 70% reacts with water to form carbonic acid; the carbonic acid then dissociates to release hydrogen ions and bicarbonate ions	Carbon dioxide Carbaminohemoglobin Bicarbonate ions

to raise the pH of blood. Table 13-4 lists oxygen and carbon dioxide transported in the blood.

U check

How is the majority of oxygen transported in the blood?

Life Span

Because air from the environment is constantly brought into the lungs, the lungs are exposed to many pollutants. Even the lungs of nonsmokers who have lived in metropolitan areas their whole life are affected. Instead of a pinkish color, the lungs are darkened due to carbon deposits from air pollution. As we get older, our body slows down. This is also true of the respiratory system. It contains fewer cilia and they are less active. This means particles are not removed as easily from the respiratory system, making the person more susceptible to infections.

learning outcome

Explain how aging of the respiratory system affects life span.

In elderly individuals, mucus thickens and reflexes such as the cough, swallowing, and gag reflexes are diminished. This further increases the risk of infection. Breathing becomes more difficult for several reasons. The thoracic cage becomes less pliable, affecting pressure changes and breathing. The number of alveoli remains relatively the same, but the actual surface area for gas exchange decreases. The amount of collagen compared to the amount of elastin increases, which also impairs breathing. By the time someone has reached 80 years of age, he or she has half the ventilation of a 20-year-old. This means there is more stagnant air in the lungs—good breeding ground for bacteria and the onset of pneumonia.

summary

chapter 13

learning outcomes	key points
13.1 Describe the components of the respiratory system and their functions.	The respiratory system moves oxygen and carbon dioxide into and out of the body. The upper respiratory tract consists of the nose, nasal cavity, paranasal sinuses, and pharynx. The lower respiratory tract consists of the larynx, trachea, bronchi, and the lungs. Components of the respiratory system moisten and warm the air that is inhaled and transport gases between the lungs and nasal cavity. The lungs exchange oxygen and carbon dioxide.
13.2 Compare internal and external respiration.	Internal respiration is the exchange of carbon dioxide and oxygen at the cellular level, and external respiration is their exchange in the lungs.
13.3 Describe the respiratory air volumes and capacities.	Respiratory volumes and capacities are used to measure the volume of air moving into and out of the lungs. They may be used to assess the health of the respiratory system.
13.4 Describe the factors that influence breathing.	Pathologies can influence breathing, but there are also many normal everyday situations that alter our breathing patterns. For example, coughing, sneezing, laughing, crying, yawning and even speaking can have an affect on breathing patterns.
13.5 Explain how gases are exchanged in the lungs and tissues.	Gas exchange occurs in the alveoli of the lungs and at the capillary-tissue interface in the body. The concentration of oxygen is greater in the lungs than in the blood, so it moves down its concentration gradient to enter the blood. Conversely, the concentration of carbon dioxide is greater in the blood than in the lungs, so carbon dioxide leaves the capillaries to enter the alveoli and then be exhaled out of the body.

learning outcomes	key points
13.6 Describe how oxygen is transported from the lungs to body cells, and how carbon dioxide is transported from body cells to the lungs.	The majority of the oxygen transported in the plasma is bound to hemoglobin. Only about 23 percent of the carbon dioxide in the plasma is transported bound to hemoglobin. The majority travels as the bicarbonate ion.
13.7 Explain how aging of the respiratory system affects life span.	An aging respiratory system causes us to be more susceptible to infections for various reasons.

case study 1 **questions**

Can you answer the following questions that pertain to Emil's case study presented earlier in this chapter?

1. Which of the following is *not* a COPD?
   - a. Emphysema
   - b. Asthma
   - c. Lung cancer
   - d. Chronic bronchitis
2. Because of their physical appearance, individuals with emphysema are sometimes called
   - a. Pink puffers
   - b. Blue bloaters
   - c. Pink bloaters
   - d. Blue puffers
3. What is the most common cause of emphysema?
   - a. Cigarette smoking
   - b. Genetic enzyme deficiency
   - c. Infection by *Streptococcus pyogenes*
   - d. Lung trauma

review **questions**

1. Which of the following are known as the pulmonary parenchyma?
   - a. Nares
   - b. Bronchi
   - c. Bronchioles
   - d. Alveoli
2. What is responsible for raising and lowering the rib cage during respiration?
   - a. Diaphragm
   - b. Lungs
   - c. Intercostal muscles
   - d. Respiratory center
3. Lack of surfactant is responsible for which of the following conditions in premature infants?
   - a. SARS
   - b. RDS
   - c. COPD
   - d. pneumonia
4. Which structure covers the larynx during swallowing?
   - a. Epiglottis
   - b. Glottis
   - c. Conchae
   - d. Hyoid
5. Which condition is more commonly known as a collapsed lung?
   - a. Bronchitis
   - b. Pleuritis
   - c. Coryza
   - d. Atelectasis

critical **thinking questions**

1. A person has a genetic disease (for example, Kartagener's syndrome) that prevents the cilia throughout the body from functioning properly. What effect does this have on the respiratory system?
2. Trace how oxygen and carbon dioxide are transported in the blood.
3. Compare external and internal respiration.
4. An 82-year-old man has to have hip replacement surgery. He will be nonambulatory for a period of time. What risks does the surgery present to the respiratory system? Does being immobilized present a risk for any complications?

patient **education**

Your friend has a terrible snoring problem that wakes you up at night when you stay in the same room. You are concerned and would like to help. Explain that in many cases snoring will not just interfere with your sleep but will be harmful to your friend as well. Write up a list of things he or she can do to reduce the snoring.

applying **what you know**

1. What is respiration?
2. What are the organs of respiration?
3. What organ systems does the pharynx belong to?
4. What is the arrangement of the cartilage in the larynx?
5. What are the branches of the bronchial tree?
6. What are the anatomical differences between the right and left main stem bronchi?
7. What is surfactant and what is its function?
8. What is the difference between tidal volume and vital capacity?
9. What is the difference between the true and false vocal cords?
10. What are the pleura?
11. How does the right lung differ from the left lung anatomically?
12. How does the etiology of COPD compare to that of lung cancer?

CASE STUDY 2 *Asthma*

Robin Kennicker, a five-year-old boy, is brought to the pediatrician's office. He has been coughing at night for the past week. Today he presents with shortness of breath, tightness in his chest, and wheezing. The doctor recognizes the child's symptoms as those of asthma and orders a bronchodilator, a drug that relaxes the smooth muscles in the airways, to be delivered using a nebulizer. He also refers the child for allergy testing.

1. What is causing Robin to wheeze?
   a. Vasoconstriction
   b. Constriction of the bronchioles
   c. Vasodilation
   d. Dilation of the bronchioles
2. Why did the doctor refer Robin for allergy testing?
   a. To find out what drugs may safely be prescribed
   b. All five-year-olds should be routinely tested for allergies
   c. Asthma may have an allergic basis
   d. To find out if the asthma has a genetic cause

CASE STUDY 3 *Tuberculosis (TB)*

Janet Lieb, a 24-year-old graduate student, plans to spend her summer doing field work in a foreign country. She was told that this country has a high rate of drug-resistant tuberculosis.

1. What is the cause of tuberculosis?
   a. A viral infection
   b. A bacterial infection
   c. Carcinogens
   d. A genetic mutation
2. Is there anything Janet can do to protect herself?
   a. Mosquito repellant
   b. Vaccination
   c. High doses of vitamin C
   d. No, because it is a genetic disease
3. Which of the following is *not* a sign or symptom of TB?
   a. Night sweats
   b. Persistent cough
   c. Chest pain
   d. Leukemia

14

The Nervous System

chapter outline

Functions of the Nervous System

Cells of the Nervous System

The Synapse and Nerve Transmission

Central Nervous System Components and Their Functions

Meninges, Ventricles, and Cerebrospinal Fluid

Peripheral Nervous System Components and Their Functions

Neurological Testing

Life Span

outcomes

AFTER COMPLETING THIS CHAPTER, YOU WILL BE ABLE TO:

14.1 Recognize the functions of the nervous system.

14.2 Compare the structure and function of a neuron and a glial cell.

14.3 Explain the synapse and nerve transmission.

14.4 Identify the components of the central nervous system and their functions.

14.5 Describe the structure and functions of the meninges, ventricles, and cerebrospinal fluid.

14.6 Identify the components of the peripheral nervous system and their functions.

14.7 Relate neurological testing procedures to their uses.

14.8 Infer how aging affects the nervous system and life span.

case study

Use the case study to focus on as you go through the chapter. The questions will guide you as you learn the anatomy and pathophysiology of the nervous system.

CASE STUDY 1 *Stroke*

Mary Ann Nicks, 38 years of age, has a severe headache on the right side of her head that started 30 minutes ago. She also has numbness in her left arm and is slurring her speech. She has no previous history of these symptoms, but she has taken birth control pills since she was 17 years old and has smoked cigarettes for about the same length of time. She is taking medication for high blood pressure.

As you go through the chapter, keep the following questions in mind:

1. What are the two major categories of strokes?
2. What risk factors for stroke does this patient have?
3. Are her symptoms typical for a stroke?
4. What are additional warning signs that a person is experiencing a stroke?
5. What is the treatment for a stroke?
6. What is the prognosis for a stroke?

essential terms

axon (AK-son)
cerebellum (ser-e-BEL-um)
cerebrum (ser-EE-brum)
corpus callosum (KOR-pus kal-LO-sum)
cortex (KOR-teks)
dendrite (DEN-drite)
diencephalon (di-en-SEF-a-lon)
ganglion (GANG-gle-on)
gyri (JI-ri)
medulla oblongata (me-DOOL-la ob-long-GAH-tah)
meninges (me-NIN-jez)
mesencephalon (mes-en-SEF-a-lon)
myelin (MI-e-lin)
neuroglia (noo-ROG-le-a)
neuron (NOO-ron)
neurotransmitter (noo-ro-TRANS-mit-ter)
node of Ranvier (NODE of RON-ve-a)
pons (PONZ)
sulcus (SUL-kus)
ventricles (VEN-tri-kulz)

Additional key terms in the chapter are italicized and defined in the glossary.

study tips

1. Outline the chapter. After each section, question yourself about what you just read.
2. Draw and label a neuron and explain the function of each component.
3. Meet with a study group; assign each member to be the expert on one concept and have members explain their concept to the rest of the group.

Introduction

The nervous system, in conjunction with the endocrine system, is a major regulator of the body's various organ systems and is essential in maintaining homeostasis. It is a fast-acting system, whereas the endocrine system (see Chapter 21, The Endocrine System) is much slower. The nervous system is very complex, so much so that scientists are still learning new things about it. It is almost ironic that we are using a part of our nervous system (the brain) to study the nervous system.

The nervous system is divided into several different anatomical and functional categories. Anatomically, the nervous system can be divided into the central nervous system (CNS) and the peripheral nervous system (PNS). The CNS is composed of the brain and spinal cord and the PNS consists of the cranial and spinal nerves (Figure 14-1). Functionally, the nervous system can be separated into somatic and autonomic divisions. The nervous system is responsible to a great extent for determining who we are by forming our thoughts, emotions, and memories. Numerous diseases of the nervous system exist that often require extensive testing to arrive at a proper diagnosis and treatment. Several medical specialties focus on specific aspects of the nervous system; for example, neurosurgery, psychiatry, and neurology.

> **U check**
> What are the two organ systems that help regulate homeostasis in the body?

Recognize the functions of the nervous system.

Functions of the Nervous System

The nervous system has many functions. One of the characteristics of nervous tissue is excitability, which allows nerves to conduct an electrical impulse, also known as an "action potential," to communicate with other nerves, muscles, or glands. For instance, the nervous system, working with the musculoskeletal system, produces movement.

Our brain is capable of thought, emotions, behavior, and memory. The spinal cord and nerves process stimuli that may be harmful, and the body

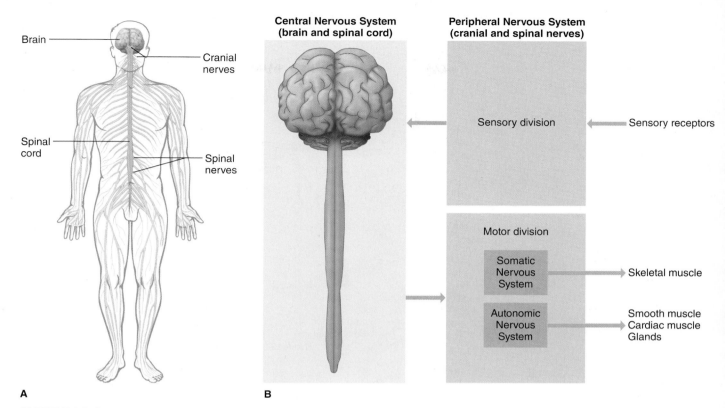

FIGURE 14-1 A diagrammatic representation of the nervous system: (A) The nervous system includes the central nervous system (brain and spinal cord) and the peripheral nervous system (cranial nerves and spinal nerves). (B) The nervous system receives information from sensory receptors and initiates responses in muscles and glands.

responds with a protective reflex action. Sensory receptors for smell, touch, sight, and hearing are part of the nervous system. The nervous system also is involved in regulating appetite and body temperature.

> **U check**
> What is an action potential?

Cells of the Nervous System

There are two categories of cells in the nervous system: neurons and neuroglia. **Neurons** are the functional cells of the nervous system (Figures 14-2 and 14-3.) They transmit electrochemical messages called nerve impulses, or action potentials, to other neurons and *effectors* (muscles or glands). Because neurons do not have the ability to divide, they are called *permanent* cells. If a person has a stroke, Alzheimer's disease, or some other disease that destroys neurons, the neurons cannot be replaced.

Neurons have a cell body (or soma), a cell membrane, a nucleus, and many of the same organelles found in other types of cells. For example, the cell body contains ribosomes and mitochondria to produce proteins and energy needed by the cell.

One unique feature of neurons is that they have two types of nerve fibers, called dendrites and axons, which are extensions from the cell body.

learning outcome

Compare the structure and function of a neuron and a glial cell.

neuron A nerve cell, consisting of a cell body, dendrites, and an axon.

FIGURE 14-2
A common neuron.

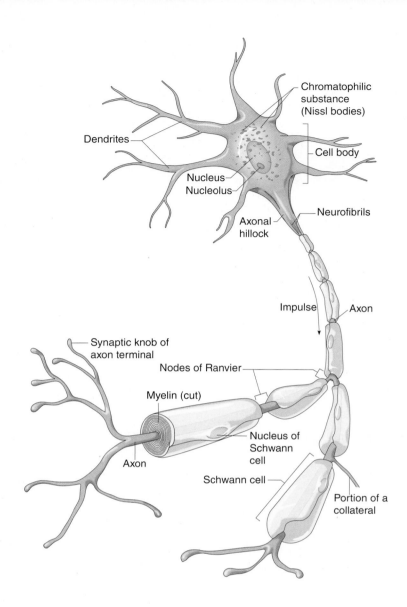

dendrite A neuronal cell process, or nerve fiber, that carries electrical signals toward the cell body.

axon Long process, or nerve fiber, of a nerve cell that sends an action potential toward the axon terminals.

When located in the central nervous system (CNS), these nerve fibers form *tracts*, and when in the peripheral nervous system (PNS), they form *nerves*. There are usually many **dendrites** from a single cell body. Their function is to transmit an electrical signal toward the cell body. Conversely, there is usually only one **axon** that carries an action potential away from the cell body.

The nervous system contains three types of neurons based on their structure (Figure 14-4 and Table 14-1):

1. Multipolar neurons are the most numerous; most of the neurons in the brain and spinal cord are of this type. The neuron has one axon and many dendrites attached to the cell body.

2. Bipolar neurons are the fewest in number. They have only one dendrite and one axon and are found in the special sense organs such as the retina of the eye, the olfactory area of the brain, and the inner ear.

3. Unipolar neurons have a single extension that divides into two separate nerve fibers. These neurons accumulate in specialized structures called ganglia. We will learn more about ganglia later in the chapter.

Neurons can also be classified based on their function:

1. *Afferent (sensory) neurons* carry impulses from the peripheral nervous system (PNS) to the central nervous system. They carry messages from receptors for touch, pain, temperature, and vibration to the CNS for interpretation. Most are unipolar in design.
2. *Efferent (motor) neurons* carry messages from the CNS to the muscles and glands in the peripheral nervous system. They are multipolar neurons.
3. *Interneurons,* or association neurons, connect sensory and motor neurons and direct the impulse to other areas of the brain or spinal cord. They are also multipolar neurons.

Neuroglia, also referred to as neuroglial cells or simply "glia," are the second major category of cells in the nervous system. "Glia" means glue and it was once thought that they held nervous tissue together. Unlike neurons, they do not transmit impulses. Instead, they act as supportive cells for neurons. Also unlike neurons, neuroglial cells are able to reproduce and replace themselves. Neuroglia are typically smaller than neurons and are up to 50 times more numerous. There are four types of glial cells in the CNS (Figure 14-5) and two types in the PNS.

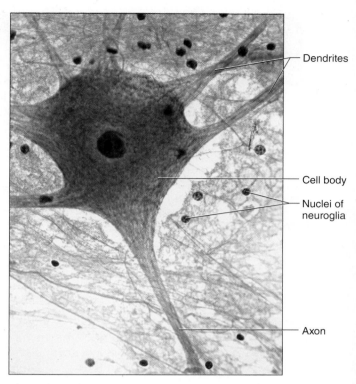

FIGURE 14-3 Neurons are the structural and functional unit of the nervous system (600x). Neuroglia are supportive and nutritive cells that surround a neuron. Neuroglia appear here as dark dots.

neuroglia Cells of the nervous system that perform various supportive functions.

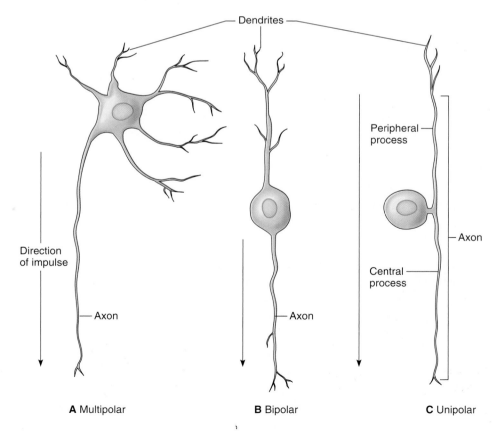

FIGURE 14-4 Structural types of neurons include (A) the multipolar neuron, (B) the bipolar neuron, and (C) the unipolar neuron.

CHAPTER 14 The Nervous System

TABLE 14-1 Types of Neurons

A. Classified by Structure

Type	Structural Characteristics	Location
1. Multipolar neuron	Cell body with many processes: one axon, and the rest dendrites	Most common type of neuron in the brain and spinal cord
2. Bipolar neuron	Cell body with a process arising from each end: one axon and one dendrite	In specialized parts of the eyes, nose, and ears
3. Unipolar neuron	Cell body with a single process that divides into two branches and functions as an axon	Found in ganglia outside the brain or spinal cord

B. Classified by Function

Type	Functional Characteristics	Structural Characteristics
1. Sensory neuron	Conducts nerve impulses from receptors in peripheral body parts into the brain or spinal cord	Most unipolar; some bipolar
2. Interneuron	Transmits nerve impulses between neurons in the brain and spinal cord	Multipolar
3. Motor neuron	Conducts nerve impulses from the brain or spinal cord to effectors—muscles and glands	Multipolar

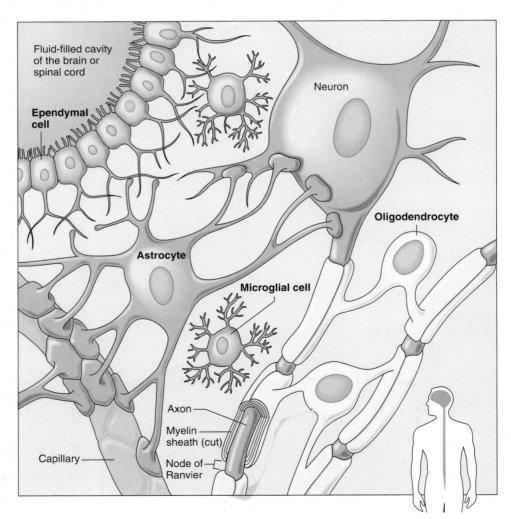

FIGURE 14-5
Types of neuroglia in the central nervous system include the astrocyte, oligodendrocyte, microglial cell, and ependymal cell.

Table 14-2 summarizes the types of neuroglia. In the CNS you will find the following glial cells:

1. *Astrocytes* are star-shaped cells that anchor blood vessels to nerve cells. They are involved in maintaining the chemical environment around neurons. They can form scar tissue when nervous tissue is damaged, they help metabolize certain substances, and they aid in providing nutrition to neurons. Astrocytes also regulate the blood-brain barrier to restrict movement of substances between the brain and blood vessels as a protective mechanism.
2. *Microglia* are small cells within the CNS that function as phagocytes that engulf bacteria and debris. They proliferate when there is injury or disease in the CNS.
3. *Oligodendrocytes* are responsible for producing the myelin sheath that is found around some axons. We will talk more about myelin later in the chapter. A single oligodendrocyte may myelinate several axons.
4. *Ependymal cells* are ciliated and are cuboidal or columnar in shape. They help regulate the cerebrospinal fluid of the CNS.

> **U check**
> What are the two categories of cells in the nervous system?

TABLE 14-2 Types of Neuroglia

Type	Characteristics	Functions
CNS		
Astrocytes	Star-shaped cells between neurons and blood vessels	Structural support, formation of scar tissue, transport of substances between blood vessels and neurons, communication with one another and with neurons, mop up excess ions and neurotransmitters, induce synapse formation
Oligodendrocytes	Shaped like astrocytes, but with fewer cellular processes; occur in rows along axons	Form myelin sheaths in the brain and spinal cord, produce nerve growth factors
Microglia	Small cells with few cellular processes; found throughout the CNS	Structural support and phagocytosis (immune protection)
Ependyma	Cuboidal and columnar cells in the inner lining of the ventricles of the brain and the central canal of the spinal cord	Form a porous layer through which substances diffuse between the interstitial fluid of the brain and spinal cord and the cerebrospinal fluid
PNS		
Schwann cells	Cells with abundant, lipid-rich membranes that wrap tightly around the axons of peripheral neurons	Speed neurotransmission
Satellite cells	Small, cuboidal cells that surround cell bodies of neurons in ganglia	Support ganglia in the PNS

In the peripheral nervous system, there are two types of neuroglial cells:

1. *Schwann cells* produce myelin in the PNS. The Schwann cells wrap themselves around axons, which are then coated by the cell membranes of the Schwann cells.
2. *Satellite cells* are flat cells that surround neurons to provide support as well as to regulate the exchange of substances between the neuron and the interstitial fluid.

myelin Multilayered covering around the axons of many neurons; composed of lipid and protein.

Myelin is a fatty substance that insulates axons and allows them to send nerve impulses more rapidly than axons of unmyelinated nerves (Figure 14-6). Note that although the axon is myelinated, the Node of Ranvier is not.

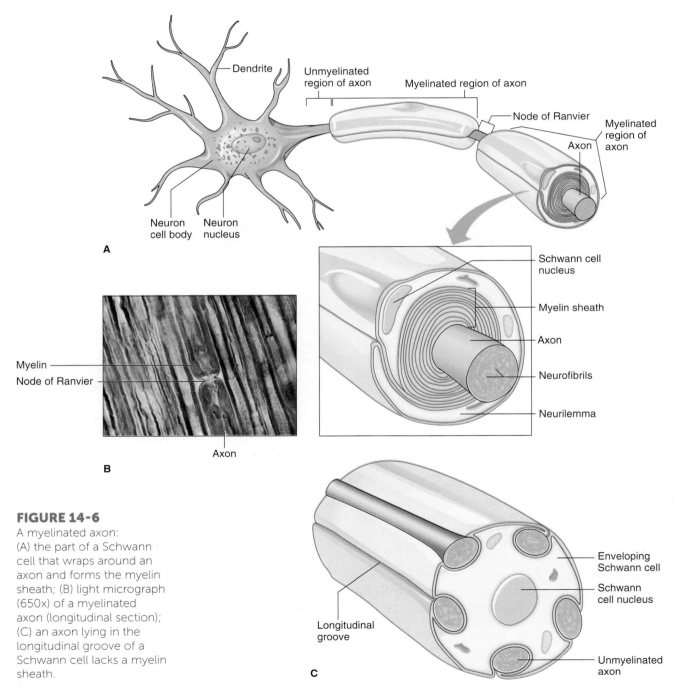

FIGURE 14-6
A myelinated axon: (A) the part of a Schwann cell that wraps around an axon and forms the myelin sheath; (B) light micrograph (650x) of a myelinated axon (longitudinal section); (C) an axon lying in the longitudinal groove of a Schwann cell lacks a myelin sheath.

When coated, these axons are referred to as "white matter." The unmyelinated nerve cell bodies make up the "gray matter" of the nervous system.

> **U check**
> What is the function of satellite cells?

The Synapse and Nerve Transmission

An electrical impulse or action potential travels in only one direction—down an axon. When traveling down an axon, the nerve impulse eventually reaches the *synaptic knob* at the end of each axon branch (Figure 14-7). These synaptic knobs contain small sacs called vesicles which contain chemicals called **neurotransmitters.** Table 14-3 lists some neurotransmitters and representative actions. A series of events cause the vesicles to fuse with the presynaptic membrane of the axon terminal. This in turn causes the neurotransmitters to be released by the synaptic knob into a space called the *synaptic cleft* (Figure 14-8). The chemical neurotransmitters then combine with receptors on the postsynaptic membrane which allows the impulse to continue on the structures on the other side of the synaptic cleft: the dendrites, cell bodies, and axons of other neurons.

There are about 50 different neurotransmitters. Although most neurons release only one type of neurotransmitter, some release more than one type. Neurotransmitters are either excitatory or inhibitory, depending on where in

14.3 learning outcome

Explain the synapse and nerve transmission.

neurotransmitter One of many varieties of molecules released by the axon terminal in response to an action potential, which then causes a change in the postsynaptic membrane potential.

Nerve impulse

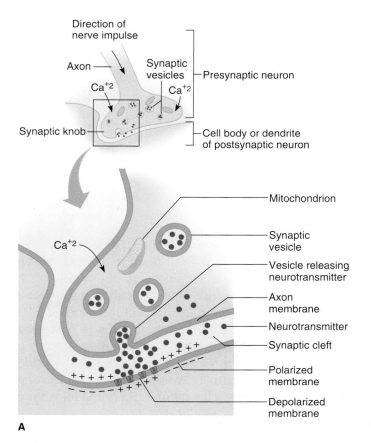

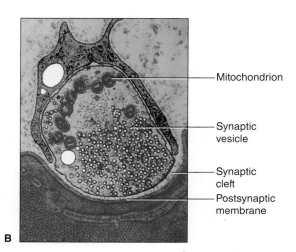

FIGURE 14-7 The synapse: (A) when a nerve impulse reaches the synaptic knob, vesicles release a neurotransmitter that crosses the synaptic cleft; (B) a transmission electromicrograph of a synaptic vesicle filled with vesicles (magnified 37,500x).

TABLE 14-3 Some Types of Neurotransmitters and Representative Actions

Neurotransmitter	Location	Major Actions
Acetylcholine	CNS	Controls skeletal muscle actions
	PNS	Stimulates skeletal muscle contraction at neuromuscular junctions; may excite or inhibit at autonomic nervous system synapses
Biogenic Amines		
Norepinephrine	CNS	Creates a sense of well-being; low levels may lead to depression
	PNS	May excite or inhibit autonomic nervous system actions, depending on receptors
Dopamine	CNS	Creates a sense of well-being; deficiency in some brain areas associated with Parkinson's disease
	PNS	Limited actions in autonomic nervous system; may excite or inhibit, depending on receptors
Serotonin	CNS	Primarily inhibitory; leads to sleepiness; action is blocked by LSD, enhanced by selective serotonin reuptake inhibitor antidepressant drugs
Histamine	CNS	Release in hypothalamus promotes alertness
Amino Acids		
GABA	CNS	Generally inhibitory
Glutamate	CNS	Generally excitatory
Neuropeptides		
Enkephalins, endorphins	CNS	Generally inhibitory; reduce pain by inhibiting substance P release
Substance P	PNS	Excitatory; pain perception
Gases		
Nitric oxide	CNS	May play a role in memory
	PNS	Vasodilation

the nervous system the neurotransmitter is found. They can cause muscles to contract or relax, glands to secrete products, and neurons to send nerve impulses. They can also inhibit neurons from sending nerve impulses.

Neuron cell membranes have an electrical charge called *membrane potential*. One side of the membrane is more negative and the other side is more positive. This difference in charge means the membrane is *polarized* and the cell is said to be resting. In a resting cell, there are more positively charged sodium ions outside the cell than there are potassium ions inside the cell. This accounts for the relative negativity inside the cell. The resting potential of most neurons is −70 millivolts. This membrane potential is very important for the generation of an action potential (Figure 14-9).

A neuron can respond to stimuli such as heat, pressure, and chemicals by changing the potential across its membrane. For example, it can respond to a stimulus by making the inside of the cell membrane more negative. This

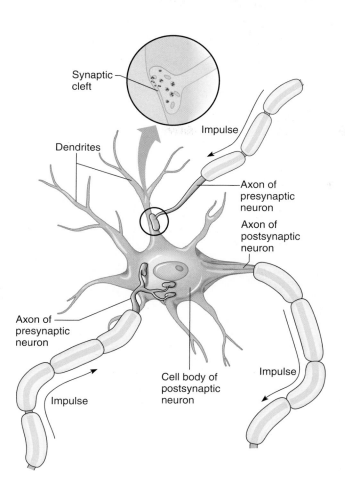

FIGURE 14-8
An action potential crosses the synaptic cleft from the presynaptic neuron to the postsynaptic neuron.

is called *hyperpolarization.* If sodium channels are opened to allow more sodium to enter the cell, then the inside of the cell becomes less negative and may reach the threshold of −55 millivolts. At this point the cell membrane is considered to be *depolarized* and an action potential is initiated. An action potential is an electrical impulse that travels in only one direction—down an axon—and it is considered an "all or none" response because it has only one magnitude (Figures 14-10 and 14-11). In other words, once the threshold is reached, an action potential is elicited. A greater stimulus does not cause a larger action potential.

Sodium channels eventually close and potassium channels open, allowing potassium ions to diffuse out of the cell and reestablishing the resting state. The return to the original polar (resting) state is called *repolarization.*

Certain factors influence the speed of conductance down an axon. A larger-diameter axon conducts an impulse faster than an axon with a smaller diameter. A myelinated nerve conducts an action potential faster than does an unmyelinated axon. In myelinated nerves, there are bare areas on the axon called **Nodes of Ranvier** where the action potential jumps from one node to the next. This is called *saltatory conductance.* In unmyelinated nerves, the action potential moves in a linear fashion down the membrane.

Node of Ranvier A bare area on an axon between myelinated areas.

U check

What is meant by an action potential being an "all or none" response?

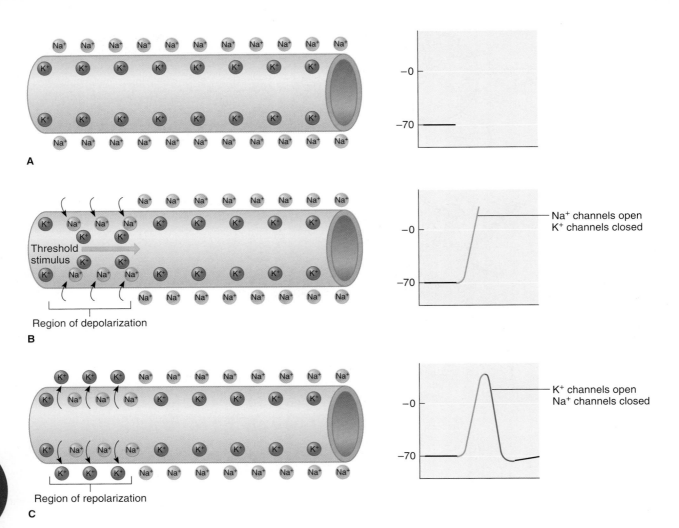

FIGURE 14-9 At rest, (A) the membrane potential is about −70 millivolts. When the membrane reaches threshold (B), voltage-sensitive sodium channels open, some sodium ions diffuse into the cell, and the membrane is depolarized. Shortly thereafter, (C) voltage-sensitive potassium channels open and some potassium ions diffuse out and the membrane is repolarized.

14.4 learning outcome

Identify the components of the central nervous system and their functions.

Central Nervous System Components and Their Functions

Recall that the nervous system is divided into the *central nervous system (CNS)* and the *peripheral nervous system (PNS)*. Table 14-4 lists the subdivisions of the nervous system. The CNS consists of the brain and the spinal cord, and the PNS consists of peripheral nerves, which are located throughout the rest of the body. The PNS is discussed in section 14.6 of this chapter.

Brain

The brain is the part of the central nervous system located within the skull. It is responsible for making decisions, solving problems, memory, and much more. The spinal cord carries electrical impulses through the body. These delicate tissues of the CNS are protected by two different structures. The blood-brain barrier (BBB) protects the brain by not allowing harmful substances to enter it from the blood. This is accomplished by tight junctions between endothelial cells of the brain capillaries. Astrocytes help maintain the tight junctions. The BBB restricts most large molecules and hydrophilic

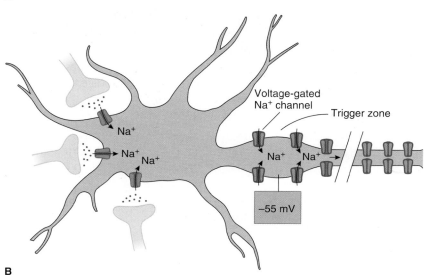

FIGURE 14-10 (A) A subthreshold depolarization will not result in an action potential. (B) Multiple stimulation by presynaptic neurons may reach threshold, opening voltage-gated channels at the trigger zone.

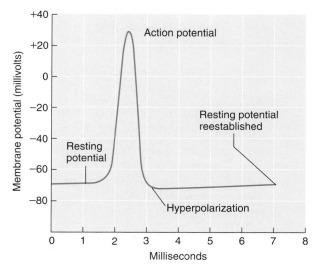

FIGURE 14-11 An oscilloscope records an action potential.

TABLE 14-4 Subdivisions of the Nervous System

1. Central nervous system (CNS)
   a. Brain
   b. Spinal cord
2. Peripheral nervous system (PNS)
   a. Cranial nerves arising from the brain
      (1) Somatic fibers connecting to the skin and skeletal muscles
      (2) Autonomic fibers connecting to viscera
   b. Spinal nerves arising from the spinal cord
      (1) Somatic fibers connecting to the skin and skeletal muscles
      (2) Autonomic fibers connecting to viscera

CHAPTER 14 The Nervous System

substances from entering the cerebrospinal fluid (CSF). It allows oxygen and lipid-soluble substances such as anesthetics to cross and it actively transports essential nutrients such as glucose. Unfortunately, some harmful substances, such as alcohol and nicotine, can cross the BBB and inflammation can make this barrier more permeable. The brain and spinal cord are also protected by the meninges, which will be discussed later in the chapter.

The brain begins its development early in embryonic life. The CNS begins as a simple tubelike structure called the neural tube. By week four of development, the brain has started developing an expansion of the anterior end of the tube. Shortly thereafter, three regions of the brain can be identified: the forebrain, midbrain, and hindbrain. They are the forerunners of the adult brain.

> **U check**
> What is the function of the blood-brain barrier?

TABLE 14-5 Major Parts of the Brain

Part	Characteristics	Functions
1. Cerebrum	Largest part of the brain; two hemispheres connected by the corpus callosum	Controls higher brain functions, including interpreting sensory impulses, initiating muscular movements, storing memory, reasoning, and determining intelligence
2. Basal nuclei (ganglia)	Masses of gray matter deep within the cerebral hemispheres	Relay stations for motor impulses originating in the cerebral cortex and passing into the brainstem and spinal cord
3. Diencephalon	Includes masses of gray matter (thalamus and hypothalamus)	The thalamus is a relay station for sensory impulses ascending from other parts of the nervous system to the cerebral cortex; the hypothalamus helps maintain homeostasis by regulating visceral activities and by linking the nervous and endocrine systems
4. Brainstem	Connects the cerebrum to the spinal cord	
a. Midbrain	Contains masses of gray matter and bundles of nerve fibers that join the spinal cord to higher regions of the brain	Contains reflex centers that move the eyes and head, and maintains posture
b. Pons	A bulge on the underside of the brainstem that contains masses of gray matter and nerve fibers	Relays nerve impulses to and from the medulla oblongata and cerebrum; helps regulate rate and depth of breathing
c. Medulla oblongata	An enlarged continuation of the spinal cord that extends from the foramen magnum to the pons and contains masses of gray matter and nerve fibers	Conducts ascending and descending impulses between the brain and spinal cord; contains cardiac, vasomotor, and respiratory control centers and various nonvital reflex control centers
5. Cerebellum	A large mass of tissue inferior to the cerebrum and posterior to the brainstem; includes two lateral hemispheres connected by the vermis	Communicates with other parts of the CNS by nerve tracts, integrates sensory information concerning the position of body parts, and coordinates muscle activities and maintains posture

The adult brain can be divided into four major areas: the cerebrum, the diencephalon, the brainstem, and the cerebellum (Figure 14-12). Table 14-5 lists the major parts of the brain.

Cerebrum: The **cerebrum** is the largest division of the brain and is what makes us uniquely human. The cerebrum allows us to think in the present, remember the past, and plan for the future. The cerebrum is divided into

cerebrum The two hemispheres of the forebrain making up the largest part of the brain.

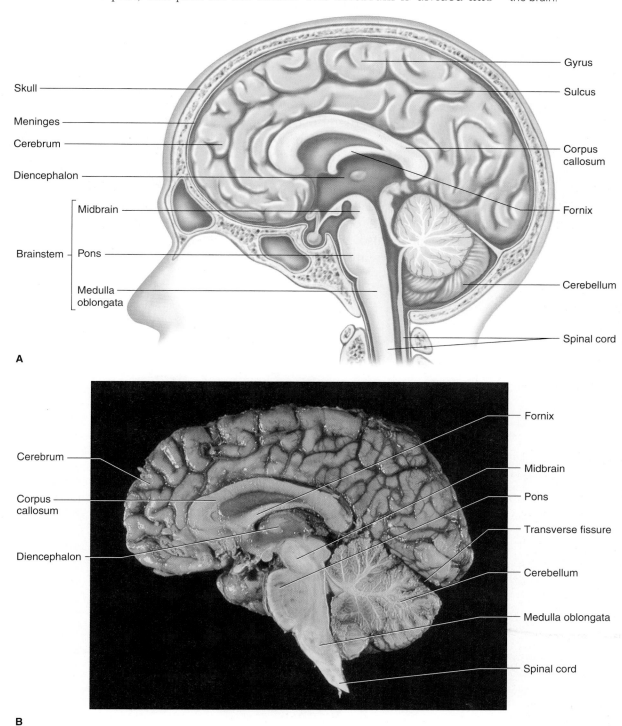

FIGURE 14-12 Sagittal section of a brain: (A) The major portions of the brain include the cerebrum, the diencephalon, the cerebellum, and the brainstem. (B) Photo of a sagittal section of a human brain.

CHAPTER 14   The Nervous System   387

corpus callosum The large is the connection between the right and left hemispheres of the brain.

sulcus A groove between the convolutions of the brain

gyri Folds of gray matter on the surface of the brain.

cortex The convoluted layer of gray matter over the cerebral hemispheres.

right and left halves called cerebral hemispheres by a deep groove known as the longitudinal fissure. A thick bundle of fibers called the **corpus callosum** connects the two hemispheres. The grooves on the surface of the cerebrum are called **sulci.** The folds of brain matter between the sulci are called **gyri** or convolutions. The gyri increase the surface area or **cortex** of the brain. The cerebral cortex also houses the limbic system, which controls emotion, mood, and memory.

Lobes: Each cerebral hemisphere is divided into four lobes—frontal, parietal, temporal, and occipital—named after the cranial bones that cover them (Figure 14-13). Table 14-6 summarizes the functions of the cerebral lobes.

The central sulcus separates the *frontal lobe* from the parietal lobes. The frontal lobe contains the primary motor area located in the precentral gyrus. The cortex is the outer portion of the brain and it is made up of nerve cell bodies. This layer contains nearly 75 percent of all neurons in the entire nervous system. It allows a person to have conscious thought and make decisions regarding movement of specific muscles or muscle groups. Each hemisphere controls muscles on the opposite or contralateral side of the body (the right side of the brain controls the left side of the body and vice versa). There is also an area of the frontal lobe close to the central sulcus called Broca's area that is responsible for the articulation of speech.

FIGURE 14-13
Colors in this figure distinguish the lobes of the cerebral hemispheres: (A) Lateral view of the left hemisphere. (B) Hemispheres viewed from above.

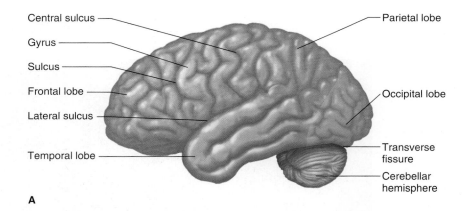

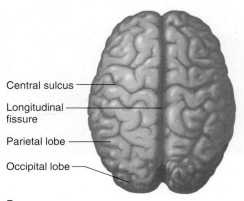

TABLE 14-6 Functions of the Cerebral Lobes

Lobe	Functions
Frontal lobes	Association areas carry on higher intellectual processes for concentrating, planning, complex problem solving, and judging the consequences of behavior.
	Motor areas control movements of voluntary skeletal muscles.
Parietal lobes	Sensory areas provide sensations of temperature, touch, pressure, and pain involving the skin.
	Association areas function in understanding speech and in using words to express thoughts and feelings.
Temporal lobes	Sensory areas are responsible for hearing.
	Association areas interpret sensory experiences and remember visual scenes, music, and other complex sensory patterns.
Occipital lobes	Sensory areas are responsible for vision.
	Association areas combine visual images with other sensory experiences.

In most people, Broca's area is in the left hemisphere (Figure 14-14). A CVA or stroke in this area prevents a person from articulating thoughts, but does not affect the person's ability to think clearly.

Other functions of the frontal lobe include intelligence, personality, learning, abstract reasoning, and recall. If the prefrontal cortex is severely damaged, a person may have a complete personality change, exhibiting a variety of unacceptable and/or inappropriate behaviors.

Somatosensory areas are located in the *parietal lobes* just posterior to the central sulcus in the postcentral gyrus. These areas interpret sensations felt

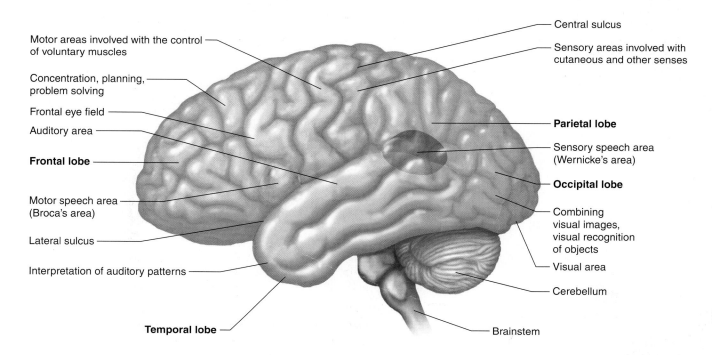

FIGURE 14-14 Some sensory association and motor areas of the left cerebral cortex. Note the motor speech area (Broca's area).

on or within the body, including touch, pressure, vibration, hot, cold, pain, and body position in space (also known as proprioception). For example, if you feel a light touch on your right hand, the somatosensory area interprets the sensation and where it is occurring.

The *temporal lobes* contain auditory areas that interpret sounds as well as olfactory areas that identify different aromas or smells. Visual areas for interpretation of what is seen are located in the *occipital lobes*.

> **U check**
> What are the four lobes of the cerebrum?

diencephalon A part of the brain consisting of the thalamus and hypothalamus.

Diencephalon: The **diencephalon** is located between the cerebral hemispheres and is superior to the brainstem. The diencephalon contains the thalamus and hypothalamus.

The *thalamus* serves as the relay station for sensory information that is conveyed to the cerebral cortex for interpretation (Figure 14-15). If sensory information does not pass through the thalamus before it reaches the cerebral cortex, it cannot be interpreted correctly. For example, if you are experiencing a painful stimulus in your left forearm, this information goes up the spinal cord, through the thalamus, and then to the cerebral cortex for interpretation. If the information does not go through the thalamus for any reason, the cerebral cortex may interpret the sensation as cold or some other sensation instead of pain in your left forearm.

The *hypothalamus* maintains homeostasis by regulating many vital activities such as heart rate (pulse), blood pressure (BP), and respiration (breathing) rate.

> **U check**
> What is the function of the thalamus?

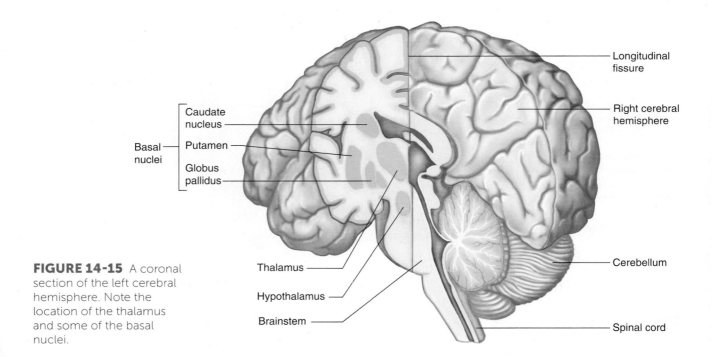

FIGURE 14-15 A coronal section of the left cerebral hemisphere. Note the location of the thalamus and some of the basal nuclei.

Brainstem: The brainstem is the structure that connects the cerebrum to the spinal cord. The three parts of the brainstem are the midbrain, the pons, and the medulla oblongata.

The *midbrain* or **mesencephalon** lies just beneath the diencephalon. It controls both visual and auditory reflexes. An example of a visual reflex is when you see something in your peripheral vision and you automatically turn your head to view it more clearly.

The **pons** is the rounded bulge on the underside of the brainstem situated between the midbrain and the medulla oblongata. It contains nerve tracts to connect the cerebrum to the cerebellum. The pons also regulates respiration.

The **medulla oblongata** is the most inferior portion of the brainstem and is directly connected to the spinal cord. It controls many vital activities such as heart rate, blood pressure, and respiration. It also controls reflexes associated with coughing, sneezing, and vomiting.

mesencephalon The part of the brain between the pons and diencephalon; the midbrain.

pons The part of the brainstem that forms a "bridge" between the medulla oblongata and the midbrain.

medulla oblongata The most inferior part of the brainstem.

> **U check**
> What are the components of the brainstem?

Cerebellum: The **cerebellum** is inferior to the occipital lobes of the cerebrum and posterior to the pons and medulla oblongata. Although much smaller than the cerebrum, nearly half the neurons of the brain are located in the cerebellum. It coordinates the complex skeletal muscle activity that is needed for body movements. For example, when you walk, many muscles have to contract and relax at appropriate times. Your cerebellum coordinates these activities. The cerebellum also coordinates fine-motor movements such as threading a needle, playing an instrument, and writing.

cerebellum The part of the brain that maintains balance and coordinates skilled movements.

PATHOPHYSIOLOGY

Alzheimer's Disease

Alzheimer's disease is a progressive, degenerative disease of the gray matter of the brain. It is one of the two most common disorders causing dementia in elderly people. (The other is cerebrovascular accident (CVA).)

Etiology: There is no single "cause" of Alzheimer's. It is likely due to a combination of factors, including genetics, head trauma, educational level, and age. Dietary habits, hypertension, and hypercholesterolemia are also being investigated as contributing factors.

Signs and Symptoms: Common symptoms include a loss of memory, confusion, personality changes, language deterioration, impaired judgment, and restlessness.

Treatment: There is no cure, but with medications and proper nutrition, physical and mental exercise, social activity, and calm environments, the disease progression may be slowed and managed.

(continued)

Alzheimer's

Parkinson's Disease

Parkinson's disease (PD) is a motor system disorder. It is slowly progressive and degenerative.

Etiology: PD is a progressive deterioration of neurons in the *substantia nigra*. These neurons produce a neurotransmitter called *dopamine*. Dopamine serves as a chemical messenger allowing communication between the *substantia nigra* and the *basal ganglia* of the brain. Most cases of PD are *idiopathic* (of unknown cause), but some may be caused by brain tumors, certain drugs, carbon monoxide, or repeated head trauma.

Signs and Symptoms: A lack of dopamine results in abnormal nerve functioning, causing a loss in the ability to control body movements. Common signs and symptoms include a resting tremor (referred to as a "pill rolling tremor" because of the movement of the fingers and thumb) that disappears with voluntary movement, muscle rigidity, and a lack of coordination and balance. The face is said to be masklike because it is expressionless (Figure 14-16). The patient may also develop a stooped posture and a shuffling gait. Additionally, patients also have a higher incidence of depression and dementia.

Treatment: There is no cure, but medications such as dopamine and levodopa alleviate some symptoms and slow the progression of this disease. Surgery is useful in some cases.

Cerebrovascular Accident

Stroke

Cerebrovascular accident (CVA), or stroke, occurs when brain cells die because of inadequate (blood) perfusion of the brain. CVA was the third leading cause of death in the United States until recently, when it was surpassed by deaths due to chronic obstructive disease. *Transient ischemic attacks (TIAs)* or "ministrokes" are not really strokes at all but are caused by temporary or transient interruptions of the blood supply to the brain. They

FIGURE 14-16 Individuals with repeated head trauma are at increased risk for Parkinson's disease (PD). Muhammed Ali has PD from many years of boxing. Michael J. Fox also has PD—his symptoms were first recognized at 29 years of age, which is an unusually young age of onset.

Cerebrovascular Accident (concluded)

do not cause the death of neurons, but are considered to be a warning of a future stroke.

Etiology: The two major categories of stroke are occlusive and hemorrhagic. *Occlusive strokes* are sometimes referred to as white strokes because of the lack of blood to an area of the brain. The occlusion (blockage) can be caused by a thrombus (blood clot) in a blood vessel of the neck or vein or an embolus that has traveled to the brain from elsewhere in the body. *Hemorrhagic strokes* are sometimes called red strokes because blood seeps into the surrounding brain tissue. These are often due to a ruptured aneurysm in a blood vessel to the brain. Of the two types, occlusive strokes are much more common than hemorrhagic strokes. Risk factors for stroke include hypertension, cigarette smoking, diabetes, oral contraceptives, and a history of TIAs or heart disease.

Signs and Symptoms: Signs and symptoms may include numbness or loss of function on one side of the body, paralysis of one side of the body, speech problems, memory and reasoning deficits, coma, and possibly death. Symptoms vary depending on the location of the stroke within the brain.

Treatment: Because neurons in the brain cannot be replaced, the effects of a stroke can be permanent. However, the sooner treatment is initiated (preferably if "clot-buster drugs" such as TPA are used within three hours of the first symptoms) the better the treatment outcome. Physical, occupational, and speech therapy are often very useful in lessening the effects of a stroke.

> REMEMBER MARY ANN from our case study? What risk factors does she have for stroke?

Headaches

Headaches affect almost everyone at some point in life. There are numerous types and causes of headaches. Although most headaches do not require medical attention, a physician should evaluate repetitive and severe headaches. Headache types include tension headaches, migraines, and cluster headaches. Tension headaches are classified as either episodic (random) or chronic (frequent).

- *Episodic tension headaches* are the most common type of tension headache.

Etiology: This type of headache is often the result of temporary stress or muscle tension.

Signs and Symptoms: Symptoms include pain or soreness in the temples and the contraction of head and neck muscles.

(continued)

Headaches (continued)

Treatment: Most of these headaches can be managed by taking an over-the-counter (OTC) medicine, and relief usually occurs in one or two hours.

- *Chronic tension headaches* occur almost daily and persist for weeks or months.

Etiology: This type of headache may be the result of stress or fatigue, but it may also be associated with physical problems, psychological issues, or depression.

Signs and Symptoms: As with episodic tension headaches, symptoms include pain or soreness in the temples and the contraction of head and neck muscles.

Treatment: People who suffer from chronic headaches should seek medical treatment to determine the underlying cause.

- *Migraines* are the most severe type of headache. They are responsible for more "sick days" than any other headache type. Almost 30 million people in the United States suffer from migraines, and more women than men suffer from migraine headaches.

Etiology: Hormones may influence migraines, which may explain why women experience migraines at least three times more often than men do. Migraine headaches are considered vascular headaches because they are associated with the dilation of the arteries of the brain.

Signs and Symptoms: Migraines often begin as dull pains that develop into throbbing pains accompanied by nausea and sensitivity to light and noise. There are many types of migraines, but the two most common are migraine with aura and migraine without aura. Auras may include the appearance of jagged lines or flashing lights, tunnel vision, hallucinations, or the detection of strange odors. The auras may last up to an hour and usually go away as the headache begins. Most migraine headaches last about four hours, but in rare cases can last up to a week.

Treatment: When treating migraines, a physician will prescribe a drug to relieve the pain but will also try to identify the factors that trigger it. Many OTC and prescription medicines are available to treat migraines.

- *Cluster headaches* are so named because the attacks come in groups for a period of time and often appear seasonal. More men than women experience these types of headaches.

Etiology: Because of their seasonal relationship, it is thought the hypothalamus may be involved. Some research indicates that alcohol consumption can bring on attacks of cluster headaches.

Headaches (concluded)

Signs and Symptoms: Common symptoms include a runny nose, watery eyes, and swelling below the eyes. Cluster headaches normally last about 45 minutes to an hour, although they can last longer. It is common for a patient with this disorder to experience one to four headaches a day during a cluster time span. Cluster time spans can last weeks or months.

Treatment: Various drugs are available for the treatment of these headaches.

Brain Tumors

Brain tumors and cancers are abnormal growths in the brain. Malignant tumors that start in any tissue of the brain are called primary brain cancers. Those that start elsewhere in the body and then metastasize (spread) to the brain are classified as secondary brain cancers. The most common primary brain tumors are *gliomas* that arise from neuroglial cells such as astrocytes. It is uncommon to have primary cancers of neurons because a neuron is a more permanent cell type that does not go through normal cell division.

Etiology: Like most cancers, the causes of brain tumors are gene mutations. Factors associated with gene mutations include exposure to carcinogens, an impaired immune system, and genetics.

Signs and Symptoms: The signs and symptoms depend on the size and location of the tumor. Common symptoms include headaches, seizures, nausea, weakness in the arms or legs, fatigue, changes in speech patterns, and a loss of memory.

Treatment: Treatment often includes surgery, radiation therapy, chemotherapy, and gene therapy. The success of the treatment depends on the type of tumor, the extent of metastasis, the location of the tumor, the tumor's response to treatment, and the overall health of the patient.

Epilepsy

Epilepsy is a condition in which the brain experiences repeated spontaneous seizures due to abnormal electrical activity of the brain. A *seizure* ("fit" or convulsion) is an episode of disturbed brain function. There are two different types of seizures: petit mal (partial) or grand mal (generalized). Petit mal seizures are sometimes referred to as "absence seizures" where the individual may have a loss of awareness of the present or appears to "space out," whereas a grand mal seizure results in the classic tonic-clonic seizure.

(continued)

Epilepsy (concluded)

Etiology: Causes vary but include birth trauma, high fevers, alcohol and drug withdrawal, head trauma, infections, brain tumors, and certain medications, but often the etiology is not known.

Signs and Symptoms: The signs and symptoms may include visual disturbances, nausea, generalized abnormal feelings, a loss of consciousness, and uncontrolled muscle contractions and tremors.

Treatment: The primary treatment is medication to prevent seizures. Surgery is sometimes an option for some patients.

> **U check**
> What are the two major categories of stroke?

Spinal Cord

The spinal cord is a slender structure that is continuous with the brain. The spinal cord descends through the vertebral canal and ends around the level of the first or second lumbar vertebra. The spinal cord is divided into 31 spinal segments (discussed later). The thickening of the spinal cord in the neck is called the cervical enlargement and it contains the motor neurons that control the muscles of the upper extremities. Another thickening of the spinal cord occurs in the lumbar region. This thickening is called the lumbar enlargement and it contains the motor neurons that control the muscles of the lower extremities (Figure 14-17).

The spinal cord stops growing by about age five, but the vertebral column continues to grow. Consequently, the vertebral column extends beyond the terminus (end) of the spinal cord. Because of this, the spinal nerves from the lumbar, sacral, and coccygeal regions of the spinal cord exit the vertebral column at a different level than from where they come off the spinal cord. The roots of these nerves are called the cauda equina ("horse's tail") because of their appearance.

> **U check**
> What is the purpose of the two enlargements in the spinal cord?

Gray and White Matter: When you view a cross section of the spinal cord, you will see what looks like the letter *H* (Figure 14-18). The *H* is a darker color than the surrounding tissue. This is the gray matter, and the surrounding lighter tissue is the

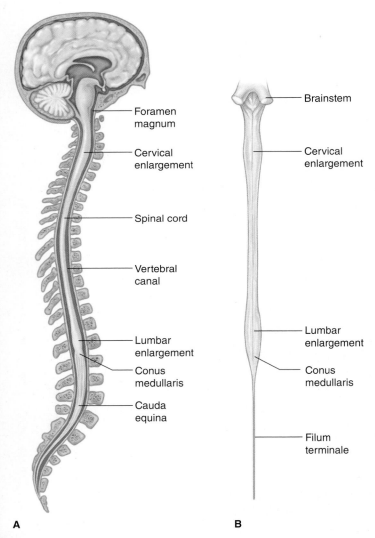

FIGURE 14-17 (A) The spinal cord begins at the level of the foramen magnum. (B) Posterior view of the spinal cord showing the cervical enlargement and the lumbar enlargement. The spinal nerves have been removed.

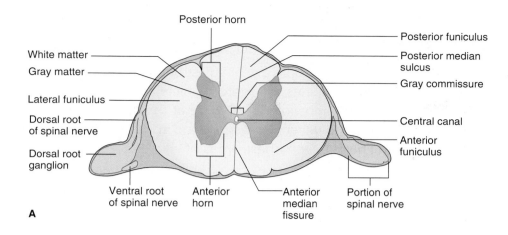

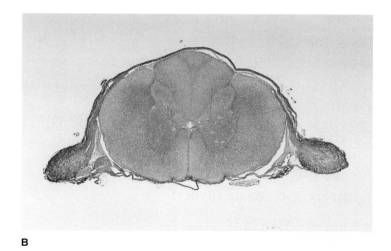

FIGURE 14-18
(A) Cross section of the spinal cord. (B) Identify the parts of the spinal cord in this micrograph (7.5x).

white matter. As noted earlier in the chapter, the gray matter consists of the cell bodies of neurons and their dendrites and the white matter is made up of myelinated axons. The central portion of the *H* is called the gray commissure and connects the right and left sides. In the center of the gray matter is the central canal through which cerebrospinal fluid (CSF) circulates.

The gray matter making up the *H* is divided into four "horns": two ventral and two dorsal. The two dorsal horns receive afferent (sensory) information from the body, whereas the two ventral horns carry efferent (motor) information to the body. The *H* separates the white matter of the spinal cord into three regions called columns—anterior (ventral), lateral, and posterior (dorsal) columns. These columns contain the groups of axons called nerve tracts.

Ascending and Descending Tracts: One function of the spinal cord is to carry sensory information up to the brain. These tracts that carry this sensory information are called ascending or afferent tracts. The spinal cord also transmits motor information from the brain to muscles and glands. These tracts are called descending or efferent tracts.

Reflexes: The spinal cord participates in reflexes. A *reflex* is a predictable and automatic response to a stimulus. For example, if you touch something

> **focus on** Wellness: Preventing Brain and Spinal Cord Injuries
>
> In the United States alone almost half a million people a year suffer brain and spinal cord injuries. The most common causes of these injuries are motor vehicle accidents, sports and recreational accidents (especially diving), and violence. People at the highest risk for spinal cord injuries are children and teens. However, most brain and spinal cord injuries can be prevented. The following tips will help prevent these types of injuries.
>
> - Know the depth of water into which you are diving. More than 90 percent of diving injuries occur in five feet of water or less.
> - Explore diving areas before diving. For example, know where rocks are located before you dive.
> - Do not drive or do any recreational activity while under the influence of alcohol or drugs. Both affect good judgment and control. Alcohol-related traffic crashes are the leading cause of disabling brain and spinal cord injuries.
> - Always wear a helmet when riding a bike or motorcycle. If you are in an accident without a helmet, your risk of brain injury is 85 percent.
> - Wear a helmet when biking and make sure your helmet fits properly.
> - Always wear appropriate protective gear while playing any sport.
> - Avoid surfing headfirst.
> - Always wear your safety belt.
> - Make sure children use car seats that are appropriate for their age and weight.
> - Be familiar with ways to get help quickly in emergencies.
> - Follow traffic rules and signs while walking, biking, or driving.
> - Follow safety rules on playgrounds.
> - Store firearms and ammunition in separate and locked places.
> - Teach children the safety rules to follow if they find a gun.

very hot, the predictable response is that you will unconsciously pull your hand away from the hot object; this type of reflex is called a withdrawal reflex. A reflex arc has several components (Figure 14-19). Table 14-7 lists the components of a reflex arc. The first component is the sensory receptors. When you touch a hot object, sensory receptors in the skin are stimulated. The sensory information is then passed along an afferent neuron to the spinal cord. An interneuron (association neuron) connects the sensory neuron that has entered the dorsal horn of the spinal cord to a motor neuron in the ventral horn of the spinal cord. The motor neuron carries efferent information to the appropriate muscles (effector organs), causing you to withdraw your hand. Reflexes can be consciously inhibited because information also goes to the cerebral cortex where conscious decisions are made.

> **U check**
>
> What are the components of a reflex arc?

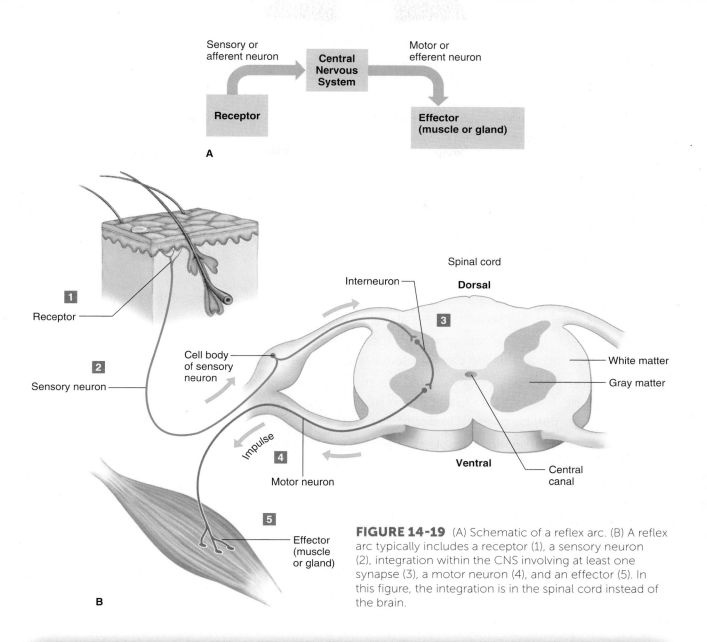

FIGURE 14-19 (A) Schematic of a reflex arc. (B) A reflex arc typically includes a receptor (1), a sensory neuron (2), integration within the CNS involving at least one synapse (3), a motor neuron (4), and an effector (5). In this figure, the integration is in the spinal cord instead of the brain.

TABLE 14-7 Parts of a Reflex Arc

Part	Description	Function
Receptor	The receptor end of a dendrite or a specialized receptor cell in a sensory organ	Sensitive to a specific type of internal or external change
Sensory neuron *afferent neuron*	Dendrite, cell body, and axon of a sensory neuron	Transmits nerve impulse from the receptor into the brain or spinal cord
Interneuron	Dendrite, cell body, and axon of a neuron within the brain or spinal cord	Serves as processing center; conducts nerve impulse from the sensory neuron to a motor neuron
Motor neuron *efferent neuron*	Dendrite, cell body, and axon of a motor neuron	Transmits nerve impulse from the brain or spinal cord out to an effector
Effector	A muscle or gland	Responds to stimulation by the motor neuron and produces the reflex or behavioral action

focus on Wellness: Signs of CVA and FAST

In our fast-paced, high-stress world, combined with our increasing life expectancy, more and more of us are witness to someone who has had a stroke (CVA). What should you look for and how should you react?

Remember that a CVA is an injury to the brain—either there is too much blood, in the form of a hemorrhage, or not enough blood, due to an occlusion (blockage) of an artery to or in the brain. Remembering the signs of stroke can be nerve-racking, but the acronym FAST can help:

F = Face. Ask the patient to smile. Look at the patient's face. Is there symmetry or is one side drooping? Is the patient experiencing visual issues? If the answer to any of these is yes, waste no time and call for an ambulance.

A = Arms and Legs. Does the patient have trouble moving both sides of the body equally? Is there weakness, numbness, or inability to walk? Call for help if you can answer yes.

S = Speech. Can the patient speak? If so, is the speech slurred? Can you understand what is being said? If the patient cannot speak clearly, call for medical aid immediately.

T = Time. Immediate reaction when a stroke is suspected is imperative. Many drug treatments have a short span of time from the onset of symptoms (three hours or less) when they can be given. If in doubt, dial 911.

Remember: Recognizing potential symptoms of stroke and telling a patient that the symptoms were a false alarm is much better than having to tell a family member "We were too late."

LO 14.5

14.5 learning outcome

Describe the structure and functions of the meninges, ventricles, and cerebrospinal fluid.

meninges The connective tissue membranes that protect the brain and spinal cord.

Meninges, Ventricles, and Cerebrospinal Fluid

Meninges are the connective tissue membranes that protect the brain and spinal cord. The three layers of meninges are dura mater, arachnoid mater, and pia mater (Figures 14-20 and 14-21). Dura mater is the toughest and outermost layer of the meninges. In fact, *dura mater* means "tough mother." The space above the dura mater is called the epidural space; below it is the subdural space. The middle meningeal layer, named for its spider-web-like appearance, is the *arachnoid mater*. The pia mater is the innermost and most delicate layer. *Pia mater* means "tender mother." It sits directly on top of the brain and spinal cord and holds blood vessels onto the surface of these structures. It is so closely attached to the brain and spinal cord that is difficult if not impossible to separate it from the CNS when doing a dissection. Between the arachnoid mater and pia mater is the *subarachnoid space*. It contains *cerebrospinal fluid (CSF)*, which cushions the CNS.

U check
What are the three meninges?

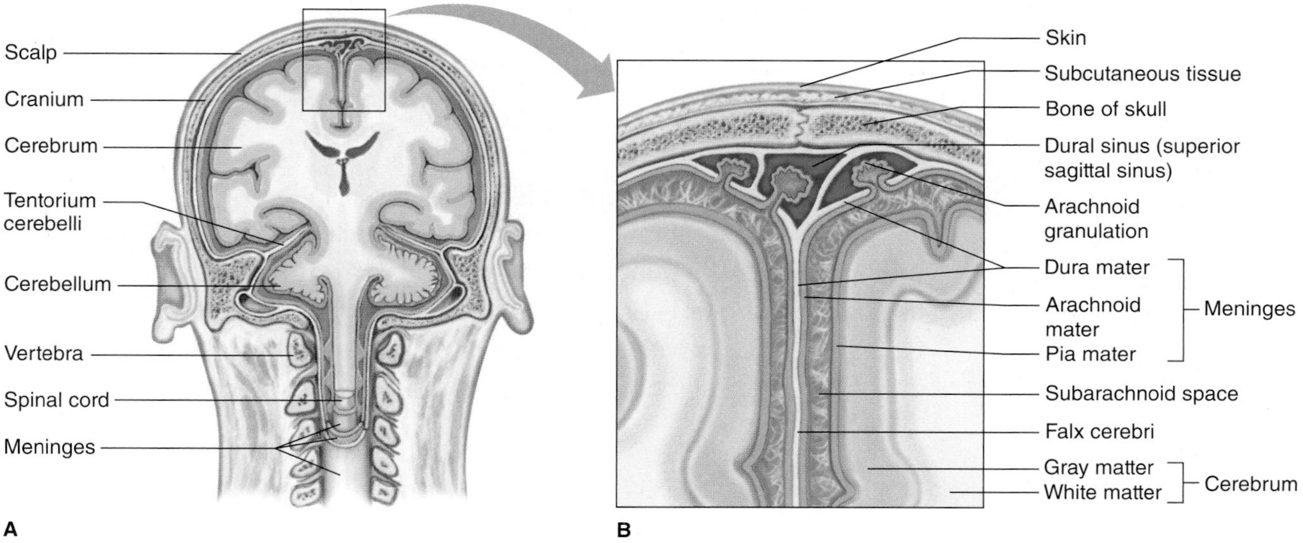

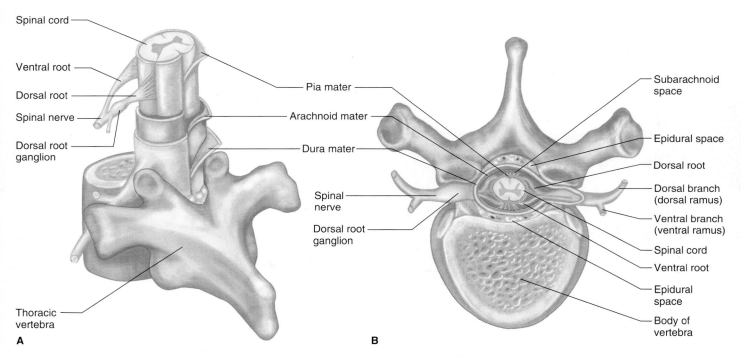

FIGURE 14-20 (A) Membranes called meninges enclose the brain and spinal cord. (B) The meninges consist of three layers: dura mater, arachnoid mater, and pia mater.

FIGURE 14-21 Meninges of the spinal cord: (A) The dura mater encloses the spinal cord. (B) Tissues forming a protective pad around the cord fill the epidural space between the dural sheath and vertebra.

PATHOPHYSIOLOGY

Meningitis

Meningitis is an inflammation of the meninges.

Etiology: Causes may include bacterial, viral, and fungal infections. Some types of meningitis can be prevented with vaccines.

(continued)

401

Meningitis (concluded)

Signs and Symptoms: Fever, headache, vomiting, *nuchal rigidity* (stiffness in the neck), and sensitivity to light, drowsiness, and joint pain usually accompany this disorder.

Treatment: The treatment varies depending on the type of meningitis. Intravenous antibiotics are used for bacterial meningitis, supportive therapy for viral meningitis, and antifungal drugs for fungal meningitis. Viral meningitis is usually self-limiting, which means the individual typically improves with supportive or palliative therapy. Bacterial meningitis can be fatal if not treated quickly soon after symptoms begin.

> **U check**
> What is nuchal rigidity?

ventricles Interconnected cavities within the brain.

Ventricles are interconnected cavities within the brain (Figure 14-22) and are filled with cerebrospinal fluid (CSF). There are two lateral ventricles, one in each cerebral hemisphere. A third ventricle is superior to the

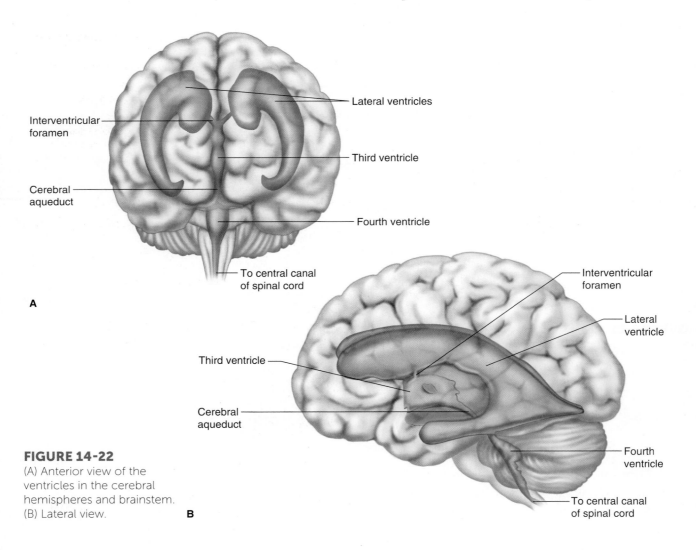

FIGURE 14-22
(A) Anterior view of the ventricles in the cerebral hemispheres and brainstem. (B) Lateral view.

hypothalamus and is located between the right and left halves of the thalamus. The lateral ventricles communicate with the third ventricle via the interventricular foramina. CSF travels from the third ventricle through the cerebral aqueduct into the fourth ventricle, which lies between the brainstem and the cerebellum. The CSF also circulates through the central canal of the spinal cord. CSF is a clear, colorless fluid that cushions the brain and spinal cord and protects them from chemical injury. It also carries oxygen and nutrients to neurons and neuroglia. CSF is produced by the ependymal cells of the choroid plexuses (Figure 14-23), a network of capillaries located with

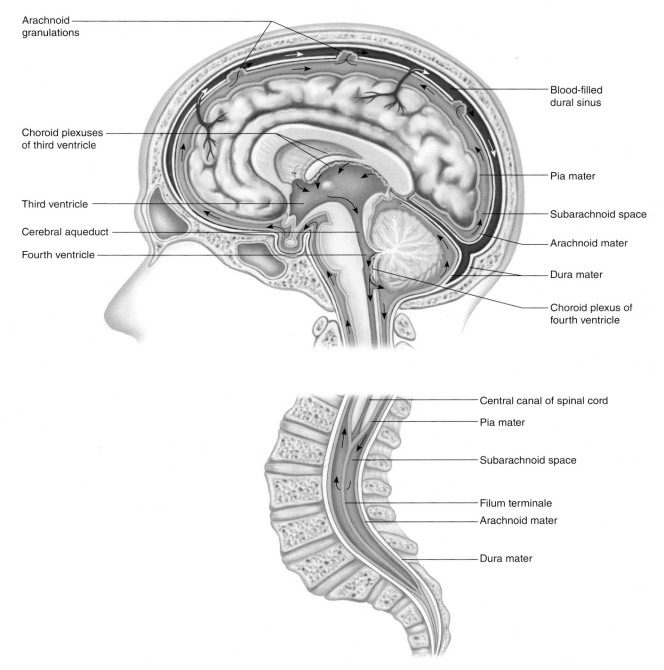

FIGURE 14-23 Choroid plexuses in the ventricles secrete cerebrospinal fluid. The CSF circulates through the ventricles and central canal of the spinal cord and is eventually reabsorbed into the blood.

the ventricles of the brain. The normal amount of CSF in the average adult is about 80 to 150 milliliters. It is constantly being formed and then reabsorbed by the venous system.

> **U check**
> What cells are responsible for the production of CSF?

Peripheral Nervous System Components and Their Functions

learning outcome 14.6
Identify the components of the peripheral nervous system and their functions.

The peripheral nervous system consists of all the nerves outside the CNS as well as sensory receptors. These peripheral nerves are classified as cranial and spinal nerves.

Cranial Nerves

Cranial nerves are peripheral nerves that originate from the brain. Roman numerals and names designate the 12 pairs of cranial nerves (Figure 14-24). They are designated as sensory, motor, or mixed. Cranial and spinal nerves can be classified as somatic if they go to the skin or skeletal muscles and autonomic if the fibers connect to viscera. Table 14-8 lists the functions of the cranial nerves. Here is a brief description of each cranial nerve:

Cranial Nerve I Olfactory: This cranial nerve is purely sensory and is responsible for transmitting aromas to the brain for interpretation.

Cranial Nerve II Optic: The optic nerve carries visual information to the brain for interpretation. It is purely sensory.

Cranial Nerve III Oculomotor: This nerve is purely motor. It innervates several eye muscles (superior rectus, inferior rectus, medial rectus,

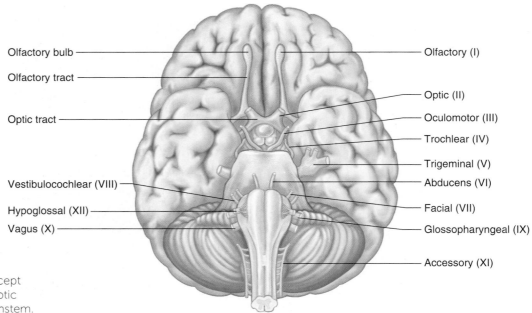

FIGURE 14-24
The cranial nerves except for olfactory (I) and optic (II) arise from the brainstem.

TABLE 14-8 Functions of the Cranial Nerves

Nerve	Type	Function
I Olfactory	Sensory	Sensory fibers transmit impulses associated with the sense of smell.
II Optic	Sensory	Sensory fibers transmit impulses associated with the sense of vision.
III Oculomotor	Primarily motor	Motor fibers transmit impulses to muscles that raise the eyelids, move the eyes, adjust the amount of light entering the eyes, and focus the lenses. Some sensory fibers transmit impulses associated with proprioceptors.
IV Trochlear	Primarily motor	Motor fibers transmit impulses to muscles that move the eyes. Some sensory fibers transmit impulses associated with proprioceptors.
V Trigeminal	Mixed	
Ophthalmic division		Sensory fibers transmit impulses from the surface of the eyes, tear glands, scalp, forehead, and upper eyelids.
Maxillary division		Sensory fibers transmit impulses from the upper teeth, upper gum, upper lip, lining of the palate, and skin of the face.
Mandibular division		Sensory fibers transmit impulses from the scalp, skin of the jaw, lower teeth, lower gum, and lower lip. Motor fibers transmit impulses to muscles of mastication and to muscles in the floor of the mouth.
VI Abducens	Primarily motor	Motor fibers transmit impulses to muscles that move the eyes. Some sensory fibers transmit impulses associated with proprioceptors.
VII Facial	Mixed	Sensory fibers transmit impulses associated with taste receptors of the anterior tongue. Motor fibers transmit impulses to muscles of facial expression, tear glands, and salivary glands.
VIII Vestibulocochlear	Sensory	
Vestibular branch		Sensory fibers transmit impulses associated with the sense of equilibrium.
Cochlear branch		Sensory fibers transmit impulses associated with the sense of hearing.
IX Glossopharyngeal	Mixed	Sensory fibers transmit impulses from the pharynx, tonsils, posterior tongue, and carotid arteries. Motor fibers transmit impulses to salivary glands and to muscles of the pharynx used in swallowing.

(continued)

TABLE 14-8 (concluded)

Nerve	Type	Function
X Vagus	Mixed	Somatic motor fibers transmit impulses to muscles associated with speech and swallowing; autonomic motor fibers transmit impulses to the viscera of the thorax and abdomen.
		Sensory fibers transmit impulses from the pharynx, larynx, esophagus, and viscera of the thorax and abdomen.
XI Accessory	Primarily motor	
Cranial branch		Motor fibers transmit impulses to muscles of the soft palate, pharynx, and larynx.
Spinal branch		Motor fibers transmit impulses to muscles of the neck and back; some proprioceptor input.
XII Hypoglossal	Primarily motor	Motor fibers transmit impulses to muscles that move the tongue; some proprioceptor input.

and inferior oblique) to move the eyeball. It also innervates the eyelid and iris.

Cranial Nerve IV Trochlear: This motor nerve innervates the superior oblique muscle of the eye to move the eye down and out. It is the smallest cranial nerve.

Cranial Nerve V Trigeminal: This is the largest of the cranial nerves. It has three divisions: the ophthalmic, maxillary, and mandibular divisions (Figure 14-25). It carries sensory information from the surface of the eye, the scalp, facial skin, the lining of the gums, and the palate to the brain for interpretation. Motor fibers innervate the muscles of mastication (chewing).

Cranial Nerve VI Abducens: As you may recall, the term "abduct" means to move away from the midline. This motor nerve innervates the lateral rectus muscle of the eyeball. When the right abducens is stimulated, the right eyeball moves laterally to look to the right. When the left abducens is stimulated, the left eyeball turns to the left.

Cranial Nerve VII Facial: This nerve has both sensory and motor components. It innervates the muscles of facial expression as well as the salivary and lacrimal (tear) glands (Figure 14-26). It also carries sensory information from the tongue.

Cranial Nerve VIII Vestibulocochlear: This sensory nerve carries hearing and equilibrium information from the inner ear to the brain for interpretation.

Cranial Nerve IX Glossopharyngeal: This nerve carries sensory information from the throat and tongue to the brain for interpretation. It also sends

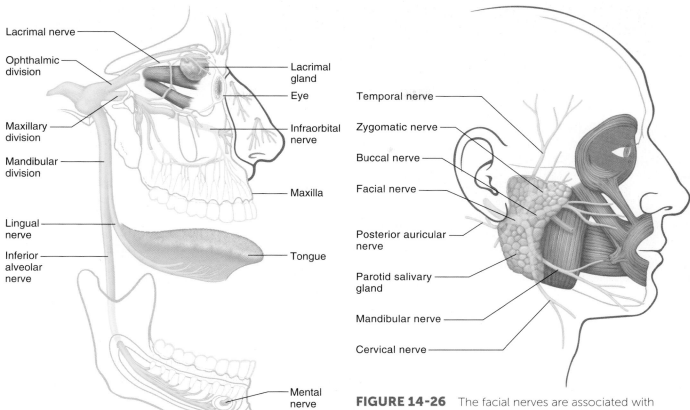

FIGURE 14-25 Each trigeminal nerve has three divisions that innervate different regions of the head and face: the ophthalmic division (sensory), the maxillary division (sensory), and the mandibular division (sensory as well as motor to the muscles of mastication).

FIGURE 14-26 The facial nerves are associated with taste receptors on the tongue and with muscles of facial expression.

motor fibers to the muscles of the throat. Along with the vagus nerve (cranial nerve X), it is responsible for the gag reflex.

Cranial Nerve X Vagus: This is the longest of the cranial nerves. It transmits sensory information from the thoracic and abdominal organs to the brain for interpretation. It also has motor fibers that innervate the muscles of the throat, stomach, and intestines, as well as the heart (Figure 14-27).

Cranial Nerve XI Accessory: This motor nerve innervates muscles of the throat, neck, back, and voice box. The trapezius muscle is innervated by this nerve.

Cranial Nerve XII Hypoglossal: This motor nerve is responsible for innervating the muscles of the tongue.

U check
What cranial nerves are involved in the gag reflex?

FIGURE 14-27
The vagus nerves (the left vagus nerve is shown here) extend from the medulla oblongata into the chest and abdomen to innervate many organs.

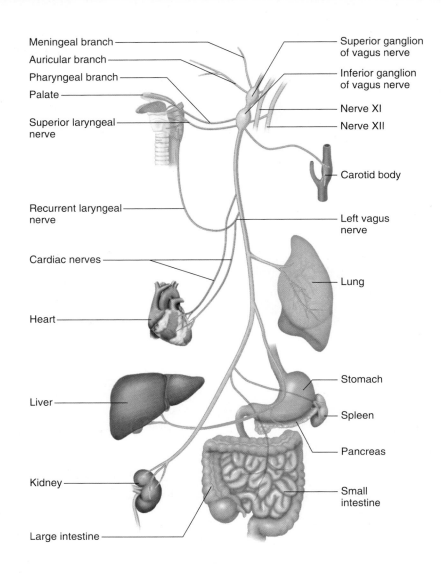

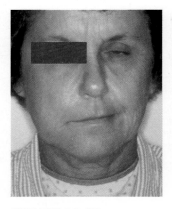

FIGURE 14-28
Individual with Bell's palsy. Note the drooping on the right side of the face and mouth.

PATHOPHYSIOLOGY

Bell's palsy

Bell's palsy is a disorder in which facial muscles are very weak or temporarily totally paralyzed (Figure 14-28).

Etiology: This condition can result from damage to cranial nerve VII (the facial nerve), but many times the cause is unknown. It is more common in people with diabetes, the flu, or a cold.

Signs and Symptoms: The most common signs and symptoms are a loss of feeling in the face, the inability to produce facial expressions, headache, and excessive tearing or drooling.

Treatment: Treatments include the use of eyedrops, pain relievers, and anti-inflammatory medications. Symptoms usually diminish or go away within 5 to 10 days.

Neuralgia

Neuralgias are a group of disorders commonly referred to as nerve pain. They most frequently occur in the nerves of the face.

Etiology: There are many causes of neuralgia, including trauma, chemical irritation of the nerves, bacterial infections, and diabetes. Many times the causes are unknown.

Signs and Symptoms: Sudden and severe skin pain is the most common symptom. The pain repeatedly occurs in the same body area. Numbness of skin areas is also common.

Treatment: Many times the disorder goes away spontaneously and treatment, other than pain medication, is not needed. Other treatments include injections of anesthetics or surgery to remove the affected nerves.

focus on Clinical Applications: Testing the Cranial Nerves

Tests of the cranial nerves are completed to assess for disorders. Knowing the function of each nerve helps the health care professional understand these tests.

- Olfactory nerves (I) are tested by asking a patient to smell various substances.
- Cranial nerves III, IV, and VI are tested by asking a patient to track the movement of the physician's finger. If a patient cannot move her eyeballs properly, there may be damage to one of these nerves. Recall that these nerves control the muscles that move the eyeballs.
- Cranial nerve V controls the muscles needed for chewing. To assess this nerve, a patient is asked to clench his teeth. The physician then feels the jaw muscles. If they feel limp or weak, this nerve may be damaged.
- Cranial nerve VII may be damaged if a person can no longer make facial expressions.
- Cranial nerve VIII is tested with a tuning fork placed on the mastoid process behind the ear and placed on the top of the head. Or you can determine the distance from which a person can hear the ticking of a clock or watch to determine hearing.
- Glossopharyngeal nerve IX and vagus nerve X are tested by having the individual drink some water and observing the swallowing reflex.
- Spinal accessory nerve XI is tested by having the individual shrug her shoulders. This nerve innervates the muscles responsible for this action.
- Hypoglossal nerve XII is tested by having the patient stick out his tongue and move it from side to side. This nerve controls tongue movement.

U check

How would you test the spinal accessory nerve (cranial nerve XI)?

Spinal Nerves

Spinal nerves are peripheral nerves that originate from the spinal cord (Figure 14-29). The segments of the spinal cord are named after the region of the spinal column they lie within. There are 31 pairs of spinal nerves: 8 pairs of cervical nerves (numbered C1 through C8), 12 pairs of thoracic nerves (numbered T1 through T12), 5 pairs of lumbar nerves (numbered L1

FIGURE 14-29
Thirty-one pairs of spinal nerves are grouped according to the level from which they arise and are numbered sequentially.

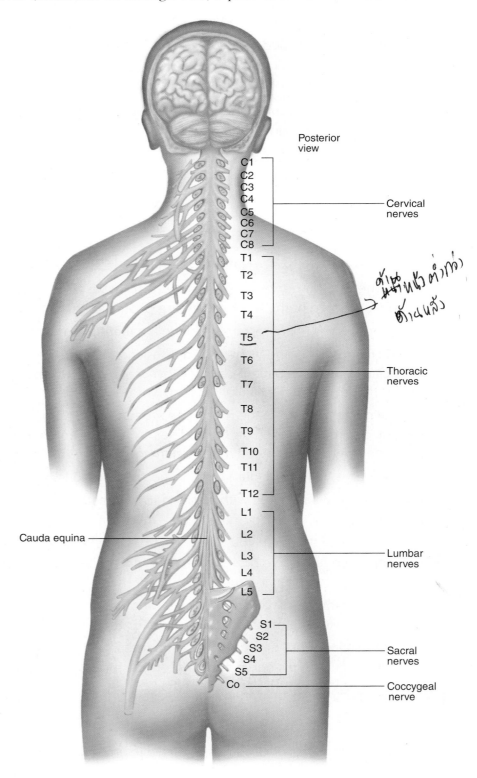

through L5), 5 pairs of sacral nerves (numbered S1 through S5), and 1 pair of coccygeal nerves (CO). Except for C1, each spinal nerve innervates a skin segment known as a *dermatome*. A map of the dermatomes with the spinal nerve responsible for the skin area is shown in Figure 14-30.

Dorsal and ventral spinal roots come off the spinal cord's dorsal and ventral horns, respectively. The ventral root contains axons of motor neurons only, and the dorsal root contains axons of sensory neurons only. Just off the spinal cord in the dorsal root is an enlargement called the dorsal root **ganglion.** It is a collection of nerve cell bodies outside the CNS. The dorsal

ganglion A group of nerve cell bodies outside the central nervous system.

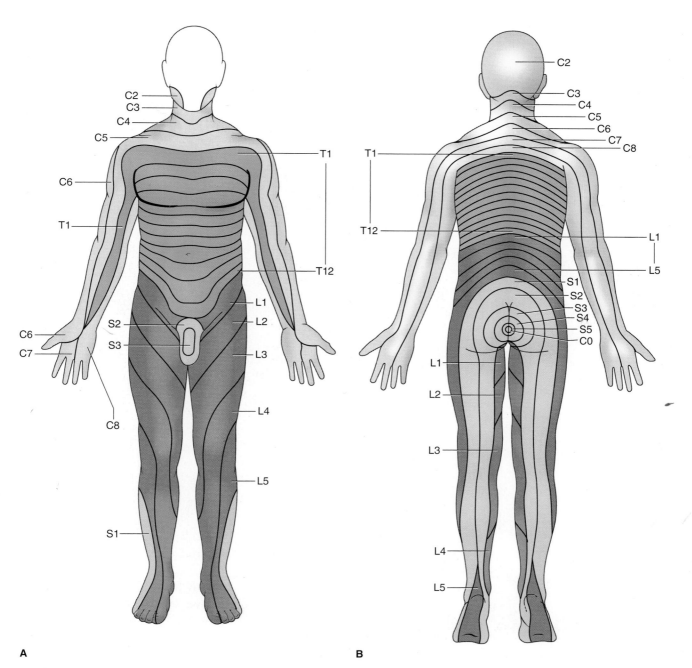

FIGURE 14-30 Dermatomes (A) on the anterior surface of the body and (B) on the posterior surface. Spinal nerve C1 does not innervate any skin area.

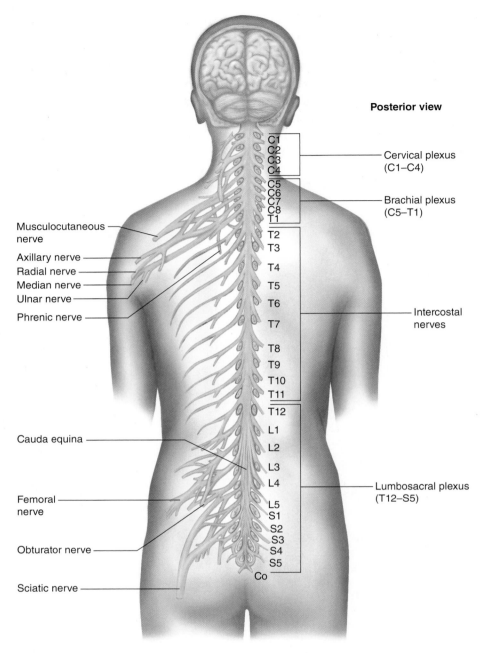

FIGURE 14-31 Dorsal view of the body illustrating the cervical plexus and the lumbosacral plexus.

and ventral roots come together to form a spinal nerve that contains both sensory and motor fibers.

A collection of nerves may come together to form a network of nerves called a plexus (Figure 14-31). The major nerve plexuses are the cervical, brachial (Figure 14-32), and lumbosacral (Figure 14-33). Nerves of the cervical plexus innervate the skin and the muscles of the neck. The phrenic nerve, which innervates the diaphragm, comes off the cervical plexus. Remember from Chapter 13, The Respiratory System, that the diaphragm is the main

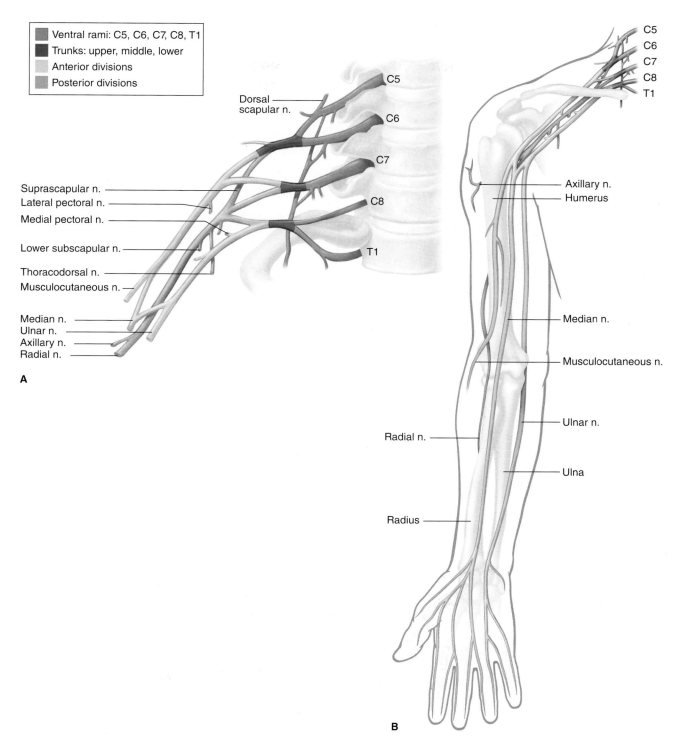

FIGURE 14-32 Nerves of the brachial plexus: (A) a close-up view and (B) an anterior view (n. stands for nerve).

muscle of respiration. That is why a spinal cord injury involving the neck can cause a person to stop breathing. The brachial plexus consists of nerves that control muscles in the arms. The *lumbosacral plexus* innervates the lower abdominal wall, external genitalia, buttocks, thighs, legs, and feet. The largest nerve of the body, the sciatic nerve, comes off this plexus.

Spinal cord injury

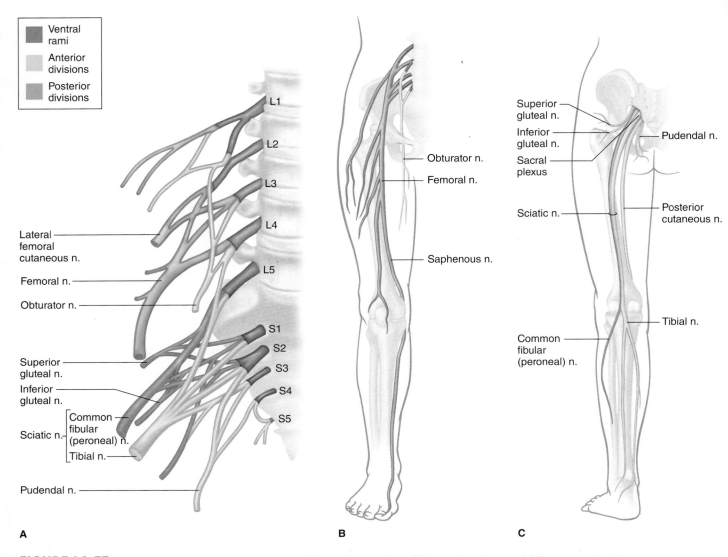

FIGURE 14-33 Nerves of the lumbosacral plexus: (A) a close-up view, (B) anterior view, and (C) a posterior view.

PATHOPHYSIOLOGY

Guillain-Barré Syndrome

Guillain-Barré syndrome is a disorder in which the body's immune system attacks part of the peripheral nervous system. It usually has a sudden and unexpected onset.

Etiology: The destruction of myelin by the body's immune system produces the signs and symptoms. Viral infections, immunizations, and pregnancy sometimes trigger the disease.

Signs and Symptoms: Symptoms may include weakness or tingling sensations in the legs or arms that can progress to paralysis. Dyspnea (difficulty breathing) and an abnormal heart rate are the more dangerous signs and symptoms. With appropriate medical treatment, the disease normally runs its course and is not fatal.

Guillain-Barré Syndrome (concluded)

Treatment: Various supportive therapies, including IV gamma globulin treatments, rest, nutritional support including NG tubes (if swallowing is affected), and respiratory and cardiac monitoring. If respiration becomes a severe issue, intubation may also be required until the disease subsides. Physical therapy is used to keep muscles strong.

Sciatica

Sciatica occurs when the sciatic nerve is damaged.

Etiology: The sciatic nerve is commonly damaged by excessive pressure on the nerve from prolonged sitting or lying down. It is also easily damaged from trauma to the pelvis, buttocks, or thighs.

Signs and Symptoms: The most usual symptoms include numbness, pain, or tingling sensations on the back of a leg or foot. Weakness of leg and foot muscles can also develop.

Treatment: This disorder is usually treated with pain and anti-inflammatory medication or steroids. Physical therapy is also needed following trauma to the nerve.

Multiple Sclerosis

Multiple sclerosis (MS) is a chronic disease of the CNS in which myelin is destroyed.

Etiology: The causes are mostly unknown, but some known causes are viruses, genetic factors, and immune system abnormalities.

Signs and Symptoms: Depending on the type of MS, symptoms can range from mild to severe. In severe cases, a person loses the ability to walk or speak.

Treatment: There is no cure for MS, but supportive treatments may lessen the symptoms. Some medications, including interferon, are available to treat and slow the progression of symptoms.

Amyotrophic Lateral Sclerosis

Amyotrophic lateral sclerosis (ALS), commonly known as Lou Gehrig's disease, is a fatal disorder characterized by the degeneration of neurons in the spinal cord and brain.

Etiology: Most causes are unknown but they are likely to involve genetic and environmental factors.

Signs and Symptoms: Early symptoms include cramping of hand and feet muscles, persistent tripping and falling, chronic fatigue, and slurred speech. Signs and symptoms that appear in later stages include dyspnea and muscle paralysis.

Treatment: There is no cure for this disorder; however, physical, speech, and respiratory therapies help manage the symptoms. Some medications relieve muscle cramping.

Somatic and Autonomic Nervous Systems: The peripheral nervous system is divided into the somatic nervous system (SNS), which governs the body's skeletal (voluntary) muscles, and the autonomic nervous system (ANS), which is in charge of the body's automatic functions, such as the respiratory and gastrointestinal systems. Table 14-9 lists some effects of autonomic stimulation on various visceral effectors. As noted earlier, there are three separate types of nerve cells or neurons that carry out the actual functions of the nervous system: sensory (afferent nerves), motor (efferent nerves), and interneurons (association neurons).

The interneurons work within the CNS (brain or spinal cord) and are the interpretive neurons between the sensory and motor nerves. These neurons act as connections between the afferent and efferent nerves. An example of this process would be noting a red light while driving. The sensory neurons within your eyes note the color and send the information to the cerebral cortex of your brain, where the interpretation takes place. The interneurons pick up the signal and send the information to the motor neurons that you are supposed to stop your vehicle, which in turn send the instructions to your right foot to step on the brake pedal of your vehicle. This entire transaction, of course, takes place in milliseconds, allowing you to stop in time.

In the autonomic nervous system, motor neurons from the brain and spinal cord communicate to other motor neurons that are located in ganglia. The motor neurons of ganglia then communicate to various organs and blood vessels. The two divisions of the autonomic nervous system are the sympathetic (Figure 14-34) and the parasympathetic nervous systems (Figure 14-35). Most organs receive innervation from both the sympathetic and parasympathetic nervous systems. The sympathetic nervous system is sometimes called the "fight or flight" system.

Many neurons of the sympathetic division are located in the thoracic and lumbar regions of the spinal cord. For this reason, this division is also called the thoracolumbar division. The sympathetic neurons usually release the neurotransmitter norepinephrine.

When sympathetic nerve fibers are stimulated, several responses are seen. The pupils of the eyes dilate, heart rate and blood pressure increase, and the airways dilate. In addition, blood vessels to organs such as the heart and skeletal muscle dilate to allow more blood and oxygen to reach the tissues. The breakdown of glycogen to glucose by the liver increases and peristalsis (the movement of food through the digestive tract) slows down.

The parasympathetic system is responsible for "rest and digest." Many neurons of the parasympathetic division are located in the brainstem and the sacral regions of the spinal cord. For this reason, this division is also referred to as the craniosacral division. Although all parasympathetic neurons release acetylcholine, most blood vessels in the body do not receive communication from parasympathetic nerves. The acronym *SLUDD* is helpful in remembering what the parasympathetic nervous system does when stimulated: **s**alivation, **l**acrimation, **u**rination, **d**igestion, and **d**efecation all increase.

> **U check**
> What are the two divisions of the ANS?

TABLE 14-9 Effects of Autonomic Stimulation on Various Visceral Effectors

Effector Location	Response to Sympathetic Stimulation	Response to Parasympathetic Stimulation
Integumentary System		
Apocrine glands	Increased secretion	No action
Eccrine glands	Increased secretion (cholinergic effect)	No action
Special Senses		
Iris of eye	Dilation	Constriction
Tear gland	Slightly increased secretion	Greatly increased secretion
Endocrine System		
Adrenal cortex	Increased secretion	No action
Adrenal medulla	Increased secretion	No action
Digestive System		
Muscle of gallbladder wall	Relaxation	Contraction
Muscle of intestinal wall	Decreased peristaltic action	Increased peristaltic action
Muscle of internal anal sphincter	Contraction	Relaxation
Pancreatic glands	Reduced secretion	Greatly increased secretion
Salivary glands	Reduced secretion	Greatly increased secretion
Respiratory System		
Muscles in walls of bronchioles	Dilation	Constriction
Cardiovascular System		
Blood vessels supplying muscles	Constriction (alpha adrenergic) Dilation (beta adrenergic)	No action
Blood vessels supplying skin	Constriction	No action
Blood vessels supplying heart (coronary arteries)	Constriction (alpha adrenergic) Dilation (beta adrenergic)	No action
Muscles in wall of heart	Increased contraction rate	Decreased contraction rate
Urinary System		
Muscle of bladder wall	Relaxation	Contraction
Muscle of internal urethral sphincter	Contraction	Relaxation
Reproductive Systems		
Blood vessels to penis and clitoris	No action	Dilatation leading to erection of penis and clitoris
Muscles associated with internal reproductive organs	Male ejaculation, female orgasm	

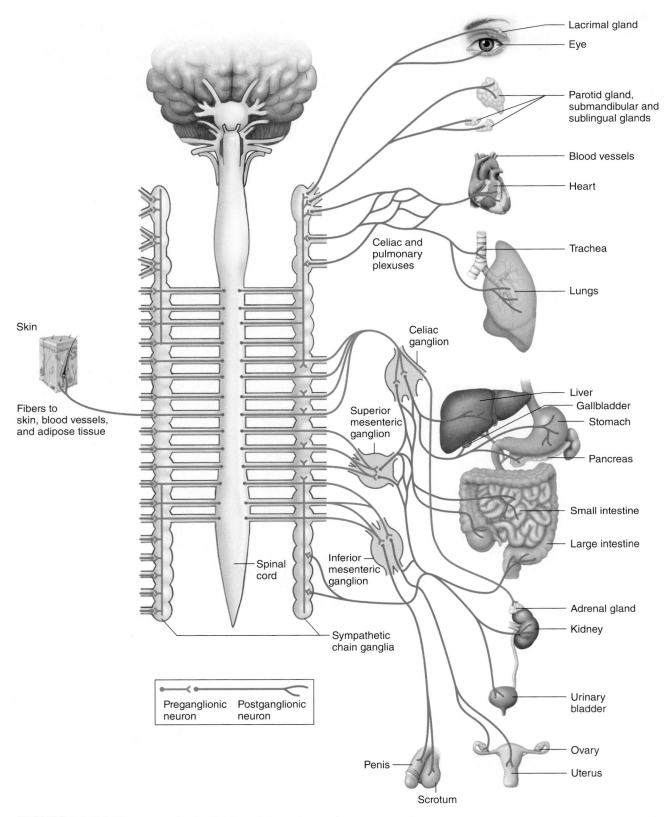

FIGURE 14-34 The sympathetic division of the autonomic nervous system.

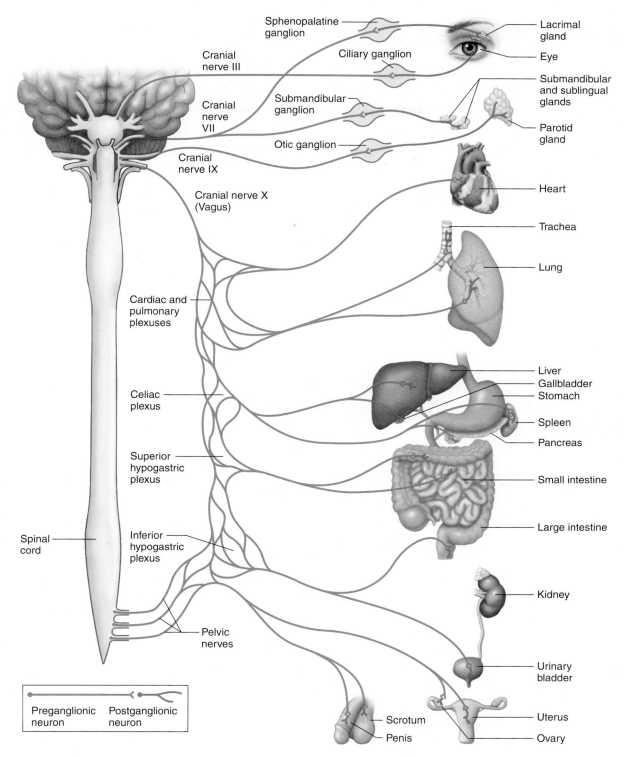

FIGURE 14-35 The parasympathetic division of the autonomic nervous system.

learning outcome 14.7

Relate neurological testing procedures to their uses.

Neurological Testing

Patients with nervous system disorders may have a wide variety of signs and symptoms, but the most common are headache, muscle weakness, and *paresthesias* (a sensation of tingling, pricking, or numbness of the skin). A typical neurologic examination can determine the following:

State of Consciousness: This state can vary from normal to a state of coma. A patient in a coma cannot respond to stimuli and cannot be awakened. Other terms used to describe states of consciousness include stupor (difficulty being awakened), delirium (being confused or having hallucinations), vegetative (having no cortical function), and asleep (can be aroused with normal stimulation).

Reflex Activity: Reflex tests primarily determine the health of the peripheral nervous system.

Speech Patterns: Abnormal speech patterns include a loss of the ability to form words correctly or to form sentences that make sense.

Motor Patterns: Abnormal motor patterns include the loss of balance, abnormal posture, or inappropriate movements of the body. For example, chorea is an exaggerated and sudden jerking of a body part.

Diagnostic Procedures

Common diagnostic procedures to determine neurologic disorders include the following specialized tests:

Lumbar Puncture: Whenever a physician needs to examine CSF, a lumbar puncture (Figure 14-36), sometimes called a "spinal tap," is performed. A needle is used to remove CSF from the subarachnoid space, usually from the space below the third lumbar vertebra of the spinal column. Analysis of this fluid provides a great deal of information about the health of a patient. For example, cancer cells in CSF may indicate a brain or spinal cord tumor. White blood cells in this fluid indicate infections such as meningitis. Red blood cells indicate abnormal bleeding.

Magnetic Resonance Imaging (MRI): This procedure allows for the brain and spinal cord to be visualized from many angles. It uses powerful magnets to generate images and is useful for detecting tumors, bleeding, or other abnormalities.

Positron Emission Tomography (PET) Scan: This procedure uses radioactive chemicals that collect in specific areas of the brain. These chemicals allow images of those specific areas to be generated. This test is useful in detecting blood flow to areas of the brain, brain tumors, and the diagnosis of such diseases as Parkinson's and Alzheimer's.

Cerebral Angiography: This procedure uses contrast material so that the blood vessels in the brain can be visualized. It is also useful in detecting aneurysms (abnormally dilated blood vessels).

Computerized Tomography (CT) Scan: This very common procedure produces images that provide more information than a standard x-ray. It is useful in detecting tumors and other abnormal structures.

Electroencephalogram (EEG): This test detects electrical activity in the brain. It is useful in diagnosing various states of consciousness and seizures.

X-ray: This procedure is useful in detecting skull or vertebral fractures.

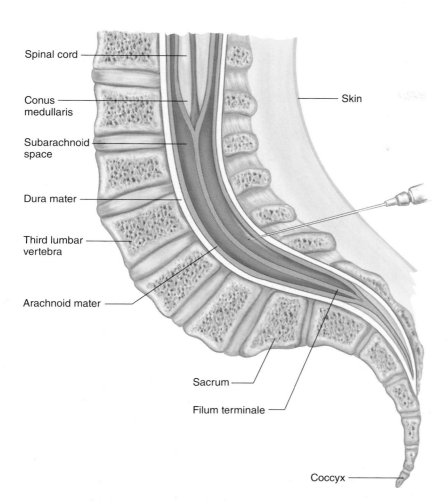

FIGURE 14-36
A lumbar puncture is performed by inserting a needle below the fourth lumbar vertebra and withdrawing a sample of cerebrospinal fluid.

Reflex Testing: Testing a patient's reflexes allows a physician to evaluate the components of a reflex as well as the overall health of the individual's nervous system. The absence of a reflex is called *areflexia*. *Hyporeflexia* is a decreased reflex, and *hyperreflexia* is a stronger than normal reflex. The following are common reflex tests:

Biceps Reflex: The absence of this reflex may indicate spinal cord damage in the cervical region.

Knee Reflex: The absence of this reflex may indicate damage to lumbar or femoral nerves (Figure 14-37).

Abdominal Reflexes: These reflexes are used to evaluate damage to thoracic spinal nerves.

Life Span

As we get older there are changes in the nervous system that can have dramatic impact on our quality of life and even life itself. For example, the brain starts to atrophy and we may see changes in memory and even cognitive and reasoning skills. We are more susceptible to diseases such as stroke, Alzheimer's, and Parkinson's disease. We may see an increase in neuropathies due to other diseases such as diabetes mellitus. There may be a slowing of reflexes. As more people are living to an older age, we will inevitably see an increase in many of these conditions. Hopefully medical research and discovery will help with treatment and prevention. We can also keep physically and mentally active to slow and perhaps prevent the onset of some of these disorders and diseases.

14.8 learning outcome
Infer how aging affects the nervous system and life span.

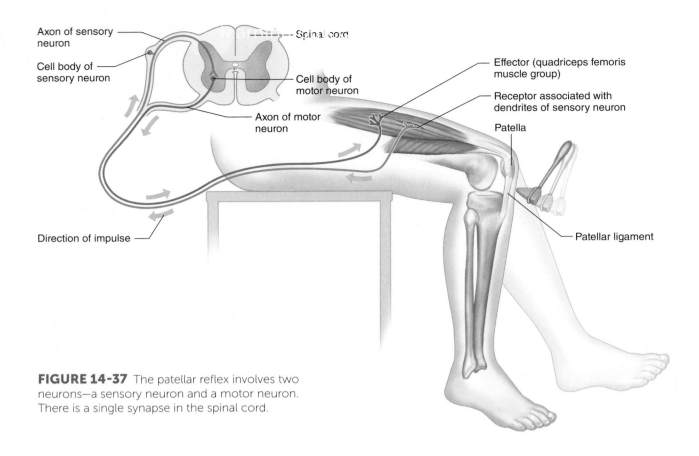

FIGURE 14-37 The patellar reflex involves two neurons—a sensory neuron and a motor neuron. There is a single synapse in the spinal cord.

summary

learning outcomes	key points
14.1 Recognize the functions of the nervous system.	The nervous system is one of the two major regulating organ systems in the body. It communicates messages between the brain and the rest of our body.
14.2 Compare the structure and function of a neuron and a glial cell.	The neuron is the functional unit of the nervous system. It has a cell body, a nucleus, organelles, and extensions called dendrites and axons. It is responsible for conscious thought, memory, and many other activities. The four major types of glial cells are astrocytes, ependymal cells, microglia, and oligodendrocytes. Astrocytes are star-shaped and ependymal cells are cuboidal. Their functions include maintaining the chemical environment of nervous tissue and producing myelin and CSF and phagocytosis.

continued

learning outcomes	key points
14.3 Explain the synapse and nerve transmission.	A synapse includes the presynaptic cell membrane of a neuron, the synaptic cleft, and the postsynaptic membrane of another nerve, muscle cell, or glandular epithelium. It is involved in the transmission of an action potential.
14.4 Identify the components of the central nervous system and their functions.	The brain is divided into the cerebrum, cerebellum, and brainstem. The cerebrum controls and integrates motor, sensory, and higher mental functions. The cerebellum is involved in the coordination of voluntary motor movement, balance and equilibrium, and muscle tone. The brainstem is involved in regulating activities such as breathing, heart rate, and blood pressure. The spinal cord is part of the CNS and is somewhat oval in cross section containing both gray and white matter. It is responsible for communication between the brain and the remainder of the body.
14.5 Describe the structure and functions of the meninges, ventricles, and cerebrospinal fluid.	The meninges are protective connective tissue coverings of the brain and spinal cord. They are the dura mater, arachnoid, and pia mater. The ventricles of the brain are four CSF-filled cavities. CSF is a clear, colorless liquid that protects the brain and spinal cord from injury.
14.6 Identify the components of the peripheral nervous system and their functions.	The cranial nerves are involved with control of the five senses, pain and touch, and the control of some muscles (especially of the head and neck). The spinal nerves are involved in the transmission of neuronal signals.
14.7 Relate neurological testing procedures to their uses.	Neurological testing procedures are performed to verify the integrity of the brain and cranial and spinal nerves.
14.8 Infer how aging affects the nervous system and life span.	Cells of the nervous system begin to die even before birth. Although diseases of the nervous system such as Alzheimer's disease and Parkinson's disease have become more common, most people can live healthy, productive lives into their later years.

chapter 14 review

case study 1 questions

Can you answer the following questions that pertain to the case (Mary Ann) presented in this chapter?

1. What are the two major categories of stroke?
   a. Hemorrhagic and occlusive
   b. Hemorrhagic and transient
   c. Transient and occlusive
   d. Transient and permanent
2. Which of the following is a risk factor this patient has for stroke?
   a. Birth control pills
   b. Hypotension
   c. Female
   d. Age
3. Are her symptoms typical for stroke?
   a. Yes
   b. No
4. What are additional warnings that a person is experiencing a stroke?
   a. Slurred speech
   b. Hyperreflexia
   c. Increased salivation
   d. Decreased salivation
5. What is the treatment for occlusive strokes?
   a. Physical therapy and antiplatelet medications
   b. Tylenol
   c. Chemotherapy
   d. Radiation therapy
6. What is the prognosis for strokes?
   a. Good
   b. Excellent
   c. Poor
   d. Depends on the location and severity of the stroke

review questions

1. A person with a history of cluster headaches should avoid
   a. Chocolate
   b. Alcohol
   c. Nuts
   d. Dark rooms
2. A myelinated nerve will _____ an action potential.
   a. Increase the rate of transmission of
   b. Decrease the rate of transmission of
   c. Have no affect on the rate of transmission of
   d. Inhibit the transmission of
3. The conductance of an action potential is called
   a. Excitatory conductance
   b. Saltatory conductance
   c. Inhibitory conductance
   d. Synaptic transmission
4. The functional unit of the nervous system is the
   a. Schwann cell
   b. Oligodendrocyte
   c. Glial cell
   d. Neuron
5. Which of the following is responsible for producing myelin in the CNS?
   a. Schwann cell
   b. Neuron
   c. Microglial cell
   d. Oligodendrocyte
6. Which of the following is (are) part of the central nervous system?
   a. Cranial nerves
   b. Sensory receptors
   c. Thalamus
   d. Spinal nerves

critical thinking questions

1. In some diseases, such as multiple sclerosis and Guillain-Barré, myelin is destroyed. What is the function of myelin in the nervous system, and how can the destruction of myelin contribute to the signs and symptoms of both of these diseases?
2. Meningitis is an inflammation of the protective covering of the central nervous system. Why is it important for physicians to know whether the cause of meningitis is viral, bacterial, or fungal when treating a patient with meningitis?

patient education

Elevated blood pressure as well as other factors increase your risk for stroke. Review each of the risk factors, then create an action plan for yourself or for someone you know to help prevent stroke. Write each risk factor in a list or in a table and for each one determine what can be done to reduce each risk. Create specific actions for each risk factor.

applying what you know

1. Explain the relationship between the central nervous system and the afferent and efferent nerves, and what role the interneurons play in this relationship.
2. What is the function of neuroglia? Give the names and functions of each of the three types of neuroglial cells.
3. What are the two divisions of the autonomic nervous system? How do these two divisions differ?

CASE STUDY 2 *Neurological Disorders*

A 70-year-old man comes to the family practice office where you are the medical assistant. He complains that he is having trouble with his sense of balance and that it is interfering with his tennis game. He also states that his right hand seems to "have a mind of its own" and he can't seem to keep it totally still. His wife observes that his posture is becoming stooped and his reactions seem slowed.

1. What is the most likely diagnosis for this individual?
   a. Alzheimer's disease
   b. Stroke
   c. Parkinson's disease
   d. Lou Gehrig's disease
2. Which of the following is *not* a typical symptom of this disorder?
   a. Masklike face
   b. Atrial fibrillation
   c. Pill rolling tremor
   d. Shuffling gait
3. Which of the following is used to treat this disorder?
   a. Levodopa
   b. Nitroglycerine
   c. Aspirin
   d. Diuretics

15 The Urinary System

chapter outline

Kidneys
Urine Formation
Urine Elimination
Life Span

outcomes

AFTER COMPLETING THIS CHAPTER, YOU WILL BE ABLE TO:

15.1 Describe the structure, location, function, and pathophysiology of the kidneys.

15.2 Identify substances normally found in urine and how urine is formed.

15.3 Compare the structures, locations, functions, and pathophysiology of the ureters, urinary bladder, and urethra.

15.4 Infer the effects of aging on the urinary system.

case study

Use the case study to focus on as you go through the chapter. The questions will guide you as you learn the anatomy, physiology, and pathology of the urinary system.

CASE STUDY 1 *Renal Stones*

Michael Huckaby, a 60-year-old patient, comes to the office complaining of severe low back pain, nausea, frequent urination, and hematuria (blood in the urine). Your physician orders an immediate urinalysis and sends the patient for a renal ultrasound.

As you go through the chapter, keep the following questions in mind:

1. What is the cause of renal stones?
2. What is kidney failure?
3. Will surgery be needed?

essential terms

afferent arteriole (AF-fer-ent ar-TEE-ree-ole)
Bowman's capsule (BO-manz CAP-sul)
calyces (KA-lih-seez)
efferent arteriole (EF-fer-ent ar-TEE-ree-ole)
hilum (HIGH-lum)
loop of Henle (loop uv HEN-lee)
micturition (mik-tuh-RISH-un)
nephrons (NEF-ronz)
pyelonephritis (pie-eh-lo-neh-FRI-tis)
renal calculi (REE-nul KAL-ku-lie)
renal column (REE-nul KOL-um)
renal corpuscle (REE-nul KOR-puss-ul)
renal cortex (REE-nul KOR-tex)
renal medulla (REE-nul meh-DUL-lah)
renal pelvis (REE-nul PEL-vis)
renal pyramid (REE-nul PEER-uh-MID)
renal tubule (REE-nul TOO-byool)
renal vein (REE-nul VANE)
retroperitoneal (reh-tro-pair-ih-toe-NEE-al)
trigone (TRI-gohn)
ureter (YU-reh-tur)
urethra (yu-REE-thrah)

Additional key terms in the chapter are italicized and defined in the glossary.

study tips

1. Draw and label the structures of a nephron and describe the function of each structure.
2. Make flash cards for the essential terms of the chapter.
3. Outline the chapter. After each section, ask yourself what you just read.
4. Write down one to three questions to ask your instructor.

Homeostasis

Introduction

The organs of the urinary system are the kidneys, ureters, urinary bladder, and urethra (Figure 15-1). The urinary system contributes significantly to homeostasis. The system removes metabolic waste products from the bloodstream and helps maintain water and pH balance in the body. Waste products are excreted from the body in the form of urine. The kidneys also produce hormones that are important in maintaining homeostasis and participate in gluconeogenesis. Nephrons are the microscopic functional units of the kidneys that filter blood, remove waste products, and form urine.

U check
What are the components of the urinary system?

FIGURE 15-1
The urinary system includes the kidneys, ureters, urinary bladder, and urethra.

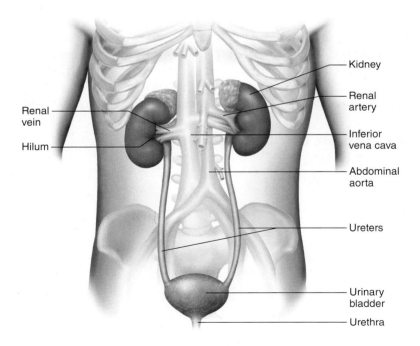

UNIT 2   Concepts of Common Illness by System

Kidneys

learning outcome

Describe the structure, location, function, and pathophysiology of the kidneys.

The kidneys are bean-shaped organs that are reddish brown in color and covered by a tough fibrous capsule. The kidneys are **retroperitoneal** in position (Figure 15-2), meaning that they lie behind the abdominal peritoneum. The kidneys lie on either side of the vertebral column between the levels of the last thoracic vertebra and the third lumbar vertebra. The 11th and 12th pairs of ribs give some protection to the kidneys. The left kidney is slightly higher than the right, which is displaced inferiorly by the liver. An adult kidney is about 4 to 5 inches long, 2 to 3 inches wide, and 1 inch thick.

retroperitoneal
In a position behind the abdominal peritoneum.

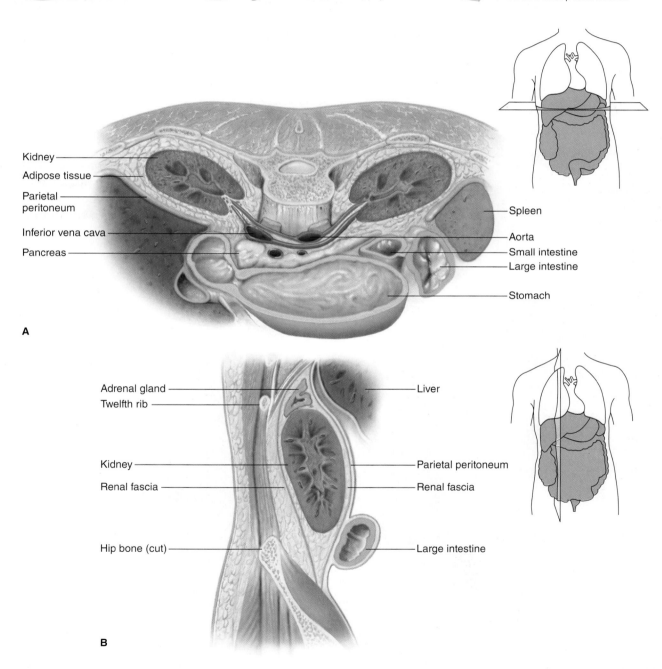

FIGURE 15-2 The kidneys are located retroperitoneally: (A) transverse section through the posterior abdominal wall; (B) sagittal section through the posterior abdominal wall.

CHAPTER 15 The Urinary System

hilum The concave depression of the kidney where the renal artery, renal vein, and ureter enter.

The surface area of the concave depression of the kidney faces medially and is called the renal **hilum** (Figure 15-3). The renal artery, renal vein, and ureter enter the kidney here.

There are three layers of tissue surrounding each kidney. The deepest layer is the renal capsule. It protects the kidney against trauma and helps give shape to the kidney. It is continuous with the connective tissue that covers the ureter. The middle layer is the adipose capsule which, as its name suggests, is made of fat. It protects the kidney against trauma and helps hold the kidney in place. If someone is starving or severely underweight, he or she can lose this fat and the kidneys may fall into the abdominal cavity in a process known as *ptosis* (drooping). The outermost layer protecting the kidney is the renal fascia consisting of dense, irregular connective tissue that anchors the kidney to the abdominal wall.

> **U check**
> What is the name of the indentation on the medial surface of the kidney where blood vessels enter and leave the kidney?

Renal function

Function of the Kidneys

The kidneys are responsible for removing metabolic waste products from the blood. These metabolic wastes are combined with water and ions to form urine, which is excreted from the body. You may remember from Chapter 10, Blood and Circulation, that the kidneys also secrete the hormone

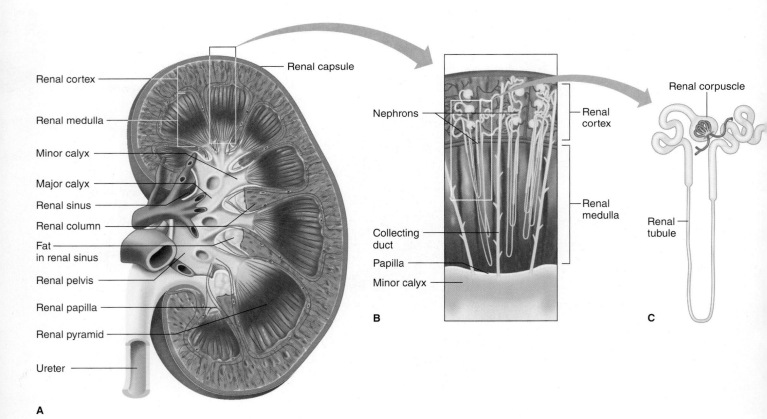

FIGURE 15-3 (A) Longitudinal section of a kidney. (B) Renal pyramid containing nephrons. (C) A single nephron.

UNIT 2   Concepts of Common Illness by System

erythropoietin, which stimulates the red bone marrow to produce red blood cells, and the enzyme *renin,* which helps regulate blood pressure and blood volume. Because the kidneys are capable of excreting hydrogen ions, they also help regulate blood pH. In addition, the kidneys participate in the synthesis of vitamin D and are involved in *gluconeogenesis,* or the synthesis of new glucose. All of these functions are important in maintaining homeostasis, which is a balanced, stable state of the body's internal environment.

Blood and Nerve Supply of the Kidneys

The kidneys make up less than 1 percent of the mass of the body, but receive almost 25 percent of the cardiac output. The combined volume of blood flow through both kidneys is about 1,200 milliliters per minute. Blood enters each kidney through its renal artery, which comes directly off the abdominal aorta (Figure 15-4). Because the aorta is slightly to the left of midline, the right renal artery is longer than the left. As the renal artery enters the kidney it divides into several segmental arteries. There it branches further into *interlobar, arcuate,* and then *interlobular* arteries. The interlobular arteries give off branches called **afferent arterioles** (Figure 15-5). Each nephron receives one afferent arteriole that, once inside the nephron, forms the glomerulus (Figure 15-6). The capillaries then form an **efferent arteriole** that exits the glomerulus and form peritubular capillaries and veins that eventually form the **renal vein.** The renal veins drain into the inferior vena cava.

Renal nerves originate in the renal ganglion and have sympathetic fibers that are vasomotor. These regulate blood flow through the kidneys by

afferent arterioles Branches of the interlobar arcuate arteries that come off the renal artery and supply blood to the nephrons.

efferent arteriole A renal vessel that exits the glomerulus to form a peritubular capillary.

renal vein The vein formed by the peritubular capillaries and veins that, in turn, drains into the inferior vena cava.

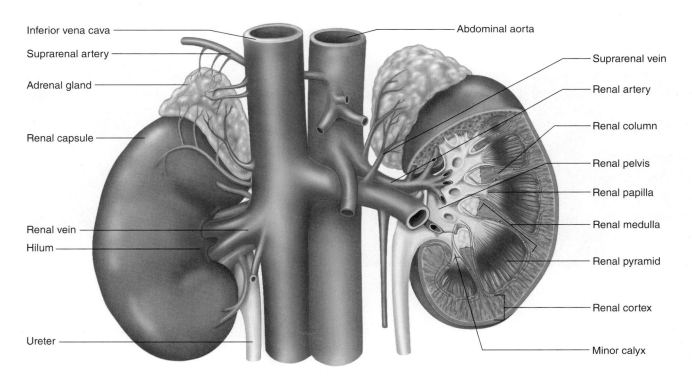

FIGURE 15-4 Blood vessels associated with the kidney and adrenal glands. Note the relationship of the aorta and inferior vena cava to the kidneys.

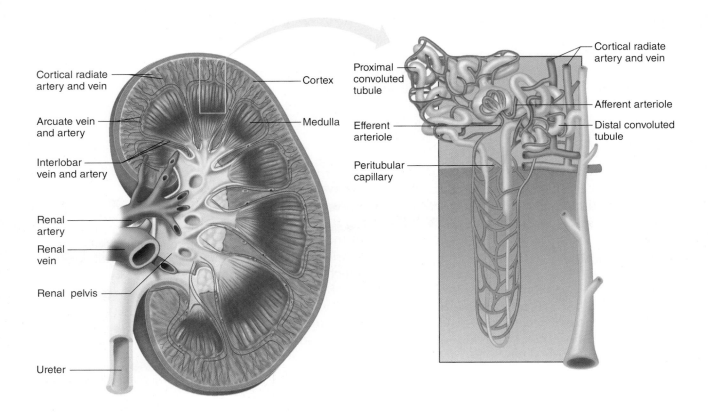

FIGURE 15-5 Renal blood vessels. One afferent arteriole enters the nephron and forms the glomerulus. An efferent arteriole exits the glomerulus.

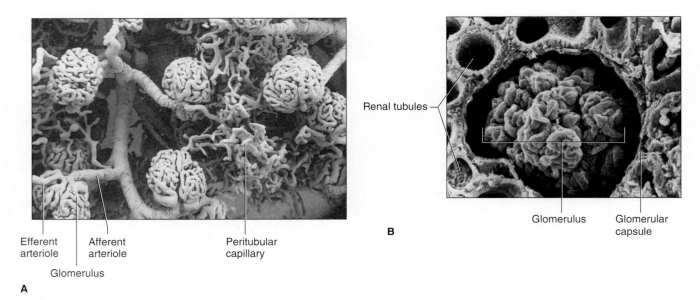

FIGURE 15-6 Blood vessels associated with nephrons: (A) a scanning electron micrograph of a cast of the renal blood vessels associated with the glomeruli (magnified 200x); (B) a scanning electromicrograph of a glomerular capsule surrounding a glomerulus (magnified 480x).

causing vasodilation and vasoconstriction. Many renal nerves enter the kidney along with the renal artery. These nerves are part of the sympathetic division of the autonomic nervous system.

> **U check**
> Why is the right renal artery longer than the left renal artery?

Nephrons

The internal anatomy of the kidney consists of two regions: the outermost **renal cortex** and the inner **renal medulla.** The renal medulla is divided into triangular-shaped areas known as **renal pyramids.** The renal cortex covers the pyramids and also extends down between them. The portion of the cortex between pyramids is called a **renal column.** Together the cortex and medulla make up the functioning part of the kidney called the *parenchyma*. The parenchyma of each kidney contains about 1 million nephrons. Approximately 85 percent of the kidney's nephrons are found in the outer renal cortex.

Nephrons are the functional units of the kidney and are made up of a **renal corpuscle** and a tubular system (Figure 15-7). Waste products are filtered from the blood by the nephrons. The renal corpuscle is composed

renal cortex The outermost layer of the kidney that covers and extends between the pyramids.

renal medulla The inner layer of the kidney that contains the renal pyramids.

renal pyramid A triangular-shaped area in the renal medulla containing the straight segments of the renal tubules.

renal column The portion of the cortex between pyramids.

nephrons The functional unit of the kidney, which comprises a renal corpuscle and a tubular system.

renal corpuscle The part of a nephron that comprises a glomerulus and Bowman's capsule and which filters blood.

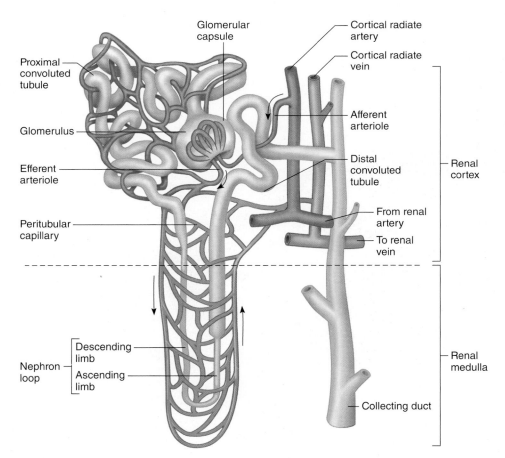

FIGURE 15-7 Structure of a nephron and the associated blood vessels.

Bowman's capsule
The glomerular capsule; a membrane that surrounds the glomerulus.

renal tubule The structure that extends from Bowman's capsule to collect the filtrate.

loop of Henle The middle section (nephron loop) of the renal tubule in which it curves toward the renal corpuscle.

of a mass or tuft of capillaries called the glomerulus, and the capsule that surrounds it is the glomerular capsule, also called **Bowman's capsule** (Figure 15-8). The renal corpuscle is the site of blood filtration. **Renal tubules** extend from the Bowman's capsule of a nephron to collect the filtrate.

The three parts of a renal tubule are the proximal convoluted tubule, the loop of Henle, and the distal convoluted tubule (Figure 15-9). The *proximal convoluted tubule* is directly attached to the Bowman's capsule and eventually straightens out to become the nephron loop, commonly called the **loop of Henle.** The loop of Henle curves back toward the renal corpuscle and starts to twist again, becoming the distal convoluted tubule. *Distal convoluted tubules* from several nephrons merge together to form *collecting ducts*.

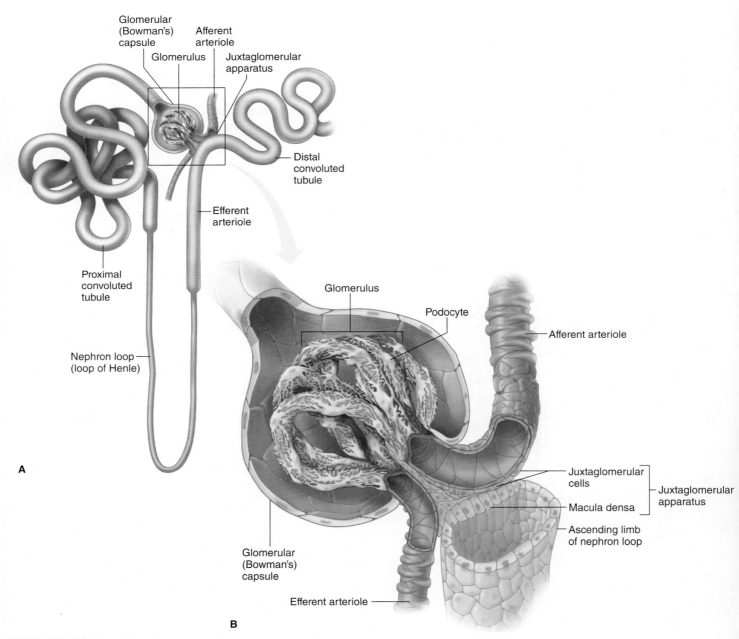

FIGURE 15-8 (A) Tubular system of the nephron. (B) A glomerulus.

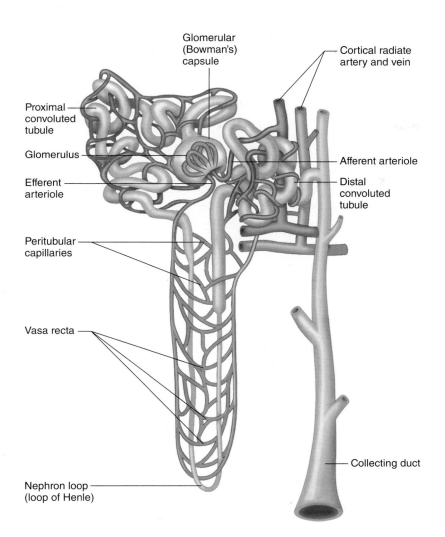

FIGURE 15-9
The capillary loop in close association with a nephron.

These ducts collect urine that is transported through cuplike structures called minor and major **calyces** (calyx is singular) to be delivered to the **renal pelvis,** which in turn empties urine into the ureters.

U check
What are the components of the renal corpuscle?

calyces Cuplike structures that transport urine from the collecting ducts to the renal pelvis.

renal pelvis Cavity in the center of the kidney into which the major calyces open.

PATHOPHYSIOLOGY

Acute Renal Failure

Acute renal (kidney) failure (ARF) is a sudden loss of kidney function.

Etiology: There are many causes and risk factors of kidney failure, including burns, dehydration, low blood pressure, hemorrhage, allergic reactions, obstruction of the renal artery, various poisons, alcohol abuse, trauma to the kidneys or skeletal muscles, blood

(continued)

Renal failure

Acute Renal Failure (concluded)

disorders, blood transfusion reactions, kidney stones (renal calculi), urinary tract infections (UTI), enlarged prostate, immune system disorders, childbirth, and food poisoning involving the bacterium *E. coli*.

Signs and Symptoms: The signs and symptoms of ARF include decreased or no urine production, swelling (edema) of the extremities, bloating, mental confusion, coma, seizures, hand tremors, nosebleeds, easy bruising, pain in the back or abdomen, hypertension, abnormal heart or lung sounds, abnormal urinalysis, and an increase in potassium levels (hyperkalemia).

Treatment: The first treatment measure is modifying the diet to decrease the amount of protein consumed. Controlling fluid intake and potassium levels is also recommended. Antibiotics and dialysis may be needed as well. If the underlying cause can be treated, acute renal failure may be reversed and kidney function returned to normal.

Chronic Renal Failure

Chronic renal (kidney) failure (CRF) is a condition in which the kidneys slowly lose their ability to function. The patient may be asymptomatic until the kidneys have lost about 90 percent of their function.

Etiology: This disorder results from diabetes, hypertension, glomerulonephritis, polycystic kidney disease, renal (kidney) stones, obstruction of the ureters, heart failure, and acute kidney failure.

Signs and Symptoms: The list of signs and symptoms is extensive and includes headache, mental confusion, coma, seizures, fatigue, itching, easy bruising, abnormal bleeding, anemia, excessive thirst, fluid retention, nausea, hypertension, abnormal heart or lung sounds, weight loss, white spots on the skin or increased pigmentation, hyperkalemia (high potassium levels), an increased or decreased urine output, urinary tract infections, and abnormal urinalysis results.

Treatment: This disorder can be treated with antibiotics; blood transfusions; medications to control anemia; restriction of fluids, electrolytes, and protein; control of hypertension; and dialysis. The most serious cases may require surgery to repair a ureteral obstruction or the patient may be a candidate for renal (kidney) transplant.

Polycystic Kidney Disease

Polycystic kidney disease (PKD) is a disorder in which the kidneys enlarge because of the presence of many cysts within them. The

Polycystic Kidney Disease (concluded)

disease develops relatively slowly, with symptoms worsening over time. Only diabetes mellitus and hypertension are responsible for more cases of renal failure.

Etiology: The cause is hereditary (both autosomal dominant and autosomal variants are seen). Autosomal-dominant (adult) polycystic kidney disease (ADPKD) is common and affects between 1 of every 400 to 1,000 live births. It is responsible for about 5 to 10 percent of chronic renal failure. Autosomal-recessive polycystic kidney disease (ARPKD) is a rare developmental anomaly.

Signs and Symptoms: Fatigue, hypertension, anemia, pain in the back or abdomen, joint pain, heart murmurs, formation of kidney stones, hematuria (blood in the urine), kidney failure, and liver disease are the symptoms of this disorder.

Treatment: Treatment includes medications to control anemia and hypertension, blood transfusions, draining of the cysts, dialysis, and surgery to remove one or both kidneys (nephrectomy)

focus on Clinical Applications: Dialysis

When the kidneys fail to clean the blood, patients may need to have their blood cleansed through an artificial means. This is known as dialysis or, more specifically, renal dialysis. There are two types of dialysis: hemodialysis and peritoneal dialysis. No matter which type of dialysis is used, the patient should be on a restricted diet and very conscious of his or her fluid intake.

Hemodialysis requires the use of a machine. A patient has two catheters, or small tubes, inserted. One catheter is placed in a vein and the other in an artery. The arterial catheter removes the blood from the patient and transports it to a dialysis machine. The dialysis machine filters the blood to rid it of waste products, similar to the way the kidneys normally do. The filtered blood then returns to the patient through the venous catheter. This type of dialysis is done by specially trained health care professionals in a health care facility.

Peritoneal dialysis uses the lining of the abdomen to filter the blood. During peritoneal dialysis, blood vessels in the abdominal lining (peritoneum) fill in for the kidneys, with the help of a special fluid (dialysate) that flows into and out of the peritoneal space. Patients require assistance from health care professionals. In some cases, after proper education, patients can give themselves treatments in their home or while traveling.

PATHOPHYSIOLOGY

Pyelonephritis

pyelonephritis A type of complicated urinary tract infection that begins as a bladder infection and spreads up one or both ureters into the kidneys.

Pyelonephritis is a type of complicated urinary tract infection (UTI). It begins as a bladder infection that spreads up one or both ureters to the respective kidney(s). This condition can develop suddenly or it may be a chronic, slowly developing condition.

Etiology: This disorder is caused by bacteria, bladder infection, renal calculi (kidney stones), or an obstruction of the urinary system ducts.

Signs and Symptoms: Signs and symptoms include fatigue, mental confusion, fever, nausea, pain in the back or abdomen, enlarged kidneys, painful urination (dysuria), and cloudy or bloody urine (hematuria).

Treatment: Treatment includes intravenous (IV) fluids, analgesics, and antibiotics.

Kidney Stones

renal calculi Kidney stones that obstruct the ducts within the kidneys or ureters.

Renal calculi are more commonly called kidney or renal stones. Besides being painful, these stones can become lodged in the ducts within the kidneys or ureters, causing obstruction.

Etiology: This condition is caused by gouty arthritis (also known as gout), defects of the ureters, overly concentrated urine, and urinary tract infections (UTI).

Signs and Symptoms: The signs and symptoms include fever, nausea, severe back or abdominal pain, a frequent urge to urinate, hematuria (blood in the urine), and abnormal urinalysis results.

REMEMBER MICHAEL, our 60-year-old patient? What do you think is the cause of his hematuria?

Treatment: Treatment includes pain medication, intravenous (IV) fluids, medications to decrease stone formation, surgery to remove kidney stones, and lithotripsy (a procedure that uses shock waves to break up stones).

Glomerulonephritis

Glomerulonephritis is an inflammation of the glomeruli of the kidneys. Chronic glomerulonephritis is one of the causes of chronic renal disease.

Etiology: This disorder is caused by bacterial infections, renal diseases, and immune disorders.

Signs and Symptoms: The signs and symptoms are drowsiness, coma, seizures, nausea, anemia, hypertension, increased skin pigmentation, abnormal heart sounds, abnormal urinalysis results, *hematuria*, and decreased or increased urine output.

Treatment: Treatment begins with a low-sodium, low-protein diet. Medications to control hypertension, corticosteroids to reduce inflammation, and dialysis are other treatment options.

Urine Formation

The three processes of urine formation are glomerular filtration, tubular reabsorption, and tubular secretion.

Glomerular Filtration

Glomerular filtration takes place in the renal corpuscles of nephrons (Figure 15-10). In this process, the fluid part of blood is forced from the glomerulus into Bowman's capsule. The fluid in the Bowman's capsule is called glomerular filtrate. Glomerular filtration occurs as a result of filtration pressures that force substances (filtrate) out of the glomerulus into Bowman's capsule. Glomerular filtration depends on the following three different pressures:

1. Glomerular filtration pressure, which is the blood pressure in the glomerular capillaries.
2. Capsular hydrostatic pressure, which actually acts in opposition to the glomerular pressure.
3. Blood osmotic pressure, which also works in opposition to the glomerular pressure.

If a person's blood pressure is too low, glomerular filtrate will not form. If filtration pressure increases, the rate of filtration and the amount of glomerular filtrate also increases. The sympathetic nervous system (SNS) largely controls the rate of filtration. If blood pressure or blood volume drops, the SNS causes the afferent arterioles in the kidneys to constrict. When constriction occurs, glomerular filtration pressure decreases and less glomerular filtrate is formed. When less glomerular filtrate is formed, less urine is ultimately formed. This allows the body to retain fluids that are needed to raise blood pressure and blood volume.

15.2 learning outcome

Identify substances normally found in urine and how urine is formed.

> **U check**
> Glomerular filtration depends on what three pressures?

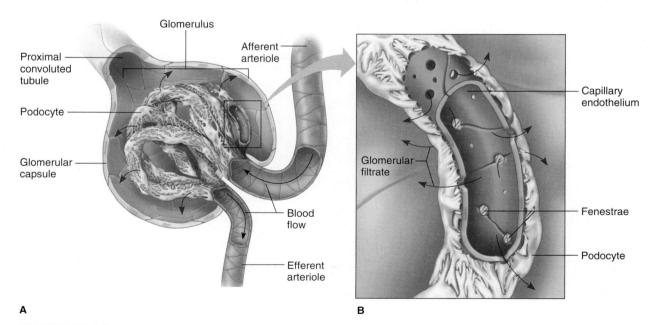

FIGURE 15-10 Glomerular filtration: (A) the first step in urine formation is filtration; (B) glomerular filtrate passes through the capillary endothelium.

Tubular Reabsorption

Tubular reabsorption is the second process in urine formation (Figure 15-11). In this process, the glomerular filtrate flows into the proximal convoluted tubule. The body needs to retain many of the substances (nutrients, water, and ions) that are found in glomerular filtrate. In tubular reabsorption, all the necessary substances in the glomerular filtrate pass through the wall of the renal tubule into the blood of the peritubular capillaries. Water reabsorption varies depending on the presence of two hormones—*antidiuretic hormone (ADH)* and *aldosterone*. Both of these hormones increase water reabsorption, which decreases urine production (Figure 15-12). About 99 percent of filtered water is reabsorbed. This retention of fluid and the resultant increase in blood pressure is one of the reasons why diuretics, which rid the body of excess fluid, are successful in treating some forms of hypertension.

Tubular Secretion

Tubular secretion is the third step in urine formation. In tubular secretion, substances move out of the blood in the peritubular capillaries and into the renal tubules. Substances that are secreted include drugs, hydrogen ions, and waste products. All of these secreted substances will be excreted in the urine.

The final solution that reaches the collecting ducts of the kidneys is urine. Urine is mostly made of water. One hundred eighty liters of water are filtered each day, but only 1 to 2 liters are excreted. Only about 5 percent or 0.1 gram of filtered protein is excreted.

Approximately 160 grams of glucose are filtered each day, but glucose is not normally found in the urine. Urine also contains *urea* and *uric acid*,

Hypertension

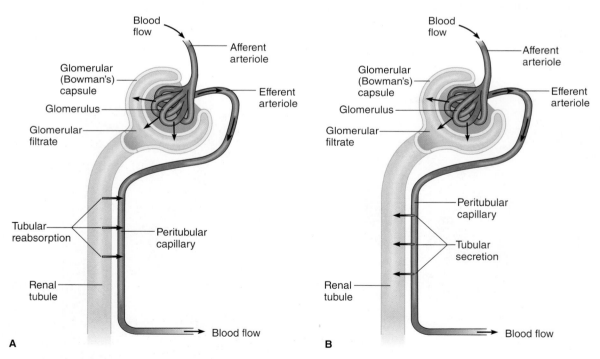

FIGURE 15-11 (A) Tubular reabsorption transports substances into the peritubular capillary from the glomerular filtrate. (B) Tubular secretion transports substances from the peritubular capillary into the renal tubule.

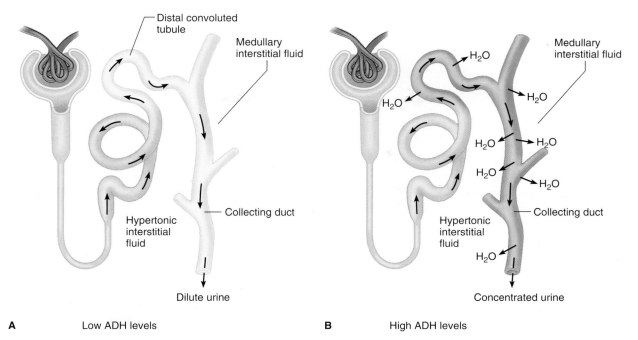

FIGURE 15-12 Urine concentrating mechanism: (A) the distal convoluted tubule and collecting duct are impermeable to water so dilute urine is excreted; (B) if ADH is present, these segments become permeable and water is reabsorbed, resulting in a more concentrated urine.

and various ions. Urea and uric acid are waste products formed by the breakdown of proteins and nucleic acids. The secretion of these waste materials assists in maintaining the body's acid–base balance.

U check
How much urine is formed and excreted each day?

from the perspective of . . .

A CLINICAL MEDICAL ASSISTANT A clinical medical assistant assists the physician with patients performing clinical and laboratory procedures. Why would knowing the composition of urine be important?

Urine Elimination

Elimination of urine is also known as micturition or urination. Approximately 125 milliliters per minute or 180 liters per day of plasma is filtered, but only one to two liters of urine is excreted each day in healthy individuals. Waste products such as urea, creatinine, and uric acid are eliminated, while important molecules such as water, sodium, and chloride are conserved.

learning outcome

Compare the structures, locations, functions, and pathophysiology of the ureters, urinary bladder, and urethra.

Ureters, Urinary Bladder, and Urethra

ureter The tube that drains urine from each kidney to the bladder.

trigone The triangle formed by the three openings of the urethra and the two ureters.

The **ureters** are the tubes that drain urine from the kidneys, carrying the urine to the bladder. Ureters are long, muscular tubes that propel urine through rhythmic muscular contractions called *peristalsis*. Urine is stored in the bladder until it is expelled via the urethra.

The *urinary bladder* is a distensible (expandable) organ that is located in the pelvic cavity just behind the pubic symphysis (Figure 15-13). The interior of the bladder has folds in the mucosa called *rugae* that allow it to expand. When full, the bladder can hold up to about 800 milliliters of urine. The internal floor of the bladder contains three openings—one for the urethra and one for each of the ureters. These three openings form a triangle called the **trigone** of the bladder (Figure 15-14). The wall of the bladder has three

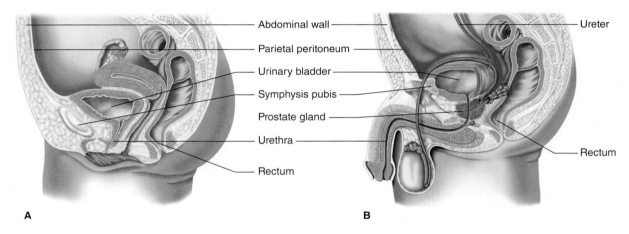

FIGURE 15-13 The urinary bladder is in the abdominal cavity behind the pubic symphysis: (A) female and (B) male.

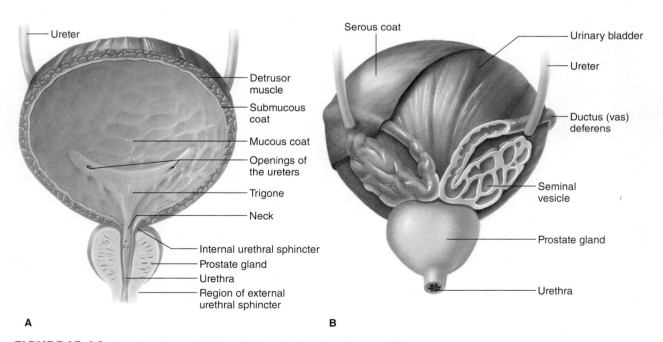

FIGURE 15-14 A male urinary bladder: (A) longitudinal section and (B) posterior view.

layers. The innermost layer is mucosa made up of transitional epithelium. The middle layer is the muscular layer, and the superficial layer is composed of connective tissue. The muscular layer is composed of smooth muscle and is known as the *detrusor muscle*. This muscle contracts to push urine from the bladder into the urethra.

The process of urination is called **micturition.** The stretching of the bladder triggers this process—usually when the bladder reaches 200 to 400 milliliters of urine. The major events of micturition include the following:

micturition The process of urination.

1. The urinary bladder distends as it fills with urine.
2. The distension stimulates stretch receptors in the bladder wall, signaling the micturition center in the spinal cord.
3. Parasympathetic nerves stimulate the detrusor muscle, which begins rhythmic contractions that trigger the sense of the need to urinate.
4. The brainstem and cerebral cortex send impulses to voluntarily contract the external urethral sphincter and to inhibit the micturition impulse.
5. Upon the decision to urinate, the external urethral sphincter is relaxed and impulses from the pons and hypothalamus start the micturition reflex.
6. Contraction of the detrusor muscle occurs and urine is expelled through the urethra.

The **urethra** is a tube that expels urine from the bladder to the outside environment. In females, the urethra is about 4 inches in length and in males it is about 8 inches in length (Figure 15-15). This anatomic difference, combined

urethra The tube that expels urine from the bladder out of the body.

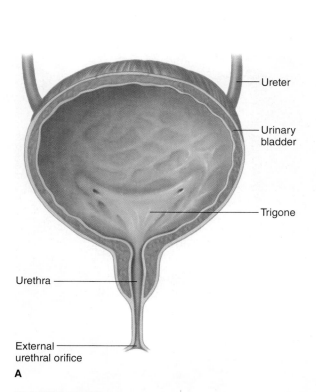
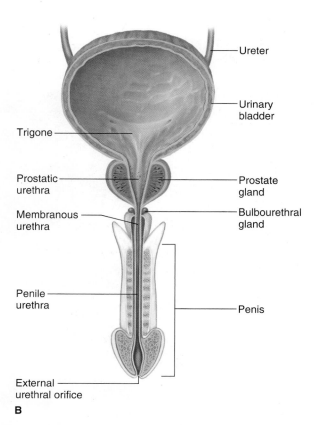

FIGURE 15-15 Urinary bladder and urethra of the (A) female (longitudinal section) and (B) male (longitudinal section).

CHAPTER 15 The Urinary System 443

> ### focus on Clinical Applications: The Urinalysis
>
> One of the most common screening tests ordered by and performed in a physician office is the urine dipstick. For this test a fresh (or refrigerated) urine specimen is supplied by the patient. A health care professional uses a reagent strip to test for the presence of the following substances: leukocytes, nitrate, urobilinogen, protein, pH, blood, specific gravity, ketones, bilirubin, and glucose.
>
> To perform the test, the reagent strip is dipped into the specimen, making sure that each reagent pad is covered completely. The strip is removed and tapped gently on the side of the urine specimen cup. Each chemical pad, when activated by the urine, turns different shades of its original color indicating the presence or absence of each substance (in the case of specific gravity, each color indicates the "weight," or concentration, of the urine). The color of each pad is carefully compared to the comparable chart located on the reagent strip bottle and the results are documented in the patient's chart. Abnormal results may require further tests to assist to diagnose a patient's complaints (Figure 15-16).

with the fact that the anus, vagina, and urethra are in close proximity in females, makes them much more susceptible to urinary tract infections. In males the urethra is part of both the urinary system and the reproductive system.

> ### U check
> What component of the autonomic nervous system is involved with micturition?

FIGURE 15-16
When performing a reagent strip urinalysis you must read the strip at the time interval recommended by the manufacturer's instructions.

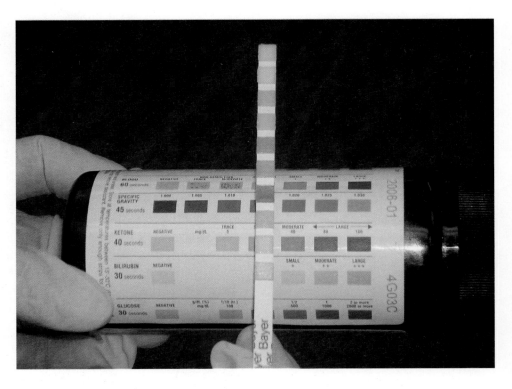

PATHOPHYSIOLOGY

Incontinence

Incontinence is a condition in which an adult cannot control urination. This condition can be either temporary or long lasting. Because of their anatomic makeup, women are more likely to develop incontinence than are men.

Etiology: This condition can be caused by various medications, excessive coughing (for example, in smokers), UTIs, nervous system disorders, and bladder cancer. In men, prostate problems can lead to the development of this disorder. The weakness of the urinary sphincters from surgery, trauma, or pregnancy can also cause incontinence. It may be prevented by avoiding urinary bladder irritants such as coffee, cigarettes, diuretics, and various medications.

Signs and Symptoms: The primary symptom is the involuntary leakage of urine.

Treatment: Treatment includes medications, incontinence pads, removal of the prostate, Kegel exercises for bladder control, and surgery to repair damaged bladders and urethral sphincters.

Cystitis

Cystitis is a urinary bladder infection. As mentioned earlier, women are much more likely to develop this disorder than men because of the shorter length of their urethras.

Etiology: Cystitis is caused by different types of bacteria (especially those that are found in the GI tract) or may be caused by the placement of a catheter in the bladder. Practicing good hygiene, and urinating promptly when the urge occurs, can help prevent this infection.

Signs and Symptoms: Common symptoms include fatigue, chills, fever, and a painful, frequent need to urinate, often with only small amounts of urine produced. Urine is often cloudy and blood may be present in the urine.

Treatment: This infection is treated with antibiotics and pain medication when needed. The patient should also be urged to drink lots of clear liquids. Drinking cranberry juice can help with these infections by reestablishing the correct (more acidic) pH in the bladder.

focus on Clinical Applications: Preventing Urinary Cystitis in Women

In females, the openings for the digestive (anus), urinary (urethra), and reproductive systems (vagina) are all within close proximity to each other in a highly functional design. However, when you combine this design with the female's shorter urethra (relative to men), women are much more likely than men to suffer from urinary tract infections (UTIs) and cystitis, more commonly known as bladder infections.

As a health care professional, if you understand why this is so you will not only be able to assist your female patients when they have a UTI with its pain (dysuria), urgency, and frequency, but you will also be able to give them helpful tips to prevent these painful episodes. Educate female patients to take these steps:

1. *Urinate when the urge occurs.* "Holding it," causes the urine to stay in the bladder and the upper urethra longer, allowing bacteria to grow and causing infection.

2. *Drink lots of clear fluids.* This is known as pushing fluids. The more clear liquids you drink, the more urine is created, flushing the system on a regular basis. Drinking cranberry juice is highly recommended as a preventive measure because it can reestablish the proper pH in the bladder and reduce the ability of microorganisms to multiply.

3. *Wipe front to back.* Teach your female patients that after a bowel movement, they should wipe front to back. Doing so prevents contamination of both the vagina and urethra by gastrointestinal (GI) bacteria (*E. coli*). Maintaining excellent hygiene in the perineal area is a very important component of UTI and cystitis prevention.

4. *Urinate after sexual intercourse.* Explain to your sexually active patients that urinating immediately after sexual intercourse will also help prevent episodes of cystitis.

5. *Fight an existing infection with antibiotics and clear fluids.* If despite preventive methods cystitis or UTI does strike, reassure your patient that antibiotics and lots of fluids should take care of the problem. If the patient suffers from more than three infections during a year, a urologic workup may be advisable to look for underlying anatomical anomalies.

15.4 learning outcome

Infer the effects of aging on the urinary system.

Medication, metabolism, and/or excretion

Life Span

The kidneys become smaller with age and are less efficient at filtering blood plasma. Blood flow to the kidneys also decreases. By 80 years of age, the kidneys have lost more than 50 percent of their filtering capability. Infections, incontinence, renal calculi, and renal failure become more common as do *dysuria* (painful urination), *nocturia* (awakening at night to urinate), and even *hematuria* (blood in the urine).

focus on Wellness: Stress Incontinence Prevention and Treatment

Many women, particularly those who have been pregnant and/or have had abdominal surgery, will secretly admit to the embarrassment of having had stress incontinence—usually while laughing, coughing, sneezing, lifting, or even exercising. The cause is a weakening of the pelvic muscles, allowing the bladder to drop, which results in an incomplete closure of the urethra and allows for urinary leakage.

The best treatment for stress incontinence is prevention. The best prevention is Kegel exercises, which strengthen the pelvic floor muscles. Performing Kegel exercises is easy and can be done anywhere because no one will know you are doing them. Simply pretend you are trying to stop urinating or avoid passing gas. When you do this you are contracting the muscles of the pelvic floor, strengthening them. Simply contract the muscles, count slowly to five, and then relax. To perform Kegels correctly, do not move your legs, buttocks, or abdominal muscles. To be effective, Kegel exercises should be performed 5 times a day, with 10 repetitions performed for each set.

If it is too late for Kegel exercises, a device called a pessary ring may be inserted to keep the bladder and urethra in place to control leakage. Unfortunately, however, one of the side effects of a pessary is increased risk of bladder and vaginal infections. Some patients have also reported success with injections of bulking agents such as collagen to increase the size of the urethral lining, which in turn increases resistance against urine flow. Repeat injections may be necessary if symptoms return. Other treatment options include surgery to suspend the bladder and tighten the urethral sphincters.

summary

learning outcomes	key points
15.1 Describe the structure, location, function, and pathophysiology of the kidneys.	The kidneys are located in the abdominal cavity in a retroperitoneal location. The kidneys remove waste products from the bloodstream and help maintain water and pH balance in the body. Waste products are excreted from the body in the form of urine. The kidneys also produce hormones that are important in maintaining homeostasis.
15.2 Identify substances normally found in urine and how urine is formed.	Urine is mostly water, but it also contains ions, small amounts of protein, and other solutes.

continued

chapter 15

learning outcomes	key points
15.3 Compare the structures, locations, functions, and pathophysiology of the ureters, urinary bladder, and urethra.	Ureters connect the kidneys to the urinary bladder. The bladder stores the urine. The urethra leads from the bladder to the exterior and voids the urine.
15.4 Infer the effects of aging on the urinary system.	The kidneys become smaller with age and are less efficient. Blood flow decreases and elderly people are more susceptible to a spectrum of kidney diseases.

chapter 15 review

case study 1 questions

Can you answer the following questions that pertain to the case (Michael Huckaby) presented earlier in this chapter?

1. Which of the following is *not* a cause of kidney stones?
   a. Gout
   b. Rheumatoid arthritis
   c. Urinary tract infections
   d. Overly concentrated urine

2. What is another name for kidney stones?
   a. Cholecystitis
   b. Renal calculi
   c. Polycystic kidney disease
   d. Renal cell carcinoma

3. A procedure that uses shock waves to break up kidney stones is called
   a. Urinalysis
   b. Lithotripsy
   c. Angioplasty
   d. Angiography

review questions

1. The renal arteries come off the
   a. Abdominal aorta
   b. Ascending aorta
   c. Inferior vena cava
   d. Superior vena cava

2. The renal veins drain into the
   a. Abdominal aorta
   b. Ascending aorta
   c. Inferior vena cava
   d. Superior vena cava

3. Which of the following are the three basic functions performed by the nephrons?
   a. Filtration, excretion, and absorption
   b. Absorption, secretion, and excretion
   c. Reabsorption, filtration, and secretion
   d. Secretion, absorption, and reabsorption

4. Which of the following is a hormone secreted by the kidneys?
   a. Aldosterone
   b. ADH
   c. Erythropoietin
   d. Insulin

5. Which of the following is *not* normally found in the urine?
   a. Water
   b. Glucose
   c. Sodium
   d. Protein

6. Why are women more likely than men to develop urinary tract infections?
   a. Women have a shorter urethra than men.
   b. Women have a longer urethra than men.
   c. Women have larger bladders than men.
   d. Women have smaller bladders than men.

critical thinking questions

1. Why would it not be surprising for a patient suffering from kidney failure to also be anemic?
2. The position of the kidneys is retroperitoneal. What would be the easiest way for a surgeon to reach a kidney?
3. Explain why patients with polycystic kidney disease tend to be less successful with kidney transplants than patients who have chronic renal failure related to other diagnoses.

patient education

A friend of yours recently had a baby and is experiencing periodic urine incontinence. What things could you tell your friend to do to treat and prevent her problem?

applying what you know

1. Normal urine contains only trace amounts of protein. Explain why a urinalysis showing a large amount of protein would be of concern.
2. Why are diuretics successful in treating some forms of hypertension?
3. Explain the functions of the kidneys.

CASE STUDY 2 *Renal Function and Aging*

A 77-year-old man has been prescribed an antibiotic for a skin infection. Based on what you know about the kidneys and aging, answer the following question:

1. The dosage of medication prescribed for this elderly man compared to the dosage prescribed for a much younger individual with the same infection would be
   a. Greater because of decreased functioning of the elderly man's kidneys
   b. Greater because of increased functioning of the elderly man's kidneys
   c. Less because of decreased functioning of the elderly man's kidneys
   d. Less because of increased functioning of the elderly man's kidneys

CHAPTER 15 The Urinary System

16

The Male Reproductive System

chapter outline

External Male Reproductive Structures

Internal Male Reproductive Structures

Male Reproductive Hormones

Spermatogenesis

Erection, Orgasm, and Ejaculation

Life Span

outcomes

AFTER COMPLETING THIS CHAPTER, YOU WILL BE ABLE TO:

16.1 Identify the external organs of the male reproductive system including the location, structure, function, and disorders of each.

16.2 Identify the internal organs of the male reproductive system including the location, structure, function, and disorders of each.

16.3 Identify the male reproductive hormones and the function of testosterone.

16.4 Describe how sperm cells are formed.

16.5 Explain the processes of erection, orgasm, and ejaculation.

16.6 Infer the effects of aging on the male reproductive system.

essential terms

acrosome (AK-ro-some)
cryptorchidism (kript-OR-kid-izm)
epididymis (ep-ih-DID-ih-mis)
penis (PE-nis)
prepuce (PREE-puse)
prostate gland (PRAH-state gland)
scrotum (SKRO-tum)
semen (SEE-men)
seminal vesicles (SEM-ih-nul VES-ih-klz)
seminiferous tubules (seh-mih-NIF-er-us TOO-byoolz)
spermatogenesis (sper-mah-toe-JEN-eh-sis)
testes (TES-teez)
testosterone (tes-TOS-ter-one)
vas deferens (vas DEF-er-enz)
vasectomy (vah-SEK-toe-mee)

Additional key terms in the chapter are italicized and defined in the glossary.

case study

Use the case study to focus on as you go through the chapter. The questions will guide you as you learn the anatomy, physiology, and pathology of the male reproductive system.

CASE STUDY 1 *Benign Prostatic Hypertrophy (BPH)*

Mark Haas is a 54-year-old banker who in the last few weeks has found himself getting up at night to urinate at least twice, sometimes more often. His wife complains because he wakes her up every time he goes. In the last few weeks he gets to the bathroom but just cannot seem to get started. He has no pain, so he does not see the need to go to the doctor. He thinks it is just part of getting older. After a particularly bad night—four times to the bathroom and he cannot get things flowing—his wife finally convinces him to see a doctor.

As you go through the chapter, keep the following questions in mind:

1. What organ of the male reproductive system might be the problem?
2. Why do you think Mark has frequency and urgency?
3. What tests might be done for Mark?
4. What treatment might help Mark?
5. What is the relationship of hormones to Mark's problem?

study tips

1. Make flash cards for each of the organs of the male reproductive system. Place the name of the organ on the front and the function on the back of each card.

2. Review your flash cards, then shuffle them and put them in the order in which sperm travels through the male reproductive system.

Introduction

The male reproductive system (Figure 16-1) functions together with the female reproductive system to produce offspring. The male reproductive system consists of external and internal organs and internal structures. The male reproductive system also produces a number of important hormones.

learning outcome 16.1

Identify the external organs of the male reproductive system including the location, structure, function, and disorders of each.

External Male Reproductive Structures

The male external reproductive organs are the scrotum, testes, and penis. These organs are described in detail in the following text.

Scrotum

The **scrotum** is the pouch of skin that holds the testes (testicles). It is lined with a serous membrane that secretes serous fluid to ensure that the testes move freely within it. The scrotum holds the testes away from the rest of the body, keeping their temperature about one degree lower than that of the core of the body, which is necessary for the sperm to live.

scrotum The pouch of skin that holds the testes.

> **U check**
> Why are the testes located outside the abdominopelvic cavity?

Testes

The **testes** are the primary organs of the male reproductive system. They produce the sex cells (sperm) of the male. They also produce the male hormone **testosterone.** Most males have two testes that are held just below the pelvic cavity in the scrotum. During the fetal stage (Figure 16-2), the testes develop in the abdominopelvic cavity of the fetus. Shortly before or soon after birth, the testes descend into the scrotal sac located just below the pelvic cavity. A fibrous capsule encloses each testis and also "invades" each testis to divide it into lobules. Each lobule contains the **seminiferous tubules,** which are filled with *spermatogenic cells* that give rise to sperm cells. Between the seminiferous tubules are the *interstitial cells*, which produce the hormone testosterone.

testes Testicles; the male gonads responsible for producing sperm and testosterone.

testosterone A male sex hormone needed for the production of sperm and responsible for male secondary sex characteristics.

seminiferous tubules Tightly coiled ducts located in the testes where sperm are produced.

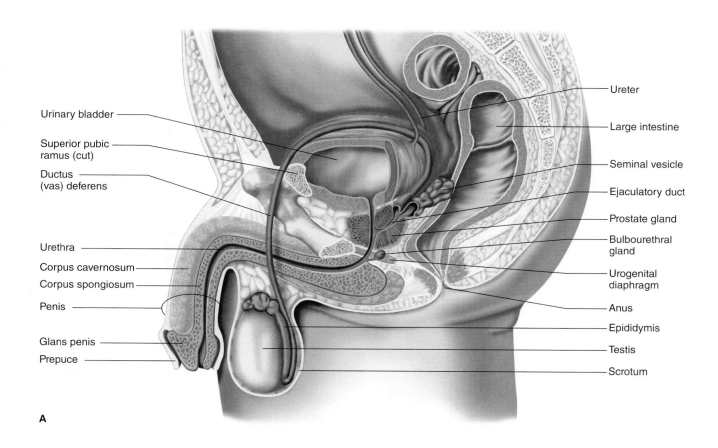

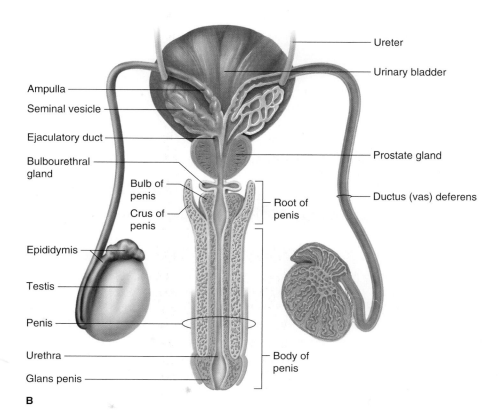

FIGURE 16-1
Male reproductive organs: (A) sagittal view and (B) posterior view.

CHAPTER 16 The Male Reproductive System

FIGURE 16-2
Fetal development of the testes. Each testis begins to develop near a kidney (A) and then descends through the inguinal canal (B) and enters the scrotum by the eighth gestational month (C).

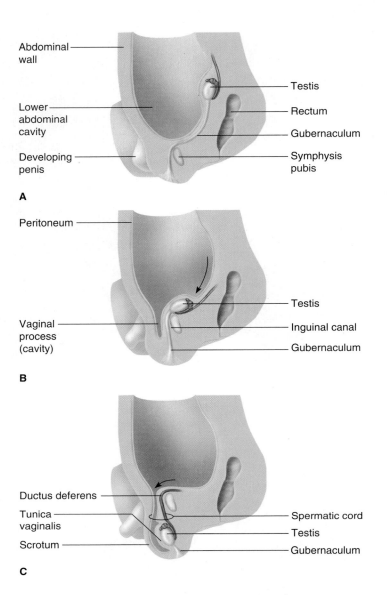

cryptorchidism
A condition in which the testicles did not descend during infancy.

PATHOPHYSIOLOGY

Testicular Cancer

Testicular cancer is a malignant growth of one or both testicles. Unlike prostate cancer, which tends to occur in older males, testicular cancer occurs in more commonly in males 15 to 30 years of age and is a much more aggressive type of malignancy.

Etiology: Predisposing factors include **cryptorchidism** (undescended testicles during infancy). Other risk factors include family history, abnormal development of the testes, Klinefelter's syndrome, certain dyes, and HIV infection.

Signs and Symptoms: A hard, painless lump in the affected testicle is a common early symptom. Patients may complain of groin or abdominal pain as the disease progresses.

Testicular Cancer (concluded)

Treatment: Orchiectomy (removal of the involved testis) is usually performed, followed by radiation therapy and chemotherapy. Caught in the early stages, testicular cancer treatment has up to a 95 percent success rate, verifying the need for males to perform testicular self-exams on a monthly basis.

Penis

The **penis** (Figure 16-3) is a highly vascularized cylindrical organ that propels urine and semen outside the body. The shaft, or body, of the penis contains specialized erectile tissue that surrounds the urethra and runs the length of the penis. The end of the penis enlarges into a cone-shaped structure called the glans penis. If a male has not been circumcised, a piece of skin, called the **prepuce** or foreskin, covers the glans penis. The function of the penis is to deliver sperm to the female reproductive tract. The *motility* of the sperm allows it to move spontaneously through the female reproductive tract towards the egg. As discussed in Chapter 15, The Urinary System, the penis, which contains the urethra, also functions during urination by draining urine from the bladder.

penis A highly vascularized cylindrical organ that propels urine and semen outside the body.

prepuce The foreskin that covers the glans penis in an uncircumcised penis.

U check
To what two organ systems does the male urethra belong?

focus on Clinical Applications: Testicular Self-Exam

As with most cancers, the earlier testicular cancer is discovered, the more likely it is that treatment outcome will be positive. Testicular self-examination, known as TSE, is something every male should do for himself. The steps of TSE are outlined as follows:

1. The best time to perform TSE is during or right after a hot shower. The scrotum is most relaxed at that time, making it easier to examine the testicles. Be aware that it is not unusual for one testicle (often the right one) to be slightly larger than the other one.
2. Examining one testicle at a time, gently roll each testicle between your fingers firmly, but gently. The easiest way to do this is to place your thumbs over the top of the testicle and with your index and middle fingers of each hand behind the testicle, roll it between your fingers.
3. The epididymis, located at the top of the back of each testicle, is not a lump—it normally should feel soft and ropelike and be somewhat tender to touch.
4. Feel for any lumps or unusual bumps on or within the testicles. You should consider a lump to be anything unusual—even as small as a pea or a grain of rice.
5. If at any time you notice any swelling, lump, or change in the size, color, or temperature of your testicles or experience any pain or achiness in your groin, you should let your physician know immediately.

   Remember, it is better to hear "it's nothing to worry about" than "why did you wait?" If something does not feel or seem right, left your physician know.

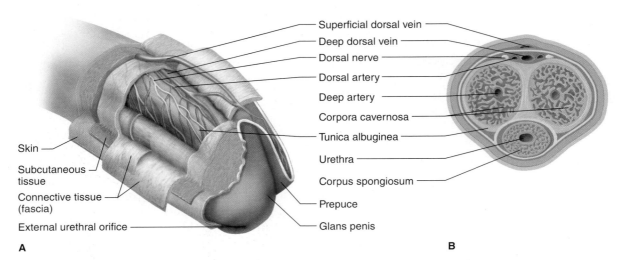

FIGURE 16-3 Structure of the penis: (a) interior and (b) cross section of the penis.

learning outcome 16.2

Identify the internal organs of the male reproductive system including the location, structure, function, and disorders of each.

epididymis A coiled tube at the top of each testis where spermatids become mature sperm cells.

Internal Male Reproductive Structures

The internal accessory organs of the male reproductive system are the epididymis, vas deferens, and the accessory sex glands, which include the seminal vesicles, prostate gland, and bulbourethral, or Cowper's, glands. We will discuss each organ in more detail in the following sections.

Epididymis

An **epididymis** (*didymus* refers to "twin," referencing the fact that there are two testicles) sits on top of each testis (Figure 16-4). It is a highly coiled tube that receives spermatids as they are formed from the seminiferous tubules. Inside the epididymis, spermatids mature to become mature sperm cells.

PATHOPHYSIOLOGY

Epididymitis

Epididymitis is inflammation of the epididymis. Most cases start as an infection of the urinary tract that spreads to the epididymis.

Etiology: Causes include the use of certain medications, placement of a catheter in the urethra, and bacterial infections—particularly those that cause gonorrhea and chlamydia.

Signs and Symptoms: Signs and symptoms include fever, pain in the testes, a lump in the testes, swelling of the scrotum, painful ejaculation, blood in the semen, dysuria (pain during urination), urethral discharge, and enlarged pelvic lymph nodes.

Treatment: Treatment includes pain medication, antibiotics to treat the infection for both the patient *and* his sexual partner, elevation of the scrotum, and ice packs applied to the scrotum.

Vas Deferens

A tubular structure called the **vas deferens** is connected to each epididymis. These tubes carry sperm cells from each epididymis to the urethra. When a male has a **vasectomy,** these tubes are cut and tied, or *fulgurated* (burned and sealed), to prevent sperm from reaching the ovum.

Accessory Sex Glands

The accessory sex glands of the male reproductive system include the seminal vesicles, the prostate gland, and the bulbourethral, or Cowper's, glands.

Seminal Vesicle: The **seminal vesicles** are saclike organs that secrete an alkaline fluid rich in sugars and prostaglandins (tissue hormones) that makes up approximately 60 percent of the semen volume. The sugars are used by sperm cells for energy, and the prostaglandins stimulate muscular contractions in the female reproductive system. These muscular contractions (peristalsis) help propel the sperm forward into the female reproductive tract. The seminal vesicles release their seminal fluid into the vas deferens prior to ejaculation.

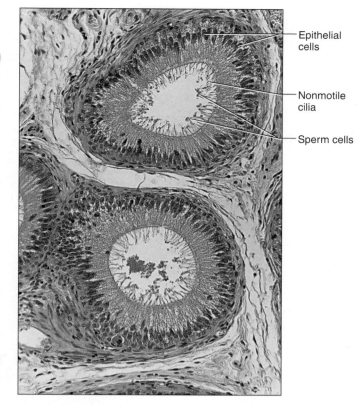

FIGURE 16-4 Cross section of a human epididymis (magnified 200x).

Prostate Gland: The **prostate gland** surrounds the proximal portion of the urethra. It produces a milky, alkaline fluid that is secreted into the urethra just before ejaculation. The alkaline nature of this fluid helps protect the sperm when they enter the more acidic environment of the female vagina. Prostatic fluid makes up approximately 40 percent of semen volume. During ejaculation, the muscular contractions of the prostate assist with the expulsion of semen from the penis.

U check
What is the role of prostaglandins in fertilization?

vas deferens A tubular structure connected to each epididymis that carries sperm cells to the urethra.

vasectomy The cutting and tying or fulguration of the vas deferens to prevent sperm from reaching and fertilizing the ovum.

seminal vesicles Saclike organs that secrete an alkaline fluid rich in sugars and prostaglandins.

prostate gland An organ that surrounds the proximal portion of the urethra and secretes a milky, alkaline fluid just before ejaculation to protect the sperm in the acidic environment of the vagina.

Prostate cancer

PATHOPHYSIOLOGY

Prostate Cancer

Prostate cancer is one of the most common cancers in men older than age 40. The risk of developing prostate cancer increases with age. Awareness of this malignancy is spreading, and access to screenings such as digital rectal exam (DRE) and prostate-specific antigen (PSA) are widely available. Therefore, many cases are being diagnosed even before symptoms present, increasing the possibility of a successful treatment outcome.

(continued)

Prostate Cancer (concluded)

Etiology: The exact cause of prostate cancer is not known, but there are several contributing factors. A high-fat diet, increased age, and genetic predisposition all increase the risk of developing prostate cancer and although testosterone does not cause prostate cancer, at higher levels it can spur its growth.

Signs and Symptoms: Common symptoms include anemia, weight loss, incontinence, difficulty starting or stopping urination, dysuria, pain in the lower back or abdomen, pain during bowel movements, high levels of PSA in the blood, and hematuria. In advanced cases, there may be bone pain caused by cancer cells metastasizing (spreading) to the bone.

Treatment: Treatments include hormone therapy, chemotherapy, radiation therapy to shrink or destroy the tumor, as well as prostatectomy (removal of the prostate).

Prostatitis

Prostatitis is an inflammation of the prostate gland. When this develops suddenly it is called acute prostatitis. Slower development of this condition is called chronic prostatitis.

Etiology: Prostatitis may be caused by bacterial infections, catheterization, trauma to the urethra or urinary bladder, excessive alcohol consumption, and scarring of the urethra or prostate because of frequent infections. Urinating frequently can help to prevent this infection because bacteria are flushed out during urination.

Signs and Symptoms: Signs and symptoms include fever; pain in the scrotum, pelvic area, or abdomen; difficult, frequent, and/or painful urination; hematuria; painful ejaculation; blood in the semen; urethral discharge; a low sperm count (oligospermia); and white blood cells in urine or semen.

Treatment: This condition is often treated with antibiotics. Surgery may also be required to repair any damage to the urethra.

Benign Prostatic Hypertrophy

Benign prostatic hypertrophy (BPH) refers to the nonmalignant enlargement of the prostate gland.

Etiology: At least 50 percent of men in their fifties have BPH, and that percentage rises to 70 percent for men in their seventies. Males produce both testosterone and estrogen, although normally the amount of testosterone is much greater than the amount of estrogen. (Estrogen is predominant in females.) Because testosterone production decreases with age, the ratio of estrogen to testosterone increases. This ratio change may increase the production of

Benign Prostatic Hypertrophy (concluded)

dihydrotestosterone (DHT). DHT in turn promotes growth of the prostate. BPH does not typically transform into prostate cancer.

Signs and Symptoms: Men with BPH often complain of frequent urination (frequency), especially at night (known as nocturia), as well as painful urination (dysuria) and difficulty starting (hesitancy) or stopping the urinary stream, including "dribbling" at the end of urination.

Treatment: Diagnosis is often confirmed by digital rectal exam (DRE), when the physician inserts a gloved finger into the rectum and palpates the prostate. Blood tests (PSA) and a biopsy may be done to rule out cancer. Once malignancy is ruled out, medications such as Avodart® or Flomax® may manage the problem, or a *transurethral resection of the prostate*, commonly abbreviated as *TURP*, may be performed to remove the enlarged tissue.

> **REMEMBER**
> Mark, our banker? What type of treatments might he receive if he is diagnosed with BPH?

Bulbourethral Glands: The *bulbourethral glands*, or *Cowper's glands*, are inferior to the prostate gland. They produce a mucus-like fluid that is secreted before ejaculation into the urethra. This fluid lubricates the end of the penis in preparation for sexual intercourse.

Semen: **Semen** is a mixture of sperm cells and fluids from the seminal vesicles, prostate gland, and bulbourethral glands. This mixture is alkaline to counteract the acidic environment of the vagina and contains nutrients and

> **semen** A mixture of sperm cells and fluids from the seminal vesicles, prostate gland, and bulbourethral glands.

---

focus on Wellness: Prostate Cancer Screening and Risks

Prostate cancer is the second leading cause of death due to cancer in men. Screening starting at age 50 with a DRE is simple and painless. If you are at high risk, African American, or have a family history of prostate cancer, screening should start earlier. The American Urological Association recommends a first-time test at age 40. A PSA is a simple blood test that can identify prostate cancer in the early stages.

Since there may be no symptoms, regular screening is essential to catch the problem early and avoid complications. Know your risk factors. Some are uncontrollable and some are able to be regulated. Advancing age is one risk factor: 80 percent of men who reach age 80 will have prostate cancer. Two-thirds of those diagnosed with prostate cancer are over 65 years of age. Men with relatives who have had prostate cancer are at increased risk. Certain genes may play a role (this accounts for less than 10 percent of all prostate cancers). African American men have a 60 percent greater chance of developing prostate cancer than white men. Smoking and high-fat diets are additional risk factors. Eating a diet high in the antioxidant lycopene (found in high levels in some fruits and vegetables, such as tomatoes and pink grapefruit) may decrease the risk of getting prostate cancer.

prostaglandins. Total semen volume is between 1.5 and 5.0 milliliters per ejaculate, with a sperm count of between 15 million and 250 million per milliliter. A normal sperm count is considered to be more than 80 million per milliliter.

> **U check**
> What are two procedures used to diagnose prostate cancer?

Male Reproductive Hormones

learning outcome
Identify the male reproductive hormones and the function of testosterone.

The hypothalamus, anterior pituitary gland, and testes secrete hormones that regulate male reproductive functions. At the onset of puberty and thereafter throughout a man's life, the hypothalamus (Figure 16-5) releases a hormone called gonadotropin releasing hormone (GnRH). You will learn more about the hypothalamus in Chapter 21, The Endocrine System. GnRH stimulates the anterior pituitary gland to release follicle-stimulating hormone (FSH).

Follicle-Stimulating Hormone and Luteinizing Hormone

Follicle-stimulating hormone (FSH) causes spermatogenesis to begin, and luteinizing hormone (LH) stimulates interstitial cells to produce testosterone. We will discuss the anterior pituitary gland in more detail in Chapter 21, The Endocrine System. Testosterone is responsible for the development of male secondary sex characteristics—those characteristics that are typically unique to males. Examples of male secondary sex characteristics include chest hair, thick facial hair, a thickening and strengthening of muscles and bones, and the thickening of vocal cords that produces a deeper voice. Testosterone also stimulates the maturation of male reproductive organs. Testosterone levels are regulated by negative feedback in the following cycle: Blood testosterone

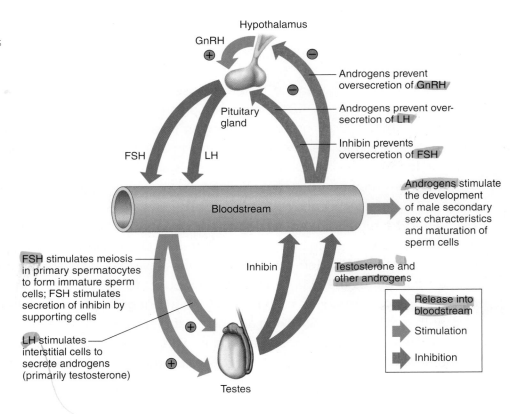

FIGURE 16-5
The hypothalamus controls maturation of sperm cells and development of male secondary sex characteristics. Negative feedback controls the concentration of androgens (steroidal hormones responsible for male secondary sex characteristics, including testosterone) in the body.

levels increase to above normal levels, which cause the hypothalamus to release GnRH. In response, the anterior pituitary ceases the secretion of LH and FSH, which in turn causes the testosterone level to fall. When the testosterone level falls below normal, GnRH is again secreted by the hypothalamus, triggering the release of LH and FSH by the anterior pituitary, and the cycle begins again.

> **U check**
> What hormone stimulates spermatogenesis?

Spermatogenesis

Spermatogenic cells of the seminiferous tubules (Figure 16-6) begin the process of making sperm cells, but the sperm cells do not mature until they travel to the epididymis. **Spermatogenesis** is the process of sperm cell formation. At the beginning of spermatogenesis, the cells are called *spermatogonia*. Spermatogonia contain 46 chromosomes. These cells undergo mitosis, which was discussed in Chapter 3, Concepts of Cells and Tissues, and the resulting cells are called primary spermatocytes. Primary spermatocytes also contain 46 chromosomes. At about the time of puberty, primary spermatocytes undergo meiosis, also discussed in Chapter 3. In meiosis, each primary spermatocyte divides to make two secondary spermatocytes. Each secondary spermatocyte divides to make two spermatids. Therefore, from one primary spermatocyte, four spermatids are formed. Spermatids develop flagella to become mature sperm cells that contain only 23 chromosomes (Figure 16-7).

16.4 learning outcome

Describe how sperm cells are formed.

spermatogenesis
The process of sperm cell formation from spermatogonia.

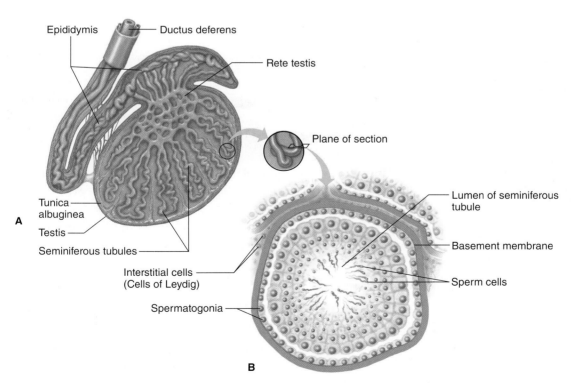

FIGURE 16-6 Structure of the testis: (A) sagittal section of a testis and (B) cross section of a seminiferous tubule.

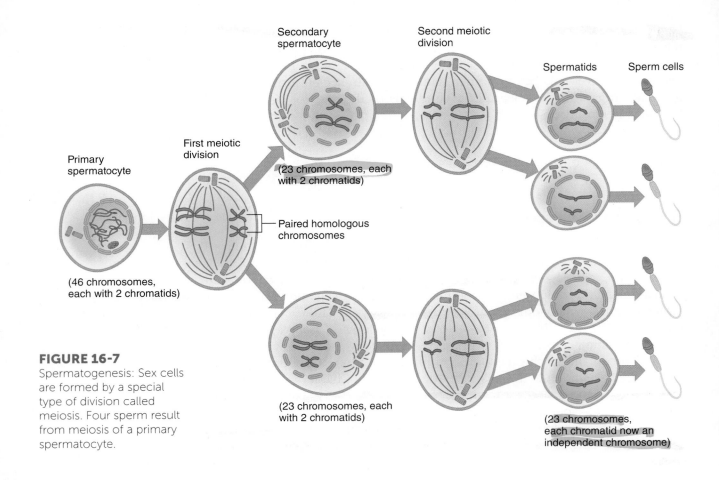

FIGURE 16-7
Spermatogenesis: Sex cells are formed by a special type of division called meiosis. Four sperm result from meiosis of a primary spermatocyte.

acrosome An enzyme-filled sac covering the head of the sperm that helps it penetrate the ovum for fertilization.

Structure of Sperm Cells

A mature sperm has the following three parts: the head, the midpiece, and the tail (Figures 16-8 and 16-9).

The Head: The sperm head is oval in structure and holds a nucleus with 23 chromosomes. The head is covered with an enzyme-filled sac called an **acrosome,** which helps the sperm penetrate an ovum at the time of fertilization.

The Midpiece: This portion of the sperm is between the head and tail. It is filled with mitochondria that generate the energy needed by the sperm cell to propel itself.

The Tail: The tail is a flagellum whose movement propels the sperm forward in the female reproductive tract.

> **U check**
> How many spermatids are produced from each primary spermatocyte?

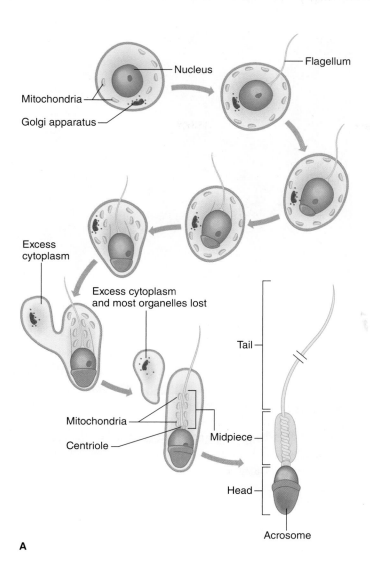
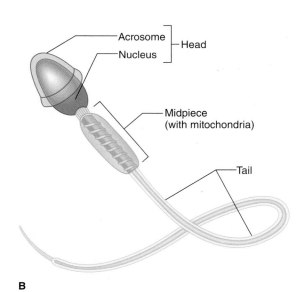

FIGURE 16-8 Spermatogenesis: (A) assembly of the parts of a sperm cell and (B) a mature sperm cell.

FIGURE 16-9 Falsely colored scanning electron micrograph of human sperm cells (magnified 100x).

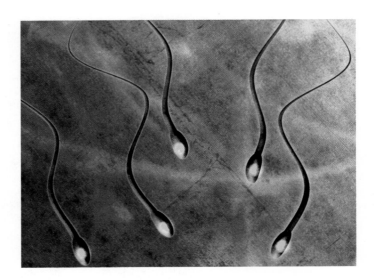

Erection, Orgasm, and Ejaculation

 learning outcome

Explain the processes of erection, orgasm, and ejaculation.

During sexual arousal, the parasympathetic nervous system causes the erectile tissue of the penis to become engorged with blood, producing erection of the penis (Figure 16-10). During orgasm, sperm cells are propelled out of the testes toward the urethra. The secretions of the prostate, seminal vesicles, and bulbourethral glands are also released into the urethra. The combination of the sperm and secretions into the urethra is called emission. The process of ejaculation occurs when semen is forced out of the urethra (Figure 16-11). After ejaculation, sympathetic nerve fibers cause the erectile tissue to release the engorged blood, and the penis gradually returns to a flaccid or nonerect state.

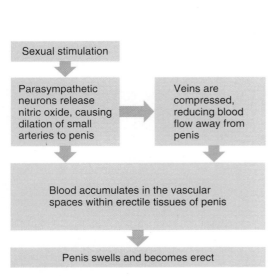

FIGURE 16-10 Mechanism of penile erection.

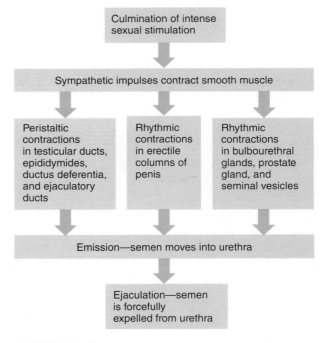

FIGURE 16-11 Mechanism of emission and ejaculation.

PATHOPHYSIOLOGY

Impotence

Impotence, or erectile dysfunction (ED), is a disorder in which a male cannot achieve or maintain an erection to complete sexual intercourse. It is estimated that half of all men between the ages of 40 and 70 years have some degree of impotence. Most causes are physical and not psychological.

Etiology: Reduced blood flow to the penis and/or nerve damage are the two most common physical causes of ED. Other common physical causes include diabetes mellitus, hypertension, anemia, coronary artery disease (CAD), peripheral vascular disease (PVD), low testosterone production, various medications, smoking, excessive alcohol consumption, and use of drugs such as cocaine, marijuana, and heroin. Psychological causes include anxiety, stress, and depression.

Signs and Symptoms: Inability to achieve an erection or maintain an erection long enough to complete sexual intercourse.

Treatment: The first treatment step should be lifestyle changes to quit smoking and stop using alcohol or drugs. Counseling to reduce anxiety and depression may also be helpful. Other treatment options include oral medications such as Viagra or Cialis, and penile injections of medications. Penile implants may be considered if oral medications are not effective.

Life Span

Males may retain their reproductive capabilities very late in their lives if they are healthy. In the mid-fifties, a decline in testosterone production leads to diminished muscle strength, fewer viable sperm, and a decreased sex drive. Diseases such as benign prostatic hypertrophy and prostate cancer also become more common.

learning outcome

Infer the effects of aging on the male reproductive system.

summary

chapter 16

learning outcomes	key points
16.1 Identify the external organs of the male reproductive system including the location, structure, function, and disorders.	The penis, testes, and scrotum are the main external organs of the male reproductive system. The scrotum holds the testicles and the ducts that produce and deliver the sperm to the urethra. The penis delivers the sperm into the vagina.
16.2 Identify the internal organs of the male reproductive system including the location, structure, function, and disorders.	The internal accessory organs of the male reproductive system are the epididymis; vas deferens; and the accessory sex glands, which include the seminal vesicles, prostate gland, and bulbourethral, or Cowper's, glands. Various secretions from these glands help in the production and nutrition of sperm.
16.3 Identify the male reproductive hormones and the function of testosterone.	Testosterone is responsible for the development of male secondary sex characteristics. Testosterone also stimulates the maturation of male reproductive organs. FSH, LS, and GnRH are other male reproductive hormones of importance.
16.4 Describe how sperm cells are formed.	Spermatogenesis starts with spermatogonia, which undergo mitosis to form primary spermatocytes. Through meiosis, four spermatids are produced for each primary spermatocyte.
16.5 Explain the processes of erection, orgasm, and ejaculation.	During sexual arousal, the parasympathetic nervous system causes erectile tissue of the penis to become engorged with blood, which produces erection of the penis. The process of ejaculation occurs when semen is forced out of the urethra. After ejaculation, sympathetic nerve fibers cause the erectile tissue to release blood, and the penis gradually returns to a flaccid or nonerect state.
16.6 Infer the effects of aging on the male reproductive system.	Although men may retain the ability to reproduce into their later years, there is a decrease in testosterone production. Diseases such as benign prostatic hypertrophy and prostate cancer also become more common.

case study 1 questions

Can you answer the following questions that pertain to the case study (Mark Haas) presented earlier in this chapter?

1. Which of the following organs of the male reproductive system is the most likely cause of Mark's problem?
   a. Epididymis
   b. Urethra
   c. Prostate gland
   d. Bulbourethral gland
2. Which is the most likely cause of Mark's urinary urgency and frequency?
   a. Excessive fluid intake
   b. Growth of the prostate gland
   c. Growth of the epididymis
   d. A tumor on the prostate gland
3. Which tests would *least* likely be done on Mark?
   a. DRE
   b. PSA
   c. Biopsy
   d. Pelvic x-ray
4. Which is the *least* likely treatment for Mark?
   a. Avodart®
   b. TURP
   c. Chemotherapy
   d. Flomax®
5. Which statement is accurate regarding the hormones related to BPH?
   a. Decreased estrogen
   b. Increased testosterone
   c. Increased DHT
   d. Decreased DHT

review questions

1. Cryptorchidism is
   a. Undescended testicle
   b. Testicular cancer
   c. Prostatic hypertrophy
   d. Prostate cancer
2. All of the following are true of semen *except*
   a. It is acidic
   b. It is alkaline
   c. It is a mixture of sperm cells and fluids from various glands
   d. It is rich in nutrients
3. All of the following secrete hormones that regulate male reproductive functions *except*
   a. Pineal gland
   b. Hypothalamus
   c. Anterior pituitary gland
   d. Testes
4. Which one of the following stimulates interstitial cells to produce testosterone?
   a. Lutenizing hormone
   b. Growth hormone
   c. Antidiuretic hormone
   d. Prostaglandin
5. The process of making sperm cells begins in the
   a. Seminiferous tubules
   b. Vas deferens
   c. Prostate gland
   d. Cowper glands

critical thinking questions

1. What are some complications that may occur when the testes do not descend into the scrotum?
2. What is the impact on male sterility when the interstitial cells are not functioning properly?
3. How are the head, midpiece, and tail of a mature sperm uniquely suited to their functions?

patient **education**

As a new health care professional, you are assigned to teach young male athletes the importance of and how to perform a testicular exam. Role-play explaining the purpose of a testicular exam, then demonstrate the steps of the self-examination using a manikin or large drawing.

applying **what you know**

1. A 58-year-old man has been diagnosed with BPH; what does this mean?
   a. He has cancer of the prostate gland.
   b. He has nonmalignant enlargement of the prostate gland.
   c. He has an undescended testicle.
   d. He has erectile dysfunction.

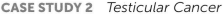

CASE STUDY 2 *Testicular Cancer*

Matt Garcia is a 24-year-old hockey player. Three years ago he was struck in the groin by an opponent's stick. At that time he was told he had testicular torsion (twisting) and a hydrocele (swelling of the scrotum). Recently he has noticed swelling of his glands in his groin with abdominal pain. The team physician finds a small mass on his right testicle. He asks Matt if the testicle has been giving him any trouble. Matt said it seemed swollen but he thought it was because of the injury he received three years ago. Matt undergoes a CT scan and blood test and is diagnosed with testicular cancer. Can you answer the following questions about Matt's condition?

1. What type of treatments will Matt undergo?
2. What self-diagnostic test could Matt have performed for this type of cancer?
3. What condition makes someone have a greater risk of developing testicular cancer?

External Female Reproductive Structures

The female external organs, or genitalia, collectively known as the **vulva**, include the following: mons pubis, labia majora, labia minora, clitoris, urethral meatus, vaginal orifice, Bartholin's glands, and perineum (Figure 17-2).

Structures of the Vulva

Mons pubis: The **mons pubis** is a fatty area that overlies the pubic symphysis.

Labia majora: The **labia majora** are rounded folds of adipose tissue and skin that protect the other external female reproductive organs. At their anterior ends, the labia majora form the mons pubis. The labia majora and mons pubis are typically covered in pubic hair in postpubescent females.

Labia minora: The **labia minora** are the folds of skin between the labia majora. They are pinkish in color because of their high degree of vascularity. They merge together anteriorly to form a hood over the clitoris. The space enclosed by the labia minora is called the **vestibule.** The *Bartholin's glands,* sometimes referred to as the vestibular glands, secrete mucus into this area during sexual arousal. This mucus eases insertion of the penis into the vagina.

Clitoris: The **clitoris** is anterior to the urethral meatus. It contains the female erectile tissue and is rich in sensory nerves.

Urethral meatus: The urethral meatus is the opening in the urethra through which urine is secreted.

Vaginal orifice: The vaginal orifice is the opening to the vagina.

Bartholin's glands: The Bartholin's glands are on either side of the vaginal orifice. These glands secrete mucus for lubrication.

17.1 learning outcome

Identify the external organs of the female reproductive system including the location, structure, function, and disorders.

vulva The external genitalia of the female.

mons pubis The rounded fatty prominence over the pubic symphysis.

labia majora Rounded folds of adipose tissue and skin that protect the other external female reproductive organs.

labia minora The folds of skin between the labia majora.

vestibule The small space at the beginning of the vagina.

clitoris An erectile organ of the female, homologous to the male penis.

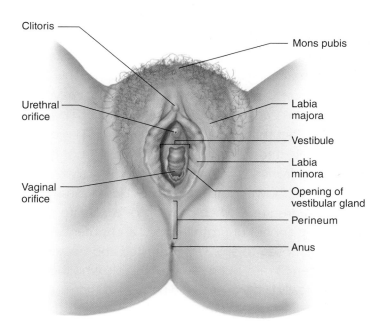

FIGURE 17-2
External female genitalia.

perineum The pelvic floor; the space between the anus and the vulva in the female and the anus and the scrotum in the male.

episiotomy A surgical incision to avoid tearing of the perineum during labor.

Perineum: The **perineum** is the area between the vagina and anus. This is the area that is sometimes "clipped" during the birth process in a procedure known as an **episiotomy**. An episiotomy is done to avoid tearing of the perineum.

Table 17-1 summarizes the major female reproductive organs and their functions.

> **U check**
> What is the purpose of an episiotomy?

Erection, Lubrication, and Orgasm: During sexual arousal, nervous system stimulation causes the clitoris to become erect and the Bartholin's

---

focus on Wellness: STI Prevention and Treatment

Sexually transmitted infections (STI) occur equally in both sexes but do appear in many cases to "play favorites," with women receiving the bulk of the symptoms and the worst of the sequelae (long-term effects) in the form of pelvic inflammatory disease (PID), infertility, and cervical and other gynecological-related cancers. Some of the most well-known STIs include AIDS, caused by HIV (human immunodeficiency virus); genital warts caused by HPV (human papillomavirus); gonorrhea, resulting from exposure to *Neisseria gonorrhea*; herpes simplex 1 (generally oral herpes) and 2 (generally enital herpes); pubic lice (parasitic infection); and syphilis caused by the bacterium *Treponema pallidum*.

Keep in mind that there are many, many more STIs out there—these are just a few of the commonly known infections. The signs and symptoms vary depending on the disease contracted, but for women, commonly there is a vaginal discharge (which may be greenish or yellowish in color or frothy in consistency and contain a strong "fishy" odor), dysuria, dyspareunia, abdominal pain, vulvovaginal itching and redness, and intermenstrual bleeding. Men often have few symptoms. When symptoms do appear, commonly they are in the form of a whitish, yellowish, or greenish penile discharge and/or dysuria.

In both male and female patients, herpes infections result in a vesicle (blister) appearing at the site of infection; HPV infection results in a characteristic wart in the genital area; pubic lice are apparent with visible eggs and crawling lice in the pubic hair; the only symptom of early syphilis infection is a painless chancre (sore) in the genital area that may not even be noticed. Untreated or maltreated infections can all result in spread of the STI, and many are spread before the patient is aware of the fact that an STI exists. When an STI is diagnosed, all exposed partners must be treated and retested so that re-exposure does not occur.

All STIs can be prevented by abstaining from sex, although we all know that is a challenge that is difficult to meet. The best treatment, however, is prevention. Limit sexual partners, use safer sex practices, and limit alcohol (and drug use) when partying or on a date. Studies show that drugs and alcohol often lead to (unplanned) unprotected sex and potential exposure to STIs. If you believe you may have been exposed to any STI, get tested and treated when necessary. People often believe "it can't happen to me," but millions of people who thought it could not happen to them are infertile or chronically ill because "it did happen to them," and diagnosis and treatment, when it did occur, was too late to avoid the long-term effects of the STI.

TABLE 17-1 Functions of the Female Reproductive Organs

Organ	Function
Ovary	Produces oocytes and female sex hormones
Uterine tube	Conveys secondary oocyte toward uterus; site of fertilization; conveys developing embryo to uterus
Uterus	Protects and sustains embryo during pregnancy
Vagina	Conveys uterine secretions to outside of body; receives erect penis during sexual intercourse; provides open channel for offspring during birth process
Labia majora	Enclose and protect other external reproductive organs
Labia minora	Form margins of vestibule; protect opening of vagina and urethra
Clitoris	Produces feelings of pleasure during sexual stimulation due to abundant sensory nerve endings
Vestibule	Space between labia minora that contains the vaginal and urethral openings
Vestibular glands	Secrete fluid that moistens and lubricates the vestibule

glands to become active. At the same time, the vagina elongates. If the clitoris is sufficiently stimulated, an orgasm occurs. During orgasm, the walls of the uterus and fallopian tubes contract to help propel sperm toward the upper ends of the fallopian tubes.

External Accessory Organs of the Female Reproductive System: *Mammary glands* are the accessory organs of the female reproductive and integumentary systems (Figure 17-3). Their reproductive function is

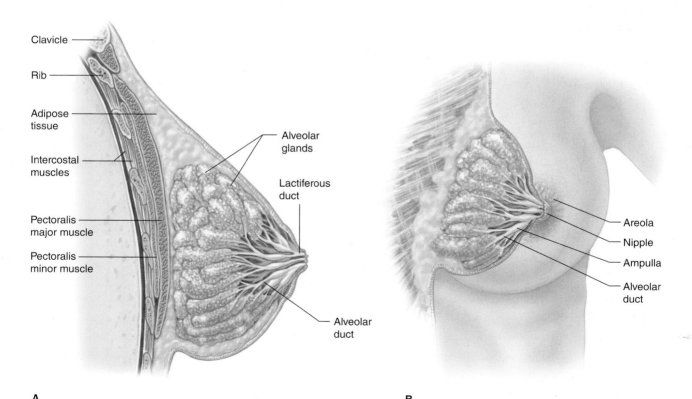

FIGURE 17-3 Female breast and mammary glands: (A) sagittal section and (B) anterior view.

the secretion of milk for newborn offspring. Mammary glands are located beneath the skin of the chest. A nipple is located near the center of each breast. The pigmented area that surrounds the nipple is called the *areola*. Each gland is made of 15 to 20 lobes and contains alveolar glands that make milk under the influence of the hormone prolactin. The hormone oxytocin (OT), which is released from the posterior pituitary, induces lactiferous ducts to deliver milk through openings in the nipples. To breast-feed, a woman must produce adequate amounts of prolactin and oxytocin. We will discuss these hormones and the posterior pituitary in more detail in Chapter 21, The Endocrine System.

Breast cancer

PATHOPHYSIOLOGY

Breast Cancer

According to the American Cancer Society, breast cancer is the second leading cause of cancer deaths in women, after lung cancer. As with all cancers, breast cancers are evaluated and graded based on the tumor size and how far cells have traveled from the site of origin. In stage 0, there is no evidence of cancer cells beyond the origin of the cancer in the breast (known as *in situ*). Stage IV describes invasive (breast) cancer that has spread to other organs such as lungs, liver, bone, and/or brain. Early diagnosis through regular mammograms, breast ultrasound (which is more effective in finding early cancers in dense or fibrous breasts), and breast self-exams greatly increases the success of treatment for breast cancer.

Etiology: Although the exact causes of breast cancer are largely unknown, several risk factors have been identified. These include increased age, genetics, high-fat diet, longer than "normal" exposure to estrogen (birth control pills), family history of breast cancer, alcohol use, and cigarette smoking.

Signs and Symptoms: Signs and symptoms include a lump in the breast that is usually painless and firm, a lump in the armpit, discharge from the nipples, dimpled skin on the breast or nipple, and breast pain. Swelling of the breast into the adjacent arm and bone pain may be present in advanced cases. Inflammatory breast cancer may present only as a painless rash of the affected breast with none of the other typical symptoms.

Treatment: Nonsurgical treatment methods include hormone therapy, radiation therapy, and chemotherapy. Surgical options include surgery to remove affected lymph nodes, lumpectomy (surgery to remove a lump), and mastectomy (surgery to remove a breast).

Fibrocystic Breast Change

Fibrocystic breast change was formerly known as *fibrocystic breast disease.* However, because it is so common—more than 60 percent of women in the United States between the ages of 30 and 50 have

REMEMBER LARA, our 47-year-old who discovered a lump in her breast? What are some risk factors for breast cancer?

Fibrocystic Breast Change (concluded)

the condition—it is now considered a "change" rather than a "disease." It consists of the presence of abnormal cysts in the breasts that vary in size related to the menstrual cycle. Because it is related to hormonal changes occurring during the menstrual cycle, it is rare in women who have gone through menopause. Thus, it is not considered to be precancerous.

Etiology: This disorder is caused by hormonal changes associated with the menstrual cycle and ingestion of various dietary substances including caffeine, nicotine, chocolate, and sugar. Birth control pills may also aggravate this condition.

Signs and Symptoms: Common symptoms include breasts that feel "lumpy"; breast tenderness or pain, particularly premenstrually; itchy nipples; and dense tissues as seen on a mammogram. The masses are usually firm and vary in size related to the stage of the menstrual cycle.

Treatment: Treatments include changing one's diet and taking vitamin supplements such as vitamin E, B complex, and magnesium. Pain may be controlled by wearing support bras. Premenopausal women should also be advised to schedule mammograms after their **menses** so that an uncomfortable procedure is not made more unpleasant because of breast tenderness.

menses A woman's menstrual flow.

Internal Female Reproductive Structures

The major internal female reproductive organs are the ovaries, fallopian tubes, uterus, cervix, and vagina, which are described in detail in the following sections. Not only are these organs important for reproductive purposes, but they are responsible for much female illness and causes of death.

 17.2 learning outcome

Identify the internal organs of the female reproductive system including the location, structure, function, and disorders.

Ovaries and Ovum Formation

The ovaries are the primary sex organs of the female. They produce the female sex cells, called *ova*. They also produce *estrogen* and *progesterone*, the female hormones. Most females have two ovaries. They are oval in shape and are located in the pelvic cavity (Figure 17-4). Each ovary is divided into an inner region called the medulla and an outer region called the cortex. The medulla contains nerves, lymphatic vessels, and many blood vessels. The cortex contains small masses of cells called ovarian follicles. Epithelial tissue and dense connective tissue cover each ovary. Before a female is born, *primordial follicles* develop in her ovarian cortex. Each primordial follicle contains a large cell called a primary *oocyte* (immature ovum) and smaller cells called *follicular cells*. Unlike males, who make sperm cells throughout their entire life, a female is born with the maximum number of primary oocytes she will ever produce.

> **U check**
> What is the difference between a primordial follicle and a primary oocyte?

FIGURE 17-4
The ovaries are located on each side against the lateral walls of the pelvic cavity. The right fallopian tube is removed to illustrate the ovarian ligament.

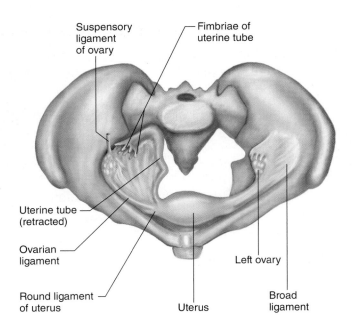

PATHOPHYSIOLOGY

Ovarian Cancer

Ovarian cancer is considered more deadly than the other types of gynecological cancers because its signs and symptoms are usually mild or indistinct until the disease has metastasized (spread) to other organs. Current statistics suggest that about 1 woman in 67 will develop ovarian cancer. Ovarian cancer primarily affects middle- and upper-class women in industrialized nations probably in large part because of diet and obesity. In the United States, it is the fifth most common cancer in women and most often appears in women who are older than 60.

Etiology: The causes are unknown, although the presence of certain genes has been indicated as a risk factor. Other possible risk factors include being white or postmenopausal; having a family history of colon, breast, or ovarian cancer; and using hormone replacement therapy. Some oral contraceptives may lower the risk of developing this disease.

Signs and Symptoms: Ovarian cancer is sometimes described as "the silent killer" because there may be no early signs and symptoms. Some of the signs and symptoms that may be seen are abdominal and pelvic discomfort, unusual menstrual cycles, indigestion, bloating, nausea, and excessive hair growth. Diagnosis is usually made after pelvic ultrasound, a CA 125 blood test, and ovarian biopsy are performed. CA 125 is a protein tumor marker that is elevated in 80 percent of women with advancing

> **Ovarian Cancer** (concluded)
>
> ovarian cancer. A biopsy is the definitive method of making the diagnosis.
>
> **Treatment:** Treatment options include surgery to remove the ovaries and reproductive organs followed by radiation therapy and chemotherapy. Currently, the five-year survival rate for all stages of ovarian cancer is only about 45 percent.

Oogenesis is the process of ovum formation. At the onset of puberty, some primary oocytes are stimulated to continue meiosis (Figure 17-5). When a primary oocyte divides, it becomes one *polar body* (a nonfunctional cell) and a secondary oocyte. It is the secondary oocyte that is released from an ovary each month during *ovulation* (Figure 17-6). When the secondary oocyte is fertilized, it divides to form a mature, fertilized ovum. Therefore,

oogenesis Formation of the female gametes.

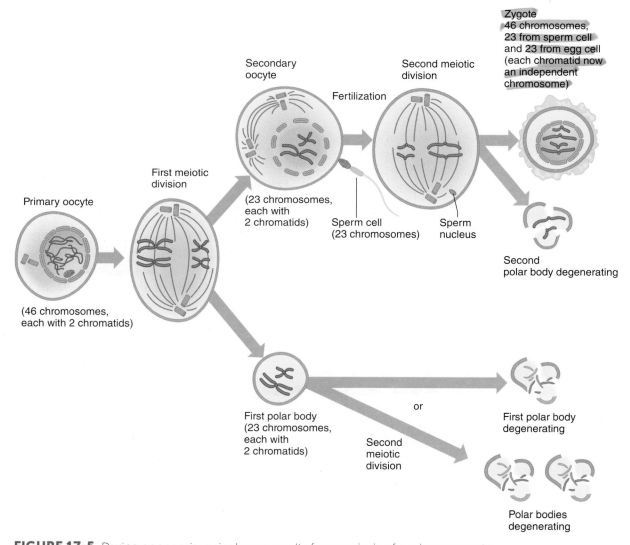

FIGURE 17-5 During oogenesis, a single egg results from meiosis of a primary oocyte.

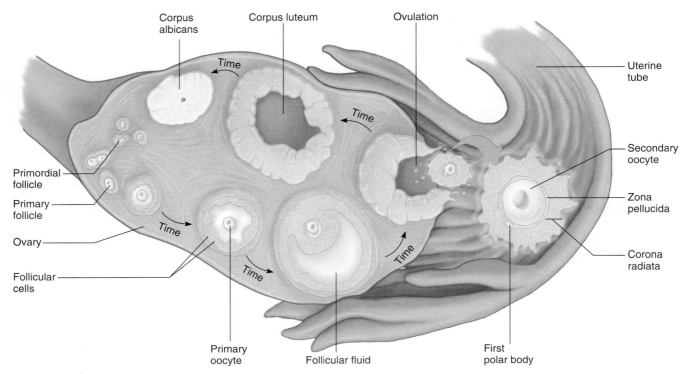

FIGURE 17-6 A developing oocyte becomes surrounded by follicular cells and fluid. The mature follicle will rupture, releasing the secondary oocyte.

the process of meiosis begins before a female is born and is completed only if a secondary oocyte is fertilized. The mature ovum contains 23 chromosomes; when it combines with a sperm cell, the resulting cell contains 46 chromosomes and is called a **zygote.**

zygote The fertilized ovum.

fallopian tube An oviduct, the tube that runs from the uterus to the ovary.

Fallopian Tubes

A **fallopian tube**, or *oviduct,* opens near each ovary and the other end connects to the uterus. The fringed, expanded end of a fallopian tube near an ovary is called the *infundibulum* (Figure 17-7) and **fimbriae.** The infundibulum and its fimbriae function to "catch" an ovum as it leaves an ovary. Fallopian tubes are muscular tubes that are lined with a mucous membrane and cilia. This construction allows the tube to propel the ovum toward the uterus using peristalsis and the sweeping motions of cilia.

fimbriae Finger-like structures near the lateral ends of the fallopian tubes.

uterus The hollow, muscular organ in the female that is the site of menstruation, implantation of the embryo, and development of the fetus.

Uterus and Cervix

The **uterus** is the hollow, muscular organ that receives a developing embryo and sustains its development. The upper domed portion of the uterus is called the **fundus;** the main portion is called the body; and the narrow, lower portion that extends into the vagina is called the **cervix.** The opening of the cervix is called the cervical orifice, or cervical os.

fundus The part of the uterus farthest from the os.

cervix The inferior, constricted part of the uterus.

U check
What is a polar body?

endometrium The mucous membrane lining the uterus.

The wall of the uterus has three layers—the endometrium, myometrium, and perimetrium (Figure 17-8). The **endometrium** is the innermost lining of

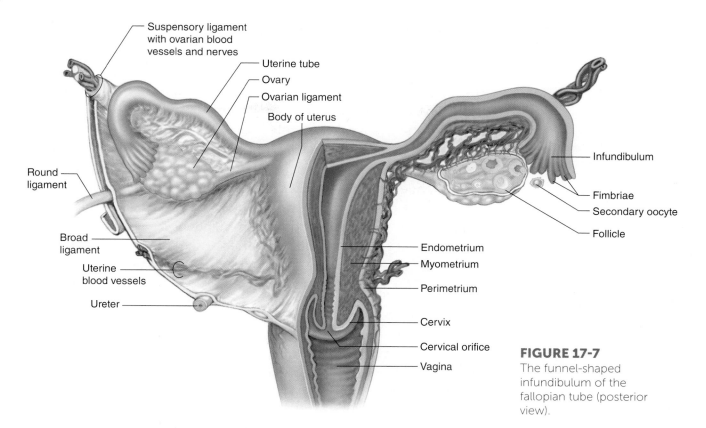

FIGURE 17-7 The funnel-shaped infundibulum of the fallopian tube (posterior view).

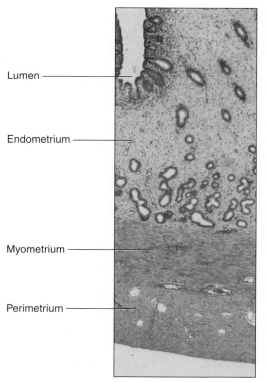

FIGURE 17-8 Light micrograph of the uterine wall (magnified 10x).

the uterus. It is vascular with a rich blood supply and also contains numerous tubular glands that secrete mucus. The **myometrium** is the middle, thick, muscular layer. The **perimetrium** is a thin, serous layer that covers the myometrium. It secretes serous fluid that coats and protects the uterus.

myometrium The smooth muscle layer of the uterus.

perimetrium The serosa of the uterus.

CHAPTER 17 The Female Reproductive System

PATHOPHYSIOLOGY

Cervical Cancer

Most cervical cancers generally develop slowly. With early detection by a yearly "Pap" (Papanicolaou) smear (a microscopic test that looks for abnormal cervical cells that have been *smeared* onto a microscope slide), treatment is often successful. In 2009, the American College of Obstetricians and Gynecologists (ACOG) revised its recommended cervical cancer screening schedule for women. It now recommends that women should have their first Pap smear at age 21. The old guidelines recommended women have their first Pap three years after they become sexually active or at age 21, whichever came first. It should also be noted that the new recommendation for Pap smear screenings is now every other year if a woman's previous screenings have been negative for five years and if the woman is in a monogamous relationship and not on birth control pills.

Etiology: Infection with human papillomavirus (HPV) that is transmitted sexually is the greatest single risk factor for cervical cancer. Risk factors also include sexual intercourse early in life and multiple sexual partners. In addition, if a woman has had very few sexual partners, but her partner has had multiple partners, this may also increase her risk of developing cervical cancer. A weak immune system may also be a factor in the development of this cancer.

Signs and Symptoms: Primary symptoms include frequent vaginal discharge, sporadic vaginal bleeding, vaginal bleeding after sexual intercourse, and abnormal cells in the cervix. Patients who are in later stages of this disease may experience pain in the pelvic area or legs, or bone fractures.

Treatment: Radiation therapy, chemotherapy, the removal or destruction of diseased tissue with cryosurgery or laser surgery, and *hysterectomy* (the removal of the uterus) are the treatments for this disease.

Cervicitis

Cervicitis is defined as an inflammation of the cervix, which is usually caused by an infection.

Etiology: Causes include bacterial or viral infections and allergic reactions to spermicidal creams and latex condoms.

Signs and Symptoms: Frequent vaginal discharge, pain during intercourse, and vaginal bleeding after intercourse are common signs and symptoms.

Treatment: This condition is treated with antibiotics and by changing the method of contraception if it is a causative factor.

Endometriosis

Endometriosis is a condition in which tissues that make up the lining of the uterus grow outside the uterus.

Etiology: The cause of this disorder is unknown, but there are several theories. These include *retrograde menstruation* (bleeding back up into the fallopian tubes), metaplasia (the change of one adult cell type to another adult cell type), genetic predisposition, and environmental factors.

Signs and Symptoms: Signs and symptoms include infertility, heavy bleeding from the uterus, pain in the abdomen or pelvis, painful periods, spotting between periods, and *dyspareunia* (pain during sexual intercourse).

Treatment: Oral contraceptives, pain medications, and various hormone therapies may be prescribed. Surgical treatments include laser surgery to remove endometrial tissue outside the uterus and hysterectomy.

Fibroids

Fibroids are benign (noncancerous) tumors that grow in the uterine wall. Their medical name is *leiomyomas* (benign tumors of smooth muscle). They are known to affect one out of four women in their thirties and forties and appear to be more common in women of African descent.

Etiology: The causes are mostly unknown, although it has been found that tumors enlarge as estrogen levels increase. Heredity does appear to play a role.

Signs and Symptoms: The signs and symptoms are pressure in the abdomen, severe menstrual cramps, abdominal gas, and heavy menstrual and intermenstrual bleeding. Back and leg pain may also occur.

Treatment: Treatment includes pain medications, hormone treatments to shrink tumors, surgery to remove tumors, hysterectomy, and surgery to decrease the blood supply to the uterus.

Uterine Cancer

Uterine (endometrial) cancer is most common in postmenopausal women. In the United States, it accounts for approximately 6 percent of cancer deaths in women.

Etiology: The causes are mostly unknown, although endometrial cancer may be related to increased levels of estrogen.

Signs and Symptoms: Signs and symptoms include abdominal pain, abnormal bleeding from the uterus, pelvic pain, and a thin, white vaginal discharge in postmenopausal women.

Treatment: Treatment includes radiation therapy, chemotherapy, and surgery to remove the uterus, fallopian tubes, and ovaries.

Vagina

vagina The muscular, tubular organ that extends from the uterus to the vestibule.

The **vagina** is a tubular, muscular organ that extends from the uterus to outside the body. The muscular folds of the vagina, called rugae, allow it to expand to receive an erect penis during sexual intercourse, and provide a passageway for delivery of offspring as well as for uterine secretions. The opening of the vagina is posterior to the urinary opening and anterior to the anal opening. The wall of the vagina has three layers—an inner mucosal layer that secretes mucus, a middle muscular layer, and an outer fibrous layer. The opening of the vagina to the outside is known as the vaginal os, vaginal orifice, or vaginal introitus.

> ## PATHOPHYSIOLOGY
> ### Vaginitis
>
> *Vaginitis* (inflammation of the vagina) and *vulvovaginitis* (inflammation of the vulva and vagina) are usually associated with an abnormal vaginal discharge. Some vaginal discharge is normal for all women, and it varies throughout the menstrual cycle. Normal vaginal discharge is clear, whitish, or yellowish in color.
>
> **Etiology:** This condition can be caused by yeast infections, tampon use, poor hygiene, bacteria, antibiotics, and STIs (sexually transmitted infections). Many cases of vaginitis may be prevented through good hygiene and safer sex practices.
>
> **Signs and Symptoms:** Common symptoms include fever, vulvar and vaginal itching and swelling, an abnormal increase or decrease in the amount of vaginal discharge, an abnormal color of vaginal discharge (brown, green, or pinkish), a change in the consistency of vaginal discharge (frothy or cheeselike), and vaginal discharge that has an abnormal odor often described as "fishy."
>
> **Treatment:** The patient may be given medications for fungal or bacterial infections, or the patient and her sexual partner may be treated for STIs.

Female Reproductive Hormones

17.3 learning outcome

Compare the actions of the hormones of the female reproductive system.

Beginning at puberty, the hypothalamus secretes increasing amounts of GnRH (Figure 17-9). This causes the anterior pituitary gland to release FSH and LH, which stimulate the ovary to produce estrogen and progesterone, and also to mature the ovarian follicles. The estrogen and progesterone are responsible for the female secondary sex characteristics: breast development, increased vascularization of the skin, and increased fat deposits in the breasts, thighs, and hips. The hypothalamus and pituitary glands will be discussed further in Chapter 21, The Endocrine System.

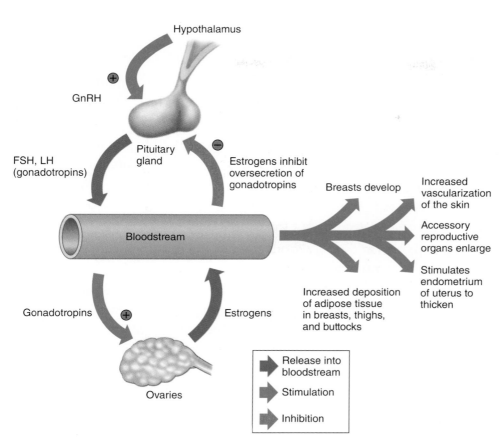

FIGURE 17-9
Control of female secondary sexual development: LH and FSH during most of the reproductive cycle except during egg cell production and ovulation.

Female Reproductive Cycle

The female reproductive cycle is also called the menstrual cycle. It consists of regular changes in the uterine lining that lead to a monthly "period," or bleeding. The first menstrual period is known as **menarche. Menopause** is the termination of the menstrual cycle related to the normal aging of the ovaries. The female reproductive cycle is divided into four phases: the menstrual phase, the preovulatory phase, ovulation, and the postovulatory phase. The typical cycle is between 24 and 35 days. We will use the average of 28 days for simplicity in this discussion.

> **U check**
> What is the difference between menarche and menopause?

17.4 learning outcome

Explain the purpose and events of the menstrual cycle.

menarche The first menses.

menopause The termination of the menstrual cycles.

1. *Menstrual phase:* The menstrual phase is also called the *menses* (which means "month"), or menstruation. This phase lasts the first 5 days of the cycle and includes the following changes in the ovaries and the uterus. In the ovaries, primordial follicles develop into primary and secondary follicles under the influence of FSH secreted by the anterior pituitary gland. Progesterone and estrogen levels decline and the uterine endometrium is shed. This causes the bleeding that is seen during this time.

Uterine endometrium is shed during the Menstrual phase

CHAPTER 17 The Female Reproductive System

> **focus on** Clinical Applications: Breast Self-Exam
>
> It is a well-known fact that the earlier breast cancer is diagnosed and treatment started, the more successful the outcome. This important technique is outlined as follows, and is shown in Figure 17-10:
>
> 1. Standing in front of a mirror with your arms at your sides, inspect each breast for anything unusual: nipple discharge or puckering, dimpling, or scaling of the skin.
> 2. Still standing in front of the mirror, clasp your hands behind your head, pressing them forward, so that your chest muscles tighten. Look for any changes in your breasts, particularly changes that occur in one breast but not the other.
> 3. Place your hands firmly on your hips and bow slightly toward the mirror, pulling your shoulders and elbows forward—any changes now?
> 4. (Some women prefer to do this step in the shower with soapy fingers.) Raise your left arm and, using three or four fingers of your right hand, explore your left breast firmly, carefully, and thoroughly. Start at the outer edge and press the flat part of your fingers in small circles moving around the breast and into and including the armpit. Feel for any unusual lumps or thickening.
> 5. Squeeze the nipple gently and look for any discharge. Repeat steps 4 and 5 on the right breast using your left hand.
> 6. Now lie on your back with your left arm over your head and a pillow or folded towel under your left shoulder. Repeat steps 4 and 5 first on the left breast and then on the right.
> 7. Repeat this exam monthly at the same time every month. Women who still have their periods often do the exam as the menstrual period ends because the breasts are less tender at that time. Menopausal women should pick the same date every month to perform the exam so it is not forgotten. Any changes or "suspicions" should be reported to your physician as soon as possible.

2. *Preovulatory phase*: This is the time between the end of the menses and ovulation, which can last from 6 to 13 days. Secondary follicles secrete estrogen and a dominant follicle appears. This follicle becomes the mature follicle. The first two phases are grouped as the follicular phase because of the growth of the follicles. The uterine endometrium thickens considerably during this time, giving rise to the term *proliferative phase* of the uterus.

3. *Ovulation*: Ovulation occurs at the middle of the cycle or about on day 14. Estrogen causes the anterior pituitary to secrete LH, which causes the ovary to expel the mature follicle.

4. *Postovulatory phase*: This is the longest phase of the cycle, lasting from day 15 to day 28. After ovulation, the mature follicle starts to break down and forms the **corpus luteum,** which secretes estrogen and progesterone as well as other hormones such as relaxin and inhibin. The corpus luteum lasts only two weeks if fertilization does not occur. It is then transformed into the corpus albicans, a white fibrous scar tissue. This phase of the ovarian cycle is also called the luteal phase. If fertilization takes place, the corpus luteum does not degenerate due

corpus luteum A yellow body in the ovary formed when a secondary follicle has been discharged from the follicle.

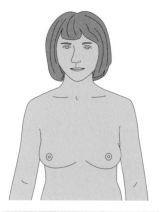
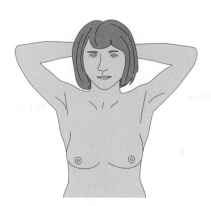
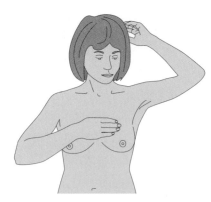
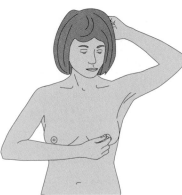
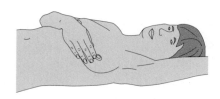

① Stand before a mirror. Inspect both breasts for anything unusual such as any discharge from the nipples or puckering, dimpling, or scaling of the skin.

The next two steps are designed to emphasize any change in the shape or contour of your breasts. As you do them, you should be able to feel your chest muscles tighten.

② Watching closely in the mirror, clasp your hands behind your head, and press your hands forward.

③ Next, press your hands firmly on your hips, and bow slightly toward your mirror as you pull your shoulders and elbows forward.

Some women do the next part of the exam in the shower because fingers glide over soapy skin, making it easy to concentrate on the texture underneath.

④ Raise your left arm. Use three or four fingers of your right hand to explore your left breast firmly, carefully, and thoroughly. Beginning at the outer edge, press the flat part of your fingers in small circles, moving the circles slowly around the breast and the underarm, including the underarm itself. Feel for any unusual lump or mass under the skin.

⑤ Gently squeeze the nipple and look for a discharge. (If you have any discharge during the month—whether or not it is during BSE—see your doctor.) Repeat steps 4 and 5 on your right breast.

⑥ Steps 4 and 5 should be repeated lying down. Lie flat on your back with your left arm over your head and a pillow or folded towel under your left shoulder. This position flattens the breast and makes it easier to examine. Use the same circular motion described earlier. Repeat the examination on your right breast.

FIGURE 17-10 The National Cancer Institute includes these instructions and illustrations in the brochure *Breast Exams: What You Should Know* (NIH Publication No. 90-2000).

to **human chorionic gonadotropin (hCG)** that is secreted by the developing embryo. The presence of hCG is one method used to determine pregnancy. During this time, the uterus thickens and glycogen is secreted by endometrial glands. This time of the uterine cycle is therefore called the secretory phase.

human chorionic gonadotropin (hCG) A hormone produced by the placenta that maintains the corpus luteum.

U check
What are the phases of the menstrual cycle?

CHAPTER 17 The Female Reproductive System

Here is a summary of the major hormonal changes seen during the female reproductive cycle (Figure 17-11):

1. The anterior pituitary gland releases FSH, which stimulates an ovarian follicle to mature.
2. The maturing follicle secretes estrogen. Estrogen causes the uterine lining to thicken.
3. The anterior pituitary gland releases a sudden surge of LH, which triggers ovulation.
4. Following ovulation, follicular cells of the follicle become a corpus luteum.
5. The corpus luteum secretes progesterone, which causes the uterine lining to become more vascular and glandular.
6. If the released oocyte is not fertilized, the corpus luteum degenerates, causing estrogen and progesterone levels to fall.
7. The decline in estrogen and progesterone levels causes the uterine lining to break down, and menses (bleeding) starts.
8. When the anterior pituitary releases FSH, the reproductive cycle begins again.

PATHOPHYSIOLOGY

Dysmenorrhea

Dysmenorrhea is the condition of experiencing severe menstrual cramps that limit normal daily activities.

Etiology: Causes include anxiety, endometriosis, pelvic inflammatory disease (PID), fibroid tumors in the uterus, ovarian cysts, abnormally high levels of prostaglandins, and multiple sexual partners.

Signs and Symptoms: Common symptoms are abdominal pain, including sharp or dull pain in the pelvic area just prior to and during the menstrual period.

Treatment: Nonsurgical treatments include pain medication, anti-inflammatory drugs, medications that inhibit prostaglandin formation, oral contraceptives, and antibiotics in the case of PID. Surgical treatments include surgery to remove cysts or fibroids and hysterectomy.

Premenstrual Syndrome

Premenstrual syndrome (PMS) is a collection of symptoms that occur just before a menstrual period.

Etiology: The causes are mostly unknown, although hormone fluctuations during the menstrual cycle are implicated.

Premenstrual Syndrome (concluded)

Signs and Symptoms: The signs and symptoms include anxiety, depression, irritability, acne, fatigue, food cravings, bloating, aches in the head or back, abdominal pain, breast tenderness, muscle spasms, diarrhea, weight gain, and loss of sex drive.

Treatment: PMS is commonly treated with pain medications, diuretics, medications to treat depression or anxiety, and oral contraceptives. Many women have also found changes in diet, including limiting caffeine, sugar, and sodium, to be helpful. The addition of B complex vitamins may also be helpful.

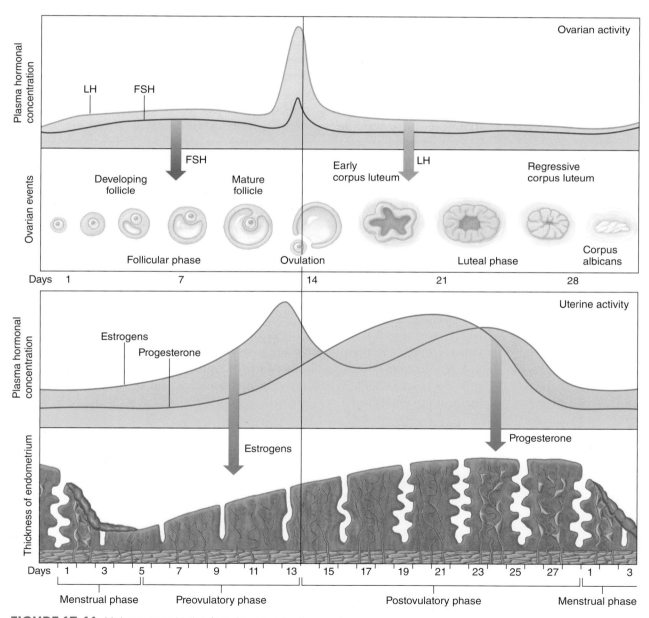

FIGURE 17-11 Major events in the female reproductive cycle.

learning outcome

Explain how and where fertilization occurs.

Fertilization

Pregnancy is defined as the condition of having a developing offspring in the uterus. Pregnancy results when a sperm cell unites with an ovum in a process called *fertilization*. Prior to fertilization, an ovum is released from an ovary and travels through a fallopian tube. During sexual intercourse, the male deposits semen into the vagina. Sperm cells must travel up through the uterus to the fallopian tubes to fertilize the ovum. Prostaglandins in semen stimulate the flagella of sperm cells to undulate, causing the swimming action of sperm. Prostaglandins also stimulate the smooth muscle in the uterus and fallopian tubes to contract. These contractions (peristalsis) help the sperm reach the ovum. Normally about 10 to 14 days after ovulation, high estrogen levels stimulate the uterus and cervix to secrete a thin, watery fluid that also promotes the movement of sperm toward the ovum.

Although many sperm cells normally reach an ovum, only one sperm cell unites with the ovum, penetrating the follicular cells and a layer called the *zona pellucida*, which surround the cell membrane of the ovum. The acrosome of this sperm releases lytic enzymes to help the sperm penetrate the membrane of the ovum. Once a sperm unites with an ovum, enzymes are released by the ovum, causing the zona pellucida to become hard and therefore impenetrable to other sperm. The nucleus of the ovum (with 23 chromosomes) and the nucleus of the sperm (with 23 chromosomes) fuse to make one nucleus that contains 46 chromosomes. The cell that is formed by this union is called a *zygote*.

> **U check**
> Where does fertilization normally take place?

learning outcome

Recall birth control methods and reasons for infertility.

Birth Control and Infertility

Birth Control

Birth control is not a new concept; it has been practiced for centuries. Birth control methods, also referred to as contraception, reduce the risk of pregnancy. Although many methods of birth control are available, some are more reliable than others. The following are some of the more commonly used birth control methods:

- *Coitus interruptus:* The penis is withdrawn from the vagina before ejaculation. This method is not very reliable because small amounts of semen may enter the vagina before ejaculation.
- *Rhythm method:* The rhythm method requires abstinence from sexual intercourse around the time a female is ovulating. However, predicting ovulation can be difficult; therefore, this type of contraception can be unreliable.
- *Mechanical barriers:* Mechanical barriers prevent sperm from entering the female reproductive tract. They include condoms, diaphragms, and cervical caps.
- *Spermicides:* These are often used in conjunction with barrier methods, particularly condoms and diaphragms.

- *Chemical barriers:* Chemical barriers destroy sperm in the female reproductive tract. They primarily include spermicides.
- *Oral contraceptives:* Birth control pills are oral contraceptives. These pills normally include low doses of estrogen or progesterone that prevent the LH surge necessary for ovulation. These pills therefore prevent ovulation. Newer oral contraceptives are being developed in which the woman takes the pill daily for three months and then is off for a week, so that a period occurs only four times a year.
- *Injectable contraceptives:* Depo-Provera® is one brand of injectable contraceptive. It prevents ovulation and alters the lining of the uterus so that implantation of a **blastocyst** is not likely.
- *Insertable contraceptives:* NuvaRing® is one of the newest forms of contraception. The ring is inserted vaginally by the patient and is left in for three weeks. It is removed at the beginning of the fourth week to allow for menstruation, in the same schedule as most oral contraceptives.
- *Contraceptive implants:* Contraceptive implants are small rods of progesterone that are implanted beneath the skin to prevent ovulation.
- *Transdermal contraceptives:* Commonly called "the Patch," transdermal contraceptives are applied to the skin once a week and removed on the seventh day for a three-week cycle. No patch is applied during the fourth week to allow for the menstrual period.
- *Intrauterine devices:* An intrauterine device (IUD) is a small, solid device that a physician places in the uterus. It prevents the implantation of a blastocyst.
- *Surgical methods:* Tubal ligation is a surgical method used in females to prevent pregnancy. In this process, each fallopian tube is cut and tied or *fulgurated* to prevent sperm from reaching the oocyte. Vasectomy is a surgical method used in males to prevent pregnancy. In this process, each vas deferens is cut and tied or fulgurated to prevent sperm from being ejaculated. Figure 17-12 illustrates these methods.

blastocyst A hollow ball of cells in the development of an embryo.

> **U check**
> What is the most effective method of contraception?

Infertility

Infertility is the inability to conceive a child. If a couple has never been pregnant and has tried for 12 months to achieve pregnancy, they are said to have "primary" infertility. If a couple has had at least one pregnancy but has not been able to get pregnant after 1 year, they are said to have "secondary" infertility. About one-third of infertility is due to problems in the man and a third to problems in the female. The remainder is either a combination of the two or due to unknown causes.

> **U check**
> What is the difference between primary and secondary infertility?

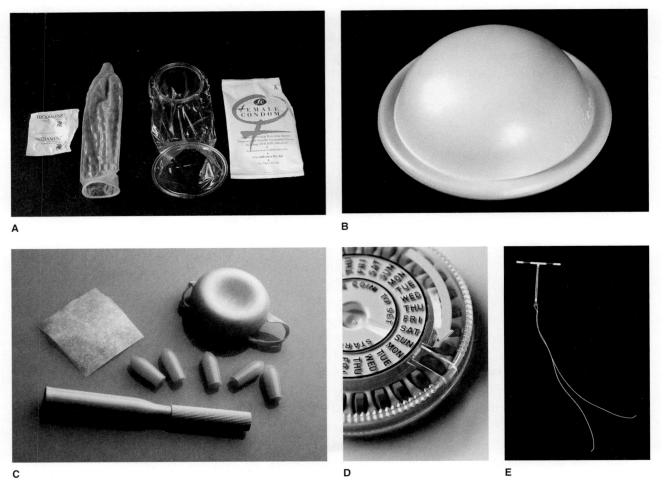

FIGURE 17-12 Methods used for birth control include (A) male and female condoms, (B) a diaphragm, (C) spermicides, (D) oral contraceptives, and (E) an IUD (intrauterine device).

Common causes of infertility as a result of male factors include the following:

- Impotence
- Retrograde ejaculation
- Low or absent sperm count
- Use of various medications or drugs
- Decreased testosterone production
- Scarring of the male reproductive tract from STIs
- Previous mumps infection that infected the testes
- Inflammation of the epididymis or testes

Infertility because of female factors includes these common causes:

- Scarring of fallopian tubes from STIs
- Pelvic inflammatory disease (PID)
- Inadequate diet
- Lack of ovulation
- Lack of menstrual cycles

- Endometriosis
- Abnormal shape of the uterus or cervix
- Hormone imbalances
- Cysts in ovaries
- Being older than age 40

Women are most likely to get pregnant in their early twenties. By the time a woman reaches the age of 40, her chance of conceiving a child is less than 10 percent each month. In general, infertility in men is not age-related.

Infertility Tests: A number of tests are used to diagnose infertility. They include the following:

- *Monitoring of morning body temperature.* If a woman's body temperature does not rise slightly once a month, which is best determined by taking her temperature first thing in the morning, a woman may not be ovulating.
- *Blood hormone measurements.* In females, various hormone levels can be monitored to predict ovulation and the general health of ovaries. In males, testosterone levels are primarily measured.
- *Endometrial biopsy.* This test determines the health of the uterine lining.
- *Urinary analysis for luteinizing hormone.* The absence of this hormone in urine may indicate a lack of ovulation.
- *Hysterosalpingogram.* This type of x-ray uses contrast media to visualize the shape of the uterus and the fallopian tubes. If a woman has excess scar tissue in her fallopian tubes, the contrast cannot run through them, nor can an ovum or blastocyst. Surgery may be required to open the oviduct(s).
- *Laparoscopy.* Laparoscopy is a procedure that is used to visualize the pelvic organs.
- *Ultrasound.* This procedure checks for the placement and appearance of reproductive structures.
- *Postcoital test.* The cervix is examined after unprotected sex to determine if the mucus is thin enough for sperm to pass through.

Treatment of Infertility: Many treatments are available for infertility, but often there is no cure for this condition. Common treatments include surgery to repair abnormal or scarred fallopian tubes, fertility drugs to increase ovulation, and hormone therapies. In cases where infertility cannot be cured, procedures such as artificial insemination, in vitro fertilization, or the use of a surrogate may help a couple conceive a child.

Life Span

17.7 learning outcome

Infer the effects of aging on the female reproductive system.

When a female experiences menarche, she has reached her reproductive years. Menarche is occurring at a younger age and menopause at an older age. This means that females now have a greater number of years when they can become pregnant. The number of follicles decreases with age and the production of estrogen declines. One of the more common symptoms of menopause is hot flashes. Other symptoms include mood swings, depression, muscle aches, vaginal dryness, and weight gain. Sexual desire does not necessarily decline with menopause.

summary

chapter 17

learning outcomes	key points
17.1 Identify the external organs of the female reproductive system including the location, structure, function, and disorders.	The female external organs or genitalia, collectively known as the vulva, include the following: mons pubis, labia majora, labia minora, clitoris, urethral meatus, vaginal orifice, Bartholin's glands, and perineum.
17.2 Identify the internal organs of the female reproductive system including the location, structure, function, and disorders.	The major internal female reproductive organs are the ovaries, fallopian tubes, uterus, cervix, and vagina. The ovaries produce the ova, which are transported along the fallopian tubes for implantation of the fertilized egg or zygote in the uterus. The vagina is the female organ for sexual intercourse.
17.3 Compare the actions of hormones of the female reproductive system.	Estrogen and progesterone are responsible for developing the ova and preparing the uterus for implantation of the zygote. They are also responsible for secondary sex characteristics of the female.
17.4 Explain the purpose and events of the menstrual cycle.	The menstrual cycle consists of regular changes in the uterine lining that leads to a monthly "period," or bleeding. The first menstrual period is known as menarche. The typical cycle is from 24 to 35 days
17.5 Explain how and where fertilization occurs.	Pregnancy results when a sperm cell unites with an ovum in a process called fertilization. This occurs in the fallopian tube.
17.6 Recall birth control methods and reasons for infertility.	There are many types and methods of birth control. These include coitus interruptus, rhythm method, mechanical barriers, spermicides, chemical barriers, oral contraceptives, injectable contraceptives, insertable contraceptives, contraceptive implants, transdermal contraceptives, intrauterine devices, and surgical methods. Only total abstinence is a foolproof method of contraception. Infertility can be caused by multiple factors affecting either the male or the female.
17.7 Infer the effects of aging on the female reproductive system.	Due to an earlier onset of menarche and a later occurrence of menopause, women now have more years during which they can become pregnant. Symptoms of menopause include hot flashes, mood swings, depression, muscle aches, vaginal dryness, and weight gain. Sexual desire does not necessarily decline with menopause.

case study 1 questions
Can you answer the following questions that pertain to the case (Lara Kempton) presented earlier in this chapter?

1. What symptom did Lara have that was consistent with breast cancer?
    a. Swelling of the breast
    b. Painless and firm lump in the armpit
    c. Dimpled skin on the breast
    d. Discharge from the nipples
2. What is another term used for stage 0 breast cancer?
    a. Invasive
    b. Benign
    c. In situ
    d. Noninvasive
3. What could Lara have done to possibly detect the lump sooner?
    a. Monthly breast self-exams
    b. Monthly appointments with her gynecologist
    c. Monthly mammograms
    d. Monthly breast ultrasounds
4. What types of treatment are used to treat breast cancer?
    a. Lumpectomy
    b. Chemotherapy
    c. Hormone therapy
    d. All of the above

review questions

1. What two female hormones are needed for a woman to produce breast milk?
    a. Oxytocin and prolactin
    b. Estrogen and progesterone
    c. Oxytocin and colostrum
    d. Prolactin and lutenizing hormone
2. Which hormones are responsible for secondary sex characteristics of the female?
    a. FSH and LH
    b. Estrogen and progesterone
    c. Relaxin and inhibin
    d. HCG and testosterone
3. The end result of meiosis in the female is
    a. One polar body and three functional eggs
    b. One polar body and two functional eggs
    c. Three polar bodies and one functional egg
    d. Two polar bodies and two functional eggs
4. Which of the following is *not* a reproductive structure in the female?
    a. Ovary
    b. Fallopian tube
    c. Prostate gland
    d. Uterus
5. If a person has a tumor in the hypothalamus, which of the following is *least* likely to be affected?
    a. If the individual is a male, testosterone levels may be affected.
    b. If the individual is a female, ovarian follicles may not mature.
    c. If the individual is a male, spermatogenesis may be affected.
    d. The person is at an increased risk of thyroid cancer.

6. A young woman was recently diagnosed with a tubal pregnancy. What does this mean?
   a. The fertilized egg was implanted in the fallopian tubes.
   b. The fertilized egg was implanted in the myometrium of the uterus.
   c. The fertilized egg was implanted in the endometrium of the uterus.
   d. The fertilized egg will degenerate with no health risk to the woman.

critical **thinking questions**

1. Discuss the pros and cons of the following birth control methods.
   a. Coitus interruptus and rhythm method
   b. Mechanical methods, such as the diaphragm and cervical cap
   c. Oral contraceptives
   d. Insertable and implantable contraceptives
2. Trace the path of the ovum through the female reproductive system. Describe each structure the ovum travels through and what happens it is are fertilized.

patient **education**

Your best friend's mother was recently diagnosed with breast cancer and you realize that your friend's chances of breast cancer are increased. Create a short teaching plan for your friend that includes:

- Why a breast self-exam should be done.
- When a breast self-exam should be done.
- How often a breast self-exam should be done.
- The correct steps to perform a breast self-exam.
- What to do if a lump is found during a breast self-exam.

applying **what you know**

1. Outline the four phases of the female menstrual cycle.
2. Explain the functions of GnRH, FSH, and LH in the female reproductive system.

CASE STUDY 2 *Sexually Transmitted Infection*

Jennifer Pierce is a 23-year-old marketing assistant who is sexually active. She recently had unprotected sexual relations with a new partner. Jennifer has only had one other partner in her lifetime. However, her new partner has had multiple partners in the past. Two days ago Jennifer noticed she had a yellow, foul-smelling vaginal discharge and vulvovaginal itching. She went to a local urgent care center where she was tested and diagnosed with an STI. Both Jennifer and her partner were treated and retested and re-exposure did not occur.

1. How can STIs be prevented?
   a. Abstain from sex
   b. Limit sexual partners
   c. Practice safe sex
   d. All of the above
2. What is a common symptom of multiple STIs?
   a. Fever
   b. Warts
   c. Abnormal vaginal or urethral discharge
   d. Parasites

18 Human Development and Genetics

chapter outline

- Human Development and Inheritance
- Hormonal Changes During Pregnancy
- The Birth Process
- The Postnatal Period
- Human Genetics
- Life Span

learning outcomes

AFTER COMPLETING THIS CHAPTER, YOU WILL BE ABLE TO:

18.1 Describe the periods of development from conception through the neonatal period.

18.2 Describe the changes that occur in a woman during pregnancy.

18.3 Describe the stages of the birth process.

18.4 Explain the major events of the postnatal period.

18.5 Explain the relationship of genotype and phenotype to genetic disorders.

18.6 Explain the effects of genetics on aging.

essential terms

allele (al-LEEL)
amniocentesis (am-nee-oh-sen-TEE-sis)
amnion (AM-nee-ahn)
chorion (KO-re-eahn)
complex inheritance (KOM-plex in-HAIR-ih-tants)
ductus arteriosus (DUK-tus ar-tee-ree-OH-sus)
ductus venosus (DUK-tus vee-NO-sus)
ectoderm (EK-toe-derm)
effacement (ee-FACE-ment)
endoderm (EN-do-derm)
foramen ovale (fo-RA-men o-VA-lee)
gene (jeen)
genotype (JEEN-oh-tipe)
homologous chromosomes (ho-MAH-luh-gus KRO-mo-sohmz)
mesoderm (MEZ-oh-derm)
morula (MOR-yu-luh)
neonate (NEE-oh-nate)
parturition (par-tu-RISH-un)
phenotype (FEE-noh-tipe)
placenta (plah-SEN-tah)
primary germ layer (PRIME-air-ee jerm LAY-er)
yolk sac (yoke sak)

Additional key terms in the chapter are italicized and defined in the glossary.

case study

Use the case study to focus on as you go through the chapter. The questions will guide you as you learn about human development and genetics.

CASE STUDY 1 *Genetic Counseling*

Dan and Lauren Lang have come to the physician's office where you are employed. Dan is 36 years of age and Lauren is 35. They want to have children and they have several questions. Other than an uncle on Dan's side who is an achondroplastic dwarf, neither knows of any unusual or abnormal conditions in either family. They do know this disorder is inherited in an autosomal dominant manner.

As you go through the chapter, keep the following questions in mind:

1. Is there an increased risk of having a child who will be a dwarf?
2. Are there any increased risks of having a child with Down syndrome?
3. Is there any prenatal testing that can be done to uncover any abnormalities?
4. They really would like a girl; what are the chances of that happening?

> **study tips**
> 1. Make flash cards for the essential terms.
> 2. Outline the chapter.
> 3. Write a one- to two-page paper on a genetic disorders discussed in the chapter.

Introduction

Pregnancy occurs when an ovum and sperm cell unite. Once this takes place, multiple phases of development occur before an infant is born. Whereas each of the periods of development is unique, each person is also unique based on his or her genetics. This chapter will discuss human development, genetics, and some common genetic disorders.

18.1 learning outcome

Describe the periods of development from conception through the neonatal period.

Human Development and Inheritance

Human development begins at the moment of conception. The time period prior to birth is the prenatal period and is divided into the zygote, embryonic, and fetal periods. The prenatal period is also divided into three periods of three months each: the first, second, and third trimester. The first trimester is crucial to the development of the major organ systems. The second trimester sees further development of the organ systems, and the third trimester is a period of rapid growth of the fetus. The neonatal period is the first 28 days after birth.

The Prenatal Period

Cells and tissues

morula A solid sphere of cells produced by successive divisions of the fertilized egg.

The prenatal period is the time before the child is born. As stated, the prenatal period is divided into a zygote period (the fertilized egg is called a zygote), an embryonic period (weeks 2 through 8 of pregnancy), and a fetal period (week 9 to the delivery of the offspring). The prenatal period can also be divided into three periods known as trimesters, which consist of three calendar months each. About one day after the zygote forms, it begins to undergo mitosis at a relatively rapid rate. This rapid cell division is called *cleavage*. The resulting ball of cells is called a **morula,** which means "berry," because of the appearance of the mass of rapidly dividing cells (Figure 18-1). The morula travels down the fallopian tube to the uterus. Fluid then invades the morula, forming and filling a cavity surrounded by cells. The morula is then called a blastocyst and will implant in the endometrial wall of the uterus (Figure 18-2). The cells that form the walls of the blastocyst are called the *trophoblast*.

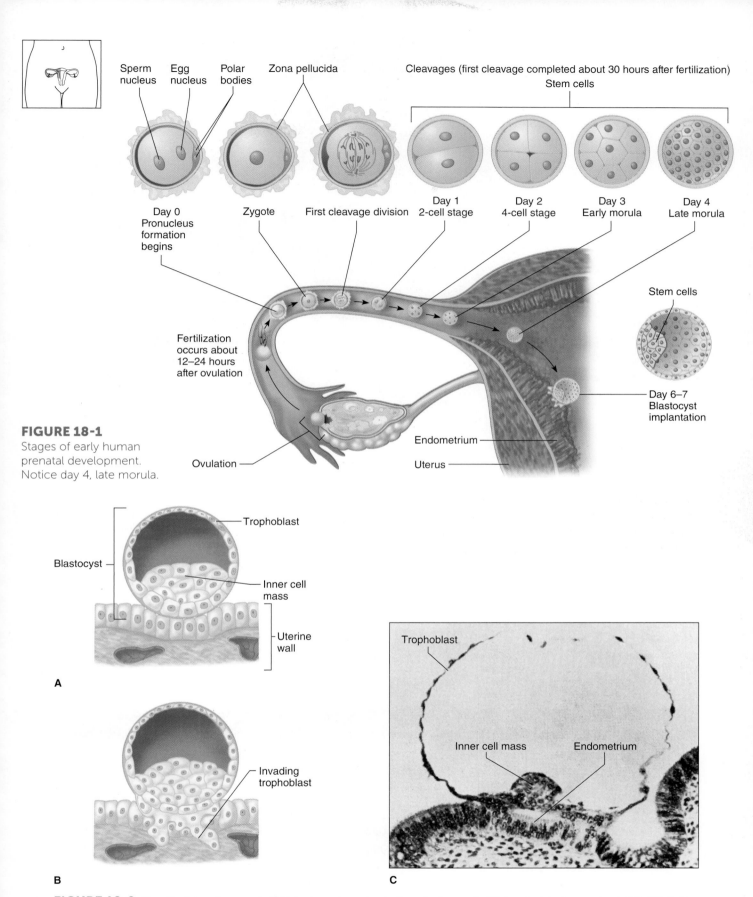

FIGURE 18-1 Stages of early human prenatal development. Notice day 4, late morula.

FIGURE 18-2 Day 6 of development: (A) blastocyst contacts the uterine wall; (B) trophoblast secretes hCG; (C) light micrograph of a monkey trophoblast (150x).

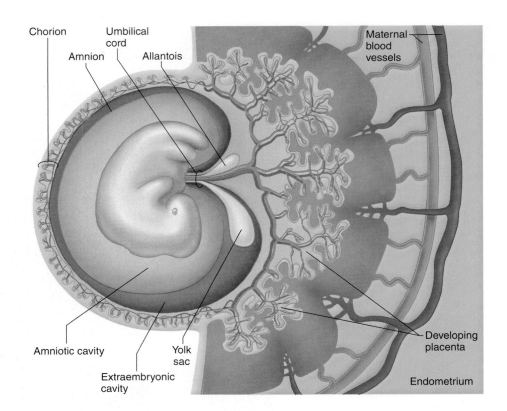

FIGURE 18-3
Development of the placenta and amnion and associated structures.

placenta A structure that allows nutrients and oxygen from maternal blood to pass to embryonic blood.

Meiosis versus mitosis

The process of moving from zygote formation to implantation into the uterine lining of the blastocyst takes about one week. Once the blastocyst implants, a group of cells in the blastocyst, called the inner cell mass, gives rise to the embryo proper. Other cells in the blastocyst, called the outer cell mass, along with cells of the uterus form the **placenta.** The placenta is a highly vascular structure that attaches the embryo to the uterine wall and exchanges gases, nutrients, and wastes between the maternal and fetal circulations (Figure 18-3). Keep in mind that the fetal and maternal blood do not mix in normal pregnancies. Table 18-1 summarizes these early stages.

U check
When is the fetal period?

TABLE 18-1	Stages and Events of Early Human Prenatal Development	
Stage	**Time Period**	**Principal Events**
Zygote	12–24 hours following ovulation	Secondary oocyte fertilized, meiosis is completed; zygote has 46 chromosomes and is genetically distinct
Cleavage	30 hours to third day	Mitosis increases cell number
Morula	Third to fourth day	Solid ball of cells
Blastocyst	Fifth day through second week	Hollowed ball forms trophoblast (outside) and inner cell mass, which implants and flattens to form embryonic disc
Gastrula	End of second week	Primary germ layers form

focus on Genetics: Down Syndrome

Human cells normally contain 23 pairs of chromosomes. One chromosome in each pair comes from your father, the other from your mother. In Down syndrome, there is abnormal cell division involving chromosome 21, which is responsible for the characteristic features and developmental problems of Down syndrome (Figure 18-4). Women over 35 are at a greater risk of having a child with Down syndrome.

Three genetic variations can cause Down syndrome; most cases are not inherited. Trisomy 21 is the most common, accounting for approximately 90 percent of the cases. In mosaic Down syndrome, a very rare type of Down syndrome, some cells have 46 chromosomes and some have 47. Another type, translocation Down syndrome, is also uncommon and is the only type that can be transmitted from a parent to the child. In this type of Down syndrome, the extra #21 chromosome is attached to another chromosome, such as #14.

You may know someone with Down syndrome. Though not all children with Down syndrome have the same features, some of the more common physical features are:

- Flattened facial features
- Protruding tongue
- Small head
- Upward slanting eyes, unusual for the child's ethnic group
- Unusually shaped ears

Children with Down syndrome may also have poor muscle tone, broad, short hands with a single crease in the palm, relatively short fingers, excessive flexibility, and some form of mental retardation.

REMEMBER DAN AND LAUREN, our couple that has some questions regarding birth defects? Do they have any risk factors for having a child with Down syndrome?

FIGURE 18-4
Child with Down syndrome.

The Embryonic Period: The embryonic period extends from the second week of pregnancy to the end of the eighth week of development (Figures 18-5 and 18-6). The cells of the inner cell mass organize into layers

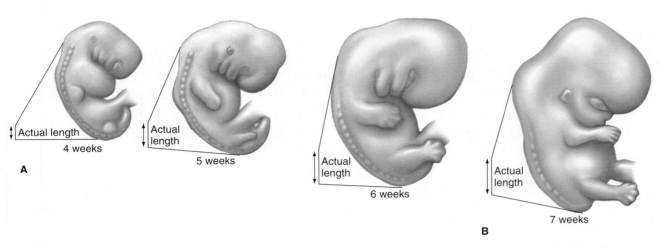

FIGURE 18-5 Development of an embryo: (A) from week 4 through week 7 of gestation; (B) At about 7 weeks of development the embryonic body and face become more distinct.

primary germ Layers of cells formed in the inner cell mass of a blastocyst that form all of the embryo's organs.

endoderm A primary germ layer of the developing embryo that gives rise to the gastrointestinal tract, the urinary tract, and the respiratory tract.

mesoderm The middle primary germ layer that gives rise to connective tissue, blood vessels, and muscle.

ectoderm The primary germ layer that gives rise to the nervous system, epidermis, and its derivatives.

amnion The protective, fluid-filled sac that surrounds an embryo.

yolk sac An extraembryonic membrane that transfers nutrients to the embryo and is the site of blood cell formation for the embryo.

chorion The structure that develops during the embryonic stage that allows the exchange of substances between the fetus and mother and acts as part of the immune system.

called **primary germ layers** (Figures 18-7 and 18-8). The primary germ layers are the **endoderm, mesoderm,** and **ectoderm.** The first two layers to form are the endoderm and ectoderm, with the mesoderm forming later. At this point, the embryo is called a *gastrula*. All organs are formed from the primary germ layers. The endoderm develops into tissues such as epithelium and glands, whereas the mesoderm becomes connective tissues, muscle, and the dermis. The epidermis, nails, hair, and nervous system develop from the ectoderm.

During this stage, the **amnion,** umbilical cord, and **yolk sac** form, along with most of the internal organs and external structures of the embryo (Figure 18-9). The umbilical cord contains three blood vessels—one umbilical vein that carries oxygenated blood from the placenta to the embryo and two umbilical arteries that carry deoxygenated blood from the embryo back to the placenta. The yolk sac makes new blood cells for the fetus as well as cells that eventually become sex cells of the baby. In addition to the placenta and yolk sac, the **chorion** develops. It acts as the primary structure for the exchange of substances between the fetus and mother. It also acts as part of the fetus's immune system. By the end of the embryonic stage, arms, hands, legs, feet, as well as head structures are recognizable.

> **U check**
> What is the purpose of the yolk sac?

The Fetal Period: The fetal period begins at the end of the eighth week and ends at birth (Figure 18-10). During this period, the growth of the offspring, which is now called a fetus, is rapid. During development, body proportions change considerably (Figure 18-11). By the twelfth week bones have begun to harden and the external reproductive organs are distinguishable as male or female. The growth rate of the fetus slows down in the fifth month but skeletal muscles become active. In the sixth month, the fetus starts to gain substantial weight. In the seventh month, the eyelids open. In the last three months of pregnancy, fetal brain cells divide rapidly and organs continue to grow. The testes of the male descend into

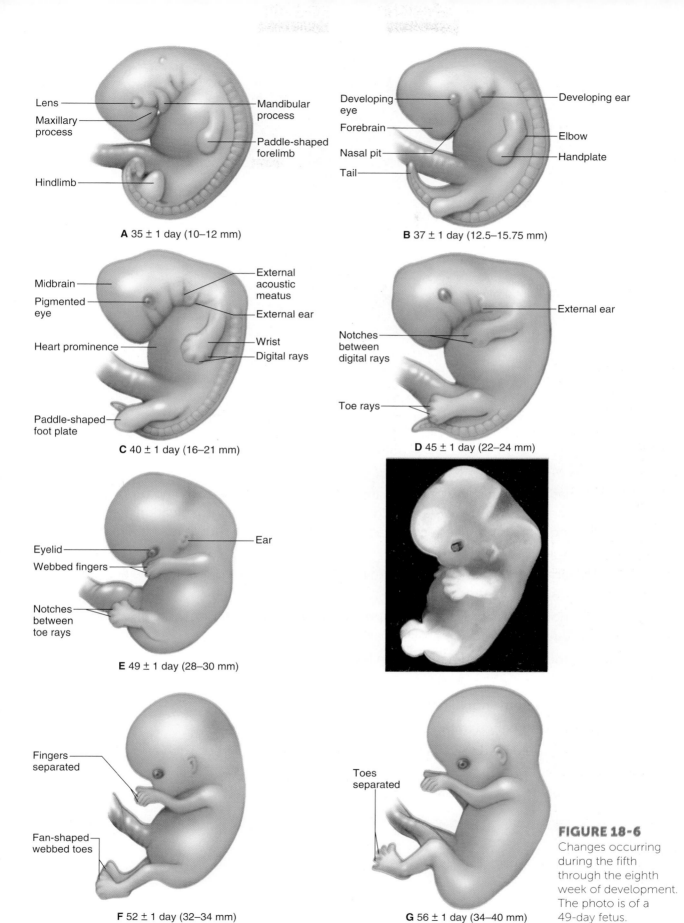

FIGURE 18-6 Changes occurring during the fifth through the eighth week of development. The photo is of a 49-day fetus.

CHAPTER 18 Human Development and Genetics

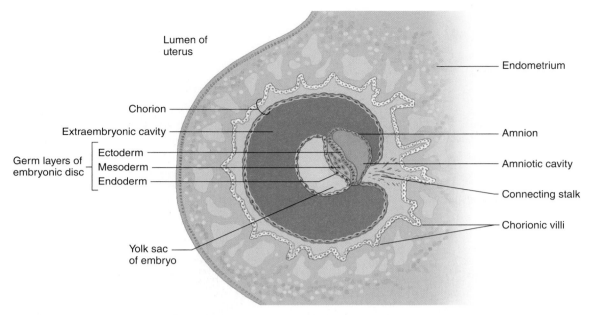

FIGURE 18-7 The three primary germ layers: endoderm, mesoderm, and ectoderm.

focus on Wellness: Routine Well-Woman Care During Pregnancy

The ideal situation is that the woman be in good health prior to becoming pregnant. Any chronic illnesses such as diabetes should be controlled and any medications should be reviewed for possible harmful effects. In addition, the following should be performed on a regular basis during the pregnancy.

1. A Pap smear and breast and pelvic examination should be performed on the first visit and as indicated thereafter.
2. Nutrition should be discussed with emphasis placed on the intake of calcium, folate, and iron. A goal should be established as to weight gain.
3. Screening (hematocrit and hemoglobin levels) for anemia should be done.
4. If necessary, hygiene should be discussed.
5. An exercise program should be recommended.
6. STIs should be screened for and treated if present.
7. Counseling should be done against the use of harmful substances such as cigarettes, drugs, and alcohol.
8. Weight, blood pressure, reflexes, and urine should be checked regularly.

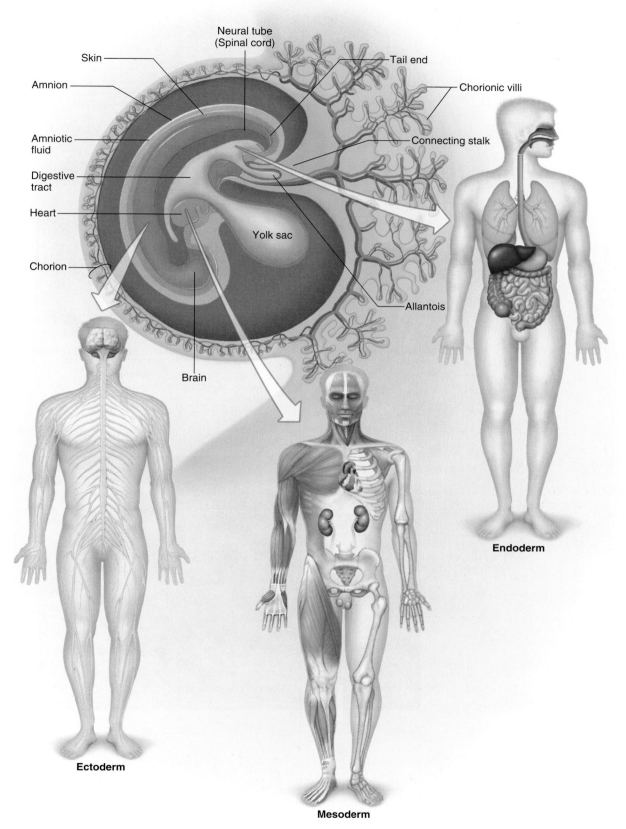

FIGURE 18-8 Each of the primary germ layers are the original tissue from which our organs are developed.

CHAPTER 18 Human Development and Genetics

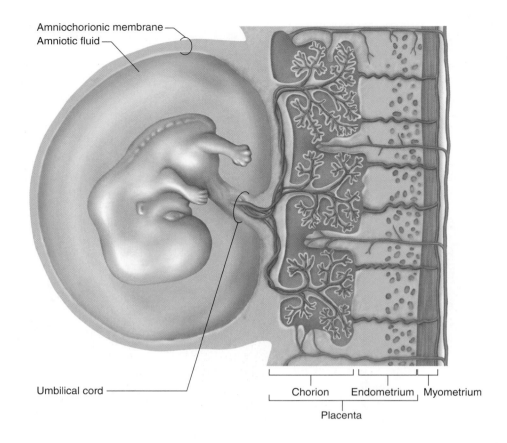

FIGURE 18-9
The amniochorionic membrane forms from the fusion of the amnion and chorion.

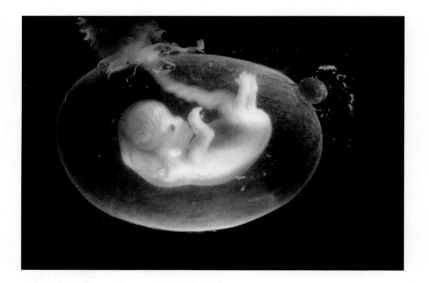

FIGURE 18-10
The embryo is distinctly human in appearance by the beginning of week 8 of gestation.

the scrotum. The last organ systems to completely develop are the digestive and respiratory systems. By the end of the ninth month, the fetus is usually positioned upside down in the uterus in preparation for delivery. Table 18-2 summarizes the pre-embryonic, embryonic, and fetal stages of development.

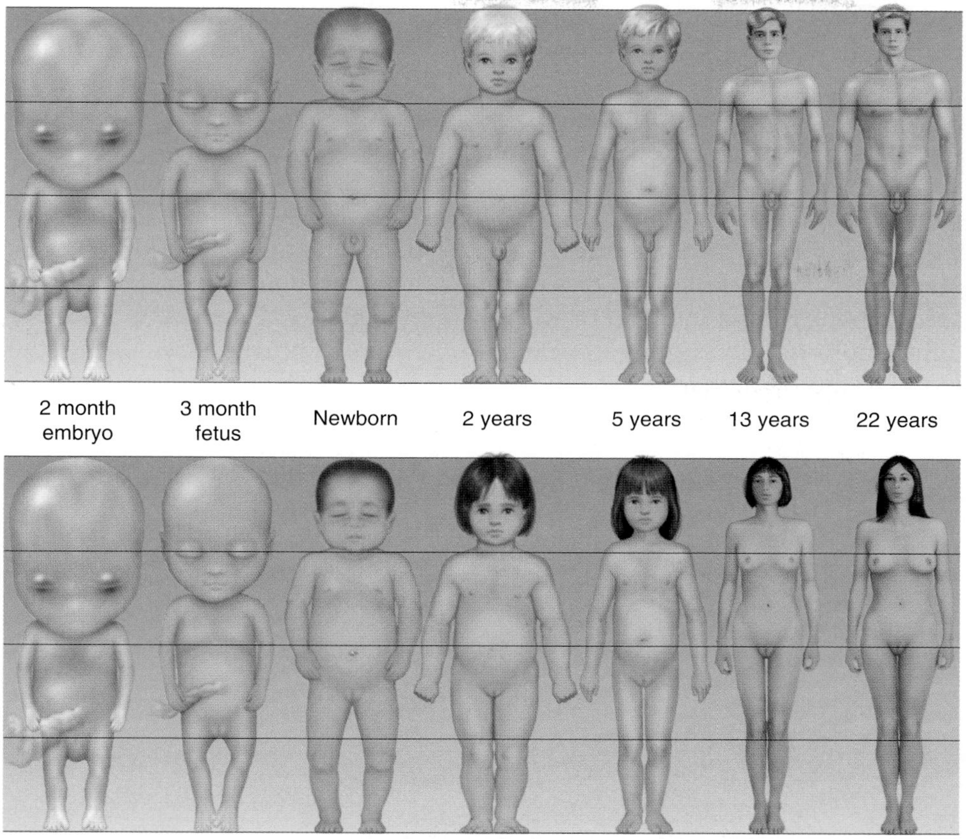

FIGURE 18-11
During development, body proportions change considerably.

2 month embryo | 3 month fetus | Newborn | 2 years | 5 years | 13 years | 22 years

TABLE 18-2	Stages of Prenatal Development	
Stage	**Time Period**	**Major Events**
Preembryonic stage	First week	Cells undergo mitosis, blastocyst forms; inner cell mass appears; blastocyst implants in uterine wall Size: ¼ inch (0.93 centimeter), weight: 1/30 ounce (0.8 gram)
Embryonic stage	Second through eighth week	Inner cell mass becomes embryonic disc; primary germ layers form, embryo proper becomes cylindrical; main internal organs and external body structures appear; placenta and umbilical cord form, embryo proper is suspended in amniotic fluid Size: 1 inch (2.5 centimeters), weight: 1/30 ounce (0.8 gram)
Fetal stage	Ninth through twelfth week	Ossification centers appear in bones, sex organs differentiate, nerves and muscles coordinate so that the fetus can move its limbs Size: 4 inches (10 centimeters), weight: 1 ounce (28 grams)
	Thirteenth through sixteenth week	Body grows rapidly; ossification continues Size: 8 inches (20 centimeters), weight: 6 ounces (170 grams)
	Seventeenth through twentieth week	Muscle movements are stronger, and woman may be aware of slight flutterings; skin is covered with fine downy hair (lanugo) and coated with sebum mixed with dead epidermal cells (vernix caseosa) Size: 12 inches (30.5 centimeters), weight: 1 pound (454 grams)
	Twenty-first through thirty-eighth week	Body gains weight, subcutaneous fat is deposited; eyebrows and lashes appear; eyelids reopen; testes descend Size: 21 inches (53 centimeters), weight: 6 to 10 pounds (2.7 to 4.5 kilograms)

Fetal Circulation: Throughout prenatal development, the placenta and umbilical blood vessels carry out the exchange of nutrients, oxygen, and waste products between maternal and fetal blood. Therefore, the fetus does not need to send blood to the lungs to pick up oxygen, nor does it need to send blood to the liver to process nutrients.

Fetal circulation has some important differences from adult circulation, which are illustrated in Figures 18-12 and 18-13. In the fetal heart, an opening called the **foramen ovale** is located between the right and left atria. This allows most of the fetal blood to flow from the right atrium into the left atrium, bypassing the lungs. However, some fetal blood does flow from the right atrium into the right ventricle, and the right ventricle then delivers the blood to the pulmonary trunk. In the adult heart, blood flows from the right

foramen ovale An opening in the fetal heart that is located between the right and left atria, allowing blood to bypass the lungs.

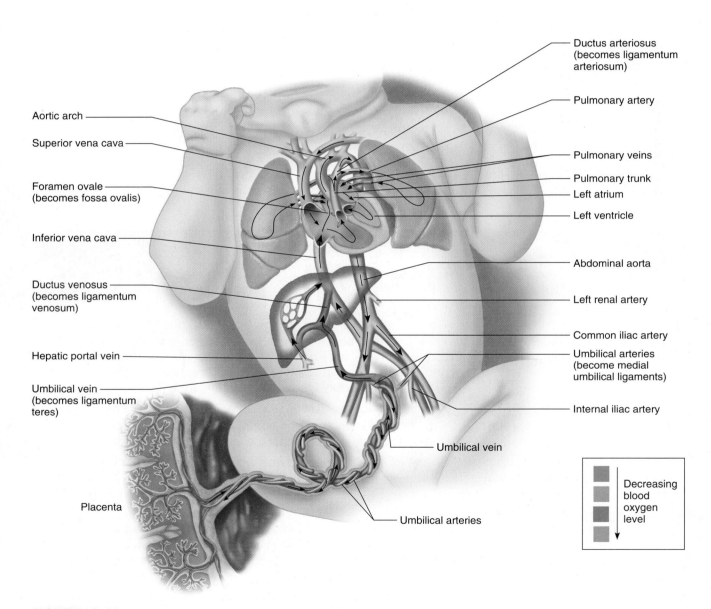

FIGURE 18-12 Anatomical representation of fetal circulation.

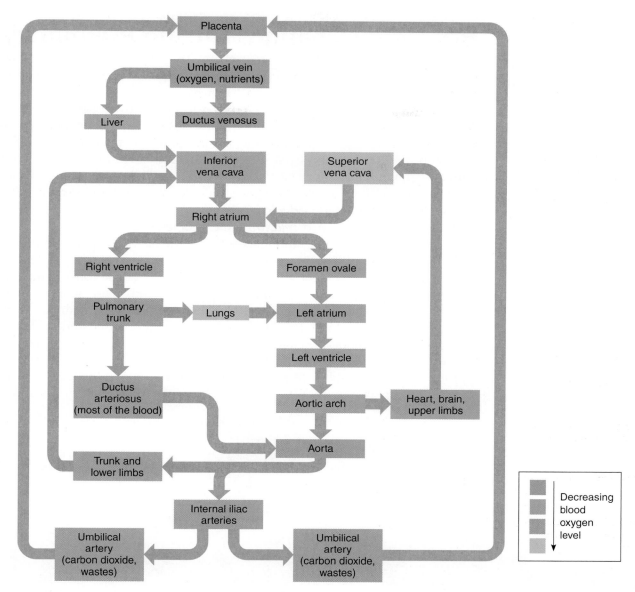

FIGURE 18-13 Schematic representation of fetal circulation.

atrium into the right ventricle so it can be pumped to the lungs. After birth, the foramen ovale closes in most individuals and becomes the *fossa ovalis*.

In the fetus, there is also a connection between the pulmonary trunk and the aorta called the **ductus arteriosus.** This connection allows blood to flow from the pulmonary trunk into the aorta again, bypassing the lungs. In the adult, this connection does not exist and blood flows from the pulmonary trunk to the lungs. After birth, the ductus arteriosus becomes the *ligamentum arteriosum*. The fetus also contains a blood vessel that allows most of the blood to bypass the liver. This vessel is called the **ductus venosus.** After a baby is born, the ductus venosus normally closes to form the *ligamentum venosum*. Hemoglobin within the fetus has a much higher affinity for oxygen than does the normal hemoglobin that is found after birth. Therefore, the fetus's blood is adapted to carry more oxygen.

ductus arteriosus An opening in the fetal heart between the pulmonary trunk and the aorta that allows blood to bypass the lungs.

ductus venosus A fetal blood vessel that allows blood to bypass the liver.

18.2 learning outcome
Describe the changes that occur in a woman during pregnancy.

Hormonal Changes During Pregnancy

Many hormonal changes take place when a woman is pregnant. Following implantation of the embryo, the cells of the placenta begin to secrete human chorionic gonadotropin (hCG) that maintains the corpus luteum in the ovary so it will continue to secrete estrogen and progesterone. The placenta also secretes large amounts of progesterone and estrogen. Progesterone and estrogen stimulate the uterine lining to thicken and inhibit the anterior pituitary gland from secreting FSH and LH to prevent ovulation during pregnancy. Estrogen and progesterone also stimulate the development of the mammary glands, inhibit uterine contractions, and stimulate the enlargement of female reproductive organs.

Another hormone called *relaxin*, which comes from the corpus luteum, inhibits uterine contractions and relaxes the ligaments of the pelvis in preparation for childbirth. The placenta also secretes *lactogen*, a hormone that stimulates the enlargement of mammary glands. Aldosterone, which is secreted from the adrenal gland, increases sodium and water retention. The secretion of parathyroid hormone (PTH) increases, helping maintain high calcium levels in the blood. PTH and aldosterone will be discussed further in Chapter 21, The Endocrine System. During pregnancy, the pregnant female may experience calcium loss, increased cardiac output, increased urination, and low back pain. Table 18-3 summarizes the hormonal changes that are seen during pregnancy.

18.3 learning outcome
Describe the stages of the birth process.

The Birth Process

The birth process, which ends pregnancy, begins when progesterone levels fall and uterine contractions are no longer inhibited. The uterus secretes prostaglandins that stimulate uterine contractions, which causes the posterior

TABLE 18-3 Hormonal Changes During Pregnancy

1. Following implantation, cells of the trophoblast begin to secrete hCG.
2. hCG maintains the corpus luteum, which continues to secrete estrogens and progesterone.
3. As the placenta develops, it secretes abundant estrogens and progesterone.
4. Placental estrogens and progesterone
   a. Stimulate the uterine lining to continue development.
   b. Maintain the uterine lining.
   c. Inhibit secretion of FSH and LH from the anterior pituitary gland.
   d. Stimulate development of the mammary glands.
   e. Inhibit uterine contractions (progesterone).
   f. Enlarge the reproductive organs (estrogens).
5. Relaxin from the corpus luteum also inhibits uterine contractions and relaxes the pelvic ligaments.
6. The placenta secretes placental lactogen that stimulates breast development.
7. Aldosterone from the adrenal cortex promotes reabsorption of sodium.
8. Parathyroid hormone from the parathyroid glands helps maintain a high concentration of maternal blood calcium.

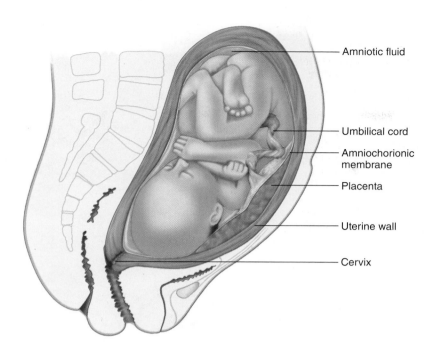

FIGURE 18-14
A full-term fetus is positioned with its head near the cervix.

pituitary gland to release oxytocin. Oxytocin stimulates stronger uterine contractions in a positive feedback manner until the birth process ends. The birth process itself occurs in three stages after the fetus settles into position in the mother's pelvis (Figure 18-14). Figure 18-15 summarizes the stages of the birth process.

1. *Dilation stage.* The cervix thins and softens in a process known as **effacement,** dilating to approximately 10 centimeters. Regular contractions occur at this stage and the amniotic sac ruptures. If rupture does not occur, the sac may be surgically punctured. This stage normally lasts 8 to 24 hours.

2. *Expulsion stage, also known as* **parturition**. This is the actual childbirth stage. Forceful contractions and abdominal compressions force the fetus from the uterus into the vagina. This stage may take up to 30 minutes or longer or may be completed in only a few minutes. If the fetus cannot be turned manually, forceps may be used to assist in the birth. Alternately, a cesarean section (C-section) may be performed to deliver the infant through the abdominal wall.

3. *Placental stage.* This stage is also referred to as the afterbirth. Approximately 10 to 15 minutes after the birth, the placenta separates from the uterine wall and is expelled. Uterine contractions continue during this stage and the blood vessels constrict to prevent hemorrhage. Normal blood loss is less than 12 ounces (350 milliliters). If the fetus is not in the usual head-down position, the child is said to be *breech,* and usually the buttocks or feet present first.

effacement The thinning and softening of the cervix during the birth process.

parturition The actual childbirth stage of the birth process.

U check
What are the stages of the birth process?

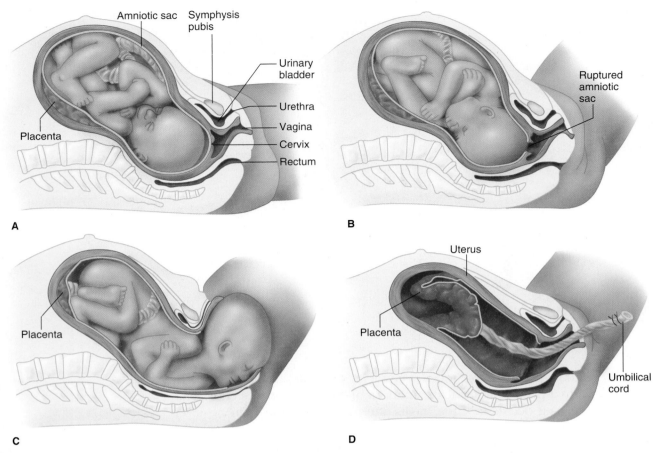

FIGURE 18-15 Stages in birth: (A) Fetal position before labor. (B) Stage 1: Dilation of the cervix. (C) Stage 2: Expulsion of the fetus. (D) Stage 3: Placenta stage.

learning outcome

Explain the major events of the postnatal period.

neonate An infant the first four weeks after birth.

The Postnatal Period

The *postnatal period* is the six-week period following birth. The first four weeks of the postnatal period are called the *neonatal period,* and the offspring is called a **neonate.** The neonatal period is marked by adjustment to life outside the uterus. The lungs of the neonate must expand, which is why the first breath of a baby is forceful. The liver of the newborn is immature, so the baby must obtain most of its glucose from fat stores in the skin. The newborn urinates a lot because the kidneys are too immature to concentrate the urine. In addition, body temperature tends to be unstable. The umbilical vessels of the newborn constrict, and the foramen ovale, ductus arteriosus, and ductus venosus close.

Milk Production and Secretion

During pregnancy, hormones stimulate the breasts to enlarge. After childbirth, prolactin causes the mammary glands to produce milk. The hormone oxytocin stimulates the ejection of milk from mammary gland ducts. As long as milk is removed from the mammary glands, milk production will continue. Once a female stops breast-feeding, the hypothalamus will inhibit the release of prolactin and oxytocin so that milk production will stop.

514 UNIT 2 Concepts of Common Illness by System

Infancy

Infancy is from the end of the first four weeks after birth to one year of age. This is a period of rapid growth. The infant gains weight rapidly, his or her vision improves, and he or she is able to grasp objects and is learning to sit, stand, crawl, and even walk.

Childhood

Childhood extends from one year of age to puberty. The primary teeth erupt and begin to be replaced by the permanent teeth. The child becomes more coordinated and has control over his or her bladder and bowel movements. The child learns to communicate both verbally and in writing.

Adolescence

Adolescence is the period from childhood to adulthood and encompasses many changes, including the ability to reproduce. Hormonal changes take place that are responsible for secondary sex characteristics. There is an increase in motor skills as well as intellectual accomplishment.

Adulthood

Adulthood, or maturity, continues from the end of adolescence until death. Physical and mental functioning increase, then later decline. Depending on genetics, personal circumstances, and environment, decline may be rapid or slow. People are living into their nineties and even past one hundred and many are able to participate in a full and active lifestyle.

Human Genetics

Human genetics is the transfer of genetic traits from parent to child. A person's **genotype** is his or her total genetic makeup. A person's **phenotype** is any observable characteristic, including morphological and biochemical properties. As discussed previously, the joining of the male sperm and female ovum produces a zygote with 23 chromosomal pairs (half from Mom and half from Dad). The first 22 pairs are known as autosomes. The 23rd pair consists of sex chromosomes. If the sex chromosomes consist of an X chromosome and a Y chromosome, the child is a male. If they are both X chromosomes, the child is a female. Since a female has two X chromosomes to donate to the child and the male has one X and one Y chromosome to donate, it is the male that "determines" the sex of the child.

In addition to determining the sex of an individual, these chromosomes are also involved in the secondary sex characteristics, such as a deeper voice in males. However, it is the autosomes that determine most body traits. Among them, the 46 total chromosomes contain between 20,000 and 25,000 **genes.** In 2003, the Human Genome Project (HGP) was completed. HGP was a 13-year effort coordinated by the U.S. Department of Energy and the National Institutes of Health (NIH). One of the goals of the project was to identify all the approximately 20,000 to 25,000 genes

18.5 learning outcome

Explain the relationship of genotype and phenotype to genetic disorders.

genotype The genetic makeup of an individual.

phenotype The observable expression of the genotype.

gene A segment of DNA that determines a body trait found in the human body.

in human DNA. Besides the United States, several countries including the United Kingdom, Japan, France, Germany, and China participated in the project.

> **U check**
> What is the difference between genotype and phenotype?

homologous chromosomes Chromosomes that carry all the genes that code for a particular trait.

allele One of two or more different forms of a gene that may or may not influence a particular trait.

Homologous chromosomes carry all the genes that code for a particular trait, but there may be different variations of the genes called **alleles**. Sometimes only one allele is actually expressed as a trait even if another allele is present. An allele that is always expressed over the other is called a "dominant" allele. The one that is not expressed is called the "recessive" allele. A recessive allele can be expressed only if no dominant allele is present.

Detached earlobes are an example of a trait that is determined by a dominant allele. If a child inherits a dominant allele for this trait from one parent but inherits the recessive allele from the other parent, the child will have detached earlobes. If the child inherits recessive alleles from both parents, then he or she will have attached earlobes. Most traits in the body are determined by multiple alleles. For example, hair color, height, skin tone, eye color, and body build are each determined by many different genes as well as by environmental factors.

> **U check**
> What is an allele?

complex inheritance Inherited traits that are determined by multiple genes.

Complex inheritance is the term used to describe inherited traits that are determined by multiple genes. It explains why different children within the same family often have different characteristics.

Sex-linked traits are carried on the sex chromosomes. The Y chromosome is much smaller than the X chromosome and does not carry many genes. Therefore, if the X chromosome carries a recessive allele, it is likely to be expressed because there is usually no corresponding allele on the Y chromosome. For example, red–green color blindness is determined by the presence of a recessive allele that is always found on the X chromosome. This disorder (like most sex-linked disorders) primarily affects males because the corresponding Y chromosome does not have any allele to prevent the expression of the recessive allele. As you have seen in the Pathophysiology and Focus On . . . features throughout the text, genetics plays a role in many health conditions.

Genetic Techniques

Deoxyribonucleic acid (DNA) is the primary component of genes and is found in the nucleus of most cells in the body. The chemical structure of every person's DNA is the same. The DNA molecule is made up of nucleotides and is a double-stranded molecule that is shaped like a double helix or stairway. The differences in DNA between individuals are found in the order of the nitrogen bases of the nucleotides. It is the unique sequence of the nucleotides that determines the characteristics of an individual. One DNA molecule contains hundreds or thousands of genes. Each gene occupies a particular location on the DNA molecule, making it possible to compare the same gene in a number of different samples. Genetic techniques involve using or manipulating genes. Two widely used genetic techniques in the clinical setting are the polymerase chain reaction (PCR) and DNA fingerprinting.

> **U check**
> What accounts for the differences in DNA between individuals?

Polymerase Chain Reaction (PCR): The polymerase chain reaction (PCR) is a quick, easy method for making millions of copies of any fragment of DNA. This technique has been revolutionary in the study of genetics and has quickly become a necessary tool for improving human health. PCR can produce millions of gene copies from tiny amounts of DNA, even from just one cell. This method is especially useful for detecting disease-causing organisms that are impossible to culture, such as many kinds of bacteria, fungi, and viruses. It can, for example, detect the AIDS virus sooner than other tests—during the first few weeks after infection. PCR is also more accurate than standard tests. The technique can detect bacterial DNA in children's middle ear fluid, which indicates an infection, even when culture methods fail to detect bacteria.

Other diseases diagnosed through PCR include Lyme disease, stomach ulcers, viral meningitis, hepatitis, tuberculosis, and many sexually transmitted infections (STIs) including herpes and chlamydia. PCR is also leading to new kinds of genetic testing because it can easily distinguish among the tiny variations in DNA among individuals. PCR can diagnose people with inherited disorders or identify those who carry mutations that could be passed to their children. PCR is also used to determine who may develop common disorders such as heart disease and various types of cancer. This knowledge can help at-risk individuals take steps to prevent those diseases.

DNA Fingerprinting: A DNA "fingerprint" refers to the unique sequences of nucleotides in a person's DNA and is the same for every cell, tissue, and organ of that person. Consequently, DNA fingerprinting is a reliable method for identifying and distinguishing among human beings to establish paternity and identify suspects in criminal cases. It is also used to diagnose genetic disorders such as cystic fibrosis, hemophilia, Huntington's disease, familial Alzheimer's, sickle cell anemia, thalassemia, and many other conditions. Detecting genetic diseases early, even in utero, allows patients and medical staff to prepare for proper treatment. Researchers also use this information to identify DNA patterns associated with genetic diseases.

PATHOPHYSIOLOGY

Albinism

Albinism is a condition in which a person is born with little or no pigmentation in the skin, eyes, or hair. Albinism affects all races, and in most cases there is no family history of the condition.

Etiology: At least six different genes are involved with pigment production. This condition develops when a person inherits one or more faulty genes that do not produce the usual amounts of a pigment.

(continued)

Albinism (concluded)

Signs and Symptoms: People with the condition experience visual problems, including severe photophobia and sun-sensitive skin.

Treatment: Although there is no cure, treatments are available to help with the symptoms. Prenatal testing for the condition is available.

Cystic Fibrosis

Cystic fibrosis is a life-threatening disease that mainly affects the lungs and pancreas. This disease is one of the most common inherited life-threatening disorders among Caucasians in the United States.

Etiology: Inheritance is autosomal recessive, so if both parents are carriers, there is a 25 percent chance that each child born to them will develop cystic fibrosis. In this condition, there is a problem with chloride transport across cell membranes.

Signs and Symptoms: One of the first signs is production of thick, sticky mucus and saltier than normal sweat. Individuals with this disorder have increasing problems with breathing and digestion. Thick secretions eventually block passage in the airways, and these secretions may become infected.

Treatment: There is no cure, but treatments are available to help patients live with the complications associated with this disorder. Newborn babies are commonly screened for the disease because the sooner treatment begins, the healthier the child can be. Parents are also commonly screened for the gene to determine the likelihood of having a child with cystic fibrosis. Parents can perform a "kiss test," kissing the child and tasting the skin. If the child "tastes salty," more sophisticated tests may be indicated to make a definitive diagnosis.

Down Syndrome

Down syndrome, also called Trisomy 21, is a disorder that causes mental retardation and physical abnormalities. It is the single most common type of birth defect.

Etiology: This disorder occurs when a person has three copies of chromosome 21 instead of two. This is the result of "nondysjunction" during meiosis—two chromosomes do not separate. This condition can be diagnosed through prenatal testing by performing **amniocentesis,** which involves withdrawing some of the amniotic fluid surrounding the fetus and then analyzing the fluid and fetal cells. The test is done early in the second trimester. The risk of having a child with Down syndrome increases with the age of the mother.

amniocentesis A prenatal test that may be used to diagnose Down syndrome. It is performed early in the second trimester by withdrawing and testing some of the amniotic fluid surrounding a fetus.

Down Syndrome (concluded)

Signs and Symptoms: The signs of Down syndrome include a flat facial profile; protruding tongue; oblique, slanting eyes; abundant neck skin; short, broad hands; and poor muscle tone. Heart, digestive, hearing, and visual problems are also common in people with this condition. Learning difficulties are common in Down syndrome and can range from moderate to severe.

Treatment: There is no cure, but support programs and the treatment of health problems allow many individuals with Down syndrome to live relatively normal lives.

Fragile X Syndrome

Fragile X syndrome is the most common inherited cause of learning disability. All races and ethnic groups seem to be affected equally by this syndrome.

Etiology: In this disorder, one of the genes on the X chromosome is defective and makes the chromosome susceptible to breakage. This sex-linked disorder affects males more severely than females. It is estimated that approximately 1 in 300 females is a carrier for this disorder.

Signs and Symptoms: Mental impairment, learning disabilities, attention deficit disorder (ADD), a long face, large ears, and flat feet are some of the signs and symptoms. Fragile X syndrome can be easily diagnosed prior to birth using prenatal tests such as amniocentesis.

Treatment: There is no cure, but some treatments and support groups are available for patients (and their families) with this disorder.

Hemophilia

Hemophilia is a group of inheritable bleeding disorders. The effects of each disorder vary from mild to very severe symptoms.

Etiology: In each type of hemophilia, an essential clotting factor is low or missing. Most types of hemophilia are X-linked recessive disorders; therefore, this disorder primarily affects males. Carriers of the gene can be identified with a blood test, and prenatal tests can diagnose the condition in the fetus.

Signs and Symptoms: Symptoms include easy bruising, spontaneous bleeding, and prolonged bleeding. Repeated bleeding in the joints leads to arthritis and permanent joint damage.

Treatment: Treatment includes injections or infusions of the missing clotting factors, often Factor VIII.

(continued)

Klinefelter's Syndrome

Klinefelter's syndrome is a chromosomal abnormality that affects males.

Etiology: Males with this disorder have an extra X chromosome: XXY.

Signs and Symptoms: Tall stature, pear-shaped fat distribution, small testes, sparse body hair, and infertility are the most common signs and symptoms. Thyroid problems, diabetes, and osteoporosis are also common in individuals with this syndrome. Their intelligence may be slightly lower than normal.

Treatment: There is no cure, but treatments such as testosterone replacement therapy can decrease the risk of osteoporosis and produce more distinctive male characteristics.

Muscular Dystrophy

Muscular dystrophy (MD) is a group of genetic disorders that primarily affect the muscular and nervous systems. It most often affects males. The two most common types are Duchene's and Becker's MD.

Etiology: Most types involve mutations in the genes responsible for producing muscle proteins. Some types of muscular dystrophy are inherited as an X-linked disorder, but some are caused by gene mutations.

Signs and Symptoms: In this disorder, muscle cells gradually break down, causing progressive muscle wasting and weakness.

Treatment: There is no cure, and few treatments are available to slow down the loss of muscle cells. Prenatal genetic tests are available for some types of muscular dystrophy.

Phenylketonuria

Phenylketonuria (PKU) develops when a person cannot synthesize the enzyme that converts phenylalanine to tyrosine. Phenylalanine is an essential amino acid, but too much of it can be harmful, so the body regularly converts it to tyrosine.

Etiology: This condition is inherited as an autosomal recessive disorder.

Signs and Symptoms: If phenylalanine builds up in the blood, it can lead to irreversible organ and brain damage.

Treatment: Phenylalanine is found in many proteins, so meats and other protein-rich foods must be avoided. Early detection of PKU is important to prevent developmental delays. There is no cure for PKU, but special diets allow a person to lead a normal life. Most newborns are tested for PKU, and prenatal testing is also available.

Turner's syndrome

Turner's syndrome is a disorder that results when females have a single X chromosome (XO).

Turner's syndrome (concluded)

Etiology: This disease results when an X chromosome is completely or partially missing.

Signs and Symptoms: The signs and symptoms may include webbed neck, broad chest, widely spaced nipples, low hairline, short stature, and infertility. Intelligence is normal. Prenatal tests can diagnose the condition, but most girls are diagnosed in late childhood when they fail to start menstruating.

Treatment: There is no cure for Turner's syndrome, but treatments with growth hormone replacements can increase the height of the patient.

Life Span

18.6 learning outcome

Explain the effects of genetics on aging.

Our genetics play a role in our life span. We have genes that regulate the number of cell divisions that different cells can undergo. Some cell types have a very limited number of cell divisions while others have the capability of many more divisions. DNA is eliminated from our chromosomes as a normal process and contributes to aging of cells and of the individual.

A test to determine the biological age of person is in development. This test is based upon the length of the telomeres (structures on the tips of the chromosomes) Short telomeres are linked to age-related diseases and premature death.

summary

learning outcomes	key points
18.1 Describe the periods of development from conception through the neonatal period.	The prenatal period is the time before the individual is born. The fertilized egg is the zygote. The prenatal period is then divided into an embryonic period (weeks 2 through 8 of pregnancy) and a fetal period (week 9 to the delivery of the offspring). It is further divided into three periods known as trimesters, which consist of three calendar months each. The neonatal period is the first 28 weeks after birth.
18.2 Describe the changes that occur in a woman during pregnancy.	Estrogen and progesterone stimulate the development of the mammary glands, inhibit uterine contractions, and stimulate the enlargement of female reproductive organs. Relaxin inhibits uterine contractions and relaxes the ligaments of the pelvis in preparation for childbirth. During pregnancy, the pregnant female may experience calcium loss, increased cardiac output, increased urination, and low back pain.

continued

chapter 18

learning outcomes	key points
18.3 Describe the stages of the birth process.	The three stages of the birth process are dilation, expulsion of the fetus, and expulsion of the placenta.
18.4 Explain the major events of the postnatal period.	The postnatal period is the six weeks following birth and the neonate period is the very first four weeks.
18.5 Explain the relationship of genotype and phenotype to genetic disorders.	Genotype is a person's total genetic makeup. Phenotype is any observable characteristic including morphological and biochemical properties. These distinguish human genetics. Various disorders may occur if a person's genotype includes faulty genes. The most common is Down syndrome, or Trisomy 21.
18.6 Explain the effects of genetics on aging.	There are numerous genetic disorders. Some that were discussed include albinism, cystic fibrosis, Down syndrome, Fragile X syndrome, and muscular dystrophy.

chapter 18 review

case study 1 questions

Can you answer the following questions that pertain to Dan and Lauren Lang's case study presented earlier in this chapter?

1. Is there an increased risk of having a child who will be a dwarf?
   a. Yes, because the condition is inherited in an autosomal dominant manner
   b. Yes, because it is inherited in an autosomal recessive manner
   c. No, because it is a sex-linked disorder
   d. No, because it is inherited in an autosomal dominant manner

2. Are there any increased risks of this couple having a child with Down syndrome?
   a. Yes, because of the man's age
   b. Yes, because of the woman's age
   c. Yes, because the man's uncle is an achondroplastic dwarf
   d. No, because dwarfism has no effect on having a child with Down syndrome

3. Is there any prenatal testing that can be done to uncover any abnormalities?
   a. Yes, a PSA can be done on the man.
   b. Yes, a DRE can be done on the man.
   c. No, there are no tests available to discover this abnormality.
   d. Yes, amniocentesis can be done in the second trimester of the pregnancy.

4. Dan and Lauren really would like a girl; what are the chances of that happening?
   a. Zero likelihood
   b. 1 out of 2
   c. 1 out of 4
   d. 100 percent since they really want a girl

review questions

1. How does human chorionic gonadotropin aid in pregnancy?
   a. It allows nutrients and oxygen from maternal blood to pass to embryonic blood.
   b. It maintains the corpus luteum in the ovary so it will continue to secrete estrogen and progesterone.

c. It helps maintain high calcium levels in the blood.
   d. It increases sodium and water retention.
2. What is the haploid number of chromosomes in humans?
   a. 13
   b. 23
   c. 46
   d. 96
3. The end result of meiosis in the female is
   a. One polar body and three functional eggs
   b. One polar body and two functional eggs
   c. Three polar bodies and one functional egg
   d. Two polar bodies and two functional eggs
4. Which of the following is *not* a common physical feature of Down syndrome?
   a. Small head
   b. Protruding tongue
   c. Enlarged mouth
   d. Flattened facial features
5. In Down syndrome, there is abnormal cell division involving which chromosome?
   a. 21
   b. 17
   c. 23
   d. 8

critical **thinking questions**

1. Why does maternal age play a part in Down syndrome?
2. Why is the genetic disorder of color blindness seen in males?
3. What is the difference between your genotype and your phenotype?

patient **education**

Giving birth can be painful and traumatic. Research shows that when a woman has a better understanding about the birth process before she gives birth, the outcome improves. Research and review the stages of birth and create a teaching brochure to guide a pregnant woman through the process.

applying **what you know**

1. Turner's syndrome is a disorder that results when females have a single X chromosome; what does this mean?
   a. A Y chromosome is completely or partially missing.
   b. There are three X chromosomes.
   c. An X chromosome is completely or partially missing.
   d. All chromosomes are missing.
2. A newborn is have a test for a genetic disorder. Which of the following is most likely being tested?
   a. Fragile X syndrome
   b. PKU
   c. Muscular dystrophy
   d. Cystic fibrosis

19 The Digestive System

chapter outline

Overview of Digestive System Functions

Digestive System Organs and Their Functions

Accessory Organs and Their Functions

Phases of Digestion

Life Span

learning outcomes

AFTER COMPLETING THIS CHAPTER, YOU WILL BE ABLE TO:

19.1 Identify the general functions of the digestive system.

19.2 Describe each digestive system organ's structure, specific function(s), and related disorders.

19.3 Describe each accessory organ's structure, specific function(s), and related disorders.

19.4 Explain the phases of digestion.

19.5 Relate the effects of aging on the digestive system.

essential terms

alimentary canal (al-ih-MENT-ah-ry ka-NAL)
cecum (SEE-kum)
deglutition (deg-lu-TISH-un)
duodenum (du-o-DEE-num)
frenulum (FREN-you-lum)
gingiva (JIN-jih-vah)
ileocecal valve (il-ee-oh-SEE-kul valv)
ileum (IL-ee-um)
jejunum (je-JU-num)
mastication (mas-tih-KAY-shun)
mesentery (MES-en-ter-ee)
peritoneum (per-ih-to-NEE-um)
pharynx (FAIR-inx)
pyloric sphincter (pie-LOR-ik SFINK-ter)
rugae (ROOG-eye)
uvula (YOO-vyoo-lah)
vermiform appendix (VER-mih-form ah-PEN-dix)

Additional key terms in the chapter are italicized and defined in the glossary.

case study

Use the case study to focus on as you go through the chapter. The questions will guide you as you learn the anatomy and pathophysiology of the digestive system.

CASE STUDY 1 *Cholelithiasis*

Yesterday afternoon, Sandy Wolbers, a 55-year-old female, came to the gastroenterologist's office complaining of severe pain in her right upper abdomen. She complained of nausea and stated that for several months—and especially following meals—she had been having periodic abdominal pain. After several tests, she was diagnosed as having gallstones and was scheduled for the cholecystecomy (surgical removal of her gallbladder).

As you read this chapter, consider the following questions:

1. What is the function of the gallbladder?
2. How does the gallbladder empty bile into the small intestine?
3. What conditions can result if gallstones are not removed?
4. How will this patient have to change her diet once her gallbladder is removed?

study tips

1. Trace a bite of food through the digestive system. Identify each organ it enters and exits and what happens to it in that organ.

2. Assign each person in a study group the role of an organ in the digestive system to explain and describe to the rest of the group.

Introduction

The digestive system is involved with maintaining homeostasis by taking in food and water and then eliminating the waste products. Food contains the building blocks of proteins that the body needs to survive. Food also contains vitamins and minerals that are essential to homeostasis. The functions of the digestive system are carried out by the organs of the **alimentary canal** or gastrointestinal tract (GI tract) and accessory organs (Figure 19-1). Organs of the alimentary canal extend from the mouth to the anus. In addition to the mouth, the organs include the pharynx, esophagus, stomach, small intestine, large intestine, rectum, and anal canal (Figure 19-2). The accessory organs of digestion include the teeth, tongue, salivary glands, liver, gallbladder, and pancreas. You may find it helpful to review the abdominal regions and quadrants discussed in Chapter 1, Concepts of the Human Body, while studying the organs of this chapter.

alimentary canal The gastrointestinal tract and accessory organs.

LO 19.1

19.1 learning outcome
Identify the general functions of the digestive system.

Overview of Digestive System Functions

The functions of the digestive system can be divided into six separate processes:

1. *Ingestion,* the process of eating and drinking.
2. *Secretion,* which occurs when saliva, water, acids, and enzymes enter the

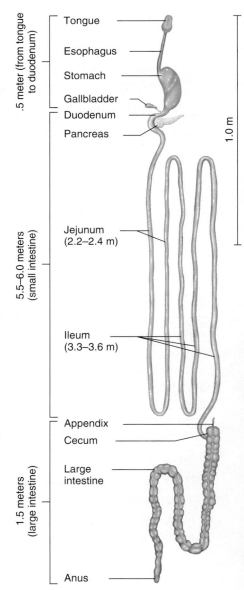

FIGURE 19-1 The alimentary canal is a muscular tube about 8 meters (26 feet) long.

526 UNIT 2 Concepts of Common Illness by System

mouth and gastrointestinal tract to help with the breakdown and absorption of foods.

3. *Mixing and propulsion,* which occur throughout the GI tract to move food along its way to the anal canal.
4. *Digestion,* which consists of the mechanical and chemical processes that break down food into small molecules that can be absorbed. Mechanical

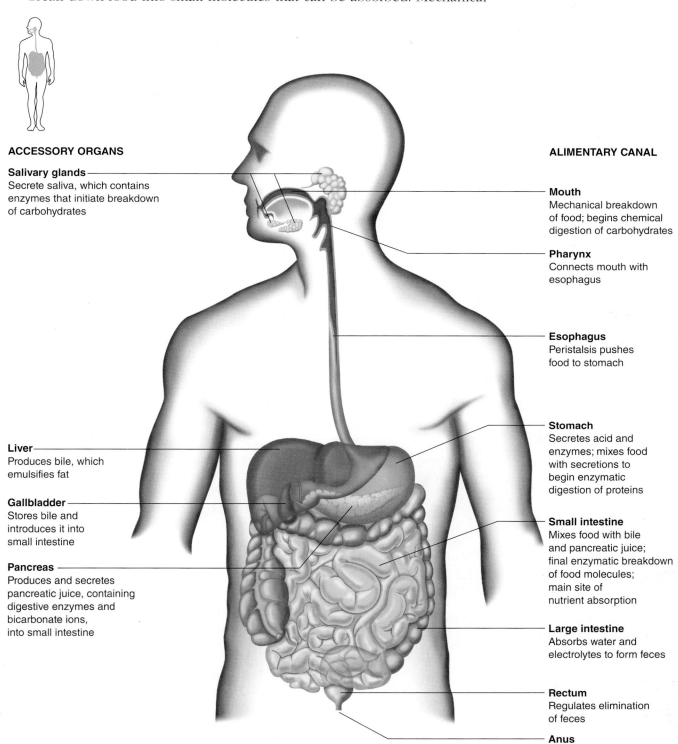

FIGURE 19-2 Organs of the digestive system.

CHAPTER 19   The Digestive System   527

mastication The act of chewing.

digestion begins in the mouth with the **mastication** (chewing) of food. Chemical digestion involves the further breakdown of food by the enzymes secreted by salivary glands, stomach, pancreas, and small intestine.

5. *Absorption into the blood and lymph,* which takes place within the GI tract lumen by the epithelial surface layer.
6. *Defecation,* the elimination of wastes, indigestible substances, unabsorbed substances, water, some cells, and bacteria by the GI tract.

> **U check**
> What is the difference between ingestion and digestion?

19.2 learning outcome
Describe each digestive system organ's structure, specific function(s), and related disorders.

frenulum A fold of the mucous membrane that attaches two structures together.

gingiva Gums.

Digestive System Organs and Their Functions

Mouth

The mouth, which is also known as the oral or buccal cavity, takes in food and reduces its size through mastication (mechanical digestion). The mouth also starts the process of chemical digestion when saliva, which contains the enzyme *amylase,* begins to break down carbohydrates (Figure 19-3). The boundaries of the oral cavity are the cheeks, lips, hard and soft palates, and tongue. The vestibule of the oral cavity is the space located between the lips and cheeks and the teeth. The oral cavity proper is the space interior to (behind) the teeth. The cheeks, which act to hold food in the mouth, consist of skin, adipose tissue, skeletal muscles, and an inner lining of moist nonkeratinized stratified squamous epithelium. The buccinator muscle (refer to Chapter 9, The Muscular System) is the muscle that performs this function. The lips (also referred to as labia) are formed by the orbicularis muscle, which closes the mouth, and are covered externally by skin and internally by a moist mucous membrane. Both the upper and lower lips have a fold of the mucous membrane called the labial **frenulum** that attaches the lips to the **gingiva** (gums). The tongue is made of skeletal muscle and is covered by a mucous membrane.

The tongue has both extrinsic muscles that are attached to various bones such as the mandible and hyoid bone, and intrinsic muscles that change the shape and size of the tongue to assist with speech and swallowing. The body of the tongue is held to the floor of the oral cavity by a fold of mucous membrane called the lingual frenulum. If the lingual frenulum is too short, the person is "tongue tied" or has the condition known as *ankyloglossia,* which may cause difficulty with speech. Treatment for

FIGURE 19-3 The mouth ingests food and prepares it for digestion, both mechanically and chemically.

Labels: Lip, Hard palate, Soft palate, Uvula, Palatine tonsils, Tongue, Lingual frenulum, Vestibule, Lip

528   UNIT 2   Concepts of Common Illness by System

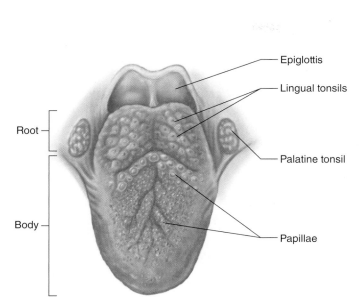

FIGURE 19-4 The surface of the tongue, superior view.

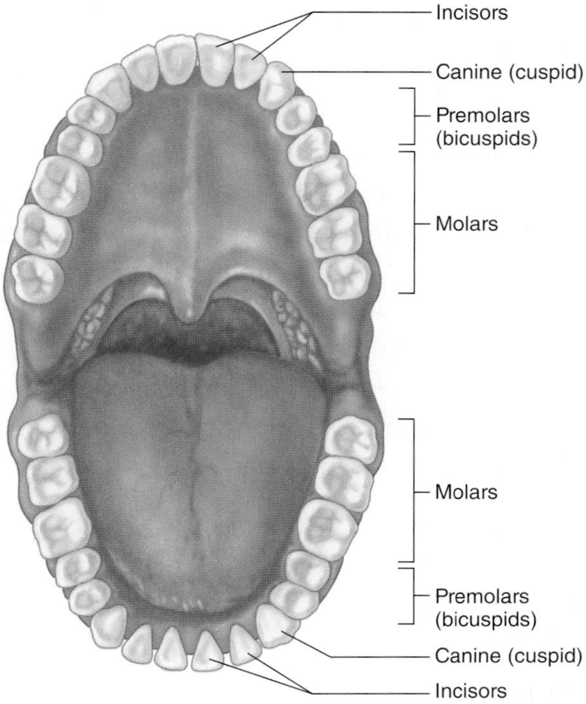

FIGURE 19-5 The permanent teeth of the upper and lower jaws.

this condition consists of clipping the frenulum to free up the tongue. On the dorsum (upper surface) of the tongue are many small projections called *papillae* (Figure 19-4). Some papillae contain taste receptors (taste buds) and others contain touch receptors. The tongue helps mix food and holds it between the teeth. The back of the tongue contains lymphatic tissue, called lingual tonsils, which destroy bacteria and viruses on the back of the tongue (refer to Chapter 12, The Lymphatic and Immune Systems).

The roof of the mouth includes the hard and soft palates. The hard palate is formed by the maxillary and palatine bones and is covered by a mucous membrane and stratified squamous epithelium. The soft palate is just posterior to the hard palate. It is formed by muscle and is also covered by a mucous membrane. The palate separates the oral cavity from the nasal cavity. Projecting off the posterior aspect of the soft palate is the muscular **uvula.** The uvula prevents food and liquids from entering the nasal cavity during swallowing. Just lateral to the uvula are the palatine tonsils. The lingual tonsils, palatine tonsils, and pharyngeal tonsils protect the area from bacteria and viruses. Refer to Chapter 12, The Lymphatic and Immune Systems, for a more in-depth discussion of the tonsils.

uvula The projection off the posterior aspect of the soft palate that prevents food and liquids from entering the nasal cavity during swallowing.

U check
What is the difference between the vestibule and the oral cavity proper?

As you should know now, the teeth break up food and each tooth type is adapted to handle food in different ways (Figure 19-5). The most medial teeth, called *incisors,* cut off food pieces. The cuspids, also known as the "canines," are the sharpest teeth and they tear tough food. The back teeth, called premolars (bicuspids) and molars, are flatter. Both are designed to grind food. During

> ### focus on Wellness: Oral Cancer
>
> Oral cancer has a strong association with tobacco use. Cigarette smoking is well recognized as a cause of lung cancer. Other types of tobacco use impact the health of the oral cavity. For example, pipe smoking "burns hotter" than cigarette smoking. It therefore can cause damage to the palate and over a long period of time can cause cancer. Long-term cigar smoking is associated with lip cancer. Chewing tobacco is related to cancer of the mouth including the tongue. Alcohol use can increase the risk of some of these cancers. Many of these cancers of the tongue and mouth spread quickly because of the highly vascular nature of this area.
>
> Prevention—not using tobacco products—is the best way of not getting oral cancer. The disease is more common in young men. Regular examinations by a dentist can help detect these cancers in the early stages and therefore improve the prognosis.

our lifetime, humans have two sets of teeth: the primary or *deciduous* dentition ("baby" teeth) and the secondary or permanent dentition (adult teeth). Most children have 20 primary teeth, and adults have 32 secondary teeth.

> **U check**
> Which dentition is considered the deciduous dentition?

Salivary glands secrete saliva, which is a mixture of water, enzymes, and mucus (Figure 19-6). Salivary glands are made of two types of cells—serous

FIGURE 19-6
Location of the major salivary glands.

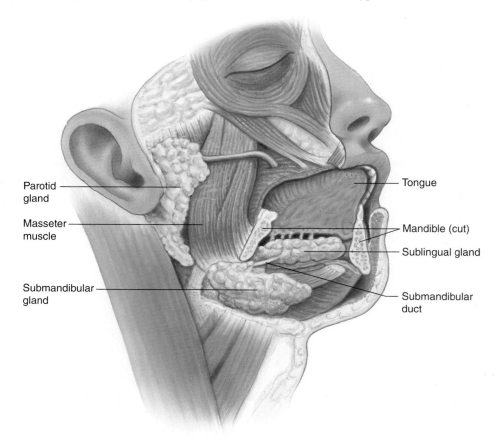

cells and mucous cells. Serous cells secrete a fluid made up mostly of water and also amylase. As their name implies, mucous cells secrete mucus. The mass created by food mixed with the saliva and mucus mixture is called a *bolus*. All major salivary glands are paired as follows:

- The parotid glands are the largest of the salivary glands and are located beneath the skin just in front of the ears. They secrete serous saliva and their ducts open adjacent to the maxillary second molars.
- The submandibular glands are located in the floor of the mouth just inside the surface of the mandibles. They secrete a mixture of serous and mucous saliva.
- The sublingual glands are the smallest of the major salivary glands and are located in the floor of the mouth beneath the tongue. As we get older, the amount of saliva we produce lessens. Some diseases and drugs can also cause us to secrete less saliva. A person with the symptom of "dry mouth" is said to have xerostomia. The underlying cause should be discussed with the patient's dentist or physician.

> **U check**
> What are some causes of xerostomia?

Pharynx

The **pharynx** is more commonly called the throat. It is a long, muscular structure that extends from the area behind the nose to the esophagus (Figure 19-7). It connects the nasal cavity with the oral cavity for breathing through the nose. The pharynx is composed of skeletal muscle and is lined with a mucous membrane. It pushes food into the esophagus (Figure 19-8).

pharynx The throat.

PATHOPHYSIOLOGY

Oral Cancer

Oral cancer usually involves the lips or tongue but can occur anywhere in the mouth. This type of cancer tends to spread rapidly to other organs because of the high vascularity of this area.

Etiology: The cause of 80 to 90 percent of oral cancers is smoking cigarettes, cigars, or pipes, and chewing tobacco. Alcohol use is another major contributor. Some oral cancers have no known cause.

Signs and Symptoms: Signs and symptoms include difficulty tasting, dysphagia (problems swallowing), and ulcers on the tongue, lip, or other oral structures. Leukoplakia, or hardened white patches in the mucous membrane of the mouth, are considered precancerous lesions. They should be examined by a health care professional and biopsied if there is a question as to their malignancy.

Treatment: Radiation therapy, chemotherapy, and surgical removal of the malignant area are the treatment options.

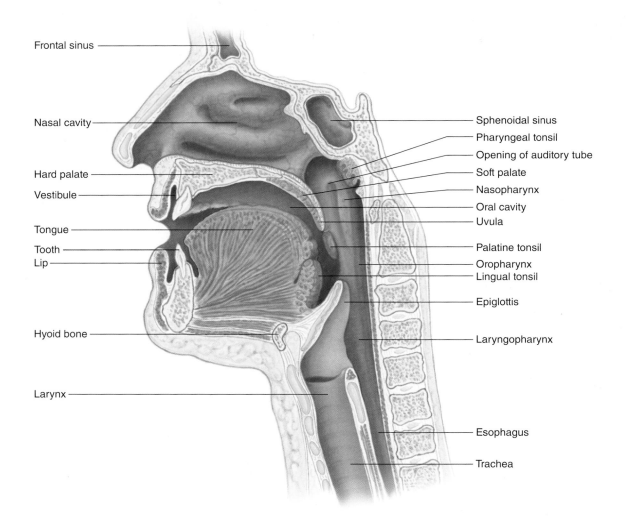

FIGURE 19-7 Sagittal section of the mouth, nasal cavity, and pharynx.

deglutition The act of swallowing.

The divisions of the pharynx are the *nasopharynx*, the portion behind the nasal cavity; the *oropharynx*, the portion behind the oral cavity; and the *laryngopharynx*, the portion behind the larynx. The laryngopharynx continues on as the esophagus. The oropharynx and laryngopharynx are considered to be parts of both the digestive and respiratory systems.

Deglutition (swallowing) is largely a reflex. In other words, it is an automatic response that does not require much thought. The following events occur during deglutition:

1. The soft palate rises, causing the uvula to cover the opening between the nasal cavity and the oral cavity.
2. The epiglottis covers the opening of the larynx so that food does not enter it.
3. The tongue presses against the roof of the mouth, forcing food into the oropharynx.

FIGURE 19-8 The esophagus is the passageway between the mouth and the stomach.

4. The muscles in the pharynx contract, forcing food toward the esophagus.
5. The esophagus opens.
6. Food is pushed into the esophagus by the muscles of the pharynx.

> **U check**
> What structure of the digestive tract is also part of the respiratory system?

Esophagus

The esophagus is a muscular tube that connects the pharynx to the stomach. It is about 10 inches in length and lies posterior to the trachea. It descends through the mediastinum in the thoracic cavity, through the diaphragm, and into the abdominal cavity where it joins the stomach. The opening in the diaphragm through which the esophagus passes is called the *esophageal hiatus*. This hiatus is a common location for hernias to occur. A *hernia* develops when an organ pushes through a wall that contains it. The esophagus has two sphincters that control food entering and exiting it. The *upper esophageal sphincter (UES)* is made up of skeletal muscle and controls food entering the esophagus from the laryngopharynx. The *lower esophageal sphincter (LES)* is made up of smooth muscle and is sometimes called the "cardiac sphincter." It controls food entering the stomach from the esophagus.

Starting with the esophagus and ending at the anal canal, the GI tract is composed of the same four layers, although some are more or less developed depending on the area of region of the GI tract. The four layers are the (1) mucosa or inner lining, (2) submucosa, (3) muscularis, and (4) serosa. The *mucosa* is the GI tract inner lining and, as its name suggests, is a mucous membrane. The *submucosa* is the next layer and it is made up of areolar connective tissue, blood vessels, and a network of nerves called the submucosa plexus. The *muscularis* layer is composed mostly of smooth muscle,

focus on Genetics: Cystic Fibrosis

Mutation in the cystic fibrosis transmembrane conductance regulator (CFTR) gene causes cystic fibrosis (CF). Cystic fibrosis occurs 1 in 2,500 to 3,500 Caucasian births. This condition is inherited in an autosomal recessive pattern, which means both copies of the gene in each cell have mutations. The parents of an individual with an autosomal recessive condition each carry one copy of the mutated gene, but they typically do not show signs and symptoms of the condition. The inherited CF gene directs the body's epithelial cells to produce a defective form of a protein called CFTR found in cells that line the lungs, digestive tract, sweat glands, and genitourinary system. When the CFTR protein is defective, epithelial cells can't regulate the way chloride (part of the salt called sodium chloride) passes across cell membranes. This disrupts the essential balance of salt and water needed to maintain a normal thin coating of fluid and mucus inside the lungs, pancreas, and passageways in other organs. The mucus becomes thick, sticky, and hard to move.

some areas also have skeletal muscle. The *serosa* is the outermost layer and is a serous membrane with areolar connective tissue and simple squamous epithelium. The upper third of the esophagus is skeletal muscle; the middle third is a combination of skeletal and smooth muscle; and the lower third of the esophagus is smooth muscle. Rhythmic muscle movements known as *peristalsis* move food through the esophagus.

> **U check**
> What are the four layers of the digestive tract?

Stomach

The stomach lies below the diaphragm in the left upper quadrant (LUQ) of the abdominal cavity (Figure 19-9). The stomach receives the food bolus from the esophagus. It then mixes the food with *gastric juice* (secretions of the stomach lining), starts protein digestion, and moves food into the small intestine. The beginning portion of the stomach that is attached to the esophagus is known as the *cardiac region*. The portion of the stomach that is superior to the cardiac region is the *fundus*. The main part of the stomach

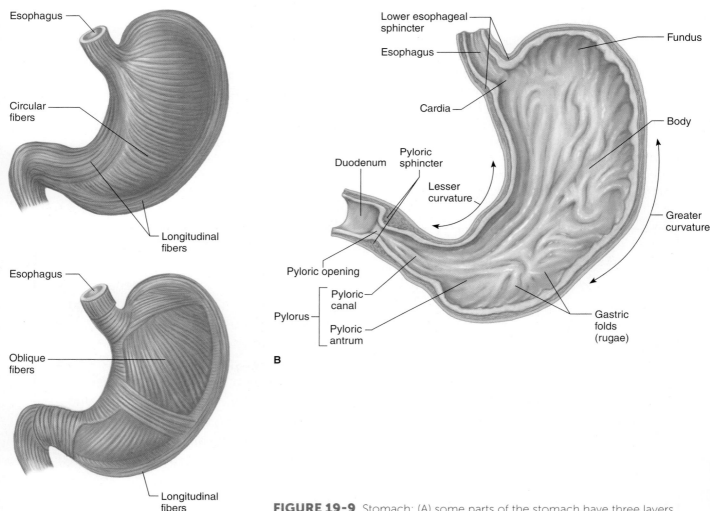

FIGURE 19-9 Stomach: (A) some parts of the stomach have three layers of muscle; (B) major regions of the stomach and its associated structures.

is the body, and the narrow portion that is connected to the small intestine is the pylorus. The **pyloric sphincter** controls the movement of substances from the pylorus of the stomach into the small intestine. Numerous folds on the inner lining of the stomach, called **rugae,** help churn and mix the gastric contents. The lining of the stomach contains gastric glands made of several cell types. Mucous cells secrete mucus to protect the lining of the stomach against the acidic pH found in it. Chief cells secrete *pepsinogen*, which is changed to pepsin in the presence of acid. Pepsin digests proteins.

Parietal cells secrete hydrochloric acid, which is necessary to convert pepsinogen to pepsin (Figure 19-10). They also secrete *intrinsic factor*, which is necessary for vitamin B_{12} absorption. When a person smells, tastes, or sees appetizing food, the parasympathetic nervous system stimulates the gastric glands to secrete their products. The hormone gastrin produced by the stomach stimulates the gastric glands to become active, and the hormone cholecystokinin (CCK) made by the small intestine inhibits gastric glands.

The stomach does not absorb many substances (most absorption takes place in the small intestine) but it can absorb alcohol, water, and some fat-soluble drugs. The mixture of food and gastric juices in the stomach is called *chyme*. Once chyme is well mixed, stomach contractions push small amounts into the small intestine a little at a time. It takes four to eight hours for the stomach to empty following a meal. If a patient is unable to swallow for any reason (aphagia), a tube called a gastrostomy tube, or "G tube," may be inserted into the patient's stomach. He or she can be fed liquid meals, such as Ensure®, through this tube.

pyloric sphincter The opening between the pylorus of the stomach and the small intestine that controls movement of substances from the stomach to the small intestine.

rugae Folds on the inner lining of the stomach that help mix gastric contents.

Food absorption

> **U check**
> What is the purpose of intrinsic factor?

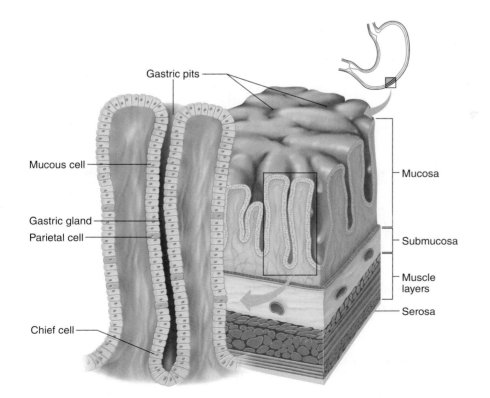

FIGURE 19-10

Lining of the stomach: Gastric glands include mucous cells, parietal cells, and chief cells.

PATHOPHYSIOLOGY

Gastritis

Gastritis is an inflammation of the stomach lining. It is often referred to as an "upset stomach."

Etiology: Gastritis can be caused by bacteria or viruses, some medications, the use of alcohol, caffeine, spicy foods, excessive eating, poisons, and stress. Cooking food properly to kill harmful bacteria and viruses and avoiding stress can help prevent or minimize this condition.

Signs and Symptoms: Symptoms include nausea, lack of appetite, heartburn, vomiting, and abdominal cramps. An upper endoscopy may be done to confirm the diagnosis and rule out more serious conditions, such as an ulcer or cancer.

Treatment: Lifestyle changes should be implemented to avoid foods or medications that irritate the stomach lining. Treatment with medications such as Pepcid® and Nexium® to reduce the production of stomach acids can provide relief from the symptoms of this disorder.

Gastroesophageal Reflux Disease

Heartburn is also called *gastroesophageal reflux disease (GERD)*. This occurs when stomach acids are pushed into the esophagus. If not treated, GERD can cause erosion of the esophagus and even esophageal cancer.

Etiology: Alcohol, some foods, a defective cardiac sphincter, pregnancy, obesity, hiatal hernia, and repeated vomiting can contribute to the development of this disease.

Signs and Symptoms: Common symptoms include frequent burping, difficulty swallowing, pharyngitis, a burning sensation in the chest following meals and when lying down, nausea, and blood in the vomit.

Treatment: Treatment includes losing weight, making dietary changes, reducing the consumption of alcohol, taking medications such as Pepcid® or Nexium®, and elevating the head, neck, and chest when lying down.

Hiatal Hernia

Hiatal hernias occurs when a portion of the stomach protrudes into the thoracic cavity through an opening (esophageal hiatus) in the diaphragm.

Etiology: The causes are mostly unknown, although obesity and smoking are considered risk factors. Eating smaller meals can be an effective preventive measure.

Hiatal Hernia (concluded)

Signs and Symptoms: Signs and symptoms include excessive burping, difficulty swallowing, chest pain, and heartburn.

Treatment: Treatments are weight reduction, medications to reduce the production of stomach acid, and surgical repair of the hernia.

Stomach Cancer

Stomach cancer most commonly occurs in the uppermost (cardiac portion) of the stomach. It appears to occur more frequently in Japan, Chile, and Iceland than in the United States. This may be due to diets of foods high in nitrates that are known to be carcinogenic.

Etiology: Although no single cause has been identified as the cause of stomach cancer, there are several risk factors. For example, the organism *Helicobacter pylori* (*H. pylori*) has been implicated as being responsible for many ulcers, including gastric ulcers, which is a risk factor. Also associated with increased risk of stomach cancers are polyps, some types of atrophic gastritis, and diets high in salt or nitrates.

Signs and Symptoms: Signs and symptoms include frequent bloating, loss of appetite, early satiety (feeling full after eating small amounts), nausea, vomiting (with or without blood), abdominal cramps, excessive gas, and blood in the feces.

Treatment: Treatment includes radiation therapy, chemotherapy, and surgical removal of the tumor.

Gastric Ulcers

Gastric, or *stomach*, *ulcers* occur when the lining of the stomach breaks down.

Etiology: Stomach ulcers can be caused by bacteria such as *Helicobacter pylori* (*H. pylori*), smoking, alcohol, excessive aspirin use, and hypersecretion of stomach acid. Sometimes ulcers may be prevented by stopping smoking and avoiding aspirin, certain foods, and alcohol.

Signs and Symptoms: Symptoms include nausea, abdominal pain, vomiting (with or without blood), and weight loss. Diagnosis can be confirmed by upper endoscopy.

Treatment: Treatment options include antibiotics, medications to reduce stomach acid production, surgery to remove the affected portion of the stomach (partial gastrectomy), and a vagotomy (cutting the vagus nerve) to reduce the production of stomach acid.

U check
What are some risk factors for gastric cancer?

Small Intestine

The small intestine is a coiled, tubular organ that extends from the stomach to the large intestine, filling most of the abdominal cavity. The small intestine carries out most of the digestion in the body and is also responsible for absorbing most of the nutrients into the bloodstream (Figure 19-11).

The first section of the small intestine is called the **duodenum.** It is C-shaped and measures about 10 inches in length. It receives secretions from the pancreas that aid in the digestive process. The middle portion of the small intestine is called the **jejunum** and is about 8 feet in length. If a patient's stomach is diseased or removed, a jejunostomy, or "J-tube," may be inserted into the jejunum to allow the patient to receive nutrition directly into the jejunum. The longest part of the small intestine is the third and final region, the **ileum.** It measures about 12 feet in length and joins the large intestine at the **ileocecal valve.** The jejunum and ileum are held in the abdominal cavity by a broad, fanlike tissue called the **mesentery** that attaches to the posterior wall of the abdomen (Figure 19-12).

In the small intestine, the four layers of the GI wall take on specialized characteristics. The mucosa, composed mostly of enzyme and mucus-secreting epithelial tissue, secretes these substances into the lumen of the canal. This layer also is very active in absorbing nutrients. The blood vessels of the submucosa carry away absorbed nutrients. The muscular layer contracts to move materials through the canal. The serosa, or **peritoneum,** is the double-walled outermost layer of the canal. The innermost wall of the serosa is known as the *visceral peritoneum.* It secretes serous fluid to keep the outside of the canal moist, preventing it from sticking to other organs or to its outer layer, the *parietal peritoneum,* which is also called the abdominal lining.

Smooth muscle in the wall of the canal can contract to produce two basic types of movements—churning (or segmentation) and peristalsis. Churning mixes the substances in the canal and peristalsis propels substances through the tract (Figure 19-13). The cells of the lining of the small intestine have

> **duodenum** The first section of the small intestine that receives secretions from the pancreas to aid in digestion.
>
> **jejunum** The middle portion of the small intestine.
>
> **ileum** The last and longest section of the small intestine.
>
> **ileocecal valve** The valve that joins the last segment of the small intestine (ileum) with the first segment of the large intestine (cecum).
>
> **mesentery** Broad, fanlike tissue that attaches to the posterior wall of the abdomen and holds the jejunum and ileum in the abdominal cavity.
>
> **peritoneum** The double-walled outermost layer of the small intestine.

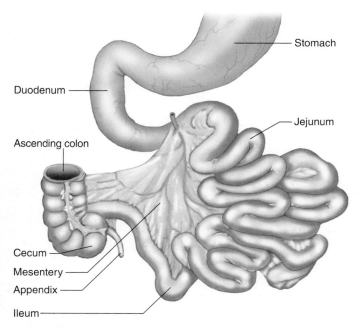

FIGURE 19-11 The three parts of the small intestine are the duodenum, jejunum, and ileum.

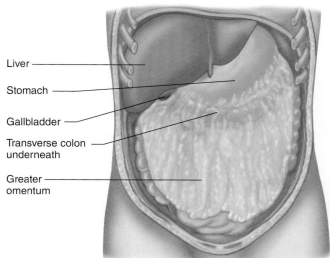

FIGURE 19-12 The mesentery hangs like a curtain over the abdominal organs.

focus on Wellness: GI Cancer Prevention and Screening

As the incidence of GI cancers in the United States continue to rise, there is a renewed interest in prevention and early screenings. It is suspected that the high-fat fast-food diet of many Americans may be an underlying cause of this increase. Physicians and other health practitioners are urging people to cut down on high-fat foods and greatly increase the intake of fresh fruits and vegetables, particularly in their raw state. In addition, increased activity and exercise and their resulting weight loss are also encouraged.

Early screenings beginning at age 50 (or younger if there is a family history), including routine colonoscopy to view the colon for polyps and other lesions, is recommended. For many people the initial reaction is "Ugh, I am not having *that* done." However, celebrity Katie Couric, whose husband died of colon cancer, believes in early detection so strongly that she underwent a colonoscopy on television so the public could see that it is not such a "big deal." Most people who have undergone colonoscopy say the preparation is the worst part, as you "clean yourself out" so that the lower GI tract can be viewed without fecal obstruction, and that once the conscious sedation is administered you enter "the twilight zone" and remember nothing.

In addition to colonoscopy, another test done to view the lower GI tract is the barium enema, although colonoscopy is the screening exam of choice. An upper GI series may include a barium swallow and upper endoscopy, which examines from the pharynx through to the stomach.

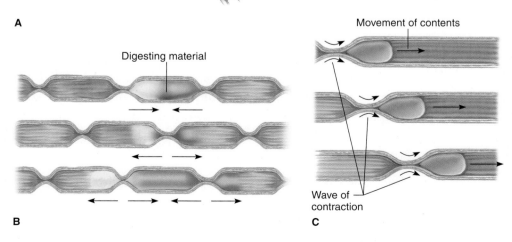

FIGURE 19-13
Mixing movements through the alimentary canal: (A) mixing movements, (B) segmentation, and (C) peristalsis.

CHAPTER 19   The Digestive System

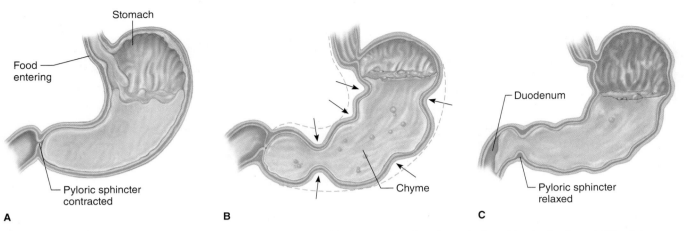

FIGURE 19-14 Stomach movements: (A) as the stomach fills, it stretches but the pyloric sphincter is closed; (B) mixing movements produce chyme; (C) peristalsis moves the chyme toward the pyloric sphincter which relaxes, allowing some chime to enter the duodenum.

Lactose intolerance

cecum The first portion of the large intestine.

microvilli, small fingerlike projections, that greatly increase the surface area of the small intestine so that it can absorb many nutrients. The lining of the small intestine also contains intestinal glands that secrete various substances including mucus and water. Water aids in digestion, but some toxins cause the secretion of too much water and this leads to diarrhea which in turn aids the body in eliminating the toxins.

Mucus protects the lining of the small intestine. Enzymes secreted by the small intestine include peptidase, sucrase, maltase, lactase, and intestinal lipase. Peptidase digests proteins, whereas sucrase, maltase, and lactase digest sugars. A person who cannot produce lactase will not be able to digest lactose, which is the sugar in dairy products—a condition called lactose intolerance. Lactose intolerance is most common among people of Asian, African, Native American, and Hispanic descent and for most diagnosed with this condition, it is a lifelong problem.

The small intestine ends at the ileocecal valve which, as stated earlier, controls the movement of chyme from the ileum to the **cecum,** the first portion of the large intestine (Figure 19-14).

> **U check**
> What is the difference between churning (segmentation) and peristalsis?

PATHOPHYSIOLOGY

Crohn's Disease

Crohn's disease is a type of inflammatory bowel disease. It can affect any region of the digestive tract from the mouth to the anus, but most often affects the ileum in the small intestine. For this reason it is sometimes called "terminal ileitis."

Etiology: Because the cause has not yet been identified, Crohn's is described as an "idiopathic" disease. There is some evidence that Crohn's disease has a genetic component since it tends to run in

Crohn's Disease (concluded)

families. Twenty to 25 percent of patients with Crohn's disease have a relative with the disease or ulcerative colitis. There is evidence that Crohn's has an autoimmune component and, although it can occur at any age, it is most common in young adults. Some environmental factors have been linked to initial episodes or relapses and the condition is more common in Western, industrialized societies.

Signs and Symptoms: Symptoms include abdominal pain and diarrhea. Rectal bleeding, weight loss due to malnutrition, joint pain, skin problems, and febrile episodes may be present. Some people have long periods of remission when they are free of symptoms.

Treatment: The first treatment is to change the patient's diet. Other treatments include medications to reduce inflammation, including steroids, as well as antibiotics and bowel rest where IV (intravenous) feedings are given so the patient's digestive system is not used and so "rests." In some situations, surgery to remove the affected part of the intestine (enterectomy) may be needed.

Diarrhea

Diarrhea is the condition of watery and frequent feces. Many cases of diarrhea do not require treatment because they are usually self-limiting and stop within a day or two.

Etiology: The causes of diarrhea include bacterial, viral, or parasitic infections of the digestive system. It may also be caused by the ingestion of toxins; food allergies, including lactose intolerance; ulcers; Crohn's disease; laxative use; antibiotics; chemotherapy; and radiation therapy. Diarrhea related to infections may be prevented by thoroughly washing hands and food preparation utensils, and cooking food properly.

Signs and Symptoms: The symptoms include abdominal cramps, watery feces, and the frequent passage of feces.

Treatment: Patients should drink clear fluids to prevent dehydration. The underlying cause, if known, should be treated. Medications and dietary changes are the primary treatment options. In severe cases, antidiarrheal medications such as Lomotil® may be prescribed.

Large Intestine (colon)

The large intestine extends from the ileocecal valve, which joins the small intestine to the large intestine, to where it opens to the outside of the body as the anus (Figure 19-15). The large intestine measures about 5 feet in length and 2.5 inches in diameter (the diameter of the small intestine averages about 1 inch). The beginning of the large intestine is the cecum. Projecting off the cecum is the **vermiform appendix.**

vermiform appendix
The wormlike projection off the cecum that contains lymphoid tissue and has a role in immunity.

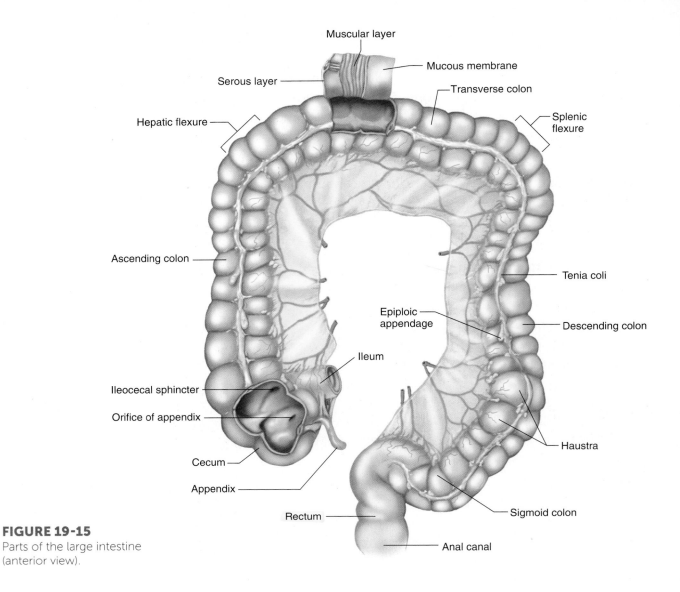

FIGURE 19-15
Parts of the large intestine (anterior view).

Vermiform means "wormlike," which describes the appearance of the appendix. The appendix contains lymphoid tissue and has a role in immunity. The cecum gives rise to the *ascending colon,* which is the portion of the large intestine that runs up the right side of the abdominal cavity. Remember that the appendix is in the right lower quadrant (RLQ); it will then be easy to remember that the ascending colon also goes up the right side of the abdomen. The ascending colon makes a turn toward the midline. Because this is near the liver, this bend or turn of the colon is called the hepatic flexure. The ascending colon becomes the *transverse colon* as it horizontally crosses the abdominal cavity. Near the spleen, it takes another turn and becomes the *descending colon.* This turn is called the splenic flexure. In the pelvic cavity, the descending colon then forms the S-shaped tube called the *sigmoid colon.* The function of the large intestine is to absorb water, produce certain vitamins (for example, vitamin K), and form and expel the feces from the body.

U check
What are the components of the large intestine?

PATHOPHYSIOLOGY

Colitis

Colitis is defined as inflammation of the large intestine. This condition can be chronic or short-lived, depending on the cause.

Etiology: Colitis can be caused by a viral or bacterial infection or the use of antibiotics. Ulcers in the large intestine, Crohn's disease, various other diseases, and stress may also contribute to the development of this disorder.

Signs and Symptoms: The primary symptoms are abdominal pain, bloating, and diarrhea.

Treatment: The first goal of therapy is to treat any underlying causes. Changing antibiotics, treating existing ulcers, and drinking plenty of fluids are additional treatment options. In advanced cases, surgery to remove the affected area of the colon, known as a *colectomy*, may be recommended. If too much of the colon is affected, a *colostomy* may be performed. In this procedure, the majority of the colon is removed and the opening to the outside of the body is moved to the abdomen where an appliance known as an ostomy is placed. Often a collecting device, commonly called a bag, attaches to the ostomy and collects the fecal material.

Colorectal Cancer

Colorectal cancer usually arises from the lining of the rectum or colon. This type of cancer is curable if diagnosed and treated early. It is the third most common cause of death due to cancer in both men and women.

Etiology: The cause of colon cancer is really not certain, but there are definite risk factors. The biggest risk factor is age. Most colorectal cancer is diagnosed in those over 60 years of age. Having a first-degree relative who had colorectal cancer is a risk factor. Certain ethnicities have a higher risk. For example, Jews of Eastern European descent are at greater risk. A diet high in fat and low in fiber, as well as smoking and alcohol put people at risk. Another "danger signal" of increased risk is the finding of multiple colonic polyps.

Signs and Symptoms: Anemia, unintended weight loss, abdominal pain, blood in the feces, narrow "pencil-like" feces, or changes in bowel habits are common symptoms.

Treatment: Chemotherapy is the first line of treatment. Surgery to remove a cancerous tumor or the affected portions of the colon or rectum (colectomy or colostomy) may be needed in more advanced cases. Prevention through regular screenings (colonoscopy) will help with early diagnosis.

(continued)

Constipation

Constipation is the condition of difficult defecation or elimination of feces.

Etiology: The primary causes are a lack of physical activity, lack of adequate fiber and water in the diet, use of certain medications, and thyroid and colon disorders.

Signs and Symptoms: Common signs and symptoms include infrequent bowel movements (for example, no bowel movement for three days), bloating, abdominal pain and pain during bowel movements, hard feces, and blood on the surface of feces.

Treatment: Treatment includes an increase in dietary fiber, adequate fluid intake, regular exercise, and the use of stool softeners, laxatives, and enemas (for extreme cases only).

Diverticulitis

Diverticulitis is inflammation of diverticuli in the intestine. Diverticuli are abnormal dilations or pouches in the intestinal wall. When the diverticuli are not inflamed, the condition is known as *diverticulosis*.

Etiology: The causes are mostly unknown, but a lack of fiber in the diet and bacterial infection of the diverticuli are associated with this disorder. Both of these conditions seem to be a problem of industrialized countries that have low-fiber diets. It is not common in countries that have high-fiber diets. Patients have found that certain foods, such as peanuts and seeds, may aggravate this disorder.

Signs and Symptoms: Signs and symptoms include fever, nausea, abdominal pain, constipation or diarrhea, blood in the feces, and a high white blood cell count.

Treatment: Treatments include a diet high in fiber, antibiotics, and keeping a food diary to track foods that cause flare-ups. Colectomy to remove the affected portion of the intestine may be necessary in severe cases.

Inguinal Hernias

Inguinal hernias occur when a portion of the large intestine protrudes into the inguinal canal, which is located where the thigh and the body trunk meet. In males, the hernia can also protrude into the scrotum.

Etiology: The causes are mostly unknown, although these hernias may be caused by weak muscles in the abdominal walls.

Signs and Symptoms: A lump in the groin or scrotum, or pain in the groin area that gets worse when bending or straining are the common symptoms.

Inguinal Hernias (concluded)

Treatment: Pain medications may be prescribed. Surgery to repair the hernia consists of pushing the large intestine back into the abdominal cavity.

Appendicitis

Appendicitis is an inflammation of the appendix. If not treated promptly, it can be life threatening.

Etiology: This disorder may be caused by blockage of the appendix with feces or tumor, infection, or other idiopathic (unknown) cause.

Signs and Symptoms: The signs and symptoms include lack of appetite, pain in the RLQ that may radiate throughout the abdomen and even down the right leg, nausea, slight fever, and an increased white blood cell count. Undiagnosed abdominal pain of 24 hours should be considered appendicitis until proven otherwise.

Treatment: The primary treatments are antibiotics to prevent infection and appendectomy (surgery to remove the appendix).

Rectum and Anal Canal

Eventually the sigmoid colon straightens out to become the *rectum* which measures about 8 inches in length. The last inch of the rectum is known as the anal canal, and the opening of the anal canal to the outside of the body is called the anus (Figure 19-16).

The lining of the large intestine secretes mucus to aid in the movement of substances. As chyme leaves the small intestine and enters the large intestine, the proximal portion of the large intestine absorbs water and a few electrolytes from it. The leftover chyme is then called feces. Feces consist of undigested solid materials, a little water, ions, mucus, cells of the intestinal lining, and bacteria. The contractions of the large intestine propel feces forward, but these contractions normally occur periodically and as mass movements. Mass movements trigger the defecation reflex, which allows the anal sphincters to relax and feces to move through the anus in the process of defecation (elimination). The squeezing actions of the abdominal wall muscles also aid in the emptying of the large intestine. The anal sphincters include an internal sphincter made up of smooth muscle and an external anal sphincter made up of skeletal muscle.

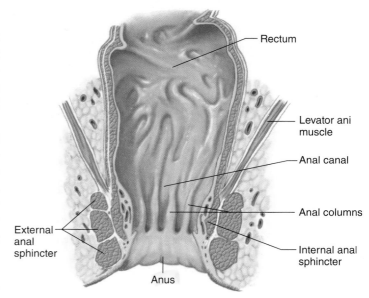

FIGURE 19-16 The rectum and anal canal are at the end of the alimentary canal.

focus on Wellness: Stool Specimens

With the number of colon cancer diagnoses on the rise in the United States, there is a new focus on early screening to increase the number of early diagnoses and do successful treatment. One of the most common early screening tests done in the physician office is the *Hemoccult*® test. Specimens are obtained in two ways: During the patient's physical exam, the practitioner often performs a digital rectal exam (DRE) with a gloved finger. A small amount of stool from the examining finger is deposited on a chemically treated *Hemoccult* slide. When a drop or two of developing fluid is deposited on the specimen, it will turn blue if blood is present, which may indicate that further testing is required.

The same test can be done by sending the patient home with a three-day "kit" in which he or she deposits a small amount of stool on specialized cards for three days in a row and sends them back to the office for development. Larger specimens of stool may also be checked in laboratories when looking for worms or other parasites that may cause a myriad of other GI tract infections.

PATHOPHYSIOLOGY
Hemorrhoids

Hemorrhoids are *varicosities* (varicose veins) of the rectum or anus.

Etiology: Hemorrhoids are caused by constipation, excessive straining during bowel movements, liver disease, pregnancy, and obesity.

Signs and Symptoms: Signs and symptoms include itching of the anal area, painful bowel movements, bright red blood on feces, and varicosities (veins) that protrude from the anus.

Treatment: Constipation can be avoided or improved by eating a high-fiber diet and drinking adequate water during the day. Other treatments include stool softeners, medications to reduce the inflammation of hemorrhoids, and banding or surgical removal of hemorrhoids (hemorrhoidectomy).

Describe each accessory organ's structure, specific function(s), and related disorders.

Accessory Organs and Their Functions

Certain organs, such as the pancreas, liver, and gallbladder, contribute to digestion but are not part of the alimentary canal and food does not travel through them. Thus, they are called accessory organs. In some case, such as with the pancreas and liver, these organs also perform functions not related to digestion.

Pancreas

The pancreas is located behind the stomach. It is composed of a head, a body, and a tail. The head makes up about 50 percent of the pancreas and fits in the "C" of the duodenum. The tail is positioned toward the left with the body in between (Figure 19-17). The pancreas is both endocrine and exocrine in function. Only 1 percent of the pancreas is endocrine and 99 percent is exocrine. The "endocrine" side of the pancreas is discussed in Chapter 21, The Endocrine System. Its exocrine function consists of pancreatic acinar cells that produce pancreatic juice, which ultimately flows through the pancreatic duct to the duodenum. Pancreatic juice contains amylase, lipase, nucleases, trypsin, chymotrypsin, and carboxypeptidase. Pancreatic amylase (*amyl* = starch) digests carbohydrates. Remember that the salivary glands also secrete amylase, but the acids in the stomach inactivate the amylase, so much of the carbohydrate that enters the duodenum is undigested. Pancreatic lipase digests lipids (*lipo* = fats). Nucleases digest nucleic acids, while trypsin, chymotrypsin, and carboxypeptidase digest proteins. When stimulated by the parasympathetic nervous system, the pancreas also secretes bicarbonate ions into the duodenum which neutralize the acidic

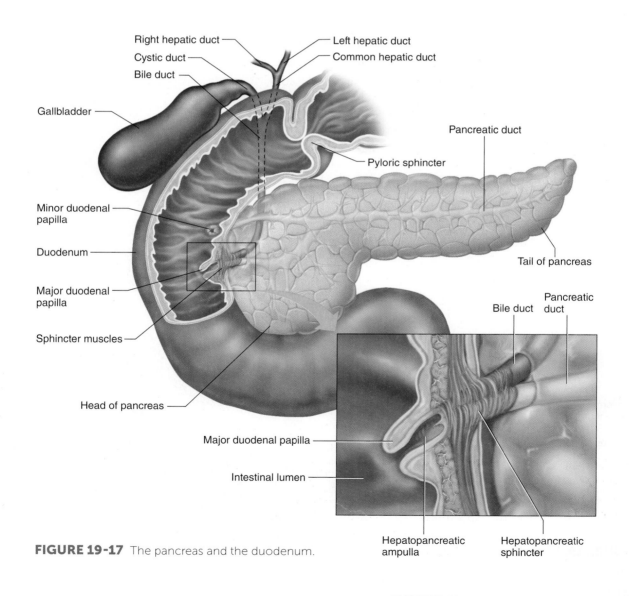

FIGURE 19-17 The pancreas and the duodenum.

chyme arriving from the stomach. The hormones secretin and cholecystokinin that come from the small intestine also stimulate the pancreas to release digestive enzymes.

> **U check**
> What are the two main functional parts of the pancreas?

PATHOPHYSIOLOGY
Pancreatic Cancer

Pancreatic cancer is the fourth leading cause of cancer death in the United States. The poor prognosis and five-year survival rate of only 5 percent is largely due to the late diagnosis.

Etiology: Causes are mostly unknown, although smoking and alcohol consumption are considered risk factors. Additional risk factors are increased age, race (African Americans are more at risk), being male, and a history of pancreatitis.

Signs and Symptoms: Common signs and symptoms include depression, fatigue, lack of appetite, nausea or vomiting, abdominal pain, constipation or diarrhea, jaundice, and unintended weight loss.

Treatment: Treatment includes radiation therapy, chemotherapy, and surgical removal of the tumor.

Liver and Gallbladder

The liver is the largest visceral organ in the body. It weighs about 3 pounds and occupies most of the right upper quadrant (RUQ). It is reddish-brown in color and is enclosed by a tough fibrous capsule (Figure 19-18). This capsule divides the liver into a large right lobe and a smaller left lobe by a fold of mesentery called the *falciform ligament*. Each lobe is separated into

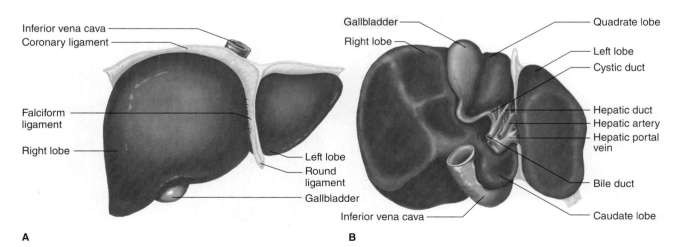

FIGURE 19-18 Lobes of the liver viewed (A) anteriorly and (B) inferiorly.

UNIT 2 Concepts of Common Illness by System

smaller divisions called hepatic lobules. Branches of the hepatic portal vein carry blood from the digestive organs to the hepatic lobules. The hepatic lobules contain macrophages that destroy bacteria and viruses in the blood. Each lobule contains many cells called *hepatocytes*. Hepatocytes process the nutrients in blood and make *bile*, which is used in the emulsification (digestion) of fats. Bile leaves the liver through the *hepatic duct*. The hepatic duct merges with the *cystic duct* (the duct from the gallbladder) to form the *common bile duct*, which delivers bile to the duodenum.

The functions of the liver include carbohydrate, lipid, and protein metabolism; storage of excess glucose as glycogen; synthesization of cholesterol; storage of triglycerides; and breakdown of fatty acids. In addition, the liver produces many proteins, including albumin and fibrinogen, and metabolizes and detoxifies many substances such as alcohol and drugs. It also produces bile, which is stored in the gallbladder. The liver also stores vitamins A, B_{12}, D, E, and K and helps activate vitamin D (assisting the skin and kidneys). Finally, the phagocytes in the liver, called Kupffer cells, phagocytize red and white blood cells and bacteria.

Medication metabolism

U check
What are the functions of the liver?

PATHOPHYSIOLOGY

Hepatitis

Hepatitis (*hepat* = liver; *itis* = inflammation) is defined as inflammation of the liver. There are many different types of hepatitis (see the Focus on Clinical Applications feature), but because they all involve inflammation of the liver, they share many signs and symptoms.

Etiology: The two major causes of hepatitis are alcohol and viruses, but other causes include bacteria, parasites, immune disorders, and an overdose of acetaminophen. If the cause is viral, depending on the virus involved, hepatitis can be contracted through contaminated water or food, sex, and blood products. Preventive measures include getting the hepatitis B virus (HBV) vaccinations, practicing safer sex, avoiding undercooked food (especially seafood), and using prescription or over-the-counter drugs (especially those containing acetaminophen) at their recommended dosages or as prescribed by a physician.

Signs and Symptoms: Symptoms include mild fever, bloating, lack of appetite, nausea, vomiting, abdominal pain, weakness, and jaundice, the itching of various body parts, an enlarged liver, dark urine, and *gynecomastia*—breast development in males.

Treatment: Patients should avoid using alcohol and intravenous drugs. Various medications may be prescribed.

focus on Clinical Applications: Hepatitis

Liver failure

There are several variations of hepatitis:

- *Hepatitis A.* Hepatitis A is caused by the hepatitis A virus (HAV). Hepatitis A is spread mainly through the fecal-oral route. For example, HAV can spread in daycare or health care settings when an attendant changes the diaper of an infected child, then helps another child with feeding before performing adequate hand washing. The disease is rarely fatal (the recovery rate is 99 percent), and there is a vaccine that prevents infection.

- *Hepatitis B.* Hepatitis B is the most common bloodborne hazard that health care professionals face. It is spread through contact with contaminated blood or body fluids and through sexual contact. Most patients recover fully from HBV infection, but some patients develop chronic infection or remain carriers of the pathogen for the rest of their lives. Adults and children with hepatitis B who develop lifelong infections may experience serious health problems including cirrhosis (scarring of the liver), liver cancer, liver failure, and death.

    Preventing the spread of the infection is the most effective means of combating the disease. Following standard precautions and receiving the HBV vaccination are the most effective ways to control the spread of the infection. The Occupational Safety and Health Administration (OSHA) has established guidelines that require employers to offer this vaccine to employees who are at risk of exposure to HBV while performing their job. The offer must be made within 10 days of employment and at no cost to the employee. If they decline the offer, they must sign a waiver. Current standard medical practice recommends against declining the vaccine. If employees are exposed to hepatitis B and have not been vaccinated, they can receive a postexposure inoculation of hepatitis B immune globulin (HBIG). HBIG is given in large doses during the seven-day period after exposure. Shortly after beginning the treatment, they also receive the HBV vaccination. The HBIG inoculation also is used for infants born to HBV-infected mothers.

- *Hepatitis C.* Also referred to as non-A/non-B, hepatitis C is spread through contact with contaminated blood or body fluids and through sexual contact, so it is a risk for health care professionals as well. There is no cure for this variant, which has resulted in more deaths than hepatitis A and hepatitis B combined. Many people become carriers of hepatitis C without knowing it because they do not experience any symptoms of the virus. If the infection causes immediate symptoms, they often resemble the flu. Although treatment exists to suppress the virus, nothing can prevent or stop the virus from replicating. Over time it is likely to damage the liver, causing cirrhosis, liver failure, and cancer. As with hepatitis B, preventing the spread of the infection is the best way to combat the disease.

- *Hepatitis D.* Hepatitis D (delta agent hepatitis) occurs only in people infected with HBV. Delta agent infection may make the symptoms of hepatitis B more severe, and it is associated with liver cancer. The HBV vaccine also prevents delta agent infection.

- *Hepatitis E.* Hepatitis E is caused by hepatitis E virus (HEV). Hepatitis E is transmitted by the fecal-oral route, usually through contaminated water. Chronic infection does not occur, but acute hepatitis E may be fatal in pregnant women. Proper hand hygiene is a must.

PATHOPHYSIOLOGY
Cirrhosis

Cirrhosis is a chronic liver disease in which normal liver tissue is replaced with nonfunctional scar tissue.

Etiology: Chronic alcoholism is the number one cause of cirrhosis of the liver. Men and women respond differently to alcohol. It takes much less alcohol to damage a woman's liver than a man's liver. People who drink more have a higher risk of developing cirrhosis than those who drink less. Other causes of cirrhosis include hepatitis B and C infections, certain immune disorders, and exposure to toxic metals such as copper and iron.

Signs and Symptoms: The many symptoms of cirrhosis include anemia, fatigue, mental confusion, fever, vomiting, blood in the vomit, *hepatomegaly* (enlarged liver), jaundice, unintended weight loss, edema (swelling of the legs), *ascites* (fluid buildup in the abdomen), abdominal pain, decreased urine output, and pale feces.

Treatment: Alcohol consumption should be discontinued. A patient with cirrhosis may be given various medications, including antibiotics and diuretics. A liver transplant may be needed for the most seriously ill patients.

The gallbladder is a small, saclike or pear-shaped structure located beneath the liver (Figure 19-19). Its function is to store bile. Bile leaves the gallbladder through the cystic duct. The hormone cholecystokinin causes the gallbladder to release bile. The salts in bile emulsify (break up) large fat globules into smaller ones so that they can be more quickly digested by various enzymes. Bile salts also increase the absorption of fatty acids, cholesterol, and fat-soluble vitamins into the bloodstream.

PATHOPHYSIOLOGY
Cholelithiasis

Cholelithiasis (*chole* = gall; *lith* = stone; *iasis* = condition) is a condition in which hardened deposits of bile form in the gallbladder. Two types of calculi are cholesterol and pigment stones.

Etiology: Causes of gallstone formation may be due to too much cholesterol in the bile, too much bilirubin in the bile, or inadequate emptying of the gallbladder. Risk factors for gallstones include being female, over 40 years of age, obese, Mexican American, Native American, or pregnant and having diets high in fat and cholesterol and a family history of cholelithiasis.

(continued)

REMEMBER SANDY, who came to the gastroenterologist's office complaining of severe pain in her right upper abdomen? Why does cholelithiasis cause abdominal pain?

Cholelithiasis (concluded)

Signs and Symptoms: Patients may be asymptomatic or may have terrible pain in the abdomen if the gallstones become lodged in a duct. Pain may radiate to the right shoulder or right scapular (shoulder blade) region.

Treatment: Cholecystecomy (surgery to remove the gallbladder) is the most common treatment, but medications may be used on those who are unable to undergo surgery. Prevention of attacks involves changing to a low-fat, low-cholesterol diet; increasing fiber in the diet; and adding certain vitamins such as vitamin C and vitamin E on a daily basis.

U check
What is cholelithiasis?

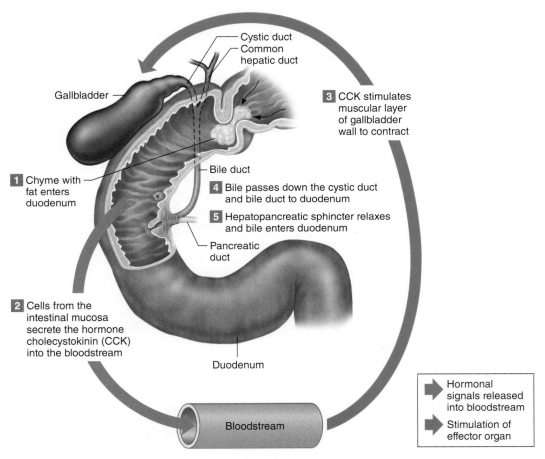

FIGURE 19-19 Fatty chyme enters the duodenum, stimulating the release of bile from the gallbladder.

Phases of Digestion

Explain the phases of digestion.

There are three phases of digestion: the cephalic phase, gastric phase, and intestinal phase. The *cephalic phase* starts in the cerebral cortex. The mere thought, sight, or smell of food as well as the taste of food can activate the cortex, hypothalamus, and brain stem. Cranial nerve VII, the facial nerve, and cranial nerve IX, the glossopharyngeal nerve, stimulate the secretion of saliva. Cranial nerve X, the vagus nerve, stimulates the stomach to secrete gastric juices. (Refer to Chapter 14, The Nervous System, for a review of the cranial nerves.)

The *gastric phase* is under the regulation of neural and hormonal control. Stretch and chemical receptors in the stomach activate the parasympathetic system to increase gastric secretions and increase peristalsis and emptying of the stomach. Gastrin is a hormone secreted by the stomach in response to stretching of the stomach or the presence of protein and caffeine, to mention a few stimuli. This increases the secretions of gastric acid and aids in gastric emptying. The *intestinal phase* also involves neural and hormonal regulation. It inhibits gastric emptying to allow the small intestine time to absorb the nutrients that enter it.

U check
What are the three phases of digestion?

Life Span

Relate the effects of aging on the digestive system.

Changes related to aging are evident throughout all body systems, and the digestive system is no exception. In general, the mucous lining thins and the blood supply and number of smooth muscle cells in the alimentary canal decrease, which leads to decreased motility. The decreasing motility often causes gastroesophageal reflux disease (GERD) and constipation. The decreasing blood supply also affects the patient's ability to absorb medications and nutrients. Secretions from the stomach, liver, and pancreas decrease as well, and the liver's ability to detoxify the blood lessens, making the prescribing of medication dosages more difficult.

Individuals who drink alcohol may notice a difference in the way alcohol affects them. Advancing age also makes it more likely for the person to develop ulcerations and cancers of the GI system, including colorectal cancer. Many elderly people say that their sense of taste is altered and that food loses some of its appeal, and eating, its enjoyment. Bowel obstruction and impaction are both more common with age, as is divertculosis and diverticulitis.

If loneliness and/or isolation is also present, the individual's diet will likely change because "it is no fun or too much work to cook just for myself." The person's diet may also change if he or she is experiencing depression. As a health care professional you should be aware of these possibilities, looking for clues such as weight loss, depression, or patient comments, and should relate information to the health care provider with a goal of securing social service assistance for the individual if needed.

summary

chapter 19

learning outcomes	key points
19.1 Identify the general functions of the digestive system.	The overall functions of the digestive system include ingestion, secretion, mixing, propulsion, digestion, absorption, and defecation.
19.2 Describe each digestive system organ's structure, specific function(s), and related disorders.	The cheeks, lips, palate, teeth, tongue, and salivary glands make up the mouth. The structures of the mouth are responsible for the ingestion, mastication, dissolution, and lubrication of food. The pharynx is a hollow muscular tube that connects the mouth to the esophagus. The esophagus is a muscular tube; the upper third is skeletal muscle, the middle third is a mixture of skeletal and smooth muscle, and the final third is smooth muscle. It moves food into the stomach through peristalsis. The stomach has several regions: the cardiac, fundus, body, and pylorus. It mixes food and adds acids and enzymes for digestion. The small intestine, like most of the GI tract, has four layers—the mucosa, submucosa, muscularis, and serosa. It is about 10 feet in length and is responsible for absorbing the majority of nutrients. The large intestine is made up the ascending, transverse, descending, and sigmoid portions. It absorbs water and produces feces. The anal canal and rectum are the terminus for digestion and the elimination of feces.
19.3 Describe each accessory organ's structure, specific function(s), and related disorders.	The pancreas is made up of acinar cells (exocrine) and islets of Langerhans (endocrine) that secrete enzymes and hormones for the digestion of food and the regulation of glucose. The liver has two major lobes and is highly vascularized. The liver is involved in metabolism, detoxification, synthesis, and activation of vitamins and blood clotting. The liver synthesizes bile, proteins, and cholesterol. It also stores vitamins such as vitamins A, D, E, K, and B_{12}. The gallbladder is a pear-shaped structure on the underside of the liver. The gallbladder stores and releases bile for the emulsification of fats.
19.4 Explain the phases of digestion.	The phases of digestion are the cephalic phase, which involes the cerebral cortex and several cranial nerves; the gastric phase, which involves gastric secretions and the emptying of the stomach; and the intestinal phase, which involves nutrient absorption.
19.5 Relate the effects of aging on the digestive system.	As a person ages, the digestive system becomes less efficient. The person may have a greater risk of GERD, constipation, ulcers, and even cancer. Medication dosages may need adjusting due to the liver's decreased ability to detoxify the blood.

case study 1 questions
Can you answer the following questions that pertain to the case (Sandy Wolbers) presented in this chapter?

1. What is the function of the gallbladder?
   a. Produces bile
   b. Produces gastrin
   c. Stores bile
   d. Stores glucose
2. The gallbladder secretes bile into the
   a. Pancreas
   b. Stomach
   c. Duodenum
   d. Ileum
3. Sandy would benefit from a diet
   a. High in cholesterol
   b. Low in fiber
   c. High in sodium
   d. High in fiber

review questions

1. Which of the following is the correct order of the layers of the gastrointestinal tract from the lumen side to that farthest from the lumen?
   a. Submucosa, muscularis, serosa, mucosa
   b. Mucosa, muscularis, serosa, submucosa
   c. Mucosa, submucosa, muscularis, serosa
   d. Muscularis, mucosa, submucosa, serosa
2. Which one of the following is *not* an accessory organ of digestion?
   a. Liver
   b. Pancreas
   c. Gallbladder
   d. Spleen
3. Motility of the GI tract includes
   a. Segmentation and propulsion
   b. Segmentation and defecation
   c. Propulsion and defecation
   d. Ingestion and defecation
4. Which one of the following is *not* a function of the digestive system?
   a. Absorption
   b. Ingestion
   c. Digestion
   d. Reabsorption
5. Salivary amylase is involved in the chemical breakdown of
   a. Carbohydrates
   b. Fats
   c. Nucleic acids
   d. Proteins
6. The upper esophageal sphincter consists of
   a. Smooth muscle
   b. Lymphoid tissue
   c. Skeletal muscle
   d. A combination of smooth and skeletal muscle
7. Intrinsic factor is necessary
   a. For the absorption of fats
   b. For the absorption of vitamin B_{12}
   c. For the digestion of proteins
   d. For the emulsification of fats
8. Which of the following is a cause of duodenal ulcers?
   a. *Helicobacter pylori*
   b. *Staphylococcus aureus*
   c. *Streptococcus pyogenes*
   d. Herpes simplex 1

CHAPTER 19 The Digestive System

9. If a person has his or her gallbladder removed, which of the following would be the dietary recommendations for the individual?
   a. Eat a diet low in fiber.
   b. Eliminate protein from the diet.
   c. Eat smaller amounts at each meal with limited fats.
   d. Eliminate carbohydrates from the diet.
10. An 82-year-old man has suffered a stroke. Which of the following may be a complication that affects the digestive system?
    a. He is at increased risk of diabetes mellitus.
    b. He may have trouble swallowing because the epiglottis may not be closing off the passageway to the lungs.
    c. He is likely to suffer chronic diarrhea.
    d. He is more at risk of developing Crohn's disease.

critical **thinking questions**
1. Why may a person with chronic gastritis suffer from pernicious anemia?
2. What complications might a person encounter after the removal of his gallbladder?

patient **education**
Early screening is the best way to avoid colon cancer. Patients need to understand that having a colonoscopy and having their stool tested for blood, however disgusted they may be about it, can make the difference between early detection and cure and severe long-term illness or death. Research the Internet about colon cancer prevention and come up with at least three reasons to *convince* a patient to participate in colon cancer screening.

applying **what you know**
1. List the path of the alimentary canal, beginning with the mouth and ending with the anus. Don't forget to include the sphincters and the individual parts of organs where applicable.
2. What substances are normally digested in the small intestine? What substances are normally absorbed through the wall of the small intestine?

20
Metabolic Function and Nutrition

chapter outline

- Carbohydrate Metabolism
- Lipid Metabolism
- Protein Metabolism
- Homeostasis
- Nutrition
- Malnutrition
- Life Span

outcomes

AFTER COMPLETING THIS CHAPTER, YOU WILL BE ABLE TO:

20.1 Describe carbohydrate metabolism.

20.2 Describe lipid metabolism and disorders of lipid metabolism.

20.3 Describe protein metabolism.

20.4 Relate the role of metabolism to homeostasis.

20.5 Relate the major vitamins and minerals to their functions.

20.6 Compare types of malnutrition.

20.7 Explain the relationship between aging and nutrition.

essential terms

acetyl coenzyme A (uh-SEE-tul koh-EN-zime ay)
anabolic reaction (a-nuh-BOL-ik ree-AK-shun)
anaerobic reaction (an-air-OH-bik ree-AK-shun)
basal metabolic rate (BMR) (BAY-zal met-uh-BAH-lik rate)
calorie (KAL-or-ee)
carbohydrates (kar-boh-HY-drayts)
catabolic reaction (kat-uh-BOL-ik ree-AK-shun)
decarboxylation (dee-kar-bok-seh-LAY-shun)
gluconeogenesis (glu-koh-nee-oh-JEH-neh-sis)
glycogenolysis (gli-ko-jeh-NAH-lih-sis)
glycolysis (gli-KOH-lih-sis)
inorganic (in-or-GAN-ik)
kwashiorkor (kwash-ee-OR-kor)
lipid (LIP-id)
lipoprotein (ly-poh-PRO-teen)
malnutrition (mal-nu-TRISH-un)
marasmus (mah-RAZ-mus)
triglycerides (try-GLIS-er-idez)

Additional key terms in the chapter are italicized and defined in the glossary.

case study

Use the case study to focus on as you go through the chapter. The questions will guide you as you learn the concepts of metabolism and nutrition.

CASE STUDY 1 *Nutrition*

The physician you work for has asked you to visit with one of his elderly patients, Tessa White. Your employer would like you to discuss nutrition with the patient. The patient has a 15-year history of type 2 diabetes mellitus and has recently been diagnosed with psoriasis. Other than these two conditions, the octogenarian is in good health and takes no medications except daily insulin injections and a medication for her psoriasis.

As you go through the chapter, keep the following questions in mind:

1. What are the six main types of nutrients that you must discuss?
2. Are there any supplements (minerals or vitamins) that may help Tessa with her diabetes and insulin injections?
3. Are there any age-specific considerations that must be taken into account in discussing nutrition with this patient?

study tips

1. Flowcharts are a great way to learn the different types of cellular respiration.
2. Look at food packaging labels to see what vitamins, minerals, fats, proteins, and carbohydrates are found in some "everyday" foods.
3. Design a healthy diet for Tessa White.

Introduction

anabolic reaction
A metabolic reaction that involves the synthesis of substances.

catabolic reaction
A metabolic reaction that involves the breaking down of larger molecules into smaller molecules.

carbohydrates A category of food that includes starches, simple sugars, and cellulose.

Metabolism is the sum of all chemical reactions in the body. These reactions are classified as anabolic or catabolic. **Anabolic reactions** involve the synthesis of substances; more complex molecules are made from simpler molecules. For example, large, complex proteins are made from smaller, simpler amino acids. Anabolic reactions require energy. **Catabolic reactions** involve the breaking down larger molecules into smaller molecules. These reactions give off more energy than they consume. Examples of important catabolic reactions are glycolysis, the Krebs cycle, and the electron transport chain. Anabolic (synthesis) and catabolic (decomposition) reactions must be in balance for the body to be in homeostasis.

ATP (adenosine triphosphate) is the main "energy currency" of the cell (Figure 20-1). If you had a one-dollar bill for each ATP molecule in a cell, you would be a billionaire. ATP is made up of adenine, ribose (a 5-carbon sugar), and three phosphate groups. The energy of the ATP molecule is in these bonds. Energy and ADP (adenosine diphosphate) result when a phosphate group is broken off from ATP.

U check
What is the difference in terms of energy requirement between anabolic and catabolic reactions?

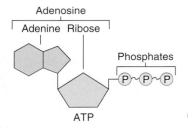

FIGURE 20-1
ATP provides energy for the cell. An ATP (adenosine triphosphate) molecule consists of an adenine, a ribose, and three phosphates. The wavy lines between phosphorus molecules represent high-energy bonds.

Three major food categories are required for balanced metabolism: carbohydrates, lipids, and proteins.

Carbohydrates ingested by humans include starches (polysaccharides), simple sugars (monosaccharides and disaccharides), and cellulose. Starches come from foods such as pasta, potatoes, rice, and breads. Monosaccharides and disaccharides are obtained from sweet foods and fruits. Cellulose is a type of carbohydrate found in many vegetables that cannot be digested by humans. Therefore, cellulose provides fiber or bulk for the large intestine, which helps the large intestine empty more regularly. A connection has been made between higher-fiber diets and a decrease in colon diseases, including cancer. This may be because fiber increases water absorption and bulk, causing more rapid emptying of the colon. This decreases the production of benign growths such as adenomas or polyps, which have been shown to increase the risk of cancer. Fiber may also neutralize toxins produced by bacteria in the gut.

> **U check**
> What are some foods that we eat that contain cellulose?

Lipids (fats) are obtained through various foods. The most abundant dietary lipids are **triglycerides,** which are formed from one molecule of glycerol and three fatty acid molecules. They are found in meats, eggs, milk, and butter. Cholesterol is another common dietary lipid and is also found in eggs, whole milk, butter, and cheeses. Lipids are used by the body primarily to make energy when glucose levels are low. Excess triglycerides are stored in adipose tissue. Cholesterol is essential to cell growth and function; cells use it to make cell membranes and some hormones such as estrogen and testosterone. People should have the essential fatty acid linoleic acid in their diet, because the body cannot make it. This fatty acid is found in corn and sunflower oils. The body also needs a certain amount of fat to absorb fat-soluble vitamins.

lipids Fats.

triglyceride A molecule formed from one molecule of glycerol and three fatty acid molecules.

> **U check**
> What is the meaning of the term *essential* as used in "essential fatty acid"?

Protein is found in meats, eggs, milk, fish, chicken, turkey, nuts, cheese, and beans. Protein requirements vary from individual to individual, but all people must take in proteins that contain certain amino acids (called essential amino acids) because the body cannot make them. Proteins are used by the body for growth and the repair of tissues.

Carbohydrate Metabolism

Carbohydrate metabolism is basically sugar metabolism. Complex sugars are broken down into simpler sugars such as glucose, fructose, and galactose. The most abundant of these simpler sugars is glucose (Figure 20-2). Glucose is the preferred substrate for the synthesis of ATP. A normal blood glucose level is 80 to 100 milligrams per 100 milliliters of plasma. In addition to the production of ATP, glucose can be used in several other ways. Glucose can be used to make some amino acids, which are the building blocks of proteins. Excess glucose can be converted into glycogen, which consists of long chains of glucose molecules linked together and then stored in the liver and skeletal muscle.

20.1 learning outcome
Describe carbohydrate metabolism.

It makes sense that you would expect to find glycogen in skeletal muscle since, when a burst of energy is needed, it is good to have glycogen readily available. Skeletal muscle can store about three times as much glycogen as the liver. When glycogen storage capacity is met, excess glucose can be converted into glycerol and fatty acids for the synthesis of triglycerides, which can then be stored in adipose tissue. To be used by the body, glucose in the plasma must enter the cells. It does this by facilitated diffusion. This means that energy is not required for the entry of glucose into cells. However, there are molecules in the cell

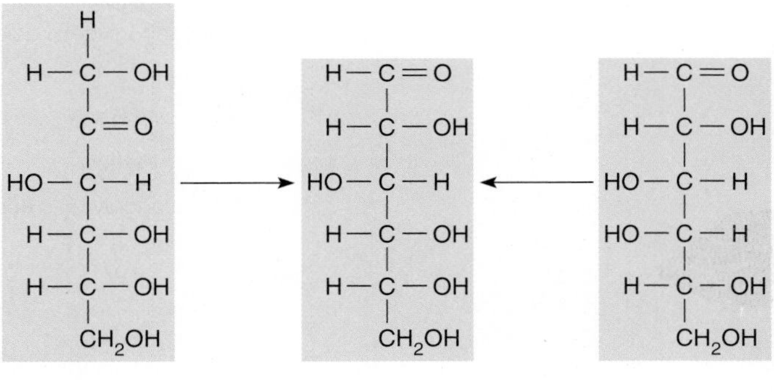

FIGURE 20-2 Liver enzymes catalyze reactions that convert the monosaccharides fructose, and galactose into glucose.

membrane that attach to the glucose molecule. This "facilitates" the passage of glucose into the cell. Insulin causes more of these facilitator molecules to be present in the cell membrane, allowing a more rapid entry of glucose into the cell. Once in the cell, the glucose becomes trapped and cannot leave the cell. There the glucose is oxidized, or "burned," to produce ATP.

Four sets of reactions involve oxidation of glucose, also known as *cellular respiration*: glycolysis, acetyl coenzyme A formation, the Krebs cycle, and the electron transport chain.

glycolysis The net production of two ATP molecules through an anaerobic pathway.

anaerobic reaction A chemical reaction that does not require oxygen.

> **U check**
> What are the two main tissues where glycogen is stored?

Glycolysis, which occurs in the cytoplasm of a cell, does not require oxygen and so is called an **anaerobic reaction** (Figure 20-3). For each glucose

FIGURE 20-3
Chemical reactions of glycolysis. There is a net gain of two ATP molecules from each glucose molecule.

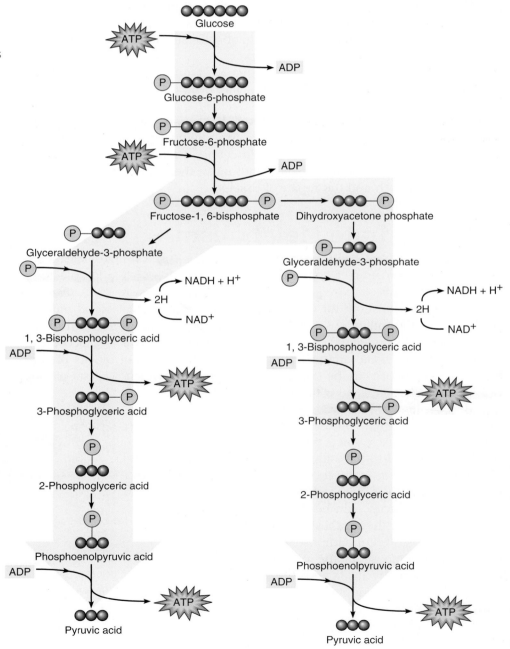

molecule (a 6-carbon sugar), two molecules of pyruvic acid are produced (each pyruvic acid molecule has three carbons). In addition, four ATP molecules are produced, but the reaction requires the input of two ATP molecules, so the net gain is two ATP molecules. If oxygen is not readily available to the cell, the pyruvic acid is converted to lactic acid. When oxygen is available, the pyruvic acid is converted to acetyl coenzyme A.

Acetyl coenzyme A is formed in the matrix of mitochondria. In a process known as **decarboxylation,** each pyruvic acid molecule loses a carbon dioxide (CO_2) molecule and is then called an acetyl group. This acetyl group is combined with coenzyme A (derived from vitamin B_5) in the matrix of mitochondria to form **acetyl coenzyme A (acetyl CoA)** (Figure 20-4). No energy is produced in this process, but two carbon dioxide molecules and two acetyl coenzyme A molecules are formed for every molecule of glucose that is oxidized. Each acetyl CoA molecule is then ready for the entry into the Krebs cycle.

The **Krebs cycle,** also known as the citric acid cycle, or tricarboxylic acid cycle (TCA cycle), is a series of oxidation-reduction reactions that produces carbon dioxide and ATP (Figure 20-5). Hans Krebs won the Nobel Prize in physiology in 1953 for his identification of the citric acid cycle in 1937. The Krebs cycle occurs in the mitochondria and since oxygen is required, it is classified as an aerobic process. There are eight reactions in the Krebs cycle and each requires a different enzyme. For each acetyl CoA that enters the cycle, two carbon dioxide molecules and two ATP molecules are produced.

decarboxylation A process in which a pyruvic acid molecule loses a CO_2 molecule.

acetyl coenzyme A (acetyl CoA) A molecule formed in the matrix of the mitochondria that is the substrate that enters the Krebs cycle.

Krebs cycle An aerobic process of eight reactions that occurs in the mitochondria and produces CO_2, ATP, NADH, and $FADH_2$.

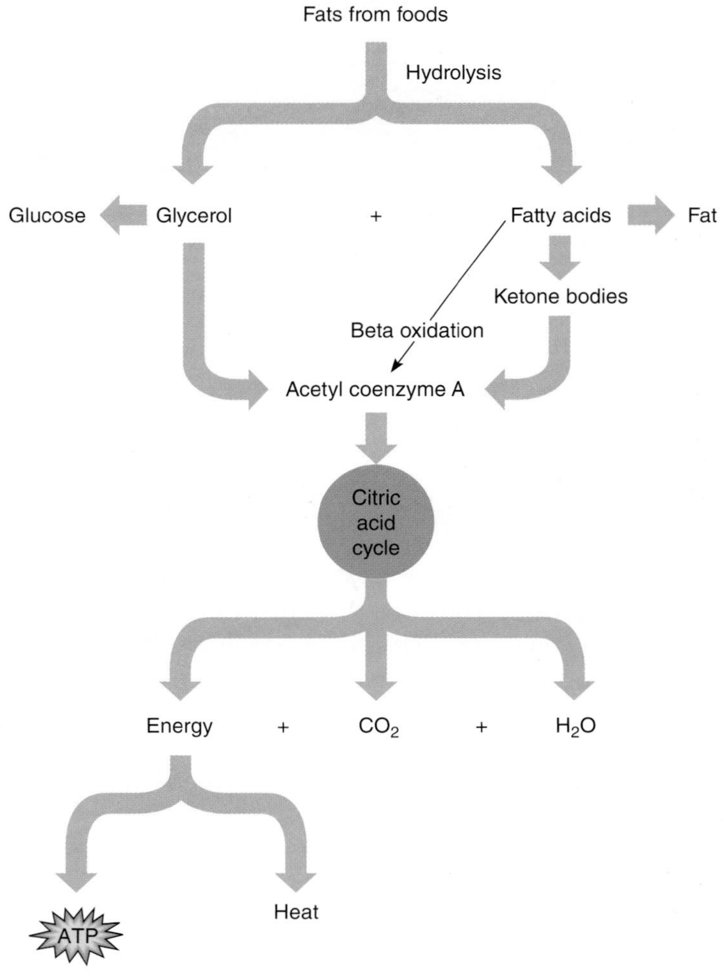

FIGURE 20-4
The body digests fats from foods into glycerol and fatty acids that can enter metabolic reactions to produce energy for the body.

FIGURE 20-5
Chemical reactions of the Krebs cycle.

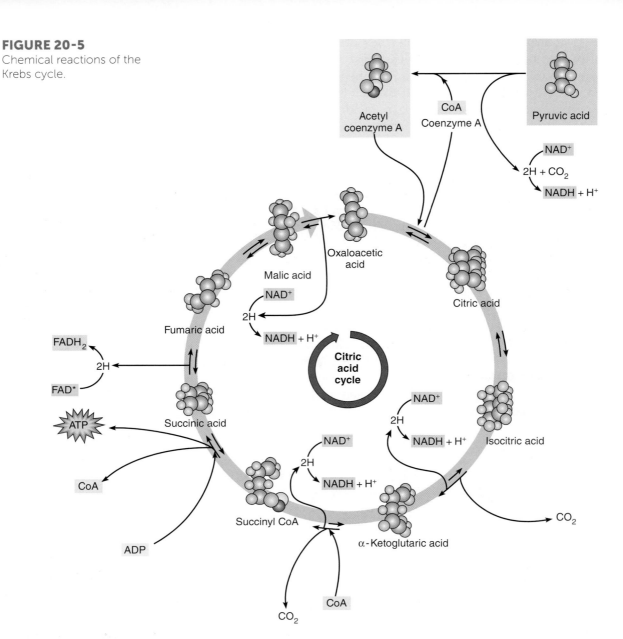

electron transport chain (ETC) An aerobic series of electron transfers that occurs on the inner membrane of the mitochondria, resulting in the production of 32 ATP molecules for every glucose molecule that is oxidized.

glycogenolysis The conversion of glycogen stored in the liver and skeletal muscle into glucose.

gluconeogenesis The formation of glucose from proteins and fats.

In addition, NADH and $FADH_2$ are produced. These molecules enter the electron transport chain and function as electron carrier molecules.

The **electron transport chain (ETC)** is an aerobic series of electron transfers that occurs on the inner membrane of mitochondria (Figure 20-6). The net result is the production of 32 ATP molecules for every glucose molecule that is oxidized. The total overall production of ATP from cellular respiration—glycolysis (2 ATP), Krebs cycle (2 ATP), and the electron transport chain (32 ATP)—is 36 ATP per molecule of glucose. In addition to obtaining glucose from the breakdown of more complex sugars, glucose can be converted from the glycogen stored in the liver and skeletal muscle in a process called **glycogenolysis,** or can be formed from proteins and fats in a process known as **gluconeogenesis.**

> **U check**
> Which metabolic reaction accounts for most of the ATP produced by the cell?

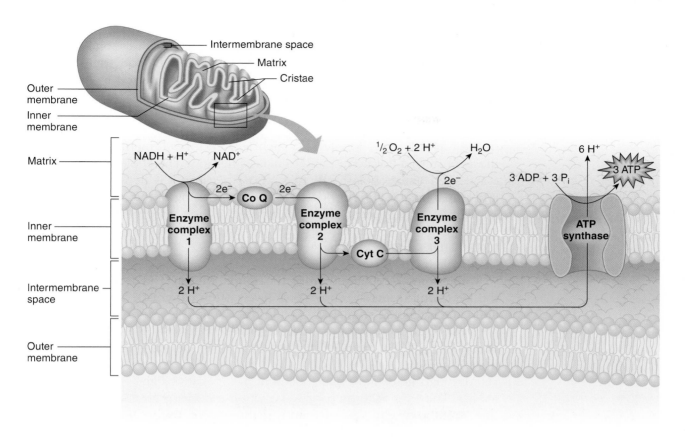

FIGURE 20-6 High-energy electrons moving down the electron transport chain located on the inner membrane of a mitochondrion.

LO 20.2

Lipid Metabolism

The liver uses fatty acids to synthesize a variety of lipids (Figure 20-7). Because lipids are not water soluble, they must be combined with proteins for transport. This makes them more water soluble and allows them to be transported in the plasma. This combination of lipid and protein is termed **lipoprotein.**

20.2 learning outcome

Describe lipid metabolism and disorders of lipid metabolism.

lipoprotein A combination of lipid and protein that makes the lipid more water soluble and able to be transported in the plasma.

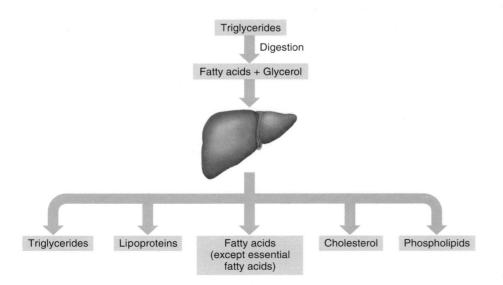

FIGURE 20-7
The liver uses fatty acids to synthesize a variety of lipids.

CHAPTER 20 Metabolic Function and Nutrition

Lipoproteins are assigned to various classes based on their density. *Very low-density lipoproteins* (VLDLs) consist of proteins, triglycerides, phospholipids, and cholesterol and are made in the liver. VLDLs transport triglycerides to adipose tissue for storage. Eventually, VLDLs are converted to low-density lipoproteins (LDLs).

LDLs are made up of the same four substances as VLDLs, but have a higher percentage of proteins and cholesterol. They carry roughly 75 percent of the total plasma cholesterol, which is used for cell membranes, the synthesis of steroid hormones, and to make bile salts. The cholesterol in LDLs is considered "bad" cholesterol (as is VLDL) because it is deposited in arteries and is responsible for arteriosclerosis or "hardening of the arteries."

High-density lipoproteins (HDLs) have the highest percentage of proteins (40 to 45 percent compared to 10 percent for VLDLs and 25 percent for LDLs). HDLs transport cholesterol to the liver for elimination and are therefore considered "good" cholesterol.

Cholesterol is obtained from the foods we eat and from synthesis by the liver. Unfortunately, cholesterol levels often cannot be managed by diet and exercise alone, so medications known as lipid-lowering agents are often needed. Total cholesterol in an adult should be less than 200 milligrams per deciliter (mg/dL) of blood, LDL cholesterol less than 130 mg/dL, and HDL at least 40 mg/dL.

Although lipids are often looked at in a negative light, they are essential for life. They are necessary for cell membranes, many hormones, blood clotting, and the myelin that surrounds many nerves. Excess lipids are stored in adipose tissue. Adipose tissue helps cushion organs and hold them in place, and acts as an insulator as well. Fat should account for 30 percent or less of the calories we consume each day, and saturated fats should be no more

focus on Clinical Applications: Metabolic Syndrome

A *syndrome* is defined as a group or cluster of symptoms or conditions. Patients who experience the following conditions together may be considered to have metabolic syndrome: hypertension, hyperinsulinism, excess body fat around the waist (apple-shaped body), and/or hypercholesterolemia. Being diagnosed with even one of these conditions increases your risk of developing heart disease, diabetes, or stroke. Occurring in concert, the risk increases exponentially.

Patients should make and keep regular appointments with their health care provider and should be compliant with all recommended medication regimens. As a health care professional, it may be your duty to assist patients diagnosed with any or all of the components of metabolic syndrome to begin aggressive changes in lifestyle. These changes could delay or possibly even prevent serious health problems down the road. The changes will include:

- Weight loss with a fiber-rich diet including a base of fruits and vegetables
- Increased physical activity
- Smoking cessation (if applicable)
- Decreased alcohol intake (if applicable)

than 10 percent of our total fat intake. One gram of fat is equivalent to 9 calories of energy. One gram of carbohydrate or 1 gram of protein is each equal to 4 calories of energy.

> # PATHOPHYSIOLOGY
> ## Hypercholesterolemia
>
> Hypercholesterolemia is an excess of cholesterol in the blood.
>
> **Etiology:** The cause may be dietary, but more often there is a genetic component. Familial hypercholesterolemia is an autosomal dominant disorder that affects 1 in 500 individuals and runs in families.
>
> **Signs and Symptoms:** Although high cholesterol levels normally do not cause any symptoms, cholesterol may be deposited around the eyelids (xanthelasma palpebrarum), the iris (arcus senilis corneae) of the eye, and as lumps in the tendons of the hands, elbows, knees, and feet. More importantly, the condition puts individuals at risk for heart disease even at a very young age.
>
> **Treatment:** A diet low in fat is important, but usually is not sufficient by itself. Drugs classified as statins and bile acid sequestrants are typically necessary.
>
> **U check**
> Which lipoprotein is the "good" cholesterol?

Protein Metabolism

learning outcome 20.3
Describe protein metabolism.

Proteins are catabolized or broken down into amino acids, which are then used for the production of ATP or to make new proteins needed by the body. Whereas carbohydrates and lipids can be stored if there is an excess, proteins are not stored. If there is an excess of proteins, the amino acids are converted into glucose or lipids.

Proteins can be categorized into structural and functional types. An example of a structural protein is collagen, which is the most abundant protein in our body and is a component of many structures such as membranes, bone, and blood vessels.

Hemoglobin is an example of a functional protein which, as we learned in Chapter 10, Blood and Circulation, is involved in the transport of oxygen by the red blood cells. Proteins continually undergo both catabolism and anabolism. As proteins are broken down, amino acids may be recycled or converted into glucose or lipids, or used to synthesize ATP (Figure 20-8). Anabolism of amino acids involves the formation of new chemical (peptide) bonds. Anabolism can be stimulated by thyroid hormone, insulin, testosterone, and estrogen. Protein synthesis occurs in the cytoplasm at the site of ribosomes.

FIGURE 20-8
The body hydrolyzes (breaks down) proteins from foods to obtain the amino acids to build other proteins.

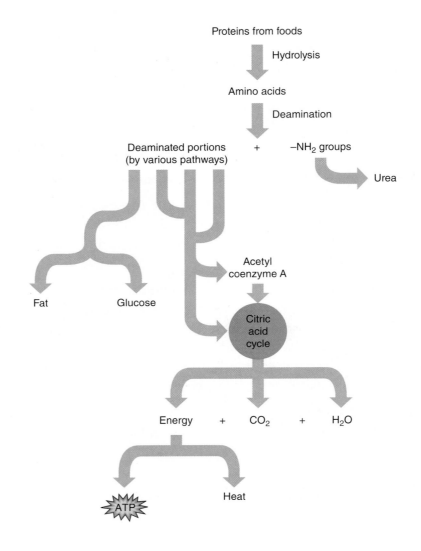

The process of protein synthesis is called "translation." There are 20 amino acids in the body. Eight of them are considered "essential" because we cannot synthesize them; they must be obtained from the food we eat. The essential amino acids include histidine, isoleucine, leucine, methionine, phenylalanine, threonine, tryptophan, and valine (see Table 20-1).

TABLE 20-1 Amino Acids in Foods

Alanine	Glycine	Proline
Arginine (ch)	Histidine (ch)	Serine
Asparagine	Isoleucine (e)	Theonine (e)
Aspartic acid	Leucine (e)	Tryptophan (e)
Cysteine	Lysine (e)	Tyrosine
Glutamic acid	Methionine (e)	Valine (e)
Glutamine	Phenylalanine (e)	

NOTE: Eight essential amino acids (e) cannot be synthesized by human cells and must be provided in the diet. Two additional amino acids (ch) are essential in growing children.

All of the essential amino acids are found in complete proteins such as beef, pork, fish, and poultry, milk, and eggs. The only plant protein that is considered a complete protein with all eight essential amino acids is soybeans. Other foods that contain some but not all of the essential amino acids are called incomplete proteins. Examples of incomplete proteins include nuts, dried beans, dried peas, lentils, peanut butter, seeds, cereals, and grains.

> **U check**
> Where does protein synthesis (translation) take place?

focus on Wellness: Eating Healthy

Guidelines for eating healthy have been modified over the years. In June 2011, the United States Department of Agriculture (USDA) introduced *ChooseMyPlate,* a personal approach to making dietary choices and exercising regularly (Figure 20-9). MyPlate includes the five basic food groups (grains, vegetables, fruits, milk, meat, and beans) and oils. Individuals should consume foods from each group daily. The size of the portion of the plate indicates the relative portion to be eaten. Typical recommendations are that 50 to 60 percent of the diet should be carbohydrates (with an emphasis on low glycemic carbohydrates, which generally come from natural sources) with less than 15 percent from simple sugars; less than 30 percent from fats, and no more than 10 percent as saturated fats; and 10 to 15 percent from proteins. Individuals should also watch the amount of salt they use; no more than 2,300 milligrams of sodium should be used daily.

FIGURE 20-9
Food guide pyramids and the newly released MyPlate illustrate the relative amounts of food that should make up a healthy diet. The USDA MyPlate program consists of an individual approach to healthy eating and exercise. To formulate an individual plan, visit the website http://choosemyplate.gov.

20.4 learning outcome

Relate the role of metabolism to homeostasis.

Homeostasis

calorie The amount of heat required to raise 1 gram of water 1°C.

basal metabolic rate (BMR) The amount of energy used by the body when it is in a resting state.

Homeostasis

Homeostasis as it relates to metabolism and nutrition means obtaining and using the right amount of nutrients as needed. Too little or too much of certain substances can interfere with the body's balance and in some cases can result in serious illness or even death. We refer to metabolic reactions and metabolic rate when talking about the way the body uses the energy it gets from food. Energy can be measured in units of heat known as calories. One **calorie** (cal) is the amount of heat required to raise 1 gram of water 1 degree Celsius. Because this is such a small amount of heat, we more often speak in kilocalories (kcal) or Calories (Cal). One thousand calories is equivalent to 1 kilocalorie or 1 Calorie.

When the body is in a quiet resting state, the **basal metabolic rate (BMR)** is the amount of energy being used. This is determined by measuring the amount of oxygen used per kilocalorie of food utilized. Earlier, we said that 1 gram of fat is equivalent to 9 calories and proteins and carbohydrates each are equivalent to 4 calories. A typical combination of these foods in the body gives approximately 4.8 Cal of energy for every liter of oxygen used. This means that over a 24-hour period, the average adult BMR is 1,200 to 1,900 Cal per day. In men, this is 22 to 26 Calories per kilogram (Cal/kg) of body mass and for women, 20 to 24 Cal/kg of body mass.

Homeostasis allows the body to maintain a relatively stable body temperature of 37°C (98.6°F). If the body temperature is too high (hyperthermia) or too low (hypothermia), severe consequences may result. The body has several mechanisms for generating or losing heat.

The hormones of the thyroid gland are the main regulators of BMR and heat generation. Other hormones such as insulin and testosterone can also elevate BMR (and heat). The sympathetic nervous system and epinephrine and norepinephrine can also raise BMR. Other influences on heat generation include eating (raises BMR), exercise, age (elderly people have a lower BMR than younger individuals), gender (lower BMR in females), and sleep (lower BMR).

Heat is lost from the body through evaporation, conduction, radiation, and convection. *Evaporation* is the loss of water (and heat) through the skin and mucous membranes. *Conduction* is the exchange of heat between two objects that are in direct contact with each other. *Radiation* is the transfer of heat in the form of infrared rays between objects that are not in direct contact. *Convection* is the transfer of heat by the movement of air or liquids with different temperatures. The hypothalamus is the control center for temperature regulation. By receiving signals from the body, the hypothalamus can raise or lower body temperature as necessary.

> **U check**
> What is the difference between a calorie and a Calorie?

20.5 learning outcome

Relate the major vitamins and minerals to their functions.

Nutrition

Most of us should have a basic understanding of nutrition. We know there are certain substances our body needs and others that are best avoided. We have talked about carbohydrates, lipids, and proteins. One gram of protein or carbohydrate provides 4.1 calories of energy, whereas 1 gram of lipid produces 9.5 calories. See Table 20-2 for carbohydrate, lipid, and protein nutrients. We know that all three are necessary and should have a balance of them in our diet. Another essential nutrient that is often not addressed appropriately is

TABLE 20-2 Carbohydrate, Lipid, and Protein Nutrients

Nutrient	Sources and RDA* for Adults	Calories per Gram	Use	Conditions Associated with Excesses	Conditions Associated with Deficiencies
Carbohydrate	Primarily from starch and sugars in foods of plant origin and from glycogen in meats 125–175 g	4.1	Oxidized for energy; used in production of ribose, deoxyribose, and lactose; stored in liver and muscles as glycogen; converted to fats and stored in adipose tissue	Obesity, dental caries, nutritional deficits	Metabolic acidosis
Lipid	Meats, eggs, milk, lard, plant oils 80–100 g	9.5	Oxidized for energy; production of triglycerides, phospholipids, lipoproteins, and cholesterol, stored in adipose tissue; glycerol portions of fat molecules may be used to synthesize glucose	Obesity, increased serum cholesterol, increased risk of heart disease	Weight loss, skin lesions
Protein	Meats, cheese, nuts, milk, eggs, cereals, legumes 0.8 g/kg body weight	4.1	Production of protein molecules used to build cell structure and to function as enzymes or hormones; used in the transport of oxygen, regulation of water balance, control of pH, formation of antibodies; amino acids may be broken down and oxidized for energy or converted to carbohydrates or fats for storage	Obesity	Extreme weight loss, wasting, anemia, growth retardation

*RDA = Recommended dietary allowance.

water. We must have sufficient water to live a healthy life. Although rare, we can have too much water (water intoxication) which also can be harmful. The other two types of nutrients we need are vitamins and minerals.

Vitamins

Vitamins are organic substances that are required in small amounts by the body. Most vitamins act as coenzymes in metabolic reactions. The majority are obtained by eating a balanced diet. Vitamin K is actually produced by bacteria (normal flora) in the gastrointestinal tract. Vitamins can be categorized as fat soluble (vitamins D, A, K, and E) and water soluble (vitamin C and the B vitamins). Fat-soluble vitamins can be stored in cells (especially the liver), whereas water-soluble vitamins are easily excreted in the urine if there is an excess of the vitamin (see Tables 20-3 and 20-4).

Food absorption

- Vitamin D is essential for the absorption of calcium and phosphorus from the GI tract. A deficiency leads to rickets in children and osteomalacia in adults.
- Vitamin A is essential for epithelial cells, acts as an antioxidant, participates in the formation of photoreceptors in the eye, and helps with bone and tooth development. A deficiency (hypovitaminosis A) can lead to skin problems, increase in infections, night blindness, and poor bone and tooth development. Too much vitamin A (hypervitaminosis A) during pregnancy may cause birth defects.
- Vitamin K is involved in blood clotting. A deficiency may result in excessive bleeding.
- Vitamin E is an antioxidant and also participates in the formation of red blood cells, DNA, and RNA. It is also involved in normal functioning of the nervous system. A deficiency may result in abnormal functioning of cell organelles and anemia.
- Vitamin B_2 (riboflavin) is a coenzyme in metabolic reactions and carbohydrate and protein metabolism. A deficiency may result in cataracts, blurred vision, and skin problems.
- Niacin is essential in many metabolic reactions. A deficiency can lead to a disease known as pellagra. Pellagra is characterized by "4 Ds"—diarrhea, dermatitis, dementia, and death.
- Vitamin B_6 (pyridoxine) is an important coenzyme. A deficiency can cause dermatitis.
- Vitamin B_{12} (cobalamin) is an important coenzyme in the Krebs cycle and red blood cell formation. A deficiency can lead to neurological problems and pernicious anemia. Deficiency of this vitamin is often treated with cyanocobalamin, a man-made form of vitamin B_{12}.
- Folic acid is important for many enzyme systems. It is also essential for development of the nervous system. A deficiency can lead to neural tube defects in infants.
- Vitamin C is an antioxidant and is involved in wound healing, is important for the synthesis of collagen, and acts as a coenzyme in many reactions. A deficiency can cause scurvy, anemia, poor wound healing, and inadequate collagen synthesis.

> **U check**
> What is the difference between a fat-soluble vitamin and a water-soluble vitamin?

TABLE 20-3 Fat-Soluble Vitamins

Vitamin	Characteristics	Functions	Sources and RDA* for Adults	Conditions Associated with Excesses	Conditions Associated with Deficiencies
Vitamin A	Exists in several forms; synthesized from carotenes; stored in liver; stable in heat, acids, and bases; unstable in light	An antioxidant necessary for synthesis of visual pigments, mucoproteins, and mucopolysaccharides; for normal development of bones and teeth; and for maintenance of epithelial cells	Liver, fish, whole milk, butter, eggs, leafy green vegetables, yellow and orange vegetables and fruits 4,000–5,000 IU**	Nausea, headache, dizziness, hair loss, birth defects	Night blindness, degeneration of epithelial tissues
Vitamin D	A group of steroids; resistant to heat, oxidation, acids, and bases; stored in liver, skin, brain, spleen, and bones	Promotes absorption of calcium and phosphorus; promotes development of teeth and bones	Produced in skin exposed to ultraviolet light; in milk, egg yolk, fish liver oils, fortified foods 400 IU	Diarrhea, calcification of soft tissues, renal damage	Rickets, bone decalcification and weakening
Vitamin E	A group of compounds; resistant to heat and visible light; unstable in presence of oxygen and ultraviolet light; stored in muscles and adipose tissue	An antioxidant; prevents oxidation of vitamin A and polyunsaturated fatty acids; may help maintain stability of cell membranes	Oils from cereal seeds, salad oils, margarine, shortenings, fruits, nuts, and vegetables 30 IU	Nausea, headache, fatigue, easy bruising and bleeding	Rare, uncertain effects
Vitamin K	Exists in several forms; resistant to heat but destroyed by acids, bases, and light; stored in liver	Required for synthesis of prothrombin, which functions in blood clotting	Leafy green vegetables, egg yolk, pork liver, soy oil, tomatoes, cauliflower 55–70 mg	Jaundice in formula-fed newborns	Prolonged clotting time

*RDA = Recommended dietary allowance.
**IU = International unit.

TABLE 20-4 Water-Soluble Vitamins

Vitamin	Characteristics	Functions	Sources and RDA* for Adults	Conditions Associated with Excesses	Conditions Associated with Deficiencies
Thiamine (Vitamin B_1)	Destroyed by heat and oxygen, especially in alkaline environment	Part of coenzyme required for oxidation of carbohydrates; coenzyme required for ribose synthesis	Lean meats, liver, eggs, whole-grain cereals, leafy green vegetables, legumes 1.5 mg	Uncommon; vasodilation, cardiac dysrhythmias	Beriberi, muscular weakness, enlarged heart
Riboflavin (Vitamin B_2)	Stable to heat, acids, and oxidation; destroyed by bases and ultraviolet light	Part of enzymes and coenzymes, such as FAD, required for oxidation of glucose and fatty acids and for cellular growth	Meats, dairy products, leafy green vegetables, whole-grain cereals 1.7 mg	None known	Dermatitis, blurred vision
Niacin (Nicotinic acid, Vitamin B_3)	Stable to heat, acids, and bases; converted to niacinamide by cells; synthesized from tryptophan	Part of coenzyme NAD and NADP required for oxidation of glucose and synthesis of proteins, fats, and nucleic acids	Liver, lean meats, peanuts, legumes 20 mg	Flushing, vasodilation, wheezing, liver problems	Pellagra, photosensitive dermatitis, diarrhea, mental disorders
Pantothenic acid (Vitamin B_5)	Destroyed by heat, acids, and bases	Part of coenzyme A required for oxidation of carbohydrates and fats	Meats, whole-grain cereals, legumes, milk, fruits, vegetables 10 mg	None known	Rare; loss of appetite, mental depression, muscle spasms
Vitamin B_6	Group of three compounds; stable to heat and acids; destroyed by oxidation, bases, and ultraviolet light	Coenzyme required for synthesis of proteins and various amino acids, for conversion of tryptophan to niacin, for production of antibodies, and for nucleic acid synthesis	Liver, meats, bananas, avocados, beans, peanuts, whole-grain cereals, egg yolk 2 mg	Numbness, clumsiness, paralysis	Rare; convulsions, vomiting, seborrhea lesions

TABLE 20-4 (concluded)

Vitamin	Characteristics	Functions	Sources and RDA* for Adults	Conditions Associated with Excesses	Conditions Associated with Deficiencies
Cobalamin (Vitamin B$_{12}$)	Complex, cobalt-containing compound; stable to heat; inactivated by light, strong acids, and strong bases; absorption regulated by intrinsic factor from gastric glands; stored in liver	Part of coenzyme required for synthesis of nucleic acids and for metabolism of carbohydrates; plays role in myelin synthesis; required for normal red blood cell production	Liver, meats, milk, cheese, eggs 3–6 mg	None known	Pernicious anemia
Folacin (Folic acid)	Occurs in several forms; destroyed by oxidation in acid environment or by heat in alkaline environment; stored in liver where it is converted into folinic acid	Coenzyme required for metabolism of certain amino acids and for DNA synthesis; promotes production of normal red blood cells	Liver, leafy green vegetables, whole-grain cereals, legumes 0.4 mg	None known	Megaloblastic anemia
Biotin	Stable to heat, acids, and light; destroyed by oxidation and bases	Coenzyme required for metabolism of amino acids and fatty acids and for nucleic acid synthesis	Liver, egg yolk, nuts, legumes, mushrooms 0.3 mg	None known	Rare; elevated blood cholesterol, nausea, fatigue, anorexia
Ascorbic acid (Vitamin C)	Chemically similar to monosaccharides; stable in acids but destroyed by oxidation, heat, light, and bases	Required for collagen production, conversion of folacin to folinic acid, and metabolism of certain amino acids; promotes absorption of iron and synthesis of hormones from cholesterol	Citrus fruits, tomatoes, potatoes, leafy green vegetables 60 mg	Exacerbates gout and kidney stone formation	Scurvy, lowered resistance to infection, wounds heal slowly

*RDA = Recommended dietary allowance.

inorganic Consisting of nonliving materials.

REMEMBER TESSA, our elderly patient? Can you think of any vitamins that would be especially helpful to improving and maintaining her health?

Minerals

Minerals are **inorganic** (from nonliving materials) elements that are found in the earth's crust. The more important minerals found in our body include calcium, phosphorus, potassium, sulfur, sodium, chloride, iron, iodide, cobalt, chromium, copper, fluoride, magnesium, zinc, manganese, and selenium. You may be familiar with some of these, such as calcium and phosphorus, while others like manganese and chromium may be new to you. (See Table 20-5 for the major minerals and Table 20-6 for trace elements.) Bone acts as a reservoir for many of these elements. Let's talk about some of these elements.

- Calcium is the most abundant mineral in the body. Approximately 99 percent of the body's calcium is stored in bones and teeth. It gives these two tissues their hardness. Calcium is also important in nerve transmission, muscle activity, blood clotting, and cellular activity.
- Phosphorus is found in bones and teeth. It is involved in many metabolic reactions, buffering blood pH, part of DNA and RNA, and production of ATP.
- Potassium is the most abundant cation in the extracellular fluid. It is necessary for the generation of nerve impulses (action potentials).
- Sulfur is needed for the production of ATP and is part of several hormones and vitamins.

focus on Wellness: Maintaining Healthy Nutrition and Metabolism

Maintaining a healthy diet literally affects every body system, allowing for growth and development in utero and during childhood. As adults, as our bodies change and age, our metabolism also changes. You may have heard of the "freshman ten" referring to the weight gain that often occurs in college as students study more, party more, and exercise less. Watching the diet more carefully, by decreasing fats and highly processed foods and increasing vegetables and fruits at each meal, as well as increasing physical activity, is key for many people. As adults "hit middle age," again the metabolism often takes a hit and the pounds begin to slowly (or maybe not so slowly) add up.

We now know that the body needs consistent nutrition throughout the day. Remember, we were created as "hunters and gatherers," rarely being sure where our next meal was coming from or when. Even though today that is not a problem for many of us, our bodies do not know that, which is why "crash diets" and "quicky" liquid programs do not work. The body goes into "starvation mode," so the next time you do eat, it "holds on" to those calories as fat. It is far better to eat frequent, small, low-fat meals and/or snacks during the day so the body does not get hungry, using the calories being eaten on an as-needed basis. Boosting your metabolism throughout the day and increasing physical activity does not have to be hard: Take the stairs instead of the elevator, park at the end of the parking lot, take a brisk walk at lunch, get off the bus a stop earlier than needed and walk to your destination, dance to the radio—just keep moving. Your body will thank you for it.

TABLE 20.5 Major Minerals

Mineral	Distribution	Functions	Sources and RDA* for Adults	Conditions Associated with Excesses	Conditions Associated with Deficiencies
Calcium (Ca)	Mostly in the inorganic salts of bones and teeth	Structure of bones and teeth; essential for neurotransmitter release, muscle fiber contraction, and blood coagulation; increases permeability of cell membranes; activates certain enzymes	Milk, milk products, leafy green vegetables 800 mg	Kidney stones, deposition of calcium phosphate in soft tissues	Stunted growth, misshapen bones, fragile bones, tetany
Phosphorus (P)	Mostly in the inorganic salts of bones and teeth	Structure of bones and teeth; in nearly all metabolic reactions; in nucleic acids, many proteins, some enzymes, and some vitamins; in cell membrane, ATP, and phosphates of body fluids	Meats, cheese, nuts, whole-grain cereals, milk, legumes 800 mg	None known	Stunted growth
Potassium (K)	Widely distributed; tends to be concentrated inside cells	Helps maintain intracellular osmotic pressure and regulate pH; required for nerve impulse conduction	Avocados, dried apricots, meats, nuts, potatoes, bananas 2,500 mg	Uncommon	Muscular weakness, cardiac abnormalities, edema
Sulfur (S)	Widely distributed; abundant in skin, hair, and nails	Essential part of certain amino acids, thiamine, insulin, biotin, and mucopolysaccharides	Meats, milk, eggs, legumes No RDA established	None known	None known
Sodium (Na)	Widely distributed; mostly in extracellular fluids and bound to inorganic salts of bone	Helps maintain osmotic pressure of extracellular fluids; regulates water movement; plays a role in nerve impulse conduction; regulates pH and transport of substances across cell membranes	Table salt, cured ham, sauerkraut, cheese, graham crackers 2,500 mg	Hypertension, edema, body cells shrink	Nausea, cramps, convulsions
Chlorine (Cl)	Closely associated with sodium; most highly concentrated in cerebrospinal fluid and gastric juice	Helps maintain osmotic pressure of extracellular fluids, regulates pH; maintains electrolyte balance; forms hydrochloric acid; aids transport of carbon dioxide by red blood cells	Same as for sodium No RDA established	Vomiting	Cramps
Magnesium (Mg)	Abundant in bones	Required in metabolic reactions in mitochondria that produce ATP; plays a role in the breakdown of ATP to ADP	Milk, dairy products, legumes, nuts, leafy green vegetables 300–350 mg	Diarrhea	Neuromuscular disturbances

*RDA = Recommended dietary allowance.

TABLE 20-6 Trace Elements

Mineral	Distribution	Functions	Sources and RDA* for Adults	Conditions Associated with Excesses	Conditions Associated with Deficiencies
Iron (Fe)	Primarily in blood; stored in liver, spleen, and bone marrow	Part of hemoglobin molecule; catalyzed formation of vitamin A; incorporated into a number of enzymes	Liver, lean meats, dried apricots, raisins, enriched whole-grain cereals, legumes, molasses 10–18 mg	Liver damage	Anemia
Manganese (Mn)	Most highly concentrated in liver, kidneys, and pancreas	Activates enzymes required for fatty acids and cholesterol synthesis, formation of urea, and normal functioning of the nervous system	Nuts, legumes, whole-grain cereals, leafy green vegetables, fruits 2.5–5 mg	None known	None known
Copper (Cu)	Most highly concentrated in liver, heart, and brain	Essential for hemoglobin synthesis, bone development, melanin production, and myelin formation	Liver, oysters, crabmeat, nuts, whole-grain cereals, legumes 2–3 mg	Rare	Rare
Iodine (I)	Concentrated in thyroid gland	Essential component for synthesis of thyroid hormones	Food content varies with soil content in different geographic regions; iodized table salt 0.15 mg	Decreased uptake by the thyroid gland	Decreased synthesis of thyroid hormones
Cobalt (Co)	Widely distributed	Component of cobalamin; required for synthesis of several enzymes	Liver, lean meats, milk No RDA established	Heart disease	Pernicious anemia
Zinc (Zn)	Most concentrated in liver, kidneys, and brain	Component of enzymes involved in digestion, respiration, bone metabolism, liver metabolism; necessary for normal wound healing and maintaining integrity of the skin	Meats, cereals, legumes, nuts, vegetables 15 mg	Slurred speech, problems walking	Depressed immunity, loss of taste and smell, learning difficulties
Fluorine (F)	Primarily in bones and teeth	Component of tooth structure	Fluoridated water 1.5–4 mg	Mottled teeth	None known
Selenium (Se)	Concentrated in liver and kidneys	Components of certain enzymes	Lean meats, fish, cereals 0.05–2 mg	Vomiting, fatigue	None known
Chromium (Cr)	Widely distributed	Essential for use of carbohydrates	Liver, lean meats, wine 0.05–2 mg	None known	None known

- **Sodium** is essential for the generation of action potentials and helps in water regulation.
- **Chloride** is part of hydrochloric acid, which is necessary for digestion. It is also involved in buffering blood pH and water balance in the body.
- **Iron** is an important component of hemoglobin, the molecule responsible for transporting oxygen. About two-thirds of the body's iron is found in hemoglobin. Iron is also involved in the electron transport chain (Figure 20-10).
- **Iodide** is required for the synthesis of thyroid hormones.
- **Cobalt** is an important component of vitamin B_{12}.
- **Chromium** helps with insulin activity.
- **Copper** is a coenzyme in several metabolic reactions including the synthesis of hemoglobin.
- **Fluoride** is an important component of bone and teeth as well as other structures.
- **Magnesium** is an important coenzyme and is required for the functioning of muscle and nervous tissue.
- **Zinc** is involved in the sensation of taste, wound healing, and various metabolic reactions.
- **Manganese** is involved in many metabolic reactions, such as hemoglobin synthesis and urea formation.
- **Selenium** is an antioxidant and so may be involved in the prevention of diseases such as coronary artery disease and prostate cancer.

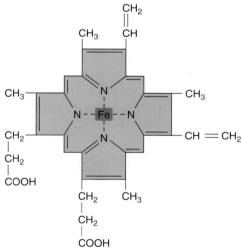

A Heme group

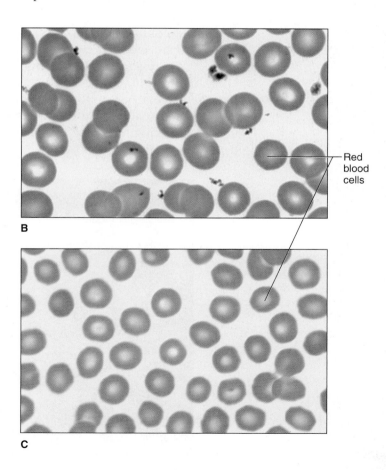

FIGURE 20-10 Iron in hemoglobin: (A) A hemoglobin molecule consists of four heme groups, each with a single iron atom that can combine with oxygen. Iron deficiency anemia can result from a diet lacking in iron or from blood loss such as occurs with gastric ulcers or a heavy menstrual flow. The red blood cells in (B) are normal (1,250x), but many of those in (C) are small and pale (1,250x) due to a deficiency of hemoglobin.

20.6 learning outcome

Compare types of malnutrition.

malnutrition Inadequate or excessive caloric intake.

kwashiorkor A type of starvation in which there is too little protein in the diet.

marasmus A type of starvation that entails both protein and calorie insufficiency.

Malnutrition

Malnutrition comes in two forms: starvation and obesity. You might think obesity is not malnutrition, because obesity usually indicates overeating. However, just like starvation, obesity is a form of malnutrition because **malnutrition** comprises both inadequate or excessive caloric intake.

Starvation

Starvation can be due to inadequate resources, dietary imbalances, or illness, or it can be self-imposed. Starvation can take the form of too little protein or a combination of too little protein and too few calories. Starvation is more common in third-world countries, but is becoming more common even in wealthy industrialized countries during difficult economic times. Some starvation is also self-imposed, such as different eating disorders. **Kwashiorkor** is a type of starvation where there is too little protein in the diet. This may occur even if caloric intake is normal or near normal. Kwashiorkor is a common disorder seen in Africa where there may be sufficient calories in the diet but there is inadequate protein. This disease often shows up in the firstborn who had been breast-fed, but who is displaced by a second infant that is now being breast-fed. Signs of kwashiorkor include ascites (edema of the abdomen), *hepatomegaly,* hypotension, and delayed mental and physical development (Figure 20-11).

The second major type of starvation is **marasmus,** which entails both protein and calorie insufficiency. This is a more severe type of starvation than kwashiorkor. It may cause emaciation, muscle wasting, delayed growth, and death. Even in the United States, there are areas of hunger and starvation when people are faced with limited resources. In the United States, as in most countries where this is a problem, it is the very young and very aged who are most often and most severely affected.

Anorexia nervosa and bulimia are two eating disorders. Anorexia nervosa affects up to 1 out of 250 adolescents and most are females. Many individuals have a perception of being overweight regardless of their actual weight. They are always "too fat." Anorexia nervosa is a serious illness and is life threatening. Medical intervention is necessary and counseling is essential.

FIGURE 20-11
Two types of starvation: (A) kwashiorkor; the children received an adequate diet early in life while breast-feeding, but then suffered malnutrition when they switched to a protein-poor diet, and (B) marasmus due to an insufficient diet as an infant.

A

B

Anorexia can cause musculoskeletal complications as well as gastrointestinal effects. It can impact the endocrine system resulting in amenorrhea or decreased thyroid function. The skin may become dry and scaly and bone density is decreased. It can also cause serious heart damage that may result in death.

The person with bulimia may be of normal weight and may eat normal or even excessive amounts of food, but then vomits to rid herself or himself of the calories. This is called "binge and purge" eating and bulimia poses certain medical risks. Teeth and the esophagus may become eroded due to the acidic contents of the vomitus. In addition, there may be menstrual irregularities, electrolyte imbalances and cardiac arrhythmias. Medical treatment and psychological counseling are typically required for this disorder as well.

> **U check**
> What is the major difference between kwashiorkor and marasmus?

focus on Wellness: Eating Disorders

Eating disorders can occur in all types of individuals of any age. These disorders can cause serious illness, long-term effects, and even death. Recognizing the following signs and symptoms of eating disorders is the first step in obtaining assistance.

Anorexia Nervosa
- Unexplained weight loss of at least 15 percent
- Self-starvation
- Excessive fear of gaining weight
- Malnourishment
- Cessation of menstruation in women
- Drastic reduction in food consumption
- Denial of feeling hungry
- Ritualistic eating habits
- Overexercising
- Unrealistic self-image as being obese
- Extremely controlled behavior

Bulimia
- Eating large quantities of food in a short period, followed by purging
- Pretexts for going to the bathroom after meals
- Using laxatives or diuretics to control weight
- Buying and consuming large quantities of food
- Feeling out of control while eating
- Maintaining a constant weight while eating a large amount of fattening foods
- Mood swings
- Awareness of having a disorder, but fear of not being able to stop
- Depression, self-deprecation, and guilt following the episodes

Obesity

body mass index (BMI)
A tool used to measure obesity, based on weight and height.

Diabetes Type 2

 learning outcome

Explain the relationship between aging and nutrition.

Obesity

The other major type of malnutrition is obesity, which can be "plain" obesity or "morbid" obesity. Although starvation does occur in industrial countries, obesity due to poor eating habits is much more common. A body weight greater than 20 percent above the standard is considered obesity and 50 percent over the standard is considered morbid obesity. Morbid obesity can result in a person's health becoming seriously and dangerously affected.

Body mass index (BMI) is a tool used to measure obesity. BMI is determined using weight and height. A BMI of 30 to 39.9 is considered obese, whereas a BMI of 40 or more is considered morbidly obese. Serious health consequences of obesity include, but are not limited to, adult-onset diabetes mellitus (type 2 diabetes mellitus), cardiovascular disease, hypertension, joint problems, and cancer (such as colon, breast, and uterine cancer, among others). If you are curious about your own BMI, you can visit the National Heart, Lung, and Blood Institutes website at www.nhlbisupport.com/bmi/ to calculate your BMI. All you need to know is your height and your weight.

Life Span

It is hard to think of a single factor that can affect life span as dramatically as diet. The adage "we are what we eat" holds a lot of truth. By watching what we eat and exercising regularly, we not only can live longer lives, but the lives we live will be more enjoyable and healthier. As we get older, our bodies change in many ways. For example, food may not taste as good or the drive for thirst may not be as great as when we were younger. Being alert to these changes, the individual can consciously make decisions to eat properly and take in plenty of healthy fluids. Inviting others over to eat occasionally (potluck if resources are limited) may make eating more enjoyable.

summary

learning outcomes	key points
20.1 Describe carbohydrate metabolism.	Carbohydrates, especially glucose, are the body's main source of nutrients for the production of ATP.
20.2 Describe lipid metabolism and disorders of lipid metabolism.	Lipids are one of the major nutrients used by the body and are used to make cell membranes and hormones. They can also be stored for future use. Hypercholesterolemia is one type of lipid disorder.
20.3 Describe protein metabolism.	Proteins are broken down or catabolized. The resulting amino acids are then used to make new proteins or are used for the production of ATP.
20.4 Relate the role of metabolism in homeostasis.	Metabolism is essential to maintaining homeostasis. One way it does this is by maintaining a stable body temperature.

learning outcomes	key points
20.5 Relate the major vitamins and minerals to their functions.	Vitamins can be categorized as water soluble or fat soluble. They are organic substances that help maintain growth and normal metabolism. Minerals are inorganic substances that are needed by the body for many metabolic processes.
20.6 Compare types of malnutrition.	Obesity and starvation are both malnutrition. Kwashiorkor and marasmus are two types of starvation. Kwashiorkor is a starvation state due to inadequate protein in the diet. Marasmus is starvation in which there is too little protein and too few calories.
20.7 Explain the relationship between aging and nutrition.	As we age, our nutritional needs change. In addition, the attention we give to our diet may accelerate or slow the aging process.

case study 1 questions

Can you answer the following questions that pertain to Tessa White's case study presented earlier in this chapter?

1. Which of the following is *not* a major nutrient needed by the body?
   - a. Lipids
   - b. Nucleic acids
   - c. Protein
   - d. Water
2. Which if the following may help with the activity of insulin?
   - a. Chromium
   - b. Potassium
   - c. Chloride
   - d. Phosphorus
3. Which of the following would be most helpful in maintaining the overall health of the epithelium in the skin?
   - a. Vitamin A
   - b. Vitamin D
   - c. Vitamin K
   - d. Vitamin E

review questions

1. Which of the following is responsible for the greatest production of ATP?
   - a. Formation of acetyl coenzyme A
   - b. Krebs cycle
   - c. Electron transport chain
   - d. Glycolysis
2. Which one of the following produces the most ATP per molecule of glucose oxidized?
   - a. Krebs cycle
   - b. Electron transport chain
   - c. Glycolysis
   - d. Formation of acetyl coenzyme A
3. Which of the following is the breakdown of glycogen?
   - a. Gluconeogenesis
   - b. Glycogenolysis
   - c. Glycogenesis
   - d. Glucogenesis

CHAPTER 20 Metabolic Function and Nutrition

4. Which of the following are the building blocks of proteins?
   a. Lipids
   b. VLDLs
   c. Amino acids
   d. Carbohydrates
5. How many carbon atoms are in each glucose molecule?
   a. 2
   b. 4
   c. 6
   d. 8
6. The body's "thermostat" is located in the
   a. Hypothalamus
   b. Thalamus
   c. Pituitary gland
   d. Cerebellum
7. Which of the following would *not* increase metabolic rate?
   a. Fever
   b. Stimulation of the parasympathetic nervous system
   c. Eating
   d. Increased secretion of thyroid hormones
8. Vitamins that are quickly excreted in the urine are less likely to build up to toxic levels. Which one of the following vitamins is rapidly excreted in the urine?
   a. Vitamin C
   b. Vitamin A
   c. Vitamin D
   d. Vitamin E

critical **thinking questions**

1. If a person has a chronic lung disease such as emphysema or chronic bronchitis, which metabolic reaction would be *least* affected: Formation of acetyl coenzyme A, Krebs cycle, electron transport chain, or glycolysis? Explain why.
2. Why would eating disorders such as kwashiorkor and marasmus cause heart problems?

patient **education**

Whenever you teach patients about nutrition and diet, you help them take steps to improve their health. You should teach patients about the role of nutrition in helping prevent specific medical conditions. You should also teach patients how to be wise consumers when they shop by reading food package labels. You will be better equipped to educate patients and answer their questions if you have solid knowledge of diet and nutrition, and if you stay current with recent research findings. What are some current trends in diet and nutrition that you and your patients should be aware of?

applying **what you know**

1. For each of the following vitamins, name one disease that can occur as a result of a deficiency and one disease that can occur from toxicity.
   a. Vitamin A
   b. Vitamin D
   c. Vitamin C
   d. Folate
2. An adult patient reads you his food journal for the previous week. His average daily intake of calcium is 540 milligrams. Is he getting enough calcium to supply his body's needs?

21 The Endocrine System

chapter outline

Major Endocrine Glands and Hormones

Additional Endocrine Glands and Tissues

Regulatory Mechanisms

Life Span

outcomes

AFTER COMPLETING THIS CHAPTER, YOU WILL BE ABLE TO:

21.1 Recognize the major endocrine glands, their location in the body, and their hormones, target tissues, functions, and disorders.

21.2 Identify additional endocrine glands and tissues that secrete hormones and their functions.

21.3 Describe the regulatory mechanisms involved in maintaining homeostasis.

21.4 Infer the effects of aging on the endocrine system.

case study

Use the case study to focus on as you go through the chapter. The questions will guide you as you learn the anatomy, physiology, and pathology of the endocrine system.

CASE STUDY 1 *Diabetes Mellitus*

Manny Washington, a 58-year-old marketing manager, reports that he is always hungry and thirsty. He says he is going to the bathroom more frequently including two to three times per night. He also says he sometimes feels "shaky." He is about 75 pounds overweight and often skips meals and eats a lot of candy since he does not take time to eat lunch while at work.

As you go through the chapter, keep the following questions in mind:

1. What organs are possibly involved in this disorder?
2. What are the causes of this disorder?
3. How is diabetes insipidus similar to diabetes mellitus? How is it different?
4. What tests would the physician run to confirm the diagnosis?
5. What treatments would you expect the doctor to prescribe?
6. How will this condition affect Manny's daily activities and life expectancy?

essential terms

acromegaly (ak-ro-MEG-uh-lee)

adrenocorticotropic hormone (ACTH) (ah-dree-no-kor-tih-ko-TRO-pik HORE-mone)

antidiuretic hormone (ADH) (an-te-dy-you-REH-tik HORE-mone)

calcitonin (kal-sih-TOE-nin)

cortisol (KOR-tih-sal)

diabetes mellitus (die-uh-BEE-teez MEL-lit-tus)

epinephrine (ep-ih-NEF-rin)

exophthalmos (ek-sof-THAL-mos)

gigantism (jie-GAN-tizm)

glucagon (GLU-kuh-gon)

hypercalcemia (hy-per-kal-SEE-mee-uh)

hyperthyroidism (hy-per-THY-roy-dizm)

hypocalcemia (hy-poh-kal-SEE-mee-uh)

hypothyroidism (hy-poh-THY-roy-dizm)

islets of Langerhans (EYE-lets uv LAHNG-er-hanz)

prolactin (PRL) (pro-LAK-tin)

prostaglandin (pros-tah-GLAN-din)

protein (PRO-teen)

steroid (STAIR-oyd)

Additional key terms in the chapter are italicized and defined in the glossary.

study tips

1. As you study the endocrine glands, if you remember five key points about each gland you will have a pretty good understanding of each gland and the hormone(s) it secretes. For each gland, you should know
   a. The location in the body of the endocrine gland.
   b. The hormones secreted by the gland (name and type).
   c. The target organ.
   d. The effect of the hormone.
   e. The disorders of the gland.

2. Make flash cards for the various endocrine glands with the information above and use the cards to review.

3. Read the chapter. After each section, formulate a question or two as though you were writing an exam. Exchange questions within a study group for review.

Introduction

The nervous and endocrine systems are the two regulatory systems of the body. (Figure 21-1). As you learned in Chapter 14, The Nervous System, the nervous system acts very rapidly through nerves and nerve impulses, or action potentials. The endocrine system is slower acting, but is very important in terms of health and disease for its longer acting effects. It works through chemical messengers called hormones.

FIGURE 21-1
Chemical communication: (A) Neurons release neurotransmitters into synaptic clefts. (B) Endocrine glands secrete hormones into the bloodstream.

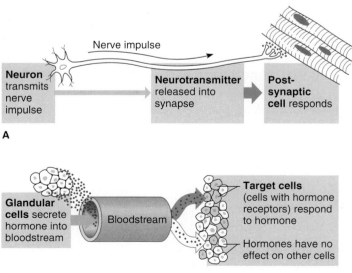

Major Endocrine Glands and Hormones

The endocrine system consists of glands and hormones (Figure 21-2). As seen in Table 21-1, each hormone has a particular function in the body.

The endocrine system is ductless. The glands of the endocrine system are therefore often referred to as ductless glands. This means the substances secreted by endocrine glands directly enter the circulatory system or act directly on specific cells. In contrast, glands such as the salivary glands secrete their chemicals into ducts. These glands are discussed in other chapters and are called exocrine glands.

Each endocrine gland secretes one or more *hormones*. These are chemical messengers that are typically steroids, proteins, amino acids, or lipids. The different types of hormones are summarized in Table 21-2.

Steroids are hormones that are built from cholesterol. Steroid hormones are bound loosely to plasma proteins as they circulate in the blood. They are lipid-soluble, and like all lipid-soluble substances they can easily cross cell membranes. Once they cross the cell membrane, they target receptors on the nuclear membrane (Figure 21-3). Once inside the nucleus, the steroid hormones turn specific genes on or off. This results in new products being formed or new functions occurring. Examples of steroid hormones are aldosterone, estrogen, progesterone, and testosterone.

Another class of hormones is made up of **protein.** One major difference between steroid hormones and protein hormones is that proteins cannot

> **21.1 learning outcome**
> Recognize the major endocrine glands, their location in the body, and their hormones, target tissues, functions, and disorders.

steroid A lipid-soluble hormone built of cholesterol that turns specific genes on or off.

proteins Contain essential amino acids and are used by the body for growth and the repair of tissues.

FIGURE 21-2
Location of major endocrine glands.

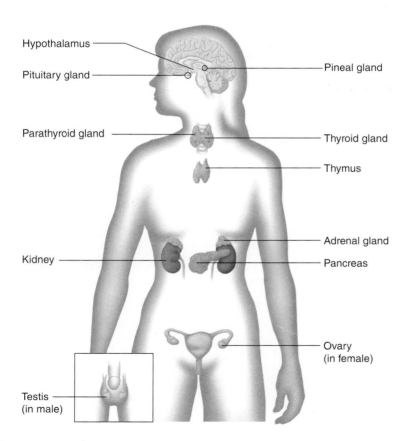

CHAPTER 21 The Endocrine System

TABLE 21-1 Endocrine Glands: Their Hormones and Actions

Gland	Hormone	Action Produced
Hypothalamus (produces)	Antidiuretic hormone (ADH)	Is stored and released by posterior pituitary
	Oxytocin (OT)	Is stored and released by posterior pituitary
Anterior pituitary	Growth hormone (GH)	Promotes growth and tissue maintenance
	Melanocyte stimulating hormone (MSH)	Stimulates pigment regulation in epidermis
	Adrenocorticotropic hormone (ACTH)	Stimulates adrenal cortex to produce its hormones
	Thyroid stimulating hormone (TSH)	Stimulates the thyroid to produce its hormones
	Follicle stimulating hormone (FSH)	(F) Stimulates ovaries to produce ova and estrogen; (M) Stimulates testes to produce sperm and testosterone
	Luteinizing hormone (LH)	(F) Stimulates ovaries for ovulation and estrogen production; (M) Stimulates testes to produce testosterone
	Prolactin (PRL)	(F) Stimulates the breasts to produce milk; (M) Works with and complements LH
Posterior pituitary (releases)	Antidiuretic hormone (ADH)	Stimulates the kidneys to retain water
	Oxytocin (OT)	Stimulates uterine contractions for labor and delivery and milk ejection from breast
Pineal body	Melatonin	Regulates biological clock; linked to onset of puberty
Thyroid	T_3 and T_4	Synthesize protein and increase energy production for all cells
	Calcitonin	Increases bone calcium and decreases blood calcium
Parathyroid	Parathyroid hormone (PTH)	Acts as antagonist to calcitonin; decreases bone calcium and increases blood calcium
Thymus	Thymosin and thymopoietin	Both hormones stimulate the production of T lymphocytes
Adrenal cortex	Aldosterone	Stimulates the body to retain sodium and water
	Cortisol	Decreases protein synthesis; decreases inflammation
Adrenal medulla	Epinephrine and norepinephrine	Prepare body for stress; increase heart rate, respiration, and blood pressure
Pancreas (islet of Langerhans)	Alpha cells—glucagon	Increases blood sugar; decreases protein synthesis
	Beta cells—insulin	Decreases blood sugar; increases protein synthesis
Gonads: ovaries (female)	Estrogen and progesterone	Develop secondary sex characteristics; female sex hormone
Testes (male)	Testosterone	Develops secondary sex characteristics; male reproductive hormone

TABLE 21-2 Types of Hormones

Types of Compound	Formed from	Examples
Amines	Amino acids	Norepinephrine, epinephrine
Peptides	Amino acids	ADH, OT, TRH, SS, GnRH
Proteins	Amino acids	PTH, GH, PRL
Glycoproteins	Proteins and carbohydrates	FSH, LH, TSH
Steroids	Cholesterol	Estrogens, testosterone, aldosterone, cortisol

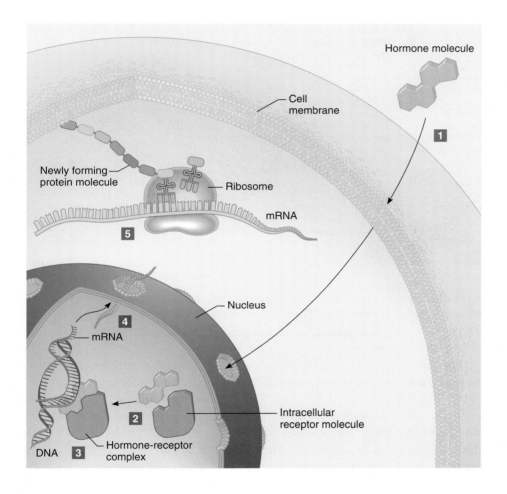

FIGURE 21-3
Steroid hormones: (1) A steroid hormone crosses a cell membrane and (2) combines with a receptor protein. (3) Activation of transcription. (4) mRNA leaves the nucleus to enter the cytoplasm. (5) Protein synthesis.

easily cross the cell membrane. This means that these hormones must have assistance to exert their effects. *Receptors* are structures in the cell membranes of a target tissue that combine with a chemical such as a hormone or drug to alter the functioning of the cell. Once the protein hormone binds to receptors on the cell membrane, a "second messenger" is activated (Figure 21-4). The best example of a second messenger is G-protein, which is the second messenger that causes the needed changes in the cell. Examples of hormones that are proteins include growth hormone secreted by the anterior pituitary gland, calcitonin from the thyroid gland, and antidiuretic (ADH) hormone from the posterior pituitary gland. A more in-depth discussion on each of these hormones is found later in the chapter.

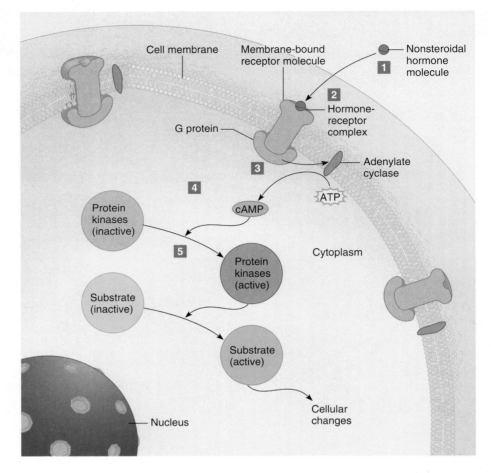

FIGURE 21-4
Nonsteroidal hormone action: (1) Nonsteroidal hormone travels to the cell membrane where (2) it binds to a receptor molecule. (3) This activates adenylate cyclase which (4) converts ATP into cAMP. (5) cAMP promotes a series of reactions.

prostaglandins Lipid hormones that are produced by a number of organs that have nearby receptors and thus do not travel through the blood; they have a role in inflammation and intensifying pain, among other effects.

Yet another type of hormone is made up of *amino acids,* which are the building blocks of proteins. These hormones also require a second messenger. An example of an amino acid hormone is melatonin, which is secreted by the pineal gland.

Lipids are the final category of hormones. As lipids, hormones are fat soluble and easily cross the cell membrane, making a second messenger unnecessary. Examples of lipid hormones are **prostaglandins.** They received their name because they were first discovered in the prostate gland. We now know prostaglandins are produced by many cells in the body including the kidneys, heart, uterus, stomach, and even the brain. They normally target receptors on nearby organs, so they do not travel through the blood. The have a role in promoting inflammation, intensifying pain, and increasing fever. They also are involved in smooth muscle contraction, glandular secretions, and altering the immune response.

Endocrine Disorders

Endocrine disorders are usually the result of "hypo" (too little) or "hyper" (too much) secretion of hormones. Other endocrine disorders include benign and malignant tumors. Tumors that are benign are nonmalignant, which means they are not recurrent or progressive. Tumors that are malignant are cancerous, harmful, or tend to produce death. Table 21-3 is a quick reference to endocrine disorders.

An example of a hyposecretion of a hormone is hypopituitarism. When the anterior pituitary gland secretes too little growth hormone in childhood,

TABLE 21-3 Endocrine System Diseases and Conditions: Quick Reference Guide

Hormone	Hypo- or Hyper-Secretion	Disease or Condition
GH (somatotropin)	Hyposecretion (children)	Dwarfism
	Hypersecretion (children)	Gigantism
	Hypersecretion (adults)	Acromegaly
ACTH	Hyposecretion (adrenal cortex—cortisol)	Addison's disease
	Hypersecretion (adrenal cortex—cortisol)	Cushing's syndrome
ADH	Hyposecretion	Diabetes insipidus (dehydration)
	Hypersecretion	Edema, hypertension
T_3 and T_4	Hyposecretion (congenital/children)	Cretinism
	Hyposecretion (adults)—severe cases	Hypothyroidism and *myxedema*
	Hypersecretion	Graves' disease
Glucagon	Hyposecretion	Hypoglycemia
	Hypersecretion	Hyperglycemia
Insulin	Hyposecretion	Hyperglycemia (diabetes mellitus)
	Hypersecretion	Hypoglycemia (Hyperinsulinism)

dwarfism may result. An example of hypersecretion of a hormone is **hyperthyroidism.** When the thyroid secretes too much thyroid hormone it may cause a condition called Graves' disease. Pheochromocytoma is a benign tumor of the adrenal medulla that can cause an increase in epinephrine and a corresponding rise in blood pressure.

hyperthyroidism An excess secretion of thyroid hormone.

Hyperthyroidism

> **Ʊ check**
> What are the major categories of hormones?

Pituitary gland

The pituitary gland is sometimes called the "master gland" because it produces or secretes several different hormones and has an effect on nearly every organ in the body (Figure 21-5). It also is called the hypophysis (Latin for "beneath-grow") because it is located at the base of the brain. It about the size of a pea and is found in a bony structure of the sphenoid bone called the *sella turcica* (Latin for "Turkish saddle") because of its shape. The pituitary gland is located inferior to the optic chiasm, which as we discovered in Chapter 14, The Nervous System, is a visual pathway from the optic nerve of the eyes to the occipital lobe of the brain. This has clinical significance that we will talk about later.

The pituitary gland is divided into anterior and posterior lobes (Figure 21-6). The anterior lobe is called the adenohypophysis; it develops from oral ectoderm in the roof of the mouth called Rathke's pouch and secretes seven different hormones. The posterior lobe is called the neurohypophysis. It develops from nervous tissue in the brain and secretes

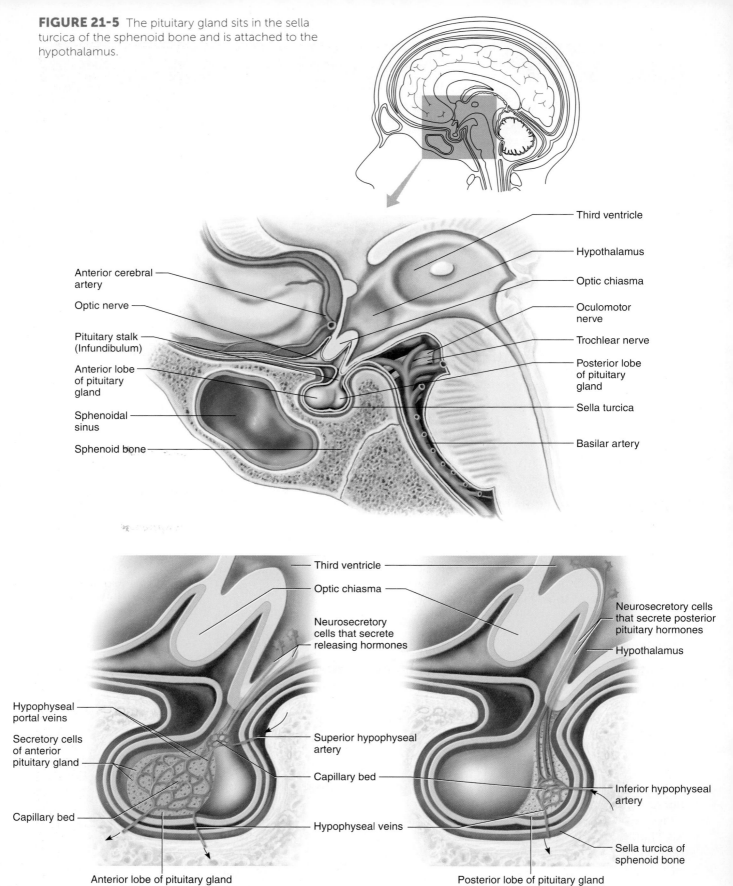

FIGURE 21-5 The pituitary gland sits in the sella turcica of the sphenoid bone and is attached to the hypothalamus.

FIGURE 21-6 Hormones from the hypothalamus stimulate cells of the anterior lobe of the pituitary to secrete hormones. Nerve impulses from the hypothalamus stimulate cells of the posterior lobe of the pituitary to secrete hormones.

two hormones. We will now discuss each hormone in more detail and include the conditions that can result when the gland is not functioning correctly.

> **U check**
> Why is the pituitary gland called the master gland?

Anterior Pituitary (Adenohypophysis): Growth hormone (GH) is a protein hormone that targets virtually every cell and tissue in the body. It is responsible for the growth seen in childhood and an increase in mass of skeletal muscle and bones. It is also involved in repair of injured tissues. GH is sometimes referred to as somatotropin (*soma* = body; *tropin* = growth).

FIGURE 21-7
Undersecretion of human GH produces pituitary dwarfism, in which a person has short stature. Notice that the individual has a normal head and torso and short extremities.

PATHOPHYSIOLOGY
Dwarfism

Hyposecretion of growth hormone (GH) can result in a form of dwarfism. Dwarfism is an abnormal underdevelopment of the body (Figure 21-7). In cases of GH hyposecretion, the individual may be treated with injections of a growth hormone preparation such as Humatrope (somatropin). Growth hormone hypersecretion can result in gigantism or acromegaly. **Gigantism** is the result of too much growth hormone being secreted during childhood and causes an increase in the length of the long bones. In some cases there may be a tumor that causes too much growth hormone to be secreted by the anterior pituitary after puberty resulting in a condition known as **acromegaly** which causes the hands, feet, and face to take on an unusual enlargement (Figure 21-8).

gigantism An abnormal increase in the length of long bones due to hypersecretion of growth hormone during childhood.

acromegaly A condition caused by the secretion of too much growth hormone after puberty that causes the hands, feet, and face to take on an unusual enlargement.

A

B

C

D

FIGURE 21-8 Oversecretion of growth hormone in adulthood causes acromegaly: Same individual at (A) 9 years of age, (B) 16 years, (C) 33 years, and (d) 52 years.

focus on Genetics: Dwarfism

Dwarfism (medical term is achondroplasia) is generally defined as an adult height of 4 feet 10 inches tall or less (Figure 21-7). In most cases achondroplasia results from a random genetic mutation in either the father's sperm or the mother's egg. Interestingly, about 80 percent of those diagnosed with dwarfism are born to parents of average height. A person with dwarfism may or may not pass the genetic mutation to his or her own child.

There is no treatment for congenital dwarfism to increase the person's height, although rods and staples may be placed in the long bones to keep the bones growing in the correct direction. Rods may also be placed in the spine to keep it straight. Some patients elect to undergo limb-lengthening procedures, but many more do not because there is "nothing wrong with being short."

Melanocyte Stimulating Hormone (MSH): Melanocyte stimulating hormone (MSH) is a peptide hormone that targets the melanocytes and their organelles, melanosomes, which are responsible for producing melanin, the protective pigment found in the epidermis of the skin.

adrenocorticotropic hormone (ACTH) An amino acid hormone that targets the adrenal cortex to secrete aldosterone and cortisol and other adrenal cortical hormones.

Adrenocorticotropic Hormone (ACTH): As the name suggests, **adrenocorticotropic hormone (ACTH)** has the adrenal cortex as its target. It causes this portion of the adrenal gland to secrete its hormones. Aldosterone and cortisol are the two major adrenal cortical hormones.

PATHOPHYSIOLOGY

Addison's Disease

Hyposecretion of ACTH is known as Addison's disease. People with this disorder experience anorexia (loss of appetite), fatigue, weight loss, and gastrointestinal problems. They often have "bronzing of the skin" that resembles a perpetual tan. President Kennedy (Figure 21-9) is thought to have had Addison's disease. Treatment includes administration of the missing glucocorticoids and mineralcorticoids normally secreted by the adrenal cortex. Salt and potassium levels must also be monitored.

Cushing's Syndrome

Hypersecretion of ACTH is known as Cushing's syndrome. These individuals will have fatigue, muscle weakness, hyperlipidemia, osteoporosis, and atherosclerosis. Treatment will include cessation of any steroid hormone use. Radiation therapy and/or surgery may be necessary to treat any tumor responsible for the hypersecretion.

FIGURE 21-9
President Kennedy may have had Addison's disease.

Thyroid Stimulating Hormone (TSH): Thyroid stimulating hormone (TSH) is a peptide hormone. It causes the release of thyroxine from the thyroid gland.

> ## PATHOPHYSIOLOGY
> ### Hypothyroidism and Hyperthyroidism
> Hyposecretion of TSH may result in too little thyroid hormone being secreted, causing hypothyroidism. Hypersecretion of TSH may result in too much thyroid hormone being secreted, causing hyperthyroidism.

Follicle Stimulating Hormone (FSH): Although the name *follicle stimulating hormone* (FSH) may indicate that this is a hormone found in females, it is actually found in both men and women. FSH stimulates the maturation of the ova in the female and production of sperm in the male. It also causes increased production of estrogen by the ovaries. FSH is a glycoprotein.

Luteinizing Hormone (LH): Luteinizing hormone (LH) is a glycoprotein that is important in stimulating ovulation (release of mature ova from the ovary). It also increases the production of estrogen in the female and testosterone in the male.

Prolactin (PRL): **Prolactin (PRL)** is the peptide responsible for the production of milk by the mammary glands in the female. In males, prolactin helps with the function of luteinizing hormone.

prolactin The peptide responsible for milk production in the mammary glands in females and for helping with the function of luteinizing hormone in males.

> ⋓ check
> What is the function of FSH in women? What is the function of FSH in men?

Antidiuretic Hormone (ADH): **Antidiuretic hormone (ADH)** is another peptide hormone and is secreted at the posterior pituitary (neurohypophysis). It helps the body retain water (think "against diuresis or urination") by increasing water absorption by the kidneys. By increasing fluid volume, it increases blood pressure. Depending on the situation, this may be a good or bad thing. When blood pressure increases too much (hypertension), a category of drugs called diuretics may be administered. They have the opposite effect of ADH, causing an increase in excretion of urine and a decrease in blood pressure.

antidiuretic hormone A peptide hormone that causes the body to retain water by increasing water absorption in the kidneys.

> ## PATHOPHYSIOLOGY
> ### Diabetes Insipidus
>
> *Diabetes insipidus* is a condition in which a person does not have enough ADH. This may be due to hyposecretion of ADH or a problem with the receptors for ADH. Common symptoms include excessive urination, thirst, and dehydration. Treatment often includes hormone replacement therapy.

Oxytocin (OT): Oxytocin (OT) is a protein hormone that assists in childbirth by increasing the force of contractions. It also causes milk ejection from the breasts. Although it might seem that OT is strictly a "female" hormone, like FSH it also has a role in the male body. It causes contraction of the prostate and the vas deferens during sexual arousal and possibly facilitates sperm motility.

> **U check**
> What do diuretics do?

Thyroid Gland

The thyroid gland has two lobes and an isthmus that connects them. The thyroid is located in the front of the neck just inferior to the larynx (voice box) and is easily palpable on a routine physical examination (Figure 21-10). It is responsible for important functions that help maintain homeostasis. It is also involved in several common diseases. Some of these diseases have an autoimmune component and are seen more frequently in women. When examining thyroid tissues under a microscope, you will see follicles that appear as large spaces filled with a pinkish substance (when stained with a special dye), which are the thyroid hormones T_3 and T_4. Additional cells adjacent to the follicles, called extrafollicular cells, are responsible for producing another thyroid hormone known as calcitonin (Figure 21-11).

Thyroid Hormones: Thyroid hormones are amino acid hormones. There are actually three different thyroid hormones: triiodothyronine (T_3), which has three iodine molecules attached to it, tetraiodothyronine (T_4), which has four iodine molecules, and calcitonin. T_3 is more active than T_4. **Calcitonin** is responsible for moving calcium from the blood to the bone (the major reservoir for calcium). It stimulates osteoblasts (immature bone cells) to produce new bone tissue. Thyroid hormones are very important in homeostasis and are involved in energy production, protein synthesis, and tissue repair.

Hypothyroidism: With hyposecretion of thyroid hormones the individual often experiences slowing of the metabolism. Other symptoms may include dry skin, fatigue, cold intolerance, constipation, and difficulty getting pregnant. A childhood (congenital) form of **hypothyroidism** is called *cretinism* (Figure 21-12). These children often experience both mental and physical growth retardation.

Hyperthyroidism: Hypersecretion of thyroid hormones causes an overall increase in metabolism. Symptoms include an increase in heart rate, heart

---

calcitonin A protein hormone that moves calcium from the blood to the bone.

hypothydroidism A condition in which there is an undersecretion of thyroid hormones.

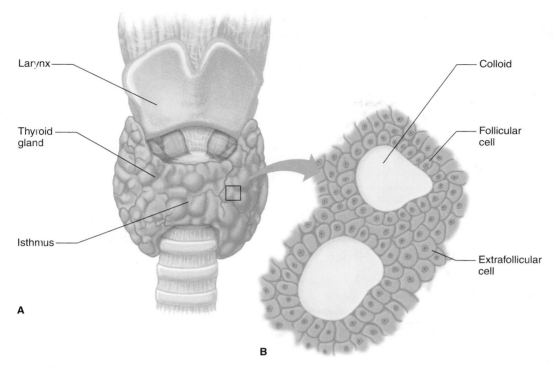

FIGURE 21-10 The thyroid gland: (A) the thyroid consists of two lobes connected by an isthmus; (B) follicular cells secrete thyroid hormones.

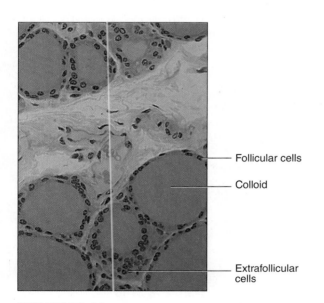

FIGURE 21-11 A light micrograph of thyroid gland tissue (240x).

FIGURE 21-12 Cretinism is due to a hypoactive thyroid gland during infancy and childhood.

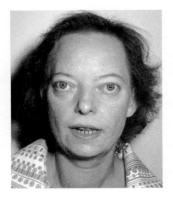

FIGURE 21-13
Graves' disease may cause protrusion of the eyeballs (exophthalmos).

exophthalmos A condition in which the eyeballs protrude, often caused by hyperthyroidism.

hypercalcemia An increase in blood calcium levels.

palpitations, nervousness, and insomnia (inability to sleep). The individual may also be heat intolerant and have protruding eyeballs, a condition known as **exophthalmos** (Figure 21-13). One form of hyperthyroidism is called Graves' disease. Former First Lady Barbara Bush suffers from Graves' disease.

Thyroid Tumors: The thyroid gland may develop both benign and malignant (cancerous) tumors. The type of tumor determines the treatment chosen.

PATHOPHYSIOLOGY

Hypercalcemia

Calcitonin hyposecretion results in less calcium going into bone; possibly resulting in **hypercalcemia** (an increase in blood calcium levels). Hypersecretion of calcitonin results in more calcium going into bone.

Parathyroid Glands

The human body typically has four parathyroid glands, although some people may have six. These glands are rather small and embedded in the posterior aspect of the thyroid gland (Figure 21-14). They secrete parathyroid hormone, which is the antagonist to calcitonin.

Parathyroid Hormone (PTH): Parathyroid hormone (PTH) is a polypeptide that stimulates osteoclasts to help "break down" bone. When the bone is broken down, calcium leaves the bone to enter the blood circulation. PTH also causes the kidneys to conserve more calcium by stimulating the

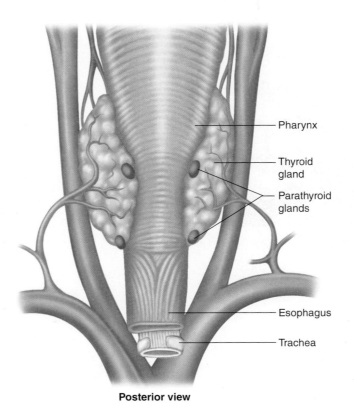

Posterior view

FIGURE 21-14
The parathyroid glands are embedded in the posterior surface of the thyroid gland.

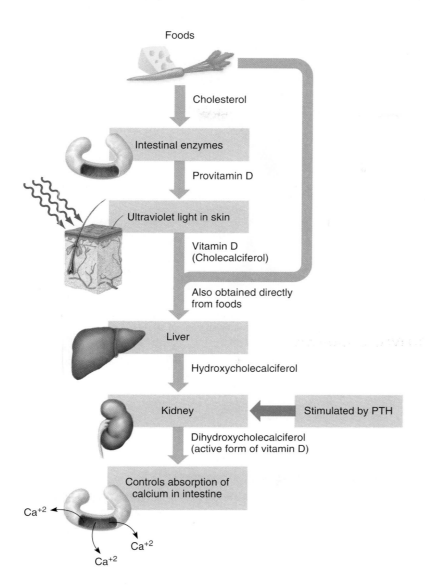

FIGURE 21-15
Mechanism by which PTH regulates calcium absorption in the intestine.

production of the active form of Vitamin D. Vitamin D then induces synthesis of a calcium-binding protein in intestinal epithelial cells that facilitates efficient absorption of calcium into blood (Figures 21-15 and 21-16).

PATHOPHYSIOLOGY

Hypocalcemia

With hyposecretion of PTH there is less osteoclast activity and calcium does not readily leave the bone to enter the blood. This may contribute to **hypocalcemia** (too little calcium in the blood). With hypersecretion of PTH, increased calcitonin activates the osteoclasts, causing breakdown of bone and an increase in the blood calcium levels (hypercalcemia).

U check
What are the two hormones that regulate calcium levels in the blood and bone?

hypocalcemia A condition in which there is too little calcium in the blood.

CHAPTER 21 The Endocrine System

FIGURE 21-16 Parathyroid hormone (PTH) stimulates osteoclasts so that calcium is released from bone, stimulates the kidneys to conserve calcium, and indirectly stimulates the intestines to absorb calcium. PTH secretion is then inhibited by negative feedback.

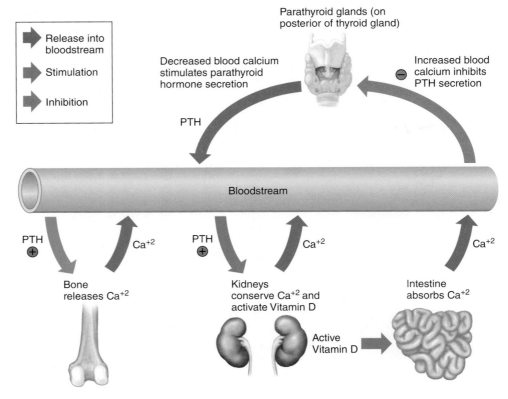

Adrenal Glands

There are two adrenal glands in the body (Figure 21-17). Each is located on the superior surface of one of the kidneys. That is why the adrenal ("next to" the kidneys) glands are sometimes called the suprarenal glands ("above" the kidneys). Each adrenal gland has an inner portion called the medulla (a term that you see in several different structures which means "inner") and an outer portion called the cortex (Figure 21-18). The medulla secretes

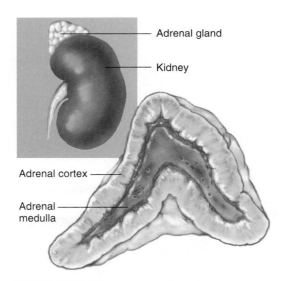

FIGURE 21-17 Adrenal glands sit on top of the kidneys and are composed of an outer cortex and an inner medulla.

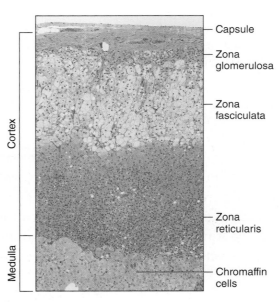

FIGURE 21-18 Light micrograph of the outer cortex and inner medulla of an adrenal gland (75x).

hormones that are considered neurotransmitters (refer to Chapter 14, The Nervous System) that mimic the "flight or fight" response of the sympathetic nervous system. The hormones of the adrenal medulla are referred to as catecholamines. The adrenal cortex secretes two categories of hormones—mineralcorticoids and glucocorticoids—that are important in the regulation of minerals such as sodium and glucose, respectively.

Adrenal Medulla: The inner portion of the adrenal gland is the medulla and is responsible for secreting two vey important hormones. **Epinephrine** and norepinephrine (adrenalin and noradrenalin, respectively) are released by the adrenal medulla. These are amino acid hormones that prepare the body for emergency situations. They are also released during times of physical or emotional stress. They increase heart rate, breathing rate, and blood pressure. The bronchioles are dilated to allow more air into the lungs and peristalsis slows, so some of the target organs are the heart, lungs, and arterioles (small arteries).

epinephrine An amino acid hormone secreted by the adrenal medulla that increases heart rate, breathing rate, and blood pressure in response to physical or emotional stress.

PATHOPHYSIOLOGY
Epinephrine and Norepinephrine Imbalances

Hyposecretion conditions are uncommon in the adrenal medulla. Hypersecretion of epinephrine and/or norepinephrine causes an increase in blood pressure, tachycardia (increased heart rate), and tachypnea (faster breathing rate), among other symptoms.

Tumors

Pheochromocytoma is a benign tumor, but it can have serious and even fatal consequences. It can dramatically increase the secretion of epinephrine which causes hypertension, sometimes extreme.

Adrenal Cortex: Several hormones are secreted by the adrenal cortex. *Aldosterone* is one of two that we will mention here. It is a steroid hormone classified as a mineralcorticoid because it helps regulate the amount of sodium in the body. It targets the tubules of the kidneys to increase retention of sodium. There is a saying "where sodium goes, water follows." When sodium is retained by the body, water is also retained. This causes an increase in fluid volume and an increase in blood pressure. You may have noticed that when you eat very salty foods, for example, your rings or shoes suddenly become too tight.

Cortisol: **Cortisol** is a glucocorticoid that is released from the adrenal cortex and it is also a steroid hormone. When a person is stressed, cortisol levels are increased in the blood. This helps decrease inflammation and pain in the body, but it also decreases protein synthesis and tissue repair. People who are on long-term steroid therapy are susceptible to other complications such as joint pain, infections, headaches, weight gain, and increased risk of developing diabetes mellitus.

cortisol A glucocorticoid released by the adrenal cortex in response to stress to decrease inflammation, pain, protein synthesis, and tissue repair.

focus on Wellness: Stress

We have all experienced stress. Both pleasant (for example, buying a new home) and unpleasant (for example, having a minor car accident) experiences can produce stress. Those events or stimuli that produce stress are called *stressors*. Additional negative stressors may be physical factors such as extreme heat or cold, infections, injuries, heavy exercise, and loud sounds. They can also include psychological or emotional factors such as personal loss, grief, anxiety, depression, and guilt. Some additional positive stressors include sexual arousal, joy, and happiness.

The body's physiologic response to stress consists of a collection of reactions called the *general stress syndrome*, which is primarily caused by the release of specific hormones. This syndrome results in an increase in the heart rate, breathing rate, and blood pressure. Glucose and fatty acid concentrations also increase in the blood, which leads to weight loss. If overwhelming or negative stress persists for a long duration, cortisol is released. Cortisol interferes with protein synthesis and inhibits tissue repair. The immune response also becomes compromised and the individual is more prone to infection and other illnesses. Some stress is okay, but it is best to control our level of stress to maintain our health.

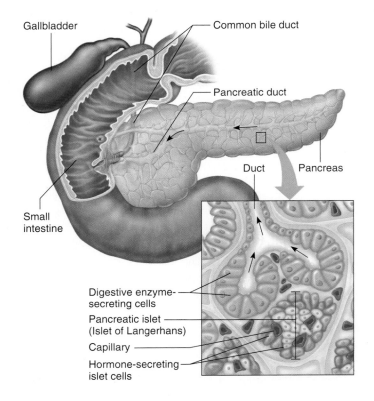

FIGURE 21-19
Pancreas. The hormone-secreting cells of the pancreas are grouped in islets and release their hormones into nearby blood vessels. Digestive enzymes are released into ducts.

Pancreas

The pancreas is located behind the stomach, so it is not palpable (Figure 21-19). The pancreas is a "mixed" gland which means that it is both exocrine (about 80 to 90 percent) and endocrine (10 to 20 percent). The exocrine portion secretes digestive enzymes such as amylase, protease, and lipase. These enzymes are secreted into the duodenum of the small intestine (refer to Chapter 19, The Digestive System). Several hormones are secreted by the endocrine portion. We will focus on two—insulin and **glucagon.**

The endocrine pancreas is made up of groups of cells called the **islets of Langerhans** (Figure 21-20). The two major cell types making up the islets are alpha cells, which secrete glucagon and beta cells, which secrete insulin. These two hormones act as antagonists to each other, meaning they create opposite effects. Glucagon increases blood glucose levels and insulin does the opposite. In fact, this concept of two hormones working as antagonists to each other along with negative feedback is common in the body and is important in maintaining homeostasis. As stated earlier, another example of this concept is calcitonin, which increases bone calcium, and parathyroid hormone, which decreases bone calcium.

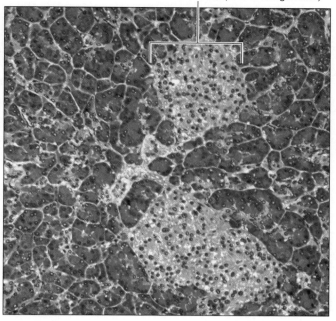

FIGURE 21-20 Light micrograph of the islets of Langerhans (200x).

Insulin: Insulin is a peptide hormone that lowers blood glucose levels. Because it also helps take up amino acids into cells, it facilitates protein synthesis. Insulin is secreted by specialized cells called beta cells.

glucagon A hormone excreted by the endocrine pancreas that increases blood glucose levels.

> ### focus on Clinical Applications: Glucose Testing
>
> With the increase of cases of diabetes mellitus, glucose testing—both at home or in the medical office or lab—is becoming more common. As a health care professional, it is important that you be able to advise patients on preparation for glucose testing in the office or lab, as well as on the use of home glucose monitors.
>
> To prepare a patient for glucose testing in the office or lab, you must first know whether the physician is ordering a fasting or "random" glucose test. Random tests are just that: The patient goes to the office or lab at the appointed time having had meals and snacks as usual. Fasting tests are usually performed in the morning because the patient must be advised to take nothing by mouth (NPO) for at least eight hours (usually midnight the night before). The patient may have water, but must otherwise be in a fasting state.
>
> Home glucose monitors come in many shapes, styles, and designs, but all have one thing in common—they are user-friendly and use only a small amount of blood to test blood sugar "on the spot." Many diabetic patients test their sugar levels at least three times a day and keep a log which they bring to their medical appointments. These logs allow the patient and medical care provider to track and trend the patient's blood sugar to provide an individualized plan of care to keep the blood sugar level stable and avoid the "peaks and valleys" found in uncontrolled diabetes mellitus.

islets of Langerhans
Groups of cells that form the endocrine pancreas and are made of glucagon-secreting alpha cells and insulin-secreting beta cells.

diabetes mellitus
A condition caused by hyposecretion of insulin that is categorized into type 1, or insulin-dependent diabetes mellitus, and type 2, or non-insulin-dependent diabetes mellitus.

Type 1 Diabetes and Type 2 Diabetes

REMEMBER MANNY
Washington, our 58-year-old marketing manager? Which type of diabetes do you think he has?

PATHOPHYSIOLOGY

Diabetes Mellitus

Hyposecretion of insulin (hypoinsulinism) can result in the very common disease **diabetes mellitus.** There are two primary forms of diabetes mellitus. Type 1 diabetes mellitus, also called juvenile diabetes because of its age of onset (usually before 30 years of age), is also known as insulin-dependent diabetes mellitus (IDDM) because of the individual's dependence on insulin therapy. Symptoms include *polydipsia* (excessive thirst), *polyphagia* (excessive hunger), and *polyuria* (excessive urination). Individuals suffer from neuropathies, infections, fatigue, and weight loss. Type 2 diabetes mellitus is referred to as adult-onset or non-insulin-dependent diabetes mellitus (NIDDM). The term *adult onset* is losing favor as more and more children and teens in the United States are being diagnosed with NIDDM, believed to be related to our high-fat, sedentary lifestyles. Many of the same symptoms occur in type 2 diabetes as in type 1, but type 2 individuals are typically obese. Diabetes is the number one cause of preventable blindness and amputation of extremities (the toes, feet, and legs in particular) because of circulatory issues related to diabetes. Hypersecretion of insulin (hyperinsulinism) may occur when cells become more resistant to insulin.

Pancreatic Tumors

Pancreatic tumors may occur in the exocrine or endocrine portion of the pancreas. Malignant tumors of the pancreas have some of the poorest prognoses (outcomes) of all cancers. They are difficult to diagnose, which means the cancer often spreads before treatment can be started.

Glucagon: Glucagon is a peptide hormone that raises blood glucose levels. Therefore, it has the opposite effect of insulin. Glucagon is secreted by specialized cells called alpha cells.

U check
What is the difference between diabetes mellitus and diabetes insipidus?

PATHOPHYSIOLOGY

Glucagon Imbalances

Hyposecretion of glucagon results in chronic hypoglycemia (a low level of glucose in the blood). Hypersecretion of glucagon results in chronic hyperglycemia (elevated glucose levels in the blood), also known as diabetes mellitus.

> ### focus on Wellness: Prevention of NIDDM
>
> The increasingly sedentary lifestyle found in the United States is taking a toll on our health and well-being. This is true not only of adults, but increasingly we are seeing an increase in chronically ill children and teens as well. The major culprit—obesity leading to diabetes mellitus type 2 or NIDDM (noninsulin-dependent diabetes mellitus). More children and adults spend their days sitting in classrooms or at office desks, arrive home, have something to eat, and then return to their chairs to sit at the computer, play a video game, or watch television. To add insult to injury, while doing these sedentary activities, a high-fat, high-calorie snack and beverage is often added to the mix.
>
> NIDDM can usually be prevented—the recipe is simple, yet so hard to follow:
>
> 1. Take one individual and get him or her out of the chair.
> 2. Actively move the individual to increase the heart rate and burn calories.
> 3. Repeat this activity or a new one, 20 to 30 minutes a day, at least three to five times a week.
> 4. Consume fruits, vegetables, and water instead of sugary, high-calorie snacks and beverages.
>
> Simply put: Decrease calories and increase exercise for a decrease in diagnosis of NIDDM!

Thymus

The thymus lies in the space between the lungs called the mediastinum. It reaches its maximum size by adolescence and then atrophies or involutes ("turns in on itself") as the body ages. It matures or processes the T lymphocytes that are very important in immunity. It produces the hormone thymosin.

Thymosin: Thymosin is a peptide hormone that is involved in the maturation of T lymphocytes. Refer to Chapter 12, The Lymphatic and Immune Systems, to read more about T cells and immunity.

PATHOPHYSIOLOGY
Thymosin Imbalances

Hyposecretion of thymosin may result in lack of maturation of T lymphocytes and a decrease in immunity. Tumors are rare, but when they do occur they are often very aggressive.

Pineal Gland

The pineal gland (pineal body) is located in the brain between the cerebral hemispheres. At one time it was thought that this was a vestigial organ, which means over time it has lost its function. We now know that is not true because we have identified the function of this gland: It produces melatonin.

Melatonin is an amino acid hormone that regulates circadian rhythms or the body's biological clock. In other words, it helps with the awake–sleep cycle. Melatonin is also involved with the onset of puberty.

> ## PATHOPHYSIOLOGY
> ### Melatonin Imbalances
>
> Hyposecretion of melatonin may disturb the sleep cycle and contribute to depression. In addition, because of its involvement with the onset of puberty, a deficiency of melatonin may contribute to delay of maturation.

Gonads

The general term for the male and female reproductive organs is *gonads*. As we learned in Chapters 16 and 17 on the male and female reproductive systems, the female gonads are the ovaries and the male gonads are the testes.

Female Hormones: The ovaries of the female are located in the pelvic cavity and secrete two important hormones, estrogen and progesterone. Estrogen is a steroid hormone that is the main female sex hormone. It regulates the menstrual cycle and is important in the development of female secondary sex characteristics. Progesterone is also a steroid hormone with many functions. It helps reduce anxiety and induce sleep, build and maintain bone tissue, and it maintains the embryo and fetus during pregnancy.

> ## PATHOPHYSIOLOGY
> ### Estrogen Imbalances
>
> Hyposecretion of estrogen, either before or during menopause, may result in hot flashes, vaginal dryness, increased urinary tract infections, mood changes, depression, and loss of bone density.
>
> ### Ovarian Tumors
>
> Ovarian cancer is the second most common type of gynecological cancer. Symptoms, especially early in the disease, may be mild and include bloating and pelvic pain. Typically the diagnosis of the disease occurs in the later stages causing a poor survival rate. Although the five-year survival rate has improved, it remains under 50 percent.

Male Hormones: Testosterone is a steroid hormone responsible for the development of the male reproductive organs. It is also responsible for male secondary sex characteristics. These include more facial and chest hair, increased muscle mass and bone density, and a deeper voice than females.

Additional Endocrine Glands and Tissues

Heart

The heart is not usually considered an endocrine gland. However, it does secrete a hormone that helps regulate blood pressure.

Atrial natriuretic peptide (factor) is a protein hormone that is intimately involved in the regulation of blood pressure. *Natrium* is the Latin term for sodium, which accounts for sodium's chemical symbol Na. Atrial natriuretic factor (ANF) is secreted by specialized cells in the right atrium in response to high blood pressure. It helps eliminate sodium and consequently water and is important in maintaining homeostasis of the body.

Kidneys

Erythropoietin is an amino acid hormone that is responsible for stimulating the synthesis of erythrocytes (red blood cells). It is sometimes used as a "blood doping agent" by athletes to increase their endurance. Too many red blood cells (erythrocytosis) can increase blood viscosity (thickness) and put the individual at risk for a myocardial infarction or stroke.

21.2 learning outcome

Identify additional endocrine glands and tissues that secrete hormones and their functions.

Regulatory Mechanisms

To achieve homeostasis there has to be some method to regulate the endocrine system. Feedback loops are the way the body does this. There are two basic types of feedback loops, negative feedback and positive feedback. Figure 21-21 illustrates negative feedback regulation of cortisol.

21.3 learning outcome

Describe the regulatory mechanisms involved in maintaining homeostasis.

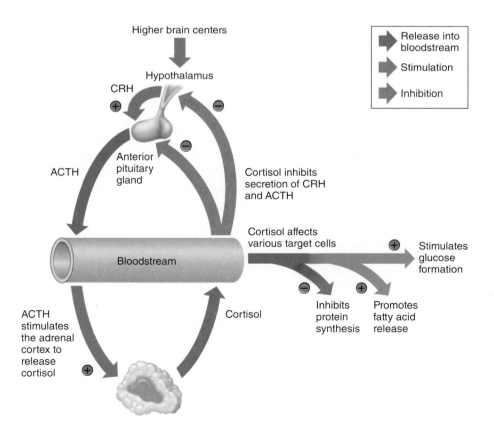

FIGURE 21-21
Negative feedback regulates cortisol secretion.

CHAPTER 21 The Endocrine System

You could possibly think of these as two highway systems. A negative feedback loop is like a highway that goes both north and south. If you take the highway north, you can also take it south back to your original destination. For example, the pituitary gland secretes several hormones. One of these hormones is growth hormone. When enough growth hormone is produced and secreted, there is a message or "feedback" to the pituitary gland to stop secreting growth hormone. This returns the system back closer to its original state. You may recall also that two other hormone antagonists that use negative feedback are glucose and insulin, and calcitonin and PTH.

Most feedback loops are negative feedback systems. A positive feedback loop is like a highway going in only one direction. It does not provide much opportunity to return to the initial state. The two examples of positive feedback loops are the birth process and blood clotting.

Let's look at the birth process. With the onset of labor, oxytocin is released from the posterior pituitary (neurohypophysis). This stimulates uterine contractions which then cause even more oxytocin to be secreted. This cycle continues until the birth of the infant and then the release of oxytocin is stopped.

> **U check**
> What are the two types of feedback mechanisms?

21.4 learning outcome

Infer the effects of aging on the endocrine system.

Life Span

As individuals age, some glands get smaller or decrease their secretions while other glands actually increase their secretions. For example, the anterior pituitary gland secretes less growth hormone, which is responsible in part for the decrease in muscle mass as we get older. The thyroid commonly decreases its output of thyroid hormones, resulting in decreased metabolism and increased fat deposits in the body. Parathyroid hormone levels increase possibly due to the decrease in dietary calcium intake, whereas calcitonin levels fall. The combination of rising PTH with declining calcitonin leads to increased osteoporosis. The adrenal glands produce less aldosterone but relatively the same amount of epinephrine and norepinephrine. The thymus begins to atrophy at puberty and thymus tissue is replaced by fat and areolar connective tissue. However, the thymus can still participate in the immune response. Estrogen in female and testosterone in males decrease with age, although the effects are more dramatic and are seen at an earlier age in women.

summary

chapter 21

learning outcomes	key points
21.1 Recognize the major endocrine glands, their location in the body, and their hormones, target tissues, functions, and disorders.	The endocrine system consists of a number of ductless glands distributed throughout the body. They secrete substances called hormones that help regulate many body functions and maintain homeostasis.
21.2 Identify additional endocrine glands and tissues that secrete hormones and their functions.	The heart, although not considered an endocrine gland, secretes atrial natriuretic peptide (factor), a hormone that helps to regulate blood pressure. The kidneys secrete erythropoietin, a hormone that stimulates the synthesis of red blood cells.
21.3 Describe the regulatory mechanisms involved in maintaining homeostasis.	Negative and positive feedback loops are mechanisms that help to regulate homeostasis.
21.4 Infer the effects of aging on the endocrine system.	Some endocrine glands decrease their output, while others actually increase theirs. The overall effect is a general slowing down of metabolism.

case study 1 questions

Can you answer the following questions that pertain to Manny Washington's case study presented earlier in this chapter?

1. The endocrine gland involved in the etiology of diabetes mellitus is
   - a. Pituitary
   - b. Thyroid
   - c. Adrenal
   - d. Pancreas
2. Which cells are responsible for secreting insulin?
   - a. Alpha cells
   - b. Beta cells
   - c. Goblet cells
   - d. Extrafollicular cells
3. Which one of the following is also known as insulin-dependent diabetes?
   - a. Type 1
   - b. Type 2
   - c. Diabetes insipidus
   - d. Gestational diabetes

review questions

1. All of the following are true regarding the endocrine system *except*
   - a. It is a ductless system.
   - b. It is faster than the nervous system.
   - c. It is one of the two major regulating systems in the body.
   - d. It secretes chemical substances called hormones.
2. Which of the following glands has a major regulatory effect on the pituitary gland?
   - a. Thalamus
   - b. Epithalamus
   - c. Hypothalamus
   - d. Adrenal gland
3. What gland secretes growth hormone (GH)?
   - a. Posterior pituitary
   - b. Anterior pituitary
   - c. Adrenal gland
   - d. Pancreas

4. What gland secretes ACTH?
   a. Posterior pituitary
   b. Anterior pituitary
   c. Adrenal gland
   d. Pancreas
5. What gland secretes melatonin?
   a. Posterior pituitary
   b. Anterior pituitary
   c. Pineal gland
   d. Pancreas
6. What gland secretes erythropoietin?
   a. Posterior pituitary
   b. Anterior pituitary
   c. Kidneys
   d. Pancreas
7. What gland secretes glucagon?
   a. Posterior pituitary
   b. Anterior pituitary
   c. Adrenal gland
   d. Pancreas
8. What gland secretes vasopressin?
   a. Posterior pituitary
   b. Anterior pituitary
   c. Adrenal gland
   d. Pancreas
9. What gland secretes insulin?
   a. Posterior pituitary
   b. Anterior pituitary
   c. Adrenal gland
   d. Pancreas
10. What gland secretes glucocorticoids?
    a. Posterior pituitary
    b. Anterior pituitary
    c. Adrenal gland
    d. Pancreas
11. A tumor in the anterior pituitary gland may cause
    a. Diabetes insipidus
    b. Acromegaly
    c. Cushing's syndrome
    d. Addison's disease

critical **thinking questions**

1. Patients who are on hemodialysis due to end-stage renal disease are often anemic. Explain how renal disease can cause or worsen this problem.
2. List the organs and hormones that cause the unrelated diseases of diabetes mellitus and diabetes insipidus. What is at least one consequence of the progression of each of these diseases?
3. Why is hyposecretion (insufficient secretion) of thyroid hormone in newborns more serious than hyposecretion in adults?

patient **education**

Diabetes mellitus is an epidemic in the United States, with children being at risk of having shorter life spans than their parents. You have been asked to present a talk to a high school class of students and their parents regarding type 2 diabetes mellitus. What would you say to this audience? You want to address prevention, risk factors, detection, complications, and treatment of the disease.

applying what **you know**

1. Use your knowledge of erythropoietin and its effects on blood viscosity to explain the risks of a 17-year-old athlete taking erythropoietin to increase his endurance?
2. What are three to seven things that you can personally do to improve your overall "endocrine health"?

CASE STUDY 2 *Hypothyroidism*

Tara McDonald, a 28-year-old nurse's assistant, has complained about feeling cold all the time. She also has felt sluggish and is concerned that she has not been able to conceive, although she and her husband have been trying to have a family for some time.

1. Which of the following is *not* typical of hypothyroidism?
   a. Problems with conception
   b. Cold intolerance
   c. Lethargy (sluggishness)
   d. Rapid heart rate
2. Which of the following hormones is *not* secreted by the thyroid gland?
   a. Thymosin
   b. Triiodothyronine
   c. Tetraiodothyronine
   d. Calcitonin
3. What is cretinism?
   a. Uncontrolled diabetes mellitus
   b. A form of hyperthyroidism
   c. A form of hypothyroidism
   d. Hyperaldosteronism

CASE STUDY 3 *Cushing's Syndrome*

A 62-year-old dog groomer, Ellie Weis, is concerned because when she last checked her blood pressure at the mall it was much higher than usual. Her face has also taken on a very round appearance and the back of her neck "seems fatter" to her. Ellie also mentions that she has developed facial hair and she is feeling depressed frequently.

1. What gland is responsible for this disorder?
   a. Adrenal   c. Pancreas
   b. Pituitary   d. Pineal
2. What is the main cause of this disorder?
   a. Hypersecretion of cortisol
   b. Hyposecretion of cortisol
   c. Hypersecretion of insulin
   d. Hypersecretion of glucagon
3. The opposite of Addison's disease is
   a. Cushing's syndrome
   b. Gestational diabetes
   c. Diabetes insipidus
   d. Graves' disease

22

The Special Senses

chapter outline

- Olfaction
- Taste
- Vision
- Hearing and Equilibrium
- Life Span

outcomes

AFTER COMPLETING THIS CHAPTER, YOU WILL BE ABLE TO:

22.1 Relate the anatomy of the nose to the sensation of olfaction (smell).

22.2 Relate the anatomy of the tongue to the sensation of taste.

22.3 Relate the anatomy and pathophysiology of the eye and the accessory structures to their functions.

22.4 Explain how the anatomy of the ear relates to the sense of hearing and to equilibrium.

22.5 Describe the changes that are seen in the special senses as we get older.

case study

Use the case study to focus on as you go through the chapter. The questions will guide you as you learn the anatomy, physiology, and pathology of the special senses.

CASE STUDY 1 *Night Blindness*

Matthew Garcia is a 68-year-old, over-the-road truck driver. He has been having difficulty with his night driving. The "eye doctor" says that he also has hyperopia and a cataract in his left eye.

As you go through the chapter, keep the following questions in mind:

1. What is the medical term for "night blindness"?
2. What type of "eye doctor" can treat all conditions of the eye?
3. What is hyperopia?
4. What is the cause of cataracts?
5. What is the cause of glaucoma?

essential terms

aqueous humor (AH-kwee-us HYOO-mer)
auricle (AW-rih-kul)
choroid (KO-royd)
cochlea (KOK-lee-uh)
cones (kohnz)
conjunctiva (kon-junk-TIE-vah)
cornea (KOR-nee-uh)
eustachian tube (yoo-STAY-shun toob)
fovea centralis (FOH-vee-uh sen-TRAL-is)
incus (ING-kus)
malleus (MAL-lee-us)
olfaction (ol-FAK-shun)
optic disk (blind spot) (OP-tik disk)
ossicles (OS-ih-kulz)
papillae (pah-PIL-ee)
retina (REH-tih-nah)
rods (RODZ)
sclera (SKLEH-rah)
semicircular canals (seh-mih-SIR-kyoo-lar kah-NALZ)
stapes (STAY-peez)
vestibule (VES-tih-bule)
vitreous humor (VIH-tree-us HU-mor)

Additional key terms in the chapter are italicized and defined in the glossary.

> **study tips**
>
> 1. Draw and label the eyeball.
> 2. Make your own "Snellen" chart and see how your vision compares to that of one or two of your friends.
> 3. Make flash cards for the outer ear, middle ear, and inner ear, listing the structures found in each. Make flash cards for the essential terms of the chapter.

Introduction

Special senses include smell, taste, vision, hearing, and equilibrium. They are called special senses because their sensory receptors are located within relatively large sensory organs in the head—the nose, tongue, eyes, and ears. The skin, which provides us the special sense of touch, is considered a generalized sense organ and was discussed in Chapter 7, The Integumentary System. Regardless of the nature of the stimulus—smell, taste, light, touch, or sound—the signal is sent via the nervous system to the cerebral cortex for interpretation and a response. This chapter introduces the structure and function of the special sense organs. If you choose a career in health care, it is likely that you will be asked to assist with or perform examinations and treatments for common disorders of the eyes and ears. Therefore, you will need to understand the functioning of these important sense organs.

U check
Where are special senses interpreted?

22.1 learning outcome

Relate the anatomy of the nose to the sensation of olfaction (smell).

olfaction The sense of smell.

Olfaction

Olfaction is the sense of smell. Smell receptors, also called olfactory receptors (Figure 22-1), are *chemoreceptors*. This means that they respond to changes in chemical concentrations. Chemicals that activate smell receptors must first be dissolved in the mucus of the nose. This explains why a person with either a "dry nose" or excessive mucus related to an upper respiratory tract infection or allergies has trouble smelling. Olfactory epithelium covers the upper part of the nasal cavity and covers the inferior aspect of the cribriform plate of the ethmoid bone (Figure 22-2). Three types of cells are found in the olfactory epithelium—olfactory receptors, supporting cells, and basal cells.

Olfactory receptors are bipolar neurons that have a dendrite and an axon (refer to Chapter 14, The Nervous System, for more information on neurons). The cilia on the olfactory receptors are stimulated by odorants. The chemical message is changed (transduced) into an electrical signal. Once smell receptors are activated, they send their information to the olfactory nerves. The olfactory nerves send the information along olfactory bulbs and tracts to different areas of the cerebrum. The cerebrum interprets the information as a particular type of smell.

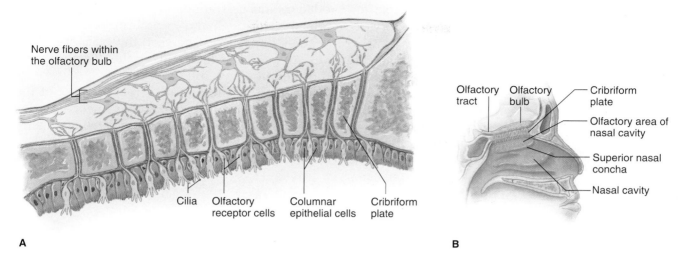

FIGURE 22-1 Olfactory receptors: (A) columnar epithelial cells support the olfactory receptors; (B) the olfactory area is associated with the superior nasal concha.

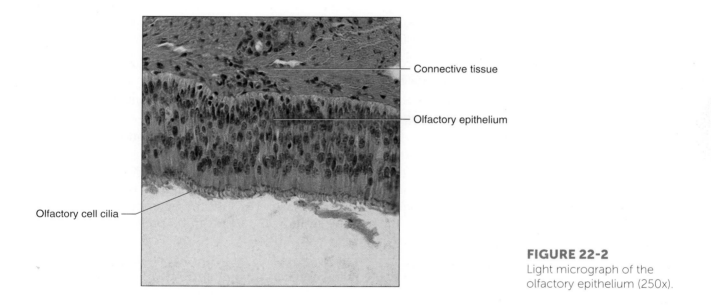

FIGURE 22-2 Light micrograph of the olfactory epithelium (250x).

Our sense of smell undergoes *sensory adaptation,* which means that the same chemical can stimulate smell receptors for only a limited amount of time. In a relatively short period of time, the smell receptors fatigue and no longer respond to the same odor, and it can no longer be smelled. Sensory adaptation explains why you smell perfume when you first encounter it, but after a few minutes you cannot smell it or may be less aware of it. *Supporting cells* are columnar in shape and provide nutrition, support, and insulation for the receptors. *Basal cells* are mitotically active cells that produce new receptor cells which live only about a month.

U check

What type of neuron is found in the olfactory receptors?

Cranial nerve VII (the facial nerve) innervates the epithelium and the glands responsible for producing mucus. Bundles of unmyelinated axons from the olfactory receptors pass through the cribriform plate to enter the cranial cavity. These bundles form cranial nerve I, the olfactory nerve (remember, all cranial nerves are paired), which ends in the olfactory bulb. From here, the axons travel posteriorly as the olfactory tract.

Axons travel to the primary olfactory area of the cerebral cortex of the temporal lobe for conscious awareness of the aroma being smelled. Axons also travel to the hypothalamus for the emotional aspect of smell and memory of smells. In addition, axons go to the frontal lobe where odors are identified. Interestingly, olfactory sensations are the only sensations that do not synapse in the thalamus.

> **U check**
> Which lobe of the brain is where conscious awareness of smell is located?

Taste

22.2 learning outcome
Relate the anatomy of the tongue to the sensation of taste.

Taste is also called gustation. Like olfaction, taste is a chemical sensation. However, whereas we can recognize perhaps 10,000 different smells, we can distinguish only five different tastes: sweet, sour, bitter, salty, and umami.

1. *Sweet*. Taste cells that respond to "sweet" chemicals are concentrated at the tip of the tongue.
2. *Sour*. Taste cells that respond to "sour" chemicals are concentrated on the sides of the tongue.
3. *Salty*. Taste cells that respond to "salty" chemicals are concentrated on the tip and sides of the tongue.
4. *Bitter*. Taste cells that respond to "bitter" chemicals are concentrated at the back of the tongue. The interesting thing about this is that many poisonous plants have a bitter taste. Because the bitter taste is detected at the back of the tongue, swallowing them can be prevented when the bitterness is detected.
5. *Umami*. In addition to the four well-known taste sensations, in the 1980s science recognized a taste known as *umami*. In 1908, a Japanese scientist discovered that glutamic acid found in kelp produced a savory taste that he named umami. This unique taste is found most notably in tomatoes and also in meats, fish, and dairy products, including breast milk. Although discovered and named in the early 1900s, it was not until the 1980s that studies found that glutamate, the substance responsible for this unique taste, created a fifth basic taste that has now become universally recognized.

Gustatory (taste) receptors (Figure 22-3) are located on taste buds. Taste buds are microscopic structures found on the **papillae** (elevations) of the tongue (Figure 22-4). There are four types of papilla in humans:

papillae The projections on the surface of the tongue where the taste buds are located.

1. Circumvallate, or vallate papillae, number 12 to 22 and are located at the back of the tongue. They form a V that points toward the throat. Each papilla contains 100 to 300 taste buds.
2. Fungiform papillae are mushroom-shaped and are dispersed over the dorsum of the tongue. Each papilla has about five taste buds.

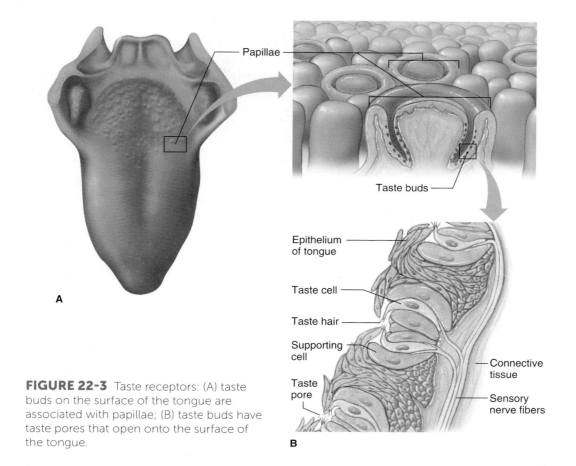

FIGURE 22-3 Taste receptors: (A) taste buds on the surface of the tongue are associated with papillae; (B) taste buds have taste pores that open onto the surface of the tongue.

3. Foliate papillae are found on the sides of the tongue. Their taste buds degenerate by adulthood. They also contain lymphoid tissue and so help fight infections.
4. Filliform papilla are the most numerous type of papilla, but they do not contain taste buds; they contain tactile receptors. Filliform papillae give the tongue its typical roughened surface.

Some taste buds are also scattered on the roof of the mouth and in the walls of the throat. Each taste bud is made of taste cells and supporting cells. The taste cells function as taste receptors. Like the olfactory cells of smell, taste cells are chemoreceptors. They are activated by chemicals found in food and drink that must be dissolved in saliva as part of the digestive process. When taste cells are activated, they send their information to several cranial nerves. The information eventually reaches the gustatory cortex in the parietal lobe of the cerebrum. The gustatory cortex interprets the information as a particular taste. Although not an actual taste sensation, eating spicy foods activates pain receptors on the tongue that are then interpreted by the brain as "spicy." We also adapt to a specific taste in as little as one to five minutes.

Three cranial nerves are directly involved in taste sensation. Cranial nerve VII (the facial nerve) innervates the anterior two-thirds of the tongue. Cranial nerve IX (the glossopharyngeal nerve) innervates the posterior one-third of

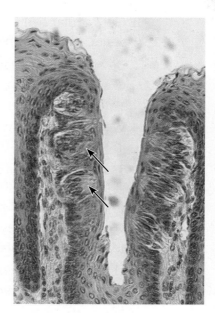

FIGURE 22-4
A light micrograph of taste buds (225x).

CHAPTER 22 The Special Senses

the tongue. Cranial nerve X (the vagus) innervates taste buds in the throat and the epiglottis. In addition, cranial nerves IX and X are responsible for the gag reflex. Nerve impulses are sent to the medulla oblongata and from there to the limbic system, the hypothalamus, and the thalamus. From the thalamus, impulses are sent to the cortex of the parietal lobe for the conscious perception of taste.

> **U check**
> What is the location and function of the circumvallate papillae?

22.3 learning outcome

Relate the anatomy and pathophysiology of the eye and the accessory structures to their functions.

Vision

Vision and the sense of sight are very important to our activities of daily living and to survival itself. This is evidenced by the number of sensory receptors that are located in the eye: more than 50 percent of the body's total number of sense receptors! A person's visual system consists of the eyes; the optic nerve (cranial nerve I), which connects the eye to the vision center of the brain; the occipital cortex; and several accessory structures. If all parts of the system are healthy and normal, the individual is able to see normally.

focus on Wellness: Protecting Your Vision

The following guidelines can help make your vision last a lifetime.

1. Get regular health checkups. There is a connection between general health and eyes. High blood pressure and diabetes can cause eye problems.

2. Get regular eye examinations. Most people need eye examinations every one to two years. People with diabetes should see their eye care specialist more frequently.

3. Be alert for the warning signs of eye disease such as floaters, narrowing of your field of vision, eye pain, cloudy or blurred vision, or blind spots. Visit your eye care specialist immediately if you experience any of the warning signs.

4. Wear sunglasses with ultraviolet protection to shield the eyes from bright sunlight, even in the winter. Have ultraviolet protection added to new distance prescription glasses. The cornea can get sunburned, which can be painful and damaging. Excessive exposure to the sun is a contributing factor in the development of cataracts and malignant melanoma of the eye.

5. Wear protective eye equipment to prevent eye injury. (See the Focus on Wellness feature on eye safety later in the chapter.)

6. Use nonprescription eye medications properly. (See the Focus on Clinical Applications feature on eye medications later in the chapter.)

7. Never share eye makeup because bacterial infections can be passed via the applicators.

PATHOPHYSIOLOGY

Amblyopia

Amblyopia is more commonly called lazy eye and occurs when a child does not use one eye regularly. A child with this disorder does not have normal depth perception and often also has concurrent *strabismus,* a condition in which the eyes do not focus in unison.

Etiology: Amblyopia can be caused by any disorder of the eyes that affects normal eye development and use, including hyperopia (farsightedness), myopia (nearsightedness), strabismus, cataracts, and astigmatism.

Signs and Symptoms: The most common symptoms are blurred vision and an eye that appears to turn inward or outward.

Treatment: Treating the underlying conditions, placing a patch over the normal eye to strengthen the lazy eye, and the use of corrective lenses are the primary methods of treatment.

Strabismus

Strabismus is a misalignment of the eyes. Convergent strabismus is more commonly referred to as crossed eyes. In this condition, the eyes do not focus together and one or both eyes turn inward. In divergent strabismus, sometimes called wall eyes, one or both eyes turn outward instead of focusing together.

Etiology: The causes are mostly unknown, although some known causes include eye and brain injuries, cerebral palsy, and various disorders of the retina.

Signs and Symptoms: Blurred vision and depth perception are the most common symptoms.

Treatment: Treatment includes eyeglasses, eye exercises, an eye patch for the stronger eye to force the weaker eye to work, and surgery to realign the eyes.

> **U check**
> What is the medical term for crossed eyes?

The Eye

The eye is a complex organ that processes light to produce images. The majority of the eye is protected by the bony orbit that is made of several different facial bones. The eyeball is made up of three layers (Table 22-1) known as tunics, two chambers, and a number of specialized parts (Figure 22-5).

TABLE 22-1 Layers of the Eye

Layer/Tunic	Posterior Portion	Function	Anterior Portion	Function
Outer layer	Sclera	Protection	Cornea	Light transmission and refraction
Middle layer	Choroid coat	Blood supply, pigment prevents reflexion	Ciliary body, iris	Accommodation; controls light intensity
Inner layer	Retina	Photoreception, impulse transmission	None	

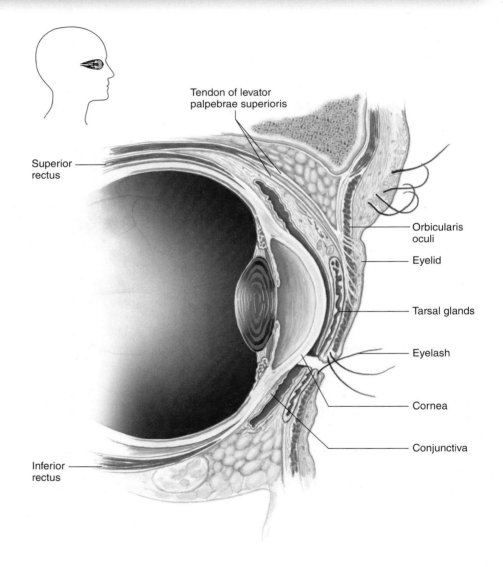

FIGURE 22-5 Sagittal section of the closed eyelids and the anterior portion of the eye.

cornea The transparent anterior portion of the fibrous tunic that helps focus light on the retina.

Fibrous Tunic: The outer fibrous tunic actually has two components. The anterior portion is the **cornea** which helps focus light on the retina. The cornea is transparent and covers the iris. It is made up of epithelium as well as collagen fibers and fibroblasts. The cornea receives oxygen through diffusion, but itself is avascular, which allows for corneal transplants without risk of rejection.

The **sclera** forms the posterior aspect of the outer fibrous tunic. The sclera is made up of tough collagen fibers and gives structure to the eyeball. It is the site of attachment of the six extrinsic eye muscles. It forms the "white "of the eye and posteriorly, it is continuous with the dura mater.

The limbus is the area where the cornea and sclera meet. In this area is the *scleral venous sinus* (canal of Schlemm) which drains aqueous humor from the eye.

sclera The outer fibrous tunic (white of the eye) that gives structure to the eyeball; the posterior aspect is the site of attachment of the eye muscles.

> **U check**
> What is the name of the area where the cornea and sclera meet?

PATHOPHYSIOLOGY
Astigmatism

Astigmatism occurs when the cornea or lens has an abnormal shape, which causes blurred images during near or distant vision.

Etiology: This is normally considered to be a congenital condition.

Signs and Symptoms: There are no symptoms with this condition other than blurred vision. However, it can be diagnosed during an ophthalmic (eye) exam.

Treatment: Treatment includes corrective lenses or surgery to reshape the cornea, such as photorefractive keratectomy (PRK), or the procedure more commonly done today called laser-assisted in situ keratomileusis (LASIK).

Uvea (Vascular Tunic): The *uvea* is the middle layer of the eyeball and consists of the **choroid,** ciliary body, and iris. The choroid is the vascular component of the eyeball and contains most of the eye's blood vessels. In the anterior part of the choroid are the iris and the ciliary body. The iris is the colored part of the eye that is made of smooth muscle arranged in both a radial and circular orientation. As these muscles contract or relax, an opening at its center, the *pupil* (Figure 22-6), grows larger or smaller. The size of the pupil regulates the amount of light that enters the eye. In bright light the pupil becomes constricted (smaller). In dim light it becomes dilated (larger). The *ciliary body* is a wedge-shaped thickening in the middle layer of the eyeball. Muscles in the ciliary body control the shape of the lens, making the lens more or less curved for viewing either near or distant objects, respectively. The choroid, ciliary body, and iris all contain melanocytes and so are subject to malignant melanoma.

choroid The vascular component of the eyeball.

> **U check**
> What is the name of the structure that controls the shape of the lens?

retina The inner layer of the eye that senses light.

optic disk The area of the retina where the optic nerve enters and where there are no sensory nerves; also known as the blind spot.

Retina: The inner layer of the eye consists of the **retina.** Nerve cells at the posterior aspect of the retina sense light. The area where the optic nerve enters the retina is known as the optic disk. Because the **optic disk**

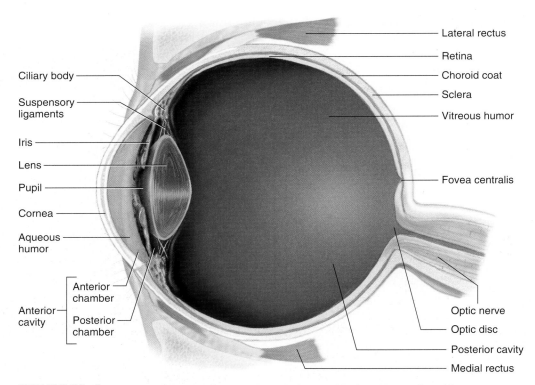

FIGURE 22-6 Transverse, superior view of the right eye. Notice the pupil and its relationship to the lens.

fovea centralis The small depression in the center of the macula lutea that contains only cones and thus is the area of greatest visual acuity.

cones Photoreceptors that detect blue, green, and red and provide sharper images than rods.

(Figure 22-7) contains no sensory nerves itself, it is referred to as the blind spot. On the posterior aspect of the retina is a flat region called the *macula lutea* which appears yellow in color. In the center of this flat yellow region is a small depression called the **fovea centralis.** This is the area of greatest visual acuity (sharpest vision) and is made up of only one type of photoreceptors known as **cones.**

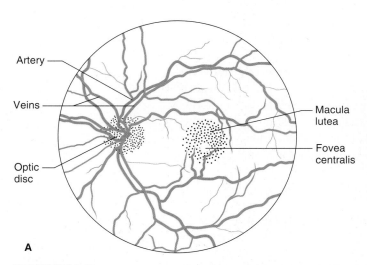
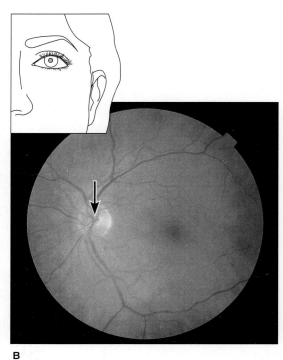

FIGURE 22-7 The retina: (A) major features of the retina; (B) fibers leave the retina in the area of the optic disk (arrow) in this magnified view (53x).

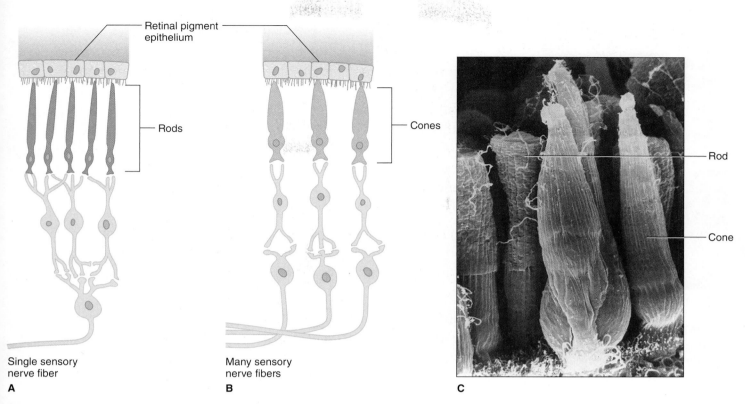

FIGURE 22-8 Rods and cones: (A) a single sensory nerve fiber transmits a message from several cones to the brain; (B) separate nerve fibers transmit impulses from cones to the brain; (C) scanning electron micrograph of rods and cones (1,350x).

The other type of photoreceptors found in the eye are called rods. **Rods** are highly sensitive to light and function well in dim light. They are responsible for night and peripheral (side) vision. Rods are more concentrated in the periphery of the eyeball and can detect only black and white and shades of gray (Figure 22-8).

Cones are much more numerous than rods and function best in bright light. They are more concentrated in the center of the retina and detect colors and provide sharper images. The three colors that cones can detect are blue, green, and red. Deficiencies in the number or types of cones are responsible for the various types of color blindness, which is generally an inherited condition that is most common in boys.

rods One of the two types of photoreceptors found in the retina; specialized for vision in dim light.

U check

What is the name of the area where the optic nerve enters the eyeball?

PATHOPHYSIOLOGY

Macular Degeneration

Macular degeneration is a progressive disease that usually affects people over the age of 50. It occurs when the retina no longer receives an adequate blood supply, and is the most common cause of vision loss in the United States.

(continued)

A   Normal vision

B   The same scene as viewed by a person with macular degeneration

FIGURE 22-9
Macular degeneration: (A) normal vision and (B) macular degeneration.

Macular Degeneration (concluded)

Etiology: Genetics, age, smoking, and exposure to ultraviolet radiation (from sunlight) are known risk factors. Non-age-related conditions such as diabetes mellitus, ocular infection, and trauma are additional causes of this condition. Nutrition may play a role in preventing this disease; diets high in fruits and vegetables that contain vitamins A, C, and E, as well as diets low in saturated fats, are recommended. However, there are no current definitive data on the effectiveness of nutrition in preventing or slowing the progression of this disease.

Signs and Symptoms: Common symptoms include loss of central vision (may be gradual or sudden), distortions in vision (straight lines begin to look wavy, for example), and difficulty seeing details (Figure 22-9).

Treatment: In many cases there is no treatment, although specialized vitamins such as Preservision® are recommended. Laser treatments may repair the damaged blood vessels of the retina.

Retinal Detachment

Retinal detachment occurs when the layers of the retina separate. It is considered a medical emergency and if not treated right away leads to permanent vision loss.

Etiology: This disorder is sometimes caused by fluids that seep between layers of the retina, occurring most commonly in myopic (nearsighted) people. In diabetics, vitreous body or scar tissue pulls the retina loose. Eye trauma is another cause of retinal detachment.

Signs and Symptoms: Signs and symptoms include light flashes, wavy vision, a sudden loss of vision, particularly of peripheral vision, and a larger number of floaters noted in the visual field.

Treatment: The treatment measures include any of the following:

- Pneumatic retinopexy, which involves injecting a gas bubble into the posterior segment of the eye. The pressure flattens the retina, and the retina is later fixed in place with a laser.
- Scleral buckle, which involves using a silicone band to hold the retina in place.
- Replacing the vitreous body with silicone oil to reattach the retina.

U check
Which photoreceptors are responsible for the detection of colors?

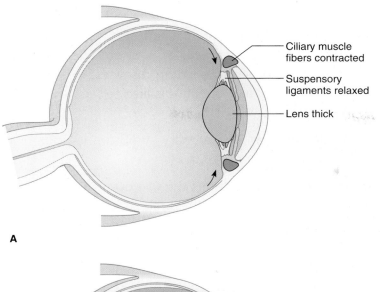

A

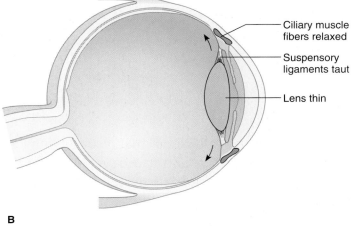

B

FIGURE 22-10 In accommodation, (A) the lens thickens as the ciliary muscle fibers contract. (B) The lens thins as the ciliary muscle fibers relax.

Lens

Just posterior to the iris is a clear, circular disk called the *lens*. Because it can change shape, it helps the eye focus images of objects that are near or far away. This process is called *accommodation* (Figure 22-10). Clouding and hardening of the lens, which often occurs with aging or some diseases, leads to visual changes in a condition known as cataracts. The lens also divides the interior of the eyeball into anterior and posterior cavities.

Eye without a cataract

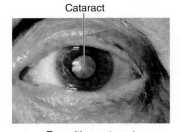

Eye with a cataract

FIGURE 22-11 Cataracts.

PATHOPHYSIOLOGY

Cataracts

Cataracts are opaque structures within the lens that prevent light from going through the lens. Over time, images begin to look fuzzy and, if left untreated, cataracts may cause blindness (Figure 22-11).

(continued)

Cataracts (concluded)

Etiology: Aging is the most significant risk factor associated with this disorder. Cataracts can also be caused by eye injuries, some medications, and certain underlying diseases. Excessive exposure to damaging UV rays from sunlight is also felt to be a risk factor for developing cataracts.

Signs and Symptoms: The primary symptom is poor or impaired vision with a "milky" spot over the lens.

Treatment: Treatment includes the use of eyeglasses, medications to dilate the pupils, or surgery—such as phakoemulsification—to remove the cataract followed by implantation of an artificial intra-ocular lens.

> **REMEMBER MATTHEW,** our 68-year-old over-the-road truck driver? What part of the eye is involved with cataracts?

Ü check
What is the cause of cataracts?

PATHOPHYSIOLOGY

Presbyopia

Presbyopia is a common eye disorder that results in the loss of lens elasticity. It develops with age and causes a person to have difficulty seeing objects that are close up (farsightedness due to aging; *presby* = age).

Etiology: Different hypotheses or possibilities exist as to the exact cause. One possibility is that fibers in the lens become stiffened and therefore are unable to change shape as easily. Another hypothesis is that the lens grows too big and is unable to change shape. Trauma, occupation, and lifestyle may be factors as well.

Signs and Symptoms: Difficulty reading or doing close work, or blurry closeup vision are the major signs and symptoms.

Treatment: Treatments include contact lenses, eyeglasses, and eye surgery such as conductive keratoplasty (CK).

Chambers of the Eye

The anterior cavity is further divided into anterior and posterior chambers that are filled with a watery fluid called the **aqueous humor** (Figure 22-12). The aqueous humor is continually produced and reabsorbed throughout an individual's life. Aqueous humor provides nutrients to and bathes the structures in the anterior chamber of the eyeball. An abnormal accumulation of aqueous humor causes a visual condition known as *glaucoma*. The *posterior cavity* or *vitreous chamber* is filled with a thick, jellylike substance called the **vitreous humor,** which is present at birth with no additional

---

aqueous humor The watery fluid that fills the anterior and posterior chambers of the eye; provides nutrients to and bathes the structures in the anterior chamber of the eyeball.

vitreous humor The thick, jellylike substance that fills the posterior cavity of the eye and helps maintain the shape of the eye.

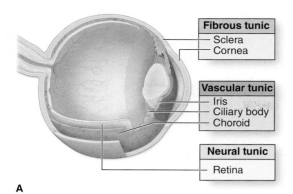
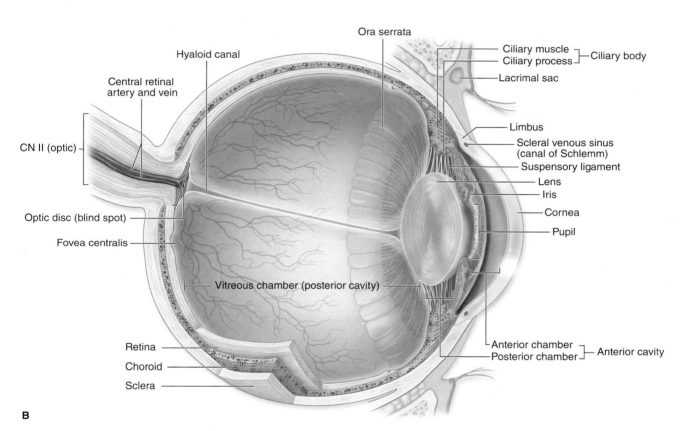

FIGURE 22-12 Anatomy of the internal eye showing the anterior and posterior chambers and associated structures.

production. The vitreous humor keeps the retina flat and helps maintain the shape of the eye.

> **U check**
> Where in the eyeball do you find the vitreous humor?

PATHOPHYSIOLOGY

Glaucoma

Glaucoma is indicated by an increase in intraocular pressure, caused by a buildup of aqueous humor in the anterior chamber.

(continued)

Glaucoma (concluded)

If untreated, this excess pressure can lead to permanent damage to the optic nerve that can result in blindness.

Etiology: *Open-angle glaucoma* progresses relatively slowly; it can be caused by slowing of the drainage of aqueous humor from the anterior chamber of the eye. *Angle closure,* or *narrow-angle glaucoma,* is a more serious type of glaucoma; it results when the space between the iris and the cornea, which is more narrow than usual, is suddenly blocked so that the humor cannot be drained. Certain medications, trauma, and tumors may all cause secondary glaucoma.

Signs and Symptoms: Open-angle glaucoma usually has no symptoms. Common symptoms of angle closure glaucoma include nausea, vomiting, extreme eye pain, headache, and a sudden loss of vision related to the sudden buildup of the aqueous humor. The vision loss is initially more peripherally located.

Treatment: Treatments include medications and eye drops such as Timoptic® and pilocarpine to control pressure in the anterior segment of the eye. In the case of angle closure glaucoma, surgery to correct the blockage may be needed.

Visual Pathways

The eye works much like a camera. Light reflected from an object, or produced by one, enters the eye from the outside and passes through the cornea, pupil, lens, and fluids in the eye. The cornea, lens, and fluids help focus the light onto the retina by bending it, in a process known as *refraction* (Figure 22-13). As in a camera, an image of an object is carried by light patterns. The image is projected upside down—on film in a camera and on the retina in the eye (Figure 22-14). The retina converts the light into nerve impulses. These impulses are transmitted along the optic nerve (cranial nerve II) to the brain (Figure 22-15). This nerve, which consists of about a million fibers, serves as a flexible cable connecting the eyeball to the brain.

Parts of the optic nerve cross at a structure called the *optic chiasm,* located at the base of the brain. The visual area in the occipital lobes of the cerebrum is responsible for

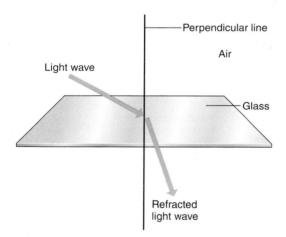

FIGURE 22-13 When light passes at an oblique angle from air into glass, light waves are bent or refracted toward a line perpendicular to the surface of the glass.

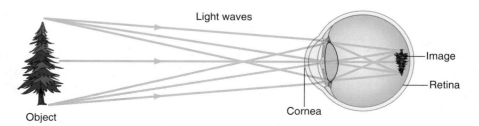

FIGURE 22-14 The image of an object forms upside down on the retina.

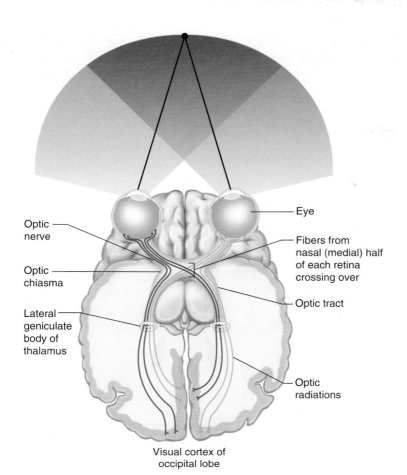

FIGURE 22-15
The visual pathway includes the optic nerve, optic chiasm, optic tract, and optic radiations.

interpreting vision. Because visual information crosses in the optic chiasm, about half of the visual information detected in each eye is interpreted on the opposite side of the brain. Therefore, half of what a person sees in the right eye is interpreted in the left side of the brain and vice versa, where it is brought together as one image. The brain interprets these impulses, turns the image right-side up, and "develops" a picture of the object from which the light originally came. Table 22-2 summarizes functions of the parts of the eye.

> **U check**
> What is refraction?

Visual Accessory Organs

Visual accessory organs assist and protect the eyeball. They include the orbits, eyebrows, eyelids and eyelashes, conjunctiva, the lacrimal apparatus, and extrinsic eye muscles. The eye sockets, or bony orbits, form a protective shell around the eyes. Eyebrows protect the eyes by reducing the chances that sweat and direct sunlight will enter the eyes. Eyelashes line both the upper and lower eyelids and have sebaceous glands located at their base that help lubricate the hair follicles.

Each eyelid is composed of skin, muscle, and dense connective tissue. The muscle in the eyelid is called the orbicularis oculi and is responsible for

TABLE 22-2 The Functions of the Parts of the Eye

Structure	Function
Aqueous humor	Nourishes and bathes structures in the anterior eye cavity
Vitreous humor	Holds the retina in place: maintains the shape of the eyeball
Sclera	Protects the eye
Cornea	Allows light to enter the eye; bends light as it enters the eye (refraction)
Choroid	Supplies nutrients and provides a blood supply to the eye
Ciliary body	Holds the lens; controls the shape of the lens for focusing
Iris	Controls the amount of light entering the eye
Lens	Focuses light onto the retina (accommodation)
Retina	Contains visual receptors
Rods	Allow vision in dim light: detect black, white, and gray images; detect broad outlines of images
Cones	Allow vision in bright light; detect colors; detect details
Optic nerve	Carries visual information (stimuli) from rods and cones toward the brain

blinking and squinting. Blinking the eyelids prevents the mucous membrane surface of the eyeball from drying out. A moist eyeball surface is much less likely to grow bacteria than a dry one is. Blinking also protects the eyes, keeping foreign material from entering the eyes, with the assistance of the eyelashes, which catch foreign substances, including perspiration and dust, and keep those substances from getting into the eye. The space between the two eyelids is called the palbebral fissure, and the corners of the eyes are called the medial and lateral commissures.

> **U check**
> Which muscle is responsible for blinking?

PATHOPHYSIOLOGY

Ectropion

Ectropion is characterized by eversion of the lower eyelid.

Etiology: Aging and skin relaxation or scar tissue may cause this condition.

Signs and Symptoms: Common signs and symptoms include redness, irritation, and drying of the conjunctiva. Poor tear drainage through the nasolacrimal system may also be present.

Treatment: Surgery to repair the defect may be needed if the condition is bothersome to the patient.

Entropion

Entropion is characterized by an inversion of the lower eyelid.

Etiology: Aging and scar tissue may cause this condition.

Signs and Symptoms: Signs include irritation of the sclera as the lashes brush against it, which may lead to corneal ulceration or scarring.

Treatment: Surgery is the only treatment option to correct this problem.

The **conjunctiva** is the mucous membrane that lines the inner surfaces of the eyelids and covers the anterior surface of each eyeball. The conjunctiva produces mucus that keeps the surface of the eyeballs moist. The conjunctiva that covers the eyelids is called the palpebral conjunctiva, and that which covers the sclera, but not the cornea, is called the bulbar conjunctiva. The highly vascular nature of the bulbar conjunctiva is responsible for bloodshot eyes.

conjunctiva The mucous membrane that lines the inner surface of the eyelids and the anterior surface of the eyeball.

U check
What is the difference between an entropion and an ectropion?

PATHOPHYSIOLOGY

Conjunctivitis

Conjunctivitis is commonly called pink eye and is highly contagious when the cause is bacterial in nature.

Etiology: This disease is caused by bacteria, viruses, or allergies.

Signs and Symptoms: The signs and symptoms are red, itchy eyes; swollen eyelids; a watery discharge when caused by viruses and allergies; and a sticky, purulent discharge when the cause is bacterial. The allergic type usually affects both eyes, whereas viral and bacterial conjunctivitis more commonly begin in one eye and spread to the other.

Treatment: Cool compresses and anti-inflammatory drugs are used to treat conjunctivitis caused by viruses and allergies, with antihistamines added for those caused by allergies. The bacterial type is best treated with antibiotics.

Dry Eye Syndrome

Dry eye syndrome (xerophthalmia) is one of the most common eye problems treated by physicians. This syndrome results from a

(continued)

Dry Eye Syndrome (concluded)

decreased production of the oil within tears, which normally occurs with age.

Etiology: Dry eye can be caused by cigarette smoke; air conditioning; staring for long hours at a computer monitor; some medications; contact lenses; hormonal changes associated with menopause; and hot, dry, or windy climates.

Signs and Symptoms: The common eye symptoms include burning, irritation, redness, itching, and excessive tearing.

Treatment: Artificial tears may provide relief to many patients, and newer drugs such as Restasis® have helped patients with this condition make more of their own tears. People with this condition should drink 8 to 10 glasses of water a day and make a conscious effort to blink more frequently and avoid rubbing their eyes. In addition, punctal plugs can be inserted to trap tears on the eyes, which prevents the tears from entering the nasolacrimal duct and being drained.

The lacrimal apparatus consists of lacrimal glands and nasolacrimal ducts (Figure 22-16). Lacrimal glands are located on the upper lateral edge of each eyeball and produce tears. Tears are mostly water, but also contain enzymes (lysozymes) that can destroy bacteria and viruses as part of the body's defense mechanism. Tears also have an oily component that prevents them from evaporating. Nasolacrimal ducts are located on the medial aspect of each eyeball and drain tears into the nose. When a person cries, tears entering the nose produce the "runny nose" associated with crying.

Extrinsic eye muscles are skeletal muscles that attach to the sclera and move the eyeball in different directions (Figure 22-17), as summarized in

FIGURE 22-16
The lacrimal apparatus.

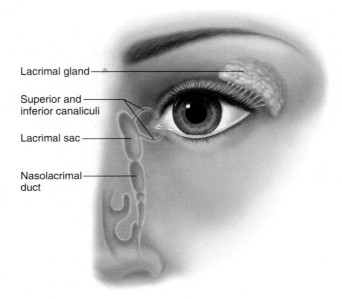

Table 22-3. Each eyeball has six extrinsic eye muscles attached to it. The superior rectus allows you to look up, and the inferior rectus allows you to look down. The lateral rectus moves the eyeball laterally, and the medial rectus moves the eyeball medially. The superior oblique muscle moves the eyeball

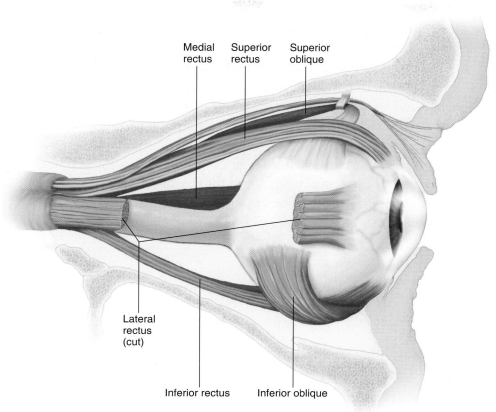

FIGURE 22-17
Extrinsic eye muscles of the right eye (lateral view).

TABLE 22-3 Muscle Associated with the Eyelids and Eyes		
Skeletal Muscles	**Innervation**	**Function**
Muscles of the eyelids		
Orbicularis oculi	Facial nerve (VII)	Closes eye
Levator palpebrae superioris	Oculomotor nerve (III)	Opens eye
Extrinsic muscles of the eyes		
Superior rectus	Oculomotor nerve (III)	Rotates eye upward and toward midline
Inferior rectus	Oculomotor nerve (III)	Rotates eye downward and toward midline
Medial rectus	Oculomotor nerve (III)	Rotates eye toward midline
Lateral rectus	Abducens nerve (VI)	Rotates eye away from midline
Superior oblique	Trochlear nerve (IV)	Rotates eye downward and away from midline
Inferior oblique	Oculomotor nerve (III)	Rotates eye upward and away from midline

> **focus on** Clinical Applications: Eye Medications
>
> As a health care professional you may administer or dispense medications and explain their use. Some medications are used to diagnose conditions, whereas others are used to treat conditions.
>
> Only medications for ophthalmic use should be used in the eye. You should teach patients to check medication labels carefully before administering them at home. *Optic* medications for use in the eye could easily be confused with *otic* medications for the ear. Medications other than optic medications may be too concentrated or may contain substances that will injure sensitive eye tissue. If you administer eye medications, always avoid touching a dropper or ointment tube tip to the eye. Such touching can injure the eye, cause an infection, and contaminate the medication.
>
> When foreign materials enter the eye, they must be flushed out. Flushing (or irrigation) should be done with a sterile solution especially formulated for this purpose. The eye may also need to be irrigated to relieve discomfort from irritating substances, such as smog, pollen, chemicals, or chlorinated water.

down and laterally, and the inferior oblique moves the eyeball up and laterally. Cranial nerves III (the oculomotor), IV (the trochlear), and VI (the abducens) innervate the extrinsic eye muscles.

U check
Which eye muscle moves the eyeball laterally?

Vision Testing

When doctors perform complete physical examinations, they usually screen patients for vision problems and for the general health of the eyes. An ophthalmologist (a medical doctor who is an eye specialist) usually performs a thorough eye examination. He or she tests the external as well as internal structures of the eyes, along with eye movement and coordination. The kinds of testing you will perform or help perform will depend on where you work. You will, however, be expected to assist the doctor and ensure that the patient is comfortable.

Other eye professionals include optometrists and opticians. The optometrist is not an MD but an OD, or doctor of optometry. An optometrist often works with or under the direction of an ophthalmologist, providing vision screenings and diagnostic testing. If disease or further workup is needed, the patient will be referred to an ophthalmologist. The optician is responsible for filling the vision prescriptions for glasses and contacts written by the ophthalmologist and optometrist.

U check
What is the difference between an ophthalmologist and an optician?

focus on Clinical Applications: Vision Screening Tests

Screening tests are used to detect a number of common visual problems. Some problems involve the ability to see clearly, known as hyperopia, presbyopia, and myopia. Others involve the ability to distinguish shades of gray or colors. When testing vision you will need to test each eye separately then both eyes together without and then with glasses. When recording results you must identify which eye, R, L, or Both. You must also identify if the test was with correction (wearing glasses) or without correction (no glasses). You will also check near and distance vision.

Distance Vision

Impairment of distance vision is known as *myopia* (Figure 22-18). In myopia, the eyeball is too "long," resulting in the light being focused anterior to the retina. To test the distance vision of adults, the Snellen letter chart is commonly used. This chart has several rows of letters of the alphabet. Letter size is the same within each row and decreases from top to bottom. Patients are asked to read the letters from larger to smaller. Another chart often used is the Landolt C chart, which is similar to the Snellen chart but uses the letter C instead of the E used in the Snellen. For children and adults who do not know the alphabet, a pictorial chart is used.

When distance vision is tested, normal vision is referred to as 20/20. This number means that at the standard testing distance of 20 feet (first number), a patient can see what a person with normal vision can see at 20 feet (second number). A patient who has 20/80 vision can see at 20 feet what a person with normal vision sees at 80 feet. The first number (20) always stays the same because a patient always stands 20 feet away from the chart. The second number, however, changes with a patient's visual acuity. If a patient misses only one or two letters on a line during vision testing, record the results with a minus sign. For example, one letter missed with the right eye would be Right Eye 20/30−1. Two letters would be Right Eye 20/30−2.

Near Vision

Refractive error in close vision is called *hyperopia*. In hyperopia, the eyeball is shorter than normal, resulting in objects being focused posterior to the retina.

To test for near vision, special handheld charts are used. These charts contain letters, numbers, or paragraphs in various sizes of print. They may be held and read at a normal reading distance or mounted in a plastic and metal frame and read through optical lenses.

Presbyopia is visual impairment as a result of aging and is caused by the loss of lens elasticity, resulting in difficulty seeing items close to you. The combination of myopia and presbyopia requires many people to wear bifocal lenses as they age.

U check
What does it mean if a person is diagnosed with 20/100 vision?

FIGURE 22-18
Point of focus: (A) Myopia—if an eye is too long, the focus point lies in front of the lens. (B) Normal vision—the focus point is on the retina. (C) Hyperopia—if the eye is too short, the focus point lies behind the retina.

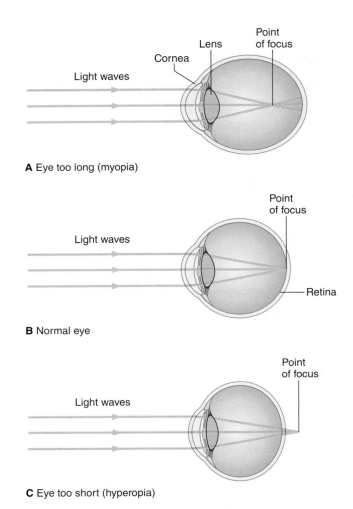

A Eye too long (myopia)

B Normal eye

C Eye too short (hyperopia)

PATHOPHYSIOLOGY

Nystagmus

Nystagmus is rapid, involuntary eye movements. The eye movements may be horizontal or vertical.

Etiology: Alcohol and some drugs may cause nystagmus. Inner ear disturbances may also result in involuntary eye movements. Brain lesions and injury (including those that may occur during birth), and cerebrovascular accidents (CVA) or strokes may also cause nystagmus.

Signs and Symptoms: Signs and symptoms include rapid, irregular eye movements that may be horizontal, vertical, or rotary, depending on the underlying cause of the nystagmus.

Treatment: Treatment is focused on the underlying cause of the disorder.

focus on Wellness: Eye Safety and Protection

Almost 90 percent of all eye injuries could be prevented by eye safety practices or proper protective eyewear.

Eye Safety in the Home

Follow these suggestions to protect your eyes in the home:

- Pad or cushion the sharp corners and edges of furniture and home fixtures.
- Make sure adequate lighting and handrails are available on stairs.
- Keep personal use items (for example, cosmetics and toiletries), kitchen utensils, and desk supplies out of the reach of children.
- Keep toys with sharp edges out of the reach of children. Also, make sure toys intended for older children are kept away from younger children.
- Before mowing the lawn, remove dangerous debris.
- Wear safely goggles when operating any type of power equipment.
- Keep dangerous solvents, paints, cleaners, fertilizers, and other chemicals out of the reach of children.
- Never mix cleaning agents.

Eye Safety at Work

Approximately 15 percent of eye injuries in the workplace lead to temporary or permanent vision loss. Eye injuries at work can be avoided by using the following precautions:

- Choose safety eyewear according to the type of work being performed and the type of eye protection that is needed.
- Wear safety eyewear whenever there is a chance of objects flying from machines.
- Wear safety eyewear whenever exposure to harmful chemicals, body fluids, or radiation is possible.

Eye Safety During Sports and Recreational Activities

Common eye injuries that occur while playing a sport include scratched corneas, inflamed irises and retinas, bleeding in the anterior chamber of the eye, traumatic cataracts, and fractures of the eye socket. Wearing sports eye guards can prevent most sports eye injuries. These guards are recommended for baseball, basketball, soccer, football, rugby, and hockey. Virtually any type of contact sport requires appropriate eye protection.

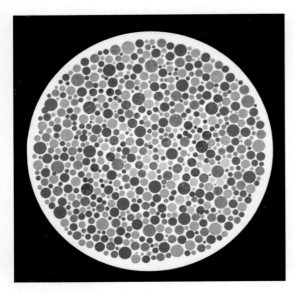

FIGURE 22-19
Ishihara color vision test.

Contrast Sensitivity: To test for the ability to distinguish shades of gray (contrast sensitivity), the Pelli-Robson contrast sensitivity chart, the Vistech Consultants vision contrast test system, or another testing system is used. Newer systems use special equipment to provide contrast variations in a projected image. These tests can detect cataracts or problems of the retina even before the sharpness of the patient's vision is impaired.

Color Vision: To test color vision, illustrations such as those of the Ishihara color system or the Richmond pseudoisochromatic color test are used (Figure 22-19). These illustrations contain numbers or symbols made up of colored dots that appear among other colored dots. The patient is asked to identify what he or she sees. A patient who is color blind cannot see the numbers or symbols. Color blindness may be inherited; it occurs more commonly in males. Changes in the ability to see colors, however, may indicate a disease of the retina or optic nerve.

Treatment of Eye Problems: Some common eye problems include conjunctivitis (inflammation of the conjunctiva), *blepharitis* (inflammation of the eyelid), and corneal abrasions (scratching of the cornea). The eye is an extremely delicate organ. Even what seems to be a minor injury or infection

focus on Wellness: LASIK

LASIK is the abbreviation for laser-assisted in situ keratomileusis. This procedure is performed to eliminate or reduce a person's dependency on corrective lenses (glasses or contacts). In LASIK, the shape of the cornea (the clear covering of the front of the eye) is permanently changed, using an excimer laser. Depending on the method chosen by the surgeon, a mechanical microkeratome (a blade) or a laser keratome (a laser device) is used to cut a flap in the cornea and a hinge is left at one end of this flap. The flap is folded back to reveal the stroma, which is the middle section of the cornea. Pulses from a computer-controlled laser vaporize a portion of the stroma and the flap is replaced. Most patients who have the procedure state there is a dramatic improvement in their vision almost immediately.

Like many surgical procedures today, LASIK is constantly evolving. Although the procedure is increasingly common, many insurance plans still do not cover LASIK—it is an out-of-pocket expense. Anyone considering LASIK should chose a practitioner with experience in the procedure—it is still surgery! Keep in mind that you are dealing with your health and vision—do not make a medical decision based solely on who will give you "the best deal." Ask about the physician's experience and success rate. Ask to see testimonials from actual patients or speak with one or two. Get all of your questions answered and feel confident and comfortable with your choice. Once the procedure is completed, as with any medical procedure, remain compliant with the physician's postoperative medical care and advice.

can have lasting consequences. Therefore, you must use the greatest caution as well as proper (sterile) techniques when treating a patient's eyes. You should also provide patients with information on how to routinely care for their eyes.

Hearing and Equilibrium

The ear is responsible for hearing and, in part, equilibrium. Like the eye, the ear is divided into three parts: the external, middle, and inner ear.

Components of the Ear

The external ear (Figure 22-20) is composed of the **auricle,** the external auditory canal, and the eardrum, or tympanic membrane. The *auricle,* also referred to as the pinna, is the structure we typically consider "the ear." The upper part of the auricle is known as the helix or the rim of the ear. The inferior part of the auricle is the lobule, more commonly known as the ear lobe. The ear is shaped by elastic cartilage and adipose tissue and is covered by stratified squamous epithelium. Hairs and ceruminous glands are found inside the external ear. The hairs and the *cerumen* (ear wax) produced by these glands prevent foreign objects from entering the ear. The *tympanic membrane* is the boundary between the external and middle ears.

The middle ear consists of the ossicles and the **eustachian** (auditory) **tube.** The **ossicles** are the smallest bones in the body and include the **malleus** (mallet or hammer), **incus** (anvil), and **stapes** (stirrup)—so named because of their shapes (Figure 22-21). The stapes is in contact with the *oval window,* which is the boundary between the middle and inner ears. The eustachian tube is a continuous, mucous membrane–lined pathway between the throat and middle ear that allows organisms to travel from the throat to the middle ear. This is the reason it is not unusual to have a sore throat and earache at the same time. Children have a shorter auditory tube than adults

22.4 learning outcome

Explain how the anatomy of the ear relates to the sense of hearing and to equilibrium.

auricle The external part of the ear; also called the *pinna.*

eustachian tube A pathway between the throat and the middle ear that allows organisms to pass between them.

ossicles The bones of the middle ear, the smallest bones in the body.

malleus The hammer-shaped bone of the middle ear.

incus The anvil-shaped bone of the middle ear.

stapes The stirrup-shaped bone of the middle ear.

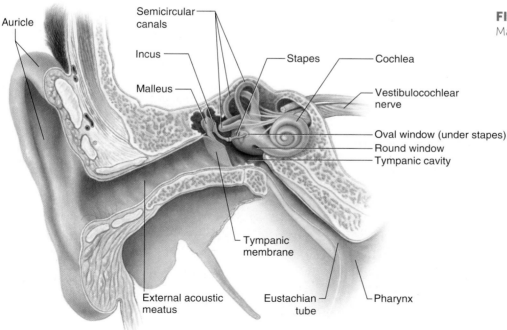

FIGURE 22-20
Major parts of the ear.

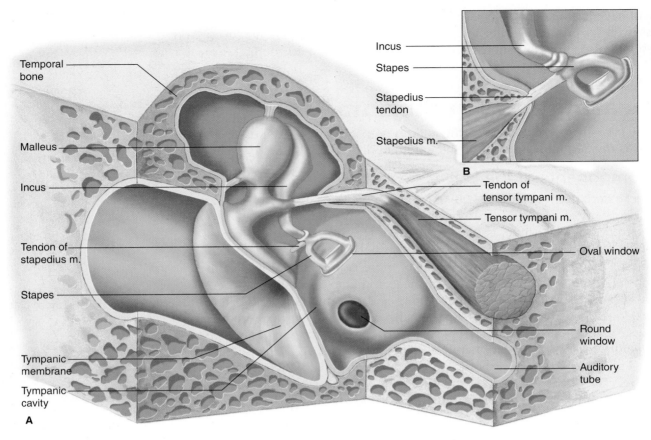

FIGURE 22-21 Ossicles of the ear—malleus, incus, and stapes—and associated structures.

and it's angled in such a way that middle ear infections, known as *otitis media*, are very common in children.

> **U check**
> What are the names of the ossicles?

PATHOPHYSIOLOGY

Otosclerosis

Otosclerosis is the immobilization of the stapes within the middle ear and is a common cause of conductive hearing loss.

Etiology: Genetics seems to play a role. This disease occurs more frequently in females than in males.

Signs and Symptoms: Signs and symptoms include a slow and progressive hearing loss that may be accompanied by tinnitus (ringing in the ear).

Treatment: Surgical removal of the stapes (stapedectomy) with insertion of a prosthetic stapes may result in at least a partial restoration of hearing.

focus on Wellness: Preventive Ear Care Tips

You can protect your ears and take care of your hearing by following these guidelines:

1. Get routine hearing exams. Screening for hearing problems is often part of a comprehensive physical exam. Have your hearing screened.
2. Avoid injury when cleaning the ears. Do not put objects in the ear that might injure the eardrum or ear canal, which includes vigorous probing with cotton-tipped swabs.
3. Avoid injury from nonprescription ear care products. Check with a doctor before using nonprescription products for softening earwax.
4. Wear ear protectors around loud equipment and avoid listening to loud music. It is especially important to keep the volume at a reasonable level when listening through earphones and through ear buds that fit directly into the external auditory canal.
5. Use all medications properly and follow instructions carefully. This applies to all medications because many, including aspirin and some antibiotics, may cause hearing loss if used improperly.
6. Be alert for warning signs. Call the doctor immediately if you experience any of these signs of ear problems: ear pain, stuffiness, discharge from the ear, and vertigo (dizziness).

Also notify the doctor if you have any of these signs of hearing problems: tinnitus (ringing in the ears), hearing others' speech as mumbled sounds, or speaking in a very loud voice without being aware of it.

The inner ear is composed of a very complex system of communicating chambers and tubes known as the *labyrinth*. It is divided into three portions—**semicircular canals** (Figure 22-22), the **vestibule,** and the **cochlea** (Figure 22-23). There are three semicircular canals per ear, and they detect the balance of the body. The cochlea is shaped like a snail's shell and contains hearing receptors, including the *organ of Corti* (Figure 22-24), which is known as the spiral organ or organ of hearing. The vestibule is the area between the semicircular canals and the cochlea. Like the semicircular canals, it also functions in equilibrium. When the head moves, fluids known as *perilymph* and *endolymph* in the semicircular canals and vestibule move, which activate equilibrium receptors as well as hearing receptors.

In addition, a thickened region called the macula contains hair cells and supporting cells. The supporting cells secrete a thick glycoprotein layer called the otolithic membrane. A dense layer of calcium carbonate crystals, called otoliths, cover the membrane. When the head is tilted, it causes the otoliths to bend the hair cells. This helps to detect the position of the head in space.

The equilibrium receptors send the information along vestibular nerves to the cerebrum for interpretation. The cerebrum can then advise the body if it needs to make any adjustments to prevent a fall. When sound waves of different volumes and frequencies activate the hearing receptors in the cochlea, they send their information to auditory nerves. Auditory nerves

semicircular canals The segment of the labyrinth that detects the balance of the body.

vestibule The area between the semicircular canals and the cochlea that functions in equilibrium.

cochlea The spiral structure of the ear that contains hearing receptors.

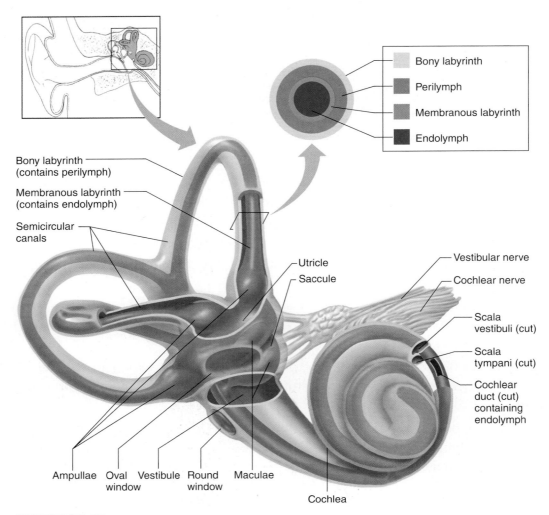

FIGURE 22-22 Semicircular canals.

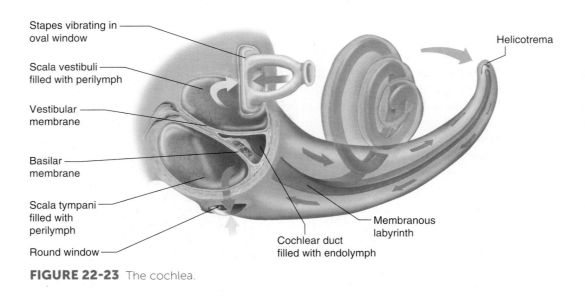

FIGURE 22-23 The cochlea.

UNIT 2 Concepts of Common Illness by System

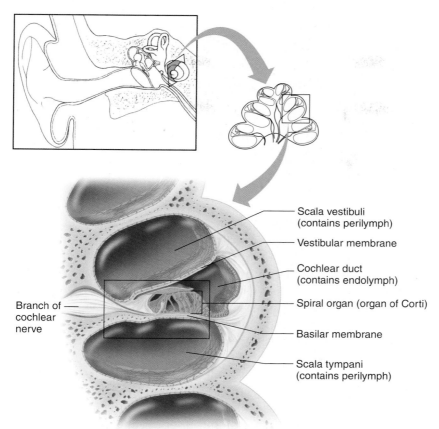

FIGURE 22-24
Cross section of the cochlea.

(vestibulocochlear nerves; cranial nerve VIII) deliver the information to the auditory cortex in the temporal lobe of the cerebrum. The auditory cortex interprets the information as sounds.

PATHOPHYSIOLOGY

Meniere's Disease

Meniere's disease is a disturbance in the equilibrium characterized by *vertigo* (dizziness), *tinnitus* (ringing in the ears), nausea, and progressive hearing loss.

Etiology: The cause is not totally understood, but it is related to an accumulation of endolymph fluid within the membranous labyrinth of the inner ear.

Signs and Symptoms: Patients often complain of episodes of tinnitus, vertigo, nausea, and balance disturbances. Hearing is affected by the tinnitus and, in the early periods of the disease, episodes of hearing loss are temporary. However, permanent hearing loss usually results as the disease progresses.

Treatment: Antihistamines and fluid restrictions are commonly used to control fluid accumulation. Antinausea and antivertigo drugs may

(continued)

> **Meniere's Disease** *(concluded)*
>
> also be prescribed. Because tinnitus is often more bothersome at night (during quiet), mild sedatives are commonly used, especially at bedtime. Patients are often advised to avoid caffeine, alcohol, and cigarettes. Fatigue and stress may also aggravate the condition.
>
> **U check**
> What is vertigo?

The Hearing Process

A sound consists of waves of different frequencies that move through the air. The external ear initiates sound conduction when it collects these waves and channels them to the tympanic membrane. Sound waves make the tympanic membrane vibrate. The vibrations, in turn, are amplified by the ossicles of the middle ear. The amplified waves enter the inner ear and the cochlea and cause tiny hairs that line the cochlea to bend. Movement of the hairs triggers nerve impulses, which are transmitted by cranial nerve VIII (auditory nerve) to the brain where they are perceived as sound.

Sound waves are also conducted through the bones of the skull directly to the inner ear, a process called bone conduction. This alternative pathway for sound bypasses the external and middle ears. When you hear your own voice, the sound has reached your inner ear mainly through bone conduction. By comparing a person's ability to sense sounds by bone conduction and through the entire ear, doctors can often identify what part of the ear is causing a hearing problem. For example, if bone conduction is normal, a hearing problem likely involves the middle or external ear rather than the inner ear.

Hearing Loss

Hearing loss is actually a symptom of a disease, not a disease in itself. Hearing loss once was accepted as part of the aging process. It is now thought that much of it is preventable. About 75 percent of hearing loss in elderly people is not due to aging alone. Therefore, hearing loss should always be evaluated for proper treatment.

Types of Hearing Loss: The two types of hearing loss are conductive and sensorineural. The types differ in the point at which the hearing process is interrupted.

Conductive Hearing Loss: This type of hearing loss is caused by an interruption in the transmission of sound waves to the inner ear. Conditions that can cause conductive hearing loss include obstruction of the ear canal, as with cerumen impaction, foreign object, tumor, infection of the middle ear, or reduced movement of the stirrup.

Sensorineural Hearing Loss: Sensorineural hearing loss occurs when there is damage to the inner ear or to the nerve that leads from the ear to the brain, or damage or injury to the brain itself. In this kind of loss, sound waves reach the inner ear but the brain does not perceive them as sound. This type of hearing loss can be hereditary, and it also can be caused by repeated exposure to loud noises, viral infections, or it can occur as a side effect of medications.

Tinnitus, as previously stated, is an abnormal ringing in the ear. In some individuals it may be caused by damage to the auditory nerve or by certain medications such as aspirin usage. Sensorineural hearing loss can be differentiated from conductive hearing loss by hearing tests. It is possible for both types of hearing loss to occur together.

Hearing loss: Sensorineural

> **U check**
> What is tinnitus?

---

focus on Wellness: Can Your Baby Hear?

When a mother has a new infant she wants to know her baby is "all right." Being aware of what a normal infant's hearing abilities are will help you know if your infant may have a problem. Here is what to expect. Any deviations from these guidelines may indicate a hearing loss and you should tell your health care provider.

Infants up to 4 months old:
- Should be startled by loud noises (barking dog, hand clap, etc.).
- Should wake up at the sound of voices when sleeping in a quiet room.

Around the 4th month of age, infants:
- Should turn their head or move their eyes to follow a sound.
- Should recognize the mother's or primary caregiver's voice better than other voices.

Infants 4 to 8 months of age:
- Should regularly turn their heads or move their eyes to follow sounds.
- Should change their facial expressions at the sound of familiar voices or loud noises.
- Should begin to enjoy certain sounds such as rattles or ringing bells.
- Should begin to babble at people who talk to them.

Babies 8 to 12 months of age:
- Should turn quickly to the sound of their name.
- Should begin to vary the pitch of the sounds they produce in their babbling.
- Should begin to respond to music.
- Should respond to the instruction "no."

PATHOPHYSIOLOGY

Presbycusis

Presbycusis is hearing loss that is due to the aging process.

Etiology: As part of the natural aging process, the auditory system deteriorates, resulting in a loss of hair cells in the organ of Corti.

Signs and Symptoms: Signs and symptoms include gradual, progressive hearing loss, usually of high-frequency sounds first. Tinnitus may accompany this loss and the patient may become depressed and frustrated at his or her developing inability to communicate well as a result of the hearing loss.

Treatment: In most cases hearing aids are prescribed to alleviate some of the hearing loss, although the condition itself is irreversible.

Noise Pollution

Prolonged exposure to loud noises is a common cause of hearing loss because of damage to the sensitive cells in the cochlea. People who work around noisy equipment, including construction workers, aircraft personnel, and machine operators, are likely to suffer from this type of hearing loss unless they protect their ears (Figure 22-25). Repeatedly listening to loud music from a stereo, iPod, or car radio set at too high a volume can also damage the ears.

> **U check**
> What sounds are the first to be lost in elderly individuals?

FIGURE 22-25 Ear protection being worn to protect against loud noises.

Hearing and Diagnostic Tests

Various tests are performed to find out whether a person hears normally. If the tests reveal a problem, follow-up tests are performed to determine the cause of the problem.

Hearing Tests: As part of a general examination, physicians may perform a simple hearing test with one or more tuning forks. Physicians use the tuning forks to determine whether there is a hearing loss. Tuning forks can also be helpful in differentiating conductive from sensorineural hearing loss.

If you have the necessary training, you may help perform a test that uses an audiometer. An audiometer is an electronic device that measures hearing acuity by producing sounds in specific frequencies and intensities (Figure 22-26). A frequency is the number of complete fluctuations of energy per second in the form of waves. Frequency is best described as the pitch of sound. High frequency is high pitched, and low frequency is low-pitched. The audiometer allows a physician or other health practitioner to test a person's hearing and to determine the nature and extent of a person's hearing loss.

focus on Clinical Applications: Working with Patients with a Hearing Impairment

Health care professionals may come in contact with patients of all ages who have hearing impairments. Many patients wear hearing aids to amplify normal speech. Some patients, however, may not admit they have a problem—out of fear, vanity, or misinformation. It is estimated that one-third of patients between the ages of 65 and 74 and one-half of those between the ages of 75 and 79 suffer from some loss of hearing. To improve communication with a patient whose hearing is impaired, you can do the following:

1. Speak at a reasonable volume. Do not shout. Shouting can actually make your words harder to understand. A hearing aid filters out loud sounds, so the patient may not hear everything you say if you shout.
2. Speak in clear, low-pitched tones. Elderly patients lose the ability to hear high-pitched sounds first.
3. Avoid speaking directly into the patient's ear. Stand 3 to 6 feet away, and face the patient so she can see your lip movements and facial expressions. Avoid covering your mouth with your hands. Speak at a normal rate.
4. Avoid overemphasizing your lip movements, which makes lip reading difficult.
5. Avoid hand gestures unless they are appropriate.
6. If the patient does not understand what you say, restate the message in short, simple sentences. Have the patient repeat the message to verify that your words were understood.
7. Treat patients who have a hearing impairment with patience and respect.

Many types of audiometers are available. Some machines automatically generate the various tones at different decibels (units for measuring the relative intensity of sounds on a scale from 0 to 130) and print out the patient's responses. Others must be manually adjusted and the results charted by hand. During the test, the patient wears a headset to hear the sounds produced by the audiometer. Depending on the particular unit, the patient indicates hearing a sound by raising a finger or by pushing a button. In the former case the person administering the test records the response. In the latter case the response may be recorded automatically or by the test giver. Adults and children who can understand directions and respond appropriately can be screened in this manner. If you work in a pediatrician's office, you may also help check an infant's response to sounds. These tests require special techniques because infants cannot understand directions.

FIGURE 22-26 Audiometer.

Diagnostic Testing: A diagnostic test called tympanometry measures the eardrum's ability to move and thus gauges pressure in the middle ear. Tympanometry is used to detect diseases and abnormalities of the middle ear. To perform the test, a small, soft-rubber cuff is placed over the external ear canal. This produces an airtight seal. The tympanometer then automatically measures air pressure and prints out a graph of the results.

Treating Ear and Hearing Problems

Some common ear problems that you may encounter in the physician office include cerumen impaction (a buildup of earwax in the ear canal), rupture of the eardrum, otitis media (inflammation of the middle ear), and otitis externa (inflammation of the outer ear). Physicians use various approaches and techniques with each problem to restore the health of a patient's ears. They employ special techniques and devices to improve hearing and maintain the health of the ears. Health care professionals provide patients with information on preventive ear care techniques.

PATHOPHYSIOLOGY
Cerumen Impaction

Cerumen impaction refers to the buildup of ear wax within the external auditory canal.

Etiology: Cerumen impaction may be caused by inadequate cleaning of cerumen from the ear canal, or by overactive ceruminous glands producing more than the normal amount of cerumen.

Signs and Symptoms: The most common sign is some degree of hearing loss because of the blockage of sound waves. Some patients may also complain of ear pain.

Treatment: Treatment includes ear irrigation to remove the blockage. Severely hardened cerumen may require an ear wax softener such as Debrox® to soften and loosen the cerumen before the impaction can be removed.

U check
What is cerumen?

FIGURE 22-27 A hearing aid worn inside the ear.

Hearing Aids: Hearing aids may be worn inside or outside the ear (Figure 22-27). If worn outside, they may be located behind the ear, mounted on eyeglasses, or worn on the body. Hearing aids consist of the following parts:

- A tiny microphone to pick up sounds
- An amplifier to increase the volume of sounds
- A tiny speaker to transmit sounds to the ear

You may need to teach patients how to obtain a hearing aid. You can also pass along tips to patients to help them take proper care of their hearing aids and to troubleshoot problems.

focus on Wellness: Obtaining and Using a Hearing Aid

Obtaining a Hearing Aid
A patient with signs of hearing loss should be referred to an otologist, a medical doctor specializing in the health of the ear, or an audiologist, a specialist who focuses on evaluating and correcting hearing problems. Audiologists are not medical doctors and do not treat diseases of the ear. Instead, they evaluate the patient's hearing, fit hearing aids, give instruction in the use of hearing aids, and provide service for hearing aids if necessary. It is important for hearing aids to fit properly.

Care and Use of Hearing Aids
Hearing aids run on batteries that typically last about two weeks. Therefore, the patient must keep a fresh supply of batteries on hand. The hearing aid itself must be routinely cleaned, or the microphone, switches, or dials may not work properly. Moisture can damage the aid, so it must not get wet. Hair sprays can clog hearing aid openings or interfere with the operation of moving parts. For these reasons spray should be applied before a hearing aid is inserted. Cerumen often builds up behind hearing aids that are worn in the ear, reducing sound transmission. If buildup does occur, the cerumen plug should be removed by ear irrigation.

Other Devices and Strategies
People whose hearing cannot be substantially improved by hearing aids may need to use other devices or strategies to overcome the problem. These devices include appliances that light up as well as ring, such as telephones, doorbells, smoke detectors, alarm clocks, and burglar alarms. Patients can purchase amplifiers for the telephone, television set, and radio. Many closed-captioned television programs are also available. To benefit from closed captioning, the patient must have a television set with a decoder that translates the captioning and displays the captions on the screen.

Life Span

22.5 learning outcome
Describe the changes that are seen in the special senses as we get older.

As individuals age, changes to their bodies are probably no more evident than those they experience concerning the special senses. Each of the special senses undergoes some changes. However, with proper attention and care, these changes can often be lessened.

Aging Changes in the Eye
With age, a number of changes occur in the structure and function of the eye:

- The amount of fat tissue diminishes; this loss may cause the eyelids to droop.
- The quality and quantity of tears decrease.
- The conjunctiva becomes thinner and may be drier because of a decrease in tear production.

- The cornea begins to appear yellow, and a ring of fat deposits may appear around it.
- The sclera may develop brown spots.
- Changes in the iris cause the pupil to become smaller, limiting the amount of light entering the eye.
- The lens becomes denser and more rigid; this trend reduces the amount of light that reaches the retina and makes focusing more difficult. Cataracts become more common.
- Yellowing of the lens causes problems in distinguishing colors.
- Changes in the retina may make vision fuzzy.
- The ability of the eye to adapt to changes in different light intensities may be reduced; glare can become painful as this ability diminishes.
- Night vision may be impaired. This is called *nyctalopia*.
- Peripheral vision is reduced, limiting the area a person can see and reducing depth perception.
- The vitreous humor breaks down, producing tiny clumps of gel or cellular material that cause floaters—dark spots or lines—that appear in a person's field of vision.
- Rubbing of the vitreous humor on the retina produces flashes of light or "sparks."
- Because of changes that impair vision—such as reductions in the field of vision, depth perception, and visual clarity—elderly people may fall more often than younger people.
- Glaucoma increases due to the buildup of aqueous humor in older people.

Aging Changes in the Ear

As a person grows older, several changes occur in the ear:

- The external ear appears larger because of continued growth of cartilage and the loss of skin elasticity.
- The earlobe gets longer and may have a wrinkled appearance.
- The glands that produce cerumen become less efficient, producing earwax that is much drier and prone to impaction.
- The ear canal becomes narrower.
- In the middle ear, changes in the eardrum cause it to shrink and appear dull and gray.
- The joints between the bones of the middle ear degenerate, so they do not move as freely.
- In the inner ear, the semicircular canals become less sensitive to changes in position, and this reduced sensitivity affects balance.
- Hearing loss (especially high-pitched sounds) becomes more common.
- Tinnitus and imbalance occur more often in elderly people.

Aging Changes in Smell and Taste

Receptors are lost at an increased rate and are replaced at a decreased rate, resulting in decreased acuity in both smell and taste.

summary

chapter 22

learning outcomes	key points
22.1 Relate the anatomy of the nose to the sensation of olfaction (smell).	The nose is made of olfactory receptors (bipolar neurons) and cilia specialized for olfaction. It has supporting cells that provide nutrition and support, and basal cells that produce new receptor cells. The bipolar neurons are chemical receptors that change odors into chemical signals that are relayed along the olfactory nerves to the cerebral cortex for interpretation and memory storage.
22.2 Relate the anatomy of the tongue to the sensation of taste.	The tongue is a muscular organ and has several different types of papilla on its surface. Some of the papilla contain taste buds for the sensation of taste. There are taste sensations for sweet, sour, salty, bitter, and umami. Taste buds within papilla detect the different taste sensations. Cranial nerves VII, IX, and X relay information to the brain for taste perception.
22.3 Relate the anatomy and pathophysiology of the eye and the accessory structures to their functions.	The eye is the most complex organ of the body after the brain. It is made up of tunics that cover the eyeball as well as a lens and retina for the transmission of light and vision. It has several accessory structures, such as eyelashes and glands, that help protect the eye. Multiple diseases and disorders affect the eye and our ability to see.
22.4 Explain how the anatomy of the ear relates to the sense of hearing and to equilibrium.	The external ear is made up of skin, adipose tissue, and elastic cartilage. The internal ear is divided into three regions. The smallest bones of the body are found in the ear. Through sound waves and vibrations of membranes and small bones, sound is transmitted to cranial nerve VIII, where it is sent to the brain for interpretation and integration. Through a complex system of channels and canals, cilia and fluid, as well as cranial nerve VIII, the ear is involved in establishing and maintaining equilibrium.
22.5 Describe the changes that are seen in the special senses as we get older.	As we get older, functioning of all the special senses decreases. We do not have the same ability to smell and taste food, for example. Our eyesight becomes dimmer and we may experience more visual problems. Our hearing also may diminish. These changes can have dramatic effects on our activities of daily living.

chapter 22 review

case study 1 questions

Can you answer the following questions that pertain to Matthew Garcia's case study presented earlier in this chapter?

1. What is the medical term for night blindness?
   - a. Hyperopia
   - b. Myopia
   - c. Diplopia
   - d. Nyctalopia
2. What type of "eye doctor" can treat all conditions of the eye?
   - a. Ophthalmologist
   - b. Optician
   - c. Optometrist
   - d. Otologist
3. What is hyperopia?
   - a. Farsightedness
   - b. Nearsightedness
   - c. Crossed eyes
   - d. Degeneration of the retina
4. What is the cause of cataracts?
   - a. Detachment of the retina
   - b. Thickening and cloudiness of the lens
   - c. A tumor in the optic chiasm
   - d. A buildup of aqueous humor in the eye
5. What is the cause of glaucoma?
   - a. Detachment of the retina
   - b. Thickening and cloudiness of the lens
   - c. A tumor in the optic chiasm
   - d. A buildup of aqueous humor in the eye

review questions

1. How does the orientation of the semicircular canals contribute to equilibrium?
   - a. Their orientation to each other helps detect small changes in head movement.
   - b. They transmit electrical impulses to the cerebral cortex for interpretation of head movement.
   - c. They transmit electrical impulses to the cerebellum for interpretation of head movement.
   - d. Their orientation to each other allows for integration of visual input.
2. If a person has a pituitary tumor, which of the following is most likely to result?
   - a. Problems with taste
   - b. Problems with vision
   - c. Problems with hearing
   - d. Problems with equilibrium

critical thinking questions

1. Explain why smell and taste are affected when a person has an upper respiratory infection (URI) with excess mucus production.
2. How are the signs and symptoms of cataracts, glaucoma, and macular degeneration different?
3. Explain why listening to music, music videos, and video games with ear buds for prolonged periods of time may be cause for concern.

patient education

Damage can occur to the cilia (tiny hairs) in our ears from prolonged loud noise. This damage will results in hearing loss. A certain amount of hearing loss occurs with aging but some hearing loss can be prevented. What are some things that you could discuss with a patient to help them that could be done to avoid hearing loss?

applying what you know

1. An 84-year-old woman was just diagnosed with presbyopia. What does this diagnosis mean?
2. If cranial nerve VII (the facial nerve) is damaged, what senses are affected?
3. Trauma to the face that damages the cribriform plate will affect what cranial nerve?

CASE STUDY 2 *Meniere's Disease*

A 52-year-old man comes to the physician office complaining of dizziness, nausea, and a loud ringing in his left ear. He is diagnosed with Meniere's disease.

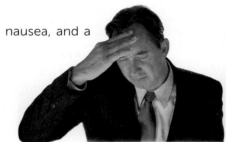

1. The doctor explains to him that this disorder is caused by
   a. The buildup of fluid in the inner ear
   b. Cerumen impaction
   c. Damage to the cochlea
   d. Damage to cranial nerve II (the optic nerve)
2. What is the clinical term for ringing in the ear?
   a. Nystagmus
   b. Tinnitus
   c. Strabismus
   d. Vertigo
3. A diuretic is a drug that decreases fluids in the body. Why might the doctor prescribe a diuretic for this patient's problem?
   a. To prevent heart failure
   b. To make the kidneys work more efficiently
   c. To decrease the buildup of fluid in the inner ear
   d. To prevent edema of the lower extremities

appendix I
diseases and disorders

Infectious Diseases Caused by Bacteria

Name	Description
Boils	Localized skin infection usually caused by the *Staphylococcus* bacterium.
Botulism	If the bacterium *Clostridium botulinum* gets on food, it produces a toxin that can lead to a type of food poisoning.
Conjunctivitis	Commonly called pink eye and is highly contagious when the cause is bacterial in nature. Can also be caused by viruses and allergies.
Gonorrhea	Highly contagious condition transmitted by sexual intercourse and caused by gonococcus bacterium.
Impetigo	A *Staphylococcus* or *Streptococcus* infection found more commonly in children that causes lesions on the face or lower leg. Impetigo is highly contagious when contact is made with the lesions. This condition is treated with antibiotics.
Legionnaire's disease	An acute type of bacterial pneumonia caused by *Legionnaire bacillus,* a gram-negative organism. It grows in standing water such as that found in commercial air conditioners, humidifiers, water heaters, and evaporative condensers.
Lyme disease	An arthropod-borne disease caused by the spirochete *Borrelia burgdorferi*. The disease is transmitted by deer ticks. The signs and symptoms include skin lesions, central nervous system involvement, arthritis, and cardiac involvement.
Meningitis	Inflammation of the meninges. Causes may include bacterial, viral, and fungal infections.
Pertussis	Also called whooping cough. Caused by bacillus bacteria.
Pneumonia	Respiratory condition in which lung tissue is inflamed. Can include difficulty breathing and cough. Can also be caused by fungus or chemicals.
Rheumatic fever	Febrile (characterized by fever) disease, usually occurring after a streptococcal infection.
Strep throat	Inflammation and infection of the throat caused by the *Streptococcus* bacterium.
Syphilis	Infectious venereal disease, usually transmitted by sexual contact. Caused by a spirochete bacterium.
Tetanus	Infectious disease produced by the toxins from the tetanus bacillus. The first sign is stiffness of the jaw, hence the common name "lockjaw."
Tuberculosis (TB)	A disease that primarily affects the lungs but can spread to other parts of the body. Caused by various strains of the bacterium *Mycobacterium tuberculosis*.

Infectious Diseases Caused by Fungi

Name	Description
Athlete's foot	Known as *Tinea pedis*. This same microorganism can affect the body (*T. corporis*) and the scalp (*T. capitis*).
Tinea (ringworm)	A fungal skin infection that gets its name because it appears serpentine like a worm. *Tinea corporis* means the location is on the trunk or body, *tinea capitis* affects the scalp, and *tinea pedis* refers to the feet, and is commonly known as athlete's foot.

Infectious Diseases Caused by Parasites

Name	Description
Giardiasis	Transmitted by the oral-fecal route and through untreated water. Causes diarrhea.
Malaria	Disease transmitted to humans from the bite of an infected anopheles mosquito. Protozoan parasites invade the red blood cells and create symptoms such as high fever and chills.
Scabies	A highly contagious skin condition as the result of a mite that burrows beneath the skin, leaving its feces behind to leave red lines of inflammation to be seen on the skin.

Infectious Diseases Caused by Protozoa

Name	Description
Amebic dysentery	Condition characterized by loose stools; caused by inflammation of the intestines.
Trichomoniasis	Affects the reproductive organs and can be transmitted through sexual intercourse.

Infectious Diseases Caused by Viruses

Name	Description
Acquired immune deficiency syndrome (AIDS)	Syndrome caused by HIV (human immunodeficiency virus), resulting in decreased resistance to infections. Transmitted by blood and body fluids.
Hepatitis	Hepatitis types A, B, C, D, and E are all caused by a virus. This disease affects the liver and can cause mild to moderate symptoms or chronic illness, and possibly death. Health care workers are encouraged to be vaccinated for hepatitis B because they are at risk for contact with client blood and body fluids.
Herpes simplex virus type 1 and 2 (HSV-1, HSV-2)	Typically HSV-1 causes cold sores, is very contagious, and is spread through contact with infected saliva. HSV-2, known as genital herpes, is sexually transmitted. However, it should be noted that there can be crossing over, with HSV-1 infecting the genitals and HSV-2 infecting the mouth.

Infectious Diseases Caused by Viruses (concluded)

Name	Description
Human immunodeficiency virus (HIV)	Virus that destroys the immune system and can result in AIDS (acquired immune deficiency syndrome).
Human papillomavirus (HPV)	The most common sexually transmitted infection. Some strains of HPV cause harmless verrucae (warts); other strains are the greatest single risk factor for cervical cancer and may also cause other, less common cancers such as cancer of the vagina, penis, and oropharynx.
Influenza	Commonly called "the flu"; is an infection of both the upper and lower respiratory tracts.
Mononucleosis	A highly contagious viral infection spread through the saliva of the infected person. Caused by either the Epstein-Barr virus or the cytomegalovirus (CMV).
Severe acute respiratory syndrome (SARS)	Highly contagious disease that causes severe flulike symptoms.
Upper respiratory (tract) infection (URI)	The term often used for the common cold, including pharyngitis (sore throat). Caused by the rhinovirus.
Varicella	Highly contagious disease caused by the varicella-zoster virus. Also called chickenpox, it is characterized by the presence of skin lesions. Shingles is also caused by varicella and is seen in patients who have previously had chickenpox.

Genetic Diseases

Disease	Description
Albinism	A genetic condition in which a person is born with little or no pigmentation in the skin, eyes, or hair.
Cystic fibrosis	A life-threatening disease that affects the lungs and pancreas. It is one of the most common inherited life-threatening disorders among Caucasians in the United States.
Down syndrome	Abnormal cell division involving chromosome 21 resulting in physical abnormalities and some form of mental retardation. It is the single most common type of birth defect.
Fragile X syndrome	The most common inherited cause of learning disability. One of the genes on the X chromosome is defective. It affects boys more severely than girls.
Hemophilia	Inheritable bleeding disorder in which an essential clotting factor is low or missing, primarily affecting males.
Klinefelter's syndrome	A disorder in which males have an extra X chromosome, resulting in tall stature, pear-shaped fat distribution, small testes, sparse body hair, infertility, and slightly lower intelligence.
Muscular dystrophy	A group of genetic disorders that affect the muscular and nervous systems, most often affecting males.
Phenylketonuria (PKU)	A genetic disorder in which the body is unable to properly eliminate phenylalanine, which is an essential amino acid. Organ damage, mental retardation, and even death are possible.
Turner's syndrome	A disorder that results when females have a single X chromosome. Symptoms include a webbed neck, broad chest, short stature, and infertility. Intelligence is normal.

Common Diseases and Disorders in the United States*

Disease	Examples	Screening and Prevention
Heart disease	Coronary artery disease, atherosclerosis, congestive heart failure	Monitor blood pressure and cholesterol and triglyceride levels, and have routine electrocardiograms (ECG).
Cancer	Skin, lung, breast, prostate, blood, colorectal, gynecologic, HPV-related	Have routine breast or prostate exam, Pap smear, and colonoscopy. Monitor size and shape of skin lesions. Maintain healthy eating habits and a good body weight, exercise, do not use tobacco products, and limit alcohol consumption.
Cerebrovascular diseases	Stroke, deep vein thrombosis, aneurysm, embolism	Maintain healthy eating habits and a good body weight, exercise, do not use tobacco products, and limit alcohol consumption. Have routine screening tests and physical exams.
Chronic lower respiratory diseases	Asthma, COPD, tuberculosis	Do not use tobacco products, avoid irritants and allergens, and monitor air quality. Have routine screening tests (TB skin test) and vaccinations.
Diabetes mellitus	Type 1, type 2, gestational	Have blood glucose tests including fasting blood sugar, and hemoglobin A1c blood test. Maintain healthy eating habits and a good body weight, exercise, and limit alcohol and sugar consumption.
Alzheimer's disease		Maintain a healthy lifestyle and have a regular physical examination.
Influenza		Maintain a healthy lifestyle and obtain a yearly flu vaccination.
Pneumonia		Maintain a healthy lifestyle and obtain a yearly pneumonia vaccination.
Nephritis/nephrosis	Chronic renal disease (CRD), end stage renal disease (ESRD)	Have routine urinalysis tests. Maintain healthy eating habits and a good body weight, exercise, do not use tobacco products, and limit alcohol consumption.

*Entries ordered from most to least common.

Skin Lesions

Name and Example	Description
Bulla	A large blister or cluster of blisters.
Cicatrix	A scar, usually inside a wound or tissue.

Skin Lesions (continued)

Name and Example	Description
Crust	Dried blood or pus on the skin.
Ecchymosis	A black and blue mark or bruise.
Erosion	A shallow area of skin worn away by friction or pressure.
Excoriation	A scratch; may be covered with dried blood.
Fissure	A crack in the skin's surface.
Keloid	An overgrowth of scar tissue.
Macule	A flat skin discoloration, such as a freckle or a flat mole.
Nodule	A large pimple or small node (larger than 6 cm).
Papule	An elevated mass similar to but smaller than a nodule.
Petechiae	Pinpoint skin hemorrhages that result from bleeding disorders.
Plaque	A small, flat, scaly area of the skin.

Skin Lesions (concluded)

Name and Example	Description
Purpura	Purple-red bruises; usually the result of clotting abnormalities.
Pustule	An elevated (infected) lesion containing pus.
Scale	Thin plaques of epithelial tissue on the skin's surface.
Tumor	A swelling of abnormal tissue growth.
Ulcer	A wound that results from tissue loss.
Vesicle	A blister.
Wheal	Another term for hive.

Classifications of Fractures

Name and Example	Description
Closed fracture (simple fracture)	Fracture in which the skin remains intact.

Classifications of Fractures (concluded)

Name and Example	Description
Comminuted fracture	Fracture in which the bone has broken into several fragments.
Complete fracture	Fracture that goes across the entire bone.
Greenstick fracture	Fracture commonly seen in children; occurs in bones that are not completely ossified, so there is a bending and only one side of the bone is fractured, rather than a complete breaking of bone.
Impacted fracture	Fracture in which one end of the fractured bone is driven into the interior of the other.
Incomplete fracture	Fracture that goes through only part of the bone.
Open fracture (compound fracture)	Fracture in which the skin is broken.

Bone Disorders

Name	Description
Arthritis	A general term meaning joint inflammation.
Bursitis	Inflammation of a bursa, which is the fluid-filled sac that cushions tendons. (Tendons attach muscles to bone.)
Carpal tunnel syndrome (CTS)	Occurs when the median nerve in the wrist is excessively compressed by an inflamed tendon (the flexor retinaculum).
Gout (gouty arthritis)	A type of arthritis associated with high uric acid levels in the blood with crystalline deposits in the joints, kidneys, and various soft tissues.
Kyphosis	An abnormal, exaggerated curvature of the spine, most often at the thoracic level. This condition is often referred to as humpback.
Lordosis	An exaggerated inward (convex) curvature of the lumbar spine. The condition is sometimes called swayback.
Osteoarthritis (OA)	Also known as degenerative joint disease (DJD) or wear-and-tear arthritis.
Osteogenesis imperfecta (brittle bone disease)	People with brittle bone disease have decreased amounts of collagen in their bones, which leads to very fragile bones.
Osteoporosis	A condition in which bones become thin (more porous) over time.
Osteosarcoma	A type of bone cancer that originates from osteoblasts, the cells that make bone tissue. It is most often seen in children, teens, and young adults, and occurs more often in males than females.
Paget's disease	Causes bones to enlarge and become deformed and weak. It usually affects people older than 40 years of age.
Rheumatoid arthritis (RA)	A chronic, systemic inflammatory disease that attacks the smaller joints such as those in the hands and feet.
Scoliosis	An abnormal, S-shaped lateral curvature of the thoracic or lumbar spine.

Classifications of Burns

Name	Description
First-degree	A superficial burn that causes pain and makes the surrounding skin turn red.

Classifications of Burns (concluded)

Name	Description
Second-degree	A partial-thickness burn that extends deeper into the skin than first-degree burns; causes blistering along with pain and redness.
Third-degree	A full-thickness burn that involves all layers of the skin and requires immediate medical assistance.

Diseases and Disorders of the Integumentary System

Name	Description
Acne vulgaris	An infection of the sebaceous gland(s), which causes pustules, papules, and scarring of the affected skin.
Alopecia	The absence or loss of hair, especially of the head (baldness).
Basal cell carcinoma	Skin cancer that originates from the basal layer of the epidermis and rarely metastasizes (spreads).
Cellulitis	An inflammation of the connective tissue in skin that is most often seen on the face and legs. A bacterial infection by *Staphylococcus aureus* or *Streptococcus* is the most common cause.
Comedos	Commonly known as blackheads; collections of bacteria, dead epithelial cells, and dried sebum.
Dermatitis	A general term describing any inflammation of the skin; can be caused by a wide range of disorders.
Eczema	A very common chronic dermatitis that often has acute phases or flareups followed by periods of remission. The rash of eczema can appear anywhere on the body and appears as a red, scaly, pruritic (itchy) rash that may be painful.
Folliculitis	Inflammation of hair follicles. When it involves a single hair follicle, the condition is called a furuncle. When more than one hair follicle is involved, it is a carbuncle.

Diseases and Disorders of the Integumentary System (concluded)

Name	Description
Jaundice	A yellow cast to the skin that often occurs with liver disease.
Lentigos	Commonly called liver spots or age spots; these are not caused by the liver but are the result of excessive melanin production due to overexposure to sunlight (UV rays).
Malignant melanoma	Skin cancer that arises from melanocytes and often metastasizes (spreads).
Psoriasis	A common chronic inflammatory skin condition that has an autoimmune basis.
Rosacea	A skin disorder that commonly appears as facial redness, predominantly over the cheeks and nose.
Squamous cell carcinoma	Skin cancer that arises from the upper cells of the epidermis and often metastasizes (spreads).

Diseases and Disorders of the Muscular System

Name	Description
Fibromyalgia	A fairly common condition that results in chronic pain primarily in joints, muscles, and tendons. It most commonly affects women between the ages of 20 and 50.
Muscular dystrophy (MD)	A group of inherited disorders characterized by muscle weakness and a loss of muscle tissue.
Myasthenia gravis	A condition in which affected persons experience muscle weakness. In this autoimmune condition, a person produces antibodies that prevent muscles from receiving neurotransmitters from neurons.
Sprains	Injuries that excessively stretch or tear ligaments at a joint.
Strains	Caused by stretching or tearing of muscles or tendons.
Tendonitis	Painful inflammation of a tendon as well as of the tendon–muscle attachment to a bone.
Torticollis	Also known as wry neck. This condition is due to abnormally contracted neck muscles. The head typically bends toward the side of the contracted muscle while the chin rotates to the opposite side.

Diseases and Disorders of the Blood and Circulatory System

Name	Description
Anemia	A symptom of an underlying disease process. Anemia occurs when the blood has less than its normal oxygen carrying capacity. Anemia is the most common blood disorder in the United States and occurs more often in women than in men.
Aneurysm	Results from a ballooned, weakened arterial wall. The most common locations of aneurysms are the aorta and the arteries in the brain, legs, intestines, and spleen. An aortic aneurysm is a bulge in the wall of the aorta.

Diseases and Disorders of the Blood and Circulatory System (concluded)

Name	Description
Leukemia	A neoplastic condition in which the bone marrow produces a large number of WBCs that are not normal.
Thrombocythemia	A condition in which there is an increase in the platelet count. This condition is the opposite of thrombocytopenia.
Thrombocytopenia	A condition in which there are too few platelets, causing abnormal bleeding. This can be caused by a variety of situations, such as leukemia, certain medications, or idiopathic (unknown) reasons. The bleeding may be mild or life threatening.
Thrombophlebitis	A condition in which a thrombus and inflammation develop in a vein. It most commonly occurs in the deeper veins of the legs.
Varicose veins (varices or varicosities)	Tortuous or twisted, dilated veins that are usually seen in the legs. They affect women more often than men.

Diseases and Disorders of the Cardiovascular System

Name	Description
Arrhythmias	Also known as dysrhythmias; abnormal heart rhythms and/or rates.
Congenital heart disease	A problem with the heart's structure and function due to abnormal heart development before birth.
Congestive heart failure (CHF)	Failure of the heart to pump effectively. The heart weakens over time and loses its ability to supply blood to the body.
Coronary artery disease (CAD)	Condition involving partial or complete blockage of major coronary arteries that supply blood to the heart.
Endocarditis	An inflammation of the innermost lining of the heart and heart valves, usually caused by bacterial infections.
Myocardial infarction (MI, heart attack)	Death of heart tissue due to deprivation of oxygen. The cardiac muscle sustains damage because of ischemia.
Myocarditis	An inflammation of the muscular layer of the heart caused by a viral infection. It leads to weakening of the heart wall.
Murmurs	Abnormal heart sounds. Not all murmurs indicate a heart disorder. Murmurs are graded from 1 to 6, with 6 being quite loud and the most serious.
Pericarditis	Inflammation of the pericardium, usually caused by complications of viral or bacterial infections, MIs, or chest injuries.

Diseases and Disorders of the Lymphatic and Immune Systems

Name	Description
Allergies	Excessive immune responses to stimuli that would not ordinarily cause a reaction.
Autoimmune disease	A disease in which the immune system targets itself.
Chronic fatigue syndrome (CFS)	Causes a person to feel severe tiredness, possibly caused by the Epstein-Barr virus (EBV).

Diseases and Disorders of the Lymphatic and Immune Systems (concluded)

Name	Description
HIV/AIDS	A viral disease spread through body fluids. AIDS is caused by the human immunodeficiency virus (HIV). HIV attacks T lymphocytes.
Systemic lupus erythematosus (SLE)	Autoimmune disorder that affects women more often than men. Affects many organ systems of the body and has numerous symptoms usually including joint pain and swelling.
Lymphadenitis	Inflammation of lymph nodes; can be viral or bacterial.
Lymphadenopathy	Any disease of lymph nodes; causes can be autoimmune disease or malignancy.
Lymphedema	Blockage of lymphatic vessels; can be caused by genetics, parasitic infections, trauma to the vessels, tumors, radiation therapy, cellulitis, and surgeries.

Diseases and Disorders of the Respiratory System

Name	Description
Allergic rhinitis	A hypersensitivity reaction to various airborne allergens.
Asthma	Hyperactivity of the bronchioles; an inflammatory response with excess mucus production.
Atelectasis	Commonly called collapsed lung; may occur after surgery or pleural effusion.
Bronchitis	Inflammation of the bronchi; can be acute or chronic. Often follows a cold. One common cause of chronic bronchitis is cigarette smoking.
Chronic obstructive pulmonary disease (COPD)	A group of lung disorders in which obstruction limits airflow to the lungs.
Emphysema	A lung disease in which the alveolar walls are destroyed; most often caused by cigarette smoking.
Laryngitis	An acute inflammation of the larynx; caused by viruses, bacteria, polyp formation, excessive use, allergies, smoking, heartburn, alcohol use, nerve damage, or stroke.
Lung cancer	The number-one cause of death in the United States. Caused by smoking or exposure to radon, asbestos, and industrial carcinogens. The three types are small cell lung cancer, squamous cell lung cancer, and adenocarcinoma.
Mesothelioma	A type of cancer that affects the pleura and is a result of asbestos exposure.
Pleuritis or pleurisy	A condition in which the pleura become inflamed.
Pneumonoconiosis	Lung diseases that result from years of exposure to different environmental or occupational types of dust.
Pulmonary edema	A condition in which fluids fill the alveoli of the lungs, most commonly occurring with left heart failure.
Pulmonary embolism	Blocked artery in the lungs, usually caused by a blood clot.
Respiratory distress syndrome (RDS)	Kills apparently healthy infants; the cause is unknown.
Sinusitis	Inflammation of the membranes lining the sinuses. It can be acute or chronic.
Sudden infant death syndrome (SIDS)	The sudden and unexpected death of an infant under one year of age. There is no explainable cause of death.

Diseases and Disorders of the Nervous System

Name	Description
Alzheimer's disease	A progressive, degenerative disease of the gray matter of the brain. Causes dementia in elderly people.
Amyotrophic lateral sclerosis (ALS)	Commonly known as Lou Gehrig's disease; a fatal disorder characterized by the degeneration of neurons in the spinal cord and brain.
Bell's palsy	A disorder in which facial muscles are very weak or temporarily totally paralyzed. Results from damage to the facial nerve; causes unknown.
Brain tumors and cancers	Abnormal growths in the brain. Malignant tumors that start in brain tissue are called primary brain cancers. Those that start elsewhere and metastasize to the brain are classified as secondary brain cancers.
Epilepsy	A condition in which the brain experiences repeated spontaneous seizures due to abnormal electrical activity of the brain.
Guillain-Barré syndrome	A disorder in which the body's immune system attacks part of the peripheral nervous system. It has a sudden and unexpected onset.
Headaches	There are numerous types and causes. Types include tension, migraine, and cluster headaches.
Multiple sclerosis (MS)	A chronic disease of the central nervous system in which myelin is destroyed. Some known causes are viruses, genetic factors, and immune system abnormalities.
Neuralgia	A group of disorders commonly referred to as nerve pain; most frequently occur in the nerves of the face.
Parkinson's disease	A motor system disorder that is slowly progressive and degenerative.
Sciatica	A condition in which the sciatic nerve is damaged, commonly due to excessive pressure on the nerve from prolonged sitting.
Stroke or cerebrovascular accident (CVA)	Occurs when brain cells die because of inadequate blood perfusion of the brain.

Diseases and Disorders of the Urinary System

Name	Description
Acute renal failure (ARF)	A sudden loss of kidney function due to burns, dehydration, low blood pressure, hemorrhage, allergic reactions, obstructions, poisons, alcohol abuse, or trauma.
Chronic renal failure (CRF)	A condition in which the kidneys slowly lose their ability to function due to diabetes, hypertension, kidney disease, or heart failure.
Cystitis	A urinary bladder infection caused by different types of bacteria.
Glomerulonephritis	Inflammation of the glomeruli of the kidney, caused by bacterial infections, renal diseases, and immune disorders.
Incontinence	A condition in which an adult cannot control urination. Can be temporary or long lasting; caused by various medications, UTIs, nervous system disorders, cancers, surgery, trauma, or pregnancy.

Diseases and Disorders of the Urinary System (concluded)

Name	Description
Polycystic kidney disease (PKD)	A disorder in which the kidneys enlarge because of the presence of many cysts within them. The cause is hereditary.
Pyelonephritis	A type of complicated UTI that begins as a bladder infection and spreads up one or both ureters into the kidneys.
Renal calculi (kidney stones)	Obstruct the ducts within the kidneys or ureters. Painful condition caused by gouty arthritis, ureter defects, overly concentrated urine, or UTIs.

Diseases and Disorders of the Male Reproductive System

Name	Description
Benign prostatic hypertrophy (BPH)	The nonmalignant enlargement of the prostate gland.
Epididymitis	Inflammation of the epididymis. Most cases start as an infection of the urinary tract.
Impotence or erectile dysfunction (ED)	A disorder in which a male cannot achieve or maintain an erection to complete sexual intercourse. Can be caused by many physical or psychological conditions.
Prostate cancer	One of the most common cancers in men older than 40. A high-fat diet, increased age, and genetic predisposition all increase the risk.
Prostatitis	An inflammation of the prostate gland; can be acute or chronic. May be caused by bacterial infections, catheterization, trauma, excessive alcohol consumption, or scarring.
Testicular cancer	A malignant growth of one or both testicles. Occurs more commonly in males 15 to 30 years of age and is very aggressive.

Diseases and Disorders of the Female Reproductive System

Name	Description
Breast cancer	Evaluated and graded based on tumor size and how far cells have traveled from the site of origin.
Cervical cancer	Most develop slowly, and with early detection by a yearly Pap smear, treatment is often successful.
Cervicitis	Inflammation of the cervix, caused by an infection.
Dysmenorrhea	Condition of experiencing severe menstrual cramps that limit normal daily activities.
Endometriosis	A condition in which tissues that make up the lining of the uterus grow outside the uterus.
Fibrocystic breast change	Common condition consisting of abnormal cysts in the breasts that vary in size related to the menstrual cycle.
Fibroids	Benign tumors that grow in the uterine wall.

Diseases and Disorders of the Female Reproductive System (concluded)

Name	Description
Infertility	The inability to conceive a child due to scarring of the fallopian tubes, STIs, PID, endometriosis, or hormone imbalance.
Ovarian cancer	Considered more deadly than the other types of gynecological cancers because its symptoms are mild and indistinct.
Pelvic inflammatory disease (PID)	Acute or chronic infection of the reproductive tract caused by untreated STIs or bacteria.
Premenstrual syndrome (PMS)	A collection of symptoms that occur just before the menstrual period. Symptoms include anxiety, depression, irritability, bloating, and diarrhea.
Uterine (endometrial) cancer	Most common in postmenopausal women. It may be related to increased levels of estrogen.
Vaginitis	Inflammation of the vagina, associated with abnormal vaginal discharge.
Vulvovaginitis	Inflammation of the vulva and vagina.

Diseases and Disorders of the Digestive System

Name	Description
Appendicitis	An inflammation of the appendix. If not treated promptly, it can be life threatening.
Cholelithiasis (gallstones)	Gallstones are hardened deposits of bile that can form in the gallbladder. Two types of calculi are cholesterol and pigment stones.
Cirrhosis	A chronic liver disease in which normal liver tissue is replaced with nonfunctional scar tissue.
Colitis	An inflammation of the large intestine. This condition can be chronic or short-lived, depending on the cause.
Colorectal cancer	Usually arises from the lining of the rectum or colon. This type of cancer is curable if diagnosed and treated early. It is the third most common cause of death due to cancer in both men and women.
Constipation	The condition of difficult defecation or elimination of feces.
Crohn's disease	A type of inflammatory bowel disease. It can affect any region of the digestive tract from the mouth to the anus, but most often affects the lower part of the small intestine called the ileum.
Diarrhea	The condition of watery and frequent feces. Many cases of diarrhea do not require treatment because they are usually self-limiting and stop within a day or two.
Diverticulitis	Inflammation of diverticuli in the intestine. Diverticuli are abnormal dilations or pouches in the intestinal wall. When the diverticuli are not inflamed, the condition is known as *diverticulosis*.
Gastric or stomach ulcers	Occur when the lining of the stomach breaks down.
Gastritis	An inflammation of the stomach lining; often referred to as an upset stomach.

Diseases and Disorders of the Digestive System (concluded)

Name	Description
Gastroesophageal reflux disorder (GERD)	This occurs when stomach acids are pushed into the esophagus (also called heartburn). If not treated, GERD can cause erosion of the esophagus and even esophageal cancer.
Helicobacter pylori (*H. pylori*)	Organism implicated as being responsible for many diseases and disorders, including gastric ulcers.
Hemorrhoids	Varicosities (varicose veins) of the rectum or anus.
Hepatitis	Inflammation of the liver. There are many different types of hepatitis, but because they all involve inflammation of the liver. Many signs and symptoms are shared, regardless of the cause of the inflammation.
Hiatal hernias	Occur when a portion of the stomach protrudes into the thoracic cavity through an opening (esophageal hiatus) in the diaphragm.
Inguinal hernias	Occur when a portion of the large intestine protrudes into the inguinal canal, which is located where the thigh and the body trunk meet. In males, the hernia can also protrude into the scrotum.
Oral cancer	Usually involves the lips or tongue but can occur anywhere in the mouth. This type of cancer tends to spread rapidly to other organs because of the high vascularity of this area.
Pancreatic cancer	The fourth leading cause of cancer death in the United States. The poor prognosis and five-year survival rate of only 5% is due to the late diagnosis in many cases.
Stomach cancer	Most commonly occurs in the uppermost (cardiac portion) of the stomach. It appears to occur more frequently in Japan, Chile, and Iceland than in the United States. This may be due to diets of foods high in nitrates that are known to be carcinogenic.

Diseases and Disorders of Metabolism and Nutrition

Name	Description
Anorexia nervosa	An eating disorder in which individuals have a perception of being overweight regardless of their actual weight.
Bulimia	An eating disorder where the person may be of normal weight and may eat normal or even excessive amounts of food, but then vomits to rid himself or herself of the calories. Also called binge and purge eating.
Hypercholesterolemia	An excess of cholesterol in the blood.
Kwashiorkor	A type of starvation in which there is too little protein in the diet.
Malnutrition	Inadequate or excessive caloric intake.
Marasmus	A type of starvation that entails both protein and calorie insufficiency.
Metabolic syndrome	A group of symptoms such as hypertension, hyperinsulinism, excess body fat around the waist, and hypercholesterolemia. Lifestyle changes, medication, and regular appointments with a health care provider are essential.
Obesity	A body weight greater than 20% above the standard is considered obesity and 50% over the standard is considered morbid obesity.
Starvation	A type of malnutrition resulting from inadequate caloric intake, inadequate resources, dietary imbalances, illness, or self-imposed starvation.

Diseases and Disorders of the Endocrine System

Name	Description
Acromegaly	A condition caused by the secretion of too much growth hormone after puberty that causes the hands, feet, and face to take on unusual enlargement.
Addison's syndrome	Hyposecretion of ACTH in which patients experience anorexia, fatigue, weight loss, GI problems, and bronzing of the skin.
Diabetes insipidus	Hyposecretion of ADH. Symptoms include excessive urination, thirst, and dehydration.
Diabetes mellitus	Hyposecretion of insulin, categorized into type 1 (insulin dependent) and type 2 (non-insulin dependent).
Dwarfism (achondroplasia)	Abnormal underdevelopment of the body due to hyposecretion of growth hormone (GH). Adult height of 4 feet 10 inches or less.
Epinephrine and norepinephrine imbalances	Cause increase in blood pressure, tachycardia, and tachypnea.
Estrogen imbalances	Hyposecretion before or during menopause results in hot flashes, vaginal dryness, mood changes, depression, and loss of bone density.
Gigantism	Abnormal increase in the length of long bones due to hypersecretion of growth hormone during childhood.
Hypercalcemia	An increase in blood calcium levels, resulting in more calcium going into bone.
Hyperthyroidism	Lower level of TSH due to overactive thyroid. Also called Grave's disease, it causes an overall increase in metabolism.
Hypocalcemia	A condition in which there is too little calcium in the blood.
Hypoglycemia	A low level of glucose in the blood.
Hypothyroidism	Excess secretion of TSH due to inactive thyroid, resulting in slowing of metabolism.
Melatonin imbalances	Hyposecretion disturbs the sleep cycle and contributes to depression.
Pancreatic tumors	Malignant tumors have a poor prognosis due to difficulty in diagnosing.
Pheochromocytoma	A benign tumor of the adrenal medulla that can cause an increase in blood pressure.
Thymosin imbalances	Hyposecretion results in lack of mature T lymphocytes and a decrease in immunity.
Thyroid tumor	Abnormal growth on the thyroid gland. May be benign or malignant.

Diseases and Disorders of the Special Senses—The Eye

Name	Description
Amblyopia	More commonly called lazy eye; occurs when a child does not use one eye regularly.
Astigmatism	Occurs when the cornea or lens has an abnormal shape, which causes blurred images in near or distant vision.
Blepharitis	Inflammation of the eyelid, as well as corneal abrasions (scratching of the cornea).

Diseases and Disorders of the Special Senses—The Eye (concluded)

Name	Description
Cataracts	Opaque structures within the lens that prevent light from going through the lens. Over time, images begin to look fuzzy; if left untreated, cataracts may cause blindness.
Color blindness	The inability to see certain colors. May be inherited; occurs more commonly in males.
Dry eye syndrome (xerophthalmia)	One of the most common eye problems treated by physicians. This syndrome results from a decreased production of the oil within tears, which normally occurs with age.
Ectropion	Eversion of the lower eyelid.
Entropion	Inversion of the lower eyelid.
Glaucoma	Indicated by an increase in intraocular pressure, caused by a buildup of aqueous humor in the anterior chamber. If untreated, this excess pressure can lead to permanent damage of the optic nerve that can result in blindness.
Hyperopia	Farsightedness.
Macular degeneration	A progressive disease that usually affects people over the age of 50. It occurs when the retina no longer receives an adequate blood supply. It is the most common cause of vision loss in the United States.
Myopia	Nearsightedness.
Nystagmus	Rapid, involuntary eye movements. The movements may be horizontal or vertical.
Presbyopia	A common eye disorder that results in the loss of lens elasticity. It develops with age and causes a person to have difficulty seeing objects that are close up.
Retinal detachment	Occurs when the layers of the retina separate. It is considered a medical emergency and if not treated right away leads to permanent vision loss.
Strabismus	A misalignment of the eyes. Convergent strabismus is commonly referred to as crossed eyes.

Diseases and Disorders of the Special Senses—The Ear

Name	Description
Cerumen impaction	Buildup of ear wax within the external auditory canal.
Meniere's disease	A disturbance in the equilibrium characterized by *vertigo* (dizziness), ringing in the ears, nausea, and progressive hearing loss.
Otitis externa	Inflammation of the outer ear.
Otitis media	Inflammation of the middle ear.
Otosclerosis	The immobilization of the stapes within the middle ear; a common cause of conductive hearing loss.
Presbycusis	Hearing loss because of the aging process.
Tinnitus	Ringing in the ears.
Vertigo	Dizziness.

appendix II

prefixes, suffixes, and word roots in commonly used medical terms

Prefixes

a-, an- without, not
ab- from, away
ad- to, toward
ambi-, amphi- both, on both sides, around
ante- before
antero- in front of
anti- against, opposing
auto- self
bi- twice, double
brachy- short
brady- slow
cata- down, lower, under
centi- hundred
chol-, chole-, cholo- gall
circum- around
co-, com-, con- together, with
contra- against
cryo- cold
de- down, from
deca- ten
deci- tenth
demi- half
dextro- to the right
di- double, twice
dia- through, apart, between
dipla-, diplo- double, twin
dis- apart, away from
dys- difficult, painful, bad, abnormal
e-, ec-, ecto- away, from, without, outside
em-, en- in, into, inside
endo- within, inside
ento- within, inner
epi- on, above
erythro- red
eu- good, true, normal
ex-, exo- outside of, beyond, without
extra- outside of, beyond, in addition
fore- before, in front of
gyn-, gyno-, gyne-, gyneco- woman, female
hemi- half
hetero- other, unlike
homeo, homo- same, like
hyper- above, over, increased, excessive
hypo- below, under, decreased
idio- personal, self-produced
im-, in-, ir- not
in- in, into
infra- beneath
inter- between, among
intra-, intro- into, within, during
iso- equal
juxta- near, nearby
leuk- white
macro- large, long
mal- bad
mega-, megalo- large, great
melan- black
mes-, meso- middle
meta- beyond
micro- small
mono- single, one
multi- many
nat- birth
neo- new
non-, not- no
nuli- none
olig-, oligo- few, less than normal
ortho- straight
oxy- sharp, acid
pachy- thick
pan- all, every
par-, para- alongside of, with; woman who has given birth
per- through, excessive
peri- around
pluri- more, several
poly- many, much
post- after, behind
pre-, pro- before, in front of
presby-, presbyo- old age
primi- first
pseudo- false
quadri- four
re- back, again
retro- backward, behind
semi- half
steno- contracted, narrow
stereo- firm, solid, three-dimensional
sub- under
super-, supra- above, upon, excess
sym-, syn- with, together

tachy- fast
tele- distant, far
tetra- four
tox- poison
trans- across
tri- three
ultra- beyond, excess
uni- one
xeno- foreign material
xero- dry
xantho- yellow

Suffixes
-ad to, toward
-aesthesia sensation
-al characterized by
-algia pain
-ase enzyme
-asthenia weakness
-cele swelling, tumor
-centesis puncture, tapping
-cidal killing
-cide causing death
-cise cut
-cyst bladder, sac
-cyte, cell, cellular
-derm skin
-dynia pain
-ectomy cutting out, surgical removal
-emesis vomiting
-emia blood
-esthesia sensation
-gene, -genic, -genetic, -genesis arising from, origin, formation
-gram recorded information
-graph instrument for recording
-graphy the process of recording
-gravida pregnant female
-ia condition
-iasis condition of
-ic, -ical pertaining to
-ism condition, process, theory
-ist one who specializes in
-itis inflammation of
-ium membrane
-ize to cause to be, to become, to treat by special method
-kinesis, -kinetic motion
-lepsis, -lepsy seizure, convulsion
-lith stone
-logy science of, study of
-lysis setting free, disintegration, decomposition
-malacia abnormal softening
-mania insanity, abnormal desire
-megaly enlargement
-meter measure
-metry process of measuring
-odynia pain
-oid resembling
-ole small, little
-oma tumor
-opia vision
-opsy to view
-orexia appetite
-osis disease, condition of
-ostomy to make a mouth, opening
-otomy incision, surgical cutting
-ous having
-oxia oxygen
-pathy disease, suffering
-penia too few, lack, decreased
-pexy surgical fixation
phagia, -phage eating, consuming, swallowing
-phobia fear, abnormal fear
-phylaxis protection
-plasia formation or development
-plastic molded
-plasty operation to reconstruct, surgical repair
-plegia paralysis
-pnea breathing
-poiesis formation
-rrhage, -rrhagia abnormal or excessive discharge, hemorrhage, flow
-rrhaphy suture of
-rrhea flow, discharge
-sclerosis hardening
-scopy examining
-sepsis poisoning, infection
-spasm cramp or twitching
-stasis stoppage
-stomy opening
-therapy treatment
-thermy heat
-tome cutting instrument
-tomy incision, section
-tripsy surgical crushing
-trophy nutrition, growth
-tropy turning, tendency
-ule small
-uria urine

Word Roots
adeno- gland, glandular
adipo- fat
aero- air
andr-, andro- man, male
angio- blood vessel
ano- anus
arterio- artery
arthro- joint
bili- bile
bio- life
blasto-, blast- developing stage, bud
blephar- eyelid
bracheo- arm
broncho- bronchial (windpipe)
bucc- cheek
carcino- cancer
cardio- heart
cephal-, cephalo- head
cerebr-, cerebro- brain
cervico- neck
chondro- cartilage
chromo- color
colo- colon
colp-, colpo- vagina
coro- body
cost-, costo- rib
crani-, cranio- skull
cry- cold
cutane- skin
cysto- bladder, bag
cyto- cell, cellular
dacry-, dacryo- tears, lacrimal apparatus
dactyl-, dactylo- finger, toe
dent-, denti-, dento- teeth
derma-, dermat-, dermato- skin
dorsi-, dorso- back
encephalo- brain
entero- intestine
esophag- esophagus
esthesio- sensation
fibro- connective tissue
galact-, galacto- milk
gastr-, gastro- stomach
gingiv- gums
glosso- tongue
gluco-, glyco- sugar, sweet
haemo-, hemato-, hem-, hemo- blood
hepa-, hepar-, hepato- liver

herni- rupture
histo- tissue
hydra-, hydro- water
hyster-, hystero- uterus
ictero- jaundice
ileo- ileum
karyo- nucleus, nut
kera-, kerato- horn, hardness, cornea
lact- milk
laparo- abdomen
latero- side
linguo- tongue
lipo- fat
lith- stone
mast-, masto- breast
med-, medi- middle
meno- month
metro-, metra- uterus
my-, myo- muscle
myel-, myelo- marrow
narco- sleep
nas-, naso- nose
necro- dead
nephr-, nephro- kidney
neu-, neuro- nerve
niter-, nitro- nitrogen

nucleo- nucleus
oculo- eye
odont- tooth
omphalo- navel, umbilicus
onco- tumor
oo- ovum, egg
oophor- ovary
ophthalmo- eye
orchid- testicle
os- mouth, opening
oste-, osteo- bone
oto- ear
paedo, pedo- child
palpebro- eyelid
path-, patho- disease, suffering
pepso- digestion
phag-, phago- eating, consuming, swallowing
pharyng- throat, pharynx
phlebo- vein
pleuro- side, rib
pneumo- air, lungs
pod foot
procto- rectum
psych- the mind
pulmon-, pulmono- lung
pyelo- pelvis (renal)

pyo- pus
pyro- fever, heat
reni-, reno- kidney
rhino- nose
sacchar- sugar
sacro- sacrum
salpingo- tube, fallopian tube
sarco- flesh
sclero- hard, sclera
septi-, septico- poison, infection
somato- body
stomato- mouth
teno-, tenoto- tendon
thermo- heat
thoraco- chest
thrombo- blood clot
thyro- thyroid gland
tonsillo- tonsil
trach- trachea
tricho- hair
urino-, uro- urine, urinary organs
utero- uterus, uterine
vaso- vessel
ven- vein
ventri-, ventro- abdomen
vesico- blister
viser- organ

appendix III
abbreviations and symbols commonly used in medical notations

Abbreviations

a before
āā, AA of each
a.c. before meals
ADD attention deficit disorder
ADL activities of daily living
ad lib as desired
ADT admission, discharge, transfer
AIDS acquired immunodeficiency syndrome
a.m.a. against medical advice
AMA American Medical Association
amp. ampule
amt amount
aq., AQ water; aqueous
ausc. auscultation
ax axis
Bib, bib drink
b.i.d., bid, BID twice a day
BM bowel movement
BP, B/P blood pressure
BPC blood pressure check
BPH benign prostatic hypertrophy
BSA body surface area
c̄ with
Ca calcium; cancer
cap, caps capsules
CBC complete blood (cell) count
C.C., CC chief complaint
CDC Centers for Disease Control and Prevention

CHF congestive heart failure
chr chronic
CNS central nervous system
Comp, comp compound
COPD chronic obstructive pulmonary disease
CP chest pain
CPE complete physical examination
CPR cardiopulmonary resuscitation
CSF cerebrospinal fluid
CT computed tomography
CV cardiovascular
d day
D&C dilation and curettage
DEA Drug Enforcement Administration
Dil, dil dilute
DM diabetes mellitus
DOB date of birth
DTP diptheria-tetanus-pertussis vaccine
Dr. doctor
DTs delirium tremens
D/W dextrose in water
Dx, dx diagnosis
ECG, EKG electrocardiogram
ED emergency department
EEG electroencephalogram
EENT eyes, ears, nose, and throat
EP established patient
ER emergency room

ESR erythrocyte sedimentation rate
FBS fasting blood sugar
FDA Food and Drug Administration
FH family history
Fl, fl, fld fluid
F/u follow-up
Fx fracture
GBS gallbladder series
GI gastrointestinal
Gm, gm gram
gr grain
gt, gtt drop, drops
GTT glucose tolerance test
GU genitourinary
GYN gynecology
HA headache
HB, Hgb hemoglobin
HEENT head, ears, eyes, nose, throat
HIV human immunodeficiency virus
HO history of
Hx history
ICU intensive care unit
I&D incision and drainage
I&O intake and output
IM intramuscular
inf. infusion; inferior
inj injection
IT inhalation therapy
IUD intrauterine device
IV intravenous

677

KUB kidneys, ureters, bladder
L1, L2, etc. lumbar vertebrae
lab laboratory
lb pound
liq liquid
LLL left lower lobe
LLQ left lower quadrant
LMP last menstrual period
LUQ left upper quadrant
M mix (Latin *misce*)
m- meta-
MI myocardial infarction
mL milliliter
MM mucous membrane
MRI magnetic resonance imaging
MS multiple sclerosis
NB newborn
NED no evidence of disease
no. number
noc, noct night
npo, NPO nothing by mouth
NPT new patient
NS normal saline
NSAID nonsteroidal anti-inflammatory drug
NTP normal temperature and pressure
N&V nausea and vomiting
NYD not yet diagnosed
o- ortho-
OB obstetrics
OC oral contraceptive
oint ointment
OOB out of bed
OPD outpatient department
OPS outpatient services
OR operating room
OTC over-the-counter
p- para-
p̄ after
P&P Pap smear (Papanicolaou smear) and pelvic examination
PA posteroanterior
Pap Pap smear
Path pathology
p.c., pc after meals
PE physical examination
per by, with
PH past history
PID pelvic inflammatory disease
p/o postoperative

POMR problem-oriented medical record
PMFSH past medical, family, social history
PMS premenstrual syndrome
p.r.n., prn, PRN whenever necessary
Pt patient
PT physical therapy
PTA prior to admission
PVC premature ventricular contraction
pulv powder
q. every
q2, q2h every 2 hours
q.a.m., qam every morning
q.h., qh every hour
q.i.d., QID four times a day
qns, QNS quantity not sufficient
qs, QS quantity sufficient
RA rheumatoid arthritis; right atrium
RBC red blood cells; red blood (cell) count
RDA recommended dietary allowance, recommended daily allowance
REM rapid eye movement
RF rheumatoid factor
RLL right lower lobe
RLQ right lower quadrant
R/O rule out
ROM range of motion
ROS/SR review of systems/systems review
RUQ right upper quadrant
RV right ventricle
Rx prescription, take
s̄ without
SAD seasonal affective disorder
SIDS sudden infant death syndrome
Sig directions
sig sigmoidoscopy
SOAP subjective, objective, assessment, plan
SOB shortness of breath
sol solution
S/R suture removal
Staph staphylococcus
stat, STAT immediately
STD sexually transmitted disease

Strep streptococcus
subling, SL sublingual
subcut subcutaneously
surg surgery
S/W saline in water
SX symptoms
T1, T2, etc. thoracic vertebrae
T & A tonsillectomy and adenoidectomy
tab tablet
TB tuberculosis
tbs., tbsp tablespoon
TIA transient ischemic attack
t.i.d., tid, TID three times a day
tinc, tinct, tr tincture
TMJ temporomandibular joint
top topically
TPR temperature, pulse, and respiration
TSH thyroid stimulating hormone
tsp teaspoon
Tx treatment
UA urinalysis
UCHD usual childhood diseases
UGI upper gastrointestinal
ung, ungt ointment
URI upper respiratory infection
US ultrasound
UTI urinary tract infection
VA visual acuity
VD venereal disease
Vf visual field
VS vital signs
WBC white blood cells; white blood (cell) count
WNL within normal limits
wt weight
y/o year old

Symbols
Weights and Measures
\# pounds
° degrees
′ foot; minute
″ inch; second
mcg micrometer
mEq milliequivalent
mL milliliter
dL deciliter
mg% milligrams percent; milligrams per 100 mL

Mathematical Functions and Terms

\# number
− minus; negative; alkaline reaction
± plus or minus; either positive or negative; indefinite
× multiply; magnification; crossed with, hybrid
÷ divided by
= equal to
≈ approximately equal to
≠ not equal to
√ square root
³√ cube root
∞ infinity
: ratio; "is to"
∴ therefore
% percent
π pi (3.14159)—the ratio of circumference of a circle to its diameter

Chemical Notations

Δ change; heat
⇌ reversible reaction
↑ increase
↓ decrease

Warnings

Ⓒ Schedule I controlled substance (CI)
Ⓒ Schedule II controlled substance (CII)
Ⓒ Schedule III controlled substance (CIII)
Ⓒ Schedule IV controlled substance (CIV)
Ⓒ Schedule V controlled substance (CV)
☠ poison
☢ radiation
☣ biohazard

Others

R_x prescription; take
□, ♂ male
○, ♀ female
† one
†† two
††† three

glossary

a

Abdominal reflexes (ab-DOM-in-al RE-fleks-es) Reflexes used to evaluate damage to thoracic spinal nerves.

Abdominopelvic cavity (ab-do-mih-no-PEL-vic KAV-i-tee) Body cavity that contains the stomach, liver, spleen, gallbladder, kidneys, intestine, urinary bladder, and internal reproductive organs.

Abduction (ab-DUK-shun) Moving a body part away from the midline of the body.

Absorption (ab-SORP-shun) The passage of digested foods from the gastrointestinal tract into the blood or lymph.

Accommodation (uh-kom-oh-DA-shun) The ability of the lens to change shape to help the eye focus on images at various distances.

Acetabulum (ass-eh-TAB-u-lum) The hip socket.

Acetylcholine (ACh) (uh-see-til-KOH-leen) A neurotransmitter involved in smooth muscle contraction.

Acetylcholinesterase (uh-see-til-koh-lih-NES-tuh-race) An enzyme that breaks down acetylcholine and causes muscle relaxation.

Acetyl coenzyme A (uh-SEE-tul koh-EN-zime ay) A molecule formed in the matrix of the mitochondria that is the substrate that enters the Krebs cycle.

Acidosis (as-ih-DO-sis) A condition in which the pH of the blood plasma is less than 7.35.

Acne vulgaris (AK-knee vul-GAIR-is) An infection of the sebaceous gland(s) that causes pustules, papules, and scarring of the affected skin.

Acquired immunodeficiency syndrome (AIDS) (a-KWIRD im-yu-no-dee-FISH-en-see SIN-drom) The most advanced stage of HIV infection; it severely weakens the immune system.

Acquired (specific) immunity (a-KWIRD i-MU-ni-te) Formation of antibodies and lymphocytes after exposure to an antigen.

Acromegaly (ak-ro-MEG-uh-lee) A condition caused by too much growth hormone after puberty that causes the hands, feet, and face to take on an unusual enlargement.

Acrosome (AK-ro-some) An enzyme-filled sac covering the head of the sperm that helps it penetrate the ovum for fertilization.

Actin (AK-tin) The thin protein filaments that, along with myosin, create the striations in cardiac and skeletal muscle.

Action potential (AK-shun po-TEN-shal) Flow of electrical current along the axon membrane.

Active transport (AK-tiv TRANS-port) Movement across cell membranes against a concentration gradient, which requires energy (ATP).

Acute (uh-CUTE) Having a rapid onset and progress, such as acute appendicitis.

Acute inflammation (uh-CUTE in-fluh-MA-shun) Inflammation that lasts from one to two days.

Acute renal failure (ARF) (uh-CUTE REE-nul FAYL-yur) A sudden loss of kidney function. Also called acute kidney failure (AKF).

Adduction (ad-DUK-shun) Moving a body part toward the midline of the body.

Adenocarcinoma (add-ih-no-kar-sih-NO-muh) A type of lung cancer that arises from the mucus-producing cells of the lungs; the most common type of lung cancer in nonsmokers.

Adenosine triphosphate (ATP) (ah-DEN-o-sene tri-FOS-fate) Compound in cells that stores energy.

Adipocyte (ADD-ih-po-site) Fat cell.

Adrenocorticotropic hormone (ACTH) (ah-dree-no-kor-tih-ko-TRO-pik HORE-mone) An amino acid hormone that targets the adrenal cortex to secrete aldosterone and cortisol and other adrenal cortical hormones.

Aerobic (air-O-bik) Requiring oxygen.

Aerobic respiration (air-O-bik res-per-A-shun) A process in which the body's store of glucose and oxygen are used to make ATP.

Afebrile (a-FEB-rile) Without fever; having a normal body temperature.

Afferent arterioles (AF-fer-ent ar-TEE-ree-ole) Branches of the interlobar arcuate arteries that come off the renal artery and supply blood to the nephrons.

Afferent (sensory) neurons (AFF-er-ent NOO-ronz) Neurons that carry impulses from the peripheral nervous system to the central nervous system.

Agglutination (uh-gloo-tih-NA-shun) The clumping of RBCs that can occur if a patient is given the wrong blood type during a blood transfusion.

Agonist (AG-uh-nist) A muscle that assists the prime mover in the muscle movement.

Agranulocytes (ay-GRAN-yoo-lo-sites) WBCs that do not have granules in their cytoplasm.

Albinism (al-BINE-ism) A condition in which a person is born with little or no pigmentation in the skin, eyes, or hair.

Albumin (al-BYOO-men) Smallest but most abundant of the plasma proteins.

Alcoholism (AL-ko-hol-izm) An addiction to alcohol that has characteristics of dependence and withdrawal symptoms.

Aldosterone (al-DOS-ter-one) Hormone secreted by the adrenal cortex for the absorption of sodium by the kidneys.

Alimentary canal (al-ih-MENT-ah-ry ka-NAL) The gastrointestinal tract and accessory organs.

Alkalosis (al-kah-LO-sis) Condition in which the pH of the blood plasma is greater than 7.45.

Allele (al-LEEL) One of two or more different forms of a gene that may or may not influence a particular trait.

Allergen (AL-er-jen) An antigen that elicits a hypersensitivity reaction.

Allergic rhinitis (uh-LER-jik ri-NI-tis) A hypersensitivity reaction to various airborne allergens.

Allograft (AL-o-graft) Type of transplant in which the tissue is taken from someone other than the individual receiving the transplant or an identical twin.

Allopathic approach (allo-PATH-ic a-PROCH) Treats disease by the use of remedies that produce effects different from those produced by the disease itself.

Alopecia (al-o-PE-she-a) Baldness.

Alveolus (al-VE-o-lus) Air sac in the lungs; the functional unit of the lungs.

Alzheimer's disease (ALZS-hi-merz dih-ZEEZ) A progressive, degenerative disease of the gray matter of the brain that causes dementia.

Amblyopia (am-blee-O-pe-uh) More commonly called lazy eye; occurs when a child does not use one eye regularly.

Amino acids (a-MEEN-o AH-sidz) Organic compounds used by the body to create protein.

Amniocentesis (am-nee-oh-sen-TEE-sis) A prenatal test that may be used to diagnose Down syndrome; it is performed early in the second trimester by withdrawing and testing the amniotic fluid surrounding a fetus.

Amnion (AM-nee-ahn) The protective, fluid-filled sac that surrounds an embryo.

Amphitrichous (am-FIH-trik-us) Having flagella on opposite ends of a bacterial cell.

Amylase (AM-ih-lase) An enzyme that starts the process of chemical digestion in saliva in the mouth.

Amyotrophic lateral sclerosis (ALS) (a-me-o-TRO-fic LAT-er-al skle-RO-sis) Commonly known as Lou Gehrig's disease, a fatal disorder characterized by the degeneration of neurons in the spinal cord and brain.

Anabolic reaction (a-nuh-BOL-ik ree-AK-shun) A metabolic reaction that involves the synthesis of substances.

Anabolism (a-NA-bo-lizm) Chemical reactions requiring energy where smaller molecules are used to build larger, more complex molecules.

Anaerobic (an-air-OH-bik) Bacteria that do not require oxygen to live.

Anaerobic reaction (an-air-OH-bik ree-AK-shun) A chemical reaction that does not require oxygen.

Anaerobic respiration (an-air-O-bik res-per-A-shun) A process in which glucose is derived from blood and from glycogen stored in muscle fibers and the liver to make ATP in the absence of oxygen.

Anaphase (AN-uh-faze) The third phase of mitosis; sister chromatids have separated and moved to opposites poles of the cell.

Anaphylactic shock (an-uh-fih-LAK-tik SHOK) An extreme hypersensitivity (allergic) reaction that leads to widespread vasodilation, with a resulting drop in blood pressure.

Anaphylaxis (an-uh-fih-LAK-sis) A condition caused by an extreme allergic reaction in which blood vessels dilate so quickly that blood pressure drops very suddenly and the body cannot compensate.

Anatomical position (anaTOM-ical po-ZIH-shun) The body is standing upright, facing forward with the palms of the hands also facing forward.

Anatomy (uh-NAT-uh-mee) The science of the study of the body structures.

Anemia (uh-NEE-mee-uh) A condition in which the blood has less than its normal oxygen-carrying capacity.

Aneurysm (AN-yur-izm) A ballooned, weakened arterial wall.

Angina pectoris (an-JI-nuh PEK-tor-iss) Pain caused by a narrowing or spasm of coronary arteries.

Angiotensin (an-je-oh-TEN-sin) A hormone involved in the regulation of water and sodium.

Angle closure glaucoma (ANG-gul CLOZ-yur glaw-KO-muh) See *Narrow-angle glaucoma*.

Ankyloglossia (ang-kih-lo-GLOSS-ee-uh) A condition in which the lingual frenulum is too short, and the person is "tongue tied."

Anorexia (an-o-REK-see-uh) Loss of appetite.

Anorexia nervosa (an-o-REK-see-uh ner-VOH-suh) A serious illness in which the person has a perception of being overweight regardless of his or her actual weight.

Antagonist (an-TAG-uh-nist) A muscle that produces a movement opposite to that of the prime mover.

Anterior (an-TER-e-or) Toward the front of the body when in anatomical position or in front of another structure.

Anthracosis (an-thruh-KO-sis) A type of pneumoconiosis that results from exposure to coal dusts; black lung disease.

Antibodies (AN-tih-bah-deez) Proteins in plasma that react to antigens on red blood cells to cause agglutination.

Antibody-mediated response (AN-ti-bah-dee MEE-dee-a-ted re-SPONS) See *Humoral immunity*.

Antidiuretic hormone (ADH) (an-ti-di-yoo-RET-ik HOR-mone) A peptide hormone that causes the body to retain water by increasing water absorption in the kidneys.

Antigens (AN-tih-jinz) Proteins on the surface of red blood cells that can react with antibodies in plasma to cause agglutination; a substance that has the ability to elicit an immune response.

Antihistamines (an-tih-HISS-tah-meenz) Medications that treat the histamine-related symptoms of allergies.

Antipyretics (an-ti-pi-RET-iks) Fever-reducing medications.

Aorta (ay-OR-tuh) The largest artery in the body.

Aortic semilunar valve (a-OR-tik sem-e-LOO-nar valve) The valve located between the left ventricle and the aorta.

Apical (A-pik-ul) The free surface of epithelial cells, which faces the lumen; slightly domed to allow stretching or expansion without tearing the epithelium.

Aplastic anemia (a-PLAS-tik uh-NEE-mee-uh) Anemia that results when bone marrow is destroyed.

Apneic (AP-nee-ik) Not breathing.

Apocrine (APP-o-krin) A type of sweat gland.

Aponeurosis (a-po-noo-RO-sis) A broad, sheetlike structure made of fibrous connective tissue that typically attaches muscles to other muscles.

Apoptosis (ap-op-TOE-sis) Programmed cell death or suicide.

Appendicitis (uh-pen-duh-SI-tis) An inflammation of the appendix.

Appendicular skeleton (ap-en-DIK-you-lar SKEL-uh-ton) Division of the adult skeleton that includes the pectoral and pelvic girdles and the upper and lower extremities.

Aqueous humor (AH-kwee-us HYOO-mer) The watery fluid that fills the anterior and posterior chambers of the eye; provides nutrients to and bathes the structures in the anterior chamber of the eyeball.

Arcuate artery (AR-kyoo-ate AR-ter-ee) A branch of the renal artery.

Areflexia (a-re-FLEK-se-a) The absence of a reflex.

Areola (uh-RHEE-o-la) The pigmented area that surrounds the nipple.

Arrector pili (ah-REK-tor PI-li) Smooth muscles that connect hair follicles to the papillary layer of the dermis.

Arrhythmia (a-RITH-me-a) An irregular heart rhythm; also called dysrhythmia.

Arterioles (ar-TEER-ee-olz) Blood vessels that carry oxygenated blood throughout the body; smaller than arteries.

Arthritis (arth-RYE-tis) A general term meaning joint inflammation.

Asbestosis (as-bes-TO-sis) A type of pneumonoconiosis that results from lung exposure to asbestos.

Ascending colon (uh-SEND-ing KO-lun) The portion of the large intestine that runs up the right side of the abdominal cavity.

ascites (uh-SI-teez) Fluid buildup in the abdomen.

Aspiration pneumonia (as-pih-RA-shun noo-MON-ee-uh) Type of pneumonia that results when foreign matter enters the lungs.

Asthma (AZ-ma) Reaction characterized by smooth muscle spasms in the bronchi and excessive production of mucus.

Astigmatism (uh-STIG-muh-tizm) A condition that occurs when the cornea or lens has an abnormal shape, which causes blurred images during near and distant vision.

Astrocytes (ASS-tro-sites) Star-shaped cells in the nervous system that anchor blood vessels to nerve cells.

Atelectasis (at-el-EK-tah-sis) Collapsed lung.

Atherosclerosis (ath-er-o-skle-RO-sis) An accumulation of cholesterol in the smooth muscle fibers of the tunica media of an artery.

Atlas (ATT-lus) The first cervical vertebra.

Atom (AH-tom) Unit of matter that makes up a chemical element.

Atomic number (uh-TOM-ic NUM-ber) The number of protons in an element.

Atomic weight (uh-TOM-ic WAIT) The sum of the number of protons plus neutrons.

Atrial diastole (A-tre-ul di-AS-to-le) Relaxation of the atria of the heart.

Atrial fibrillation (A-tre-ul fih-brih-LA-shun) Rapid but sporadic beating of the atria that results in poor blood flow through the heart.

Atrial natriuretic peptide (AY-tree-uhl na-tre-u-RET-ik PEP-tide) A hormone secreted by the right cardiac atrium to increase sodium and water excretion when blood volume increases.

Atrial systole (A-tre-ul SIS-toh-lee) Contraction of the atria of the heart.

Atrioventricular (AV) node (a-tree-o-ven-TRIK-u-lar node) A collection of cells located in the interatrial septum that is part of the conduction system of the heart, also known as the secondary pacemaker.

Atrioventricular septum (a-tree-o-ven-TRIK-yoo-lar SEP-tum) The septum or wall that separates the atria of the heart from the ventricles.

Atrium (A-tre-um) (plural, atria) A superior chamber of the heart.

Atrophy (AT-tro-fee) The reduction in size or wasting of an organ.

Auditory ossicles (AW-dih-to-ree OSS-ih-kulls) The smallest bones of the body located in the middle ear.

Auricle (AW-rih-kul) The external part of the ear; also called the pinna.

Auscultation (aw-skul-TAY-shun) Listening.

Autograft (AW-toe-graft) Tissue transplant in which the tissue is taken from another site on the patient's own body.

Autoimmune diseases (aw-to-im-MYUN di-ZEEZ) Diseases in which a person's immune system targets the person's own tissues.

Autoimmunity (auto-ih-MUN-ih-te) An individual's abnormal immune reaction of his or her own tissues.

Autosome (AW-to-some) A chromosome not involved in sex determination.

Avascular (a-VASS-ku-lar) Without a blood supply.

Axial skeleton (AK-cee-al SKEL-a-ton) Division of the adult skeleton made up of the skull, hyoid bone, auditory ossicles, vertebral column, sternum, and ribs.

Axis (AXE-iss) The second cervical vetebra.

Axon (AK-son) Long process, or nerve fiber, of a nerve cell that sends an action potential toward the axon terminals.

b

Bacillus (buh-SIL-us) A bacterium that is longer than it is wide.

Bactericidal (bak-te-re-SI-dal) Able to kill bacteria.

Bacteriostatic (bak-te-re-o-STA-tik) Unable to kill an organism but able to inhibit its multiplication

Bartholin's glands (BAR-to-linz glandz) Sometimes referred to as the vestibular glands, they secrete mucus during sexual arousal.

Baroreceptors (BAR-o-re-sep-torz) Sensors in the aorta and carotid arteries that help regulate blood pressure.

Basal cells (BAZ-ul cells) Cells in the olfactory epithelium that produce new receptor cells.

Basal cell carcinoma (BAY-sul cell kar-si-NO-ma) The most common malignancy of the skin.

Basal ganglia (BAY-zul GANG-lee-uh) Masses of gray matter deep inside the cerebrum.

Basal metabolic rate (BMR) (BAY-zul met-uh-BAH-lik rate) The amount of energy used by the body when it is in a resting state.

Basophil (BA-so-fil) A type of white blood cell with a granular cytoplasm that is involved in hypersensitivity reactions.

Bell's palsy (BELZ PALL-zi) A disorder in which facial muscles are very weak or temporarily totally paralyzed.

Benign (be-NINE) Describing a tumor that stays localized in the tissue.

Benign prostatic hypertrophy (BPH) (be-NINE pros-STAT-ik hi-PER-tro-fee) The nonmalignant enlargement of the prostate gland.

Biceps reflex (BI-ceps RE-fleks) Flexion of the forearm on percussion of the biceps brachii tendon.

Bicuspid valve (bi-KUS-pid valve) The atrioventricular valve on the left side of the heart.

Bile (BYLE) A substance created in the liver that is used in the digestion of fats.

Biliverdin (bil-ih-VER-din) A pigment released from destroyed RBCs that is converted to bilirubin in the liver.

Biochemistry (bio-KEM-is-tre) A branch of chemistry dealing with the chemistry of life.

Black piedra (BLAK pe-A-dra) An infection of the hair shaft by *Piedraia hortai*.

Blastocyst (BLAS-toe-sist) A hollow ball of cells in the development of an embryo.

Blepharitis (blef-ar-I-tis) Inflammation of the eyelid.

Blood pressure (BLUD PRE-shur) The force that blood exerts on the inner walls of blood vessels.

Blood type A Blood that has antigen A on the surface of its RBCs.

Blood type B Blood that has antigen B on the surface of its RBCs.

Blood type AB Blood that has both antigen A and antigen B on the surface of its RBCs.

Blood type O Blood that has neither antigen A nor antigen B on the surface of its RBCs.

Body (BAH-dee) The middle portion of the sternum (breastbone).

Body mass index (BMI) (BA-dee MASS IN-deks) A tool used to measure obesity, based on weight and height.

Bolus (BO-lus) Mass created by food mixed with the saliva and mucus mixture.

Botulism (BOT-u-lizm) A rare but very serious disorder caused by the bacterium *Clostridium botulinum*, which normally lives in soil and water.

Bowman's capsule (BO-manz KAP-sul) The glomerular capsule; a membrane that surrounds the glomerulus.

Brachiocephalic trunk (bray-kee-o-seh-FAL-ik trunk) A large artery that corresponds to the aorta, but actually comes off it and arches to the right side of the body.

Brachium (BRAK-ee-um) Referring to the arm.

Bradycardia (bray-dee-KAR-dee-uh) Heart rate less than 60 beats per minute.

Breech (BREECH) If the fetus is not in the usual head-down position, the child is said to be breech, and usually the buttocks or feet present first.

Bronchi (BRON-ki) Branches of the respiratory passageway.

Bronchial tree (BRON-kee-ul tree) The air passages that branch off from the trachea.

Bronchioles (BRON-kee-olz) Smaller air passages that branch off the bronchi in the lungs.

Bronchitis (bron-KI-tis) Inflammation of the bronchi; can be acute or chronic.

Bronchodilators (bron-ko-DI-lay-torz) Medications that dilate or open the bronchi, usually by reducing inflammation.

Bronchomediastinal trunk (bron-ko-mee-dee-uh-STI-nul trunk) The lymphatic trunk that drains lymph from the thorax.

Buffy coat (BUF-ee KOTE) The thin layer of white blood cells and platelets in a centrifuged blood sample.

Bulla (BULL-uh) A larger blister or a collection of small vesicles.

Bulimia (boo-LEE-mee-uh) A disorder in which a person eats normal or excessive amounts of food, but then vomits to rid herself or himself of the calories.

Bundle branches (BUN-dul branches) Pathways that transmit electrical impulses from the bundle of His to the Purkinje fibers of the heart.

Bundle of His (bundle of HISS) Part of the conduction system of the heart that extends from the AV node down the interventricular septum and splits into left and right bundle branches.

Bursitis (bur-SI-tis) Inflammation of the bursa, or fluid-filled sac that cushions tendons.

C

Cachexia (kuh-KEK-see-uh) Severe wasting of the body.

Calcitonin (kal-sih-TOE-nin) A protein hormone that moves calcium from the blood to the bone.

Calorie (KAL-or-ee) The amount of heat required to raise 1 gram of water 1°C.

Calyces (KA-lih-seez) Cuplike structures that transport urine from the collecting ducts to the renal pelvis.

Canaliculi (kan-ah-LIK-yoo-lie) Small canals in the bone that are perpendicular to the central canal and carry blood vessels and nerves to the lamellae and lacunae.

Canalize (KAN-uh-lize) The formation of a new lumen or channel through a thrombus.

Cancellous (KAN-cell-us) Describing spongy bone tissue that contains red bone marrow.

Cancer (KAN-ser) A malignant neoplasm or tumor.

Capsule (CAP-sul) A sheath or continuous enclosure around an organ or structure

Capillaries (KAP-ih-lar-eez) The smallest of the blood vessels, where gas and nutrient exchange takes place.

Carbohydrates (kar-boh-HI-drayts) A category of food that includes starches, simple sugars, and cellulose.

Carbon dioxide (CARB-un di-OX-ide) Gas that is a waste product of cellular metabolism.

Carboxyhemoglobin (kar-bok-see-HEE-mo-glo-bin) Hemoglobin that is carrying carbon dioxide.

Carbuncle (KAR-bun-kul) Inflammation of more than one hair follicle.

Carcinoma (kar-sih-NO-mah) A malignant tumor of epithelial origin.

Cardiac region (KAR-dee-ak RE-jun) Beginning portion of the stomach that is attached to the esophagus.

Cardiogenic shock (car-de-oh-GEN-ic SHOK) Shock caused by pump or heart failure.

Cardiomyopathy (kar-dee-o-mi-AH-puh-thee) Disease of the heart muscle.

Carina (kuh-RI-nuh) The point at which the trachea branches, or bifurcates, to create the left and right main stem bronchi.

Carotene (KARE-o-teen) A yellow pigment found mainly in the stratum corneum and the fat cells of the hypodermis.

Carpal tunnel syndrome (KAR-pull TUN-el SIN-drom) A condition that occurs when the median nerve in the wrist is excessively compressed by an inflamed tendon.

Cartilage (KAR-tih-lij) A rigid connective tissue.

Catabolic reaction (kat-uh-BOL-ik ree-AK-shun) A metabolic reaction that involves the breaking down of larger molecules into smaller molecules.

Catabolism (kah-TAB-o-lizm) The chemical breakdown of complex molecules into simpler molecules with the release of energy.

Cataracts (CAT-uh-racts) Opaque structures within the lens of the eye that cause clouding and hardening of the lens and lead to vision problems.

Caudal (KAW-dal) Away from the head.

Cecum (SEE-kum) The first portion of the large intestine.

Cell (SELL) The basic unit of life.

Cell cycle (SELL SI-kil) The series of events that occur in a cell that results in its replication.

Cell death (SELL DETH) The demise of a cell due to normal or pathological circumstances.

Cell-mediated response (SELL MEE-dee-a-ted re-SPONS) See *Cellular immunity*.

Cellular immunity (SELL-u-lar ih-MYOO-nih-tee) Immune response in which T lymphocytes bind to antigens on cells and attack them directly.

Cellular respiration (SELL-u-lar res-per-A-shun) Four sets of reactions that involve oxidation of glucose.

Cellulitis (sell-u-LI-tis) An inflammation of the connective tissue in skin.

Central nervous system (CNS) (SEN-tral NER-vus SIS-tem) A system that consists of the brain and spinal cord.

Centrioles (SEN-tree-olz) Cylinder-shaped organs near the cell nucleus that are essential for cell division.

Centromere (SEN-tro-meer) A region on a chromosome that joins two sister chromatids.

Cephalic phase (su-PHAL-ik FAZE) Part of digestion that starts in the cerebral cortex.

Cerebellum (ser-e-BEL-um) The part of the brain that maintains balance and coordinates skilled movements.

Cerebral angiography (su-RHEE-bral an-jee-OG-gruh-fee) A procedure that uses contrast material so that the blood vessels in the brain can be visualized.

Cerebrospinal fluid (CSF) (suh-ree-bro-SPI-nal FLU-id) A fluid that circulates in and around the brain and spinal cord.

Cerebrovascular accident (CVA) (suh-ree-bro-VAS-kyoo-lar AK-si-dent) Also known as a stroke, this condition occurs when brain cells die because of inadequate blood to the brain.

Cerebrum (ser-EE-brum) The two hemispheres of the forebrain making up the largest part of the brain.

Cerumen (seh-RU-men) Ear wax.

Cerumen impaction (seh-RU-men im-PAK-shun) The buildup of ear wax within the external auditory canal.

Cervical vertebrae (SER-vick-ul VER-tuh-bray) Located in the neck, these are the smallest and lightest of all the vertebrae.

Cervicitis (ser-vuh-SYE-tus) An inflammation of the cervix.

Cervix (SER-viks) The inferior, constricted part of the uterus.

Chancre (SHANK-ur) A hard, painless sore that is a symptom of syphilis.

Chemical bonds (KEM-a-kul bondz) An attraction between atoms that holds them together.

Chemistry (KEM-is-tre) The study of matter and how it undergoes change.

Chemoreceptors (KEY-mo-re-sept-ors) Smell receptors, also called olfactory receptors.

Chickenpox (CHIK-en-poks) A childhood disease caused by the varicella zoster virus.

***Chlamydia trachomata* (kla-MID-e-ah trah-ko-MA-ta)** Bacteria that causes the STI known as chlamydia.

Cholecystitis (koh-luh-sis-TI-tis) Inflammation of the gallbladder.

Cholelithiasis (ko-leh-lih-THI-uh-sis) Gallstones; hardened deposits of bile that form in the gallbladder.

Chordae tendineae (KOR-duh ten-DIN-ee-uh) The cordlike structures that anchor the cusps of the tricuspid valve to the papillary muscles.

Chorion (KO-rhee-ahn) The structure that develops during the embryonic stage that allows the exchange of substances between the fetus and mother and acts as part of the immune system.

Choroid (KO-royd) The vascular component of the eyeball.

Chromosomal abnormalities (kro-mo-SOM-al ab-nor-MAL-ih-teez) Abnormalities such as the deletion or translocation of chromosomes, or having a different number of chromosomes.

Chromosome (KRO-mo-some) Small structures in the cell nucleus consisting of the genetic material (DNA) and proteins (histones).

Chronic inflammation (CRON-ik in-fla-MAY-shun) A condition of inflamed tissues that develops over time or recur.

Chronic obstructive pulmonary disease (KRON-ik ob-STRUK-tiv PUL-mon-air-ee di-ZEEZ) A group of lung disorders in which obstruction limits airflow to the lungs. The most common types are chronic bronchitis and emphysema.

Chronic renal failure (CRF) (CRON-ik REE-nul FAYL-yur) A condition in which the kidneys slowly lose their ability to function. Also called chronic kidney failure (CKF).

Chronic tension headache (KRON-ik TEN-shun HED-ake) Pain in the head that occurs almost daily and persist for weeks or months.

Chylomicrons (ki-lo-MI-kronz) A type of lipoprotein transported by plasma.

Chyme (KIME) A mixture of food and gastric juices in the stomach.

Cilia (SILL-e-ah) Hairlike projections on the outside of the cell membrane.

Ciliary body (SILL-e-air-ee BOD-ee) A wedge-shaped thickening in the middle layer of the eyeball.

Circumduction (sir-kum-DUK-shun) Moving a body part in a circle.

Cirrhosis (ser-OH-sis) A chronic liver disease in which normal liver tissue is replaced with nonfunctional scar tissue.

Cisterna chyli (sis-TERN-uh KI-li) The dilated sac at the beginning of the thoracic duct that contains chyle (the milky fluid obtained from dietary lipids).

Cisternae (sis-TERN-ee) The stacked membranous sacs in a Golgi apparatus.

Clitoris (KLIT-oh-ris) An erectile organ of the female, homologous to the male penis.

Closed fracture (KLOZED FRAK-shur) A simple fracture, in which the skin remains intact.

***Clostridium* species (closs-TRID-e-um SPEE-seez)** Large, anaerobic, rod-shaped bacteria (oxygen is deadly to these bacteria).

***Clostridium tetani* (closs-TRID-e-um TET-uh-ni)** A rod-shaped, anaerobic bacterium of the genus *Clostridium* that causes tetanus.

Cluster headaches (KLUSS-tur HED-akes) Headaches that come in groups for a period of time and often appear seasonally.

Coagulation (ko-ag-yoo-LA-shun) The formation of a blood clot.

Coccus (KOK-us) A bacterial cell that is spherical or oval in shape.

Coccyx (KOK-siks) The tailbone—a small, triangularshaped bone made up of three to five fused vertebrae.

Cochlea (KOK-lee-uh) The spiral structure of the ear that contains hearing receptors.

Colectomy (ko-LEK-to-me) Surgery to remove the affected area of the colon.

Colitis (ko-LI-tus) Inflammation of the large intestine.

Collecting ducts (kuh-LEK-ting ducts) Ducts that collect urine as it is transported through the calyces.

Colorectal cancer (colo-RECK-tal KAN-sir) A tumor that arises from the lining of the rectum or colon.

Colostomy (ko-LOSS-tuh-me) Surgery in which the majority of the colon is removed and the opening to the outside of the body is moved to the abdomen.

Colostrum (kul-LOS-trum) Breast fluid (milk) that contains important antibodies and nutrients for an infant.

Comedos (KOM-e-doz) More commonly known as blackheads, collections of bacteria, dead epithelial cells, and dried sebum.

Common bile duct (KOM-mon bile duct) Delivers bile to the duodenum.

Compact bone (KOM-pakt bone) The denser of the two types of bone tissue.

Complementary and alternative medicine (CAM) (kom-pla-MEN-ta-ree and al-TER-na-tiv MED-ih-sin) A branch of medicine that uses herbal medicines, chiropractic, aromatherapy, and a variety of Eastern medicine practices.

Complex inheritance (KOM-plex in-HAIR-ih-tants) Inherited traits that are determined by multiple genes.

Compound (COM-pound) The combination of two or more atoms of more than one element.

Compound fracture (KOM-pound FRAK-shur) An open fracture; one in which the skin is broken.

Computerized tomography (CT) scan (kom-PU-ta-rized to-MAH-gruh-fee) A procedure that produces images utilizing a computer.

Conduction (con-DUCK-shun) The exchange of heat between two objects that are in direct contact with each other.

Condyloma acuminatum **(kon-dl-OH-muh uh-ku-min-AH-tum)** Sexually transmitted genital warts.

Cones (kohnz) Photoreceptors that detect blue, green, and red and provide sharper images than rods.

Confocal microscope (kon-FOK-ul MI-kro-skope) A microscope that uses laser light and computer-assisted image enhancement to give a three-dimensional image.

Congenital (kon-JEN-ih-tul) Present at birth.

Congestive heart failure (kon-JES-tiv heart failure) Failure of the heart to pump effectively.

Conjugation (kon-joo-GAY-shun) Transmission of genetic information from one bacterium to a second.

Conjunctiva (kon-junk-TIE-vah) The mucous membrane that lines the inner surface of the eyelids and the anterior surface of the eyeball.

Conjunctivitis (kon-junk-tiv-EYE-tis) Commonly called pink eye; highly contagious when the cause is bacterial in nature.

Connective tissue (kon-NECK-tiv TISS-oo) Tissue made up of living cells and a nonliving matrix.

Constipation (kon-stih-PA-shun) The condition of difficult defecation or elimination of feces.

Conventional medicine (kun-VEN-shun-al MED-ih-sin) Another term for allopathic medicine.

Convection (kon-VECK-shun) The transfer of heat by the movement of air or liquids with different temperatures.

Cornea (KOR-nee-uh) The transparent anterior portion of the fibrous tunic that helps focus light on the retina.

Coronal plane (Kuh-ROH-nul PLANE) The frontal anatomical plane; divides the body into anterior (frontal) and posterior (rear) portions.

Coronary sinus (KOR-o-nar-ee SI-nus) The collection site for blood returning to the heart from the coronary veins.

Corpus callosum (KOR-pus kal-LO-sum) The large connection between the right and left hemispheres of the brain.

Corpus luteum (KOR-pus LU-tee-um) A yellow body in the ovary formed when a secondary follicle has been discharged from the follicle.

Cortex (KOR-teks) The convoluted layer of gray matter over the cerebral hemispheres.

Cortisol (KOR-tih-sal) A glucocorticoid released by the adrenal cortex in response to stress to decrease inflammation, pain, protein synthesis, and tissue repair.

Costal (KAHS-tul) Rib cartilage.

Costochondritis (kahs-to-kon-DRI-tiss) Inflammation at the junction of the rib bone and sternum; also called Tietze's syndrome.

Covalent bonds (co-VALE-ent bondz) When two atoms share the same electrons.

CPAP (SEE-pap) Continuous positive airway pressure. A machine that uses a mask attached to a pump that forces air into the passageways while a person sleeps; may be used to treat obstructive sleep apnea.

Cramps (kramps) Painful, involuntary contractions of muscles.

Cranial (KRAY-nee-al) Above or close to the head.

Creatine phosphate (KREE-uh-teen FOSS-fate) A high-energy molecule that is stored in the muscle cells and can provide enough energy for about 15 seconds of muscle activity.

Cretinism (KRE-tin-izm) A congenital form of hypothyroidism that often causes mental and physical growth retardation.

Cribriform plate (KRIB-rih-form plate) The roof of the nasal cavity, through which the olfactory nerves pass on their way to the brain.

Cricoid cartilage (KRI-coyd KAR-tuh-lij) The third cartilage of the larynx, which forms most of the posterior wall of the larynx and part of the anterior wall.

Crista (KRIS-ta) A double membrane thrown into folds within the mitochondria.

Crohn's disease (KRONZ di-SEZ) A type of inflammatory bowel disease.

Cryosurgery (CRY-oh-sur-jur-ee) A procedure that uses freezing to kill cancer cells.

Cryptorchidism (kript-OR-kid-izm) A condition in which the testicles did not descend during infancy.

Curettage (kyur-eh-TAJ) A procedure using a sharp instrument to scoop out the cancerous spot.

Cyanosis (si-ah-NO-sis) A bluish discoloration.

Cystic duct (SIS-tic DUCT) Duct that carries bile into and out of the gallbladder.

Cystic fibrosis (SIS-tik fi-BRO-sis) A life-threatening disease that mainly affects the lungs and pancreas.

Cystitis (sis-TI-tis) A urinary bladder infection.

Cytogenetics (cy-toe-ge-NET-iks) The study of chromosomal abnormalities.

Cytokines (SI-toe-kines) Chemicals that are made by lymphocytes that aid in the destruction of antigens.

Cytokinesis (cy-toe-kin-EE-sis) Distribution of cytoplasm into two separate cells during cell division.

Cytology (cy-TOL-o-je) The branch of anatomy that studies cells.

Cytoplasm (SI-to-plazm) The portion of the cell between the cell membrane and the nucleus.

Cytosol (SI-to-sol) The fluid portion of the cytoplasm of a cell.

Cytotoxic T cells (si-toe-TOK-sik T cells) T lymphocytes that can destroy infected cells directly.

d

Decarboxylation (dee-kar-bok-seh-LAY-shun) A process in which a pyruvic acid molecule loses a CO_2 molecule.

Deciduous dentition (de-SID-yoo-us den-TI-shun) Baby teeth.

Decline phase (dee-KLINE faze) The last phase of the growth curve of bacteria caused by the death of cells exceeding the production of cells.

Decubitus ulcers (de-KU-bih-tus UL-sers) Also referred to as bedsores or pressure sores, sores seen in patients that are bedridden or wheelchair bound and are not turned or repositioned regularly.

Deep (DEEP) Away from the surface of the body or organ.

Defecation (def-uh-KAY-shun) The elimination of wastes, indigestible substances, unabsorbed substances, water, some cells, and bacteria from the GI tract.

Definitive diagnosis (duh-FIN-a-tiv di-ag-NO-sis) The diagnosis of which the clinician is most certain.

Deglutition (deg-lu-TISH-un) The act of swallowing.

Dendrite (DEN-drite) A neuronal cell process, or nerve fiber, that carries electrical signals toward the cell body.

Deoxyhemoglobin (dee-ok-see-HEE-mo-glo-bin) Hemoglobin that is not carrying oxygen.

Deoxyribonucleic acid (DNA) (de-ok-se-ri-bo-nu-KLE-ik A-sihd) A nucleic acid consisting of nucleotides made up of deoxyribose (a sugar), nitrogenous bases (adenine, guanine, cytosine, and thymine), and phosphate groups.

Depolarization (De-po-lar-i-ZA-shun) A reversal of charges at the cell membrane; a change in the resting potential of a cell that results in an action potential.

Depression (de-PREH-shun) Lowering a body part.

Dermal papillae (DER-mal pap-ILL-ee) The thickest part of the skin that is made up of dense, irregular connective tissue.

Dermatitis (der-muh-TI-tis) A general term for inflammation of the skin.

Dermatome (DER-ma-tome) A region of the skin supplied by a single spinal nerve.

Dermis (DER-mis) The bottom layer of the skin.

Descending colon (de-SEND-ing KO-lun) The part of the large intestine that runs down the left side of the abdomen.

Detrusor muscle (dee-TRU-zur MUS-ul) A muscle that contracts to push urine from the bladder into the urethra.

Diabetes mellitus (die-uh-BEE-teez mel-LIE-tus) A condition caused by hyposecretion of insulin that is categorized into type 1, insulin-dependent diabetes mellitus, and type 2, non-insulin-dependent diabetes mellitus.

Diagnosis (di-ag-NO-sis) The identification of a specific disease or condition.

Diapedesis (di-uh-peh-DEE-sis) The process of WBCs crossing the blood vessel wall into the interstitium.

Diaphoresis (di-uh-for-EE-sis) Profuse sweating.

Diaphragm (DI-a-fram) The dome-shaped muscle between the thoracic and abdominopelvic cavities; a major muscle of respiration.

Diaphysis (die-AH-fih-sis) Shaft of a long bone.

Diarrhea (di-uh-REE-uh) The condition of watery and frequent feces.

Diastolic pressure (di-uh-STOL-ik PREH-shur) Blood pressure in the arteries at its lowest point.

Diencephalon (di-en-SEF-a-lon) A part of the brain consisting of the thalamus and hypothalamus.

Differential diagnosis (dif-er-EN-shul die-ag-NO-sis) A ranked list of possible diagnoses based on the information available.

Differentiation (diff-ur-en-te-AY-shun) The process of cells becoming specialized.

Diffusion (dif-FU-zyun) Movement of molecules or ions from an area of higher concentration to an area of lower concentration with no energy required.

Digestion (di-JEST-yun) The mechanical and chemical processes that break down food into small molecules that can be absorbed.

Diplopia (dih-PLO-pe-uh) Double vision.

Dysphonia (dis-FON-ee-uh) A hoarse voice.

Distal (DISS-tal) Further from the attachment of an extremity to the trunk.

Distal convoluted tubules (DISS-tal kon-vo-LOO-ted TOO-byoolz) Tubules in a nephron that extend from the loop of Henle to the collecting ducts.

Diverticulitis (di-ver-tik-u-LI-tis) Inflammation of diverticuli in the intestine.

Diverticulosis (di-ver-tik-u-LOH-sis) Abnormal dilations or pouches in the intestinal wall.

DNA See *Deoxyribonucleic acid*.

Dopamine (DO-pah-meen) A chemical messenger allowing communication between the substantia nigra and the basal ganglia of the brain.

Dorsal (DOOR-sal) Toward the back of the body.

Down syndrome (DOWN SIN-drohm) A congenital disorder due to an extra chromosome 21.

Drug abuse The use of any substance that is not according to accepted medical, social, or legal purposes.

Dry eye syndrome (Dry eye SIN-drome) Also called xerophthalmia, a syndrome that results from a decreased production of the oil within tears.

Ductus arteriosus (DUK-tus ar-tee-ree-OH-sus) An opening in the fetal heart between the pulmonary trunk and the aorta that allows blood to bypass the lungs.

Ductus venosus (DUK-tus vee-NO-sus) A fetal blood vessel that allows blood to bypass the liver.

Duodenum (du-o-DEE-num) The first section of the small intestine that receives secretions from the pancreas to aid in digestion.

Dysmenorrhea (dis-men-or-REE-uh) The condition of experiencing severe menstrual cramps that limit normal daily activities.

Dysphagia (dis-FAJ-ee-uh) Difficulty swallowing.

Dysplasia (dys-PLAY-ze-uh) Disordered growth and maturation of cells.

Dyspnea (DISP-nee-uh) Difficulty breathing.

Dysuria (diss-YUR-ee-uh) Painful urination.

Eccrine (EK-rin) A type of sweat gland.

Ectoderm (EK-toe-derm) The primary germ layer that gives rise to the nervous system, epidermis, and its derivatives.

Ectropion (ek-TRO-pe-on) Eversion of the lower eyelid.

Eczema (EK-zih-ma) A very common chronic dermatitis that often has acute phases or flareups followed by periods of remission from flares.

Edema (eh-DEE-ma) The accumulation of excess fluid in the spaces between cells.

Effacement (ee-FACE-ment) The thinning and softening of the cervix during the birth process.

Effector (ef-FEC-tor) Muscle or gland that responds to a stimulus.

Efferent (motor) neurons (EF-fer-ent NOO-ronz) A neuron that carries nerve impulses away from the central nervous system.

Efferent arteriole (EF-fer-ent ar-TEE-ree-ole) A renal vessel that exits the glomerulus to form a peritubular capillary.

Elastic connective tissue (e-LAS-tik kon-NEK-tiv TISH-oo) Made up mostly of elastic fibers that are yellow in color.

Electrodessication (e-lek-tro-dess-uh-CAY-shun) A procedure that uses electrical currents to minimize bleeding as well as to kill cancer cells.

Electroencephalogram (EEG) (e-lek-tro-en-CEPH-ah-luh-gram) A test that detects electrical activity in the brain.

Electron (e-LECK-tron) Negatively charged subatomic particle.

Electron beam computerized tomography (EBCT) (ee-LEK-tron beam kom-PYOO-tur-izd to-MOG-ruh-fee) Similar to a CT scan of the arteries; used to find narrowing of the arteries.

Electronegative atom (ee-lek-tro-NEG-uh-tiv A-tom) An atom with an overall negative charge, such as oxygen, fluorine, or nitrogen.

Electron microscope (ee-LEK-tron MI-kro-skope) A microscope that uses electrons rather than visible light to view objects.

Electron transport chain (ee-LEK-tron TRANZ-port chane) An aerobic series of electron transfers that occurs on the inner membrane of the mitochondria, resulting in the production of 32 ATP molecules for every glucose molecule that is oxidized.

Element (EL-uh-ment) Another term for atom.

Elevation (el-uh-VA-shun) Lifting a body part.

Embolism (EM-bo-lizm) An obstruction caused by the movement of a substance through arterial or venous circulations.

Embolization (em-bo-li-ZA-shun) The process in which a thrombus breaks away from its location on the lumen of a blood vessel and moves freely through the vessel.

Emphysema (em-fih-SE-ma) A lung disease in which the alveolar walls are destroyed; often caused by cigarette smoking.

Empyema (em-pi-EE-muh) A buildup of pus in the lungs due to infective processes.

Endemic (en-DEM-ik) Very common.

Endocarditis (en-do-kar-DI-tis) Inflammation of the innermost lining of the heart and heart valves.

Endocardium (en-do-KAR-de-um) Endothelium and smooth muscle that lines the inner layer of the heart, the heart valves, and the tendons that hold the valves open.

Endochondral (en-doe-KON-dral) Describing a type of ossification in which bones start out as cartilage models.

Endocytosis (en-doe-cy-TOE-sis) A process where energy is required and where part of the cell membrane encloses a substance outside the cell forming a vesicle and then invaginates, bringing the substance into the cell interior.

Endoderm (EN-do-derm) A primary germ layer of the developing embryo that gives rise to the gastrointestinal tract, the urinary tract, and the respiratory tract.

Endolymph (EN-do-limf) Fluid in the semicircular canals and vestibule that moves, activating equilibrium receptors as well as hearing receptors.

Endometrial cancer (en-do-MEE-tree-ul KAN-ser) See *Uterine cancer*.

Endometriosis (en-do-me-tree-OH-sis) A condition in which tissues that make up the lining of the uterus grow outside the uterus.

Endometrium (en-do-MEE-tree-um) The mucous membrane lining the uterus.

Endomysium (en-do-MIZ-ee-um) The connective tissue surrounding an individual muscle cell.

Endoplasmic reticulum (en-do-PLAS-mik reh-TIK-yoo-lum) Membranes in the cellular cytoplasm.

Endosteum (en-DOS-tee-um) A membrane lining the central canal of a bone and the holes of cancellous bone containing the bone-forming osteoblasts.

Endotracheal tube (en-do-TRAY-kee-ul toob) A tube inserted through the mouth into the trachea to assist in delivering oxygen to the lungs.

Entropion (en-TRO-pe-on) Inversion of the lower eyelid.

Eosinophil (ee-o-SIN-o-fil) A type of white blood cell that has a granular cytoplasm and is involved in allergic reactions and parasitic infections.

Ependymal cells (eh-PEN-dih-mal cells) Cells lining the ventricles of the brain that produce cerebrospinal fluid.

Epicardium (ep-i-KAR-de-um) The thin outer layer of the heart wall, also known as the visceral pericardium.

Epidemic (eh-pi-DIM-ik) Affecting many people in a region simultaneously.

Epidermis (ep-ih-DER-mis) The top layer of the skin.

Epididymis (ep-ih-DID-ih-mis) A coiled tube at the top of each testis where spermatids become mature sperm cells.

Epididymitis (ep-ih-did-ih-MI-tis) Inflammation of the epididymis.

Epiglottic cartilage (eh-pi-GLOT-ik KAR-tuh-lij) The cartilage that forms the framework of the epiglottis.

Epiglottis (eh-pi-GLOT-iss) The flaplike structure that closes off the larynx during swallowing so that food and liquids do not enter the respiratory system.

Epilepsy (EP-ih-lep-see) A condition in which the brain experiences repeated spontaneous seizures due to abnormal electrical activity of the brain.

Epimysium (ep-ih-MIZ-ee-um) Dense, irregular connective tissue that surrounds an entire muscle.

Epinephrine (ep-ih-NEF-rin) An amino acid hormone secreted by the adrenal medulla that increases heart rate, breathing rate, and blood pressure in response to physical or emotional stress. Ephinephrine may also be injected to treat a person experiencing anaphylaxis to constrict the blood vessels and dilate the bronchi.

Epiphysis (eh-PIH-fih-sis) The end of a long bone.

Episiotomy (e-piz-e-OT-o-me) A surgical incision to avoid tearing of the perineum during labor.

Episodic tension headaches (eh-pih-SOD-ik TEN-shun HED-akes) A headache that is usually the result of temporary stress or muscle tension.

Equatorial plate (ek-wa-TO-re-ul PLATE) Another term for the metaphase plate.

Erectile dysfunction (ee-REK-tile dis-FUNK-shun) See *Impotence*.

Erosion (e-RO-shun) An ulcer is an example of erosion where there is an excavation of the skin.

Erythroblastosis fetalis (ee-rith-roh-blas-TOH-sis fe-TAL-iss) Another term for hemolytic disease of the newborn (HDN).

Erythrocyte (e-RITH-ro-site) A mature red blood cell.

Erythropoiesis (e-rith-ro-poy-EE-sis) The formation of red blood cells.

Erythropoietin (e-rith-ro-POY-eh-tin) The hormone responsible for regulating erythropoiesis.

Esophageal hiatus (eh-sof-uh-JEE-ul hi-ATE-us) Opening in the diaphragm through which the esophagus passes.

Essential hypertension (eh-SIN-shul hi-per-TEN-shun) Hypertension that has no identifiable cause; also called idiopathic or primary hypertension.

Estrogen (ESS-tro-jen) A female hormone responsible for female secondary sex characteristics.

Ethmoid bone (ETH-moyd bone) An unpaired bone located between the sphenoid bone and the nasal bones.

Ethmoid sinus (ETH-moyd SI-nus) The paranasal sinus located in the ethmoid bone.

Etiology (e-te-OL-o-je) Study of the cause of a disease.

Eukaryotes (you-KARE-ee-otes) An organism that contains cells that have nuclei.

Eukaryotic cells (you-kar-ee-AHT-ik cells) Cells that have nuclei.

Eustachian tube (yoo-STAY-shun toob) A pathway between the throat and the middle ear that allows organisms to pass between them.

Evaporation (e-vap-o-RA-shun) The loss of water (and heat) through the skin and mucous membranes.

Eversion (ee-VER-zyun) Turning the sole of the foot laterally (outward) so that the soles of the feet are pointing away from each other.

Excoriation (ex-kor-e-A-shun) An abrasion of the skin.

Exhalation (eks-ha-LA-shun) Breathing out.

Exocytosis (ex-o-si-TO-sis) The opposite of endocytosis.

Exophthalmos (ek-sof-THAL-mos) A condition in which the eyeballs protrude, often caused by hyperthyroidism.

Exponential phase (eks-po-NEN-shul faze) The second phase of the growth curve of bacteria in which there is very rapid growth.

Extension (ek-STEN-shun) Straightening a body part or increasing the angle of a joint.

External nares (eks-TERN-ul NAIR-eez) The external openings of the nose; the nostrils.

External respiration (eks-TERN-ul res-pih-RA-shun) The exchange of oxygen and carbon dioxide between the alveoli and the capillaries.

Exudate (EKS-yu-date) Fluid from the lesions.

f

Fascicle (FAS-ih-kul) A group of 10 to 100 muscle fibers.

Facilitated diffusion (fa-SIL-ih-ta-ted dih-FYU-zion) Diffusion process in which small molecules that are not lipid soluble cross the plasma membrane with the help of a carrier molecule.

Falciform ligament (FALS-ih-form LIG-a-mint) A fold of mesentery in the liver.

Fallopian tube (fal-LOP-e-an toob) An oviduct, the tube that runs from the uterus to the ovary.

False ribs (fals ribz) Their costal cartilages attach indirectly to the sternum or do not attach at all.

False vocal cords (FALS VO-kul cords) The upper vestibular folds in the larynx; so-called because they do not produce sound.

Fascia (FAY-shuh) Connective tissue located just below the skin that helps support and hold together muscles, bones, nerves, and blood vessels.

Febrile (FEE-brile) Having a fever or elevated body temperature.

Femoral (FEM-oh-rul) Referring to the thigh.

Femur (FEE-mur) Thigh bone, the longest and heaviest bone in the body.

Fertilization (fer-til-ih-ZA-shun) The process of a sperm cell uniting with an ovum.

Fibrinogen (fi-BRIN-o-jin) A protein that is important in blood clotting.

Fibroblasts (FI-bro-blasts) Immature cells responsible for making the fibers and matrix.

Fibrocystic breast change (fi-bro-SIS-tik brest chanj) The presence of abnormal cysts in the breasts that vary in size related to the menstrual cycle; formerly called fibrocystic breast disease.

Fibrocytes (FI-bro-sites) A mature, less active form of a fibroblast.

Fibroids (FI-broydz) Benign (noncancerous) tumors that grow in the uterine wall.

Fibromyalgia (fi-bro-mi-AL-jee-uh) A condition that results in chronic pain primarily in joints, muscles, and tendons.

Fibrosis (fi-BRO-sis) Scar formation.

Fibrous joints (FI-brus joints) The bones of fibrous joints are connected together with short connective tissue fibers abundant with collagen that do not allow much, if any, movement.

Filtration (fil-TRA-shun) The flow of a liquid through a filter due to hydrostatic pressure; no energy is required.

Fimbriae (FIM-bree-ee) Fingerlike structures near the lateral ends of the fallopian tubes.

Flagella (flah-JEL-lah) Structures on microorganisms used for locomotion; similar to cilia, but are longer.

Flexion (FLEK-shun) Bending a body part or decreasing the angle of a joint.

Floating ribs (FLOH-ting ribz) Rib pairs 11 and 12, which do not attach anteriorly to the sternum or any other structure.

Fluid loss (FLU-id LOSS) A process that results in not having enough fluid in the body.

Fluid overload (FLU-id OH-ver-lode) An abnormal condition in which the body retains too much fluid.

Flu salute (FLOO sa-LOOT) Coughing or sneezing into the upper sleeve to help prevent the spread of pathogens.

Follicle (FOL-lih-kul) A vesicle on the ovary that contains a developing egg surrounded by a covering of cells.

Follicle stimulating hormone (FSH) (FOL-lih-kul STIM-yoo-la-ting HOR-mone) A hormone that stimulates the ovaries to produce ova and estrogen in females, and stimulates the testes to produce sperm and testosterone in males.

Follicular cells (fo-LIK-yoo-lar cells) Cells in the thyroid gland that are responsible for the production and secretion of thyroid hormones.

Folliculitis (fol-lik-u-LI-tis) Inflammation of hair follicles.

Fomites (FO-mites) Contaminated surfaces and objects that can transmit viruses to a susceptible host.

Fontanelles (fon-tan-ELZ) "Soft spots" on an infant's skull that are tough membranes that have not yet ossified.

Foramen magnum (fo-RA-men MAG-nem) The large opening at the base of the occipital bone.

Foramen ovale (fo-RA-men o-VA-lee) An opening in the fetal heart that is located between the right and left atria, allowing blood to bypass the lungs.

Fossa ovalis (FOS-a o-VAL-is) A depression on the interatrial septum that is formed when the foremen ovale closes.

Fovea centralis (FOH-vee-uh sen-TRAL-is) The small depression in the center of the macula lutea that contains only cones and thus is the area of greatest visual acuity.

Fragile X syndrome (FRAJ-ul X SIN-drome) A disorder in which one of the genes on the X chromosome is defective, making the chromosome susceptible to breakage; one of the most common inherited causes of learning disability.

Frenulum (FREN-you-lum) A fold of the mucous membrane that attaches two structures together.

Frontal bone (FRON-til bone) A single bone that forms the anterior portion of the cranium; the bone of the forehead.

Frontal lobes (FRON-til lobes) The lobes of the brain that are responsible for reasoning, judgment, and problem solving and also control movements of voluntary skeletal muscles.

Frontal plane (FRON-til PLANE) Anatomic plane that divides the body into anterior (frontal) and posterior (rear) portions.

Frontal sinus (FRON-til SI-nus) The paranasal sinus located in the frontal bone.

Fulgurated (FUL-gur-a-tid) Burned and sealed, as with a vasectomy.

Fundus (FUN-dus) Portion of the stomach that is superior to the cardiac region; the part of the uterus farthest from the os.

Furuncle (FYUR-un-kul) Inflammation of a single hair follicle.

g

Ganglion (GANG-gle-on) A group of nerve cell bodies outside the central nervous system.

Gastric juice (GAS-trik JUS) Secretions of the stomach lining.

Gastric phase (GAS-trik FAZE) Part of digestion that is under the regulation of neural and hormonal control.

Gastric ulcer (GAS-trik UL-ser) A stomach ulcer.

Gastritis (gas-TRI-tis) An inflammation of the stomach lining.

Gastroesophageal reflux disease (GERD) (ga-stro-e-sof-a-JEE-al RE-flux di-ZEZ) Occurs when stomach acids are pushed into the esophagus; also called heartburn.

Gastrula (GA-stroo-la) An embryo at the stage following the blastula, consisting of a hollow, two-layered sac of ectoderm and endoderm.

Gene (jeen) A segment of DNA that determines a body trait found in the human body.

General stress syndrome (JEN-er-al STRES SIN-drome) The body's physiologic response to stress; consists of a collection of reactions primarily caused by the release of specific hormones.

Genotype (JEEN-oh-tipe) The genetic makeup of an individual.

Genus (JEEN-us) A group of species exhibiting similar characteristics.

Gigantism (jie-GAN-tizm) An abnormal increase in the length of long bones due to hypersecretion of growth hormone during childhood.

***Giardia lamblia* (JAR-dee-a LAM-blee-a)** A common pathogenic protozoan found in the intestinal tract of humans.

Gingiva (JIN-jih-vah) Gums.

Glaucoma (glaw-KO-muh) An increase in intraocular pressure, caused by a buildup of aqueous humor in the anterior chamber; if untreated, this excess pressure can lead to permanent damage of the optic nerve that can result in blindness.

Glioma (glee-OH-muh) The most common type of brain tumor, which arises from neuroglial cells such as astrocytes.

Goblet cells (GOB-let cells) Ciliated cells in the nasal cavity that secrete mucus.

Globulins (GLOB-yoo-linz) Antibodies; proteins produced by the liver, and by plasma cells in the blood, that work as part of the immune system.

Glomerulonephritis (glo-mer-yoo-lo-neh-FRI-tis) An inflammation of the glomeruli of the kidney.

Glottis (GLOT-is) The true vocal cords and the space between them.

Glucagon (GLU-kuh-gon) A hormone excreted by the endocrine pancreas that increases blood glucose levels.

Gluconeogenesis (glu-koh-nee-oh-JEH-neh-sis) The formation of glucose from proteins and fats.

Glucose (GLU-kose) The most common carbohydrate.

Glycogenolysis (gli-ko-jeh-NAH-lih-sis) The conversion of glycogen stored in the liver and skeletal muscle into glucose.

Glycolysis (gli-KOH-lih-sis) The metabolic breakdown of glucose and other sugars that releases energy in the form of ATP; the net production of two ATP molecules through an anerobic pathway.

Goblet cell (GOB-let cell) A type of epithelial cell that secretes mucus.

Gold standard (GOLD STAN-dard) Best test to diagnose a specific problem.

Golgi apparatus (GOL-je ap-pa-RAH-tus) Part of the cell that synthesizes carbohydrates and prepares and stores secretions for discharge from the cell.

Gonads (GO-nadz) The general term for the male and female reproductive organs.

Gout (gouty arthritis) (GOWT) A type of arthritis associated with high uric acid levels in the blood with crystalline deposits in the joints, kidneys, and various soft tissues.

Grading (GRA-ding) Examining the nucleus of a cancer cell to see how it differs from a normal cell.

Gram stain (gram stane) Method of staining that differentiates bacteria by the chemical composition of the cell walls.

Granulocytes (GRAN-yoo-lo-sites) White blood cells that have granules in their cytoplasm.

Greenstick fracture (GREEN-stik FRAK-shur) Fracture in a bone that is not completely ossified, as in children's bones; the bone bends instead of breaking completely.

Guillain-Barré syndrome (gue-LAN ba-RAY SIN-drome) A disorder in which the body's immune system attacks part of the peripheral nervous system.

Gynecomastia (guy-neh-ko-MASS-tee-uh) Breast development in males.

Gyri (JI-ri) Folds of gray matter on the surface of the brain.

h

***Haemophilus influenza* (he-MOF-i-lis in-flu-ENZ-uh)** A rod-shaped Gram-negative bacterium found on the mucous membranes of the upper respiratory tract; causes respiratory infections.

Haptens (HAP-tenz) Foreign substances that are too small to elicit an immune response by themselves, but can combine with larger proteins in the blood to elicit an immune response.

Headache (HED-ake) Pain in the head, caused by dilation of cerebral arteries, muscle contraction, insufficient oxygen in the cerebral blood, reaction to drugs, or other causes. The medical term is cephalgia.

***Helicobacter pylori* (he-lik-o-BAK-tur pi-LO-ree)** A spiral-shaped Gram-negative rod associated with the formation of duodenal (intestinal) and gastric (stomach) ulcers and even gastric cancer.

Helper T cells (HELP-ur T cells) T lymphocytes that increase antibody formation, memory cell formation, B-cell formation, and phagocytosis.

Hematocrit (he-MAT-o-krit) The percentage of blood made up of red blood cells.

Hematopoiesis (hee-ma-toh-poy-EE-sis) The production of blood cells.

Hematuria (hee-ma-TUR-ee-uh) A condition of blood in the urine.

Hemodynamics (he-mo-di-NAM-iks) The study of blood flow through the heart and blood vessels.

Hemoglobin (HE-mo-glo-bin) An iron-containing pigment on red blood cells responsible for transporting most of the oxygen and some of the carbon dioxide in the body.

Hemolytic anemia (hee-mo-LIT-ik uh-NEE-mee-uh) Anemia in which red blood cells are destroyed faster than they can be produced.

Hemolytic disease of the newborn (HDN) (hee-mo-LIH-tik di-ZEEZ of the NU-born) A disease that occurs when an Rh-negative female has an Rh-positive child; the Rh-positive fetal blood mixes with the mother's Rh-negative blood, producing antibodies against the fetal RBCs. If the mother has a second Rh-positive child, the mother's antibodies may cross the placenta and attack the second fetus's RBCs.

Hemophilia (hee-mo-FIL-e-a) A group of inheritable bleeding disorders.

Hemoptysis (hee-MOP-tih-sis) Coughing up blood.

Hemorrhoids (HEM-uh-roydz) Varicosities (varicose veins) of the rectum or anus.

Hemorrhagic stroke (hem-o-RAJ-ik stroke) A cerebrovascular accident in which blood seeps into the surrounding brain tissue, often due to a ruptured aneurysm in a blood vessel in the brain.

Hemostasis (he-mo-STA-sis) Stoppage of blood flow by natural or artificial means.

Hemothorax (hee-mo-THOR-aks) A buildup of blood in the pleural cavity.

Hepatic duct (ha-PAT-ik DUCT) A duct conveying the bile away from the liver that unites with the cystic duct to form the common bile duct.

hepatitis (hep-uh-TI-tis) Inflammation of the liver.

Hepatitis A (hep-uh-TI-tis A) Caused by the hepatitis A virus (HAV); hepatitis A is spread mainly through the fecal-oral route.

Hepatitis B (hep-uh-TI-tis B) The most common bloodborne hazard heath care professionals face; spread through contact with contaminated blood or body fluids and through sexual contact.

Hepatitis C (hep-uh-TI-tis C) Also referred to as non-A/non-B; spread through contact with contaminated blood or body fluids and sexual contact so is a risk for health care professionals as well; there is no cure for this variant.

Hepatitis D (delta agent hepatitis) (hep-uh-TI-tis D) Occurs only in people infected with HBV; delta agent infection may make the symptoms of hepatitis B more severe, and it is associated with liver cancer.

Hepatitis E (hep-uh-TI-tis E) Caused by hepatitis E virus (HEV); hepatitis E is transmitted by the fecal-oral route, usually through contaminated water.

Hepatocytes (he-PAT-o-sites) Cells in the liver that process the nutrients in blood and make bile.

Hepatomegaly (heh-pat-o-MEG-uh-lee) Enlarged liver.

Hernia (HER-nee-uh) Develops when an organ pushes through a wall that contains it.

Herpes simplex (HER-peez SIM-plex) A viral infection that causes oral blisters or genital blisters.

Herpes zoster (HER-peez ZAH-stir) An inflammatory condition of the skin caused by the varicella zoster virus more commonly known as shingles.

Hiatal hernia (hi-ATE-ul HER-nee-uh) Occurs when a portion of the stomach protrudes into the thoracic cavity through an opening (esophageal hiatus) in the diaphragm.

High-density lipoproteins (HDLs) (HI DEN-si-tee li-po-PRO-teens) Contain the highest percentage of proteins and transport cholesterol to the liver for elimination; called "good" cholesterol.

Hilum (HI-lum) The concave depression of the kidney where the renal artery, renal vein, and ureter enter.

Histamine (HISS-ta-men) Chemical found in mast cells, basophils, and plasma cells; when released, causes vasodilation, increased vascular permeability, and constriction of bronchioles.

Histology (his-TOL-ogy) The branch of anatomy that studies tissues.

Histones (HISS-tones) Dispersed or uncoiled DNA and protein.

***Histoplasma capsulatum* (hiss-to-PLAZ-muh kap-su-LAH-tum)** A saprophyte and the most common cause of fungal pulmonary infections in humans and animals.

Homan's sign (HO-manz sine) Pain in the calf or popliteal region when the foot is dorsiflexed with the knee flexed to 90 degrees.

Homeostasis (ho-me-o-STAY-sis) Relative consistency of the body's internal environment.

Homologous chromosomes (ho-MAH-luh-gus KRO-mo-sohmz) Chromosomes that carry all the genes that code for a particular trait.

Hormone (HORE-mone) A chemical secretion that acts as a messenger and has regulatory effects on other cells.

Host factors (HOST FAK-torz) Factors such as physical barriers, inherited traits, and age that have a role in infection.

Human chorionic gonadotropin (hCG) (HYOO-man ko-ree-ON-ik go-nah-do-TRO-pin) A hormone produced by the placenta that maintains the corpus luteum.

Human immunodeficiency virus (HIV) (HU-man ih-myu-no-de-FISH-in-see VI-russ) Either of two retroviruses that infect and destroy helper T cells of the immune system causing the marked reduction in their numbers that is diagnostic of AIDS.

Human leukocyte antigens (HLA) (HYOO-man LOO-ko-site AN-ti-jenz) Surface proteins on white blood cells and other nucleated cells used to type tissues for transplantation; encoded by genes of the major histocompatibility complex in humans.

Humoral immunity (HU-mor-al i-MU-ni-te) Immune response in which plasma cells are derived from B lymphocytes.

Humors (HU-morz) Fluids of the body.

Humerus (HU-mer-us) The upper arm bone.

Hyaline membrane disease (HI-uh-lin MEM-brane dih-ZEEZ) See *Respiratory distress syndrome (RDS)*.

Hydrogen bond (HI-dro-jin bond) An attractive intermolecular force between two partial electric charges of opposite polarity.

Hydrothorax (hi-dro-THOR-aks) A buildup of fluid in the pleural cavity.

Hypercalcemia (hi-per-kal-SEE-mee-uh) An increase in blood calcium levels.

Hypercapnia (hi-per-KAP-nee-uh) An arterial blood carbon dioxide concentration greater than normal.

Hyperchloremia (hi-per-klor-EE-me-uh) A plasma chloride concentration greater than normal.

Hyperchromatic (hi-per-kro-MAT-ik) Having a nucleus with a color that is darker than normal.

Hyperextension (hi-per-eks-TEN-shun) Extending a body part past the normal anatomical position.

Hyperkalemia (hi-per-ka-LEE-me-uh) A plasma potassium concentration that is greater than normal.

Hypermagnesemia (hi-per-mag-ne-SEE-me-uh) A plasma magnesium concentration greater than normal.

Hypernatremia (hi-per-na-TRE-me-uh) A plasma sodium concentration greater than normal.

Hyperopia (hi-per-O-pe-uh) Refractive error in close vision.

Hyperphosphatemia (hi-per-fos-fay-TEE-me-uh) A plasma phosphate concentration greater than normal.

Hyperplasia (hi-per-PLAY-zhee-a) An increase in the number of cells in an organ or tissue.

Hyperpolarization (hi-per-pol-a-ri-ZA-shun) Increase in the potential difference across a membrane.

Hyperreflexia (hi-per-re-FLEK-se-a) A stronger than normal reflex.

Hypertension (hi-per-TEN-shun) High blood pressure.

Hyperthyroidism (hy-per-THY-royd-izm) An excess secretion of thyroid hormone.

Hypertrophy (hi-PER-tro-fee) An increase in the size of an organ with physiologic or pathologic origin.

Hypocalcemia (hi-poh-kal-SEE-mee-uh) A condition in which there is too little calcium in the blood.

Hypochloremia (hi-po-klor-EE-me-uh) A plasma chloride concentration less than normal.

Hypodermis (hi-po-DERM-is) A layer of support tissue below the dermis.

Hypokalemia (hi-po-ka-LEE-me-uh) A plasma potassium concentration less than normal.

Hypomagnesemia (hi-po-mag-ne-SEE-me-uh) A plasma magnesium concentration less than normal.

Hyponatremia (hi-po-na-TRE-me-uh) A sodium plasma concentration less than 135 milliequivalent per liter.

Hypoperfusion (hi-po-per-FYU-zyun) Too little blood flow.

Hypophosphatemia (hi-po-foss-fa-TEE-me-uh) A plasma phosphate concentration less than normal.

Hyporeflexia (hi-po-re-FLEK-se-a) A decreased reflex.

Hypothalamus (hi-po-THAL-a-mus) A part of the brain that maintains homeostasis by regulating many vital activities such as heart rate (pulse), blood pressure (BP), and respiration (breathing) rate.

Hypothyroidism (hi-po-THI-royd-izm) A condition in which there is an undersecretion of thyroid hormone.

Hypovolemic shock (hi-po-vo-LE-mik SHOK) Occurs when there is a dramatic decrease in blood volume.

Hypoxia (hi-POX-ia) Inadequate oxygenation.

Hysterectomy (his-tuh-REK-to-me) The removal of the uterus.

i

Iatrogenic (i-at-ro-JEN-ik) Describing unintended injuries caused by doctors and other health care workers and by prescribed drugs or therapies.

Idiopathic (id-e-o-PATH-ik) Of unknown cause.

Ideopathic hypertension (ih-dee-o-PATH-ik hi-per-TEN-shun) Hypertension that has no identifiable cause; also called essential or primary hypertension.

Ileocecal valve (il-ee-oh-SEE-kul valv) The valve that joins the last segment of the small intestine (ileum) with the first segment of the large intestine (cecum).

Ileum (IL-ee-um) The last and longest section of the small intestine.

Immune system (ih-MYUN SIS-tem) The body system that helps protect the body from organisms in the environment, such as bacteria, viruses, fungi, toxins, parasites, and cancer.

Immunity (i-MU-ni-te) The ability to resist or overcome the effects of diseases or harmful materials; being resistant to injury and infection by pathogens.

Immunopathology (i-myu-no-path-OL-o-gee) Pathology of the immune system.

Impacted fracture (im-PAK-ted FRAK-shur) A fracture in which one end of the fractured bone is driven into the interior of the other.

Impetigo (im-peh-TI-go) A *Staphylococcus* or *Streptococcus* infection found more commonly in children that causes lesions on the face or lower leg.

Impotence (IM-po-tints) Also called erectile dysfunction (ED); a disorder in which a male cannot achieve or maintain an erection to complete sexual intercourse.

Incisors (in-SIZE-orz) The most medial teeth; used to cut off food pieces.

Incontinence (in-KON-ti-nents) A condition in which an adult cannot control urination.

Incus (ING-kus) The anvil-shaped bone of the middle ear.

Infarction (in-FARCT-shun) The death or necrosis of tissue distal to the occlusion of an artery.

Infection (in-FEK-shun) The presence of a pathogen in or on the body.

Inferior (in-FEER-e-ur) Below, lower than.

Inferior nasal conchae (in-FER-e-ur NA-zal KONG-kee) Paired bones that are separate from the ethmoid bone and its middle nasal conchae.

Inflammation (in-flah-MAY-shun) Response of tissue to injury characterized by redness, swelling, and pain.

Influenza (in-floo-IN-zuh) An infection of both the upper and lower respiratory tracts; commonly called "the flu."

Infundibulum (in-fun-DIB-u-lum) Fringed, expanded end of a fallopian tube near an ovary.

Ingestion (in-JES-chun) The process of eating and drinking.

Inguinal hernia (IN-gwi-nal HER-ne-uh) Occurs when a portion of the large intestine protrudes into the inguinal canal, which is located where the thigh and the body trunk meet.

Inhalation (in-ha-LA-shun) Breathing in.

Innate (nonspecific) immunity (in-NATE i-MU-ni-te) Immunity a person is born with.

Inorganic matter (in-or-GAN-ik MAT-ur) Generally does not contain carbon and hydrogen; these molecules tend to be smaller than organic molecules; consisting of nonliving materials.

Inorganic salts (in-or-GAN-ik SALTS) The fourth major category of inorganic substances that are important to life.

Insertion (in-SER-shun) The site of attachment of a muscle to the more movable bone during muscle contraction.

Integrative medicine (IN-te-gra-tive MED-i-sin) An approach that attempts to combine aspects of conventional medicine with complementary and alternative medicine (CAM).

Interatrial septum (in-ter-A-tre-ul SEP-tum) The septum or wall that separates the left and right atria of the heart.

Intercalated disc (in-TER-kal-a-ted DISK) A specialized structure in cardiac muscle that helps the muscle cells contract as a single unit.

Interferon (in-tur-FEER-on) A chemical in the blood that blocks viruses from infecting cells.

Interlobar artery (in-ter-LO-bar AR-ter-ee) A branch of the renal artery.

Intermittent positive-pressure breathing (IPPB) (in-ter-MIT-ent POZ-ih-tiv PRESH-ur BREETH-ing) A treatment for chronic atelectasis.

Interneuron (in-ter-NOO-ron) A nerve cell that conveys impulses from one nerve cell to another.

Interphase (IN-ter-faze) The period between cell divisions when the cell is involved in growth, metabolism, and preparation for cell division.

Interstitial cells (in-ter-STISH-al cells) Produce the hormone testosterone.

Interstitial (tissue) fluid (in-ter-STISH-al FLOO-id) The fluid found between the cells.

Interstitium (in-ter-STISH-um) Spaces between cells.

Interventricular septum (in-ter-ven-TRIK-yoo-lar SEP-tum) The septum or wall that separates the left and right ventricles of the heart.

Intestinal phase (in-TES-ti-nal FAZE) Phase of digestion that inhibits gastric emptying to allow the small intestine time to absorb the nutrients that enter it.

Intrinsic factor (in-TRIN-sik FAK-tor) A chemical that is secreted by the parietal cells of the stomach and is necessary for vitamin B_{12} absorption.

Involuntary muscle (in-VOL-un-tar-ee MUS-ul) Another term for smooth muscle.

Intrinsic factor (in-TRINS-ik FAK-tor) A substance secreted by parietal cells, which is necessary for vitamin B_{12} absorption.

Inversion (in-VER-zyun) Turning the sole of the foot medially (inward) so that the soles of the feet can touch each other.

Involuntary muscle (in-VOL-un-tar-ee MUS-ul Another term for smooth muscle.

Involute (in-voh-LOOT) Turn in on itself.

Ion (I-on) An atom or group of atoms with a net positive or net negative charge.

Iron deficiency anemia (IRN de-FISH-en-see uh-NEE-mee-uh) The most common type of anemia; people with low iron levels often have correspondingly low hemoglobin levels.

Islets of Langerhans (EYE-lets of LAHNG-er-hanz) Groups of cells that form the endocrine pancreas and are made of glucagon-secreting alpha cells and insulin-secreting beta cells.

Isograft (I-so-graft) Transplant in which the tissue or organ is taken from an identical twin.

Isotopes (I-so-topes) Atoms with the same atomic number, but different atomic weights.

j

Jaundice (JON-dis) The skin may take on a yellow cast.

Jejunum (je-JU-num) The middle portion of the small intestine.

k

Jugular trunks (JUG-yoo-lar trunks) The lymphatic trunks that drain the head and neck region.

Kaposi's sarcoma (KAP-oh-seez sar-KO-muh) A form of cancer that consists of tumors in multiple lymph nodes; mostly seen in people who have depressed immune systems.

Keloids (KEE-loydz) A thick scar resulting from excessive growth of fibrous tissue.

Keratinocyte (ker-a-TIN-o-site) The most numerous cell type in the epidermis.

Klinefelter's syndrome (KLINE-fel-turz SIN-drome) A chromosomal abnormality in which males have an extra X chromosome.

Keratoconjunctivitis sicca (ker-a-to-kon-junc-ti-VI-tis) Dry eyes.

Knee reflex (NEE RE-fleks) The absence of this reflex may indicate damage to lumbar or femoral nerves.

Krebs cycle (KREBS SI-kul) An anaerobic process of eight reactions that occurs in the mitochondria and produces CO_2, ATP, NADH, and $FADH_2$.

Kwashiorkor (kwash-ee-OR-kor) A type of starvation in which there is too little protein in the diet.

Kyphosis (ki-FO-sis) An abnormal exaggerated curvature of the spine, most often at the thoracic level (commonly known as humpback).

l

Labia majora (LA-be-a ma-JO-ra) Rounded folds of adipose tissue and skin that protect the other external female reproductive organs.

Labia minora (LA-be-a min-OR-a) The folds of skin between the labia majora.

Labile cells (LA-bil cells) Cells that have the ability to constantly divide.

Labyrinth (LAB-i-rinth) A part of the inner ear composed of a very complex system of communicating chambers and tubes.

Lacrimal bones (LAK-ri-mal bones) Paired bones that help form the bony orbit or the eye socket.

Lactogen (LAK-to-jen) A hormone that stimulates the enlargement of mammary glands.

Lacunae (lah-KU-ni) Spaces within the lamellae containing osteocytes.

Lag phase (lag faze) The first phase of the growth curve of bacteria in which there is no evident growth in numbers.

Lamella (la-MELL-ah) Layers of bone that surround the central canal.

Large cell carcinoma (large cell kar-sih-NO-ma) A type of lung cancer that arises from the peripheral areas of the lungs.

Laryngopharynx (la-rin-go-FAR-inks) The portion of the throat behind the larynx.

Larynx (LAR-inks) The voice box.

Lateral (LAT-er-al) Away from the midline of the body.

Lavage (lah-VAJ) Irrigation to cleanse an area or clear a passageway.

Legionnaire's disease (lee-jun-AIRZ disease) An acute type of bacterial pneumonia.

Leiomyomas (li-o-mi-OH-muhz) Benign tumors of smooth muscle.

Lens (LENZ) Circular disk; because it can change shape, it helps the eye focus images of objects that are near or far away.

Lentigos (len-TEE-goes) Commonly called "liver spots" or age spots.

Leukemia (loo-KEE-mee-uh) A neoplastic condition in which the bone marrow produces a large number of WBCs that are not normal.

Leukocyte (LOO-ko-site) White blood cell.

Lice (LYSE) A parasitic arthropod that feeds on its host.

Leukocytosis (loo-koh-si-TO-sis) A significantly elevated WBC count.

Leukopenia (loo-koh-PE-ne-uh) A WBC count that is below normal.

Ligament (LIG-ah-ment) Dense connective tissue fiber that attaches bones to each other.

Ligamentum arteriosum (lig-uh-MEN-tum ar-tee-ree-O-sum) A structure formed after birth when the ductus arteriosus closes.

Ligamentum venosum (LIG-a-men-tum ven-O-sum) A structure formed after birth when the ductus venosus closes.

Light microscope (LITE-Mi-kro-skope) A basic microscope that uses a light source to illuminate a specimen.

Lingula (LING-gyoo-la) A projection of the upper lobe of the left lung that is comparable to the middle lobe of the right lung.

Lipids (LIP-idz) Fats.

Lipoprotein (ly-poh-PRO-teen) A combination of lipid and protein that makes the lipid more water soluble and able to be transported in the plasma.

Lobectomy (lo-BEK-to-mee) The removal of a lung lobe or lobes.

Loop of Henle (loop of HEN-lee) The middle section (nephron loop) of the renal tubule in which it curves toward the renal corpuscle.

Lophotrichous (lo-FAH-trik-us) Having multiple flagella on one end of the cell.

Lordosis (lor-DO-sis) An exaggerated inward (convex) curvature of the lumbar spine; the condition is sometimes called swayback.

Low-density lipoproteins (LDL) (LO DEN-sih-tee li-poh-PRO-teens) Made up of the same four substances as VLDLs, but have a higher percentage of proteins and cholesterol.

Lower esophageal sphincter (LES) (LOW-er eh-sof-uh-JEE-ul SFINK-ter) The sphincter made up of smooth muscle that controls food entering the stomach from the esophagus; also called the cardiac sphincter.

Lumbar puncture (LUM-bar PUNK-shur) A local anesthetic administered with a needle in the subarachnoid space.

Lumbar trunk (LUM-bar trunk) The lymphatic trunk that drains lymph from the lower extremities.

Lumbar vertebrae (LUM-bar VER-te-bray) The most massive and sturdy of the vertebrae because they support much of the weight of the body.

Lumbosacral plexus (lum-bo-SA-kral PLEK-sus) Innervates the lower abdominal wall, external genitalia, buttocks, thighs, legs, and feet.

Lunula (LOON-yoo-luh) The crescent-shaped area of the nail just above the cuticle.

Lyse (LICE) Break up.

Lymph (LIMF) Fluid collected from the spaces between cells into the lymphatic vessels.

Lymphadenitis (lim-fad-ih-NI-tis) Inflammation of the lymph nodes.

Lymphadenopathy (lim-fad-ih-NAH-path-ee) A general term for any disease of the lymph nodes.

Lymphatic system (lim-FAT-ik SIS-tem) A body system that works with the immune system to clear the body of disease-causing agents; includes the thymus, spleen, lymphatic vessels, lymph nodes, and lymphoid tissue.

Lymphedema (lim-feh-DEE-muh) Blockage of lymphatic vessels.

Lymph node (LIMF node) An oval- or bean-shaped structure that is part of the immune system and is located along the paths of lymphatic vessels.

Lymph nodules (LIMF NAHD-yoolz) The functional unit of the lymph nodes, consisting of macrophages and lymphocytes.

Lymphocyte (LIM-fo-site) A type of white blood cell involved in cell-mediated and antibody-mediated immune responses.

Lymphokines (LIMF-o-kinez) Cytokines secreted by T lymphocytes that increase T cell production and directly kill cells that have antigenic potential.

Lysosome (LI-so-sohm) A relatively small, membrane-bound sac in the cytoplasm of a cell that contains lytic enzymes capable of destroying and digesting proteins, carbohydrates, nucleic acids, and foreign particles.

m

Macula lutea (MAK-yoo-luh LOO-te-uh) A flat region on the posterior aspect of the retina.

Macular degeneration (MAK-u-lar de-jen-uh-RA-shun) A progressive disease that usually affects people over the age of 50; occurs when the retina no longer receives an adequate blood supply.

Macule (MAK-yul) A flat lesion that is not raised above the surface of the skin.

Magnetic resonance imaging (MRI) (mag-NEH-tik REZ-o-nans IM-uh-jing) A diagnostic procedure that allows the brain and spinal cord to be visualized from many angles.

Magnification (mag-ni-fi-KA-shun) Enlarging an object, as with a microscope.

Major histocompatibility (MA-jor his-to-kum-pa-tih-BIL-ih-te) A group of genes that aid in the ability of the immune system to determine self from nonself.

Major histocompatibility complex (MHC) (MA-jor his-to-kum-pa-tih-BIL-ih-te KOM-pleks) A large protein complex in white blood cells and all other nucleated cells that is unique for every human except identical twins.

Main stem bronchi (MAIN stem BRON-ki) The primary bronchi that branch directly from the trachea.

Malignant (mah-LIG-nant) Describing a tumor that has the ability to spread to other sites of the body.

Malignant melanoma (mah-LIG-nant mel-a-NO-ma) A cancerous tumor that starts in melanocytes of normal skin or moles and metastasizes rapidly and widely.

Malleus (MAL-lee-us) The hammer-shaped bone of the middle ear.

Malnutrition (mal-nu-TRISH-un) Inadequate or excessive caloric intake.

Mammary glands (MAM-air-ee glands) Accessory organs of the female reproductive and integumentary systems, which secrete milk for newborn offspring.

Mandible (MAN-dih-bul) A single bone; it is the lower jawbone and the largest and strongest facial bone.

Manubrium (mah-NOO-bree-um) The sternum is also called the breastbone and is actually made up of three bones.

Marasmus (mah-RAZ-mus) A type of starvation that entails both protein and calorie insufficiency.

Mastication (mas-tih-KAY-shun) The act of chewing.

Matrix (MA-triks) The material between the cells made up of water, salts, proteins, fibers, and other substances.

Matter (MAT-er) Anything that takes up space and has weight.

Maxillae (mak-SILL-ee) Paired bones that have fused to form the upper jawbone of the facial skeleton.

Maxillary sinus (MAK-sill-air-ee SI-nus) The largest of the paranasal sinuses.

Media (MEE-dee-uh) A mixture of nutrients and other necessary factors used to culture or grow microorganisms.

Medial (MEE-dee-al) Near the midline of the body.

Mediastinum (me-de-as-TI-num) A space located between the two lungs laterally and the sternum in front and the vertebral column behind.

Medulla oblongata (me-DOOL-la ob-long-GA-ta) The most inferior part of the brainstem.

Megakaryocytes (meh-guh-KARE-ee-o-sites) Large cells in the bone marrow from which platelets form.

Meiosis (mi-O-sis) A type of cell division that takes place during gamete production.

Melanin (MEL-ah-nin) Black, brown, or yellow pigment found in some parts of the body such as the skin, retina, and hair.

Melanocytes (MEL-a-no-site) An epidermal cell that produces melanin, a pigment that darkens the skin.

Membrane potential (MEM-brane po-TEN-shul) An electric charge on the neuron cell membrane.

Memory B cells (MEM-o-ree B cells) B lymphocytes that form as a response to an initial exposure to an antigen; they recognize the antigen during a second exposure and provide a strong, immediate response, usually within hours of exposure.

Menarche (MEN-ar-kee) The first menses.

Meniere's disease (men-EERZ dih-ZEEZ) A disturbance in the equilibrium characterized by vertigo (dizziness), tinnitus (ringing in the ears), nausea, and progressive hearing loss.

Meninges (me-NIN-jez) The connective tissue membranes that protect the brain and spinal cord.

Meningitis (men-in-JI-tis) Inflammation of the meninges caused by bacterial, viral, or fungal infections.

Menopause (MEN-oh-pawz) The termination of the menstrual cycles.

Menses (MEN-seez) A woman's menstrual flow.

Merocrine gland (MER-o-krin GLAND) Eccrine gland.

Mesencephalon (mes-en-SEF-a-lon) The part of the brain between the pons and diencephalon; the midbrain.

Mesenchymal (mez-en-KI-mal) Mesodermal tissue that forms connective tissue and blood and smooth muscles.

Mesentery (MES-en-tar-ee) Broad, fanlike tissue that attaches to the posterior wall of the abdomen and holds the jejunum and ileum in the abdominal cavity.

Mesoderm (MEZ-oh-derm) The middle primary germ layer that gives rise to connective tissue, blood vessels, and muscle.

Mesothelioma (meh-so-thee-lee-OH-muh) A type of cancer caused by asbestos exposure that affects the pleura.

Metabolism (meh-TAB-o-lizm) The sum of all chemical reactions in the body.

Metaphase (MET-a-faze) The second phase of mitosis in which chromatids line up along the center of the cell.

Metaphase plate (MET-a-faze PLATE) The imaginary line along which the chromatids line up in the metaphase phase of mitosis.

Metaplasia (met-a-PLAY-zhee-a) The change of one adult cell from one type to another.

Metastasis (meh-TAS-tah-sis) The spread of tumor cells.

Microbiology (mi-kro-bi-OL-o-je) The science that studies organisms that are too small to study without the use of a microscope.

Microglia (mi-KROG-le-a) Glia consisting of small cells with few processes that are scattered throughout the central nervous system and have a phagocytic function.

Microvilli (mi-kro-VIL-e) Small, fingerlike projections that greatly increase the surface area of the small intestine so that it can absorb many nutrients.

Micturition (mik-tuh-RISH-un) The process of urination.

Midbrain (MID-brain) See *Mesencephalon*.

Middle nasal conchae (MID-ul NA-zal KONG-kee) The nasal conchae increase the surface area in the nose to filter and warm the air that is inhaled.

Midsagittal plane (mid-SAJ-i-tal PLANE) Runs lengthwise down the midline of the body and divides it into equal left and right halves.

Migraine (MI-grane) The most severe type of headache.

Mitochondrion (pl. mitochondria) (mi-toe-KON-dre-on) Organelle that produces energy for a cell.

Mitosis (mi-TO-sis) The division of the cell nucleus that results in two nuclei with the same number and kind of chromosomes as the original nucleus.

Mitral valve (MI-tral valve) The atrioventricular valve on the left side of the heart, also known as the bicuspid valve.

Molecule (MOL-eh-kyool) The combination of two or more atoms of the same element.

Monocyte (MON-o-site) The largest type of white blood cell with agranular cytoplasm.

Mononuclear phagocytic system (mah-no-NU-klee-ur fay-go-SIH-tik SIS-tem) A phagocytic system made up of neutrophils, monocytes, and macrophages; also called the reticuloendothelial system.

Mononucleosis (mah-no-nu-klee-O-sis) A highly contagious viral infection that spreads through the saliva of the infected person.

Monotrichous (mon-AH-trik-us) Having a single flagellum on one end of the organism.

Mons pubis (monz PYOO-bis) The rounded fatty prominence over the pubic symphysis.

Morula (MOR-yu-luh) A solid sphere of cells produced by successive divisions of the fertilized egg.

Motility (mo-TIL-ih-tee) The condition of being mobile.

Mucosa (mu-KO-sa) The GI tract inner lining.

Multifactorial inheritance (mul-ti-FAK-toe-re-al in-HER-i-tents) The inheritance of traits influenced by both genetic and nongenetic factors.

Multinucleate (mul-te-NU-kle-at) Describes a cell that has multiple nuclei.

Multiple sclerosis (MS) (MUL-tih-pul skler-O-sis) A chronic disease of the central nervous system in which myelin is destroyed.

Multiunit smooth muscle (MUL-tee-yoo-nit smooth MUS-el) A type of smooth muscle in which when one fiber is stimulated, many adjacent fibers are also stimulated.

Murmurs (MUR-murz) Abnormal heart sounds.

Muscle fatigue (MUS-el fuh-TEEG) The temporary loss of a muscle's ability to contract.

Muscular dystrophy (MD) (MUS-ku-lar DIS-tro-fee) A group of genetic disorders that primarily affect the muscular and nervous systems.

Muscularis layer (mus-kyoo-LAIR-is LAY-er) The third layer of the GI tract, between the submucosa and serous layers; composed mostly of smooth muscle, with a few areas of skeletal muscle.

Myalgia (mi-AL-jee-uh) Muscle pain.

Myasthenia gravis (mi-uhs-THEEN-ee-uh GRAH-vis) An autoimmune condition in which a person produces antibodies that prevent muscles from receiving neurotransmitters from neurons, causing muscle weakness.

***Mycobacterium tuberculosis* (mi-ko-bak-TEER-ee-um tu-ber-ku-LO-sis)** The organism that causes tuberculosis.

Mycology (my-KOL-o-je) The study of fungi.

Myelin (MI-e-lin) Multilayered covering around the axons of many neurons; composed of lipid and protein.

Myocardial infarction (MI) (mi-o-KAR-de-al in-FARK-shun) Heart attack.

Myocarditis (mi-o-kar-DI-tis) Inflammation of the muscular layer of the heart.

Myocardium (mi-o-KAR-de-um) Heart muscle.

Myocyte (MI-oh-sites) Muscle cells.

Myofibrils (mi-o-FI-brilz) Structures in the sarcoplasm of a muscle cell that contain actin and myosin.

Myoglobin (mi-o-GLO-bin) The muscle pigment from which the oxygen needed for aerobic respiration is derived.

Myosin (MI-o-sin) The thick protein filaments that, along with actin, create the striations in cardiac and skeletal muscle.

Myometrium (my-oh-MEE-tree-um) The smooth muscle layer of the uterus.

Myopia (mi-O-pe-a) Impairment of distance vision.

Myxedema (miks-eh-DEE-muh) A condition caused by severe hypothyroidism that causes edema and loss of energy.

n

Narrow-angle glaucoma (NAR-o ANG-gul glaw-KO-muh) Glaucoma that results when the space between the iris and the cornea is suddenly blocked so that the adqueous humor cannot be drained.

Nasal bones (NA-zal bones) Paired small bones that form the bridge of the nose.

Nasal conchae (NA-zal KONG-kee) Lobed structures on the lateral walls of the nasal cavity that function to increase air turbulence.

Nasal septum (NA-zal SEP-tum) The bony and cartilaginous partition between the nasal passages.

Nasopharynx (na-zo-FAR-inks) The portion of the throat behind the nasal cavity.

Natural killer (NK) cells (NACH-yur-ul KIL-ur cells) A type of lymphocyte that primarily target cancer cells but also protect the body against many types of pathogens.

Necrosis (neh-KRO-sis) Premature or pathologic cell death.

Neisseria gonorrhea (ni-SER-ee-uh go-no-REE-uh) A Gram-negative coccus that causes the common sexually transmitted infection (STI) chlamydia.

Neonatal period (ne-o-NA-tal PEER-ee-ud) The first four weeks of the postnatal period.

Neonate (NEE-oh-nate) An infant the first four weeks after birth.

Neoplasia (ne-o-PLAY-zhee-a) The new and abnormal development of cells.

Neoplasm (NEE-o-plazm) A new growth that may be benign or malignant.

Nephrons (NEF-ronz) The functional unit of the kidney, which comprises a renal corpuscle and a tubular system.

Nerves (NERVZ) Bands of nervous tissue that connect parts of the nervous system with the other organs and conduct nervous impulses throughout the body.

Nervous tissue (NER-vus TISH-oo) Tissue located in the brain, spinal cord, and peripheral nerves of the body.

Neuralgias (noo-RAL-jee-uhz) A group of disorders commonly referred to as nerve pain; they most frequently occur in the nerves of the face.

Neurogenic shock (no-ro-JEN-ik SHOK) Results from brain or spinal cord injury.

Neuroglia (noo-ROG-le-a) Cells of the nervous system that perform various supportive functions.

Neuron (NOO-ron) A nerve cell, consisting of a cell body, dendrites, and an axon.

Neurotransmitter (noo-roh-TRANZ-mih-ter) One of many varieties of molecules released by the axon terminal in response to an action potential, which then causes a change in the postsynaptic membrane potential.

Neutron (NOO-tron) An uncharged particle of matter.

Neutrophil (NOO-tro-fil) A type of white blood cell with phagocytic capabilities; characterized by a granular cytoplasm.

Nits (NITZ) Lice are often identified as eggs or nits.

Nocturia (nok-TUR-ee-uh) Frequent urination at night.

Node (node) In the heart, a collection of specialized cardiac muscle fibers that have a certain rate of activity.

Node of Ranvier (NODE of RON-ve-a) A bare area on an axon between myelinated areas.

Norepinephrine (nor-eh-pin-EFF-rihn) A neurotransmitter involved in smooth muscle contraction.

Normal flora (NOR-mal FLOR-uh) The term used for the microorganisms that inhabit the skin and mucous membranes of healthy individuals.

Nosocomial injuries (no-so-KO-me-ul IN-jur-ez) Injuries or infections caused in a hospital setting that were not present in the individual before entering the hospital.

Nuchal rigidity (NOO-kul rih-JID-ih-te) Stiffness of the neck.

Nucleus (NOO-kle-us) A spherical organelle within the cell that contains genes, the hereditary factors of the cell.

Nucleolus (noo-kle-OH-lus) An area within the nucleus, with no surrounding membrane, that produces ribosomes.

Nutritional pathology (nu-TRIH-shun-al pa-THOL-o-jee) Includes deficiencies and excesses.

Nyctalopia (nik-ta-LO-pe-a) Impaired night vision.

Nystagmus (ni-STAG-mus) Rapid, involuntary eye movements.

Oblique (o-BLEK) At an angle other than perpendicular to a sagittal, horizontal, or coronal plane.

Obstructive shock (ob-STRUK-tiv SHOK) Shock that occurs due to a blockage of an artery, causing a decrease in blood flow or perfusion to tissues.

Obstructive sleep apnea (OSA) (ob-STRUK-tiv sleep AP-nee-uh) A condition in which relaxed throat tissues cause airways to collapse, which prevents a person from breathing.

Occipital bone (ok-SIP-ih-tal bone) A single bone that forms the back of the skull.

Occipital lobes (ok-SIP-ih-tul lobes) The lobes of the brain that control visual interpretation.

Occlusive stroke (ok-KLOO-ziv stroke) A cerebrovascular accident caused by a thrombus in a blood vessel of the neck or an embolus that has traveled to the brain from elsewhere in the body.

Olfaction (ol-FAK-shun) The sense of smell.

Olfactory receptors (ol-FAK-to-ree re-SEP-turz) Bipolar neurons that have cilia that are stimulated by odorants.

Oligodendrocyte (ol-ih-go-DEN-dro-site) A glial cell resembling an astrocyte but smaller.

Oocyte (OH-oh-site) An immature ovum.

Oogenesis (oh-oh-JEN-uh-sis) Formation of the female gametes.

Open-angle glaucoma (O-pen ANG-gul glaw-KO-muh) Glaucoma caused by slowing of the drainage of aqueous humor from the anterior chamber of the eye.

Open fracture (OP-in FRAK-shur) A compound fracture; one in which the skin is broken.

Optic chiasm (OP-tik KI-azm) The opening on the undersurface of the hypothalamus through which the optic nerves are continuous with the brain.

Optic (OP-tik) For or relating to the eye.

Optic disk (OP-tik DISK) The area of the retina where the optic nerve enters and where there are no sensory nerves; also known as the blind spot.

Opposition (ah-poh-ZIH-shun) Bringing together each of the fingers and the thumb.

Organ of Corti (OR-gan of KOR-tee) The organ of hearing.

Organelles (or-gan-ELZ) Small, permanent structures within the cytoplasm that serve specific functions.

Organic matter (or-GAN-ik MAT-er) Contains carbon and hydrogen, and tends to form large molecules.

Origin (OR-ih-jin) The site of attachment of a muscle to the less movable bone during a muscle contraction.

Oropharynx (or-o-FAR-inks) The portion of the throat behind the oral cavity.

Orthopnea (or-THOP-nee-uh) Difficulty breathing that becomes worse when lying down.

Osmosis (oz-MO-sis) The net movement of water molecules through a selectively permeable membrane from an area of high water concentration to an area of low water concentration until equilibrium is reached; no energy is required.

Ossicles (OS-ih-kulz) The bones of the middle ear, the smallest in the body.

Ossification (os-ih-fi-KA-shun) A process by which bones grow.

Osteoarthritis (OA) (os-te-o-arth-RI-tis) Also known as degenerative joint disease (DJD) or "wear and tear" arthritis.

Osteoblast (OS-tee-oh-blast) An immature bone cell that causes mineralization of the bone matrix.

Osteoclast (OS-tee-oh-klast) A cell that breaks down bone for remodeling.

Osteocyte (OS-tee-oh-site) A mature bone cell.

Osteogenesis (os-te-o-JEN-uh-sis) A process by which bones grow.

Osteogenesis imperfecta (os-te-o-JEN-uh-sis im-per-FEK-tuh) Brittle-bone disease.

Osteon (OS-tee-ohn) The basic unit of structure in adult compact bone consisting of a central canal and concentrically arranged lamellae, lacunae, osteocytes, and canaliculi; also called a Haversian system.

Osteoporosis (os-tee-oh-po-RO-sis) A condition in which bones become thin and more porous.

Osteosarcoma (os-te-o-sar-KOH-ma) A type of bone cancer that originates from osteoblasts, the cells that make bone tissue.

Otic (O-tik) For or relating to the ear.

Otitis media (o-TI-tus MEE-dee-a) Middle ear infections.

Otosclerosis (o-to-skler-O-sis) The immobilization of the stapes within the middle ear; a common cause of conductive hearing loss.

Ova (O-va) The female gamete cells needed for reproduction.

Oval window (O-val WIN-doh) The boundary between the middle and inner ears.

Ovarian cancer (o-VAR-e-an KAN-sir) Considered more deadly than the other types of gynecological cancers because its signs and symptoms are usually mild or indistinct until the disease has metastasized (spread) to other organs.

Oviduct (OH-vi-dukt) See *Fallopian tube.*

Ovulation (ov-yoo-LA-shun) The discharge of a mature ovum from the ovary.

Oxygen (OKS-i-jen) An essential inorganic molecule.

Oxygen debt (OKS-i-jen DET) A condition of low oxygen that occurs when skeletal muscle is used strenuously for several minutes.

Oxyhemoglobin (ok-see-HEE-mo-glo-bin) Hemoglobin that carries oxygen.

Pack years (PAK YEERZ) The number of years a person has smoked multiplied by the number of packs per day the person smokes.

Paget's disease (PA-jets dih-ZEEZ) A disease that causes bones to enlarge and become deformed and weak; usually affects people older than 40 years of age.

Palatine bones (PAL-a-tine bones) Paired bones form the posterior aspect of the hard palate as well as the lateral wall of the nasal cavity and small part of the floor of the orbit.

Pancreatic cancer (pan-kre-AT-ik KAN-ser) Cancer of the pancreas.

Pancreatitis (pan-kree-ah-TI-tis) Inflammation of the pancreas.

Papillae (pah-PIL-ee) The projections on the surface of the tongue where the taste buds are located.

Papillary layer (pa-PIL-lar-e LAY-er) The more superficial of the two, composed of areolar or loose connective tissue with projections upward.

Papillary muscles (pa-PIL-lar-e MUS-ulz) Muscles attached to the cusps of the tricuspid valve that prevent the valve from "falling" into the right atrium during ventricular contraction and prevent the backflow of blood.

Paranasal sinus (par-a-NA-zal SI-nus) Mucus-lined air cavities in the skull.

Parasite (PAR-uh-site) An organism that needs a host to survive.

Parenchyma (par-EN-ki-ma) The functioning part of the kidney made up of the cortex and medulla.

Paresthesia (par-es-THE-ze-a) A sensation of tingling, pricking, or numbness of the skin.

Parietal bones (pa-RI-e-tal bones) Paired bones that form most of the top and sides of the skull.

Parietal cells (pa-RI-eh-tal cells) Stomach cells that secrete hydrochloric acid, which is necessary to convert pepsinogen to pepsin.

Parietal lobes (pah-RI-eh-tal lobes) The lobes of the brain responsible for sensations such as touch, temperature, and pain, as well as for understanding speech and using words to express ideas.

Parietal peritoneum (pa-RI-eh-tal per-ih-to-NEE-um) The part of the peritoneum that lines the abdominal wall.

Parietal pleura (pa-RI-e-tal PLOO-rah) The outer membrane of the pleural sac.

Parkinson's disease (PAR-kin-sonz dih-ZEEZ) A slowly progressive and degenerative motor system disorder.

Parturition (par-tu-RISH-un) The actual childbirth stage of the birth process.

Pathogenic (path-o-JEN-ik) Having the ability to cause disease.

Pathology (pah-THOL-o-je) The study of disease.

Pathogenicity (pah-tho-jin-IH-si-tee) The ability to cause a disease.

Pediculosis (peh-dik-u-LO-sis) Infestation with lice.

Penis (PEE-nis) A highly vascularized cylindrical organ that propels urine and semen outside the body.

Pepsinogen (pep-SIN-o-jen) A chemical secreted by chief cells in the stomach; becomes pepsin in the presence of acid to help digest proteins.

Pericardial cavity (per-ih-KAR-dee-ul KAV-ih-tee) The body cavity in the mediastinum in which the heart is located.

Pericarditis (per-ih-kar-DI-tis) Inflammation of the pericardium.

Pericardium (per-i-KAR-de-um) The membrane that surrounds the heart, made up of a superficial fibrous layer and a deep serous layer.

Perilymph (PER-ih-limf) Movement that activates equilibrium receptors as well as hearing receptors.

Perimetrium (per-ih-MEE-tree-um) The serosa of the uterus.

Perimysium (per-ih-MIZ-ee-um) A sheath of connective tissue surrounding a fascicle.

Perineum (per-ih-NEE-um) The pelvic floor; the space between the anus and the vulva in the female and the anus and the scrotum in the male.

Periosteum (per-ee-OS-tee-um) A membrane surrounding the shaft of a bone.

Peripheral nervous system (PNS) (per-IF-er-al NER-vus SIS-tem) The part of the nervous system that is outside the central nervous system and comprises the cranial nerves except the optic nerve, the spinal nerves, and the autonomic nervous system.

Peristalsis (per-i-STAL-sis) Rhythmic muscle movements that move food through the esophagus.

Peritoneum (per-ih-to-NEE-um) The double-walled outermost layer of the small intestine.

Peritrichous (per-IH-trik-us) Having flagella distributed over the entire perimeter of the cell.

Permanent cells (PER-ma-nent cells) Cells that do not have the ability to multiply.

Pernicious (megaloblastic) anemia (pur-NEE-shus (meh-gah-lo-BLAS-tik) uh-NEE-mee-uh) A type of anemia characterized by large red blood cells, due to a deficiency of Vitamin B_{12} or folic acid.

Peroxisomes (pe-ROKS-ih-somz) Similar to lysosomes, but contain peroxidases, enzymes involved in breaking down hydrogen peroxide that is toxic to cells.

Petechia (pe-TE-ke-a) Pinpoint hemorrhages that show as small red dots on the skin.

Phagocytosis (fay-go-cy-TOE-sis) A form of endocytosis in which larger molecules are engulfed and brought into the cell.

Phalanges (fa-LAN-jees) Bones of the digits of the hands and feet.

Pharyngitis (fair-en-JI-tis) Sore throat; inflammation of the throat.

Pharynx (FAR-inks) The throat.

Phase microscope (FAZE MI-kro-skope) A microscope that uses light waves passing through objects in different phases based on the makeup of the materials they pass through.

Phenotype (FEE-noh-tipe) The observable expression of the genotype.

Phenylketonuria (PKU) (fen-ul-kee-ton-YUR-ee-uh) Develops when a person cannot synthesize the enzyme that converts phenylalanine to tyrosine.

Phospholipids (fos-fo-LIP-id) Used to make cell membranes.

Physical barriers (FIZ-ih-kul BAR-yurz) Barriers to disease such as the skin and mucosa, as well as chemical barriers such as lysozymes in tears.

Pilus (PI-lus) A small, hairlike structure shorter than a flagellum but made of protein like a flagellum.

Pinna (PIN-a) The external part of the ear; also called the auricle.

Pinocytosis (pin-o-sih-TO-sis) Similar to phagocytosis, except liquid is engulfed by the cell.

Placenta (plah-SEN-tah) A structure that allows nutrients and oxygen from maternal blood to pass to embryonic blood.

Plantar flexion (PLAN-tar FLEK-shun) Pointing the toes down.

Plasma (PLAZ-ma) The fluid part especially of blood, lymph, or milk that is distinguished from suspended material.

Plasma cells (PLAZ-ma cells) Cells derived from B lymphocytes that produce antibodies.

Plasmapheresis (plaz-muh-fer-EE-sis) A process of removing plasma from the blood to remove harmful antibodies.

Platelets (PLATE-lets) Another term for thrombocytes; the cell fragments present in blood that play a role in stopping blood flow from an injured vessel.

Pleura (PLOO-ra) The serous membrane that covers the lungs and lines the walls of the chest and diaphragm.

Pleural cavities (PLOO-ral KAV-ih-teez) The potential space between each lung and the chest wall; normally contain a little fluid.

Pleural effusion (PLOO-ral eh-FYOO-zyun) Buildup of fluid in the pleural cavity.

Pleurisy (PLUR-uh-see) Inflammation of the membranes surrounding the lungs.

Pleuritis (ploo-RI-tis) See *Pleurisy*.

Pneumonectomy (noo-moh-NEK-toh-mee) Removal of an entire lung.

Pneumonia (noo-MOH-nee-uh) Inflammation of the lungs that may result from more than 50 different causes that range from mild to severe.

Pneumonitis (noo-mo-NI-tis) See *Pneumonia*.

Pneumoconiosis (noo-mo-no-ko-nee-OH-sis) Lung diseases that result from years of exposure to different environmental or occupational types of dust.

Pneumothorax (noo-mo-THOR-aks) Buildup of air in the pleural cavity.

Polar body (PO-lar BOD-ee) A nonfunctional cell.

Polarized (PO-la-rized) The state in which the outside of a cell membrane is positively charged and the inside is negatively charged.

Polycystic kidney disease (PKD) (pol-ee-SIS-tik KID-nee dih-ZEEZ) A disorder in which the kidneys enlarge because of the presence of many cysts within them.

Polydipsia (pol-e-DIP-se-uh) Excessive thirst.

Polyphagia (pol-e-FAJ-ee-uh) Excessive hunger.

Polyuria (pol-e-YUR-ee-uh) Excessive urination.

Pons (PONZ) The part of the brain stem that forms a "bridge" between the medulla oblongata and the midbrain.

Positron emission tomography (PET) scan (POS-ih-tron e-MISH-un toe-MOG-gruh-fee) A procedure that uses radioactive chemicals that collect in specific areas of the brain to produce images.

Posterior (pos-TER-e-or) Toward the back of the body when in anatomical position or in behind another structure.

Posterior cavity (pos-TER-e-or KAV-ih-te) See *Vitreous chamber*.

Postnatal period (post-NA-tal PEER-ee-ud) The six-week period following birth.

Pregnancy (PREG-nan-see) The condition of having a developing offspring in the uterus.

Premenstrual syndrome (PMS) (pre-MEN-stroo-ul SIN-drome) A collection of symptoms that occur just before a menstrual period.

Prepuce (PREE-puse) The foreskin that covers the glans penis in an uncircumcised penis.

Presbycusis (prez-bi-KU-sis) Hearing loss because of the aging process.

Presbyopia (prez-be-O-pe-uh) Farsightedness due to aging.

Primary antibody deficiency (PRI-mar-e AN-tih-bo-dee de-FISH-un-see) A condition in which the body does not develop antibodies to protect from recurrent bacterial or viral infections.

Primary bronchi (PRI-mair-ee BRON-ki) The main stem bronchi that branch directly from the trachea.

Primary germ layers (PRIME-air-ee jerm LAY-erz) Layers of cells formed in the inner cell mass of a blastocyst that form all of the embryo's organs.

Primary hypertension (PRI-mar-ee hi-per-TEN-shun) Hypertension that has no identifiable cause; also called idiopathic or essential hypertension.

Prime mover (PRIM MOOV-er) The muscle responsible for most of the movement of the involved joint.

Primordial follicle (pri-MOR-dee-ul FOLL-ih-kul) Vesicle in the ovary that contains a primary oocyte and follicular cells.

Progesterone (pro-JES-tur-own) A female sex hormone that is secreted by the corpus luteum to prepare the endometrium for implantation and later by the placenta during pregnancy to prevent rejection of the developing embryo or fetus.

Prognosis (prog-NO-sis) The likely outcome of a disease.

Prokaryotes (pro-KARE-ee-otes) One-celled organisms that do not have a nucleus.

Prolactin (PRL) (pro-LAK-tin) The peptide responsible for milk production in the mammary glands in females and for helping with the function of luteinizing hormone in males.

Pronation (pro-NA-shun) Turning the palm of the hand down or lying face down on the abdomen.

Prone position (PRON po-ZIH-shun) Lying on the stomach.

Propagate (PROP-a-gate) Increase in size.

Prophase (PRO-faze) The first phase of mitosis during which the nuclear envelope breaks apart and chromatin condenses to form chromosomes.

Prostaglandins (pros-tah-GLAN-dinz) Lipid hormones that are produced by a number of organs that have nearby receptors and thus do not travel through the blood; they have a role in inflammation and intensifying pain, among other effects.

Prostate cancer (PRAH-state KAN-ser) One of the most common cancers in men older than age 40. The risk of developing prostate cancer increases with age.

Prostate gland (PRAH-state gland) An organ that surrounds the proximal portion of the urethra and secretes a milky, alkaline fluid just before ejaculation to protect the sperm in the acidic environment of the vagina.

Prostatitis (pros-ta-TI-tis) An inflammation of the prostate gland.

Proteins (PRO-teenz) Contain essential amino acids and are used by the body for growth and the repair of tissues.

Proton (PRO-tahn) A particle of matter that carries a positive charge.

Protraction (pro-TRAK-shun) Moving a body part anteriorly.

Proximal (PROKS-im-al) Nearer the attachment of an extremity to the trunk or nearer to the point of attachment or origin.

Proximal convoluted tubule (PROKS-ih-mal kon-vo-LOO-ted TOO-byoolz) Tubule that is directly attached to the Bowman's capsule and eventually straightens out to become the loop of Henle.

Psoriasis (so-RI-a-sis) A common chronic inflammatory skin condition that has an autoimmune basis; silvery scaly lesions anywhere on the body, but commonly on extensor surfaces such as the elbows and knees.

Ptosis (TO-sis) Drooping.

Pulmonary circuit (PULL-mon-air-ee SIR-kut) Pulmonary circulation, in which blood travels from the heart to the lungs and back again.

Pulmonary edema (PULL-mon-air-ee eh-DEE-muh) A condition in which fluids fill the alveoli of the lungs, making gas exchange difficult or even impossible.

Pulmonary embolism (PULL-mon-air-ee EM-boh-lizm) Blood clot in an artery in the lungs.

Pulmonary semilunar valve (PUL-mo-nar-e sem-e-LOO-nar valve) The valve between the right ventricle and the pulmonary trunk.

Pupil (PU-pil) Regulates the amount of light that enters the eye.

Purkinje fibers (pur-KIN-je fibers) Muscle fibers that are part of the conduction system of the heart.

Pustule (PUS-tyul) A pus-filled lesion such as a pimple.

Pyelonephritis (pie-eh-lo-neh-FRI-tis) A type of complicated urinary tract infection (UTI) that begins as a bladder infection and spreads up one or both ureters into the kidneys.

Pyloric sphincter (pie-LOR-ik SFINK-ter) The opening between the pylorus of the stomach and the small intestine that controls movement of substances from the stomach to the small intestine.

Pyothorax (pi-o-THOR-aks) Buildup of pus in the pleural cavity.

Radiation (ra-de-A-shun) The transfer of heat in the form of infrared rays between objects that are not in direct contact.

Radioactive (ra-dee-o-AK-tiv) Spontaneously emits energetic particles by the disintegration of their atomic nuclei.

Receptor (re-SEP-ter) A cell or group of cells that receive stimuli.

Rectum (REK-tum) The last part of the intestine, extending from the sigmoid colon to the anus.

Reflex (RE-fleks) A predictable and automatic response to a stimulus.

Refraction (re-FRAK-shun) The bending of light entering the eye to focus it onto the retina.

Regeneration (re-jin-uh-RA-shun) Renewal of damaged tissue or an appendage.

Relaxin (re-LAK-sin) A hormone from the corpus luteum that inhibits uterine contractions and relaxes the ligaments of the pelvis in preparation for childbirth.

Remodeling (re-MOD-el-ing) The building up and tearing down of bone.

Renal calculi (REE-nul KAL-ku-lie) Kidney stones that obstruct the ducts within the kidneys or ureters.

Renal column (REE-nul KOL-um) The portion of the cortex between pyramids.

Renal corpuscle (REE-nul KOR-puss-ul) The part of a nephron that comprises a glomerulus and Bowman's capsule and which filters blood.

Renal cortex (REE-nul KOR-tex) The outermost layer of the kidney that covers and extends between the pyramids.

Renal medulla (REE-nul meh-DUL-lah) The inner layer of the kidney that contains the renal pyramids.

Renal pelvis (REE-nul PEL-vis) Cavity in the center of the kidney into which the major calyces open.

Renal pyramid (REE-nul PEER-uh-mid) A triangular-shaped area in the renal medulla containing the straight segments of the renal tubules.

Renal tubule (REE-nul TOO-byool) Structure that extends from Bowman's capsule to collect the filtrate.

Renal vein (REE-nul VANE) The vein formed by the peritubular capillaries and veins that, in turn, drains into the inferior vena cava.

Renin (REN-ihn) Enzyme that helps regulate blood pressure and blood volume.

Repair (re-PAIR) The replacement of the damaged tissue with connective tissue cells and fibers and new blood vessels.

Repolarization (re-po-lar-i-ZA-shun) Restoration of the difference in charge between the inside and outside of the plasma membrane of a muscle fiber or cell following depolarization; a return to the resting potential of the cell.

Resolution (re-za-LU-shun) The ability to distinguish two objects as being separate and distinct objects as with a microscope.

Respiratory distress syndrome (RDS) (RES-pir-uh-to-ree dis-STRESS SIN-drome) A syndrome of unknown etiology that kills apparently healthy infants. Also called hyaline membrane disease.

Respiratory pump (RES-pir-uh-to-ree pump) A pumping mechanism that moves lymph toward the heart using pressure changes in the thorax.

Reticular layer (reh-TIK-yu-lar LAY-er) The thickest part of the skin, made up of dense, irregular connective tissue.

Reticuloendothelial system (re-TIK-yu-lo-en-do-THEEL-ee-ul SIS-tem) See *Mononuclear phagocytic system*.

Retina (REH-tih-nah) The inner layer of the eye that senses light.

Retinal detachment (REH-tih-nal dee-TACH-ment) Occurs when the layers of the retina separate; it is considered a medical emergency and if not treated right away leads to permanent vision loss.

Retraction (re-TRAK-shun) Moving a body part posteriorly.

Retrograde menstruation (REH-tro-grade men-stroo-A-shun) Bleeding back up into the fallopian tubes.

Retroperitoneal (reh-tro-pair-ih-toe-NEE-al) In a position behind the abdominal peritoneum.

Rh antigen (R H An-tih-jin) A protein present on the red blood cells of Rh-positive individuals.

Rheumatoid arthritis (RA) (ROO-ma-toyd ar-THRI-tis) A chronic systemic inflammatory disease that attacks the smaller joints such as those in the hands and feet.

Rhinoplasty (RI-no-plas-tee) Plastic surgery on the nose.

Rhinorrhea (ri-no-REE-uh) Runny nose.

Rhinovirus (RI-no-vi-rus) A family of viruses that causes many upper respiratory infections.

Ribonucleic acid (RNA) (ri-bo-noo-KLE-ik A-sid) A single-stranded nucleic acid made of ribose (a sugar), nitrogenous bases (adenine, guanine, cytosine, and uracil), and phosphate groups.

Ribosomes (RI-bo-somz) Are made up of RNA and are responsible for the production of proteins.

RICE protocol (RICE PRO-to-kol) A protocol for treating sprains or strains with *R*est, *I*ce, *C*ompression, and *E*levation.

Right lymphatic duct (rite lim-FAT-ik dukt) The lymphatic duct that receives lymph from the upper-right side of the body and empties its contents into the right internal jugular and right subclavian veins.

Ring of Waldeyer (RING of WALL-die-ur) The ring formed by the three sets of tonsils at the junction of the oral cavity and oropharynx and the junction of the nasal cavity and nasopharynx.

RNA See *Ribonucleic acid*.

Rods (RODZ) One of the two types of photoreceptors found in the retina; specialized for vision in dim light.

Rosacea (ro-ZA-she-a) A skin disorder that commonly appears as facial redness, predominantly over the cheeks and nose.

Rotation (roh-TA-shun) Twisting or rotating a body part.

Rough endoplasmic reticulum (rER) (RUF en-do-PLAZ-mic re-TIK-u-lum) A type of endoplasmic reticulum made up of a system of membranous tubes and sacs, that is studded with ribosomes on its surface giving it a rough appearance under the microscope.

Rugae (ROOG-eye) Folds on the inner lining of the stomach that help mix gastric contents.

S

Sacrum (SA-krum) A triangular-shaped bone that actually consists of five fused vertebrae.

Sagittal (SAJ-it-al) A plane that divides the body into left and right portions.

Salivary glands (SAL-ih-ver-ee glands) Glands in the oral cavity that secrete saliva, which is a mixture of water, enzymes, and mucus.

Saltatory conductance (SAL-ta-to-re kon-DUK-tence) Action potential jumping from one node to the next.

Sarcolemma (sar-koh-LEM-uh) The cell membrane of a muscle fiber.

Sarcoma (sar-KO-mah) A malignant tumor of mesenchymal origin.

Sarcoplasm (SAR-koh-plazm) The cytoplasm of a muscle cell.

Sarcoplasmic reticulum (sar-koh-PLAZ-mik re-TIK-yoo-lum) The endoplasmic reticulum of a muscle cell.

Satellite cells (SAT-uh-lite cells) Cells surrounding a ganglion cell or a stem cell that play a role in muscle growth, repair, and regeneration.

Scabies (SKA-bez) A highly contagious skin condition as the result of a mite that burrows beneath skin leaving its feces behind to leave red lines of inflammation to be seen on the skin.

Scanning electron microscope (SKAN-ing e-LEK-tron MI-kro-skope) A type of electron microscope that gives a three-dimensional image of the object.

Schwann cells (SHWON cells) Cells that form spiral layers around a myelinated nerve fiber between two nodes of Ranvier and form the myelin sheath.

Sciatica (si-AT-ih-kuh) Numbness, pain, or tingling sensations on the back of a leg or foot that occurs when the sciatic nerve is damaged.

Sclera (SKLEH-rah) The outer fibrous tunic (white of the eye) that gives structure to the eyeball; the posterior aspect is the site of attachment of the eye muscles.

Scleral venous sinus (SKLEH-rul VEE-nus SI-nus) The sinus that drains aqueous humor from the eye.

Scleroderma (skleh-ro-DER-ma) A progressive disease affecting blood vessels; causes excessive deposition of collagen in the skin and organs such as the lungs, gastrointestinal tract, heart, and kidneys.

Scoliosis (sko-lee-OH-sis) An abnormal, S-shaped lateral curvature of the thoracic or lumbar spine.

Scrotum (SKRO-tum) The pouch of skin that holds the testes.

Sebaceous (oil) glands (se-BA-shus GLANDS) Glands on the surface of the skin, usually opening into the hair follicles, and secreting an oily material which softens and lubricates the hair and skin.

Sebum (SE-bum) An oily substance produced by the sebaceous glands.

Secondary hypertension (SEK-un-dair-ee hi-per-TEN-shun) Hypertension for which a cause has been identified.

Secondary bronchi (SEK-un-dair-ee BRON-ki) Smaller bronchi that branch off the primary or main stem bronchi.

Secretion (se-KRE-shun) Releasing a chemical into the body.

Seizure (SEE-zyur) A convulsion; an episode of disturbed brain function.

Sella turcica (SELL-uh TUR-kih-kuh) The part of the sphenoid bone that contains the pituitary gland.

Semen (SEE-men) A mixture of sperm cells and fluids from the seminal vesicles, prostate gland, and bulbourethral glands.

Semicircular canals (seh-mih-SIR-kyoo-lar kah-NALZ) The segment of the labyrinth that detects the balance of the body.

Seminal vesicles (SEM-ih-nul VES-ih-klz) Saclike organs that secrete an alkaline fluid rich in sugars and prostaglandins.

Seminiferous tubules (seh-mih-NIF-er-us TOO-byoolz) Tightly coiled ducts located in the testes where sperm are produced.

Sensory adaptation (SEN-so-re ad-ap-TA-shun) The same chemical can stimulate smell receptors for only a limited amount of time.

Septic shock (SEP-tik shok) A kind of shock caused by severe infections especially those involving Gram-negative organisms.

Serous (SEER-us) Watery.

Serosa (se-RO-sa) The outermost layer of the gastrointestinal (GI) tract; a serous membrane with areolar connective tissue and simple squamous epithelium.

Serum (SER-um) Blood plasma minus its clotting proteins.

Sex chromosomes (SEKS KRO-mo-somz) The X chromosome or the Y chromosome in humans that contains the genes to determine the inheritance of various sex-linked and sex-limited characteristics.

Sex-linked disorders (SEX-linkt dis-OR-derz) Those that are passed on through the sex chromosomes.

Shingles (SHING-gulz) An inflammatory condition of the skin caused by the varicella zoster virus.

Shock (SHOK) The inability of the circulatory system to adequately perfuse organs and tissues.

Sickle cell disease (SIK-ul sell dih-ZEEZ) A hemolytic anemia in which red blood cells deform into an abnormal shape that often resembles a sickle.

Sickle cell trait (SIK-ul sell trayt) A condition in which an individual has only one defective allele for sickle cell.

Sigmoid colon (SIG-moyd KO-lun) The S-shaped portion of the colon that connects the descending colon with the rectum.

Signs (SINES) Characteristics of disease that are evident to someone besides the patient.

Silicosis (sih-lih-KO-sis) A type of pneumonoconiosis caused by exposure to silica sand from sand blasting and ceramic manufacture.

Simple diffusion (SIM-pul dih-FYU-zion) Movement of molecules or ions from an area of higher concentration to an area of lower concentration with no energy required.

Simple fracture (SIM-pul FRAK-shur) A closed fracture; one in which the skin remains intact.

sinoatrial (SA) node (si-no-A-tre-al node) A collection of muscle fiber cells in the posterior wall of the right atrium that acts as the pacemaker of the heart.

Sinusitis (si-nyoo-SI-tis) An inflammation of the membranes lining the sinuses of the skull; can be acute or chronic in nature.

Sjogren syndrome (SHO-grenz SIN-drohm) A chronic inflammatory autoimmune disease that affects especially older women, that is characterized by dryness of mucous membranes especially of the eyes and mouth and by infiltration of the affected tissues by lymphocytes, and that is often associated with rheumatoid arthritis.

Skeletal muscle pump (SKEL-uh-tuhl MUS-ul pump) A lymphatic pump that uses skeletal muscle contractions to move lymph toward the heart.

Small cell lung cancer The most aggressive type of lung cancer; spreads readily to other organs and occurs almost exclusively in smokers.

Smear (smeer) A method of preparing cells for examination on microscopic slides.

Smooth endoplasmic reticulum (SER) (SMUTH en-do-PLAZ-mic re-TIK-u-lum) A part of endoplasmic reticulum that is tubular in form and lacks ribosomes. Its functions include lipid synthesis, carbohydrate metabolism, calcium concentration, drug detoxification, and attachment of receptors on cell membrane proteins.

Sodium-potassium pump (SO-dee-um po-TASS-ee-um PUMP) A mechanism that pumps sodium ions out of a nerve cell and also pumps potassium ions into the cell after passage of a nerve impulse.

Species (SPE-seez) A class of individuals having common attributes and designated by a common name.

Specific defense (speh-SIH-fik de-FENS) Another name for immunity; the third line of defense against disease.

Sperm (SPERM) The male cell needed for impregnating a female.

Spermatogenesis (sper-mah-toe-JEN-eh-sis) The process of sperm cell formation from immature spermatagonia.

Spermatogenic cells (sper-ma-to-JEN-ik cells) Cells that give rise to sperm cells.

Spermatogonia (sper-mat-o-GO-ne-a) Name for the cells from which sperm cells are formed, at the beginning of spermatogenesis.

Sphenoid bone (SFEE-noyd bone) An unpaired bone that is the most complex bone of the skull.

Sphenoid sinus (SFEE-noyd SI-nus) The paranasal sinus located in the sphenoid bone.

Spinal cavity (SPI-nul KAV-ih-tee) The vertebral canal, in which the spinal cord is located.

Spirochete (SPI-roh-keet) A Gram-negative bacteria that has long, helically coiled (spiral-shaped) cells. One type of spirochete causes Lyme disease.

Spleen (SPLEN) Large lymphoid organ located on the left side of the upper abdomen that is involved in blood cell formation in the fetus and filtering of the blood after birth.

Splenectomy (spleh-NEK-to-mee) Surgical removal of the spleen.

Splenomegaly (spleh-no-MEG-uh-lee) An enlarged spleen, usually caused by disease.

***Sporothrix schenckii* (SPO-ro-thriks SHANK-e-i)** A fungus associated with horticultural plants such as roses.

Squamous (SKWA-mus) Having a flattened shape.

Squamous cell carcinoma (SKWA-mus cell kar-si-NO-ma) A type of cancer made up of squamous cells that usually occurs in areas of the body exposed to strong sunlight over a period of many years.

Squamous cell lung cancer (SKWA-mus cell lung cancer) A type of lung cancer that arises from the epithelial cells that line the tubes of the lungs.

Stable angina (STAY-bul an-JI-nuh) Angina or chest pain brought about by stress or physical activity.

Stable cells (STA-bul SELLS) Cells capable of dividing when necessary.

Staging (STAY-jing) The process of determining how far a tumor has spread.

Stapes (STAY-peez) The stirrup-shaped bone of the middle ear.

***Staphylococcus* species (staf-il-o-KOK-us SPE-sees)** Gram-positive spherical bacteria that include forms occurring singly, in pairs or tetrads, or in irregular clusters; causes diseases and disorders such as food poisoning, skin infections, and endocarditis.

Stationary phase (STAY-shun-air-ee faze) The third phase of the growth curve of bacteria in which growth and death balance each other and there is neither growth nor decline.

Stem cells (STEM selz) Cells that are capable of becoming a variety of different cell types and of self-renewal or production of other stem cells.

Steroid (STAIR-oyd) A lipid-soluble hormone built of cholesterol that turns specific genes on or off.

Stomach cancer (STUM-ak KAN-ser) Most commonly occurs in the uppermost (cardiac portion) of the stomach.

Strabismus (stra-BIZ-mus) A condition in which the eyes do not focus in unison.

Stratified (STRAT-ih-fide) Stacked or arranged in several layers.

Stratum basale (STRA-tum bas-A-le) Also called the stratum germinativum; the deepest layer of the epidermis.

Stratum corneum (STRA-tum KOR-ne-um) The most superficial layer of the epidermis, sometimes called the horny layer, made of 20 to 30 layers of dead, squamous, fully keratinized cells.

Stratum germinativum (STRA-tum jer-mi-ni-TIE-vum) Also called the stratum basale; the deepest layer of the epidermis.

Stratum granulosum (STRA-tum gran-you-LO-sum) The middle layer of the epidermis, also called the granular layer, comprised of a thin layer of cells that produce granules to provide waterproofing and make keratin fibers.

Stratum lucidum (STRA-tum LU-si-dum) The epidermal layer just below the stratum corneum consisting of thin, translucent dead keratinocytes found only in thick skin.

Stratum spinosum (STRA-tum spih-NO-sum) The epidermal layer above the stratum basale.

***Streptococcus* species (strep-TO-kok-us SPE-shez)** Gram-positive bacteria that form pairs or chains.

Stressors (STRESS-orz) Events or stimuli that produce stress.

Striated muscle (STRI-a-ted MUS-el) Muscle tissue that is marked by transverse dark and light bands, that is made up of elongated fibers; includes skeletal and cardiac muscle.

Striations (stri-A-shunz) The "stripes" that appear in skeletal and cardiac muscle, due to overlapping actin and myosin muscle filaments.

Subarachnoid space (sub-uh-RAK-noyd space) The space between the arachnoid mater and pia mater.

Subclavian trunk (sub-KLAY-vee-an trunk) The lymphatic trunk that drains lymph from the upper limbs.

Subcutaneous (sub-ku-TAY-nee-us) Having to do with being under the skin.

Submucosa (sub-mu-KO-sa) The layer of the GI tract between the mucosa and muscularis layers; made up of areolar connective tissue, blood vessels, and a network of nerves called the submucosa plexus.

Substantia nigra (sub-STAN-chee-uh NI-gruh) A layer of gray matter in the midbrain that contains the cell bodies of dopamine-producing nerve cells whose secretion tends to be deficient in Parkinson's disease.

Sudden infant death syndrome (SIDS) The sudden and unexpected death of an infant under one year of age, with no explainable cause of death. Also called crib death.

Sudoriferous glands (soo-dor-RIF-or-us GLANDZ) Sweat glands.

Sulcus (SUL-kus) A groove between the convolutions of the brain.

Superficial (soop-er-FISH-al) Located on or near the surface of the body or organ.

Superior (soo-PEER-yur) Upper, above.

Supination (soo-pin-A-shun) Turning the palm of the hand up or lying face up on the back.

Supine position (SU-pine PO-si-shun) Lying on the back.

Supporting cells (suh-PORT-ing cells) Cells in the olfactory epithelium that provide nutrition, support, and insulation for the receptors.

Surfactant (sur-FAK-tant) A complex molecule of lipoproteins and phospholipids secreted by Type II alveolar cells to reduce the surface tension of the lungs.

Susceptibility (sus-sep-tuh-BIL-ih-tee) The lack of resistance or vulnerability to disease.

Suture (SU-chure) An immovable joint between the bones of the skull.

Symptoms (SIM-tumz) Subjective findings that for the most part are obvious only to the patient.

Synaptic cleft (si-NAP-tik KLEFT) The space between neurons at a nerve synapse across which a nerve impulse is transmitted by a neurotransmitter; also called a synaptic gap.

Synaptic knob (si-NAP-tik nob) The part at the end of each axon branch with vesicles that contain neurotransmitters.

Syndrome (SIN-drome) A group or cluster of symptoms or conditions.

Synergist (SIN-er-jist) A muscle that helps the prime mover by stabilizing joints.

Synovial (sin-OH-vee-al) Describing the type of joint that is the most moveable.

Syphilis (SIF-i-lis) A sexually transmitted infection caused by the bacterium *Treponema pallidum*.

Systemic circuit (sis-TEM-ik SIR-kut) Systemic circulation; blood flow from the heart through the body and back to the heart.

Systemic lupus erythematosus (SLE) (sis-TEM-ik LOO-pus er-i-thuh-ma-TO-sis) An autoimmune disease that among other things causes joint pain and swelling.

Systolic pressure (sis-TOL-ik PREH-shur) Blood pressure at its greatest in the arteries, when the ventricles of the heart contract.

t

Tachycardia (tak-ih-KAR-dee-uh) Heart rate greater than 100 beats per minute.

Tachypnea (ta-KIP-nee-uh) Rapid breathing.

Telophase (TEL-o-faze) The fourth and final phase of mitosis.

Temporal bones (TEM-por-al bones) Paired bones on each side of the head that form the lower sides of the skull.

Temporal lobes (TEM-por-ul lobes) The lobes of the brain responsible for hearing, as well as interpretation and memory of sensory experiences.

Temporomandibular joint (TMJ) (tem-po-ro-man-DIB-yoo-lar JOYNT) The joint of the jaw that allows us to open and close our mouth.

Tendon (TEN-dun) A tough, cordlike structure made of fibrous connective tissue that connects a muscle to a bone.

Tendonitis (ten-duh-NI-tis) Painful inflammation of a tendon and of the tendon-muscle attachment to a bone.

Tension headache (TEN-shun HED-ake) A diffuse, mild to moderate pain that is often described as feeling like a tight band around your head. A tension headache is the most common type of headache, and yet its causes are not well understood.

Tentative diagnosis (TEN-ta-tive di-ag-NO-sis) A "temporary" diagnosis until more information is gathered.

Teratogens (ter-AH-to-jens) Chemical, physical, and biologic agents that can cause birth defects.

Teratology (ter-uh-TOL-uh-jee) The study of developmental anomalies.

Tertiary bronchioles (TUR-shee-air-ee BRON-kee-olz) Air passages that branch off from the secondary bronchi in the lungs.

Testes (TES-teez) Testicles; the male gonads responsible for producing sperm and testosterone.

Testosterone (tes-TOS-ter-one) A male sex hormone needed for the production of sperm and responsible for male secondary sex characteristics.

Tetanus (TET-uh-nus) A disease caused by *Clostriduim tetani*; commonly called lockjaw.

Tetany (TET-uh-nee) A condition of muscle spasms, cramps, uncontrolled twitching, and sometimes seizures.

Thalamus (THAL-a-mus) The relay station for sensory information that is conveyed to the cerebral cortex for interpretation.

Thalassemia major (thal-uh-SEE-mee-uh MA-jur) The more severe form of thalassemia, which is a form of anemia in which a defective hemoglobin molecule causes small, pale, short-lived RBCs.

Thalassemia minor (thal-uh-SEE-mee-uh MI-nur) A form of thalassemia that causes only minimal symptoms, if any.

Thermoreceptors (THERM-o-re-sep-turs) Free nerve endings that send a message about warm and cold temperature variations to the hypothalamus in the brain.

Thoracic cavity (tho-RAS-ik KAV-ih-tee) Chest cavity.

Thoracic duct (tho-RAS-ik dukt) The largest lymphatic vessel, which drains lymph from all parts of the body except those that are drained by the right lymphatic duct.

Thoracic vertebrae (tho-RAS-ik VER-tuh-bray) The posterior attachments for the 12 pairs of ribs.

Thoracocentesis (thor-uh-ko-sin-TEE-sis) Needle puncture of the thorax to withdraw fluid.

Thoracostomy (thor-uh-KOS-to-mee) Insertion of a tube into the thorax to continually drain fluid.

Thrombocytes (THROM-bo-sites) The cell fragments present in blood that play a role in stopping blood flow from an injured blood vessel.

Thrombocythemia (throm-bo-si-THEE-mee-uh) A condition characterized by an increase in the platelet count.

Thrombocytopenia (throm-bo-si-to-PEEN-ee-ah) A disorder in which a person has too few platelets and is therefore at risk of increased bleeding.

Thrombosis (throm-BO-sis) The formation of a thrombus (clot) which is an aggregate of coagulated blood containing platelets, fibrin,.and other elements within a blood vessel.

Thrombus (THROM-bus) A stationary clot formed in an unbroken blood vessel.

Thymosin (THI-mo-sin) A hormone produced by the thymus that stimulates the production of mature lymphocytes.

Thymus (THI-mus) A bilobed organ of the immune system located in the mediastinum, which is responsible for the maturation of T lymphocytes (T cells).

Thyroid cartilage (THI-royd KAR-tih-lij) The largest single cartilage of the larynx.

Tinea (TIN-ee-uh) Ringworm; a fungal skin infection that gets its name because it appears serpentine like a worm.

Tinea barbae (TIN-ee-uh BAR-bee) A fungal infection of the beard.

Tinea cruris (TIN-ee-uh KRU̱ ris) A fungal infection commonly called jock itch.

Tinea pedis (TIN-ee-uh PE-dis) A fungal infection commonly called athlete's foot.

Tinea unguium (TIN-ee-uh ung-GWE-um) A fungal infection of the nails.

Tinnitus (TIN-ih-tus) Ringing in the ears.

Tissue (TISH-oo) A group of similar cells that combine to perform a specific function.

Tonsils (TON-sils) An aggregation of lymphatic nodules located in the throat.

Torticollis (tor-tih-KOL-is) A condition due to abnormally contracted neck muscles in which the head bends toward the side of the contracted muscle; also called wry neck.

Totipotent (to-TIP-o-tent) Capable of developing into virtually any cell type.

Trachea (TRA-ke-a) The airway extending from the larynx to the main stem bronchus, also known as the windpipe.

Tracheostomy (tray-kee-OS-to-mee) An emergency procedure performed when the airway is blocked; creates an opening in the trachea from outside the throat to allow air to enter the lungs, bypassing the blockage.

Transcytosis (trans-si-TO-sis) A combination of endocytosis and exocytosis going on simultaneously.

Transient ischemic attacks (TIA) (TRAN-ze-ent is-KE-mik a-TAKS) Also referred to as "mini-strokes," TIAs are not really strokes at all but are caused by temporary or transient interruptions of the blood supply to the brain.

Transmission electron microscope (TEM) (tranz-MIH-shun e-LEK-tron MI-kro-skope) Similar to a light microscope, except it uses an electron gun to shoot electrons at the specimen instead of light reflection.

Transplant rejection (TRANZ-plant re-JEK-shun) Immune reaction to transplanted tissues.

Transurethral resection of the prostate (TURP) (trans-yoo-RE-thral re-SEK-shun of the PRAH-state) Removal of enlarged tissue from the prostate.

Transverse (trans-VERSE) A plane that is also described as horizontal and divides the body into upper and lower portions.

Transverse colon (trans-VERS KO-lon) Part of the large intestine that horizontally crosses the abdominal cavity.

Trichomoniasis (trik-o-mo-NI-a-sis) A common, curable sexually transmitted infection.

Tricuspid valve (tri-KUS-pid valve) The atrioventricular valve between the right atrium and right ventricle.

Triglyceride (try-GLIS-er-ide) A molecule formed from one molecule of glycerol and three fatty acid molecules.

Trigone (TRI-gohn) The triangle formed by the three openings of the urethra and the two ureters.

Trisomy 21 (TRI-so-me 21) Also known as Down syndrome, a genetic disorder where there are three copies of chromosome 21 rather than the normal two.

Trismus (TRIZ-mus) Tetanus (lockjaw).

Trophoblast (TRO-fo-blast) The cells that form the walls of the blastocyst.

True ribs (TRU RIBZ) The first seven pairs of ribs are called true ribs because they attach directly to the sternum through pieces of cartilage called costal cartilage.

True vocal cords (Tru VO-kul cords) The lower vestibular folds in the larynx, which produce sound.

Tuberculosis (TB) (too-bur-kyoo-LO-sis) A disease caused by various strains of the bacterium *Mycobacterium tuberculosis*.

Tumor (TU-mer) A solid neoplasm.

Tunica externa (TOO-ni-kuh ek-STER-nuh) The outermost layer of an arterial wall.

Tunica intima (TOO-ni-kuh IN-tih-muh) The innermost layer of an arterial wall.

Tunica media (TOO-ni-kuh MEE-dee-uh) The middle layer of an arterial wall, made up of smooth muscle that can constrict or dilate to help control blood pressure.

Turner's syndrome (TUR-nerz SIN-drome) A disorder that results when females have a single X chromosome.

Tympanic membrane (tim-PAN-ik MEM-brane) Also called eardrum; the boundary between the external and middle ears.

Type I alveolus (type I al-VEE-oh-lus) A type of alveolus in the lungs that is composed of squamous cells; the actual site of oxygen and carbon dioxide exchange.

Type II alveolus (type II al-VEE-oh-lus) A type of alveolus in the lungs that produces surfactant to break the surface tension of the lungs, allowing them to expand.

u

Umami (oo-MAH-mee) Fifth basic savory taste that has become universally recognized.

Universal donors (yoo-ni-VER-sul DOH-nurz) People with type O blood; because their blood has neither antigen A nor antigen B, their blood can be given to most people.

Universal recipients (yoo-ni-VER-sul re-SIP-ee-ents) People with type AB blood; because their blood has neither anti-A nor anti-B antibodies, they can receive all ABO blood types.

Upper esophageal sphincter (UES) (UP-er e-sof-uh-JEE-ul SFINK-ter) Made up of skeletal muscle; controls food entering the esophagus from the laryngopharynx.

Urea (yu-RE-uh) Waste product in urine formed by the breakdown of proteins and nucleic acids.

Ureter (YU-reh-ter) The tube that drains urine from each kidney to the bladder.

Urethra (yu-REE-thrah) The tube that expels urine from the bladder out of the body.

Urethral meatus (yu-REE-thral me-A-tis) The opening in the urethra through which urine is secreted.

Uric acid (YU-rik A-sid) Waste product in urine formed by the breakdown of proteins and nucleic acids.

Urinary bladder (YU-ri-nar-e BLAD-der) A distensible (expandable) organ that is located in the pelvic cavity just behind the pubic symphysis.

Uterine (endometrial) cancer (YOO-ter-in KAN-ser) Cancer of the uterus; most common in postmenopausal women.

Uterus (YOO-ter-us) The hollow, muscular organ in the female that is the site of menstruation, implantation of the embryo, and development of the fetus.

Uvea (YOO-ve-a) The middle layer of the eyeball.

Uvula (YOO-vyoo-lah) The projection off the posterior aspect of the soft palate that prevents food and liquids from entering the nasal cavity during swallowing.

Uvulotomy (yoo-vyoo-LOT-o-mee) Surgery to remove a portion of the soft palate.

v

Vagina (vah-JIE-nuh) The muscular, tubular organ that extends from the uterus to the vestibule.

Vaginal orifice (VAJ-uh-nul OR-ih-fis) The opening to the vagina.

Vaginitis (vah-ji-NI-tis) Inflammation of the vagina.

Varicose veins (VAR-ih-kohs vaynz) Tortuous or twisted, dilated veins, usually seen in the legs.

Varicosities (var-ih-KOS-ih-teez) Varicose veins.

Vascular shock (VAS-kyoo-lar SHOK) Shock caused by vasodilation (dilation of the blood vessels).

Vas deferens (vas DEF-er-enz) A tubular structure connected to each epididymis that carries sperm cells to the urethra.

Vasectomy (vah-SEK-toe-mee) The cutting and tying or fulguration of the vas deferens to prevent sperm from reaching and fertilizing the ovum.

Vasoconstriction (va-zo-kon-STRIK-shun) Constriction of a blood vessel, which causes an increase in blood pressure.

Vasodilation (va-zo-di-LAY-shun) Relaxation or dilation of a blood vessel, which causes a decrease in blood pressure.

Vasopressin (VAS-oh-press-in) Antidiuretic hormone.

Ventral (VEN-tral) Toward the front of the body; also known as anterior.

Ventricles (VEN-tri-kulz) Interconnected cavities within the brain; also, the superior chambers of the heart.

Ventricular diastole (ven-TRIK-yoo-lar di-AS-to-lee) Relaxation of the ventricles.

Ventricular fibrillation (ven-TRIK-yoo-lar fib-rih-LA-shun) A life-threatening arrhythmia in which the ventricles receive rapid, uncoordinated electrical impulses that cause them to quiver instead of producing a coordinated contraction.

Ventricular systole (ven-TRIK-yoo-lar SIS-to-lee) Contraction of the ventricles.

Vermiform appendix (VER-mih-form ah-PEN-dix) The wormlike projection off the cecum that contains lymphoid tissue and has a role in immunity.

Verrucae warts (ver-ROO-kee WORTZ) Harmless skin growths caused by the human papillomavirus (HPV).

Vertebral canal (VER-the-bral kah-NAL) The spinal cavity, in which the spinal cord is located.

Vertebra prominens (VER-teh-brah PROM-i-nenz) The seventh cervical vertebra, so named because of its large spinal process.

Vertigo (VER-tih-go) Dizziness or the sensation of moving around in space.

Very low-density lipoproteins (VLDLs) (VAR-ee LO DEN-sih-tee li-po-PRO-teenz) Consist of proteins, triglycerides, phospholipids, and cholesterol and are made in the liver.

Vesicle (VES-i-kul) A small blister; cold sores are an example.

Vestibule (VES-tih-byool) The area between the semicircular canals and the cochlea that functions in equilibrium; the small space at the beginning of the vagina.

Viable (VI-a-bul) Living.

Vibrio (VIB-re-o) A curved, comma-shaped microorganism.

Virion (VI-re-on) The infectious part of a virus.

Virulence (VIR-yoo-lents) The degree of pathogenicity of an organism that has the ability to cause disease.

Visceral muscle (VIS-er-al MUS-el) Involuntary muscle; smooth muscle.

Visceral peritoneum (VIS-er-al per-ih-to-NEE-um) Innermost wall of the serosa.

Visceral pleura (VIS-er-al PLOO-rah) The innermost membrane of the pleural sac.

Vitreous chamber (VIH-tree-us CHAYM-ber) The posterior cavity of the eye, which is filled with vitreous humor.

Vitreous humor (VIH-tree-us HU-mor) The thick, jellylike substance that fills the posterior cavity of the eye and helps maintain the shape of the eye.

Voluntary muscle (VOL-un-tar-ee MUS-el) Another term for skeletal muscle.

Vomer (VO-mer) A single bone that forms the inferior part of the nasal septum also known as the "plow" because of its shape.

Vulva (VUL-vah) The external genitalia of the female.

Vulvovaginitis (vul-vo-vaj-ih-NI-tis) Inflammation of the vulva and vagina.

X

Xenograft (ZEE-no-graft) Transplant in which the tissue is taken from another species, such as a pig or other primate.

Xerostomia (ze-ro-STO-me-a) Dry mouth.

Xiphoid process (ZI-foyd PRAH-sess) The most inferior of the three bones that make up the sternum.

Y

Yolk sac (yoke sak) An extraembryonic membrane that transfers nutrients to the embryo and is the site of blood cell formation for the embryo.

Z

Zona pellucida (ZO-na peh-LOO-sih-duh) A layer of the ovum surrounding the cell membrane.

Zoonotic disease (zoh-uh-NOT-ik di-ZEEZ) A disease that is transmissible from an animal to humans.

Zygomatic bones (zi-go-MAT-ik bones) Paired bones that are more commonly called the cheekbones.

Zygote (ZI-gote) The fertilized ovum formed from the union of a sperm and an ovum.

credits

TEXT CREDITS

Chapter 1
Fig. 1-4: From *Hole's Human Anatomy & Physiology,* 12e, by Shier/Butler/and Lewis. Copyright © 2009. Reprinted by permission of McGraw-Hill Companies Inc.; Fig. 1-5: From *Hole's Human Anatomy & Physiology,* 12e, by Shier/Butler/and Lewis. Copyright © 2009. Reprinted by permission of McGraw-Hill Companies Inc.; Fig. 1-7: From *Hole's Human Anatomy & Physiology,* 12e, by Shier/Butler/and Lewis. Copyright © 2009. Reprinted by permission of McGraw-Hill Companies Inc.; Fig. 1-8: From *Hole's Human Anatomy & Physiology,* 12e, by Shier/Butler/and Lewis. Copyright © 2009. Reprinted by permission of McGraw-Hill Companies Inc.

Chapter 2
Fig. 2-2: From *Hole's Human Anatomy & Physiology,* 12e, by Shier/Butler/and Lewis. Copyright © 2009. Reprinted by permission of McGraw-Hill Companies Inc.

Chapter 3
Fig. 3-2 From *Hole's Human Anatomy & Physiology,* 12e, by Shier/Butler/and Lewis. Copyright © 2009. Reprinted by permission of McGraw-Hill Companies Inc.; Fig. 3-3: From *Hole's Human Anatomy & Physiology,* 12e, by Shier/Butler/and Lewis. Copyright © 2009. Reprinted by permission of McGraw-Hill Companies Inc.; Table 3-1: From *Hole's Human Anatomy & Physiology,* 12e, by Shier/Butler/and Lewis. Copyright © 2009. Reprinted by permission of McGraw-Hill Companies Inc.; Fig. 3-4: From *Hole's Human Anatomy & Physiology,* 12e, by Shier/Butler/and Lewis. Copyright © 2009. Reprinted by permission of McGraw-Hill Companies Inc.; Fig. 3-5: From *Hole's Human Anatomy & Physiology,* 12e, by Shier/Butler/and Lewis. Copyright © 2009. Reprinted by permission of McGraw-Hill Companies Inc.

Chapter 4
Fig. 4-11: From *Hole's Human Anatomy & Physiology,* 12e, by Shier/Butler/and Lewis. Copyright © 2009. Reprinted by permission of McGraw-Hill Companies Inc.

Chapter 6
Fig. 6-1: From *Hole's Human Anatomy & Physiology,* 12e, by Shier/Butler/and Lewis. Copyright © 2009. Reprinted by permission of McGraw-Hill Companies Inc.; Fig. 6-2: From *Hole's Human Anatomy & Physiology,* 12e, by Shier/Butler/and Lewis. Copyright © 2009. Reprinted by permission of McGraw-Hill Companies Inc.; Fig. 6-3: From *Hole's Human Anatomy & Physiology,* 12e, by Shier/Butler/and Lewis. Copyright © 2009. Reprinted by permission of McGraw-Hill Companies Inc.; Fig. 6-4: From *Hole's Human Anatomy & Physiology,* 12e, by Shier/Butler/and Lewis. Copyright © 2009. Reprinted by permission of McGraw-Hill Companies Inc.; Fig. 6-5: From *Hole's Human Anatomy & Physiology,* 12e, by Shier/Butler/and Lewis. Copyright © 2009. Reprinted by permission of McGraw-Hill Companies Inc.; Fig. 6-6: From *Hole's Human Anatomy & Physiology,* 12e, by Shier/Butler/and Lewis. Copyright © 2009. Reprinted by permission of McGraw-Hill Companies Inc.; Fig. 6-7: From *Hole's Human Anatomy & Physiology,* 12e, by Shier/Butler/and Lewis. Copyright © 2009. Reprinted by permission of McGraw-Hill Companies Inc.; Fig. 6-8: From *Hole's Human Anatomy & Physiology,* 12e, by Shier/Butler/and Lewis. Copyright © 2009. Reprinted by permission of McGraw-Hill Companies Inc.; Fig. 6-9: From *Hole's Human Anatomy & Physiology,* 12e, by Shier/Butler/and Lewis. Copyright © 2009. Reprinted by permission of McGraw-Hill Companies Inc.; Fig.6-10: From *Hole's Human Anatomy & Physiology,* 12e, by Shier/Butler/and Lewis. Copyright © 2009. Reprinted by permission of McGraw-Hill Companies Inc.

Chapter 7
Fig. 7-2: From *Hole's Human Anatomy & Physiology,* 12e, by Shier/Butler/and Lewis. Copyright © 2009. Reprinted by permission of McGraw-Hill Companies Inc.; Fig. 7-3: From *Hole's Human Anatomy & Physiology,* 12e, by Shier/Butler/and Lewis. Copyright © 2009. Reprinted by permission of McGraw-Hill Companies Inc.; Fig. 7-21: From *Hole's Human Anatomy & Physiology,* 12e, by Shier/Butler/and Lewis. Copyright © 2009. Reprinted by permission of McGraw-Hill Companies Inc.

Chapter 8
Fig. 8-10: From *Hole's Human Anatomy & Physiology,* 12e, by Shier/Butler/and Lewis. Copyright © 2009. Reprinted by permission of McGraw-Hill Companies Inc.; Fig. 8-11: From *Hole's Human Anatomy & Physiology,* 12e, by Shier/Butler/and Lewis. Copyright © 2009. Reprinted by permission of McGraw-Hill Companies Inc.; Fig. 8-12: From *Hole's Human Anatomy & Physiology,* 12e, by Shier/Butler/and Lewis. Copyright © 2009. Reprinted by permission of McGraw-Hill Companies Inc.; Fig. 8-13: From *Hole's Human Anatomy & Physiology,* 12e, by Shier/Butler/and Lewis. Copyright © 2009. Reprinted by permission of McGraw-Hill Companies Inc.; Fig. 8-15: From *Hole's Human Anatomy & Physiology,* 12e, by Shier/Butler/and Lewis. Copyright © 2009. Reprinted by permission of McGraw-Hill Companies Inc.; Fig. 8-16: From *Hole's Human Anatomy & Physiology,* 12e, by Shier/Butler/and Lewis. Copyright © 2009. Reprinted by permission of McGraw-Hill Companies Inc.; Fig. 8-18: From *Hole's Human Anatomy & Physiology,* 12e, by Shier/Butler/and Lewis. Copyright © 2009. Reprinted by permission of McGraw-Hill Companies Inc.; Fig. 8-19: From *Hole's Human Anatomy & Physiology,* 12e, by Shier/Butler/and Lewis. Copyright © 2009. Reprinted by permission of McGraw-Hill Companies Inc.; Fig. 8-20: From *Hole's Human Anatomy & Physiology,* 12e, by Shier/Butler/and Lewis. Copyright © 2009. Reprinted by permission of McGraw-Hill Companies Inc.; Fig. 8-21: From *Hole's Human Anatomy & Physiology,* 12e, by Shier/Butler/and Lewis. Copyright © 2009. Reprinted by permission of McGraw-Hill Companies Inc.; Fig. 8-22: From *Hole's Human Anatomy & Physiology,* 12e, by Shier/Butler/and Lewis. Copyright © 2009. Reprinted by permission of McGraw-Hill Companies Inc.; Fig. 8-23: From *Hole's Human Anatomy & Physiology,* 12e, by Shier/Butler/and Lewis. Copyright © 2009. Reprinted by permission of McGraw-Hill Companies Inc.; Fig. 8-24: From *Hole's Human Anatomy & Physiology,* 12e, by Shier/Butler/and Lewis. Copyright © 2009. Reprinted by permission of McGraw-Hill Companies Inc.; Fig. 8-25: From *Hole's Human Anatomy & Physiology,* 12e, by

Shier/Butler/and Lewis. Copyright © 2009. Reprinted by permission of McGraw-Hill Companies Inc.; Fig. 8-26: From *Hole's Human Anatomy & Physiology,* 12e, by Shier/Butler/and Lewis. Copyright © 2009. Reprinted by permission of McGraw-Hill Companies Inc.; Fig. 8-27 From *Hole's Human Anatomy & Physiology,* 12e, by Shier/Butler/and Lewis. Copyright © 2009. Reprinted by permission of McGraw-Hill Companies Inc.

Chapter 9

Fig. 9-1: From *Hole's Human Anatomy & Physiology,* 12e, by Shier/Butler/and Lewis. Copyright © 2009. Reprinted by permission of McGraw-Hill Companies Inc.; Fig. 9-2: From *Hole's Human Anatomy & Physiology,* 12e, by Shier/Butler/and Lewis. Copyright © 2009. Reprinted by permission of McGraw-Hill Companies Inc.; Fig. 9-3: From *Hole's Human Anatomy & Physiology,* 12e, by Shier/Butler/and Lewis. Copyright © 2009. Reprinted by permission of McGraw-Hill Companies Inc.; Fig. 9-12: From *Hole's Human Anatomy & Physiology,* 12e, by Shier/Butler/and Lewis. Copyright © 2009. Reprinted by permission of McGraw-Hill Companies Inc.; Fig. 9-13: From *Hole's Human Anatomy & Physiology,* 12e, by Shier/Butler/and Lewis. Copyright © 2009. Reprinted by permission of McGraw-Hill Companies Inc.; Fig. 9-14: From *Hole's Human Anatomy & Physiology,* 12e, by Shier/Butler/and Lewis. Copyright © 2009. Reprinted by permission of McGraw-Hill Companies Inc.; Fig. 9-15: From *Hole's Human Anatomy & Physiology,* 12e, by Shier/Butler/and Lewis. Copyright © 2009. Reprinted by permission of McGraw-Hill Companies Inc.; Fig. 9-16: From *Hole's Human Anatomy & Physiology,* 12e, by Shier/Butler/and Lewis. Copyright © 2009. Reprinted by permission of McGraw-Hill Companies Inc.; Fig. 9-17: From *Hole's Human Anatomy & Physiology,* 12e, by Shier/Butler/and Lewis. Copyright © 2009. Reprinted by permission of McGraw-Hill Companies Inc.; Fig. 9-18: From *Hole's Human Anatomy & Physiology,* 12e, by Shier/Butler/and Lewis. Copyright © 2009. Reprinted by permission of McGraw-Hill Companies Inc.; Fig. 9-19: From *Hole's Human Anatomy & Physiology,* 12e, by Shier/Butler/and Lewis. Copyright © 2009. Reprinted by permission of McGraw-Hill Companies Inc.; Fig. 9-20: From *Hole's Human Anatomy & Physiology,* 12e, by Shier/Butler/and Lewis. Copyright © 2009. Reprinted by permission of McGraw-Hill Companies Inc.; Fig. 9-21: From *Hole's Human Anatomy & Physiology,* 12e, by Shier/Butler/and Lewis. Copyright © 2009. Reprinted by permission of McGraw-Hill Companies Inc.; Fig. 9-22: From *Hole's Human Anatomy & Physiology,* 12e, by Shier/Butler/and Lewis. Copyright © 2009. Reprinted by permission of McGraw-Hill Companies Inc.; Fig. 9-23: From *Hole's Human Anatomy & Physiology,* 12e, by Shier/Butler/and Lewis. Copyright © 2009. Reprinted by permission of McGraw-Hill Companies Inc.

Chapter 10

Fig. 10-1: From *Hole's Human Anatomy & Physiology,* 12e, by Shier/Butler/and Lewis. Copyright © 2009. Reprinted by permission of McGraw-Hill Companies Inc.; Table 10-1: From *Hole's Human Anatomy & Physiology,* 12e, by Shier/Butler/and Lewis. Copyright © 2009. Reprinted by permission of McGraw-Hill Companies Inc.; Table 10-2: From *Hole's Human Anatomy & Physiology,* 12e, by Shier/Butler/and Lewis. Copyright © 2009. Reprinted by permission of McGraw-Hill Companies Inc.; Fig. 10-2: From *Hole's Human Anatomy & Physiology,* 12e, by Shier/Butler/and Lewis. Copyright © 2009. Reprinted by permission of McGraw-Hill Companies Inc.; Table 10-4: From *Hole's Human Anatomy & Physiology,* 12e, by Shier/Butler/and Lewis. Copyright © 2009. Reprinted by permission of McGraw-Hill Companies Inc.; Fig. 10-10: From *Hole's Human Anatomy & Physiology,* 12e, by Shier/Butler/and Lewis. Copyright © 2009. Reprinted by permission of McGraw-Hill Companies Inc.; Fig. 10-13: From *Hole's Human Anatomy & Physiology,* 12e, by Shier/Butler/and Lewis. Copyright © 2009. Reprinted by permission of McGraw-Hill Companies Inc.; Table 10-5: From *Hole's Human Anatomy & Physiology,* 12e, by Shier/Butler/and Lewis. Copyright © 2009. Reprinted by permission of McGraw-Hill Companies Inc.; Fig. 10-15: From Medical Assisting, 4/e by Kathryn Booth, et al. Copyright © 2011. Reprinted by permission of McGraw-Hill Companies, Inc.; Fig. 10-21: From *Hole's Human Anatomy & Physiology,* 12e, by Shier/Butler/and Lewis. Copyright © 2009. Reprinted by permission of McGraw-Hill Companies Inc.; Fig. 10-23: From *Hole's Human Anatomy & Physiology,* 12e, by Shier/Butler/and Lewis. Copyright © 2009. Reprinted by permission of McGraw-Hill Companies Inc.; Fig. 10-24: From *Hole's Human Anatomy & Physiology,* 12e, by Shier/Butler/and Lewis. Copyright © 2009. Reprinted by permission of McGraw-Hill Companies Inc.; Table 10-8: From *Hole's Human Anatomy & Physiology,* 12e, by Shier/Butler/and Lewis. Copyright © 2009. Reprinted by permission of McGraw-Hill Companies Inc.; Fig. 10-26: From *Hole's Human Anatomy & Physiology,* 12e, by Shier/Butler/and Lewis. Copyright © 2009. Reprinted by permission of McGraw-Hill Companies Inc.; Fig. 10-27: From *Hole's Human Anatomy & Physiology,* 12e, by Shier/Butler/and Lewis. Copyright © 2009. Reprinted by permission of McGraw-Hill Companies Inc.; Fig. 10-28: From *Hole's Human Anatomy & Physiology,* 12e, by Shier/Butler/and Lewis. Copyright © 2009. Reprinted by permission of McGraw-Hill Companies Inc.; Fig. 10-29: From *Hole's Human Anatomy & Physiology,* 12e, by Shier/Butler/and Lewis. Copyright © 2009. Reprinted by permission of McGraw-Hill Companies Inc.; Table 10-9: From *Hole's Human Anatomy & Physiology,* 12e, by Shier/Butler/and Lewis. Copyright © 2009. Reprinted by permission of McGraw-Hill Companies Inc. Fig. 10-30: From *Hole's Human Anatomy & Physiology,* 12e, by Shier/Butler/and Lewis. Copyright © 2009. Reprinted by permission of McGraw-Hill Companies Inc.; Fig. 10-31: From *Hole's Human Anatomy & Physiology,* 12e, by Shier/Butler/and Lewis. Copyright © 2009. Reprinted by permission of McGraw-Hill Companies Inc.; Fig. 10-32: From *Hole's Human Anatomy & Physiology,* 12e, by Shier/Butler/and Lewis. Copyright © 2009. Reprinted by permission of McGraw-Hill Companies Inc.; Fig. 10-33: From *Hole's Human Anatomy & Physiology,* 12e, by Shier/Butler/and Lewis. Copyright © 2009. Reprinted by permission of McGraw-Hill Companies Inc.; Fig. 10-34: From *Hole's Human Anatomy & Physiology,* 12e, by Shier/Butler/and Lewis. Copyright © 2009. Reprinted by permission of McGraw-Hill Companies Inc.

Chapter 11

Fig. 11-1: From *Hole's Human Anatomy & Physiology,* 12e, by Shier/Butler/and Lewis. Copyright © 2009. Reprinted by permission of McGraw-Hill Companies Inc.; Fig. 11-2: From *Hole's Human Anatomy & Physiology,* 12e, by Shier/Butler/and Lewis. Copyright © 2009. Reprinted by permission of McGraw-Hill Companies Inc.; Fig. 11-3: From *Hole's Human Anatomy & Physiology,* 12e, by Shier/Butler/and Lewis. Copyright © 2009. Reprinted by permission of McGraw-Hill Companies Inc.; Fig. 11-4 (A): From *Hole's Human Anatomy & Physiology,* 12e, by Shier/Butler/and Lewis. Copyright © 2009. Reprinted by permission of McGraw-Hill Companies Inc.; (B) 23-4: From *Medical Assisting,* 4/e by Kathryn Booth, et al. Copyright © 2011. Reprinted by permission of McGraw-Hill Companies, Inc.; Fig. 11-6: From *Hole's Human Anatomy & Physiology,* 12e, by Shier/Butler/and Lewis. Copyright © 2009. Reprinted by permission of McGraw-Hill Companies Inc.; Fig. 11-8: From *Hole's Human Anatomy & Physiology,* 12e, by Shier/Butler/and Lewis. Copyright © 2009. Reprinted by permission of McGraw-Hill

Companies Inc.; Table 11-1: From *Hole's Human Anatomy & Physiology,* 12e, by Shier/Butler/and Lewis. Copyright © 2009. Reprinted by permission of McGraw-Hill Companies Inc.; Figure 11-9: From *Hole's Human Anatomy & Physiology,* 12e, by Shier/Butler/and Lewis. Copyright © 2009. Reprinted by permission of McGraw-Hill Companies Inc.; Fig. 11-10: From *Hole's Human Anatomy & Physiology,* 12e, by Shier/Butler/and Lewis. Copyright © 2009. Reprinted by permission of McGraw-Hill Companies Inc.; Fig. 11-11: From *Hole's Human Anatomy & Physiology,* 12e, by Shier/Butler/and Lewis. Copyright © 2009. Reprinted by permission of McGraw-Hill Companies Inc.; Fig. 11-12: From *Hole's Human Anatomy & Physiology,* 12e, by Shier/Butler/and Lewis. Copyright © 2009. Reprinted by permission of McGraw-Hill Companies Inc.; Fig. 11-13: From *Hole's Human Anatomy & Physiology,* 12e, by Shier/Butler/and Lewis. Copyright © 2009. Reprinted by permission of McGraw-Hill Companies Inc.; Fig. 11-14: From *Hole's Human Anatomy & Physiology,* 12e, by Shier/Butler/and Lewis. Copyright © 2009. Reprinted by permission of McGraw-Hill Companies Inc.; Fig. 11-15: From *Hole's Human Anatomy & Physiology,* 12e, by Shier/Butler/and Lewis. Copyright © 2009. Reprinted by permission of McGraw-Hill Companies Inc.; Fig. 11-16: From *Hole's Human Anatomy & Physiology,* 12e, by Shier/Butler/and Lewis. Copyright © 2009. Reprinted by permission of McGraw-Hill Companies Inc.; Fig. 11-17: From *Hole's Human Anatomy & Physiology,* 12e, by Shier/Butler/and Lewis. Copyright © 2009. Reprinted by permission of McGraw-Hill Companies Inc.; Fig. 11-18: From *Hole's Human Anatomy & Physiology,* 12e, by Shier/Butler/and Lewis. Copyright © 2009. Reprinted by permission of McGraw-Hill Companies Inc.

Chapter 12

Fig. 12-1: From *Hole's Human Anatomy & Physiology,* 12e, by Shier/Butler/and Lewis. Copyright © 2009. Reprinted by permission of McGraw-Hill Companies Inc.; Fig. 12-2: From *Hole's Human Anatomy & Physiology,* 12e, by Shier/Butler/and Lewis. Copyright © 2009. Reprinted by permission of McGraw-Hill Companies Inc.; Fig. 12-3: From *Hole's Human Anatomy & Physiology,* 12e, by Shier/Butler/and Lewis. Copyright © 2009. Reprinted by permission of McGraw-Hill Companies Inc.; Fig. 12-5: From *Hole's Human Anatomy & Physiology,* 12e, by Shier/Butler/and Lewis. Copyright © 2009. Reprinted by permission of McGraw-Hill Companies Inc.; Fig. 12-6: From *Hole's Human Anatomy & Physiology,* 12e, by Shier/Butler/and Lewis. Copyright © 2009. Reprinted by permission of McGraw-Hill Companies Inc.; Fig. 12-7: From *Hole's Human Anatomy & Physiology,* 12e, by Shier/Butler/and Lewis. Copyright © 2009. Reprinted by permission of McGraw-Hill Companies Inc.; Fig. 12-8: From *Hole's Human Anatomy & Physiology,* 12e, by Shier/Butler/and Lewis. Copyright © 2009. Reprinted by permission of McGraw-Hill Companies Inc.; Fig. 12-9: From *Hole's Human Anatomy & Physiology,* 12e, by Shier/Butler/and Lewis. Copyright © 2009. Reprinted by permission of McGraw-Hill Companies Inc.; Fig.12-10: From *Hole's Human Anatomy & Physiology,* 12e, by Shier/Butler/and Lewis. Copyright © 2009. Reprinted by permission of McGraw-Hill Companies Inc.; Table 12-1: From *Hole's Human Anatomy & Physiology,* 12e, by Shier/Butler/and Lewis. Copyright © 2009. Reprinted by permission of McGraw-Hill Companies Inc.; Table 12-2: From *Hole's Human Anatomy & Physiology,* 12e, by Shier/Butler/and Lewis. Copyright © 2009. Reprinted by permission of McGraw-Hill Companies Inc.; Table 12-3: From *Hole's Human Anatomy & Physiology,* 12e, by Shier/Butler/and Lewis. Copyright © 2009. Reprinted by permission of McGraw-Hill Companies Inc.; Table 12-4: From *Hole's Human Anatomy & Physiology,* 12e, by Shier/Butler/and Lewis. Copyright © 2009. Reprinted by permission of McGraw-Hill Companies Inc.; Table 12-5: From *Hole's Human Anatomy & Physiology,* 12e, by Shier/Butler/and Lewis. Copyright © 2009. Reprinted by permission of McGraw-Hill Companies Inc.; Table 12-6: From *Hole's Human Anatomy & Physiology,* 12e, by Shier/Butler/and Lewis. Copyright © 2009. Reprinted by permission of McGraw-Hill Companies Inc.; Table 12-7: From *Hole's Human Anatomy & Physiology,* 12e, by Shier/Butler/and Lewis. Copyright © 2009. Reprinted by permission of McGraw-Hill Companies Inc.; Table 12-8: From *Hole's Human Anatomy & Physiology,* 12e, by Shier/Butler/and Lewis. Copyright © 2009. Reprinted by permission of McGraw-Hill Companies Inc.; Table 12-9: From *Hole's Human Anatomy & Physiology,* 12e, by Shier/Butler/and Lewis. Copyright © 2009. Reprinted by permission of McGraw-Hill Companies Inc.

Chapter 13

Table 13-1: From *Hole's Human Anatomy & Physiology,* 12e, by Shier/Butler/and Lewis. Copyright © 2009. Reprinted by permission of McGraw-Hill Companies Inc.; Fig. 13-2: From *Hole's Human Anatomy & Physiology,* 12e, by Shier/Butler/and Lewis. Copyright © 2009. Reprinted by permission of McGraw-Hill Companies Inc.; Fig. 13-3: From *Hole's Human Anatomy & Physiology,* 12e, by Shier/Butler/and Lewis. Copyright © 2009. Reprinted by permission of McGraw-Hill Companies Inc.; Fig. 13-4: From *Hole's Human Anatomy & Physiology,* 12e, by Shier/Butler/and Lewis. Copyright © 2009. Reprinted by permission of McGraw-Hill Companies Inc.; Fig. 13-5: From *Hole's Human Anatomy & Physiology,* 12e, by Shier/Butler/and Lewis. Copyright © 2009. Reprinted by permission of McGraw-Hill Companies Inc.; Fig. 13-6: From *Hole's Human Anatomy & Physiology,* 12e, by Shier/Butler/and Lewis. Copyright © 2009. Reprinted by permission of McGraw-Hill Companies Inc.; Fig. 13-7: From *Hole's Human Anatomy & Physiology,* 12e, by Shier/Butler/and Lewis. Copyright © 2009. Reprinted by permission of McGraw-Hill Companies Inc.; Fig. 13-9: From *Hole's Human Anatomy & Physiology,* 12e, by Shier/Butler/and Lewis. Copyright © 2009. Reprinted by permission of McGraw-Hill Companies Inc.; Fig. 13-10: From *Hole's Human Anatomy & Physiology,* 12e, by Shier/Butler/and Lewis. Copyright © 2009. Reprinted by permission of McGraw-Hill Companies Inc.; Fig. 13-13: From *Hole's Human Anatomy & Physiology,* 12e, by Shier/Butler/and Lewis. Copyright © 2009. Reprinted by permission of McGraw-Hill Companies Inc.; Fig. 13-15: From *Hole's Human Anatomy & Physiology,* 12e, by Shier/Butler/and Lewis. Copyright © 2009. Reprinted by permission of McGraw-Hill Companies Inc.; Fig. 13-16: From *Hole's Human Anatomy & Physiology,* 12e, by Shier/Butler/and Lewis. Copyright © 2009. Reprinted by permission of McGraw-Hill Companies Inc.; Fig. 13-17: From *Hole's Human Anatomy & Physiology,* 12e, by Shier/Butler/and Lewis. Copyright © 2009. Reprinted by permission of McGraw-Hill Companies Inc.; Table 13-2: From *Hole's Human Anatomy & Physiology,* 12e, by Shier/Butler/and Lewis. Copyright © 2009. Reprinted by permission of McGraw-Hill Companies Inc.; Fig. 13-18: From *Hole's Human Anatomy & Physiology,* 12e, by Shier/Butler/and Lewis. Copyright © 2009. Reprinted by permission of McGraw-Hill Companies Inc.; Table 13-3: From *Hole's Human Anatomy & Physiology,* 12e, by Shier/Butler/and Lewis. Copyright © 2009. Reprinted by permission of McGraw-Hill Companies Inc.; Fig. 13-19: From *Hole's Human Anatomy & Physiology,* 12e, by Shier/Butler/and Lewis. Copyright © 2009. Reprinted by permission of McGraw-Hill Companies Inc.; Fig. 13-20: From *Hole's Human Anatomy & Physiology,* 12e, by Shier/Butler/and Lewis. Copyright © 2009. Reprinted by permission of McGraw-Hill Companies Inc.; Fig. 13-21: From *Hole's Human Anatomy & Physiology,* 12e, by Shier/Butler/and Lewis. Copyright © 2009. Reprinted by permission of McGraw-Hill

Companies Inc.; Fig. 13-22: From *Hole's Human Anatomy & Physiology,* 12e, by Shier/Butler/and Lewis. Copyright © 2009. Reprinted by permission of McGraw-Hill Companies Inc.; Table 13-4: From *Hole's Human Anatomy & Physiology,* 12e, by Shier/Butler/and Lewis. Copyright © 2009. Reprinted by permission of McGraw-Hill Companies Inc.

Chapter 14

Fig. 14-1: From *Hole's Human Anatomy & Physiology,* 12e, by Shier/Butler/and Lewis. Copyright © 2009. Reprinted by permission of McGraw-Hill Companies Inc.; Fig. 14-2: From *Hole's Human Anatomy & Physiology,* 12e, by Shier/Butler/and Lewis. Copyright © 2009. Reprinted by permission of McGraw-Hill Companies Inc.; Fig. 14-4: From *Hole's Human Anatomy & Physiology,* 12e, by Shier/Butler/and Lewis. Copyright © 2009. Reprinted by permission of McGraw-Hill Companies Inc.; Table 14-1: From *Hole's Human Anatomy & Physiology,* 12e, by Shier/Butler/and Lewis. Copyright © 2009. Reprinted by permission of McGraw-Hill Companies Inc.; Fig. 14-5: From *Hole's Human Anatomy & Physiology,* 12e, by Shier/Butler/and Lewis. Copyright © 2009. Reprinted by permission of McGraw-Hill Companies Inc.; Table 14-2: From *Hole's Human Anatomy & Physiology,* 12e, by Shier/Butler/and Lewis. Copyright © 2009. Reprinted by permission of McGraw-Hill Companies Inc.; Fig. 14-6: From *Hole's Human Anatomy & Physiology,* 12e, by Shier/Butler/and Lewis. Copyright © 2009. Reprinted by permission of McGraw-Hill Companies Inc.; Fig. 14-7: From *Hole's Human Anatomy & Physiology,* 12e, by Shier/Butler/and Lewis. Copyright © 2009. Reprinted by permission of McGraw-Hill Companies Inc.; Table 14-3: From *Hole's Human Anatomy & Physiology,* 12e, by Shier/Butler/and Lewis. Copyright © 2009. Reprinted by permission of McGraw-Hill Companies Inc.; Fig. 14-8: From *Hole's Human Anatomy & Physiology,* 12e, by Shier/Butler/and Lewis. Copyright © 2009. Reprinted by permission of McGraw-Hill Companies Inc.; Fig. 14-9: From *Hole's Human Anatomy & Physiology,* 12e, by Shier/Butler/and Lewis. Copyright © 2009. Reprinted by permission of McGraw-Hill Companies Inc.; Fig. 14-10: From *Hole's Human Anatomy & Physiology,* 12e, by Shier/Butler/and Lewis. Copyright © 2009. Reprinted by permission of McGraw-Hill Companies Inc.; Fig. 14-11: From *Hole's Human Anatomy & Physiology,* 12e, by Shier/Butler/and Lewis. Copyright © 2009. Reprinted by permission of McGraw-Hill Companies Inc.; Table 14-4: From *Hole's Human Anatomy & Physiology,* 12e, by Shier/Butler/and Lewis. Copyright © 2009. Reprinted by permission of McGraw-Hill Companies Inc.; Table 14-5: From *Hole's Human Anatomy & Physiology,* 12e, by Shier/Butler/and Lewis. Copyright © 2009. Reprinted by permission of McGraw-Hill Companies Inc.; Fig. 14-12: From *Hole's Human Anatomy & Physiology,* 12e, by Shier/Butler/and Lewis. Copyright © 2009. Reprinted by permission of McGraw-Hill Companies Inc.; Fig. 14-13: From *Hole's Human Anatomy & Physiology,* 12e, by Shier/Butler/and Lewis. Copyright © 2009. Reprinted by permission of McGraw-Hill Companies Inc.; Table 14-6: From *Hole's Human Anatomy & Physiology,* 12e, by Shier/Butler/and Lewis. Copyright © 2009. Reprinted by permission of McGraw-Hill Companies Inc.; Fig.14-14: From *Hole's Human Anatomy & Physiology,* 12e, by Shier/Butler/and Lewis. Copyright © 2009. Reprinted by permission of McGraw-Hill Companies Inc.; Fig. 14-15: From *Hole's Human Anatomy & Physiology,* 12e, by Shier/Butler/and Lewis. Copyright © 2009. Reprinted by permission of McGraw-Hill Companies Inc.; Fig. 14-17: From *Hole's Human Anatomy & Physiology,* 12e, by Shier/Butler/and Lewis. Copyright © 2009. Reprinted by permission of McGraw-Hill Companies Inc.; Fig. 14-18: From *Hole's Human Anatomy & Physiology,* 12e, by Shier/Butler/and Lewis. Copyright © 2009. Reprinted by permission of McGraw-Hill Companies Inc.; Fig. 14-19: From *Hole's Human Anatomy & Physiology,* 12e, by Shier/Butler/and Lewis. Copyright © 2009. Reprinted by permission of McGraw-Hill Companies Inc.; Table 14-7: From *Hole's Human Anatomy & Physiology,* 12e, by Shier/Butler/and Lewis. Copyright © 2009. Reprinted by permission of McGraw-Hill Companies Inc.; Fig. 14-20: From *Hole's Human Anatomy & Physiology,* 12e, by Shier/Butler/and Lewis. Copyright © 2009. Reprinted by permission of McGraw-Hill Companies Inc.; Fig. 14-21: From *Hole's Human Anatomy & Physiology,* 12e, by Shier/Butler/and Lewis. Copyright © 2009. Reprinted by permission of McGraw-Hill Companies Inc.; Fig. 14-22: From *Hole's Human Anatomy & Physiology,* 12e, by Shier/Butler/and Lewis. Copyright © 2009. Reprinted by permission of McGraw-Hill Companies Inc.; Fig. 14-23: From *Hole's Human Anatomy & Physiology,* 12e, by Shier/Butler/and Lewis. Copyright © 2009. Reprinted by permission of McGraw-Hill Companies Inc.; Fig. 14-24: From *Hole's Human Anatomy & Physiology,* 12e, by Shier/Butler/and Lewis. Copyright © 2009. Reprinted by permission of McGraw-Hill Companies Inc.; Table 14-8: From *Hole's Human Anatomy & Physiology,* 12e, by Shier/Butler/and Lewis. Copyright © 2009. Reprinted by permission of McGraw-Hill Companies Inc.; Fig. 14-25: From *Hole's Human Anatomy & Physiology,* 12e, by Shier/Butler/and Lewis. Copyright © 2009. Reprinted by permission of McGraw-Hill Companies Inc.; Fig. 14-26: From *Hole's Human Anatomy & Physiology,* 12e, by Shier/Butler/and Lewis. Copyright © 2009. Reprinted by permission of McGraw-Hill Companies Inc.; Fig. 14-27: From *Hole's Human Anatomy & Physiology,* 12e, by Shier/Butler/and Lewis. Copyright © 2009. Reprinted by permission of McGraw-Hill Companies Inc.; Fig. 14-29: From *Hole's Human Anatomy & Physiology,* 12e, by Shier/Butler/and Lewis. Copyright © 2009. Reprinted by permission of McGraw-Hill Companies Inc.; Fig. 14-30: From *Hole's Human Anatomy & Physiology,* 12e, by Shier/Butler/and Lewis. Copyright © 2009. Reprinted by permission of McGraw-Hill Companies Inc.; Fig. 14-31: From *Hole's Human Anatomy & Physiology,* 12e, by Shier/Butler/and Lewis. Copyright © 2009. Reprinted by permission of McGraw-Hill Companies Inc.; Fig. 14-32: From *Hole's Human Anatomy & Physiology,* 12e, by Shier/Butler/and Lewis. Copyright © 2009. Reprinted by permission of McGraw-Hill Companies Inc.; Fig. 14-33: From *Hole's Human Anatomy & Physiology,* 12e, by Shier/Butler/and Lewis. Copyright © 2009. Reprinted by permission of McGraw-Hill Companies Inc.; Table 14-9: From *Hole's Human Anatomy & Physiology,* 12e, by Shier/Butler/and Lewis. Copyright © 2009. Reprinted by permission of McGraw-Hill Companies Inc.; Fig. 14-34: From *Hole's Human Anatomy & Physiology,* 12e, by Shier/Butler/and Lewis. Copyright © 2009. Reprinted by permission of McGraw-Hill Companies Inc.; Fig. 14-35: From *Hole's Human Anatomy & Physiology,* 12e, by Shier/Butler/and Lewis. Copyright © 2009. Reprinted by permission of McGraw-Hill Companies Inc.; Fig. 14-36: From *Hole's Human Anatomy & Physiology,* 12e, by Shier/Butler/and Lewis. Copyright © 2009. Reprinted by permission of McGraw-Hill Companies Inc.; Fig. 14-37: From *Hole's Human Anatomy & Physiology,* 12e, by Shier/Butler/and Lewis. Copyright © 2009. Reprinted by permission of McGraw-Hill Companies Inc.

Chapter 15

Fig. 15-1: From *Hole's Human Anatomy & Physiology,* 12e, by Shier/Butler/and Lewis. Copyright © 2009. Reprinted by permission of McGraw-Hill Companies Inc.; Fig. 15-2: From *Hole's Human Anatomy & Physiology,* 12e, by Shier/Butler/and Lewis. Copyright © 2009. Reprinted by permission of McGraw-Hill Companies Inc.; Fig. 15-3: From *Hole's Human Anatomy & Physiology,* 12e, by Shier/Butler/and Lewis. Copyright © 2009. Reprinted by permission of McGraw-Hill Companies Inc.; Fig. 15-4: From *Hole's Human Anatomy & Physiology,* 12e, by

Shier/Butler/and Lewis. Copyright © 2009. Reprinted by permission of McGraw-Hill Companies Inc.; Fig. 15-5: From *Hole's Human Anatomy & Physiology,* 12e, by Shier/Butler/and Lewis. Copyright © 2009. Reprinted by permission of McGraw-Hill Companies Inc.; Fig. 15-7: From *Hole's Human Anatomy & Physiology,* 12e, by Shier/Butler/and Lewis. Copyright © 2009. Reprinted by permission of McGraw-Hill Companies Inc.; Fig. 15-8: From *Hole's Human Anatomy & Physiology,* 12e, by Shier/Butler/and Lewis. Copyright © 2009. Reprinted by permission of McGraw-Hill Companies Inc.; Fig. 15-9: From *Hole's Human Anatomy & Physiology,* 12e, by Shier/Butler/and Lewis. Copyright © 2009. Reprinted by permission of McGraw-Hill Companies Inc.; Fig. 15-10: From *Hole's Human Anatomy & Physiology,* 12e, by Shier/Butler/and Lewis. Copyright © 2009. Reprinted by permission of McGraw-Hill Companies Inc.; Fig. 15-11: From *Hole's Human Anatomy & Physiology,* 12e, by Shier/Butler/and Lewis. Copyright © 2009. Reprinted by permission of McGraw-Hill Companies Inc.; Fig. 15-12: From *Hole's Human Anatomy & Physiology,* 12e, by Shier/Butler/and Lewis. Copyright © 2009. Reprinted by permission of McGraw-Hill Companies Inc.; Fig. 15-13: From *Hole's Human Anatomy & Physiology,* 12e, by Shier/Butler/and Lewis. Copyright © 2009. Reprinted by permission of McGraw-Hill Companies Inc.; Fig. 15-14: From *Hole's Human Anatomy & Physiology,* 12e, by Shier/Butler/and Lewis. Copyright © 2009. Reprinted by permission of McGraw-Hill Companies Inc.; Fig. 15-15: From *Hole's Human Anatomy & Physiology,* 12e, by Shier/Butler/and Lewis. Copyright © 2009. Reprinted by permission of McGraw-Hill Companies Inc.; Fig. 15-16: From *Medical Assisting,* 4/e by Kathryn Booth, et al. Copyright © 2011. Reprinted by permission of McGraw-Hill Companies, Inc.

Chapter 16

Fig. 16-1: From *Hole's Human Anatomy & Physiology,* 12e, by Shier/Butler/and Lewis. Copyright © 2009. Reprinted by permission of McGraw-Hill Companies Inc.; Fig. 16-2: From *Hole's Human Anatomy & Physiology,* 12e, by Shier/Butler/and Lewis. Copyright © 2009. Reprinted by permission of McGraw-Hill Companies Inc.; Fig. 16-3: From *Hole's Human Anatomy & Physiology,* 12e, by Shier/Butler/and Lewis. Copyright © 2009. Reprinted by permission of McGraw-Hill Companies Inc.; Fig. 16-5: From *Hole's Human Anatomy & Physiology,* 12e, by Shier/Butler/and Lewis. Copyright © 2009. Reprinted by permission of McGraw-Hill Companies Inc.; Fig. 16-6: From *Hole's Human Anatomy & Physiology,* 12e, by Shier/Butler/and Lewis. Copyright © 2009. Reprinted by permission of McGraw-Hill Companies Inc.; Fig. 16-7: From *Hole's Human Anatomy & Physiology,* 12e, by Shier/Butler/and Lewis. Copyright © 2009. Reprinted by permission of McGraw-Hill Companies Inc.; Fig. 16-10: From *Hole's Human Anatomy & Physiology,* 12e, by Shier/Butler/and Lewis. Copyright © 2009. Reprinted by permission of McGraw-Hill Companies Inc.; Fig. 16-11: From *Hole's Human Anatomy & Physiology,* 12e, by Shier/Butler/and Lewis. Copyright © 2009. Reprinted by permission of McGraw-Hill Companies Inc.

Chapter 17

Fig. 17-1: From *Hole's Human Anatomy & Physiology,* 12e, by Shier/Butler/and Lewis. Copyright © 2009. Reprinted by permission of McGraw-Hill Companies Inc.; Fig. 17-2: From *Hole's Human Anatomy & Physiology,* 12e, by Shier/Butler/and Lewis. Copyright © 2009. Reprinted by permission of McGraw-Hill Companies Inc.; Table 17-1: From *Hole's Human Anatomy & Physiology,* 12e, by Shier/Butler/and Lewis. Copyright © 2009. Reprinted by permission of McGraw-Hill Companies Inc.; Fig. 17-3: From *Hole's Human Anatomy & Physiology,* 12e, by Shier/Butler/and Lewis. Copyright © 2009. Reprinted by permission of McGraw-Hill Companies Inc.; Fig. 17-4: From *Hole's Human Anatomy & Physiology,* 12e, by Shier/Butler/and Lewis. Copyright © 2009. Reprinted by permission of McGraw-Hill Companies Inc.; Fig. 17-5: From *Hole's Human Anatomy & Physiology,* 12e, by Shier/Butler/and Lewis. Copyright © 2009. Reprinted by permission of McGraw-Hill Companies Inc.; Fig. 17-6: From *Hole's Human Anatomy & Physiology,* 12e, by Shier/Butler/and Lewis. Copyright © 2009. Reprinted by permission of McGraw-Hill Companies Inc.; Fig. 17-7: From *Hole's Human Anatomy & Physiology,* 12e, by Shier/Butler/and Lewis. Copyright © 2009. Reprinted by permission of McGraw-Hill Companies Inc.; Fig. 17-9: From *Hole's Human Anatomy & Physiology,* 12e, by Shier/Butler/and Lewis. Copyright © 2009. Reprinted by permission of McGraw-Hill Companies Inc.; Fig. 17-10: From *Medical Assisting,* 4/e by Kathryn Booth, et al. Copyright © 2011. Reprinted by permission of McGraw-Hill Companies, Inc.; Fig.17-11: From *Hole's Human Anatomy & Physiology,* 12e, by Shier/Butler/and Lewis. Copyright © 2009. Reprinted by permission of McGraw-Hill Companies Inc.

Chapter 18

Fig. 18-1: From *Hole's Human Anatomy & Physiology,* 12e, by Shier/Butler/and Lewis. Copyright © 2009. Reprinted by permission of McGraw-Hill Companies Inc.; Fig. 18-2: From *Hole's Human Anatomy & Physiology,* 12e, by Shier/Butler/and Lewis. Copyright © 2009. Reprinted by permission of McGraw-Hill Companies Inc.; Fig. 18-3: From *Hole's Human Anatomy & Physiology,* 12e, by Shier/Butler/and Lewis. Copyright © 2009. Reprinted by permission of McGraw-Hill Companies Inc.; Table 18-1: From *Hole's Human Anatomy & Physiology,* 12e, by Shier/Butler/and Lewis. Copyright © 2009. Reprinted by permission of McGraw-Hill Companies Inc.; Fig. 18-5: From *Hole's Human Anatomy & Physiology,* 12e, by Shier/Butler/and Lewis. Copyright © 2009. Reprinted by permission of McGraw-Hill Companies Inc.; Fig. 18-6: From *Hole's Human Anatomy & Physiology,* 12e, by Shier/Butler/and Lewis. Copyright © 2009. Reprinted by permission of McGraw-Hill Companies Inc.; Fig. 18-7: From *Hole's Human Anatomy & Physiology,* 12e, by Shier/Butler/and Lewis. Copyright © 2009. Reprinted by permission of McGraw-Hill Companies Inc.; Fig. 18-8: From *Hole's Human Anatomy & Physiology,* 12e, by Shier/Butler/and Lewis. Copyright © 2009. Reprinted by permission of McGraw-Hill Companies Inc.; Fig. 18-9: From *Hole's Human Anatomy & Physiology,* 12e, by Shier/Butler/and Lewis. Copyright © 2009. Reprinted by permission of McGraw-Hill Companies Inc.; Fig. 18-11: From *Hole's Human Anatomy & Physiology,* 12e, by Shier/Butler/and Lewis. Copyright © 2009. Reprinted by permission of McGraw-Hill Companies Inc.; Table 18-2: From *Hole's Human Anatomy & Physiology,* 12e, by Shier/Butler/and Lewis. Copyright © 2009. Reprinted by permission of McGraw-Hill Companies Inc.; Fig. 18-12: From *Hole's Human Anatomy & Physiology,* 12e, by Shier/Butler/and Lewis. Copyright © 2009. Reprinted by permission of McGraw-Hill Companies Inc.; Fig. 18-13: From *Hole's Human Anatomy & Physiology,* 12e, by Shier/Butler/and Lewis. Copyright © 2009. Reprinted by permission of McGraw-Hill Companies Inc.; Table 18-3: From *Hole's Human Anatomy & Physiology,* 12e, by Shier/Butler/and Lewis. Copyright © 2009. Reprinted by permission of McGraw-Hill Companies Inc.; Fig. 18-14: From *Hole's Human Anatomy & Physiology,* 12e, by Shier/Butler/and Lewis. Copyright © 2009. Reprinted by permission of McGraw-Hill Companies Inc.; Fig. 18-15: From *Hole's Human Anatomy & Physiology,* 12e, by Shier/Butler/and Lewis. Copyright © 2009. Reprinted by permission of McGraw-Hill Companies Inc.

Chapter 19

Fig. 19-1: From *Hole's Human Anatomy & Physiology,* 12e, by Shier/Butler/and Lewis. Copyright © 2009. Reprinted by permission of McGraw-Hill Companies Inc.; Fig. 19-2: From *Hole's Human Anatomy & Physiology,* 12e, by Shier/Butler/and Lewis. Copyright © 2009. Reprinted by permission of McGraw-Hill Companies Inc.; Fig. 19-3: From *Hole's Human Anatomy & Physiology,* 12e, by Shier/Butler/and Lewis. Copyright © 2009. Reprinted by permission of McGraw-Hill Companies Inc.; Fig. 19-4: From *Hole's Human Anatomy & Physiology,* 12e, by Shier/Butler/and Lewis. Copyright © 2009. Reprinted by permission of McGraw-Hill Companies Inc.; Fig. 19-5: From *Hole's Human Anatomy & Physiology,* 12e, by Shier/Butler/and Lewis. Copyright © 2009. Reprinted by permission of McGraw-Hill Companies Inc.; Fig. 19-6: From *Hole's Human Anatomy & Physiology,* 12e, by Shier/Butler/and Lewis. Copyright © 2009. Reprinted by permission of McGraw-Hill Companies Inc.; Fig. 19-7: From *Hole's Human Anatomy & Physiology,* 12e, by Shier/Butler/and Lewis. Copyright © 2009. Reprinted by permission of McGraw-Hill Companies Inc.; Fig. 19-8: From *Hole's Human Anatomy & Physiology,* 12e, by Shier/Butler/and Lewis. Copyright © 2009. Reprinted by permission of McGraw-Hill Companies Inc.; Fig. 19-9: From *Hole's Human Anatomy & Physiology,* 12e, by Shier/Butler/and Lewis. Copyright © 2009. Reprinted by permission of McGraw-Hill Companies Inc.; Fig. 19-10: From *Hole's Human Anatomy & Physiology,* 12e, by Shier/Butler/and Lewis. Copyright © 2009. Reprinted by permission of McGraw-Hill Companies Inc.; Fig. 19-11: From *Hole's Human Anatomy & Physiology,* 12e, by Shier/Butler/and Lewis. Copyright © 2009. Reprinted by permission of McGraw-Hill Companies Inc.; Fig. 19-12: From *Hole's Human Anatomy & Physiology,* 12e, by Shier/Butler/and Lewis. Copyright © 2009. Reprinted by permission of McGraw-Hill Companies Inc.; Fig. 19-13: From *Hole's Human Anatomy & Physiology,* 12e, by Shier/Butler/and Lewis. Copyright © 2009. Reprinted by permission of McGraw-Hill Companies Inc.; Fig. 19-14: From *Hole's Human Anatomy & Physiology,* 12e, by Shier/Butler/and Lewis. Copyright © 2009. Reprinted by permission of McGraw-Hill Companies Inc.; Fig. 19-15: From *Hole's Human Anatomy & Physiology,* 12e, by Shier/Butler/and Lewis. Copyright © 2009. Reprinted by permission of McGraw-Hill Companies Inc.; Fig. 19-16: From *Hole's Human Anatomy & Physiology,* 12e, by Shier/Butler/and Lewis. Copyright © 2009. Reprinted by permission of McGraw-Hill Companies Inc.; Fig. 19-17: From *Hole's Human Anatomy & Physiology,* 12e, by Shier/Butler/and Lewis. Copyright © 2009. Reprinted by permission of McGraw-Hill Companies Inc.; Fig. 19-18: From *Hole's Human Anatomy & Physiology,* 12e, by Shier/Butler/and Lewis. Copyright © 2009. Reprinted by permission of McGraw-Hill Companies Inc.; Fig. 19-19: From *Hole's Human Anatomy & Physiology,* 12e, by Shier/Butler/and Lewis. Copyright © 2009. Reprinted by permission of McGraw-Hill Companies Inc.

Chapter 20

Fig. 20-1: From *Hole's Human Anatomy & Physiology,* 12e, by Shier/Butler/and Lewis. Copyright © 2009. Reprinted by permission of McGraw-Hill Companies Inc.; Fig. 20-2: From *Hole's Human Anatomy & Physiology,* 12e, by Shier/Butler/and Lewis. Copyright © 2009. Reprinted by permission of McGraw-Hill Companies Inc.; Fig. 20-3: From *Hole's Human Anatomy & Physiology,* 12e, by Shier/Butler/and Lewis. Copyright © 2009. Reprinted by permission of McGraw-Hill Companies Inc.; Fig. 20-4: From *Hole's Human Anatomy & Physiology,* 12e, by Shier/Butler/and Lewis. Copyright © 2009. Reprinted by permission of McGraw-Hill Companies Inc.; Fig. 20-5: From *Hole's Human Anatomy & Physiology,* 12e, by Shier/Butler/and Lewis. Copyright © 2009. Reprinted by permission of McGraw-Hill Companies Inc.; Fig. 20-6: From *Hole's Human Anatomy & Physiology,* 12e, by Shier/Butler/and Lewis. Copyright © 2009. Reprinted by permission of McGraw-Hill Companies Inc.; Fig. 20-7: From *Hole's Human Anatomy & Physiology,* 12e, by Shier/Butler/and Lewis. Copyright © 2009. Reprinted by permission of McGraw-Hill Companies Inc.; Fig. 20-8: From *Hole's Human Anatomy & Physiology,* 12e, by Shier/Butler/and Lewis. Copyright © 2009. Reprinted by permission of McGraw-Hill Companies Inc.; Table 20-1: From *Hole's Human Anatomy & Physiology,* 12e, by Shier/Butler/and Lewis. Copyright © 2009. Reprinted by permission of McGraw-Hill Companies Inc.; Table 20-2: From *Hole's Human Anatomy & Physiology,* 12e, by Shier/Butler/and Lewis. Copyright © 2009. Reprinted by permission of McGraw-Hill Companies Inc.; Table 20-3: From *Hole's Human Anatomy & Physiology,* 12e, by Shier/Butler/and Lewis. Copyright © 2009. Reprinted by permission of McGraw-Hill Companies Inc.; Table 20-4: From *Hole's Human Anatomy & Physiology,* 12e, by Shier/Butler/and Lewis. Copyright © 2009. Reprinted by permission of McGraw-Hill Companies Inc.; Table 20-5: From *Hole's Human Anatomy & Physiology,* 12e, by Shier/Butler/and Lewis. Copyright © 2009. Reprinted by permission of McGraw-Hill Companies Inc.; Table 20-6: From *Hole's Human Anatomy & Physiology,* 12e, by Shier/Butler/and Lewis. Copyright © 2009. Reprinted by permission of McGraw-Hill Companies Inc.; Fig. 20-10: From *Hole's Human Anatomy & Physiology,* 12e, by Shier/Butler/and Lewis. Copyright © 2009. Reprinted by permission of McGraw-Hill Companies Inc.

Chapter 21

Fig. 21-1: From *Hole's Human Anatomy & Physiology,* 12e, by Shier/Butler/and Lewis. Copyright © 2009. Reprinted by permission of McGraw-Hill Companies Inc.; Fig. 21-2: From *Hole's Human Anatomy & Physiology,* 12e, by Shier/Butler/and Lewis. Copyright © 2009. Reprinted by permission of McGraw-Hill Companies Inc.; Table 21-2: From *Hole's Human Anatomy & Physiology,* 12e, by Shier/Butler/and Lewis. Copyright © 2009. Reprinted by permission of McGraw-Hill Companies Inc.; Fig. 21-3: From *Hole's Human Anatomy & Physiology,* 12e, by Shier/Butler/and Lewis. Copyright © 2009. Reprinted by permission of McGraw-Hill Companies Inc.; Fig. 21-4: From *Hole's Human Anatomy & Physiology,* 12e, by Shier/Butler/and Lewis. Copyright © 2009. Reprinted by permission of McGraw-Hill Companies Inc.; Fig. 21-5: From *Hole's Human Anatomy & Physiology,* 12e, by Shier/Butler/and Lewis. Copyright © 2009. Reprinted by permission of McGraw-Hill Companies Inc.; Fig. 21-6: From *Hole's Human Anatomy & Physiology,* 12e, by Shier/Butler/and Lewis. Copyright © 2009. Reprinted by permission of McGraw-Hill Companies Inc.; Fig. 21-10: From *Hole's Human Anatomy & Physiology,* 12e, by Shier/Butler/and Lewis. Copyright © 2009. Reprinted by permission of McGraw-Hill Companies Inc.; Fig. 21-14: From *Hole's Human Anatomy & Physiology,* 12e, by Shier/Butler/and Lewis. Copyright © 2009. Reprinted by permission of McGraw-Hill Companies Inc.; Fig. 21-15: From *Hole's Human Anatomy & Physiology,* 12e, by Shier/Butler/and Lewis. Copyright © 2009. Reprinted by permission of McGraw-Hill Companies Inc.; Fig. 21-16: From *Hole's Human Anatomy & Physiology,* 12e, by Shier/Butler/and Lewis. Copyright © 2009. Reprinted by permission of McGraw-Hill Companies Inc.; Fig. 21-17: From *Hole's Human Anatomy & Physiology,* 12e, by Shier/Butler/and Lewis. Copyright © 2009. Reprinted by permission of McGraw-Hill Companies Inc.; Fig. 21-19: From *Hole's Human Anatomy & Physiology,* 12e, by Shier/Butler/and Lewis. Copyright © 2009. Reprinted by permission of McGraw-Hill Companies Inc.; Fig. 21-21: From *Hole's Human Anatomy & Physiology,* 12e, by Shier/Butler/and Lewis. Copyright © 2009. Reprinted by permission of McGraw-Hill Companies Inc.

Chapter 22

Fig. 22-1: From *Hole's Human Anatomy & Physiology,* 12e, by Shier/Butler/and Lewis. Copyright © 2009. Reprinted by permission of McGraw-Hill Companies Inc.; Fig. 22-3: From *Hole's Human Anatomy & Physiology,* 12e, by Shier/Butler/and Lewis. Copyright © 2009. Reprinted by permission of McGraw-Hill Companies Inc.; Table 22-1: From *Hole's Human Anatomy & Physiology,* 12e, by Shier/Butler/and Lewis. Copyright © 2009. Reprinted by permission of McGraw-Hill Companies Inc.; Fig. 22-5: From *Hole's Human Anatomy & Physiology,* 12e, by Shier/Butler/and Lewis. Copyright © 2009. Reprinted by permission of McGraw-Hill Companies Inc.; Fig. 22-6: From *Hole's Human Anatomy & Physiology,* 12e, by Shier/Butler/and Lewis. Copyright © 2009. Reprinted by permission of McGraw-Hill Companies Inc.; Fig. 22-7: From *Hole's Human Anatomy & Physiology,* 12e, by Shier/Butler/and Lewis. Copyright © 2009. Reprinted by permission of McGraw-Hill Companies Inc.; Fig. 22-8: From *Hole's Human Anatomy & Physiology,* 12e, by Shier/Butler/and Lewis. Copyright © 2009. Reprinted by permission of McGraw-Hill Companies Inc.; Fig. 22-10: From *Hole's Human Anatomy & Physiology,* 12e, by Shier/Butler/and Lewis. Copyright © 2009. Reprinted by permission of McGraw-Hill Companies Inc.; Fig. 22-13: From *Hole's Human Anatomy & Physiology,* 12e, by Shier/Butler/and Lewis. Copyright © 2009. Reprinted by permission of McGraw-Hill Companies Inc.; Fig. 22-14: From *Hole's Human Anatomy & Physiology,* 12e, by Shier/Butler/and Lewis. Copyright © 2009. Reprinted by permission of McGraw-Hill Companies Inc.; Fig. 22-15: From *Hole's Human Anatomy & Physiology,* 12e, by Shier/Butler/and Lewis. Copyright © 2009. Reprinted by permission of McGraw-Hill Companies Inc.; Fig. 22-16: From *Hole's Human Anatomy & Physiology,* 12e, by Shier/Butler/and Lewis. Copyright © 2009. Reprinted by permission of McGraw-Hill Companies Inc.; Fig. 22-17: From *Hole's Human Anatomy & Physiology,* 12e, by Shier/Butler/and Lewis. Copyright © 2009. Reprinted by permission of McGraw-Hill Companies Inc.; Table 22-3: From *Hole's Human Anatomy & Physiology,* 12e, by Shier/Butler/and Lewis. Copyright © 2009. Reprinted by permission of McGraw-Hill Companies Inc.; Fig. 22-18: From *Hole's Human Anatomy & Physiology,* 12e, by Shier/Butler/and Lewis. Copyright © 2009. Reprinted by permission of McGraw-Hill Companies Inc.; 22-20: From *Hole's Human Anatomy & Physiology,* 12e, by Shier/Butler/and Lewis. Copyright © 2009. Reprinted by permission of McGraw-Hill Companies Inc.; 22-21: From *Hole's Human Anatomy & Physiology,* 12e, by Shier/Butler/and Lewis. Copyright © 2009. Reprinted by permission of McGraw-Hill Companies Inc.; 22-22: From *Hole's Human Anatomy & Physiology,* 12e, by Shier/Butler/and Lewis. Copyright © 2009. Reprinted by permission of McGraw-Hill Companies Inc.; 22-23: From *Hole's Human Anatomy & Physiology,* 12e, by Shier/Butler/and Lewis. Copyright © 2009. Reprinted by permission of McGraw-Hill Companies Inc.; 22-24: From *Hole's Human Anatomy & Physiology,* 12e, by Shier/Butler/and Lewis. Copyright © 2009. Reprinted by permission of McGraw-Hill Companies Inc.

PHOTO CREDITS

Design feature
Head phones icon on chapter openers: © maxuser/iStockphoto.

Chapter 1
Opener: © RubberBall Productions RF; Figure 1-6: © Comstock Images/Jupiter Images RF; p. 11: © Asia Images Group/Getty RF; Figure 1-8 (full body): © McGraw-Hill Higher Education, Inc./Joe De Grandis, photographer; (median plane): © McGraw-Hill Higher Education, Inc./Karl Ruben, photographer; (frontal plane): Courtesy Paul Reimann; (transverse plane): © McGraw-Hill Higher Education, Inc./Karl Ruben, photographer; p. 17: © Rubberball/Getty RF.

Chapter 2
Opener: © Irina Drazowa-fischer/age fotostock RF; p. 25: © McGraw-Hill Companies, Inc./Tim Fuller, photographer.

Chapter 3
Opener: © Royalty-Free/Corbis; Figure 3-10 (A): © The McGraw-Hill Companies, Inc./Al Telser, photographer; Figure 3-10 (B): © The McGraw-Hill Companies, Inc./Dennis Strete, photographer; Figure 3-10 (C): © Allen Bell/Corbis RF; Figure 3-11: © Science Photo Library/Getty RF; 3-12 (A, B) The McGraw-Hill Companies, Inc./Al Telser, photographer; Figure 3-12 (C): © PhotoLink/Getty RF; Figure 3-12 (D): © The McGraw-Hill Companies, Inc./Al Telser, photographer; Figure 3-16 (A): © Victor P. Eroschenko; Figure 3-16 (B): © The McGraw-Hill Companies, Inc./Al Telser, photographer; Figure 3-16 (C): © Victor P. Eroschenko; Figure 3-17: © The McGraw-Hill Companies, Inc; Figure 3-18: © The McGraw-Hill Companies, Inc./Jill Braaten, photographer; Figure 3-19 (A–C): The McGraw-Hill Companies, Inc./Al Telser, photographer; p. 51: © Science Photo Library/Getty Images.

Chapter 4
Opener: © Chase Jarvis/Getty RF; Figures 4-1–4-5: © The McGraw-Hill Companies, Inc.; Figure 4-6: Centers for Disease Control and Prevention; Figure 4-7: © Image Source/Getty RF; Figure 4-8: © The McGraw-Hill Companies, Inc.; Figure 4-9: © Zeva Oelbaum/Peter Arnold/PhotoLibrary Group; Figure 4-10: © Science Photo Library/Getty RF; Figure 4-12: © Stockbyte/Veer RF; Figure 4-13 (A, B): © The McGraw-Hill Companies, Inc./Al Telser, photographer; Figure 4-15: Centers for Disease Control and Prevention; Figure 4-16: © Ingram Publishing RF; Figure 4-17 (A): © Martin Diebel/Getty RF; Figure 4-17 (B): © The McGraw-Hill Companies, Inc.; Figure 4-18: © The McGraw-Hill Companies, Inc./Lars A. Niki, photographer; Figure 4-19: © Footage supplied by Goodshoot/Punchstock RF; p. 68: © Silverstock/Getty RF; Figure 4-20 (A): © C Squared Studios/Getty RF; Figure 4-20 (B): © FogStock/Alamy RF; p. 72: © Andersen Ross/Blend Images RF.

Chapter 5
Opener: © PhotoAlto/Getty RF; Figure 5-2 (A): © Richard Hutchings; Figure 5-2 (B): © Steve Allen/Brand X Pictures RF; Figure 5-3 (A, B): Centers for Disease Control; Figure 5-8: © Science Photo Library/Getty RF; Figure 5-10: © The McGraw-Hill Companies, Inc./Auburn University Photographic Services; Figures 5-11, 5-12: CDC/Janice Haney Carr; p. 87: © Andersen Ross/Blend Images RF; Figure 5-13: CDC/Janice Carr; Figure 5-14: Centers for Disease Control; Figure 5-15: Armed Forces Institute of Pathology; Figure 5-17: © Science Photo Library/Getty RF; Figure 5-18: Centers for Disease Control; p. 103: © Image Source/Veer RF.

Chapter 6
Opener: © Image Source/Getty RF.

Chapter 7
Opener: © Image Source/Getty RF; p. 128: © JGI/Blend Images RF; Figure 7-9: Corbis/SuperStock RF; Figure 7-10: Centers for Disease Control; Figures 7-12, 7-14: © The McGraw-Hill Companies; Figures 7-15–7-19 (A): Centers for Disease Control; Figure 7-19 (B): Centers for Disease Control/Dr. Lucille K. Georg; Figure 7-19 (C): Centers for Disease Control; Figure 7-22 (A): © SPL/Custom Medical Stock Photos; Figure 7-22 (B, C): © John Radcliff/Photo Researchers, Inc.; Figure 7-23 (A, B): © Biophoto Associates/Photo Researchers, Inc.; Figure 7-23 (C): © SPL/Photo Researchers, Inc.; Figure 7-24: © Ronnie Kaufman/Blend Images RF; p. 150: © Ingram Publishing RF; p. 151: © DreamPictures/Blend Images RF.

Chapter 8

Opener: © Darrin Klimek/Getty RF; Figure 8-2: © Ed Reschke; Figure 8-22 (B): © Magán-Domingo/age fotostock; Figure 8-27 (A): © Martin Rotker; p. 179: © Ingram Publishing/SuperStock RF; Figure 8-30: © Getty Images/Brand X RF; Figure 8-31: © Jim Wehtje/Getty RF; p. 190 (top): © John Lund/Sam Diephuis/Blend Images RF; (bottom): © McGraw-Hill Companies Inc./Ken Karp, photographer.

Chapter 9

Page: 192: © Jose Luis Pelaez, Inc/Getty RF; Figure 9-4: © Ingram Publishing/SuperStock RF; p. 201: © Ingram Publishing RF; Figure 9-10: © The McGraw-Hill Companies, Inc./Joe DeGrandis, photographer; Figure 9-24: © George Doyle/Getty RF; p. 225 (top): © Image Source/Getty RF; (bottom): © David Buffington/Blend Images RF.

Chapter 10

Page 226: © liquidlibrary/PictureQuest RF; Figure 10-1 (B): © The McGraw-Hill Companies, Inc./Al Telser, photographer; Figure 10-2 (B): © Bill Longcore/Photo Researchers, Inc.; Figure 10-4: CDC Sickle Cell Foundation of Georgia: Jackie George, Beverly Sinclair, photo by Jackie Haney Carr; Figures 10-5–10-7: © Ed Reschke; Figure 10-8: © R. Kessel/Visuals Unlimited; Figure 10-9: © Ed Reschke; Figure 10-11 (A): © The McGraw-Hill Companies, Inc./Al Telser, photographer; Figure 10-11 (B): © Andrew Syred/SPL/Photo Researchers, Inc.; p. 245 (top): © Digital Vision/PunchStock RF; Figure 10-14: © SPL/Photo Researchers, Inc.; Figure 10-16: © The McGraw-Hill Companies, Inc.; Figure 10-23: © The McGraw-Hill Companies, Inc./Al Telser, photographer; p. 269 (top): © Rubberball/Nicole Hill/Getty Images; (bottom): © Jose Luis Pelaez Inc./Blend Images RF.

Chapter 11

Opener: © Uppercut Images/Getty RF; Figures 11-5, 11-7: © McGraw-Hill Higher Education, Inc./University of Michigan Biomedical Communications; p. 283: © Texas Heart Institute www.texasheartinstitute.org; p. 299 (top): © Ronnie Kaufman/Blend Images RF; (bottom): © Ingram Publishing RF.

Chapter 12

Opener: © Ingram Publishing/SuperStock RF; Figure 12-2: © The McGraw-Hill Companies, Inc./Dennis Strete, photographer; Figures 12-7 (B), 12-9 (B): © The McGraw-Hill Companies, Inc./Al Telser, photographer; Figure 12-10 (B): © The McGraw-Hill Companies, Inc./Dennis Strete, photographer; p. 328 (top): © Image Source/Getty RF; (bottom): © Stockbyte/PunchStock RF.

Chapter 13

Opener: © Jack Star/PhotoLink/PhotoDisc/Getty RF; Figure 13-6 (C): © CNRI/Phototake; Figure 13-11(both): © CNRI/Phototake; Figure 13-12 (both): Used with permission of the American Cancer Society; Figure 13-14: © Dwight Kuhn; p. 362: © OJO Images/Getty RF; p. 370: © Digital Vision/SuperStock RF; p. 371: © Image Source/Getty RF.

Chapter 14

Opener: © Eric Audras/Photoalto/PictureQuest RF; Figure 14-3: © Ed Reschke; Figure 14-6 (B): © Biophoto Associates/Photo Researchers, Inc.; Figure 14-7 (B): © Don W. Fawcett/Photo Researchers, Inc.; Figure 14-12 (B): © Martin M. Rotker/Photo Researchers, Inc.; Figure 14-16: © AFP/Getty Images; Figure 14-18 (B): © Ed Reschke; Figure 14-28: Image courtesy of the Centers for Disease Control and Prevention; p. 425: © SuperStock RF.

Chapter 15

Opener: © Ingram Publishing RF; Figure 15-6 (A): © Dr. Richard Kessel & Dr. Randy Kardon/Visuals Unlimited, Inc.; Figure 15-6 (B): Courtesy of R.B. Wilson MD, Eppeley Institute for Research in Cancer, University of Nebraska Medical Center; p. 441: © Gary John Norman/Lifesize/Getty Images; Figure 15-16: © Leesa Whicker; p. 449: © Blend Images RF.

Chapter 16

Page 450: © Ronnie Kaufman/Blend Images RF; Figure 16-4: © Image Source/Getty RF; Figure 16-9: © Brand X Pictures/Corbis RF; p. 468: © Photodisc/Getty RF.

Chapter 17

Opener: © John Lund/Blend Images RF; Figure 17-8: © McGraw-Hill Higher Education, Inc./Carol D. Jacobson, PhD., Dept. of Veterinary Anatomy, Iowa State University; Figure 17-12 (A, B): © The McGraw-Hill Companies, Inc./Jill Braaten, photographer; Figure 17-12 (C): © Photolink/Getty RF; Figure 17-12 (D): © Don Farrall/Getty RF; Figure 17-12 (E): © The McGraw-Hill Companies, Inc./Jill Braaten, photographer; p. 496: © Caroline Purser/Digital Vision/Getty RF.

Chapter 18

Opener: © Moxie Productions/Blend Images RF; Figure 18-2 (C): Courtesy of Ronan O'Rahilly, M.D. Carnegie Institute of Washington; Figure 18-4: © Rhea Anna/Getty Images; Figure 18-6: © 2007 Landrum B. Shettles; Figure 18-10: © Dr. G. Moscoso/Photo Researchers, Inc.

Chapter 19

Opener: © Image Source/Getty RF.

Chapter 20

Opener: © Ariel Skelley/Blend Images RF; Figure 20-10 (B): © The McGraw-Hill Companies, Inc./Al Telser, photographer; Figure 20-10 (C): © CNRI/PhotoTake; Figure 20-11 (A): © Charles O. Cecil/Visuals Unlimited; Figure 20-11 (B): © Reuters/Corbis.

Chapter 21

Opener: © Getty RF; Figure 21-7: © The McGraw-Hill Companies, Inc./Joe DeGrandis, photographer; Figure 21-8 (A–D): Reprinted by permission of publisher from "Albert Mendeloff, "Acomegaly, diabetes, hypermetabolism, proteinura and heart failure," *American Journal of Medicine*: 20:1, 01-'56, p. 135/© by Excerpta Medica, Inc.; Figure 21-9: Library of Congress Prints and Photographs Division; Figure 21-11: © Fred Hossler/Visuals Unlimited; Figure 21-12: © Mediscan/Visuals Unlimited; Figure 21-13: © Imagingbody.com; Figure 21-18: © Ed Reschke; Figure 21-20: © Kent M. Van De Graaff; p. 613 (top): © DreamPictures/Pam Ostrow/Blend Images RF; (bottom): © Apple Tree House/Lifesize/Getty RF.

Chapter 22

Opener: © Digital Vision/Alamy RF; Figure 22-2: © Dwight Kuhn; Figure 22-4: © Victor B. Eichler, PhD; Figure 22-7 (B): © Mediscan/Corbis; Figure 22-8 (C): © Frank S. Werblin, PhD; Figure 22-9 (A, B): © Steve Mason/Getty RF; Figure 22-11 (top): © Royalty-Free/Corbis; (bottom): © Dr. P. Marazzi/Photo Researchers, Inc.; Figure 22-19: © Steve Allen/Getty RF; Figure 22-25: © Bonnie Kamin/PhotoEdit; Figure 22-26: © Royalty-Free/Corbis; Figure 22-27: © Jack Star/PhotoLink/PhotoDisc/Getty RF; p. 655: (c) Ingram Publishing RF.

index

*Note: Page numbers in **boldface** indicate figures, (t) indicates a table, and (b) boxed features.*

a

Abdomen
  muscles of, 212, **213**, 213(t)
  quadrants or regions of, 10–11, **11**
Abdominal, **15**
Abdominal aorta, 257, 258
Abdominal cavity, **10**
Abdominal reflexes, 421
Abdominopelvic cavity, 9, **10**
Abducens nerve, 405(t), 406
Abduction, 201, **202**
ABO blood group, **247**, 247–248, 249(t)
Absorption, in digestion, 528
Accessory nerve, 406(t), 407
Accommodation, 627, **627**
Acetylcholine (ACh), 195, 196, **196**, 382(t)
Acetylcholinesterase, 195
Acetyl coenzyme A, 563, **563**
Acidosis, 118, **119**
Acne, 138
  preventing, 138(b)
Acne vulgaris, 138
Acquired (specific) immunity, 310, 313–317
Acromegaly, 593(t), 595(b), **595**
Acromial, **15**
Acrosome, 462, **463**
Actin, 24, 195, 196
Action potential, 49, 374, **385**
Active transport, 36, **36**
Acute inflammation, 57
Acute renal failure, 435(b)–436(b)
Acyclovir, 97
Adam's apple, 338
Addison's disease, 593(t), 596(b)
Adduction, 201, **202**
Adductor longus, **207**, **216**, 218(t)
Adductor magnus, **207**, **215**, **216**, 218(t)
Adenine, 25, **26**
Adenocarcinoma, 352(b)
Adenohypophysis, 593, 595
Adenoids, 309
Adipocytes, 45
Adolescence, 515
ADP (adenosine diphosphate), 560
Adrenal cortex, 603
  hormones and actions of, 590(t)
Adrenal glands, **589**, **602**, 602–603
  hormones of, 602–603
Adrenalin, 603
Adrenal medulla, **602**, 603
Adrenocorticotropic hormone (ACTH), 590(t), 593(t), 596
Adulthood, 515
Aerobic respiration, 199
Afferent arterioles, 431, **432**

Afferent lymphatic vessels, 306, **306**
Afferent neurons, 377
Afterload, 285
Agglutination, 247
Agonist, 200, **200**
Agranulocytes, 236, 237, 238(t)
AIDS/HIV infection
  caring for patients with, 323(b)
  etiology, signs/symptoms, treatment, 98(b)–99(b), 324(b)–325(b)
  lymphocyte numbers, 239
  opportunistic infections and, 59
  transmission of, 58, 325(t)
Alanine, 568(t)
Albinism, 517(b)–518(b)
Albumin, 116, 230, 230(t), 549
Alcoholism, overview of, 64
Aldosterone, 590(t), 596
  deficiency of, 111(t), 112(t), 113
  during pregnancy, 512, 512(t)
  sodium level regulation and, 109, 110
  as steroid hormone, 591(t), 603
  water reabsorption, 440
Alimentary canal, 526, **526–527**
Alkalosis, 118, **119**
Alleles, 516
Allergen, 320, 321
Allergic rhinitis, 333(b)–334(b)
Allergies, 320–321
  allergic rhinitis, 333(b)–334(b)
  immune response, 320
  seasonal, 337(b)
  symptoms of, 321
Allograft, 319
Allopathic medicine, 69
Alopecia, 131, 137
Alveolar capillaries, **276**
Alveolar process, **163**
Alveoli, 344, 355, **355**, 364–365, **365–366**
Alveolus, **276**, **335**
Alzheimer's disease, 391(b)
Amblyopia, 621(b)
Amino acids, 567–569, 568(t)
  in foods, 568(t)
  hormone, 592
Amniocentesis, 518(b)
Amnion, 504, **507**, **508**
Amphitrichous, 79, **79**
Ampicillin, 88
Amylase, 528, 547
Amyotrophic lateral sclerosis (ALS), 415(b)
Anabolic reactions, 560
Anabolism, 20, 567
Anaerobic, 84
Anaerobic reaction, 562
Anaerobic respiration, 198–199
Anaphase, 38

Anaphylactic shock, 63, 320
Anatomical position, 12, **12**
Anatomical terminology, 12–15
Anatomy and physiology
  anatomical terminology, 12–15, **12–15**, 14(t)
  body cavities, regions and quadrants, 9–11
  cardiovascular system, **7**
  chemistry of, 20–26
  defined, 5
  digestive system, **6**
  endocrine system, **7**
  integumentary system, **6**
  life span, 16
  lymphatic system, **6**
  muscular system, **6**
  nervous system, **7**
  organization of body, 7–9
  organs and systems, **6–7**
  reproductive system, **7**
  respiratory system, **6**
  skeletal system, **6**
  tissue types, **8**
  urinary system, **7**
Anemia, 233(b)–236(b), 236(t)
  hemolytic, 322(t)
  pernicious, 322(t)
Aneurysm, 253(b)–254(b)
Angina pectoris, 291
Angiotensin, 108–109
Anions, 23, 110
Ankyloglossia, 528
Anorexia nervosa, 65, 580–581, 581(b)
ANP (atrial natriuretic peptide), 109, 110
Antagonist, **200**, 200–201
Antebrachial, **15**
Antecubital, **15**
Anterior, **12**, 13, 14(t)
Anterior cavity, 628, **629**
Anterior choroid artery, **258**
Anterior fontanel, **165**
Anterior pituitary, 593, 595
Anthrax, 84
Antibiotics, 82
Antibodies, **247**, 247–248, 315–317, 316(t), 317(t)
Antibody-mediated response, 313
Antidiuretic hormone (ADH), 110, 113, 590(t), 591, 593(t), 597
  water reabsorption and, 440, **441**
Antigens, **247**, 247–248, 313, 318
Antihistamines, 321
Antiseptics, 82
Antiviral medications, 97(b), 99(b)
Anus, **526**, **527**, 545, **545**
Aorta, 250, **273**, **274**, **276**
  branches of, 257–258, 257(t), **258**
Aortic arch, 257

718

Aortic semilunar valve, **277**, 278, 283
Apex of heart, 273, **274**
Aplastic anemia, 234(b)
Apocrine sweat gland, 131–132, **132**
Aponeurosis, 197, **198**
Apoptosis, 40, 56
Appendicitis, 545(b)
Appendicular skeleton, 173–177, **173–177**
    defined, 154, **155**
    lower extremities, 175, 177, **177**
    pectoral girdle, 173–174, **173–174**
    pelvic girdle, 174, **176**
Appendix, 310, **526**, 541–542, **542**
Aqueous humor, **624**, 628, **629**, 632(t)
Arcuate artery and vein, 431, **432**
Areola, **475**, 476
Areolar tissue, 45
Arginine, 568(t)
Arms
    bones of, **173**, 173–174, **175**
    muscles of, 209, **210–211**, 211(t)–212(t)
    veins of, **264**
Arrector pilli muscles, **129**, 131
Arrhythmia, 283, 293(b)–294(b)
Arteries/arterial system, 250–251, **251**, 256–260, 257(t), **258–261**, 262(t)
    aorta and branches of, 257–258, 257(t), **258**
    to brain, head and neck, **258**, 259
    disorders of, 253(b)–254(b)
    function of, 256
    major arteries of body, 262(t)
    major vessels of, **261**
    to pelvis and lower extremities, 259–260, **260**
    to shoulder and upper extremities, 259, **259**
Arterioles, 250, **252**
Arthritis, 57, **57**, 181(b)–182(b)
Arthropods, 93–94
Articular cartilage, **161**
Artificially acquired active immunity, 318(t), 319
Artificially acquired passive immunity, 318(t), 319
Arytenoid cartilage, **339**
Ascending aorta, 257, 257(t)
Ascending colon, 542, **542**
Ascending lumbar vein, **266**
Ascending tracts, 397
Ascorbic acid, 575(t)
Asparagine, 568(t)
Aspartic acid, 568(t)
Aspiration pneumonia, 338, 354(b)
Asthma, 344(b)–345(b)
Astigmatism, 623(b)
Astrocyte, **48**, 49, **378**, 379, 379(t), 384
Atelectasis, 349(b)–350(b)
Atherosclerosis, 253(b)
Atlas, 170, **170**
Atomic number, 21
Atomic weight, 21
Atoms, **5**, 7, 20, **20**, 21, **22**
ATP (adenosine triphosphate)
    carbohydrate metabolism, 560, **560**, 562, 563, 564, **565**
    defined, 32
    sources of, 198–199

Atrial fibrillation, 293(b)
Atrial flutter, 289, **290**
Atrial natriuretic factor (ANF), 609
Atrioventricular (AV) node, **282**, **286**, 286–287
Atrioventricular septum, 275
Atrium, **273**, **274**, 275–276, **276**
Atrophy, **54**, 54–55
Audiometers, 648–649, **649**
Auditory nerve, 643, 646
Auditory ossicles, 166, **166**
Auditory tube, 641, **641**, **642**
Auricle, 641, **641**
Auscultation, 283
Autograft, 319
Autoimmune diseases, 321–324, 322(t)
    defined, 321
    as genetic disease, 323(b)
    overview of, 59
    systemic lupus erythematosus, 321–322, 322(t)
Autonomic nervous system, 416, 417(t), **418–419**
Autorhythmicity, 195, 286
Autosomes, 61, 515
Axial skeleton, 162–172, **162–172**
    bones of skull, 162–166, **162–166**
    defined, 154, **155**
    facial bones, **167**, 167–168
    rib cage, 168, **168**
    spinal column, 169–171, **169–172**
Axillary, **15**
Axillary artery, 259, **259**, 262(t)
Axillary vein, 262(t)
Axis, 170, **170**
Axons, 375–376, **376**, **377**
Azithromycin, 95(b)
Azygos vein, 262(t), **264**, 266

b

Bacillus, 82
Bacillus anthracis, 84
Bacteria
    cell structure of, 78–80
    classification of, 80, **80**
    clostridium species, 85(b)–86(b)
    disease and, 84–89
    drug-resistant, 99
    gram-negative rods, 87–88
    gram-positive bacillus, 84
    gram-positive cocci, 86–87
    growth and death of, 81–82
    mycobacteria, 88
    normal flora, 82(b)
    pathogenicity and virulence, 84
    sexually transmitted infections, 94(b)–96(b)
    shapes and arrangements of, 82–83, **83**
    spirochete, 89
Bactericidal, 82
Bacteriostatic, 82
Baroreceptors, 242
Barrett's esophagus, 56, **56**
Bartholin's glands, 473
Basal cell carcinoma, **60**, 144–145, **145**
Basal cells, 617

Basal ganglia, 392(b)
Basal metabolic rate (BMR), 570
Basal nuclei, 386(t)
Base of heart, 273, **275**
Basilar artery, **258**
Basophil, **229**, 236, 237, **237**, 238(t), 239
B cells, 313, **314**, 316(t)
Becker's muscular dystrophy, 206(b)
Bedsore, **130**, 130(b)
Bell's palsy, 408(b)
Benign, 59
Benign prostatic hypertrophy (BPH), 458(b)–459(b)
Benzene, 21
Benzene hexachloride, 94
Bicarbonate ions
    concentration in extracellular and intracellular fluid, **110**, 111(t), 116
    function in body, 23(t)
Biceps brachii, 200, **200**, 207, **210**, 212(t)
Biceps femoris, **207**, **214**, **215**, 218(t)
Biceps reflex, 421
Bicuspids, 529, **529**
Bicuspid valve, **276**, **277**, 278, 283
Bile, 233, 549, 551, **552**
Bilirubin, 133, 233
Biliverdin, 233
Biochemistry, 20–21
Biotin, 575(t)
Bipolar neurons, 376, **376**, 377(t)
Birth control, 490–491, **492**
Birth control pills, 491
Birth process, 512–513, **513**, **514**
Blackheads, 138
Black piedra, 91
Bladder, urinary, 442–443, **442–443**
Blastocyst, 491, 500, **501**, 502, 502(t)
Blepharitis, 640
Blind spot, 624
Blood, **43**, 228–264
    ABO blood group, **247**, 247–248, 249(t)
    acid–base balance, 116–118
    acidosis, 118, **119**
    alkalosis, 118, **119**
    arterial system, 256–260, 257(t), **258–261**, 262(t)
    blood types, **247**, 247–249, **249–250**, 249(t)
    blood vessels, 250–253, **251–252**
    buffers, 116, 117(t)
    carbon dioxide and, 116–117
    cardiac output, 241–243, 285
    clotting, 230, 244–246, **245**
    components of, 238(t)
    as connective tissue, 43, 228
    control of bleeding, 244–247
    diseases and disorders, 233(b)–236(b), 239(b)–240(b), 253(b)–255(b)
    erythrocytes, **232**, 232–233, **233**
    fetal circulation, 510–511, **510–511**
    formation of, 228–230, **229**, 230(t)
    hematopoiesis, 228–230, **229**, 230(t)
    hemostasis, 244–247, **245**
    kidney function and, 431, **431–432**, 433
    leukocytes, 236–239, **237–239**
    life span and, 265–266
    pH of, 116

Blood *Cont.*
  plasma, 230–232, 231(t)
  platelets, 240
  pulmonary circuit, 255–256
  red blood cells, **232**, 232–233, **233**
  Rh blood group, 248–249, **249**, **250**
  shock, 246–247
  structure and function, 43
  systemic circuit, 256
  thrombocytes, 240
  venous system, 262(t), 263, **263–266**
  volume of average adult, 228
  white blood cells, 236–239, **237–239**
Blood–brain barrier, 384
Blood clotting, 244–246, **245**, **246**
Blood pressure
  cardiac control center and, 283
  cardiac output and blood pressure, 241–243
  defined, 241
  factors affecting, 241
  hypertension, 243–244
  normal ranges for, 244(t)
  taking blood pressure, 242, **242**, 243(b)
  urine formation and, 439
Blood type A, **247**, 248, 249(t)
Blood type AB, **247**, 248, 249(t)
Blood type B, **247**, 248, 249(t)
Blood type O, **247**, 248, 249(t)
Blood vessels, 250–253, **251–252**
B lymphocyte, **229**
Body mass index, 65, 582
Body movements, 201–202, **203**
Bolus, 195, 531
Bone markings, 160, 161(t)
Bones. *See also* Skeletal system
  bone fractures, 180, **180**
  bone markings, 160, 161(t)
  classification of, 158–160, **159**
  as connective tissue, 44
  diseases and disorders, 181(b)–187(b)
  evaluating, 181(b)
  facial, **167**, 167–168
  gender differences in, 178(t)
  growth of, 157–158
  joints, **178**, 178–179
  life span and, 187
  lower extremities, 175, 177, **177**
  maintaining healthy, 179
  parathyroid hormone and, 600–601, **601**, **602**
  parts of, 159, **161**
  pectoral girdle, 173–174, **173–174**
  pelvic girdle, 174, **176**
  rib cage, 168, **168**
  of skull, 162–166, **162–166**
  spinal column, 169–171, **169–172**
  structure of, 44, **44**, 155–156, **156**
  terms used to describe, 161(t)
Botox, 85(b)
Botulism, 84
  etiology, signs/symptoms, treatment, 85(b)–86(b)
Bowman's capsule, 434, 439
Brachial, **15**
Brachial artery, 259, **259**, 262(t)
Brachialis, **207**, **210**, 212(t)

Brachial plexus, 412, **412**, 413, **413**
Brachial vein, 262(t)
Brachiocephalic artery, **258**
Brachiocephalic trunk, 257
Brachiocephalic vein, 262(t), **264**
Brachioradialis, **207**, **210**, 212(t)
Bradycardia, 286, 289, **290**
Brain
  arteries to, **258**, 259
  basal nuclei, 386(t)
  brainstem, 386(t), **387**, 391
  cerebellum, 386(t), **387**, 391
  cerebrum, 386(t), **387**, 387–388
  diencephalon, 386(t), **387**, 390
  diseases and disorders, 391(b)–396(b)
  lobes of, 388–390, **388–390**
  overview of, 384–386
  tumors and cancers of, 395(b)
  veins of, 263, **263**
Brainstem, 386(t), **387**, 391
Brain tumors, 395(b)
Breast cancer, 476(b)
Breasts, **475**, 475–476
  cancer of, 476(b)
  fibrocystic breast change, 476(b)–477(b)
  milk production, 476
  milk production and secretion, 514
  self-examination, 486(b), **487**
Breathing
  common situations that alter, 362–363, 364(t)
  factors that control, 362–364, 364(t)
  inhalation and exhalation, 358–359, **360**
  respiratory volumes and capacities, 361–362, 361(t)
Breathing reflex, 362
Brittle-bone disease, 185(b)
Broca's speech area, 388–389
Bronchi, 340
Bronchial tree, 334(t), 344
Bronchioles, 344
Bronchitis, 345(b)
Bronchomediastinal trunk, 304
Buccal, **15**
Buccinator, **208**
Buffers, 116
Buffy coat, 233
Bulbourethral gland, **453**, 459
Bulimia, 65, 581, 581(b)
Bulla, 135, **136**, 137(t)
Bundle of His, **286**, 287
Burns
  as cause of death, 141
  first-degree, 142, **144**
  guidelines for treating, 142–143
  rule of nines, 142, 142(b), **143**
  second-degree, 142, **144**
  severity of, 142
  third-degree, 142, **144**
Bursitis, 182(b)–183(b)

C

Calcaneal tendon, **207**
Calcaneus, 177
Calcitonin, 114, 590(t), 591, 598, 600
Calcitriol, 115

Calcium, 22(t), 23(t), 179(b)
  bone-healthy diet and, 179(b)
  concentration in extracellular and intracellular fluid, **110**, 114
  function in body, 576, 577(t)
  hypercalcemia, 112(t), 114
  hypocalcemia, 112(t), 114
Calcium phosphate, 21
Calor, as sign of inflammation, 57
Calorie, 570
Calyces, 435
Canaliculi, 156
Canal of Schlemm, 623
Cancellous bone, 155, **156**
Cancer
  brain, 395(t)
  breast, 476(b)
  cervical, 482(b)
  colorectal, 543(b)
  grading and staging of, 60
  hyperplasia, 55
  lung, 351(b)–353(b)
  metastasis of, 60
  as neoplasia, 59–60
  oral, 530(b), 531(b)
  ovarian, 478(b)–479(b), 608(b)
  of pancreas, 548(b)
  precancerous dysplasia, 56
  prostate, 457(b)–458(b), 459(b)
  skin, 144–146
  stomach, 537(b)
  uterine, 483(b)
Candida albicans, 92
Candida infection, 65
Canines, 529, **529**
Capillaries
  alveolar, **276**
  blood vessels, 252–253
  lymphatic, 303, **303**
  systemic, **276**
Carbohydrate metabolism, 561–564, **561–565**
Carbohydrates
  conditions associated with, 571(t)
  defined, 560
  function in body, 24, 561–562, 571(t)
  metabolism of, 561–564, **561–565**
  types of, 560
Carbon, 22(t)
Carbon dioxide, **20**, 563, **563**
  blood pH and, 116–117
  breathing and, 362
  function in body, 23(t)
  gas exchange in respiration, 364, **365–366**
  transportation of, in blood, 232, 366–367, **366–367**, 367(t)
  as waste product from humans, 23
Carbonic acid-bicarbonate buffer system, 116, 117(t)
Carboxyhemoglobin, 232, 366–367, **366–367**, 367(t)
Carboxypeptidase, 547
Carbuncle, **138**, 138–139
Carcinoma, 59, **60**
Cardiac cycle, **281**, 281–283, **282**
Cardiac muscles, **8**, **47**, 48, 195(t), 196
Cardiac output, 285
Cardiac region of stomach, 534

Cardiogenic shock, 63, 246
Cardiomyocytes, 196
Cardiomyopathy, 354(b)
Cardiovascular system, 270–296
   cardiac cycle, **281**, 281–283, **282**
   cardiac output, 285
   chest pain, 291–293
   conduction system of heart, **286**, 286–287
   diseases and disorders, 278(b), 280(b)–281(b), 283(b)–284(b), 293(b)–296(b)
   electrocardiogram, 287–290, **288–290**
   function of, 9
   heart anatomy, 272–280, **273–277**
   heart disease prevention and treatment, 285(b)
   heart sounds, 283
   life span and, 296
   overview of, **7**
Carina, **342**, 344
Carotene, 133
Carotid artery, 257, **258**, 259
Carotid canal, **164**
Carotid sinus, **258**
Carpals, **15**, **155**, 174, **175**
Carpal tunnel syndrome, 183(b)
Cartilage, 44–45, **44**
Cartilaginous joints, 178
Catabolic reactions, 560
Catabolism, 20, 567
Cataracts, 627(b)–628(b)
Catecholamines, 603
Cations, 23, 110
Caudal, 13, 14(t)
CD4 cell, 315
Cecum, **526**, 540
Celiac artery, 258
Cell cycle, **37**, 37–39, **38**
Cell death, 40, 56
Cell-mediated response, 313
Cell membrane, **30**, **31**, 34, **34**
Cells, **5**, 25
   apoptosis, 56
   atrophy, 54–55
   bacteria, 78–80
   as basic unit of life, 30
   cell cycle, **37**, 37–39, **38**
   cell membrane, 34, **34**
   cell transport, 35–36
   characteristics of, 30
   components of, 31–34
   cytoplasm and organelles, 31–34, 32(t)
   death of, 40
   differentiation, 30
   dysplasia, 56
   eukaryotic, 31
   hyperplasia, 55
   hypertrophy, 55
   intracellular storage, 56
   metaplasia, 56
   mitochondria, 32
   necrosis, 57
   of nervous system, 375–381, **376–378**, 379(t), **380**
   nucleus, 31, **31**
   size of, 31
   stem cells, 39, 40
   types of, 8
Cell transport, 35–36
   active transport, 35–36
   passive transport, 35–36
Cellular immunity, 313
Cellular respiration, 562
Cellulitis, 137–138
Cellulose, 560
Central nervous system, 384–399, **385–390**
Centrioles, 32(t), 34, 37, 38
Centromere, 37
Cephalad, 13
Cephalic, **15**
Cephalic phase of digestion, 553
Cephalic vein, **264**, 266
Cerebellum, 386(t), **387**, 391
Cerebral angiography, 420
Cerebrospinal fluid (CSF), 400, 402–403, 420
   brain–blood barrier and, 386
Cerebrovascular accident (CVA), 246, 295(b), 392(b)–393(b)
   signs of and FAST, 400(b)
Cerebrum, 386(t), **387**, 387–388
Cerumen, 641
Cerumen impaction, 650(b)
Cervical, **15**
Cervical cancer, 482(b)
   HPV and, 98(b)
Cervical curve, **169**, 170
Cervical nerves, 410, **410**
Cervical plexus, 412, **412**
Cervical vertebrae, 169, **169**, 170, **171**
Cervicitis, 482(b)
Cervix, **472**, 480, **481**
   cancer of, 482(b)
   precancerous dysplasia, 56, **56**
Chancre, 89, 96(b)
Chemical barriers of birth control, 491
Chemical bond, 20
Chemistry, 20–26
   bonds, 20
   inorganic substances, 22–24
   matter and atomic structure, 21
   organic substances, 24–25
Chemoreceptors, 616
Cherry angioma, **136**
Chest pain, causes of, 291–292
Chest x-ray, 293
Chicken pox, 139
Chief cells, 535
Childbirth, 512–513, **513**, **514**
Childhood, 515
Chimera, 25
Chlamydia
   etiology, signs/symptoms, treatment, 94(b)–95(b)
   historical perspective on, 94
Chlamydia trachomatis, 94(b)
Chloride, 579
Chloride ions, 23(t)
   concentration in extracellular and intracellular fluid, **110**, 113
   hyperchloremia, 111(t), 113
   hypochloremia, 111(t), 113
Chlorine, 22(t), 577(t)
Cholecystokinin (CCK), 535, 548, 551
Cholelithiasis, 551(b)–552(b)
Cholera, 87
Cholesterol, 549
   dietary, 561
   function in body, 24, 561
   hypercholesterolemia, 567(b)
   metabolism of, 566
   normal values for, 231(b)
   testing, 231(b)
   types of, 231
ChooseMyPlate, 568(b), **569**
Chordae tendineae, **276**, **277**, 278
Chorion, 504, **507**, **508**
Choroid, 623, **624**, **629**, 632(t)
Choroid coat, 622(t)
Chromium, 22(t), 578(t), 579
Chromosomal abnormalities, 61
Chromosomes, 25, 31, 516
Chronic fatigue syndrome (CFS), 322–324
Chronic inflammation, 57, **57**
Chronic obstructive pulmonary disease (COPD), 350(b)–351(b), 362
Chronic renal failure, 436(b)
Churning, 538, **539–540**
Chylomicrons, 231
Chyme, 535, 545
Chymotrypsin, 547
Cicatrix, 137(t)
Cilia, 32(t), 33, 42
Ciliary body, 622(t), 623, **624**, **629**, 632(t)
Circulatory system
   arterial system, 256–260, 257(t), **258–261**, 262(t)
   cardiac output and blood pressure, 241–243
   fetal circulation, 510–511, **510–511**
   hemostasis, 244–247, **245**
   high blood pressure, 243–244
   improving circulation, 256(b)
   life span and, 265–266
   pulmonary circuit, 255–256
   shock, 246–247
   systemic circuit, 256
   venous system, 262(t), 263, **263–266**
Circumduction, 202, **203**
Cirrhosis, 551(b)
Cisterna chili, 304, **304**
Cisternae, 33
Citric acid cycle, 199, 563
Clavicle, **155**, 168, **168**, 173, **173**
Cleavage, 500, **501**, 502(t)
Clitoris, **472**, 473, **473**, 475(t)
Closed fracture, 180
Clostridium botulinum, 84, 85(b)–86(b)
Clostridium perfringens, 84
Clostridium tetani, 84, 85(b)
Cluster headaches, 394(b)–395(t)
Coagulation, 245
Cobalt, 22(t), 578(t), 579
Cocci, 82, 83
Coccygeal nerves, **410**, 411
Coccyx, **155**, 169, **169**, 171, **172**
   gender differences in, 178(t)
Cochlea, **641**, 643, **644**, **645**, 646
Coitus interruptus, 490
Cold sore, 96(b)–97(b), **97**, 139
Colitis, 543(b)

Collagen, 44, 45, **45**, 157, 567
Collarbones, 173, **173**
Color blindness, 625
Colorectal cancer, 543(b)
Color vision, 640, **640**
Colostrum, 319
Comedoes, 138
Common bile duct, 549
Common carotid artery, **258**
Compact bone, 155, **156**, **161**
Complementary and alternative medicine (CAM), 69
Complex inheritance, 516
Compound, 20
Compound fracture, 180
Computerized tomography scan, 420
Condoms, 490, **492**
Conduction, 570
Conduction system of heart, **286**, 286–287
Conductive hearing loss, 646
Condyle, **160**, 161(t)
Condyle canal, **164**
Condyloma acuminatum, 98(b)
Cones, 624–625, **625**, 632(t)
Confocal microscope, 78
Congenital heart disease, 278(b)
Congestive heart failure (CHF), 294(b)–295(b)
Conjugation, 81
Conjunctiva, 633
Conjunctivitis, 633(b), 640
Connective tissue, **8**, **41**, 42–46
 blood as, 43, 228
 bone as, 44
 cartilage as, 44–45
 defined, 42
 dense, 46
 loose, 45
Constipation, 544(b)
Contraceptive implants, 491
Contrast sensitivity, 640
Convection, 570
Conventional medicine, 69
Copper, 22(t), 578(t), 579
Cornea, 622, **622**, 622(t), **624**, **629**, 632(t)
Corneal abrasions, 640
Corniculate cartilage, **339**, **340**
Coronal, 13
Coronal suture, **162**, **163**, **165**
Coronary catheterization, 293
Coronary sinus, 275
Coronoid process, **163**
Corpus callosum, **387**, 388
Corpus luteum, **480**, 486, 488, **489**, 512, 512(t)
Cortex, 388
Corticosteroids, 321
Cortisol, 590(t), 596, 603, 604(b)
 negative feedback loop, **609**
Corynebacterium, 84
Corynebacterium diphtheria, 84
Costal cartilage, 168, **168**
Costochondritis, 292
Cough etiquette, **343**, 343(b)
Coughing, 362, 364(t)
Covalent bond, 20
Cowper's gland, 459

Coxa, **155**
Coxal, **15**
Coxal bones, 174, **176**
CPAP (continuous positive airway pressure), 341(b)
Cramps, muscle, 199
Cranial, 9, 13, 14(t)
Cranial cavity, **10**
Cranial nerves, **404**, 404–407, 405(t)–406(t), **407–408**, 618, 619–620
Cranium, **155**
Creatine phosphate, 198
Crenation, 35(b)
Crest, 161(t)
Cretinism, 593(t), 598, **599**
Cribriform plate, 336
Cricoid cartilage, 339, **339**, **342**, **346**
Cristae, 32
Crohn's disease, 540(b)–541(b)
Crural, **15**
Crust, **136**, 137(t)
Cryosurgery, 145
Cryptorchidism, 454(b)
Cubital, **15**
Cuneiform cartilage, **339**, **340**
Curettage, 145
Cushing's syndrome, 593(t), 596(b)
Cyanocobalamin, 575(t)
Cyanosis, 131
Cyanotic disorders, 278(b)
Cysteine, 568(t)
Cystic duct, 549
Cystic fibrosis, 518(b), 533(b)
Cystitis, 445(b), 446(b)
Cytokinesis, **37**, 38
Cytology, 30
Cytomegalovirus (CMV) infection, 99(b)
Cytoplasm, **30**, **31**, 31–34
Cytosine, 25, **26**
Cytosol, 31, 107

d

Darkfield illumination, 77–78
Decarboxylation, 563
Deciduous dentition, 530
Decline phase, 81, **81**
Decubitus ulcers, **130**, 130(b)
Deep, **14**, 14(t), 15
Deep venous thrombosis (DVT), 62, **62**, 246, **246**
Defecation, 528
Defense mechanisms against disease, 310–317
Degenerative joint disease (DJD), 181(b)
Deglutition, 532–533
Dehydration, 108
Deltoid, **207**, **209**, **210**, 211(t)
Dendrites, 375–376, **376**, **377**
Dense connective tissue, 46
Deoxyhemoglobin, 232
Deoxyribose, 24
Depolarization, 287–288, 383, **384**, **385**
Depo-Provera, 491
Depression, body movement, 202, **203**
Dermal papillae, 129–130
Dermatitis, 138

Dermatome, 139, 411, **411**
Dermatophytes, 141
Dermis, 128, **129**, 129–130
Descending aorta, 257, 257(t)
Descending colon, 542, **542**
Descending tracts, 397
Detrusor muscle, **442**, 443
Developmental diseases, 60
Diabetes insipidus, 593(t), 598(b)
Diabetes mellitus, 322(t), 606(b)
 glucose testing, 605(b)
 NIDDM (noninsulin-dependent diabetes mellitus), 607(b)
 overview of, 606(b)
 pathophysiology of, 606(b)
 preventing, 607(b)
 type 1 and 2, 606(b)
Diagnosis, of disease, 67
Dialysis, 437(b)
Diapedesis, 239, **239**
Diaphragm, 9, **10**, **273**
 respiration and, 358, **358**, 360
 as respiratory muscle, 213(t)
 spinal nerves, 412–413
Diaphragm, birth control, 490, **492**
Diaphysis, 157–158, 159, **161**
Diarrhea, 541(b)
Diastolic pressure, 241
Diencephalon, 386(t), **387**, 390
Diet. *See also* Nutrition and diet
 assisting your immune system, 318(b)
 bone health and, 179(b)
 heart disease prevention and treatment, 285(b)
 red blood cell production and, 230(t)
Differential diagnosis, 67
Differentiation, 30
Diffusion
 facilitated, 35
 simple, 35, **35**
Digestion, 527–528, 553
Digestive system, 524–553
 accessory organs of, 546–551, **547–548**
 alimentary canal characteristics, 526, **526–527**
 cystic fibrosis, 533(b)
 diseases and disorders, 531(b), 536(b)–537(b), 540(b)–541(b), 543(b)–545(b), 546(b), 548(b), 549(b), 551(b)–552(b)
 function of, overview, 526–528
 GI cancer prevention and screening, 539(b)
 hepatitis, 550(b)
 life span and, 553
 oral cancer, 530(b)
 organs of, 528–546, **528–546**
 overview of, **6**
 phases of digestion, 553
 stool specimens, 546(b)
Digital, **15**
Dilation stage, 513, **514**
Diphtheria, 84
Diplococcus pneumoniae, 87
Disaccharides, 560
Disease
 bacteria and, 84–89
 body's response to injury, 57–58

cell injury and death, 54–57
developmental and genetic
  diseases, 60–61
environmental and nutritional
  pathology, 63–65
etiology and diagnosis of, 67
hemodynamic disorders, 61–63
immunopathology, 58–59
infectious and parasitic diseases, 66
neoplasia, 59–60
prevention and treatment of, 68–69
prognosis, 68
signs and symptoms of, 66
Diseases and disorders
  blood, 233(b)–236(b), 239(b)–240(b),
    253(b)–255(b)
  cardiovascular system, 278(b),
    280(b)–281(b), 283(b)–284(b),
    293(b)–296(b)
  digestive system, 531(b), 536(b)–537(b),
    540(b)–541(b), 543(b)–545(b), 546(b),
    548(b), 549(b), 551(b)–552(b)
  of ears, 642(b), 645(b)–646(b), 648(b),
    650(b)
  endocrine system, 592–593, 593(t),
    595(b)–598(b), 600(b)–601(b), 603(b),
    606(b)–608(b)
  of eyes, 621(b), 623(b), 625(b)–630(b),
    632(b)–634(b), 638(b)
  female reproductive system,
    476(b)–479(b), 482(b)–484(b), 488(b)
  genetic disorders, 517(b)–521(b)
  immune system, 320–324, 324(b)–325(b)
  lymphatic system, 305(b)
  male reproductive system, 456(b),
    457(b)–459(b), 465(b)
  muscular system, 205(b)–206(b),
    220(b)–222(b)
  nervous system, 391(b)–396(b),
    401(b)–402(b), 408(b)–409(b),
    414(b)–415(b)
  respiratory system, 333(b)–334(b),
    336(b)–337(b), 338(b), 342(b)–345(b),
    347(b)–358(b)
  skeletal system, 181(b)–187(b)
  urinary system, 435(b)–438(b), 445(b)
Distal, **12**, **14**, 14(t), 15
Distal convoluted tubules, 434–435, **435**
Distal interphalangeal (DIP) joints, 174
Diverticulitis, 544(b)
DNA (deoxyribonucleic acid), 31, 516–517
  function in body, 24–25
  recombinant DNA, 25
DNA fingerprinting, 517
Dolor, as sign of inflammation, 57
Dopamine, 196, 382(t), 392(b)
Dorsal, 9, 13, **14**, 14(t)
Dorsiflexion, 201, **202**
Dorsum, **15**
Down syndrome, 61, **61**, 503(b),
  518(b)–519(b)
Doxycycline, 95(b)
Drug abuse, 65
Dry eye syndrome, 633(b)–634(b)
Duchenne's muscular dystrophy, 206(b)
Ductus arteriosus, **510**, 511
Ductus venosus, **510**, 511

Duodenum **526**, 538, **538**
Dura mater, 400, **401**
Dwarfism, 593(t), **595**, 595(b), 596(b)
Dysmenorrhea, 488(b)
Dysplasia, 56, **56**
Dysrhythmias, 293(b)–294(b)

e

Eardrum, 641
Ears, 641–651
  diseases and disorders, 642(b),
    645(b)–646(b), 648(b), 650(b)
  hearing aids, 650, 651(b)
  hearing loss, 646–647
  hearing process, 646
  life span and, 652
  preventive ear care tips, 643(b)
  structure and function of, **641–642**,
    641–643, **644–645**
  treating ear and hearing problems,
    650–651
Eating disorders, 581(b)
Ecchymosis, **136**, 137(t)
Eccrine sweat gland, 131–132, **132**
Echocardiogram, 293
Ectoderm, 504, **506–507**
Ectropion, 632(b)
Eczema, 138
Edema, 63, 303, 311, 593(t)
Effacement, 513
Effectors, 375
Efferent arteriole, 431, **432**
Efferent lymphatic vessels, 306, **306**
Efferent neurons, 377
Ejaculation, 464, **464**
Elastic cartilage, 44, **44**
Elastic connective tissue, 46
Elderly patient
  dehydration, 119
  fluid balance and, 119
  hydration and, 109
Electrocardiogram (ECG), 287–290,
  288–290, 292
  clinical applications of, 290(b)
Electrodessication, 145
Electroencephalogram, 420
Electrolytes, 110–116, 232
  bicarbonate, 111(t), 116
  calcium, 112(t), 114
  chloride, 111(t), 113
  function of, 110
  magnesium, 112(t), 115
  overview of, 111(t)–112(t)
  phosphate, 112(t), 115
  potassium, 111(t)–112(t), 113–114
  sodium, 110, 111(t), 113
Electron beam computerized tomography
  (EBCT), 293
Electron microscope, 77, **77**
Electrons, 21, **21**
Electron transport chain (ETC), 560,
  564, **565**
Element, 21
Elevation, 202, **203**
Embolism, 62, **62**
Embolus, 246, **246**, 254(b)

Embryonic period, 503–504,
  504–508, 509(t)
Emphysema, 350(b)
Empyema, 347(b)
Endocarditis, 280(b)
Endocardium, 279, **279**, 279(t)
Endochondral ossification, 157
Endocrine disorders, 592–593, 593(t)
Endocrine system, 587–610
  adrenal glands, 602–603
  compared to nervous system, 588, **588**
  diseases and disorders, 592–593, 593(t),
    595(b)–598(b), 600(b)–601(b), 603(b),
    606(b)–608(b)
  glucose testing, 605(b)
  gonads, 608
  hormones and actions of, 589, 590(t),
    591–592, **591–592**, 591(t)
  life span and, 610
  overview of, **7**, **589**, 589–591
  pancreas, 604–606
  parathyroid glands, 600–601
  pineal gland, 607
  pituitary gland, 593–598
  regulatory mechanisms, **609**, 609–610
  stress and, 604(b)
  thymus, 607
  thyroid gland, 598–600
Endocytosis, 36
Endoderm, 504, **506–507**
Endolymph, 643, **644**
Endometriosis, 483(b)
Endometrium, 480–481, **481**
Endomysium, 197
Endoplasmic reticulum, 32(t), 33
Endorphins, 382(t)
Endoscopy, 293
Endosteum, 156
Endothelium, 303
Energy currency, 32
Enkephalins, 382(t)
Enterobacteria, 87
Entropion, 633(b)
Environmental pathology, 63–65
Enzymes, 540
Eosinophil, **229**, 236, 237, **237**, 238(t), 239
Ependymal cell, **48**, 49, **378**, 379, 379(t)
Epicardium, 279, **279**, 279(t), 280
Epicondyle, **160**, 161(t)
Epicranial aponeurosis, **208**
Epicranius, **208**
Epidermis, 128–129, **129**
Epididymis, 456, **457**
Epididymitis, 456(b)
Epigastric region, 11, **11**
Epiglottic cartilage, 338, **339**
Epiglottis, **335**, 338, **339**, **340**, 529
Epilepsy, 395(b)–396(b)
Epimysium, 197
Epinephrine, 321, 570, 590(t), 603
  imbalance of, 603(b)
Epiphysis, 157, 159, **161**
Episiotomy, 474
Episodic tension headache, 393(b)
Epithelial tissue, 40–42, **41**
Epithelium muscles, **8**
Epstein-Barr virus (EBV), 322

Equatorial plate, 38
Erectile dysfunction (ED), 465(b)
Erection, 464, **464**, 474–475
Erosion, 135, **136**, 137(t)
Erythroblastosis fetalis, 248–249, **249–250**
Erythrocytes, 43, **43**, 159
   as cellular component of blood, 238(t)
   defined, 228
   development of, **229**
   structure and function of, **232**, 232–233, **233**
Erythropoiesis, 228
Erythropoietin, 228, 230, 431, 609
Escherichia coli, 87
Esophageal hiatus, **532**, 533
Esophagus, **335**, **526**, **527**, **532**, 533–534
Essential hypertension, 244
Estrogen, 477
   in female reproductive cycle, 488, **489**
   female secondary sexual development, 484, **485**
   imbalance of, 608(b)
   made from cholesterol, 24
   during pregnancy, 512, 512(t), 590(t), 597, 608, 608(b)
Ethambutol, 88(b)
Ethmoid bone, **162**, **163**, **164**, 166
Ethmoid sinus, 336
Etiology, 67
Eukaryote, 78
Eukaryotic cells, 31
Eustachian tube, 641
Evaporation, 570
Eversion, 202, **203**
Excoriation, 135–136, **136**, 137(t)
Exercise
   bone health and, 179(b)
   heart disease prevention and treatment, 285(b)
   heart rate and, 282
   to improve circulation, 256(b)
   Kegel exercises to prevent stress incontinence, 447(b)
   muscle health and, 216(b)
Exhalation, 344, 358–359, **360**
Exocytosis, 33, 36
Exophthalmos, 600, **600**
Expiration, 358–359, **360**
Expiratory reserve volume, 361(t)
Exponential phase, 81, **81**
Expulsion stage, 513, **514**
Extension, 201, **202**
Extensor carpi radialis brevis, 212(t), **211**
Extensor carpi radialis longus, 212(t), **211**
Extensor carpi radialis ulnaris, 212(t)
Extensor carpi ulnaris, **211**
Extensor digitorum, 212(t), **211**
Extensor digitorum longus, **207**, **217**, 219(t)
Extensor hallucis longus, **217**
Extensor retinacula, **217**
External auditory canal, 641
External auditory meatus, **163**, **164**, 166
External intercostals, 213(t)
External jugular vein, **264**
External nares, 332
External oblique, **207**, **213**, 213(t)
External respiration, 355, 364

Extracellular fluid, **106**, 107, 110–116
Extrinsic eye muscles, 634–636, **635**, 635(t)
Exudate, 139
Eyelids, 631–632
Eyes, 620–641
   accessory organs of, 631–636, **634**, **635**, 635(t)
   administering medication to, 636(b)
   chambers of, 628–629, **629**
   diseases and disorders, 621(b), 623(b), 625(b)–630(b), 632(b)–634(b), 638(b)
   LASIK, 640(b)
   lens of, 627, **627**
   life span and, 651–652
   protecting vision for lifetime, 620(b)
   safety tips for, 639(b)
   structure and function of, 621–625, **622**, **624**, **625**, 632(t)
   vision testing, 636, 637(b), 640
   visual accessory organs, 631–636, **634**, **635**, 635(t)
   visual pathways, 630–631

f

Facial artery, **258**, 262(t)
Facial bones, **163**, **164**, **167**, 167–168
Facial muscles, 208, **208**, 209(t)
Facial nerve, 405(t), 406, 618
Facilitated diffusion, 35
Falciform ligament, 548
Fallopian tubes, 480, **480**, **481**
False ribs, 168, **168**
False vocal chords, 339, **339**, **340**
Fascicle, 197
FAST, 400(b)
Fat, body, accumulation in cells, 56
Fat, dietary, lipid metabolism, **565**, 565–567
Fat tissue, 45
Fatty liver, 56
Feces, 545
Feedback loops, **609**, 609–610
Feet
   bones of, 177, **177**
   muscles of, **217**, 219(t)
Female reproductive system, 470–493
   birth control, 490–491, **492**
   breast self-examination, 486(b), **487**
   diseases and disorders, 476(b)–479(b), 482(b)–484(b), 488(b)
   erection, lubrication and orgasm, 474–475
   external structures and function of, **473**, 473–476, **475**
   fertilization, 490
   hormones of, 484, **485**, 488, **489**
   infertility, 491–493
   internal structures and function of, 477–484, **478–481**
   life span and, 493
   menstrual cycle, 485–488, **489**
   overview of, 7
   pregnancy, 498–512
   reproductive cycle, 485–488, **489**
   sexually transmitted infections (STI), 474(b)
Femoral, **15**
Femoral artery, 260, **260**, 262(t)

Femoral vein, 262(t), **265**, 266
Femur, **155**, 159, **161**, 175, **177**
Fertilization, 490
Fetal alcohol syndrome, 60, **61**
Fetal period, 504, 506, **508**, **509**
Fetus
   fetal circulation, 510–511, **510–511**
   fetal period, 504, 506, **508**, **509**, 509(t)
Fever, as body's defense mechanism, 312, 312(t)
Fiber, 560–561
Fibrinogen, 230, 230(t), 245, **245**, 549
Fibroblasts, 45, **45**
Fibrocartilage, **44**, 45
Fibrocystic breast change, 476(b)–477(b)
Fibroids, 483(b)
Fibromyalgia, 47(b), 205(b), **205**
Fibrosis, 58, **58**
Fibrous joints, 178
Fibrous tunic, **622**, 622–623, **629**
Fibula, **155**, **160**, 175, **177**, **177**
Fibularis brevis, **217**
Fibularis longus, **207**, **217**
Fibularis tertius, **217**
Fight or flight system, 416, 603
Filtration, 36
Fimbriae, 480, **481**
Fingers, muscles of, 212(t), **211**
First-degree burns, 142, **144**
Fissure, **136**, 137(t)
Flagella
   bacteria and, **79**, 79–80
   structure and function of, 32(t), 34
Flat bones, 158, **159**
Flexion, 201, **202**
Flexor carpi radialis, **210**, **211**, 212(t)
Flexor carpi ulnaris, **210**, 212(t)
Flexor digitorum longus, 219(t)
Flexor digitorum profundus, 212(t)
Flexor digitorum superficialis, **210**, **211**
Flexor retinaculum, **210**
Flight or fight response, 282
Floating ribs, 168, **168**
Fluid, body
   acid-base balance, 116–118
   electrolytes, 110–116
   extracellular fluid, **106**, 107
   hydration in elderly patient, 109
   interstitial fluid, 107
   intracellular fluid, **106**, 107
   plasma, 107
   water gain, 107
   water loss, 107–108
   water regulation and movement, 108–109
Fluoride, 579
Fluorine, 22(t), 578(t)
Folic acid, 572, 575(t)
   red blood cell production and, 230, 230(t)
Follicle-stimulating hormone (FSH), 484, **485**, 590(t), 597
   in female reproductive cycle, 488, **489**
   spermatogenesis and, **460**, 460–461
Follicular cells, 477, **480**
Folliculitis, **138**, 138–139
Fontanelles, 162

Food poisoning, bacteria and, 85–87
Foot
  bones of, 177, **177**
  muscles of, **217**, 219(t)
Foramen, 161(t)
Foramen magnum, **164**, 165
Foramen ovale, 275, 510, **510**
Fornix, **387**
Fossa, 161(t)
Fossa ovalis, 275, 511
Fovea capitis, **160**
Fovea centralis, 624, **624**, **629**
Fragile X syndrome, 519(b)
Freckle, 135
Frenulum, 528–529
Frontal, 13, **13**, **15**
Frontal bone, **162**, **163**, **164**, 165, **165**
Frontalis, **207**, **208**, 209(t)
Frontal lobe, **388**, 388–389, **389**, 389(t)
Frontal sinus, **163**, **335**, 336
Frontal suture, **165**
Fructose, 561, **561**
Functional residual capacity, 361(t)
Fundus, 480
Fundus of stomach, 534, **534**
Fungi
  infections caused by, 90–92
  uses of, 90
Furuncle, 138

g

GABA, 382(t)
Galactose, 561, **561**
Gallbladder, **526**, **527**, **548**
  function of, 551, **552**
  inflammation of, and chest pain, 292
Ganglion, 411
Gangrene, **56**, 84
Gardasil vaccine, 98(b)
Gas exchange, 364–365, **365**, **366**
Gas transport, 366–367, **366–367**
Gastric juice, 534
Gastric phase of digestion, 553
Gastric ulcers, 537(b)
Gastric vein, 262(t)
Gastrin, 535, 553
Gastritis, 536(b)
Gastrocnemius, **207**, **215**, **217**, 219(t)
Gastroesophageal reflux disorder (GERD), 56, 536(b)
Gastrula, 502(t), 504
Gender
  autoimmune diseases and, 59
  diagnosis of alcoholism and, 64
  sex-linked disorders, 61, 204(b)
  skeletal system and, 178
  urinary tract infections and, 443–444, 446(b)
General stress syndrome, 604(b)
Genes, 515
Genetic disorders
  albinism, 517(b)–518(b)
  autoimmune disease, 323(b)
  chromosomal abnormalities, 61
  congenital heart disease, 278(b)
  cystic fibrosis, 518(b), 533(b)
  Down syndrome, 61, 503(b), 518(b)–519(b)
  dwarfism, 596(b)
  Fragile X syndrome, 519(b)
  hemophilia, 519(b)
  Klinefelter's syndrome, 520(b)
  multifactorial inheritance, 61
  muscular dystrophy, 520(b)
  phenylketonuria (PKU), 520(b)
  sex-linked disorders, 61, 204(b), 205(b), 241(b)
  Trisomy 21, 61
  Turner's syndrome, 520(b)–521(b)
Genetics, 515–521
  diseases and disorders, 517(b)–521(b)
  DNA fingerprinting, 517
  polymerase chain reaction (PCR), 517
  terms and concepts of, 515–516
Genital, **15**
Genital herpes, 97(b), 139, 474(b)
Genital warts, 98(b), 474(b)
Genotype, 515
Genus, 80, **80**
Giardia lamblia, 92–93
Gigantism, 593(t), 595(b)
Gingiva, 528
Glands, 40
  ductless vs. exocrine, 589
  endocrine, 589, 590(t)
Glans penis, **453**, 455, **456**
Glaucoma, 629(b)–630(b)
Glia, 377, **378**, 379–380, 379(t)
Gliomas, 395(t)
Globulins, 230, 230(t)
Glomerular filtration, 439, **439**
Glomerulonephritis, 322(t), 438(b)
Glomerulus, 434, **434**
Glossopharyngeal nerve, 405(t), 406–407, 409(b)
Glottis, 339, **339**, **340**
Glucagon, 590(t), 593(t), 605, 606
  imbalances of, 606(b)
Glucocorticoids, 603
Gluconeogenesis, 431, 564
Glucose
  carbohydrate metabolism, 561, **561**
  energy source for muscles, 198, 199
  as most common carbohydrate, 24
Glucose testing, 605(b)
Glutamate, 382(t)
Glutamic acid, 568(t)
Glutamine, 568(t)
Gluteal, **15**
Gluteal tuberosity, **160**
Gluteus maximus, **207**, **214**, **215**, 218(t)
Gluteus medius, **207**, **214**, **215**, 218(t)
Gluteus minimus, 218(t)
Glycine, 568(t)
Glycogen, 24, 198, 549, 561
Glycogenolysis, 564
Glycolipid, 129
Glycolysis, 199, **562**, 562–563
GnRH, 484, **485**
Goblet cells, 41, 42, 336, 340
Gold standard, 67
Golgi apparatus, 32(t), 33
Gonadal artery, 258
Gonadal vein, **266**
Gonadotropin, 460
Gonads, 608
  hormones and actions of, 590(t), 608
Goniometer, 204, **204**
Gonorrhea, 474(b)
  etiology, signs/symptoms, treatment, 95(b)–96(b)
  historical perspective on, 94
Gout, 182(b)
G-protein, 591, **592**
Gracilis, **207**, **215**, **216**
Graft-versus-host disease (GVHD), 319
Gram-negative
  defined, 78
  rods, 87
Gram-positive
  bacillus, 84
  clostridium, 84, 85(b)–86(b)
  cocci, 86–87
  defined, 78
Gram stain, 78–79
Granulocytes, **229**, 236, 238(t)
Graves' disease, 322(t), 593, 593(t), **600**
Gray matter, 381, 396–397, **397**
Greenstick fracture, 180
Growth hormone, 590(t), 591, 593(t), 595, 610
Guanine, 25, **26**
Gullain-Barré syndrome, 414(b)–415(b)
Gustation, 618–620
Gyrus, **387**, 388

h

Haemophilus influenzae, 87–88
Hair, **129**, 130–131
Hair follicle, **129**, 130, **132**
Hands
  bones of, **175**
  muscles of, 212(t), **211**
  veins of, **264**
Hand washing, 91
Haptens, 313
Hard palate, **335**, **528**, 529
Haversian system, 155
HDL (high-density lipoprotein) cholesterol, 231, 231(b), 566
Head, 161(t)
  arteries to, **258**, 259
  muscles of, 208, **208**, 209(t)
  veins of, 263, **263**
Headaches, 393(b)–395(b)
Hearing, 641–651. *See also* Ears
  diseases and disorders, 642(b), 645(b)–646(b), 648(b), 650(b)
  hearing aids, 650, 651(b)
  hearing loss, 646–647
  hearing tests, 648–649
  noise pollution and, 648
  patients with hearing impairment, 649(b)
  preventive ear care tips, 643(b)
  process of, 646
  recognizing hearing problems in infants, 647(b)
Hearing aids, 650, 651(b)

Heart
    anatomy of, 272–280, **273–277**
    cardiac cycle, **281**, 281–283, **282**
    cardiac output and blood pressure, 241–243
    cardiac output formula, 285
    chambers of, 275–276, **276**
    chest pain, 291–293
    conduction system of, **286**, 286–287
    diseases and disorders, 278(b), 280(b)–281(b), 283(b)–284(b), 293(b)–296(b)
    electrocardiogram, 287–290, **288–290**
    fetal circulation, 510–511, **510–511**
    heart disease prevention and treatment, 285(b)
    heart sounds, 283
    hormone secreted by, 609
    layers of, 279, **279**, 279(t)
    life span and, 296
    location of, 273, **275**
    membranes of, 280
    pulmonary circuit, 255–256
    systemic circuit, 256
    valves of, **277**, 277–278
Heart attack, 246, 291, 295–296(b)
Heartburn, 292, 536(b)
Heart disease
    congenital, 278(b)
    prevention and treatment, 285(b)
Heart rate
    cardiac cycle, **281**, 281–283, **282**
    factors influencing, 282–283
Helicobacter pylori, 87
Helminths, 92–93
Helper T cells, 314
Hematocrit, 233, **233**
Hematopoiesis, 158, 159, 228–230, **229**, 230(t)
Hemoccult test, 546(b)
Hemocytoblasts, 228
Hemodialysis, 437(b)
Hemodynamic disorders, 61–63
Hemodynamics, 241
Hemoglobin, 116, 567
    function of, 232
    gas transport, 366–367, **366–367**, 367(t)
    skin color and, 133
Hemolytic anemia, 234(b), 322(t)
Hemolytic disease of the newborn (HDN), 248–249, **249–250**
Hemophilia, 61, 241(b), 519(b)
Hemoptysis, 352(b)
Hemorrhagic stroke, 393(b)
Hemorrhoids, 546(b)
Hemostasis, 244–247, **245**
Hemothorax, 349(b)
Heparin, 237
Hepatic duct, 549
Hepatic portal, 262(t)
Hepatic vein, 262(t), **266**
Hepatitis
    defined, 549(b)
    etiology, signs/symptoms, treatment, 89(b)–90(b), 549(b)
    types of, and transmission of, 89(b)–90(b), 550(b)
    viral, 89(b)–90(b)

Hepatitis A, 89(b)–90(b), 550(b)
Hepatitis B, 89(b)–90(b), 550(b)
Hepatitis C, 90(b), 550(b)
Hepatitis D, 550(b)
Hepatitis E, 550(b)
Hepatocytes, 38, 549
Hernia, 533, 536(b)–537(b)
    inguinal, 544(b)–545(b)
Herpes simplex virus
    etiology, signs/symptoms, treatment, 96(b)–97(b), 139
    historical perspective on, 94
    prevention of, 474(b)
Herpes zoster, 139
Hiatal hernia, 536(b)–537(b)
Hiccuping, 363, 364(t)
Hilum, 306, **306**, 430
Histamine, 237, 320, 382(t)
Histidine, 568, 568(t)
Histology, 30
Histones, 31
Histoplasma capsulatum, 92, **92**
HIV. *See* AIDS/HIV infection
Homeostasis, 6
    metabolism and nutrition, 570
Homologous chromosomes, 516
Homo sapiens, 80
Hormones. *See also specific hormones*
    actions of, 589, 590(t), 591–592, **591–592**, 591(t)
    of adrenal glands, 602–603
    of female reproductive system, 484, **485**, 488, **489**
    of gonads, 608
    of male reproductive system, **460**, 460–461
    overview of major, 589, 590(t), 591–592, **591–592**, 591(t)
    of pancreas, 605–606
    of parathyroid glands, 600–601
    of pituitary gland, 595–598
    during pregnancy, 512, 512(t)
    stress and, 604(b)
    of thymus, 607
    of thyroid gland, 598
    water regulation and, 108–109
Host factors, 66
Human chorionic gonadotropin (hCG), 487, 512, 512(t)
Human Genome Project (HGP), 515
Human Immunodeficiency Virus (HIV)/Acquired Immunodeficiency Syndrome (AIDS). *See* AIDS/HIV infection
Human leukocyte antigens (HLAs), 58, 314
Human papillomavirus (HPV)
    genital warts, 98(b)
    historical perspective on, 94
    warts and, 141, **141**
Humatrope, 595(b)
Humerus, **155**, 159, 173, **173**, 174, **175**
Humoral immunity, 313
Humors, 313
Hyaline cartilage, 44, **44**
Hyaline membrane disease, 356(b)
Hydrogen, 22(t)
Hydrogen bond, 20
Hydrothorax, 350(b)

Hyoid, **155**
Hyoid bone, **335**, **339**
Hypercalcemia, 112(t), 114, 158, 600(b)
Hypercapnia, 117
Hyperchloremia, 111(t), 113
Hypercholesterolemia, 567(b)
Hyperchromatic, 60
Hyperextension, 201, **202**
Hyperglycemia, 593(t)
Hyperkalemia, 112(t), 114
Hypermagnesemia, 112(t), 115
Hypernatremia, 111(t), 113
Hyperopia, 637(b), **638**
Hyperphosphatemia, 112(t), 115
Hyperplasia, 55, **55**
Hyperpolarization, 383
Hypertension, 243–244, 593(t)
Hyperthyroidism, 593, 597(b), 598, 600
Hypertonic, 36
Hypertrophy, 55, **55**
Hyperventilation, 362
Hypocalcemia, 112(t), 114, 158, 601(b)
Hypochloremia, 111(t), 113
Hypodermis, 128, 130
Hypogastric region, 11, **11**
Hypoglossal nerve, 406(t), 407, 409(b)
Hypoglycemia, 593(t)
Hypokalemia, 111(t), 114
Hypomagnesemia, 112(t), 115
Hyponatremia, 110, 111(t), 113
Hypoperfusion, 246
Hypophosphatemia, 112(t), 115
Hypophyseal fossa, 166
Hypophysis, 593
Hypothalamus, 390, **390**, 589
    hormones and actions of, 460, 590(t)
    regulation of body temperature, 132
    as thirst center, 108
Hypothyroidism, 593(t), 597(b), 598
Hypotonic, 36
Hypovolemic shock, 63, 246
Hypoxia, 55

i

Iatrogenic injuries, 65
Idiopathic hypertension, 244
IgA, 317, 317(t)
IgD, 317, 317(t)
IgE, 317, 317(t)
IgG, 317, 317(t), 320
IgM, 317, 317(t)
Ileocecal valve, 538
Ileum, **526**, 538, **538**
Iliac artery, 258, 259, **260**, 262(t)
Iliacus, **216**
Iliac vein, 262(t), **265**, **266**
Iliopsoas major, 218(t)
Iliotibial tract, **214**
Ilium, 174, **176**
Imidazole, 91
Imiquimod, 98(b)
Immune system, 310–326
    acquired (specific) immunity, 310, 313–319
    allergies, 320–321
    antibodies, 315–317, 316(t), 317(t)
    assisting your, 318(b)

autoimmune diseases, 59, 321–324
B cells, 313, **314**, 316(t)
chronic fatigue syndrome (CSF), 322–324
defined, 302
diseases and disorders, 320–324, 324(b)–325(b)
immunodeficiency diseases, 58–59
innate (nonspecific) immunity, 310–312, 312(t)
life span and, 326
lymphatic system components and functions, 302–310, **303–309**, 310(t)
pathophysiology, 324(b)–325(b)
protection from infectious diseases, 66
T cells, 313–315, **314**, 316(t)
transplantation and tissue rejection, 58, 319–320
Immunity
acquired (specific), 310, 313–317
artificially acquired active, 318(t), 319
artificially acquired passive immunity, 318(t), 319
cellular immunity, 313
defined, 310
humoral immunity, 313
innate (nonspecific), 310–312, 312(t)
naturally acquired active, 318–319, 318(t)
naturally acquired passive immunity, 318(t), 319
Immunization
as artificially acquired active immunity, 319
benefits of, 320(b)
Immunoglobulins, 315–317, 316(t), 317(t)
Immunopathology, 58
Impacted fracture, 180
Impetigo, 87, 139, **139**
Impotence, 465(b)
Incisive foramen, **164**
Incisors, 529, **529**
Incontinence
etiology and treatment, 445(b)
stress incontinence, 447(b)
Incus, 166, **166**, 641, **641**, **642**
Infancy, 515
Infants
erythroblastosis fetalis, 248–249, **249–250**
fluid balance and acid-base homeostasis, 118
fontanelles, 162, **165**
infectious diseases and, 66
postnatal period, 514
recognizing hearing problems, 647(b)
Infarction, 62–63
Infections
defined, 310
inflammation and, 311
upper respiratory infection, 338(b)
Infectious disease
hand washing to prevent, 91
overview of, 66
sexually transmitted infections, 94–99
Inferior, 12, **12**, 14(t)
Inferior mesenteric artery, 258
Inferior nasal conchae, 168
Inferior vena cava, 263, **265**, **266**, **274**, **276**

Infertility, 491–493
Inflammation
acute, 57
allergic reaction and, 320
as body's defense mechanism, 311
as body's response to injury, 57
chronic, 57
defined, 57
process of, 311, 312(t)
protection from infectious diseases, 66
signs of, 57, 311
Inflammatory bowel disease, 540(b)
Inflation reflex, 362
Influenza, 342(b)–343(b)
Infraorbital foramen, **162**
Infraspinatus, **207**, **209**, **210**, 211(t)
Infundibulum, 480, **481**
Ingestion, 526
Inguinal, **15**
Inguinal hernias, 544(b)–545(b)
Inhalation, 344, 358–359, **360**
Injury
body's response to, 57–58
cell injury and death, 54–57
inflammation, 57
repair, regeneration and fibrosis, 57–58
Innate (nonspecific) immunity, 310–312, 312(t)
Inner ear, 641, 643, **644**
Inorganic matter, 21, **21**
function in body, 23–24
Inorganic salts, 23–24
Inorganic substances, 22–24, 576
Insensible perspiration, 108
Insertion, 200, **200**
Inspiration, **358**, 358–359, **360**
Inspiratory capacity, 361(t)
Inspiratory reserve volume, 361(t)
Insulin, 562, 570, 590(t), 593(t), 605
Insulin-dependent diabetes mellitus, 606(b)
Integrative medicine, 69
Integumentary system, 124–147. *See also* Skin
burns, 141–144
decubitus ulcers, **130**, 130(b)
functions of, 126–128
jaundice, 133(b), **134**
overview of, **6**
regulation of body temperature, 127, 132
skin cancer, 144–146
skin color, 133
skin disorders, 137–141
skin lesions, 135–136, 137(t)
structures of, 128–132, **129–132**
wounds and healing, 134–135
Interatrial septum, 275
Intercalated disc, 48, 196
Intercostal artery, 262(t)
Intercostal muscles, 359, **360**
Intercostal vein, 262(t)
Interferon, 311
Interior jugular vein, **264**
Interlobar artery and vein, 431, **432**
Interlobular artery and vein, 431
Internal intercostals, 213(t)
Internal oblique, **213**, 213(t)
Internal respiration, 365
Interneurons, 377, 378(t), 398, **399**, 399(t)

Interphase, 37
Interstitial cells, 452
Interstitial (tissue) fluid, 107, 302
Interventricular septum, 275, **276**
Intestinal phase of digestion, 553
Intestines
large, **526**, **527**, 541–542, **542**
small, **527**, 538–540, **538–540**
Intracellular fluid, **106**, 107, 110–116
Intramembranous ossification, 157
Intrauterine devices, 491, **492**
Intrinsic factor, 235(b), 535
Inversion, 202, **203**
Involuntary muscle, 47
Iodide, 579
Iodine, 22(t), 578(t)
Ions, 23, 110
Iris, 622(t), 623, **624**, **629**, 632(t)
Iron
function in body, 578(t), 579, **579**
red blood cell production and, 230, 230(t)
Iron deficiency anemia, 234(b)
Irregular bones, 158, **159**
Ischium, 174, **176**
Islets of Langerhans, 605, **605**
Isograft, 319
Isoleucine, 568, 568(t)
Isoniazid, 88
Isotopes, 21, **22**
IV therapy, reasons for, 117

j

Jaundice, 133(b), **134**, 233
Jejunum, **526**, 538, **538**
Jenner, Edward, 319
Joints, **178**, 178–179
Jugular foramen, **163**
Jugular trunk, 303
Jugular vein, 262(t), **266**
Juvenile diabetes, 606(b)

k

Kaposi's sarcoma, 59, **59**, 99(b)
Kegel exercises, 447(b)
Keloids, 58, **58**, 135, **135**, **136**, 137(t)
Keratin, 129
Keratinocytes, 128, 129, 130–131
Keratoconjunctivitis sicca, 59
Kidneys, **428**, 429–433, **429–433**, 589
in acid–base balance, 118
blood and nerve supply of, 431, **431–432**, 433
dialysis, 437(b)
diseases and disorders of, 435(b)–438(b)
function of, 430–431
hormone secreted by, 609
nephrons, 433–435, **433–435**
structure and location of, 429–430, **429–430**
urine formation, 439–441
water loss, 107–108
water regulation and, 108–109
Kidney stones, 438(b)
Kilocalories, 570

Klinefelter's syndrome, 520(b)
Knee reflex, 421, **422**
Krebs cycle, 199, 560, **563**, 563–564
Kupffer cells, 549
Kwashiorkor, 580, **580**
Kyphosis, 183(b)–184(b), **184**

l

Labia majora, 473, **473**, 475(t)
Labia minora, 473, **473**, 475(t)
Labile cells, 38, 39
Labium majus, **472**
Labium minus, **472**
Labyrinth, 643, **644**
Lacrimal apparatus, 634
Lacrimal bone, **162**, **163**, 167
Lacrimal glands, 634, **634**
Lactase, 540
Lactic acid, 199, 563
Lactogen, 512, 512(t)
Lactose, 540
Lacunae, 156
Lag phase, 81, **81**
Lambdoidal suture, **163**, **164**
Lamellae, 156, **156**
Large cell carcinoma, 352(b)
Large intestine, **526**, **527**, 541–542, **542**
Laryngitis, 344(b)
Laryngopharynx, **335**, 337, 532
Larynx, 334(t), **335**, 338–339, **342**
LASIK, 640(b)
Lateral, **12**, 13, **14**, 14(t)
Latissimus dorsi, **207**, **210**, 211(t)
Lazy eye, 621(b)
LDL (low-density lipoprotein) cholesterol, 231, 231(b), 566
Left hypochondriac region, 11, **11**
Left iliac region, 11, **11**
Left lower quadrant (LLQ), 11, **11**
Left lumbar region, 11, **11**
Left subclavian artery, 257
Left upper quadrant (LUQ), 11, **11**
Legionnaire's disease, 351(b)
Legs
  bones of, 175, 177, **177**
  muscles of, 214, **214–217**, 218(t)–219(t)
Leiomyomas, 483(b)
Lens, 623, **624**, 627, **627**, 629, 632(t)
Leprosy, 88
Lesions, skin, 135–136, **136**, 137(t)
Leucine, 568, 568(t)
Leukemia, 239(b)–240(b), **240**
Leukocytes, 43, 228, 236–239, **237–239**, 238(t)
Leukocytosis, 238
Leukopenia, 238
Levator scapulae, **209**, **210**
Lice, 93–94, **94**, 139–140, **140**, 474(b)
Life span, 16
  blood and, 265–266
  bones and, 187
  cardiovascular system and, 296
  circulatory system and, 265–266
  digestive system and, 553
  ears and, 652

  endocrine system and, 610
  eyes and, 651–652
  female reproductive system and, 493
  heart and, 296
  human development and genetics, 521
  immune system and, 326
  life expectancy changes, 69
  male reproductive system, 465
  muscular system and, 222
  nervous system and, 421
  nutrition and metabolism, 582
  respiratory system, 367
  sense of smell and taste, 652
  skeletal system and, 187
  skin and, 146–147
  urinary system and, 446
Ligaments, 46, 179, 197, 219(b)
Ligamentum arteriosum, 511
Ligamentum venosum, 511
Light microscope, 76–77, **77**
Linea aspera, **160**
Lingual artery, **258**, 262(t)
Lingual frenulum, **528**, 528
Lingual tonsils, 309, **335**, **529**
Lingula, 346
Linoleic acid, 561
Lipase, 540, 547
Lipid metabolism, **565**, 565–567
Lipids
  as class of hormones, 592
  conditions associated with, 571(t)
  defined, 561
  function in body, 24, 566, 571(t)
  metabolism of, **565**, 565–567
Lipofuschin, 56
Lipoproteins, 231, 231(b), 565–566
Lithium, **21**
Liver, **527**, **548**, 548–549
  fatty liver, 56
Liver spots, 56
Long bones, 158–160, **159**
Loop of Henle, 434
Loose connective tissue, 45
Lophotrichous, 79, **79**
Lordosis, **184**, 184(b)
Lower esophageal sphincter (LES), 533
Lower extremities
  arteries to, 259–260, **260**
  bones of, 175, 177, **177**
  muscles of, 214, **214–217**, 218(t)–219(t)
  veins of, **265**
Lower respiratory tract, 334(t), **340**, 340–346, **341**, **346**, 355, **356**
Lumbar, **15**
Lumbar artery, 262(t)
Lumbar curve, **169**, 170
Lumbar nerves, **410**, **410**
Lumbar puncture, 420, **421**
Lumbar trunk, 303–304
Lumbar vertebrae, 169, **169**, 171, **171**
Lumbosacral plexus, **412**, 412, 413, **414**
Lung cancer, 351(b)–353(b)
Lungs, **273**, 334(t)
  alveoli, 355, **355**
  anatomy of, **335**, **342**, 345–346, **346**
  diseases and disorders, 347(b)–355(b)

  gas exchange, 364–365, **364–365**
  pulmonary circuit, 255–256
  respiration, 358–359, **360**
  respiratory volumes and capacities, 361–362, 361(t)
  systemic circuit, 256
Lunula, 131, **131**
Lupus, 59, 321–322, 322(t)
Luteinizing hormone (LH), 484, **485**, 590(t), 597
  in female reproductive cycle, 488, **489**
  testosterone production and, **460**, 460–461
Lymphadenitis, 306–307
Lymphadenopathy, 306–307
Lymphatic capillaries, 303, **303**
Lymphatic ducts, 303–305, **304**, **305**
Lymphatic nodules, 308–310
Lymphatic system, 302–310
  defined, 302
  diseases and disorders, 305(b)
  function of, 302
  interstitial fluid and lymph fluid, 302
  lymphatic vessels and lymph circulation, 302–305, **303–305**
  organs and tissues of, **306–309**, 306–310, 310(t)
  overview of, **6**
Lymphatic trunk, 303–304
Lymphedema, 305(b), 324
Lymph nodes, 305, **305–306**, 306–307, 310(t)
Lymph nodules, 306
Lymphocyte, **43**, 306
  B cells, 313, **314**
  as cellular component of blood, 238(t)
  development of, **229**
  number present and, 239
  structure and function of, 236, 237, **237**
  T cells, 313–315, **314**, 316(t)
Lymphokines, 313
Lysine, 139, 568(t)
Lysosomes, 32(t), 33
Lysozymes, 311

m

Macrophage, **229**, 306, 311
Macula lutea, 624, **624**
Macular degeneration, 625(b)–626(b)
Macule, 135, **136**, 137(t)
Magnesium, 22(t), 23(t)
  concentration in extracellular and intracellular fluid, **110**, 115
  function in body, 577(t), 579
  hypermagnesemia, 112(t), 115
  hypomagnesemia, 112(t), 115
Magnetic resonance imaging, 420
Magnification, 76–77
Main stem bronchi, 344
Major histocompatibility complex, 58, 313–314
Malaria, 99
Male reproductive system, 450–465
  diseases and disorders, 456(b), 457(b)–459(b), 465(b)
  erection, orgasm and ejaculation, 464, **464**

hormones of, **460**, 460–461
internal structures, 456–457, **457**, 459–460
life span, 465
overview of, **7**
prostate cancer screening and risks, 459(b)
sexually transmitted infections (STI), 474(b)
spermatogenesis, 461, **461–464**, 462
structure and function of organs, 452–455, **453–454**, **456**
testicular self-examination, 455(b)
Malignant, 59
Malignant melanoma, 144–146, **145**
  ABCDEs of, 145(b)
  skin cancer warning signs, 146(b)
  staging of, 146
Malleus, 166, **166**, 641, **641**, 642
Malnutrition, 580–582
Maltase, 540
Mammary, **15**
Mammary glands, **475**, 475–476
Mandible, **162**, 163, **165**, 167, **167**
Mandibular condyle, **163**
Mandibular fossa, **164**
Manganese, 22(t), 578(t), 579
Manubrium, 168, **168**
Marasmus, 580, **580**
Mask, wearing, 359(b)
Masseter, **207**, **208**, 209(t)
Mastication, 528
Mastoid fontanel, **165**
Mastoid foramen, **164**
Mastoid process, **163**, 166
Matrix, 42–43
Matter, 7, 21
Maxilla, **162**, **163**, **165**
Maxillae, 167
Maxillary artery, **258**, 262(t)
Maxillary sinus, 336
Mebendazole, 93(b)
Mechanical barriers of birth control, 490
Medial, **12**, 13, **14**, 14(t)
Mediastinum, 9, **10**
Medulla oblongata, 362, **363**, 386(t), **387**, 391
  cardiac control center, 283
Medullary cavity, 159
Megakaryocyte, **43**, 240
Megaloblastic anemia, 235(b)
Meiosis, 38, 461, **462**
Melanin, 128, **129**, 133
Melanocytes, 128, **129**, 131, 133
Melanocyte stimulating hormone (MSH), 590(t), 596
Melanosomes, 128, 133
Melatonin, 590(t), 592, 607, 608(b)
Membrane potential, 382, **384**
Memory B cells, 313
Memory T cells, 314, **315**
Menarche, 485
Meniere's disease, 645(b)–646(b)
Meninges, **387**, 400, **401**
Meningitis, 88
Menopause, 485
Menses, 485
Menstrual cycle, 485–488, **489**

Menstrual phase, 485
Mental, **15**
Mental foramen, **162**
Mesencephalon, 391
Mesenchymal, 59
Mesenteric vein, 262(t)
Mesentery, 538
Mesoderm, 504, **506–507**
Mesothelioma, 347(b)
Metabolic alkalosis, 118, **119**
Metabolic syndrome, 566(b)
Metabolism, 16, 20, 558–570
  basal metabolic rate (BMR), 570
  carbohydrate metabolism, 561–564, **561–565**
  defined, 560
  eating healthy, **569**, 569(b)
  homeostasis, 570
  hypercholesterolemia, 567(b)
  life span and, 582
  lipid metabolism, **565**, 565–567
  maintaining healthy nutrition and metabolism, 576(b)
  metabolic syndrome, 566(b)
  minerals, 576, 577(t)–578(t), 579
  overview, 560–561
  protein metabolism, 567–569, **568**, 568(t)
  vitamins, 572, 573(t)–575(t)
Metacarpals, **155**, 174, **175**
Metacarpophalangeal (MCP) joints, 174
Metaphase, 38
Metaphase plate, 38
Metaplasia, 56, **56**
Metastasis, 60, 144
Metatarsals, **155**, 175, 177, **177**
Metatarsophalangeal (MTP) joint, 177
Methicillin-resistant *S. aureus* (MRSA), 99
Methionine, 568, 568(t)
Metronidazole, 92, 97(b)
Microbiology, 74–99
  bacteria, 78–89
  defined, 76
  microscopes and, 76–78
  mycology, 90–92
  parasitology, 92–94
  sexually transmitted infections, 94–99
  virology, 89–90
Microglia cell, **378**, 379, 379(t)
Microglial cell, **48**, 49
Microscopes, 76–78, **77**
Microvilli, 42, 540
Micturition, 441, 443
Midbrain, 386(t), **387**, 391
Middle ear, 641
Middle nasal conchae, 168
Midline, **12**, **14**
Midsagittal plane, 12, **13**
Migraines, 394(b)
Milk production and secretion, 514
Mineralcorticoids, 603
Minerals, 576, 577(t)–578(t), 579
Mitochondria, 32, 32(t), 564, **565**
Mitosis, **37**, 37–38
Mitral valve, **276**, **277**, 278
Molars, 529, **529**
Molecules, 20
Monocytes, 57, 311

as cellular component of blood, 238(t)
development of, **229**
number present and, 239
structure and function of, 236, 237, **237**
Mononuclear phagocytic system, 311
Mononucleosis, 324
Monotrichous, 79, **79**
Mons pubis, 473, **473**
Morbid obesity, 65, 582
Morula, 500, **501**, 502(t)
Motor neurons, 377, 378(t), 398, **399**, 399(t)
Mouth, **527**, **528–530**, 528–531
  oral cancer, 530(b), 531(b)
Mucosa, 533
  as physical barrier for disease prevention, 311
Mucosa associated lymphoid tissue (MALT), 308
Multifactorial inheritance, 61
Multinucleate, 47
Multiple sclerosis, 322(t), 415(b)
Multipolar neurons, 376, **376**, 377(t)
Multiunit smooth muscle, 194–195
Murmur, 283(b)–284(b)
Muscle fatigue, 199
Muscle fiber, 196–197, **197**
Muscle tissue, **41**, 46–48
  cardiac, **47**, 48
  glycogen storage, 561
  skeletal, **47**, 47
  smooth, **47**, 47–48
Muscular dystrophy, 205(b)–206(b), 520(b)
Muscularis, 533–534
Muscular system, 192–222
  attachment and actions of skeletal muscles, **200**, 200–201
  body movements, 201–202, **203**
  cardiac muscles, 195(t), 196
  diseases and disorders, 205(b)–206(b), 220(b)–222(b)
  energy production for muscles, 198–199
  exercise for muscle health, 216(b)
  of eye, 634–636, **635**, 635(t)
  of face and head, 208, **208**, 209(t)
  function of muscles, 194–196, 195(t)
  life span and, 222
  of lower extremities, 214, **214–217**, 218(t)–219(t)
  major skeletal muscles, 207, **207**
  muscle fatigue and cramps, 199
  overview of, **6**
  oxygen debt, 199
  range of motion, 204(b)
  skeletal muscles, 195, 195(t), **196**
  smooth muscles, 194–195, 195(t)
  strains and sprains, 219, 220(b)
  structure and classification of muscles, 196–197, **197**
  of trunk, 212, **213**, 213(t)
  types of muscles, 194–196, 195(t)
  of upper body, 207, **209**, **210**, 211(t), **213**
Myalgia, 292
Myasthenia gravis, 206(b), 322(t)
Mycobacterium leprae, 88
Mycobacterium tuberculosis, 88, **88**, 88(b), 349(b)
Mycology, 90–92

INDEX **729**

Mycoses
  cutaneous, 91
  opportunistic, 92
  subcutaneous, 92
  superficial, 91
  systemic, 92
Myelin, **380**, 380–381
Myocardial hypertrophy, 55
Myocardial infarction, 246, 291, 295–296(b)
Myocarditis, 280(b)
Myocardium, 279, **279**, 279(t)
Myocytes, 46, 196
Myofibrils, 196
Myoglobin, 199
Myometrium, 481, **481**
Myopia, 637(b), **638**
Myosin, 24, 196
MyPlate, 568(b), **569**
Myxedema, 593(t)

n

Nail bed, 131, **131**
Nails, 131, **131**
Nasal, **15**
Nasal bone, **162**, **163**, **165**, 167
Nasal cavity, 334(t), 336
Nasal conchae, **162**, **335**, 336
Nasal septum, 166, 336
Nasolacrimal ducts, 634, **634**
Nasopharynx, **335**, 337, 532
Natural killer (NK) cells, 311, 312(t)
Naturally acquired active immunity, 318–319, 318(t)
Naturally acquired passive immunity, 318(t), 319
Neck
  arteries to, **258**, 259
  veins of, 263, **263**
Necrosis, 40, **56**, 56
Negative feedback loop, 609–610
Neisseria gonorrhea, 474(b)
Neisseria gonorrhoeae, 95(b)–96(b)
Neisseria meningitidis, 82(b)
Neonatal period, 514
Neonate, 514
Neoplasia, 59–60
Neoplasms, 59
Nephrons, 433–435, **433–435**
Nerves/nervous system, 372–422
  autonomic nervous system, 416, 417(t), **418–419**
  brain, 384–391, 385(t), **387–390**
  cells of, 375–381, **376–378**, 379(t), **380**
  central nervous system, 384–399, **385–390**
  compared to endocrine system, 588, **588**
  cranial nerves, **404**, 404–407, 405(t)–406(t), **407–408**
  diseases and disorders, 391(b)–396(b), 401(b)–402(b), 408(b)–409(b), 414(b)–415(b)
  functions of, 374–375, **375**
  to kidneys, 431, 433
  life span and, 421
  meninges, ventricles and cerebrospinal fluid, 400–404, **401–403**

neurological testing, 420–421, **421–422**
overview of, **7**, 374, **375**
parasympathetic nervous system, 416, **419**
pathophysiology, 391(b)–396(b)
peripheral nervous system, 385(t), **404**, 404–416, 405(t)–406(t), **407–414**, 417(t), **418–419**
reflexes, 397–398, **399**, 399(t)
somatic nervous system, 416, 417(t)
spinal cord, **396–397**, 396–399
spinal nerves, 410–413, **410–414**
synapse and nerve transmission, 381, 381–383, 382(t), **383–384**
Nervous tissue, **8**, **41**, **48**, 48–49
Neuralgia, 409(b)
Neural tube, 386
Neurogenic shock, 63, 247
Neuroglia, 49, 377, **378**, 379–380, 379(t)
Neuroglial cells, 377, **378**, 379–380, 379(t)
Neurohypophysis, 593, 597
Neurological testing, 420–421, **421–422**
Neurons, 39, **48**, 48–49, 375–377, **376–377**, 378(t)
Neurotransmitters, 381–382, 382(t), 603
Neutrons, 21
Neutrophils, 57, 311
  as cellular component of blood, 238(t)
  development of, **229**
  number present and, 238–239
  structure and function of, 236, 237, **237**
Niacin, 572, 574(t)
NIDDM (noninsulin-dependent diabetes mellitus), 607(b)
Nitric oxide, 382(t)
Nitrogen, 22(t)
Nits, 140, **140**
Nodes of Ranvier, **380**, 383
Nodule, **136**, 137(t)
Noise pollution, 648
Noncyanotic disorders, 278(b)
Non-insulin-dependent diabetes mellitus (NIDDM), 606(b), 607(b)
Nonspecific defense system, skin as, 126
Noradrenalin, 603
Norepinephrine, 196, 570, 603
  action of, 195, 382(t), 590(t)
  imbalance of, 603(b)
Normal flora, 82(b)
Nose, 332, **333**, 334(t)
Nosocomial injuries, 65
Nostril, **335**
Nuclear scan, 293
Nuclease, 547
Nucleic acids, 24–25
Nucleolus, 31
Nucleotide, 25
Nucleus, 21, **21**, **30**, 31, **31**, 32(t)
Nutritional pathology, 65
Nutrition and diet, 558–582
  carbohydrate metabolism, 561–564, **561–565**
  eating disorders, 581(b)
  eating healthy, **569**, 569(b)
  homeostasis, 570
  hypercholesterolemia, 567(b)
  life span and, 582

lipid metabolism, **565**, 565–567
maintaining healthy nutrition and metabolism, 576(b)
malnutrition, 580–582
metabolic syndrome, 566(b)
minerals, 576, 577(t)–578(t), 579
MyPlate, 568(b), **569**
obesity, 582
protein metabolism, 567–569, **568**, 568(t)
starvation, 580–581
vitamins, 572, 573(t)–575(t)
NuvaRing, 491
Nystagmus, 638(b)

o

Obesity, 65, 582
Oblique plane, 13
Obstructive shock, 247
Obstructive sleep apnea (OSA), 341(b)
Occipital, **15**
Occipital artery, 262(t)
Occipital bone, **163**, 165, **165**
Occipital condyle, **164**, 165
Occipitalis, **207**, **208**
Occipital lobe, **388**, **389**, 389(t)
Occlusive stroke, 393(b)
Oculomotor nerve, 404, 405(t)
Olfaction, 616–618, **617**
Olfactory nerve, 404, 405(t), 409(b)
Olfactory receptors, 616, **617**
Oligodendrocyte, **48**, **378**, 379, 379(t)
Oocyte, 477
Oogenesis, **479**, 479–480, **480**
Open fracture, 180, **180**
Ophthalmic artery, 262(t)
Opportunistic infections, AIDS patients and, 59
Opposition, 202
Optic chiasm, 593, **594**, **630**, 630–631, **631**
Optic disk, 623–624, **624**, 629
Optic nerve, **624**, 629
  function of, 404, 405(t), 620, 632(t)
  visual pathways and, **630**, 630–631, **631**
Oral, **15**
Oral cancer, 530(b), 531(b)
Oral cavity, 528–529
Oral contraceptives, 491, **492**
Orbicularis oculi, **207**, **208**, 209(t)
Orbicularis oris, **207**, **208**, 209(t)
Orbital, **15**
Orchiectomy, 455(b)
Organelles, 31–34, 32(t)
Organic matter, 21
Organic substances, 24–25
Organism, **5**
Organ of Corti, 643, **645**
Organs, **5**
Organ systems, **5**, 9
Orgasm, 464, **464**, 475
Origin, 200, **200**
Oropharynx, **335**, 337, 532
Osmolarity, 109
Osmosis, 35–36
Ossicles, 641, 646
Ossification, 157
Osteoarthritis (OA), 181(b)

Osteoblasts, 44, 156, **157**
Osteoclasts, 44, 114, 115, 156, **157**, 158
Osteocytes, 44, 156, **157**
Osteogenesis, 157
Osteogenesis imperfecta, 185(b)
Osteons, 155, **156**
Osteoporosis, 185(b)–186(b)
Osteosarcoma, 59, 186(b)
Otic, **15**
Otitis media, 642
Otosclerosis, 642(b)
Ova, 477
  meiosis and, 38
Oval window, 641, **641**, 642
Ovarian cancer, 478(b)–479(b), 608(b)
Ovaries, **472**, 475(t), **589**
  cancer of, 478(b)–479(b), 608(b)
  hormones and actions of, 590(t), 608
  menstrual cycle, 485–488, **489**
  ovum formation, **479**, 479–480, **480**
  structure and function of, 477, **478**
Overweight, BMI to define, 65
Oviduct, 480
Ovulation, 479–480, **480**, 486
Ovum, **30**
  fertilization, 490
  formation of, **479**, 479–480, **480**
Oxygen, 22(t)
  function in body, 22–23, 23(t)
  gas exchange in respiration, 364, **365–366**
  as inorganic substance, 22–23
  transportation of, in blood, 232, 366–367, **366–367**, 367(t)
Oxygen debt, 199
Oxyhemoglobin, 232, 366–367, **366–367**, 367(t)
Oxytocin, 476, 513, 514, 590(t), 598, 610

p

Paget's disease, 186(b)–187(b)
Palatine bone, **163**, 167
Palatine tonsils, 309, **335**, **528**, **529**
Palmar, **15**
Palmaris longus, **210**, 212(t)
Pancreas, **526**, **527**, **547**, 547–548, **589**, **604**, 604–606
  cancer of, 548(b)
  hormones and actions of, 590(t), 605–606
  inflammation of, and chest pain, 292
Pancreatic cancer, 548(b)
Panic attack, 292
Pantothenic acid, 574(t)
Papillae, **529**, 529, 618–619, **619**
Papillary muscles, **276**, **277**, 278
Pap smear, 482(b)
Papule, **136**, 137(t)
Paranasal sinuses, 334(t), 336
Parasites
  arthropods, 93–94
  helminths, 92–93
  protozoa, 92
Parasitology, 92–94
Parasympathetic innervation, 282
Parasympathetic nervous system, 416, **419**

Parathyroid glands, **589**
  hormones and actions of, 590(t), **600**, 600–601
Parathyroid hormone (PTH)
  action of, 590(t), 600–601, **601**
  calcium regulation and, 114
  osteoclast regulation and, 158
  phosphate regulation and, 115
  during pregnancy, 512, 512(t)
Parenchyma, **431**, 433
Paresthesias, 420
Parietal bone, **162**, **163**, 165, **165**
Parietal cells, 535
Parietal lobe, **388**, **389**, 389–390, 389(t)
Parietal pericardium, 280
Parietal peritoneum, 538
Parietal pleura, 346, **346**
Parkinson's disease, 392(b)
Parotid glands, **530**, 531
Parturition, 513
Passive transport, **35**, 35–36, **36**
Patch, **136**
Patella, **155**, 175, 177, **177**, **217**
Patellar, **15**
Patellar ligament, **216**, **217**
Patellar tendon, **216**
Pathogenicity, 66, 84
Pathology, 54
Pathophysiology
  blood, 233(b)–236(b), 239(b)–240(b), 253(b)–255(b)
  cardiovascular system, 280(b)–281(b), 283(b)–284(b), 293(b)–296(b)
  clostridium species, 85(b)–86(b)
  decubitus ulcers, **130**, 130(b)
  digestive system, 531(b), 536(b)–537(b), 540(b)–541(b), 543(b)–545(b), 546(b), 548(b), 549(b), 551(b)–552(b)
  ears, 642(b), 645(b)–646(b), 648(b), 650(b)
  endocrine system, 595(b)–598(b), 600(b)–601(b), 603(b), 606(b)–608(b)
  eyes, 621(b), 623(b), 625(b)–630(b), 632(b)–634(b), 638(b)
  female reproductive system, 476(b)–479(b), 482(b)–484(b), 488(b)
  fibromyalgia, 47(b)
  genetic disorders, 517(b)–521(b)
  immune system, 324(b)–325(b)
  jaundice, 133(b), **134**
  male reproductive system, 456(b), 457(b)–459(b), 465(b)
  muscular system, 205(b)–206(b), 220(b)–222(b)
  nervous system, 391(b)–396(b), 401(b)–402(b), 408(b)–409(b), 414(b)–415(b)
  respiratory system, 333(b)–334(b), 336(b)–337(b), 338(b), 342(b)–345(b), 347(b)–358(b)
  sexually transmitted infections, 94(b)–99(b)
  skeletal system, 181(b)–187(b)
  thrombocytopenia, 43(b)
  trichinella spiralis, 93(b)
  urinary system, 435(b)–438(b), 445(b)
  viral hepatitis, 89(b)–90(b)

Pectineus, **216**
Pectoral, **15**
Pectoral girdle
  bones of, 173–174, **173–174**
  muscles of, 212(t)
Pectoralis, 207
Pectoralis major, **207**, 211(t)
Pectoralis minor, 212(t)
Pedal, **15**
Pediculosis, 139–140
Pediculosis humanus, 140, **140**
Pediculus humanus capitis, 93
Pediculus humanus humanus, 94
Pediculus thirus pubis, 94
Pelvic cavity, **10**
  gender differences in, 178(t)
Pelvic girdle, 174, **176**
Pelvic inflammatory disease (PID)
  chlamydia and, 94(b)–95(b)
  historical perspective on, 94
Pelvis
  arteries to, 259–260, **260**
  gender differences in, 178(t)
  veins of, **265**
Penicillin, 89, 96, 313
Penis, **453**, 455, **456**
Pepsin, 535
Pepsinogen, 535
Peptidase, 540
Perforins, 311
Pericardial cavity, 9
Pericardial sac, 280
Pericarditis, 281(b), 292
Perilymph, 643, **644**
Perimetrium, **481**, **481**
Perimysium, 197
Perineal, **15**
Perineum, **473**, 474
Periosteum, 160, **161**
Peripheral nervous system, 385(t), **404**, 404–416, 405(t)–406(t), **407–414**, 417(t), **418–419**
Peristalsis, 195, 442, 534, 538, **539–540**
Peritoneal dialysis, 437(b)
Peritoneum, 538
Peritrichous, 79, **79**
Permanent cells, 38–39, 375
Pernicious anemia, 235(b), 322(t)
Peroxisomes, 32(t), 33
Perspiration, 128
  water loss through, 108
Petechia, 136, 137(t)
Petechiae, **136**
Peyer's patches, 310
pH
  acid–base balance, 116–118
  of blood, 116
Phagocytosis, 36, 311, 312(t)
Phalanges, **155**, 159, 174, 175, **175**, 177, **177**
Pharyngeal tonsils, 309, **335**
Pharynx, 334(t), 337, **527**, 531–533, **532**
Phase microscope, 78
Phenotype, 515
Phenylalanine, 26, 568, 568(t)
Phenylketonuria (PKU), 26, 520(b)
Pheochromocytoma, 593, 603(b)
Phlebotomy, 255(b)

INDEX   731

Phosphate, 23(t), 25
  concentration in extracellular and intracellular fluid, **110**, 115
  hyperphosphatemia, 112(t), 115
  hypophosphatemia, 112(t), 115
Phosphate buffer system, 116, 117(t)
Phospholipids, 24
Phosphorus, 22(t), 576, 577(t)
Phrenic arteries, 258
Phrenic nerve, **412**, 412–413
Physical barriers, 66
Physiology, 5
Pia mater, 400, **401**
Piedraia hortai, 91
Pilus, 79–80, **80**
Pineal body, 590(t)
Pineal gland, **589**, 592, 607
Pinna, 641
Pinocytosis, 36
Pituitary gland, **589**, 593–598, **594**
  hormones and actions of, 590(t), 595–598
Placenta, 502, **502**
Placental stage, 513, **514**
Plantar, **15**
Plantar flexion, 201, **202**
Plaque, **136**, 137(t)
Plasma, 43, **229**, 230–232, 231(t)
  bicarbonate level, 116
  carbon dioxide and, 116–117
  as extracellular fluid, 107
  nutrients in, 230–231
  proteins of, 230, 230(t)
Plasma cells, 313
Plasma membrane, 34
Platelets, 43, **43**, 228, 238(t), 240
  blood clotting and, 244–246, **245**
Platysma, **208**, 209(t)
Pleura, 346, **346**
Pleural cavity, 9
Pleural effusion, 347(b)–348(b)
Pleurisy, 292, 347(b)
Pleuritis, 347(b)
Pneumocystis carinii pneumonia (PCP), 99(b)
Pneumoconiosis, 353(b)
Pneumonia, 353(b)–354(b)
  aspiration pneumonia, 338
Pneumothorax, 350(b)
Polar bodies, 38, 479, **479**
Polio, 99
Polycystic kidney disease, 436(b)–437(b)
Polydipsia, 606(b)
Polymerase chain reaction (PCR), 517
Polyphagia, 606(b)
Polyuria, 606(b)
Pons, 362, **363**, 386(t), **387**, 391
Popliteal, **15**
Popliteal artery, 260, **260**, 262(t)
Popliteal vein, 262(t), 265, **266**
Positron emission tomography, 420
Posterior, **12**, 13, 14(t)
Posterior cavity, 628–629, **629**
Posterior fontanel, 162, **165**
Posterior pituitary, 593
Postherpetic neuralgia, 139
Postnatal period, 514
Postovulatory phase, 486–487

Potassium, 22(t), 23(t)
  concentration in extracellular and intracellular fluid, **110**, 113
  function in body, 576, 577(t)
  heart rate and potassium ions, 283
  hyperkalemia, 112(t), 114
  hypokalemia, 111(t), 114
  nerve transmission and, 382, 383, **384**
Pott's disease, 88
Pregnancy, 490, 498–515
  birth process, 512–513, **513**, **514**
  embryonic period, 503–504, **504–508**, 509(t)
  erythroblastosis fetalis, 248–249, **249–250**
  fetal circulation, 510–511, **510–511**
  fetal period, 504, 506, **508**, **509**, 509(t)
  hormones during, 512, 512(t)
  milk production and secretion, 514
  postnatal period, 514
  prenatal period, 500, **501–502**, 502, 502(t)
  teratogens and, 60
Prehypertension, 243, 244(t)
Preload, 285
Premenstrual syndrome, 488(b)–489(b)
Prenatal period, 500, **501–502**, 502, 502(t)
Preovulatory phase, 486
Prepuce, 455, **456**
Presbycusis, 648(b)
Presbyopia, 628(b), 637(b)
Primary antibody deficiency, 58
Primary germ layers, 504, **506–507**
Primary hypertension, 244
Primary lesions, **136**
Primary stem bronchi, 344
Prime mover, 200–201
Primordial follicles, 477, **480**
PR interval, 289(t)
Process, 161(t)
Progesterone, 477
  in female reproductive cycle, 488, **489**
  female secondary sexual development, 484, **485**
  made from cholesterol, 24
  during pregnancy, 512, 512(t), 590(t), 608
Prognosis, 68
Prokaryote, 78
Prolactin, 476, 590(t), 597
Proline, 568(t)
Pronation, 202, **203**
Pronator quadratus, **210**
Pronator teres, **210**, 212(t)
Prone position, 12
Prophase, 37
Prostaglandins, 457, 460, 490, 592
Prostate gland, **453**, 457
  cancer of, 457(b)–458(b), 459(b)
Prostatitis, 458(b)
Protein buffer system, 116, 117(t)
Protein metabolism, 567–569, **568**, 568(t)
Proteins
  as class of hormone, 589, 591–592
  conditions associated with, 571(t)
  food sources of, 569
  functional, 567
  function in body, 24, 567, 571(t)
  metabolism of, 567–569, **568**, 568(t)
  structural, 567

Protons, 21, **21**
Protozoa, 92
Protraction, 202, **203**
Proximal, **12**, 13, **14**, 14(t)
Proximal convoluted tubule, 434, **435**
Proximal interphalangeal (PIP) joints, 174
Pseudostratified columnar epithelium, 42
Psoas major, **216**
Psoas minor, **216**
Psoriasis, 140
Psoriatic arthritis, 140
Ptosis, 430
Pubis, 174, **176**
Pulmonary artery, **274**, 276
Pulmonary circuit, 255–256
Pulmonary edema, 354(b)
Pulmonary embolism, 292, 354(b)–355(b)
Pulmonary parenchyma, 355
Pulmonary semilunar valve, **277**, 278, 283
Pulmonary trunk, **273**, **274**
Pulse, locations for taking, **242**, 242–243
Pupil, 623, **624**, **629**
Purkinje fibers, **286**, 287
Purpura, **136**, 137(t)
Pustule, 135, **136**, 137(t)
P wave, 287–288, 289(t)
Pyelonephritis, 438(b)
Pyloric sphincter, **534**, 535
Pyothorax, 350(b)
Pyruvic acid, 199, **562**, 563

q

QRS complex, 287–288, 289(t)
QT interval, 289, 289(t)
Quadriceps femoris tendon, **216**
Quadriceps group, 214

r

Radial artery, 262(t)
Radial vein, 262(t)
Radiation, 570
Radioactive, 21
Radius, **155**, 173, **173**, 174, **175**
Range of motion, 204(b)
rDNA, 25
Receptors, 591
Recombinant DNA, 25
Rectum, **527**, 545, **545**
Rectus abdominis, **207**, **213**, 213(t)
Rectus femoris, **207**, **214**, **216**, 218(t)
Red blood cells, 159, 228
  blood counts of, 233
  diet and, 230(t)
  structure and function of, **232**, 232–233, **233**, 238(t)
Red bone marrow, 158, 159, 228
Red pulp, 308, **308**
Red stroke, 393(b)
Reflexes
  overview of, 397–398
  parts of reflex arc, **399**, 399(t)
  testing of, 420, 421, **422**
Refraction, 630, **630**

Regeneration, 58
Relaxin, 512, 512(t)
Remodeling, bones, 157, 158
Renal artery, 258, **428**, 431, **431**, **432**
Renal calculi, 438(b)
Renal column, **431**, 433
Renal corpuscle, 433–434, **433–434**
Renal cortex, **431**, **432**, 433, **433**
Renal medulla, **431**, **432**, 433, **433**
Renal nerves, 431, 433
Renal pelvis, 435
Renal pyramids, **431**, 433
Renal tubules, 434, **435**
Renal vein, 431, **431**
Renin, 107, 431
Repolarization, 288, 383, **384**
Reproductive system
 birth control, 490–491, **492**
 female, **7**, 470–493
 infertility, 491–493
 male, **7**, 450–465
 sexually transmitted infections (STIs), 94(b)–99(b), 474(b)
Residual volume, 361(t)
Resistance, 310
Resolution, 76–77
Respiratory acidosis, 118, **119**
Respiratory alkalosis, 118, **119**
Respiratory bronchiole, **335**
Respiratory distress syndrome (RDS), 356(b)–357(b)
Respiratory muscles, 213(t)
Respiratory pump, 304–305
Respiratory system
 caring for patients who snore, 341(b)
 cough etiquette, **343**, 343(b)
 diseases and disorders, 333(b)–334(b), 336(b)–337(b), 338(b), 342(b)–345(b), 347(b)–358(b)
 factors that control breathing, 362–364, 364(t)
 gas exchange, 364–365, **365**, **366**
 gas transport, 366–367, **366–367**
 inhalation and exhalation, 358–359, **360**
 life span, 367
 lower respiratory tract, 334(t), **340**, 340–346, **341**, **346**, 355, **356**
 nonrespiratory air movement, 364(t)
 organs of, and flow of air through, **333**, 334(t)
 overview of, **6**
 pathophysiology, 333(b)–334(b), 336(b)–337(b), 338(b), 342(b)–345(b), 347(b)–358(b)
 respiration, **358**, 358–359, **360**
 respiratory volumes and capacities, 361–362, 361(t)
 upper respiratory tract, 332–339, **333**, 334(t), **335**
 wearing a mask, 359(b)
Respiratory volumes, 361–362, 361(t)
Reticuloendothelial system, 311
Retina, 622(t), 623–625, **624**, **629**, 632(t)
Retinal detachment, 626(b)
Retinitis pigmentosa, 61
Retraction, 202, **203**

Retroperitoneal, 429, **429**
Rh antigen, 248
Rh blood group, 248–249, **249**, **250**
Rheumatic fever, 322(t)
Rheumatoid arthritis, 182(b), 322(t)
Rheumatoid scleroderma, 59
Rhinoplasty, 332
Rhomboid, **207**
Rhomboid major, **210**
Rhomboid minor, **210**
Rhythm method, 490
Rib cage
 bones of, 168, **168**
 broken ribs and chest pain, 292
Riboflavin, 574(t)
Ribonucleic acid (RNA), 24–25, 33
Ribosomes, 32(t), 33
Ribs, **155**
RICE, **219**, 220(b)
Rifampin, 88
Right hypochondriac region, 11, **11**
Right iliac region, 11, **11**
Right lower quadrant (RLQ), 11, **11**
Right lumbar region, 11, **11**
Right upper quadrant (RUC), 11, **11**
Right ventricle, **273**
Ring of Waldeyer, 309
Ringworm, 91, 140–141, **141**
RNA (ribonucleic acid), 24–25, 33
Rods (bacteria), 82
Rods (eyes), 625, **625**, 632(t)
Rogaine, 137
Rosacea, 140
Rotation, 201, **203**
Rough endoplasmic reticulum (rER), 33
Round window, **641**, **642**, **644**
Rubor, as sign of inflammation, 57
Rugae, 442, 484, 535
Rule of nines, 142, 142(b), **143**

S

Sacral, **15**
Sacral nerves, **410**, 411
Sacrum, **155**, 169, **169**, 171, **172**
 gender differences in, 178(t)
Sagittal plane, 12, **13**
Sagittal suture, 165, **165**
Salivary glands, **527**, **530**, 530–531
Saltatory conductance, 383
Saphenous vein, 262(t), **265**, **266**
Sarcolemma, 196, **197**
Sarcoma, 59, 60
Sarcoplasm, 196
Sarcoplasmic reticulum, 196
Sartorius, **207**, **214**, **215**, **216**, 218(t)
Satellite cells, 379(t), 380
Scabies, 140, **140**
Scale, **136**, 137(t)
Scanning electron microscope (SEM), 77
Scapula, **155**, **173**, **174**, **209**
Scapulae, 173
Scars, 134–135, **135**
Schwann cells, 379(t), 380, **380**
Sciatica, 415(b)
Sclera, 622(t), 623, **624**, **629**, 632(t)
Scleral venous sinus, 623

Scleroderma, 59
Scoliosis, **184**, 184(b)–185(b)
Scrotum, 452, **453**
Seasonal allergies, 337(b)
Sebaceous glands, **129**, 131
Sebum, 131
Secondary bronchi, 344
Secondary hypertension, 244
Secondary lesions, **136**
Second-degree burns, 142, **144**
Secretin, 548
Secretion, in digestion, 526–527
Selenium, 318(b), 578(t), 579
Sella turcica, **163**, 593
Semen, 457, 459–460
Semicircular canals, **641**, 643, **644**
Semimembranosus, **207**, **215**, 218(t)
Seminal fluid, 457
Seminal vesicles, **453**, 457
Seminiferous tubules, 452
Semitendinosus, **207**, **215**, 218(t)
Sensorineural hearing loss, 647
Sensory adaptation, 617
Sensory neurons, 377, 378(t), 398, **399**, 399(t)
Septic shock, 63, 246–247
Serine, 568(t)
Serosa, 280, 534
Serotonin, 382(t)
Serratus anterior, **207**
Serum, 230
Sesamoid bones, 158
Sex chromosomes, 61, 515
Sex-linked disorders, 61, 204(b), 205(b), 241(b), 516
Sex pilus, 80, **80**
Sexually transmitted infections (STI), 474(b)
 chlamydia, 94(b)–95(b)
 gonorrhea, 95(b)–96(b)
 herpes simplex virus infections, 96(b)–97(b)
 HIV/AIDS, 98(b)–99(b)
 overview of, 94
 prevention and treatment, 474(b)
 syphilis, 96(b)
 trichomonas, 97(b)–98(b)
Shingles, 139, **139**
Shock
 defined, 63
 types of, 63, 246–247
Short bones, 158, **159**
Shoulder blades, 173, **173**
Shoulders
 arteries to, 259, **259**
 veins of, **264**
Sickle cell disease, 234(b)
Sigmoid colon, 542, **542**
Signs, of disease, 66
Simple columnar epithelium, 42
Simple cuboidal epithelium, **41**, 42
Simple diffusion, 35, **35**
Simple fracture, 180
Simple squamous epithelium, **41**, 42
Simple sugars, 560
Sinoatrial (SA) node, **282**, 286, **286**
Sinuses, paranasal, **335**, 336
Sinusitis, 336(b)–337(b)

INDEX 733

Sjögren syndrome, 59
Skeletal muscle pump, 304
Skeletal muscles, **8**, **47**, 195, 195(t), **196**
   defined, 47
Skeletal system, 152–187
   appendicular skeleton, 154, **155**, 173–177, **173–177**
   axial skeleton, 154, **155**, 162–172, **162–172**
   bone classification, parts and markings, 158–160, **159–161**, 161(t)
   bone fractures, 180, **180**
   bone growth, 157–158
   bones of skull, 162–166, **162–166**
   bone structure, 155–156, **156**
   diseases and disorders, 181(b)–187(b)
   facial bones, **167**, 167–168
   functions of, 154, **155**
   gender differences in, 178(t)
   joints, **178**, 178–179
   life span and, 187
   lower extremities, 175, 177, **177**
   overview of, **6**
   pathophysiology, 181(b)–187(b)
   pectoral girdle, 173–174, **173–174**
   pelvic girdle, 174, **176**
   rib cage, 168, **168**
   spinal column, 169–171, **169–172**
   terms used to describe skeletal structures, 161(t)
Skin, 124–147. *See also* Integumentary system
   as body's defense system, 126–127, 311
   body temperature regulation and, 127, 132
   burns, 141–144
   cancer of, 144–146
   color of, 133
   decubitus ulcers, **130**, 130(b)
   disorders, 137–141
   excretion, 128
   functions of, 126–128
   jaundice, 133(b), **134**
   lesions, 135–136, 137(t)
   life span and, 146–147
   regulation of body temperature, 132
   sensory perception, 127
   structures of, 128–132, **129–132**
   vitamin D production, 127
   water loss through, 108
   wounds and healing, 134–135
Skin lesions, 135–136, **136**, 137(t)
Skull, **155**
   bones of, 162–166, **162–166**
   gender differences in, 178(t)
Small cell lung cancer, 352(b)
Small intestine, **527**, 538–540, **538–540**
Smell, sense of, 616–618, **617**, 652
Smoking, as health hazard, 63–64
Smooth endoplasmic reticulum (sER), 33
Smooth muscles, **8**, **47**, 194–195, 195(t)
   defined, 47–48
Sneezing, 362, 364(t)
Snoring, caring for patients who, 341(b)
Sodium, 22(t), 23(t)
   concentration in extracellular and intracellular fluid, 110, **110**
   function in body, 577(t), 579
   hypernatremia, 111(t), 113
   hyponatremia, 110, 111(t), 113

   nerve transmission and, 382, 383, **384**
   water regulation and, 108–109
Sodium channels, 383
Sodium-potassium pump, 36
Soft palate, **528**, 529
Soleus, **207**, **217**, 219(t)
Somatic nervous system, 416, 417(t)
Species, 80, **80**
Specific defense, 313
Sperm, **30**, **464**
   fertilization, 490
   formation of, 461, **461–462**
   meiosis and, 38
   normal sperm count, 460
   structure of mature, 462, **463**
   temperature and, 452
Spermatogenesis, **460**, 460–461
Spermatogenic cells, 452
Spermatogonia, 461
Spermicides, 490, **492**
Sphenoidal sinus, **335**
Sphenoid bone, **162**, **163**, **164**, **165**, 166
Sphenoid sinus, 336
Spinal cavity, 9
Spinal column, 169–171, **169–172**
Spinal cord, **387**, **396–397**, 396–399
   ascending and descending tract, 397
   gray matter, 396–397, **397**
   reflexes, 397–398, **399**, 399(t)
   white matter, 397, **397**
Spinal nerves, 410–413, **410–414**
Spinal tap, 420
Spirillum, 83
Spirochete, 83, 89
Spleen, 308, **308–309**, 310(t)
Splenectomy, 308
Splenic vein, 262(t)
Splenomegaly, 308
Spongy bone, 155, **156**, **161**
Sporothrix schenckii, 92
Sprains, 219, 220(b)
Squamosal suture, **162**, **163**
Squamous cell carcinoma, 144–145, **145**
Squamous cell lung cancer, 352(b)
Stable angina, 291
Stable cells, 38
Standard precautions, 83(b)
Stapes, **166**, **166**, 641, **641**, 642
Staphylococcus, **83**
Staphylococcus aureus, 86–87, 137, 139
Staphylococcus infection, 139
Starch, 24, 560
Starvation, **580**, 580–581
State of consciousness, 420
Stationary phase, 81, **81**
Stem cells, 39, 40, 228, **229**, **501**
Sterilization, 82
Sternal, **15**
Sternocleidomastoid muscle, 166, 207, **207**, **208**
Sternum, **155**, 168, **168**, 346
Steroids, 24, 589, **591**
Stomach, **526**, **527**, 534–535, **534–535**
Stomach cancer, 537(b)
Stomach ulcers, 537(b)
Stool specimens, 546(b)
Strabismus, 621(b)

Strains, muscles, 219, 220(b)
Stratified columnar epithelium, 42
Stratified squamous epithelium, 42
Stratum basale, 128, **129**
Stratum corneum, 129, **129**
Stratum germinativum, 128
Stratum granulosum, 129
Stratum lucidum, 129
Stratum spinosum, 128
Strep throat, 87
Streptococcus infection, 137, 139
Streptococcus mutans, 79
Streptococcus pneumonia, 87
Streptococcus pyogenes, 87
Stress, 604(b)
Stress incontinence, 447(b)
Stress test, 292
Striated muscle, 47, 195
Stroke, 246, 295(b), 392(b)–393(b)
ST segment, 289, 289(t)
Styloid process, **163**, **164**
Subarachnoid space, 400, **401**
Subclavian artery, **258**, 259, **259**
Subclavian trunk, 304
Subclavian vein, **266**
Subcutaneous layer, 128, **129**, 130
Sublingual glands, **530**, 531
Submandibular glands, **530**, 531
Submucosa, 533
Subscapularis, 211(t)
Substance P, 382(t)
Substantia nigra, 392(b)
Sucrase, 540
Sudden infant death syndrome (SIDS), 357(b)–358(b)
Sudoriferous glands, **131**
Sulcus, **387**, 388
Sulfate ions, 23(t)
Sulfur, 22(t), 576, 577(t)
Superficial, **14**, 14(t), 15
Superficial temporal artery, **258**
Superior, 12, **12**, 14(t)
Superior mesenteric artery, **258**
Superior thyroid artery, **258**
Superior vena cava, 263, **264**, **266**, **274**, **276**
Supination, 202, **203**
Supinator, **210**, 212(t)
Supine position, 12
Supporting cells, 617
Supraorbital foramen, **162**
Suprarenal artery, **258**
Suprarenal glands, 602
Supraspinatus, **209**, **210**
Surfactant, 355
Susceptibility, 310
Suture, 161(t), 165
Sutures, 178
Swallowing, 532–533
Sweat glands, **129**, 131–132, **132**
Sympathetic innervation, 282
Symphysis pubis, **472**
Symptoms, of disease, 66
Synapse, 381, 381–383, **383**
Synaptic cleft, 381, **383**
Synaptic knob, 381, **381**
Synergist, **200**, 200–201
Synovial joint, **178**, 178–179

Syphilis, 78, **78**, 474(b)
   etiology, signs/symptoms, treatment, 96(b)
   historical perspective on, 94
   stages of, 89
   treatment for, 89
Systematic lupus erythematosus (SLE), 59
Systemic capillaries, **276**
Systemic circuit, 256
Systemic lupus erythematosus, 321–322, 322(t)
Systolic pressure, 241

t

Tachycardia, 286, 289, **290**
Talus bone, 177
Tarsal, **15**, 177, **177**
Tarsals, **155**, 175
Taste, sense of, 618–620, **619**, 652
Taste buds, 618–619, **619**
Taxonomy, 80, **80**
TCA (trichloroacetic acid), 98(b)
T cells, 313–315, **314**, 316(t)
Tears, 634
Teeth, **529**, 529–530
Telangiectasia, **136**
Telophase, 38
Temperature, body
   control center for, 570
   heart rate and, 283
   skin and regulation of, 127, **127**, 132
Temporal bone, **162**, **163**, **164**, **165**, 166
Temporalis, **207**, **208**, 209(t)
Temporal lobe, **388**, **389**, 389(t), 390
Temporomandibular joint (TMJ), 167
Tendonitis, 220(b)–221(b)
Tendons, 46, 197, **197**, **198**, 219(b)
Tension headache, 393(b)
Tensor fasciae latae, **214**, **216**
Teratogens, 60
Teres major, **207**, **209**, 210
Teres minor, **207**, **209**, 210
Terminal ileitis, 540(b)
Tertiary bronchi, 344
Testes, 452, **453–454**, **589**
   cancer of, 454(b)–455(b)
   fetal development of, 452, **454**
   hormones and actions of, 590(t)
   testicular self-examination, 455(b)
Testicular self-examination, 455(b)
Testing
   cranial nerves, 409(b)
   hearing, 650–651
   infertility tests, 493
   neurological testing, 420–421, **421–422**
   vision, 636, 637(b)
Testosterone, 570, 590(t), 597, 608
   defined, 452
   luteinizing hormone and, **460**, 460–461
   made from cholesterol, 24
   production of, in testes, 452
   regulation of, 460–461
   secondary sex characteristics and, 460
Tetanus, 84, 85(b), 221(b)
Tetany, 221(b)
T4, 590(t), 593(t), 598
Thalamus, 390, **390**
Thalassemia, 235(b)–236(b), 236(t)

Thermoreceptors, 132
Thiamine, 574(t)
Third-degree burns, 142, **144**
Thoracic aorta, 257, 257(t), 258
Thoracic cavity, 9, **10**
Thoracic duct, 304, **304**
Thoracic mediastinum, 273
Thoracic nerves, 410, **410**
Thoracic vertebrae, 169, **169**, 170, **171**, **172**
Thoracocentesis, 348(b)
Threonine, 568, 568(t)
Throat, 531–533, **532**
Thrombocytes, 43, 43(b), 228, **229**, 238(t), 240
Thrombocythemia, 240
Thrombocytopenia, 240
Thrombophlebitis, 254(b)
Thrombosis, 61–62
Thrombus, 246, **246**
Thymine, 25, **26**
Thymopoietin, 590(t)
Thymosin, 307, 590(t), 607, 607(b)
Thymus
   hormones and actions of, 590(t), 607
   structure and function of, 307, **307**, 310(t), **589**
Thyroid cartilage, 338, **339**, **342**, 346
Thyroid gland, **589**, 598–600
   hormones and actions of, 590(t), 598
   hyperthyroidism, 598, 600
   hypothyroidism, 598
   tumors of, 600
Thyroid stimulating hormone (TSH), 590(t), 597
Thyroxine, 597
Tibia, **155**, **160**, 175, **177**, **217**
Tibial artery, 262(t)
Tibialis anterior, **207**, **217**, 219(t)
Tidal volume, 361(t)
Tinea, 140–141, **141**
Tinea barbae, 91
Tinea capitis, 91, **91**, 140–141, **141**
Tinea corporis, 140–141, **141**
Tinea cruris, 91
Tinea pedis, 91, 140–141, **141**
Tinea unguium, 91
Tinnitus, 645, 647
Tissue, **5**
   connective, **8**, **41**, 42–46
   defined, 8, 40
   epithelial, **8**, 40–42, **41**
   muscle, **8**, **41**, 46–48
   nervous, **8**, **41**, 48–49
Tissue cells, **276**
T lymphocyte, **229**
Toes
   bones of, 177, **177**
   muscles of, **217**, 219(t)
Tongue, **335**, **526**, **528**, 528-529, **529**
   sense of taste and, 618–619, **619**
Tonsils, 308–309
Torticollis, 221(b)–222(b)
Total body surface area (TBSA), 141–142
Total lung capacity, 361(t)
Totipotent, 39
Trace elements, 22(t)
Trachea, 334(t), **335**, **339**, 340–341, **342**, 346
Tracheostomy, 341

Transcytosis, 36
Transdermal contraceptives, 491
Transient ischemic attacks (TIAs), 392(b)
Transitional epithelium, **41**, 42
Translation, 568
Transmission electron microscope (TEM), 77
Transplants
   tissue rejections, 58, 319–320
   types of, 319
Transurethral resection of the prostate (TURP), 459(b)
Transverse abdominis, 213(t)
Transverse colon, 542, **542**
Transverse fissure, **387**
Transverse plane, 12, **13**
Transversus abdominis, **213**
Trapezius, **207**, **210**, 212(t)
Treponema pallidum, 78, **78**, 89, 96(b)
Tricarboxylic acid cycle, 563
Triceps brachii, **207**, **209**, 212(t)
Trichinella spiralis, 93(b)
Trichinosis, 93(b)
Trichomonas vaginalis, 97(b)
Trichomoniasis, 97(b)
Tricuspid valve, **276**, **277**, 277–278, 283
Trigeminal nerve, 405(t), 406
Triglycerides, 231(b)
   defined, 561
   function in body, 24
   metabolism of, **565**, 565–566
Trigone, 442, **442**
Trisomy 21, 61, 518(b)–519(b)
Trochanter, **160**, 161(t)
Trochlear nerve, 405(t), 406
Trophoblast, 500, **501**
True ribs, 168, **168**
True vocal cords, 339, **339**, **340**
Trypsin, 547
Tryptophan, 568, 568(t)
T3, 590(t), 593(t), 598
Tubercle, 161(t)
Tuberculosis, 88, 88(b), 348(b)–349(b)
Tuberosity, 161(t)
Tubular reabsorption, 440, **440**
Tubular secretion, **440**, 440–441
Tumor
   benign, 59
   malignant, 59
   as sign of inflammation, 57
   skin lesion, **136**, 137(t)
Tunica externa, 251, **251**
Tunica media, 251, **251**
Tunic intima, 251, **251**
Turner's syndrome, 520(b)–521(b)
T wave, 287–288, 289(t)
Tympanic cavity, **641**
Tympanic membrane, 641, **641**, **642**
Type 1 diabetes mellitus, 322(t)
Type I alveolus, 355
Type II alveolus, 355
Tyrosine, 568(t)

u

Ulcer
   skin lesion, **136**, 137(t)
   stomach, 537(b)

Ulcerative colitis, 322(t)
Ulna, **155**, 173, **173**, 174, **175**
Ulnar artery, 262(t)
Ulnar vein, 262(t)
Umbilical cord, 504, **508**
Umbilical region, 11, **11**, **15**
Unami, 618
Unipolar neurons, 376, **376**, 377(t)
Universal donors, 248
Universal recipients, 248
Upper esophageal sphincter (UES), 533
Upper extremities
  arteries to, 259, **259**
  bones of, **173**, 173–174, **175**
  muscles of, 209, **210**, **211**, 211(t)–212(t)
  veins of, **264**
Upper respiratory infection, 338(b)
Upper respiratory tract, 332–339, **333**, 334(t), **335**
Urea, 440–441
Ureters, **428**, 442, **442**
Urethra, **428**, **443**, 443–444
Urethral meatus, 473
Uric acid, 440–441
Urinalysis, 444, **444**
Urinary bladder, **428**, 442–443, **442–443**
Urinary bladder infection, 445(b)
Urinary system, 427–447
  dialysis, 437(b)
  diseases and disorders, 435(b)–438(b), 445(b)
  kidneys, 429–433, **429–433**
  life span and, 446
  nephrons, 433–435, **433–435**
  overview of, **7**, 428, **428**
  stress incontinence, 447(b)
  urine elimination, 441–444
  urine formation, 439–441
Urinary tract infections, 444
  preventing, 446(b)
Urine
  elimination of, 441–444
  formation of, 439–441, **439–441**
  water loss through, 108
Uterine cancer, 483(b)
Uterine tube, 475(t)
Uterus, 475(t)
  cancer of, 483(b)
  hyperplasia, 55, **55**
  menstrual cycle, 485–488, **489**
  structure and function, 480–481, **481**
Uvea, 623, **624**
Uvula, **335**, **528**, 529
Uvulotomy, 341(b)
U wave, 289(t)

V

Vagina, **472**, 475(t), 484
Vaginal orifice, 473, **473**
Vaginitis, 484(b)
Vagus nerve, 406(t), 407, **408**
  heart rate and, 282
Valine, 568, 568(t)
Valtrex, 97

van Leeuwenhoek, Antoine, 76, **77**
Varicella zoster virus, 139, **139**
Varicose veins, 254(b)–255(b)
Varicosities, 546(b)
Vascular lesions, **136**
Vascular shock, 246
Vascular tunic, 623, **629**
Vas deferens, **453**, 457
Vasectomy, 457
Vasoconstriction, 251
Vasodilation, 251
Vasopressin, 109
Vastus intermedius, 214, 218(t)
Vastus lateralis, **207**, 214, **215**, **216**, 218(t)
Vastus medialis, 207, **207**, 214, **216**, 218(t)
Veins/venous system, **251**, 251–252, 262(t), 263, **263-266**
  disorders of, 254(b)–255(b)
  function of, 256
  of head, neck and brain, 263, **263**
  of lower limbs and pelvis, **265**
  major veins of body, 262(t)
  major veins that drain thoracic wall, **266**
  of shoulder and upper limbs, **264**
  thoracic wall, **264**
Ventral, 9, 13, **14**, 14(t)
Ventricles
  of brain, **402–403**, 402–404
  of heart, **273**, **274**, 275
Ventricular fibrillation, 289, **289**, 293(b)
Venules, 251–252, **252**
Vermiform appendix, 310, 541–542, **542**
Verrucae, 141, **141**
Vertebral, **15**
Vertebral artery, **258**
Vertebral canal, 9, **10**
Vertebral column, **155**
Vertebra prominens, 170
Vertigo, 645
Vesicles, 32(t), 33, 135, **136**, 137(t)
Vestibular folds, 339
Vestibular glands, 473, 475(t)
Vestibule, 473, 475(t), 643, **644**
Vestibulocochlear nerve, 405(t), 406, **641**
Vibrio, 83, **83**, 87
Vibrio cholerae, 87
Vibrio vulnificus, **83**
Virchow, Rudolf, 49
Virion, 89
Virology, 89–90
Virulence, 66, 84
Viruses
  characteristics of, 89
  sexually transmitted infections, 96(b)–99(b)
  treatment of, 89
  viral hepatitis, 89
Visceral muscle, 47
Visceral pericardium, **273**, 280
Visceral peritoneum, 538
Visceral pleura, 346, **346**
Visceral smooth muscle, 195

Vision, 620–641. *See also* Eyes
  color vision, 640, **640**
  contrast sensitivity, 640
  diseases and disorders, 621(b), 623(b), 625(b)–630(b), 632(b)–634(b), 638(b)
  distance vision, 637(b)
  near vision, 637(b)
  protecting, 620(b)
  vision testing, 636, 637(b), 640
  visual pathways, 630–631
Vital capacity, 361(t)
Vitamin A, 318(b), 573(t)
Vitamin B_1, 574(t)
Vitamin B_2, 572, 574(t)
Vitamin B_3, 574(t)
Vitamin B_5, 574(t)
Vitamin B_6, 572, 574(t)
Vitamin B_{12}, 235(b), 572, 575(t)
  red blood cell production and, 230, 230(t)
Vitamin C, 318(b), 572, 575(t)
Vitamin D, 572, 573(t)
  bone-healthy diet and, 179(b)
  skin and production of, 127
Vitamin E, 318(b), 572, 573(t)
Vitamin K, 572, 573(t)
Vitamins, 572, 573(t)–575(t)
Vitreous humor, **624**, 628–629, **629**, 632(t)
VLDL (very low-density lipoprotein), 231, 566
Vocal cords, 338, 339, **339**, 340
Voice box, 338
Voluntary muscle, 47, 195
Vomer bone, **162**, **163**, **164**, 166, 167
Vulva, **473**, 473–474
Vulvovaginitis, 484(b)

W

Warts, 141, **141**
  genital, 98(b), 474(b)
Water
  extracellular fluid, **106**, 107
  function in body, 22, 23(t)
  gain of, in body, 107
  hydration in elderly, 109
  as inorganic compound, 22
  intracellular fluid, **106**, 107
  loss of, in body, 107–108
  as molecule, 20
  movement of, in body, 109
  as nutrient, 572
  percentage of body, 22
  regulation of, 108–109
  water reabsorption, 440, **441**
Wheal, **136**, 137(t)
White blood cells, 228, 236–239, **237–239**, 238(t)
  blood count, 238–239
  body's defense mechanism, 311
  diapedesis, 239, **239**
  inflammation and, 57
White coat syndrome, 242
Whiteheads, 138

White matter, 381, 397, **397**
White pulp, 308, **308**
White stroke, 393(b)
Wounds
　healing process, 134–135
　types of, **134**
Wrist, muscles of, 212(t), **211**

X

Xenograft, 319
Xerostomia, 59, 531

Xiphoid process, 168, **168**
X-linked disorders, 204(b), 205(b), 241(b)
X-ray, 420

Y

Yawning, 363, 364(t)
Yellow bone marrow, 159, **161**
Y-linked disorders, 204(b)
Yolk sac, 504, **507**

Z

Zinc, 22(t), 318(b), 578(t), 579
Zona pellucida, 490
Zoonotic disease, 310
Zygomatic bone, **162**, **163**, **164**, **165**, 167
Zygomaticus, **207**
Zygomaticus major, **208**, 209(t)
Zygomaticus minor, **208**
Zygote, 30, 480, 500, **501**, 502, 502(t), 515

Coding Corner Example Codes: CPT, ICD-9, and ICD-10

Chapter 1 Concepts of the Human Body
Disease: 50 y/o Male Physical **Body System Affected:** General/GI

Diagnostic Tests	CPT Codes	ICD-9 Diagnosis Descriptions	ICD-9 Codes	ICD-10 Diagnosis Descriptions	ICD-10 Codes
Physical exam; new patient, age 40–64	99386	General health exam	V70.0	Examination, general adult medical exam, without abnormal findings	Z00.00
Screening colonoscopy	45378	Colonoscopy, screening	V76.51	Screening, digestive tract; lower GI	Z13.811

Chapter 2 Concepts of Chemistry
Disease: Phenylketonuia (PKU) **Body System Affected:** Multiple

Diagnostic Tests	CPT Codes	ICD-9 Diagnosis Descriptions	ICD-9 Codes	ICD-10 Diagnosis Descriptions	ICD-10 Codes
Phenylalanine	84030	Phenylketonuria (PKU)	270.1	Phenylketonuria, classical	E70.0
Phenylketones (qualitative)	84035			Phenylketonuria, maternal	E70.1
				Phenylketonuria, unspecified site	C74.10

Chapter 3 Concepts of Cells and Tissues
Disease: Thrombocytopenia (low platelets) **Body System Affected:** Blood

Diagnostic Tests	CPT Codes	ICD-9 Diagnosis Descriptions	ICD-9 Codes	ICD-10 Diagnosis Descriptions	ICD-10 Codes
CBC	85025	Thrombocytopenia, primary	287.30	Thrombocytopenia, primary	D69.49
Platelet count	85049	Thrombocytopenia, unspecified	287.5	Thrombocytopenia unspecified	D69.6
Bone marrow biopsy	38221				

Chapter 4 Concepts of Disease
Disease: AIDS **Body System Affected:** Immune

Diagnostic Tests	CPT Codes	ICD-9 Diagnosis Descriptions	ICD-9 Codes	ICD-10 Diagnosis Descriptions	ICD-10 Codes
HIV (HTLV-I) testing	86687	HIV positive status	V08	HIV positive status	Z21
HIV confirmatory (Western Blot)	86689	AIDS	042	AIDS	B20

Chapter 5 Concepts of Microbiology
Disease: Strep Throat **Body System Affected:** Respiratory/Multiple

Diagnostic Tests	CPT Codes	ICD-9 Diagnosis Descriptions	ICD-9 Codes	ICD-10 Diagnosis Descriptions	ICD-10 Codes
Rapid strep throat culture	87081	Strep pharyngitis	034.0	Strep pharyngitis	J02.0
CBC	85025	Non-strep pharyngitis	462	Non-strep pharyngitis	J02.9

Chapter 6 Concepts of Fluid, Electrolyte, and Acid–Base Imbalance
Disease: Hyperkalemia **Body System Affected:** Cardiac/Multiple

Diagnostic Tests	CPT Codes	ICD-9 Diagnosis Descriptions	ICD-9 Codes	ICD-10 Diagnosis Descriptions	ICD-10 Codes
Serum potassium (OR)	84132	Hyperkalemia	276.7	Hyperkalemia	E87.5
Serum electrolytes	80051	Arrhythmia	427.9	Arrhythmia, unspecified	I49.9
ECG (EKG) with interp.	93000				

Chapter 7 The Integumentary System
Disease: Malignant Melanoma **Body System Affected:** Integumentary

Diagnostic Tests	CPT Codes	ICD-9 Diagnosis Descriptions	ICD-9 Codes	ICD-10 Diagnosis Descriptions	ICD-10 Codes
Skin biopsy (results pending)	11100	Malignant melanoma RUE	172.6	Malignant melanoma in situ RUE	D03.61
Excision malignant lesion <0.5 cm, RUE	11600				

Chapter 8 The Skeletal System
Disease: Osteosarcoma, Left Thigh **Body System Affected:** Skeletal

Diagnostic Tests	CPT Codes	ICD-9 Diagnosis Descriptions	ICD-9 Codes	ICD-10 Diagnosis Descriptions	ICD-10 Codes
X-ray left femur	73550	Osteosarcoma (malignant neoplasm) thigh	170.7	Malignant neoplasm left lower limb (thigh)	C40.22
Bone biopsy	20225				

Chapter 9 The Muscular System
Disease: Muscular Dystrophy (Duchenne's) **Body System Affected:** Muscular

Diagnostic Tests	CPT Codes	ICD-9 Diagnosis Descriptions	ICD-9 Codes	ICD-10 Diagnosis Descriptions	ICD-10 Codes
Muscle biopsy	20205	Duchenne's muscular dystrophy	359.1	Duchenne's muscular dystrophy	G71.0

Chapter 10 Blood and Circulation
Disease: Acute Myelogenous Leukemia **Body System Affected:** Blood

Diagnostic Tests	CPT Codes	ICD-9 Diagnosis Descriptions	ICD-9 Codes	ICD-10 Diagnosis Descriptions	ICD-10 Codes
CBC automated	85027	Acute myelogenous leukemia (not in remission)	205.00	Acute myeoblastic anemia, not achieving remission	C92.00
Bone marrow biopsy	38221				

Chapter 11 The Cardiovascular System
Disease: Myocardial Infarction (MI) **Body System Affected:** Cardiovascular

Diagnostic Tests	CPT Codes	ICD-9 Diagnosis Descriptions	ICD-9 Codes	ICD-10 Diagnosis Descriptions	ICD-10 Codes
EKG with interpretation	93000	Mid-sternal chest pain	786.51	Chest pain (central)	R07.9
CPK enzymes (lab)	82550	vs		vs	
Continuous cardiac monitor	93224	Acute anterior MI, unspecified site, initial episode	410.91	Acute MI with stated duration of 4 weeks or less, unspecified site	I21.3